Introduction to Mechanism Design

Introduction to Mechanism Design

With Computer Applications

Eric Constans and Karl B. Dyer

CRC Press is an imprint of the
Taylor & Francis Group, an **informa** business

MATLAB® is a trademark of The MathWorks, Inc. and is used with permission. The MathWorks does not warrant the accuracy of the text or exercises in this book. This book's use or discussion of MATLAB® software or related products does not constitute endorsement or sponsorship by The MathWorks of a particular pedagogical approach or particular use of the MATLAB® software.

CRC Press
Taylor & Francis Group
6000 Broken Sound Parkway NW, Suite 300
Boca Raton, FL 33487-2742

Printed on acid-free paper

International Standard Book Number-13: 978-1-138-74065-5 (Hardback)

Visit the Taylor & Francis Web site at
http://www.taylorandfrancis.com

and the CRC Press Web site at
http://www.crcpress.com

Contents

Preface

This book serves as an introduction to the design and analysis of mechanisms using computer-aided design tools. A mechanism is a set of components connected together in such a way as to produce a desired motion. Examples of mechanisms in everyday life are numerous, and include windshield wipers, mechanical watch movements, the piston/connecting-rod/crankshaft assembly in an automotive engine, and the fancy "European hinges" found in upscale kitchen cabinets. In each of these instances, the designer was confronted with the problem of producing a desired motion (e.g. sweeping a wiper across a windshield) in the most economical way.

Until the recent past, mechanical designers have employed drafting tools (triangle, T-square, compass) to complete their work. These tools have been entirely superseded by computer-aided design tools such as CAD software (e.g. SOLIDWORKS®) and mathematical simulation software (e.g. MATLAB®). While a mechanical engineer might use a pencil and sketch pad to help in brainstorming a design, the final result will inevitably be developed and communicated through software. This wholesale change in the mechanical design process has been largely ignored in most mechanical design textbooks, where references to compasses and dividers are still common.

With this in mind, we have written a textbook that would bring the modern practice of mechanical design into the classroom and computer lab. The book is intended to accompany a one-semester course in mechanical design at a 4-year university or technical college. The authors have used the material in this textbook to teach mechanical design to first-, second-, and third-year students for almost 20 years at our university. Some important features of the book include:

- An improved notation for conducting position, velocity, and acceleration analysis based upon the unit vector concept. This method gives the student a clearer understanding of the *meaning* of the equations, instead of the confusing jumble of trigonometric functions (or complex variables) found in most textbooks.
- A simplified (and more computationally efficient) solution to the fourbar linkage problem. This solution forms the basis for analyzing more complex linkages, such as the geared fivebar and sixbar.
- A rich set of web-based animations and simulations that are designed to be used with mobile devices, laptops, and desktop computers. QR codes interspersed through the chapters provide links to animations that illustrate the topic under discussion. Mechanical design is the study of *motion,* and students can gain a much deeper understanding of the subject by seeing and interacting with the mechanisms *as they move.*

  Web Address: http://www.mechdes.net
- A set of real-world design examples that employ the methods discussed in the text.
- Links to hands-on design projects that we have employed at our university for many years.

The target audience for this text is first-, second-, or third-year students in mechanical engineering or mechanical technology programs. The course can be taken concurrently with differential calculus, which most engineering students take during their first semester. The textbook relies to a large degree upon vector analysis, which most students learn in high school. For those who did not learn it in high school, a short "refresher" is provided in Chapter 4.

The first chapter is an introduction to kinematics, which is the study of motion. The important concept of degrees of freedom is discussed, along with a new classification scheme for the fourbar linkage that is tied to motion analysis. Next, we introduce the use of CAD software to design fourbar linkages to achieve specified motion. The techniques presented in Chapter 2 are better suited for a tutorial in a computer laboratory than in the classroom, and students will employ these design techniques for the remainder of the course. While we use SOLIDWORKS® in our example problems, the design techniques can be used with any CAD software including AutoCAD, ProEngineer, and others. Chapter 3 gives a very basic introduction to MATLAB® and is intended for students new to the software.

Chapters 4–7 form a sequence leading up to the force analysis of linkages. Chapter 4 presents techniques to conduct position analysis on a variety of linkages. Once the positions of the links have been found, the velocity analysis methods in Chapter 5 can be used to find the velocity at any point on the linkage. From here, it is a simple step to find the accelerations on a linkage, as seen in Chapter 6. The culmination of all this effort is force analysis, which is presented in Chapter 7. Here we use the prescribed motion of the linkage to solve for the forces required to produce this motion – the inverse dynamics problem. The rigid-body inverse dynamics problem is often encountered in biomechanics, and the techniques presented in Chapter 7 are also used in automotive crash safety analysis.

Speed reduction using gear trains is discussed in Chapter 8, along with some practical design examples. Chapter 9 introduces students to planetary gearsets and a variety of analysis techniques is shown. The topic of efficiency in planetary gearsets is given a more thorough treatment than is given in most textbooks.

The text concludes with a chapter on cam design and analysis. Most textbooks provide several techniques for designing a cam profile to achieve a specified motion. In the interest of keeping the book to a reasonable length, we have chosen to present only the most versatile profile functions (polynomial and sinusoidal) and have given more emphasis to the motion of the follower, which can be quite different from the cam profile in many instances. The chapter concludes with a discussion of force analysis of the cam/follower mechanism, which is often neglected in other texts.

For each of the analysis sections we have followed a four-step scheme:

1. Develop a mathematical model of the system under investigation. This results in a set of equations that can be solved for position, velocity, acceleration, or force.
2. Write a MATLAB® script to solve the equations for a set of positions of the system.
3. Verify the results of step 2 using a different mathematical technique (e.g. energy, numerical differentiation).
4. Demonstrate the results by creating a set of MATLAB® plots.

The successful engineer must be proficient at all four steps, but most textbooks emphasize only the first. In fact, verifying and plotting the results of calculations are some of the most important aspects of an engineer's role, and we have given these tasks a strong emphasis in this text.

Finally, this textbook has been written using informal, engaging language in the hope of drawing the student into the subject. The mechanical design process is rich in opportunities for creative intellectual excitement, and we hope to convey some of our own enjoyment to the student.

Acknowledgments

The authors wish to express their gratitude to their friends and colleagues at Rowan University, especially to Dr. Tirupathi Chandrupatla, who was the inspiration for writing this book. Eric Constans would like to thank his wife, Aileen, for her patient support during the years of writing this book, as well as David Mosko and Thomas Mosolovich for their assistance in preparing the problems and exercises. Karl B. Dyer would like to thank his wife, Nicole, for her encouragement to start the book writing process and Dr. Smitesh Bakrania for his help with selecting color palettes. Finally, we wish to give thanks to the students in the Mechanical Engineering Department of Rowan University – you've been our motivation all along!

Authors

Eric Constans, PhD, is a professor in Mechanical Engineering at the Rose-Hulman Institute of Technology in Terre Haute, Indiana. Prior to joining Rose-Hulman, he taught at Rowan University in southern New Jersey for 19 years and served as department chair for six years. He has taught courses in Mechanical Design for two decades, and has published over 50 articles on mechanical design and engineering education. His fields of expertise include acoustics, vibration, and mechanical design and he worked for two years at Continental, AG as an acoustician before joining academia in 1999. He lives near Philadelphia, Pennsylvania and enjoys working on old houses and building custom hi-fi equipment.

Karl B. Dyer, holds a BS in Mechanical Engineering and a MS in Electrical Engineering. He is a technologist and adjunct professor in the Mechanical Engineering Department at Rowan University. Karl's professional areas of interest include programming and mechanical design. He believes that all engineers should have hands-on experiences during their education to aide in understanding theory presented during lecture. Through use of software packages, both CAD and programming languages, students are forced to utilize theory and produce a functioning "product." As a technologist, Karl enables Rowan students to move from the software design phase to the building phase, teaching rapid prototyping and industrial machining to students so they may produce working models.

1

Introduction to Kinematics

1.1 Introduction to Mechanical Design

The subject of this textbook is mechanical design and analysis. While most people have at least a vague idea of what the word "design" means, in this text we are mainly interested in two definitions [1]:

Design:

> transitive verb 1. To make preliminary sketches of; sketch a pattern or outline for; plan. 2. To plan and carry out, esp. by artistic arrangement or in a skillful way. ...
>
> intransitive verb 8. the arrangement of parts, details, form, color, etc. so as to produce a complete and artistic unit

The goal of the text is to give the reader a set of computational tools to design and analyze *mechanisms* to achieve specific goals. A mechanism is a collection of links and joints designed in such a way as to create a desired motion output. One link of a mechanism is "grounded," that is, fixed to some reference frame, and we are commonly interested in finding the motion of the remaining links. Some examples of mechanisms are windshield wiper blades, the crankshaft/connecting rod/piston assembly in a car engine, certain types of hinges, mechanical watches and clocks, etc. Another excellent example of a mechanism, or linkage, is the human body. Each segment of the body can be modeled as a link, and the segments are connected through pin joints (the elbow) or spherical joints (the shoulder). By modeling the body in this way, biomechanical engineers can deduce the forces and moments present at the joints by analyzing the motion of the body with motion capture techniques.

Scientists, mathematicians, and engineers have studied mechanisms since the 1700s. Until very recently, all mechanism analysis was performed graphically, that is, with drafting tools. These tools have been superseded in modern times by computational tools such as CAD software, which make it possible to analyze several trial designs very quickly to find a solution. Computers have also made "linkage design optimization" possible; that is, finding the dimensions of a linkage that traces out a desired path.

The majority of the book covers the kinematic analysis of mechanisms:

> Kinematics: the study of motion without regard to forces.

The first section of the book provides an introduction to some fundamental concepts in kinematics, and the student will learn techniques for designing linkages with CAD software, such as SOLIDWORKS®. Afterwards, we discuss methods for predicting the motion of linkages using computer programs written in MATLAB®. This is actually a rather difficult task, and will take up most of the text. To begin this task, we will learn how to find the positions traced out by a given linkage. Once we are able to compute positions of linkages, it is a simple matter to compute velocities and accelerations. Accelerations must be calculated to analyze link forces, as well as to keep accelerations within limits that can be tolerated by human beings. Once the motion of a linkage has been determined, the next step is to perform kinetic (dynamic) analysis.

Kinetics: the study of forces on a system in motion.

Force analysis is necessary to keep forces (stresses) within acceptable limits. The final chapters of the book cover gear train design (both conventional and planetary) and cam design and analysis. By the end of the book, the student will have developed a set of software tools for analyzing and designing a wide variety of interesting mechanisms. The authors have endeavored to present proper programming techniques throughout the text, so that the diligent student will be well prepared for modeling and analysis in many other courses in the engineering curriculum. And so, without further ado, let us begin our journey into Mechanical Design!

1.2 Fundamentals of Kinematics

We will now introduce some fundamental concepts in the science of kinematics. The first concept is that of a *rigid body*. In a rigid body, any two points that are separated by a distance d maintain that distance regardless of the ensuing motion. In other words, a rigid body cannot stretch, twist, compress, or otherwise deform. A generic rigid body is shown in Figure 1.1; the two points A and B remain a distance d apart, even though the object has been translated and rotated.

A second fundamental concept is that of *Chasles' Theorem*, which states that any motion of a rigid body can be described by a single *rotation* and *translation* (not necessarily in that order). A *pure rotation* occurs when all points on a body describe circular arcs of constant radius about a single point (the center of rotation, see Figure 1.2).

Pure translation occurs when the motion of a point on a body describes a straight line parallel to the lines traced by every other point (Figure 1.3).

The most general type of motion, *complex motion*, occurs when we have a combination of pure rotation and pure translation. These concepts will come in handy when we describe the motion of various links in a mechanism. For a complete discussion of rigid body motion, see [2].

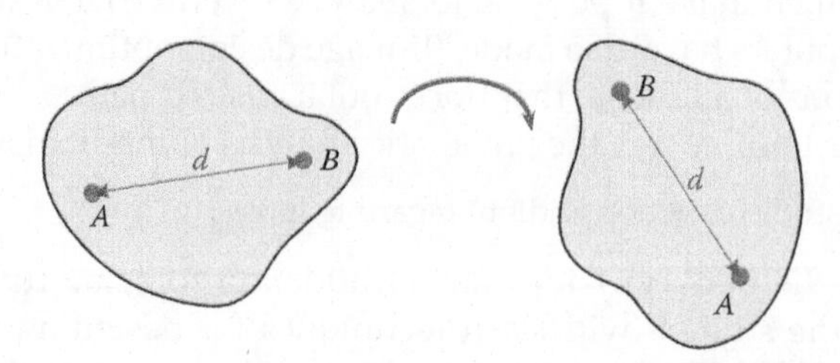

FIGURE 1.1
Any two points on a rigid body maintain a fixed distance from each other.

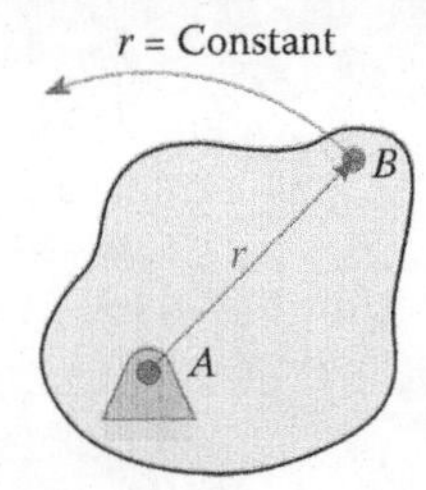

FIGURE 1.2
The body in the figure above is in pure rotation about a fixed point. The point B traces a circular arc with radius r and centered at A.

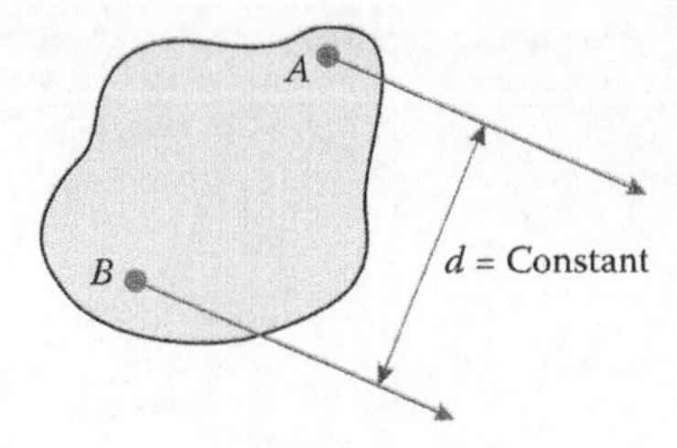

FIGURE 1.3
The points on a body in pure translation move along parallel lines.

1.3 Degrees of Freedom

In designing a mechanism, it is often critical to know its *mobility*, or the number of *degrees of freedom* (DOF) it possesses. Some mechanisms have so many links and joints that it is impossible to determine at a glance whether they are capable of movement at all. To determine the mobility of a mechanism, we define the number of *DOF* as:

> DOF = number of independent coordinates needed to completely define an object's orientation in space.

Imagine a point mass in 3D space like the one shown in Figure 1.4. The point is free to move in three directions: x, y, and z. It would take three coordinates to completely specify the position of the point mass; therefore, it has three DOF.

Now imagine a rigid body in 3D space. The body can translate in the same three directions as the point mass, but it can also rotate about its three axes as shown in Figure 1.5. To specify the configuration of the rigid body requires six coordinates: three translations and three rotations. Thus, a rigid body in 3D space has six DOF.

For the majority of this book, we will restrict ourselves to 2D (planar) space. We will be able to generalize many of the concepts that we develop to 3D, but staying in 2D will simplify the presentations considerably. In addition, a very large number of interesting mechanisms are essentially planar, rather than spatial. One important exception is the field of automotive suspension design, as we will see later in this chapter. As seen in Figure 1.6, a rigid body in 2D space has three DOF: two translations and one rotation.

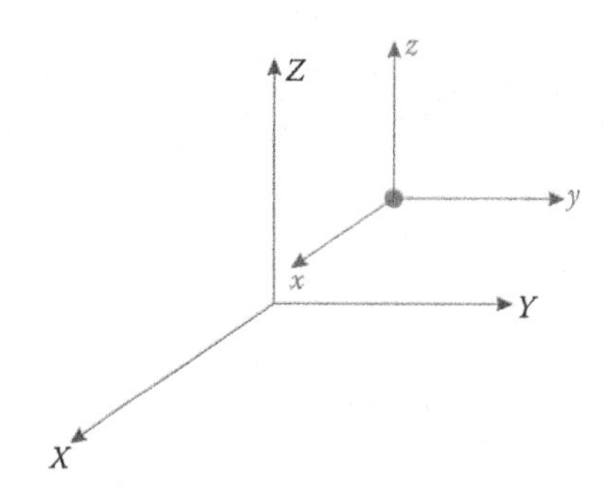

FIGURE 1.4
A point in 3D space has three DOF: three translations.

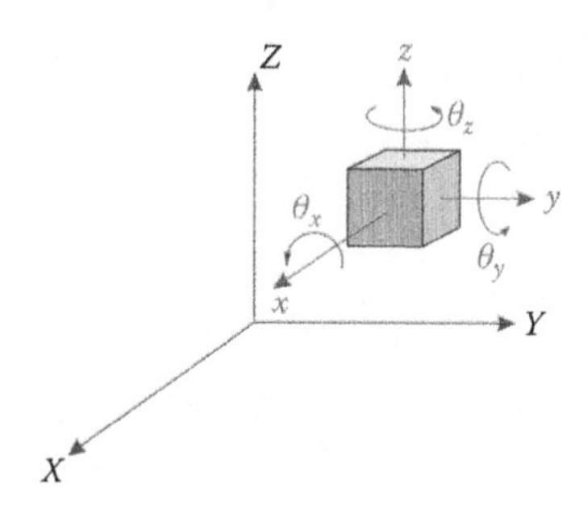

FIGURE 1.5
A rigid body in 3D space has six DOF: three translations and three rotations.

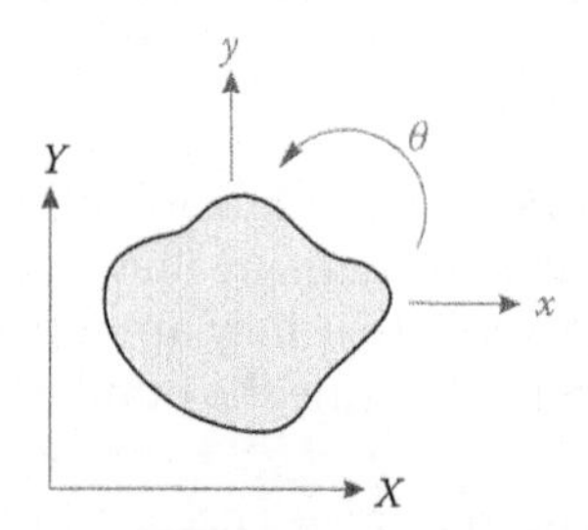

FIGURE 1.6
A 2D rigid body has three DOF.

1.3.1 Mobility of Mechanisms

With two bodies (links) we have six DOF, since each body has three of its own (Figure 1.7). We can generalize this by saying

$$\text{DOF} = 3L \tag{1.1}$$

where L is the number of links. But what happens if we connect the two links with a pin joint? (Figure 1.8).

Adding the pin constrains the translation of each link at the location of the pin without restricting rotation of each link. Denote this location (x_1, y_1) on link 1 and (x_2, y_2) on link 2. Then, for the pin joint, we have:

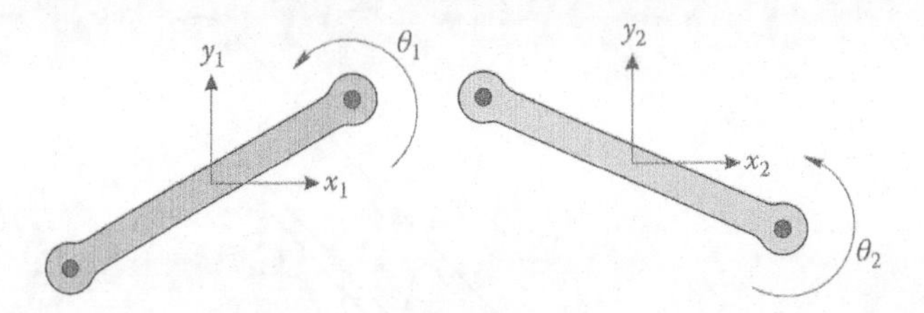

FIGURE 1.7
Two links have six DOF: four translations and two rotations.

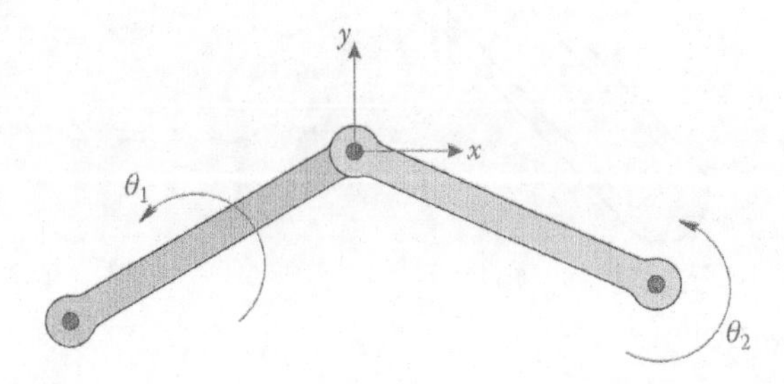

FIGURE 1.8
A pin joint removes two DOF.

$$\begin{aligned} x_1 &= x_2 = x \\ y_1 &= y_2 = y \end{aligned} \tag{1.2}$$

It now appears that coordinates x_2, y_2 (or x_1, y_1) are not independent; in fact, they have been eliminated as DOF. Thus, adding a pin joint removes two DOF from a mechanism.

$$\text{DOF} = 3L - 2J_p \tag{1.3}$$

where J_p is the number of pin joints. As we will see in a later example, a pin joint can *only* be used to pin two links together. To join three links at a point requires two pin joints (i.e., one pin to join links 1 and 2 and another pin to join links 1 and 3).

Another common type of joint is the *full-slider*, or *piston in cylinder.* Figure 1.9 shows a block that is mounted inside a slot in a link. The block is not free to rotate, and cannot move in a direction perpendicular to the slot. Thus, putting the piston in its cylinder has also removed two DOF.

$$\text{DOF} = 3L - 2J_p - 2J_{fs} \tag{1.4}$$

The "*fs*" in J_{fs} stands for "full slider." We wish to contrast this with the "half slider" joint shown below.

In the half-slider joint, shown in Figure 1.10, the pin is free to move along the slot, and the link can rotate about the pin. Thus, the half-slider, J_{hs}, only removes one DOF: it prevents the pin from moving in a direction perpendicular to the slot.

$$\text{DOF} = 3L - 2J_p - 2J_{fs} - J_{hs} \tag{1.5}$$

Another example of a half-slider joint is the cam-follower, shown in Figure 1.11 above. A cam is a rotating body with a well-defined shape that is designed to make continuous

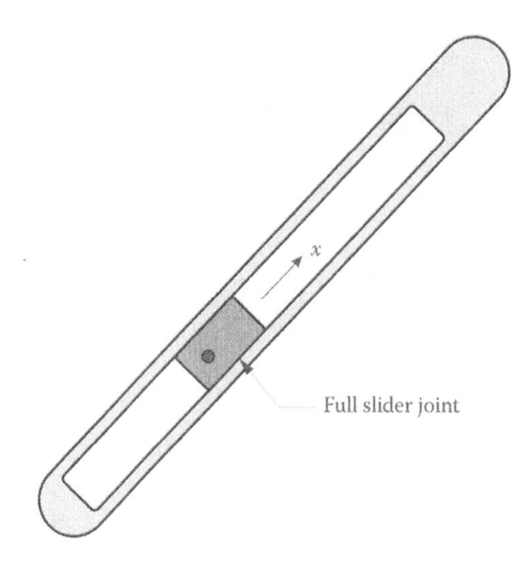

FIGURE 1.9
The full-slider removes two DOF.

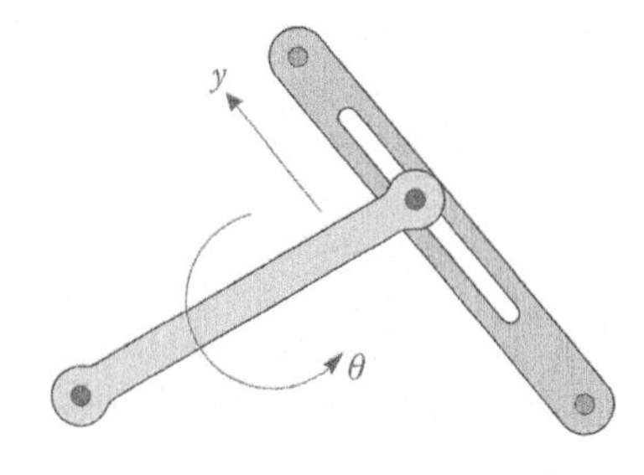

FIGURE 1.10
The half-slider removes only one DOF.

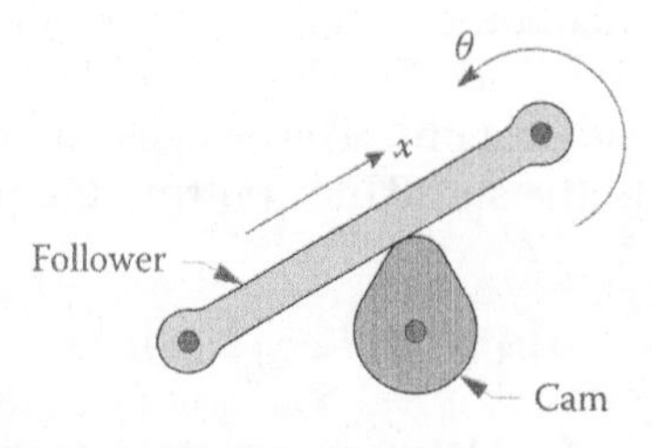

FIGURE 1.11
A cam-follower is another example of a half-slider joint.

contact with another body (the follower) and imparts a specific motion to it. In Figure 1.11, the cam is egg-shaped, and causes the follower to move up and down as it rotates. The follower can slide back and forth along the cam, and it can also rotate. The follower cannot, however, pass through the cam, which removes one DOF. Thus, the cam-follower joint is identical to the half-slider. Note that we have assumed that the cam remains in contact with the follower at all times; if contact is lost then the half-slider joint no longer exists!

FIGURE 1.12
The grounded link has zero DOF.

Finally, we note that a *grounded* link (one that is fixed in space) has all three DOF removed, as seen in Figure 1.12. Every mechanism has one grounded link, while all other links can move, completing Gruebler's equation presented in Equation (1.6)

$$\text{DOF} = 3L - 2J_p - 2J_{fs} - J_{hs} - 3G \tag{1.6}$$

where G is the number of grounded links. In the case where the entire mechanism is moving, as in the engine of a car, we choose one link (usually the engine block) to be ground, and analyze the movement of the remaining links relative to this. Since all the grounded links in a mechanism have the same movement (i.e., zero), we typically lump them together as a single link, so that $G=1$

$$\text{DOF} = 3(L-1) - 2(J_p + J_{fs}) - J_{hs} \tag{1.7}$$

For the remainder of this book we will assume that there is one, and only one grounded link. Equation (1.7) is known as the "modified Gruebler's equation," and can be used to determine the mobility of any two-dimensional linkage.

A mechanism with zero DOF is not a mechanism at all, since it cannot move. As we will see in the examples that follow, we usually wish for the mobility of a mechanism to be one. If we wish to control the position of each part of the linkage, we must provide one *actuator* (e.g. a motor, pneumatic cylinder, etc.) for each DOF. Since actuators add expense and complexity to the mechanism, our goal should be to achieve the desired motion with the lowest possible number of DOF. In some cases, such as robotic arms, a mobility of greater than 1 is unavoidable and will require multiple actuators. For most of this book, we will concentrate on mechanisms with a single DOF, but the methods presented here can easily be extended to multiple DOF systems.

1.3.2 Degrees of Freedom Example Problems

Example 1.1: Threebar Linkage

The linkage in Figure 1.13 has three links, one of which is grounded. There are three pin joints, and no full or half-sliders. Thus, we have

$$L=3 \quad J_p=3 \quad J_{fs}=0 \quad J_{hs}=0$$

Using the modified Gruebler's equation gives

$$\begin{aligned} \text{DOF} &= 3(L-1) - 2(J_p + J_{fs}) - J_{hs} \\ &= 3(3-1) - 2(3) \\ &= 0 \end{aligned}$$

Since DOF = 0, the mechanism cannot move, and is therefore called a "structure." It is interesting to note that the shape of the threebar linkage is a triangle. Since the triangle

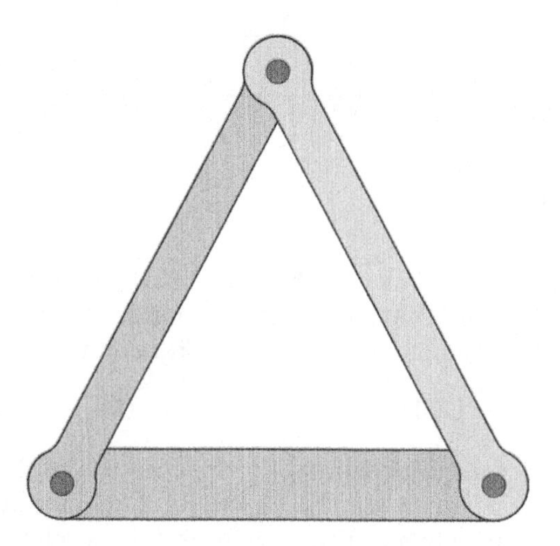

FIGURE 1.13
The threebar linkage.

has an inherently rigid structure (i.e. zero mobility), it is often used as a fundamental component in trusses and frames. For our purposes, however, the threebar is of little interest. In fact, since none of the three links can move, we may consider all three to be parts of a single "ground" link.

Example 1.2: Fourbar Linkage (Bad)

Figure 1.14 shows one method for constructing a fourbar linkage. There are four links, including the ground link. Remember that the two ground pivots count as one link, since they have the same (zero) motion! It may appear at first glance that there are four pin joints, but observe that the lower left pin connects *two* links to ground, so it counts as *two* pin joints. Therefore, there are five pin joints altogether. There are no sliders, so that we have

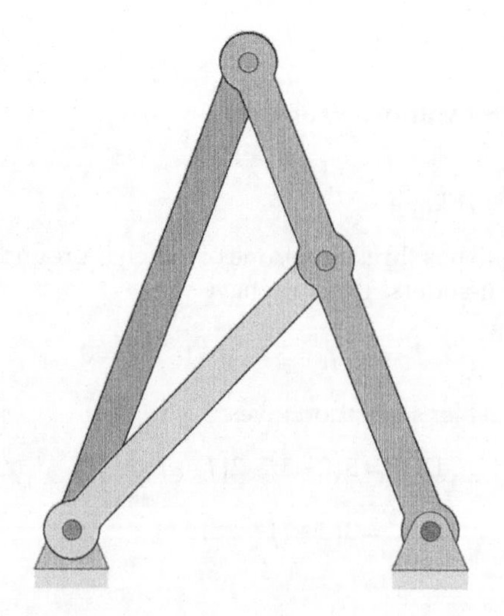

FIGURE 1.14
One way of constructing a fourbar linkage.

$$L=4 \quad J_p=5 \quad J_{fs}=0 \quad J_{hs}=0$$

$$\begin{aligned}
\text{DOF} &= 3(L-1)-2(J_p+J_{fs})-J_{hs} \\
&= 3(4-1)-2(5) \\
&= -1
\end{aligned}$$

Since DOF=−1, the mechanism cannot move, and is called a "preloaded structure." To see why, imagine trying to assemble the linkage: first, we pin the outermost links to ground, and then together at their uppermost pin holes. When we try to pin the central link between the outer links, any manufacturing error will prevent it from fitting – especially since we are assuming rigid links! That is, the distance between the two pin holes must be exactly the same as the length of the link, which is impossible in practice. If we *force* the force link to fit, we will be forced to bend (preload) the outer right link. This is the reason for the negative value of DOF. Every additional negative DOF means that another link must be forced into position, unless the pin holes are made oversize so that the pins have a "sloppy" fit.

Example 1.3: Fourbar Linkage (Good)

Figure 1.15 shows a second way of constructing a fourbar linkage. Here, we have four links (including ground) and only four pin joints. Thus,

$$L=4 \quad J_p=4 \quad J_{fs}=0 \quad J_{hs}=0$$

$$\begin{aligned}
\text{DOF} &= 3(L-1)-2(J_p+J_{fs})-J_{hs} \\
&= 3(4-1)-2(4) \\
&= 1
\end{aligned}$$

The correctly assembled fourbar linkage has one DOF, which means that one coordinate is sufficient to specify the configuration of the entire linkage. We shall have much to say about the fourbar linkage in the chapters to come.

Example 1.4: Slider-Crank

Another very common mechanism is the slider-crank, shown in Figure 1.16. This mechanism is found inside single-cylinder engines such as those found in a lawnmower or chainsaw. The slider-crank has two ordinary links, the crank and the connecting rod,

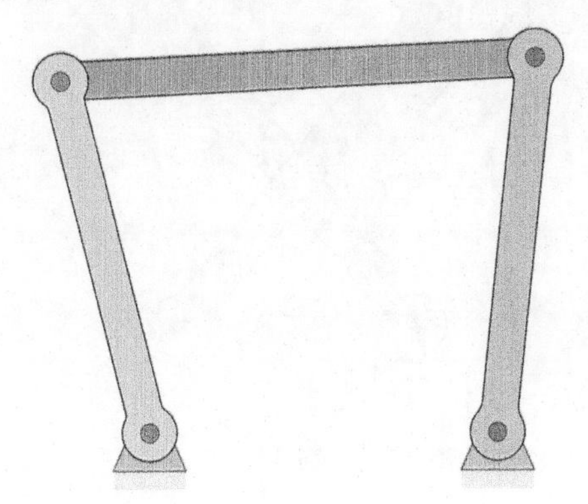

FIGURE 1.15
A better way of constructing a fourbar linkage.

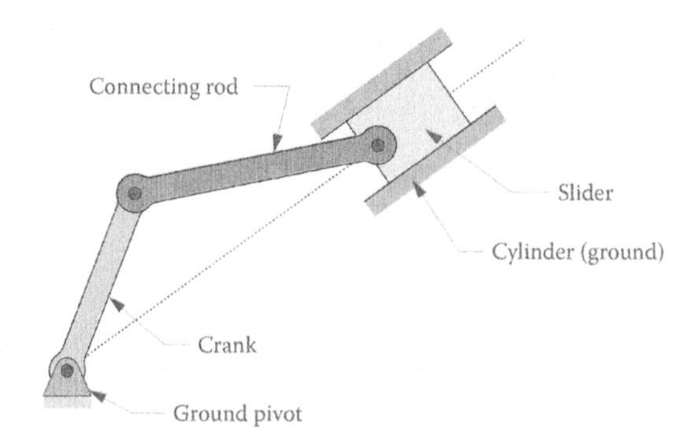

FIGURE 1.16
The slider-crank mechanism consists of a slider, crank, connecting rod, and cylinder.

and a slider, or piston. Including the ground, there are four links altogether. The slider-crank has three pin joints, and the piston rides in a full-slider joint. Thus, we have:

$$L = 4 \quad J_p = 3 \quad J_{fs} = 1 \quad J_{hs} = 0$$

$$\begin{aligned} \text{DOF} &= 3(L-1) - 2(J_p + J_{fs}) - J_{hs} \\ &= 3(4-1) - 2(3+1) \\ &= 1 \end{aligned}$$

Like the fourbar linkage, the slider-crank has one DOF. If we specify the angle of the crank, we can calculate the positions of the connecting rod and piston.

Example 1.5: Double Slider-Crank

Figure 1.17 shows a double slider-crank mechanism, as might be found in a multi-cylinder engine. It is the same as the single slider-crank of Example 1.4, but has two pistons, two

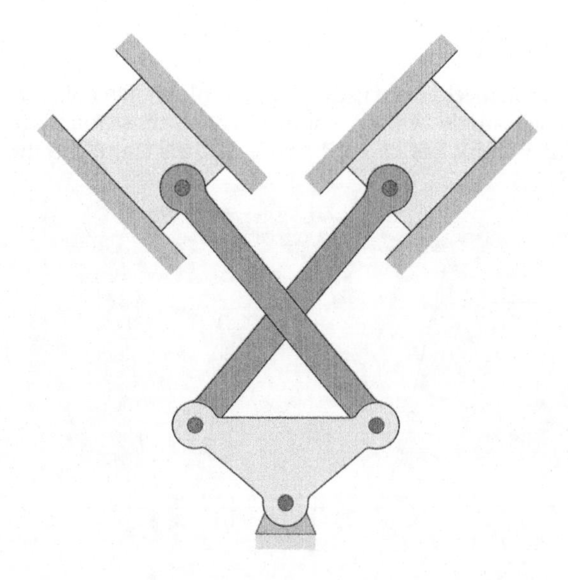

FIGURE 1.17
The double slider-crank.

connecting rods, and a crank with an additional pin hole. Including ground, the number of links is six, and there are five pin joints. Thus:

$$L=6 \quad J_p=5 \quad J_{fs}=0 \quad J_{hs}=0$$

$$\begin{aligned}\text{DOF} &= 3(L-1)-2(J_p+J_{fs})-J_{hs}\\ &= 3(6-1)-2(5+2)\\ &= 1\end{aligned}$$

The double slider-crank also has one DOF, which means that the crank angle is sufficient to determine the positions of both pistons. If this were not so, then it would not be possible to time an engine; that is, to time the firing of each spark plug when its corresponding piston reaches the desired height in the cylinder.

Example 1.6: The Inverted Slider-Crank Mechanism

The mechanism in Figure 1.18 may look strange at first, but it is surprisingly common. The inverted slider-crank, is most often seen in foot-operated bicycle pumps, but it can also be found as the "McPherson strut" suspension in many automobiles. As seen in the figure, the linkage consists of a crank, a slider, and a rocker, where the slider and rocker are connected through a full-slider joint. As with the slider-crank, there are four links (including ground), three pin joints, and one full-slider.

$$L=4 \quad J_p=3 \quad J_{fs}=1 \quad J_{hs}=0$$

$$\begin{aligned}\text{DOF} &= 3(L-1)-2(J_p+J_{fs})-J_{hs}\\ &= 3(4-1)-2(4)\\ &= 1\end{aligned}$$

Thus, like the slider-crank, the inverted slider-crank has one DOF.

Example 1.7: The Cam-Follower Mechanism

A cam-follower mechanism is shown in Figure 1.19. As the cam rotates, the roller spins on its pin joint, which is attached to the follower. There are four links in the mechanism:

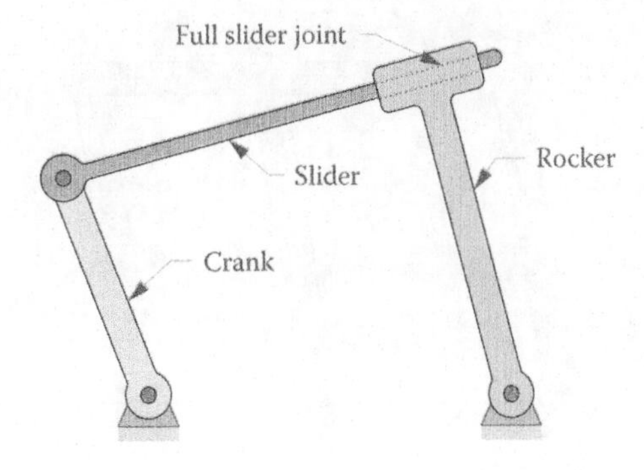

FIGURE 1.18
The inverted slider-crank mechanism.

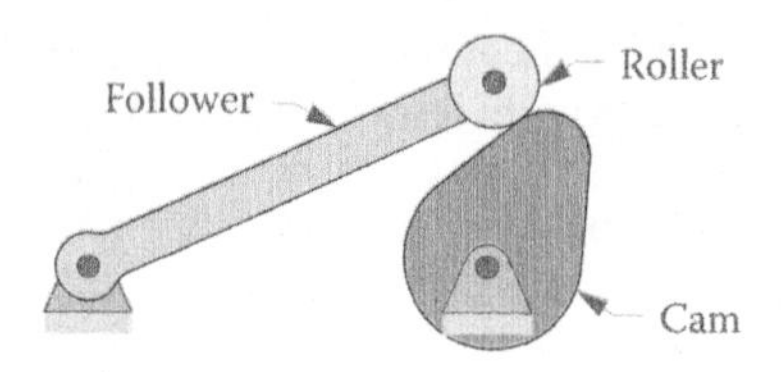

FIGURE 1.19
The follower pivots on its ground pin as the cam rotates.

cam, roller, follower, and ground. There are three pin joints, and one half-slider between the cam and roller. Conducting the DOF analysis gives

$$L = 4 \quad J_p = 3 \quad J_{fs} = 0 \quad J_{hs} = 1$$

$$\begin{aligned} \text{DOF} &= 3(L-1) - 2(J_p + J_{fs}) - J_{hs} \\ &= 3(4-1) - 2(3) - 1 \\ &= 2 \end{aligned}$$

This result may surprise you, since it appears as though the rotational position of the cam should determine the configuration of the entire mechanism. However, if we hold the cam fixed, we are still able to rotate the roller, and the roller can take on any angular coordinate without affecting the rest of the system. Hence, the two DOF are the angular coordinates of the cam and roller.

From this example, we see that Gruebler's equation is a good "sanity check" for design, but the designer should always take care to match the resulting DOF number with actual movements in the mechanism.

Example 1.8: Gruebler's Paradox

Figure 1.20 shows a fivebar linkage arranged in parallelogram form. There are five links and six pin joints so that,

$$L = 5 \quad J_p = 6 \quad J_{fs} = 0 \quad J_{hs} = 0$$

$$\begin{aligned} \text{DOF} &= 3(L-1) - 2(J_p + J_{fs}) - J_{hs} \\ &= 3(5-1) - 2(6) - 0 \\ &= 0 \end{aligned}$$

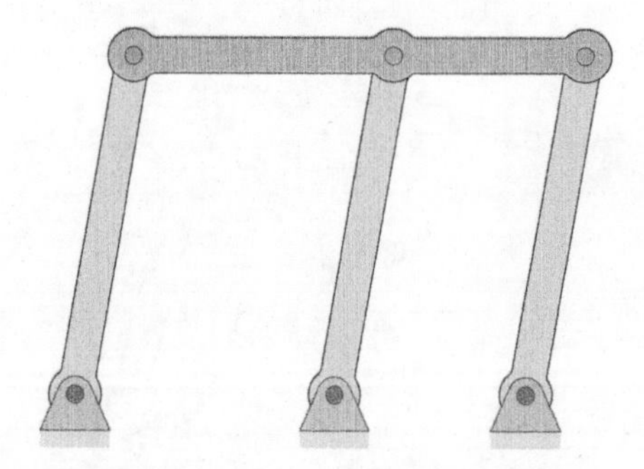

FIGURE 1.20
Another fivebar linkage arranged in a parallelogram.

Gruebler's equation predicts that the mechanism has zero DOF and cannot move, but our intuition would seem to indicate that it can move. In fact, it appears to be identical to the drive mechanism seen on steam locomotives from the 1800s. This phenomenon is known as *Gruebler's Paradox* and reinforces the notion that we should always accompany our DOF calculations with a healthy dose of skepticism and intuition. To construct the mechanism we would need to force the final link into position, unless the pin joints have sufficient slop to permit easy insertion. If the pin joints are made to a tight tolerance, the mechanism would have some difficulty moving, as the DOF equation predicts. Building some "play" into the pin joints gives us the one DOF mechanism that our intuition predicts.

Example 1.9: Stephenson's Type 1 Sixbar Linkage

Figure 1.21 shows one configuration of the Stephenson sixbar linkage. Two of the links have three pin holes, while the other links have two. There are six links in the mechanism, including ground, and seven pin joints.

$$L = 6 \quad J_p = 7 \quad J_{fs} = 0 \quad J_{hs} = 0$$

$$\begin{aligned} \text{DOF} &= 3(L-1) - 2(J_p + J_{fs}) - J_{hs} \\ &= 3(6-1) - 2(7) \\ &= 1 \end{aligned}$$

Surprisingly, this complicated linkage has only one DOF. The sixbar linkage finds many applications in function generation and in early "mechanical calculators."

Example 1.10: Fivebar Linkage

Figure 1.22 appears similar to the Stephenson sixbar, but has one less link.

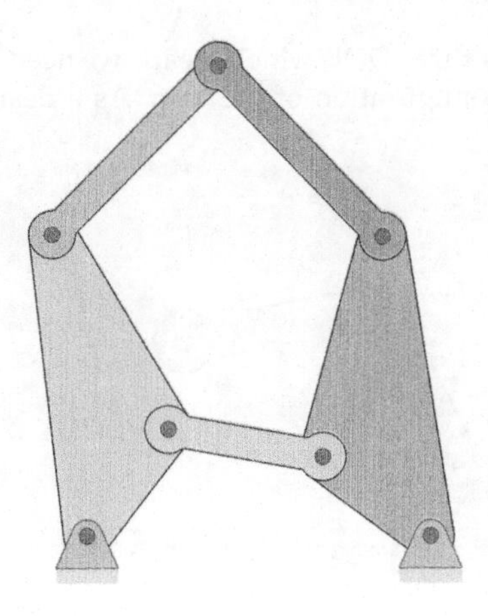

FIGURE 1.21
A Stephenson Type I sixbar linkage.

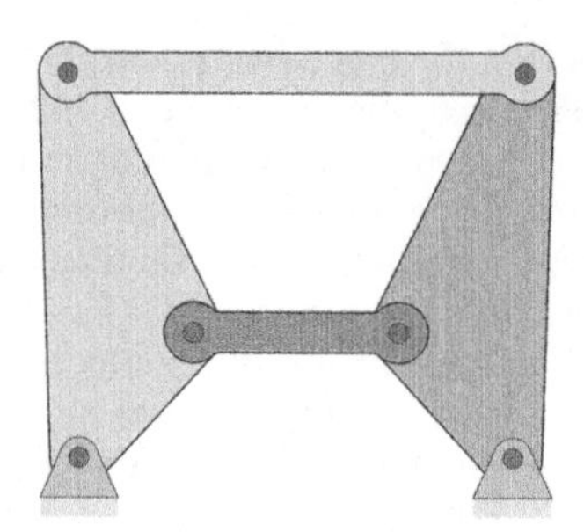

FIGURE 1.22
A fivebar linkage.

$$L=5 \quad J_p=6 \quad J_{fs}=0 \quad J_{hs}=0$$

$$\begin{aligned} \text{DOF} &= 3(L-1)-2(J_p+J_{fs})-J_{hs} \\ &= 3(5-1)-2(6) \\ &= 0 \end{aligned}$$

It appears that removing the sixth link has resulted in a structure, rather than a mechanism.

Example 1.11: Robot Arm

Figure 1.23 shows a robotic arm with three segments. There are four links (including ground) and three pin joints.

$$L=4 \quad J_p=3 \quad J_{fs}=0 \quad J_{hs}=0$$

$$\begin{aligned} \text{DOF} &= 3(L-1)-2(J_p+J_{fs})-J_{hs} \\ &= 3(4-1)-2(3) \\ &= 3 \end{aligned}$$

The robotic arm above has three DOF, which means we need three coordinates (angles) to uniquely specify the configuration of the arm. As a designer, this means that we

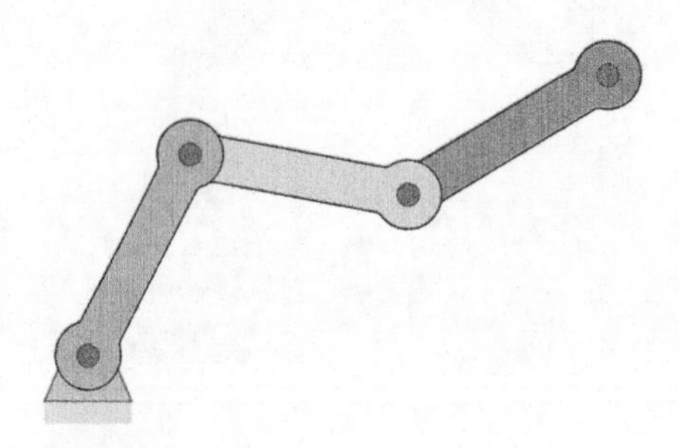

FIGURE 1.23
A robotic arm with three segments.

need at least three motors or actuators to control the arm. Some options include servo or stepper motors at each joint, hydraulic, or pneumatic cylinders at each joint, ball-screw drives, and many others.

1.4 The Fourbar Linkage and the Grashof Condition

As we saw in the previous examples, a linkage with four bars is the simplest possible mechanism; linkages with three or fewer fully connected bars cannot move, and were denoted "structures." For this reason, the *fourbar linkage* is one of the most common linkage types used in machinery today. Despite its apparent simplicity, the fourbar is capable of producing many interesting types of motion, as we will see. As it is composed only of links and pin joints, it is much simpler to fabricate than a comparable linkage with full-slider or half-slider joints. Thus, the fourbar should be the first linkage to try when designing a new mechanism.

Before we can analyze the motion of the fourbar linkage, we must first determine the type of motion it can achieve. To do this, we will make use of *Grashof's Theorem*, which is a very simple but powerful tool used in linkage design. A generic fourbar linkage is shown in Figure 1.24. Let:

S = length of the shortest link

L = length of the longest link

P = length of one remaining link

Q = length of the last link

Based upon these definitions, we can develop the classification scheme shown in Table 1.1.

If a linkage is Grashof, then at least one link (usually the shortest link) can make a full revolution without binding. If a linkage is non-Grashof, then *no* link can make a full revolution; that is, the linkage "binds up" when we try to turn a link too far. For Grashof Special Case linkages, at least one link can make a full revolution, but we must take special care with these linkages, as will be seen below.

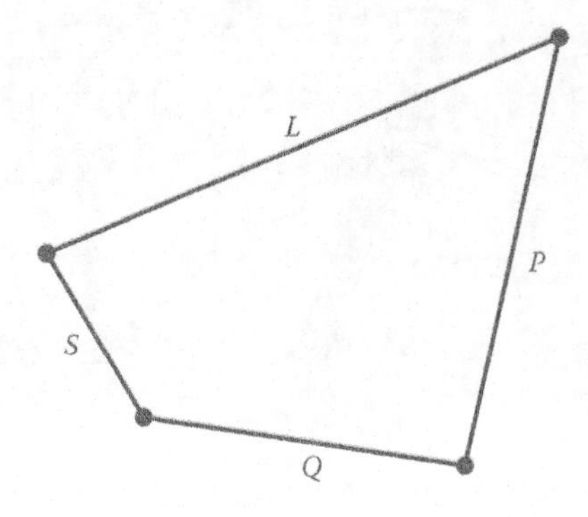

FIGURE 1.24
A generic fourbar linkage. Note that no link is "ground" at present.

TABLE 1.1
Grashof Fourbar Linkage Classification

Condition	Type
$S+L<P+Q$	Grashof
$S+L>P+Q$	Non-Grashof
$S+L=P+Q$	Grashof Special Case

In many practical situations, we will use a motor (AC or DC) to drive the linkage. This is simple in the case of a Grashof linkage – we attach the motor to the crank, and the linkage can spin forever as shown in Figure 1.25. In the non-Grashof case, we must either use a servomotor or stepper motor, or, where we desire to use a simple AC or DC motor (if we need continuous motion), we can attach a *driver dyad* as shown on the left in Figure 1.26. Note that the driver dyad must convert the left side of the linkage to a Grashof fourbar or we will be unable to drive it either! An animation of a non-Grashof linkage being driven by a driver dyad can be viewed by scanning the QRC tag or navigating to the textbook's website www.mechdes.net.

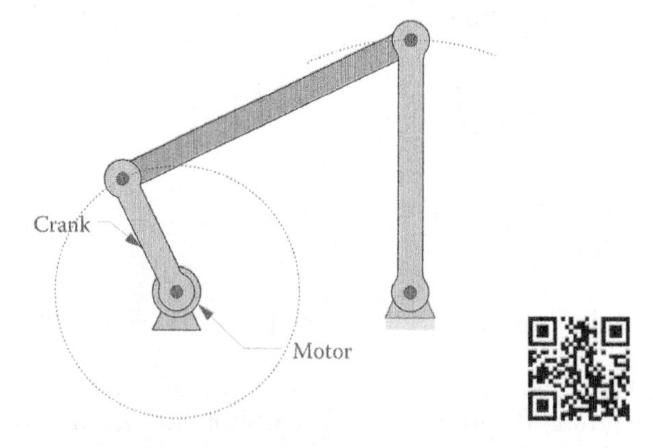

FIGURE 1.25
This linkage is Grashof, and we may attach a motor to its crank to drive it.

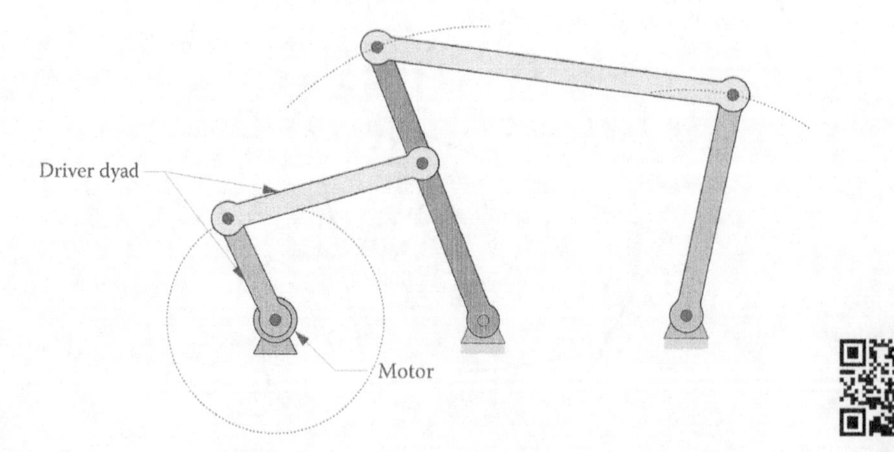

FIGURE 1.26
This linkage is non-Grashof, and we must use a driver dyad (or similar) to drive it.

1.4.1 Classifications of the Fourbar Linkage

Our goal in the next few chapters will be to develop a suite of computer programs that we can use to rapidly analyze the motion of a set of typical linkages. We will use these programs as design tools to ensure that a given linkage achieves the desired motion. To effect this, we must first create a classification scheme for the different types of fourbar (and other) linkages that can be easily implemented in software.

To begin, consider the "standard" fourbar linkage as shown in Figure 1.27. For the purposes of our discussion, we will define the *crank* as the link on the left side of the linkage and the *rocker* as the link on the right. The *coupler* is the moving link that connects the crank and the rocker, and the *ground* is the fixed link between the crank and rocker. In Figure 1.27, we have shown the ground link as two fixed pivots, but it may also be shown as an ordinary link in some illustrations.

It is important to note that much of the engineering literature defines a *crank* as any link with one fixed pin that can make a full revolution, and a *rocker* as any link with one fixed pin that cannot make a full revolution. We have deviated from this convention because the classification system proposed here permits an easier implementation into software than the ones in the literature (see e.g. [3]). As we will see, most practical linkages have cranks that make full revolutions and rockers that do not.

In most cases of practical interest, the crank receives the *input* to the linkage, either through a motor, pneumatic cylinder, or the output of a preceding linkage. Either the rocker or coupler is normally taken as the *output* of the linkage. In some cases, however, we must drive the coupler or rocker to achieve full motion of the linkage. These cases are of limited practical interest, because:

1. It is impossible to drive the coupler using a fixed motor, since all parts of the coupler move. While it is possible to attach a motor to the pin between the crank and the coupler, wiring constraints make this option rather difficult.
2. A linkage that has the rocker as the driven link can be easily "mirrored" horizontally, so that the crank is the driven link.

We present here a classification scheme that contains all possible motions of the fourbar linkage while recognizing that almost all practical cases will use the crank as the driving link.

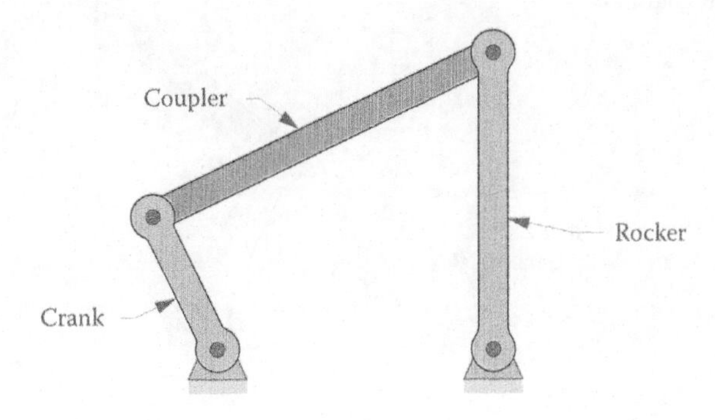

FIGURE 1.27
The fourbar linkage in its "standard" configuration. The ground link is defined as the distance between the two ground pivots.

1.4.2 Fourbar Classification: The Grashof Linkages

We begin with the set of fourbar linkages that are seen most often in practice: Grashof linkages. Recall that a linkage is considered "Grashof" if its link lengths fulfill the following condition

$$S + L \leq P + Q \tag{1.8}$$

If a linkage is Grashof, then at least one of the links can make a full revolution without the linkage binding.

Table 1.2 shows the classification scheme for all four types of Grashof linkage. The distinction between each type is the location of the shortest link, that is, if the crank is the shortest link then we have a Class 2 Grashof linkage. It is interesting to note that the crank, coupler, and rocker all make a full revolution in the Class 1 linkage.

TABLE 1.2

Classification Scheme for Grashof Linkages

Class	Driver	Notes	Illustration
1	Crank, coupler, or rocker	Short link is ground. Crank, coupler, and rocker make full revolution.	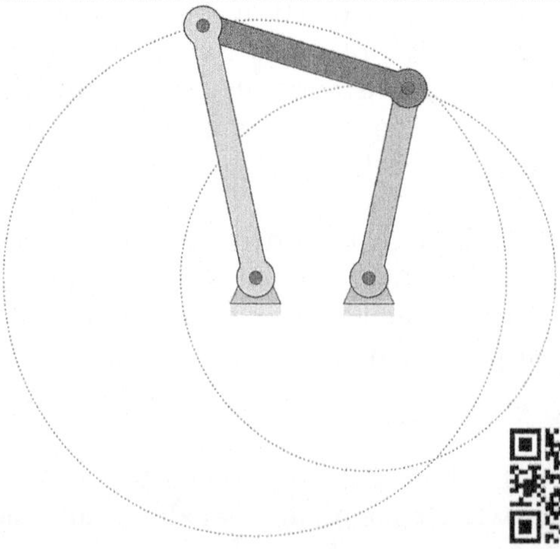
2	Crank	Short link is crank. Only crank makes full revolution.	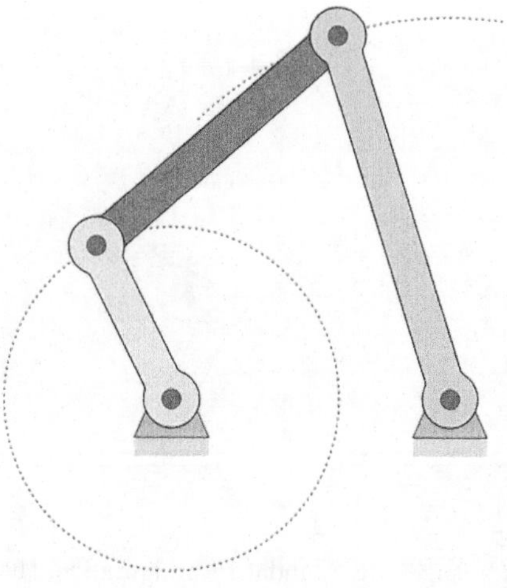

(Continued)

TABLE 1.2 (*Continued*)
Classification Scheme for Grashof Linkages

Class	Driver	Notes	Illustration
3	Coupler	Short link is coupler. Only coupler makes full revolution.	
4	Rocker	Short link is rocker. Only rocker makes full revolution.	

In the Class 3 linkage, only the coupler can make a full revolution. To drive this linkage to its full extent of motion we would need to attach a motor to one of the moving pins, which presents practical difficulties. Although the coupler makes a complete revolution, no point on the coupler traces out a circular arc. The reader will note that Class 4 is a mirror image of Class 2, and by reflecting the linkage horizontally, we arrive at the standard form of the linkage where the crank makes a full revolution.

1.4.3 Fourbar Classification: Non-Grashof Linkages

Table 1.3 shows the six types of non-Grashof linkages. All of these are similar in that they "bind up" at a certain position, and no link can make a full revolution. To visualize the non-Grashof linkages in motion scan the QRC tag or navigate to the textbook's website www.mechdes.net. Note that Class 6 is the mirror of Class 5, and Class 9 is the mirror of Class 8. The means for distinguishing the different types of non-Grashof linkages is the position of the *long* link, instead of the short link as was the case with Grashof linkages.

When a non-Grashof linkage becomes bound up at one of its extreme positions, two of the links become collinear and the linkage assumes the shape of a triangle. In some cases, the two links overlap each other (as in Classes 7–10) and in others the two links are stretched out (as in Classes 5 and 6). For either case, we will use the triangularity of the extreme positions to deduce the range of motion of the driving link.

1.4.4 Fourbar Classification – Special Cases

Recall that a Grashof Special Case linkage is defined as one whose link lengths satisfy

$$S + L = P + Q \tag{1.9}$$

TABLE 1.3

Classification Scheme for Non-Grashof Linkages

Class	Driver	Notes	Illustration
5	Crank	Long link is ground. Crank is shorter than rocker	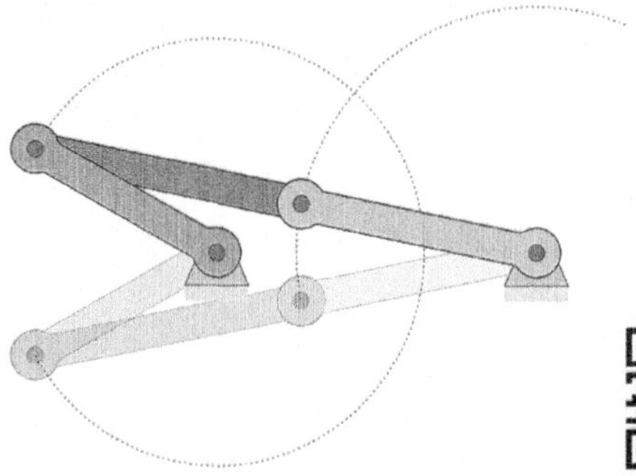
6	Rocker	Long link is ground. Rocker is shorter than crank.	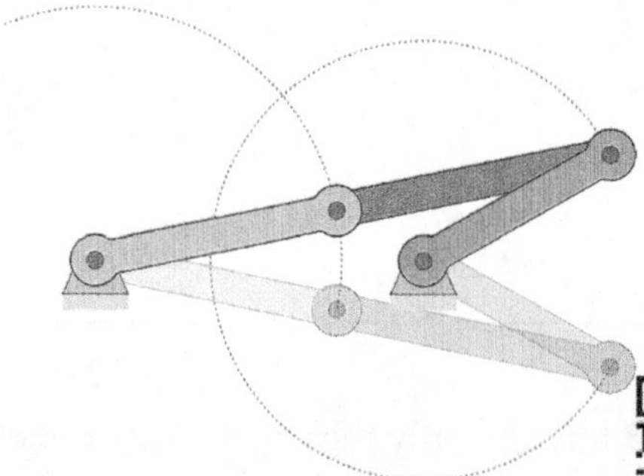
7	Rocker	Long link is crank.	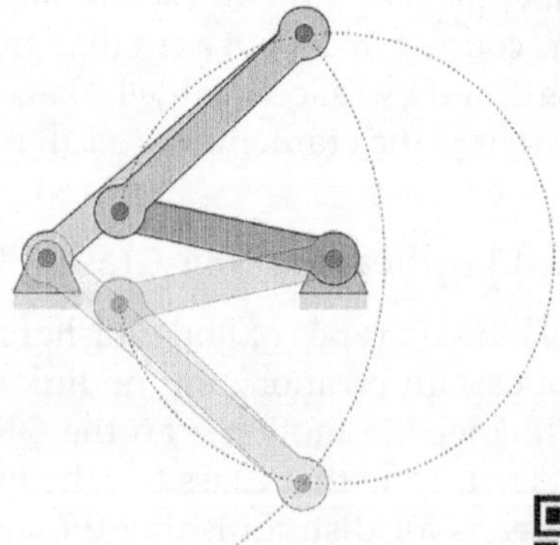

(Continued)

TABLE 1.3 (*Continued*)

Classification Scheme for Non-Grashof Linkages

Class	Driver	Notes	Illustration
8	Crank	Long link is coupler. Crank is shorter than rocker.	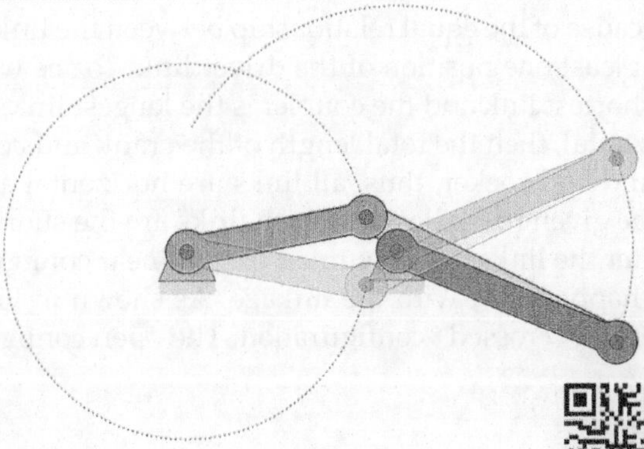
9	Rocker	Long link is coupler. Rocker is shorter than crank.	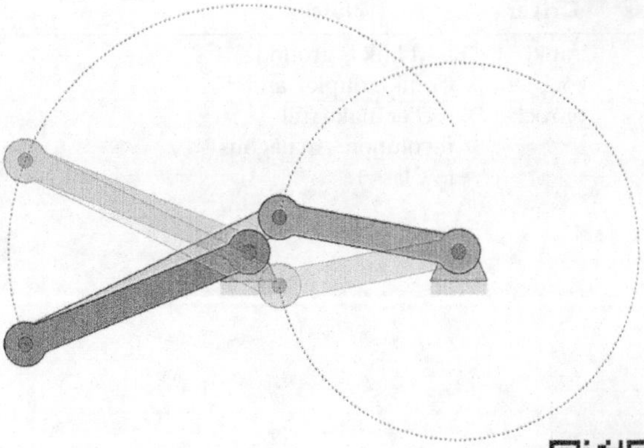
10	Crank	Long link is rocker.	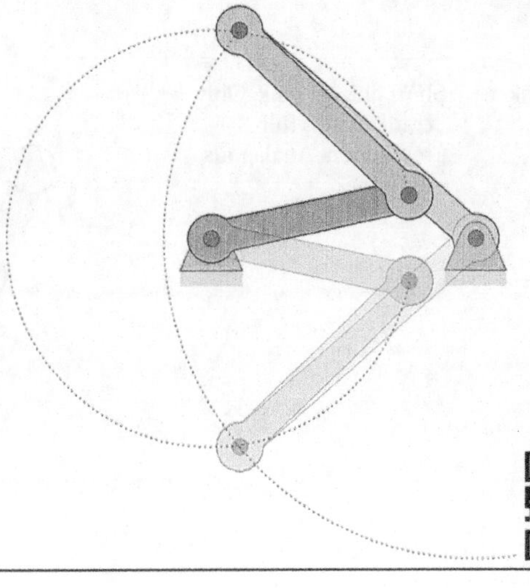

In these classes, at least one link can make a full revolution, as with the Grashof classes. The first four Grashof Special Case classes are shown in Table 1.4. Each of the classes is similar to one of the Grashof classes shown in Table 1.2, with one important difference as will be discussed below.

Because of the equal relationship between the link lengths, all four links will be collinear for at least one position of the driver link. To see why this is so, imagine that the crank is the shortest link and the coupler is the longest link, as shown in Figure 1.28. If the crank is horizontal, then the total length of the crank and coupler is equal to the total length of the ground and rocker; thus, all links are horizontal and collinear. A similar demonstration can be given regardless of which links are the shortest and longest.

After the links have assumed the collinear configuration, it is impossible to predict what will happen next with the linkage. As shown in Figure 1.29, the linkage may assume an "open" or "crossed" configuration. The open configuration takes the shape of an ordinary

TABLE 1.4

Four of the Special Case Grashof Linkages

Class	Driver	Notes	Illustration
11	Crank, coupler, or rocker	Short link is ground. Crank, coupler, and rocker make full revolution. Analagous to Class 1.	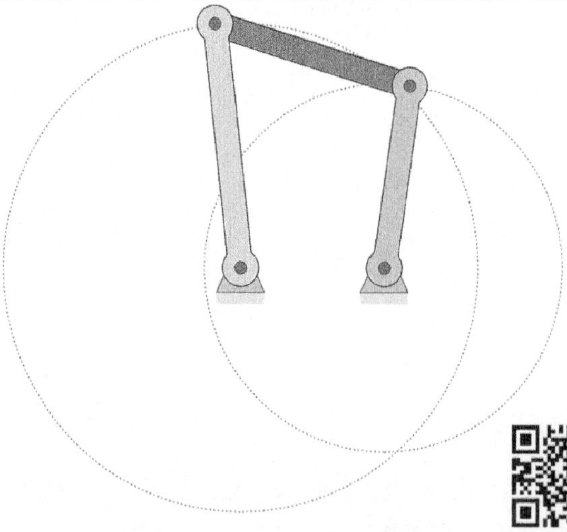
12	Crank	Short link is crank. Only crank makes full revolution. Analagous to Class 2.	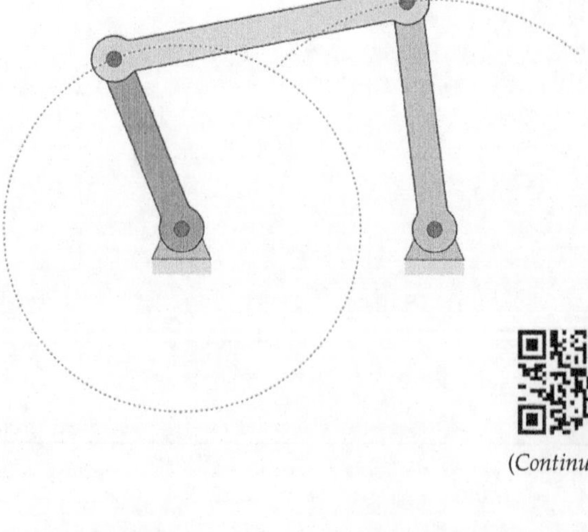

(Continued)

TABLE 1.4 (*Continued*)

Four of the Special Case Grashof Linkages

Class	Driver	Notes	Illustration
13	Coupler	Short link is coupler. Only coupler makes full revolution. Analagous to Class 3.	
14	Rocker	Short link is rocker. Only rocker makes full revolution. Analagous to Class 4.	

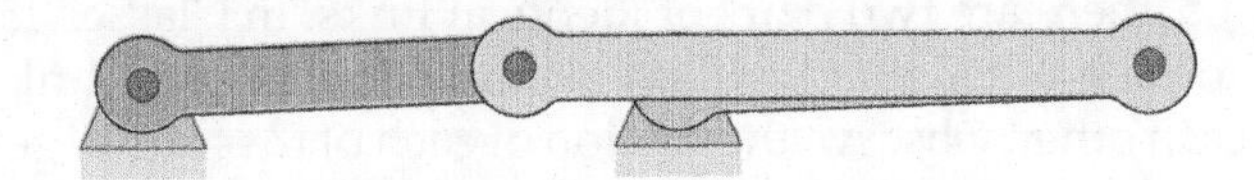

FIGURE 1.28
All links are collinear for at least one crank angle for a Special Case linkage.

quadrilateral, and the crossed configuration is so-called because the coupler *crosses* the ground link. Which configuration the linkage chooses is almost completely random, and depends upon the momentum of the links, friction in the pins, and other factors that are mostly out of the control of the designer. For this reason, a Grashof Special Case linkage should be avoided, unless absolutely necessary. If a Grashof Special Case linkage is used in a design, special care must be taken to ensure that the desired configuration (open or crossed) is achieved for each rotation of the crank. A fifth link is sometimes added to the fourbar to achieve this predictability, with the resulting linkage taking the form of the Gruebler's Paradox linkage in DOF Example 1.8 in Section 1.3. As you can see in the animation, the open and crossed configurations are mirror images of each other in

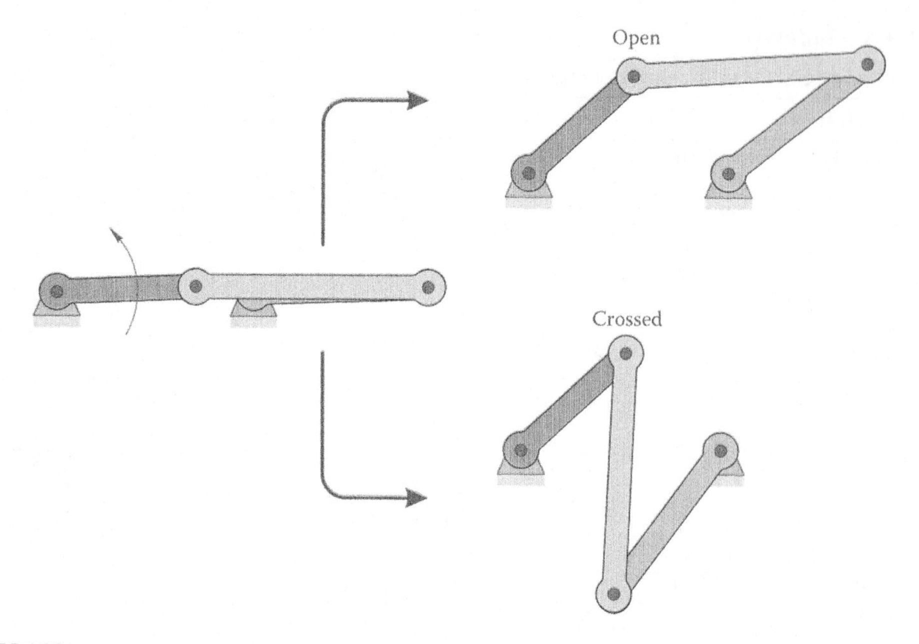

FIGURE 1.29
The links in a Special Case Grashof linkage will be collinear once per revolution of the crank. After this, the linkage may adopt the open or crossed configuration.

the vertical direction. In each case, however, the linkage "chooses" which configuration to pursue when all links are collinear, and it is impossible to predict beforehand which configuration it will choose.

1.4.5 Fourbar Classification – The Extreme Cases

We now arrive at the final set of Grashof classes: the extreme cases. In each of these classes, as shown in Table 1.5, there are two pairs of identical links. In Classes 15–18, the identical links are adjacent to each other, and in Class 19 (the parallelogram linkage) the identical pairs are opposite each other. Observe the motion of each of these linkages by scanning the QRC tags provided or by navigating to the textbook's website www.mechdes.net. Classes 15 and 18 are mirror images of each other, as are Classes 16 and 17.

The reader may wonder why these classes have been distinguished from the Grashof Special Cases shown earlier, for example, Classes 15 and 16 would seem to be special cases of Class 12, where the crank is the shortest link. There are two reasons:

1. Because there are two identical pairs in each extreme case linkage, the positions of each link are much easier to determine mathematically than for the special cases.
2. The extreme cases (especially Classes 15–18) exhibit a phenomenon not seen in the special cases.

As you can see by observing the animation of the linkages of Class 15 and 18, the coupler and rocker make a complete revolution for every *two* revolutions of the crank. In Classes 16 and 17, the coupler makes a complete revolution in the *opposite direction* of the driving link

TABLE 1.5

Extreme Classes of the Fourbar Linkage

Class	Driver	Notes	Illustration
15	Crank	Two identical, adjacent pairs. Ground and crank are shortest links. Crank, coupler, and rocker make full rotation.	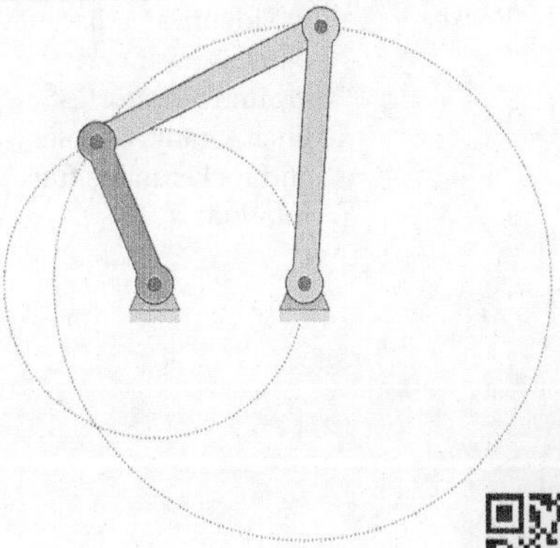
16	Crank	Two identical, adjacent pairs. Crank and coupler are shortest links. Crank and coupler make full rotation.	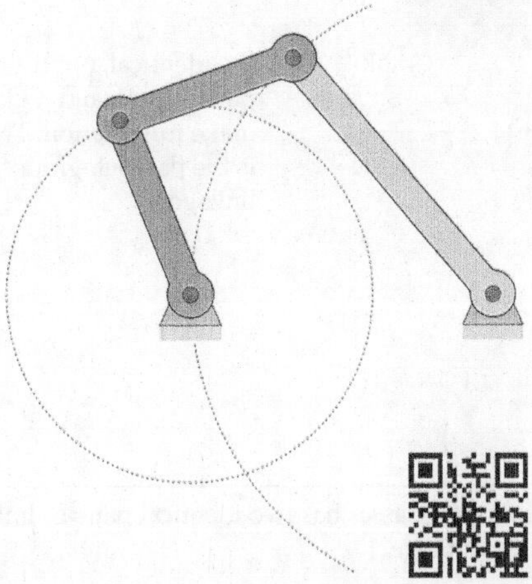
17	Rocker	Two identical, adjacent pairs. Coupler and rocker are shortest links. Coupler and rocker make full rotation.	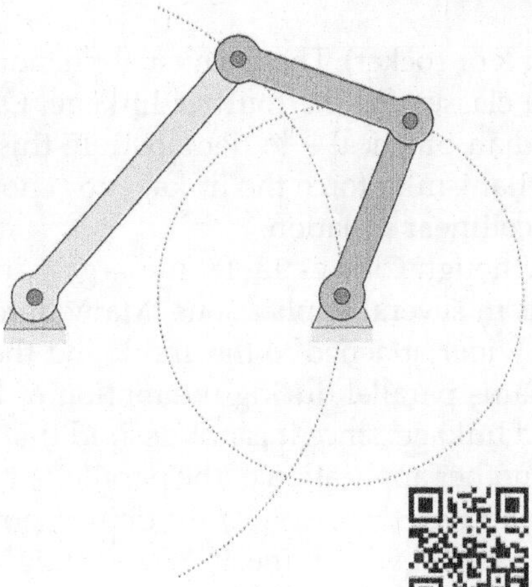

(Continued)

TABLE 1.5 (*Continued*)

Extreme Classes of the Fourbar Linkage

Class	Driver	Notes	Illustration
18	Rocker	Two identical, adjacent pairs. Rocker and ground are shortest links. Crank, coupler, and rocker make full rotation.	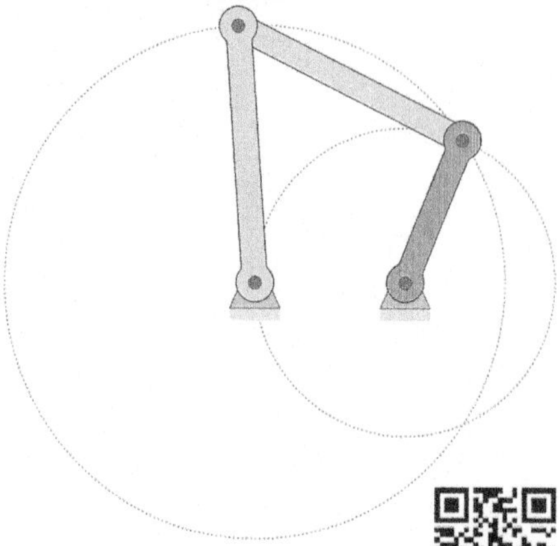
19	Crank	Two identical, opposite pairs. Crank and rocker make full rotation. This is the parallelogram linkage.	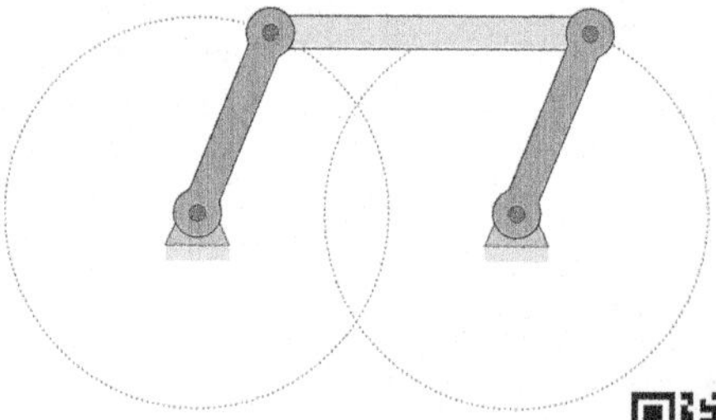

Each of these classes has two identical pairs of links. The driver is chosen to give the maximum range of motion for the linkage.

(crank or rocker). These unusual phenomena are the main reason for creating the extreme case classes for the fourbar linkage. Of course, these phenomena are more mathematical than practical – to demonstrate this in a physical linkage would require an external mechanism to force the linkage to "choose" the proper configuration when it encounters the collinear situation.

Although Classes 15–18 are largely impractical, the parallelogram linkage (Class 19) is used in several applications. Many automotive windshield wipers use this linkage with one wiper attached to the crank and the other to the rocker (see Figure 1.30). Each wiper remains parallel during the motion of the linkage, and it is normally driven by a driver dyad linkage, since it is not desired that a wiper make a complete revolution.

Another application of the parallelogram linkage was found in the drive wheels of steam locomotives (Figure 1.31). To deliver power to multiple wheels (to increase traction) each drive wheel was connected by a straight link. To ensure that the linkage did not assume the crossed configuration, the straight links were placed "out of phase" with each other

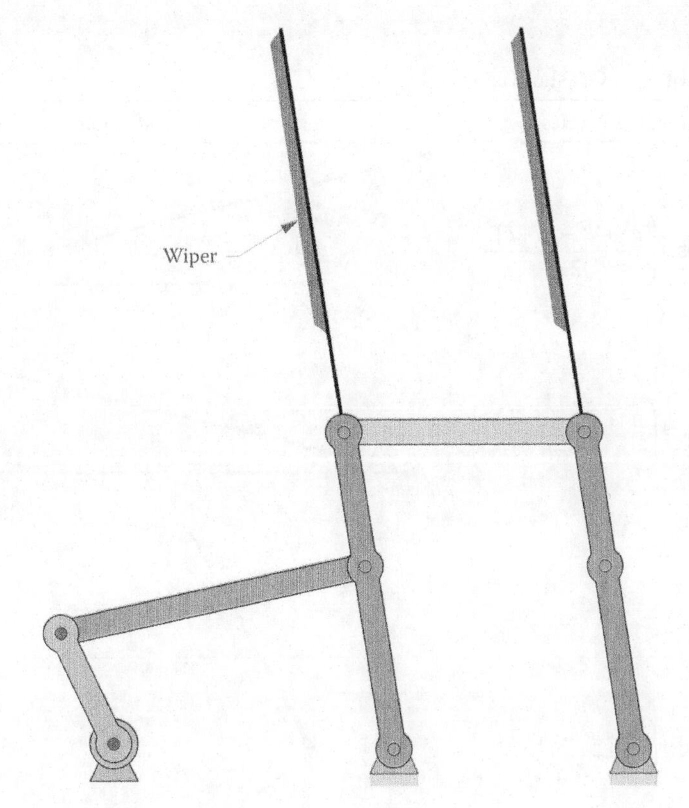

FIGURE 1.30
A set of windshield wipers is a common application of the parallelogram linkage.

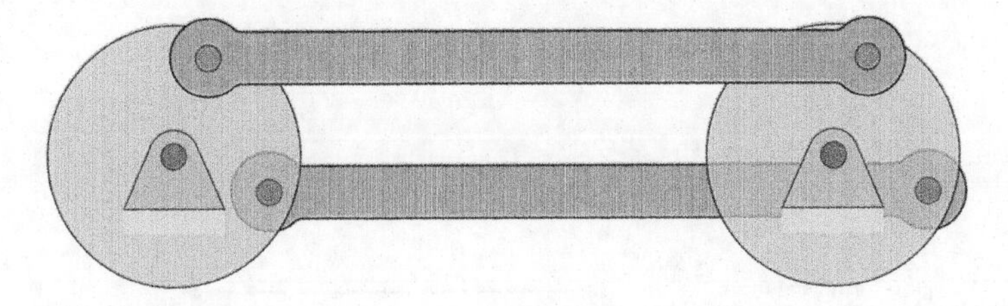

FIGURE 1.31
Steam locomotives used a parallelogram linkage to transmit power to multiple wheels.

on opposite sides of the locomotive. In this way, at least one linkage was non-collinear at all times.

1.4.6 Limiting angles for Non-Grashof Linkages

We have seen that the configuration of the linkage is undefined if the driving link is outside its limiting position for the non-Grashof linkage. Table 1.6 shows the limiting angles for all classes of non-Grashof linkages. As you can see, the limiting angle is prescribed on the *driving* link for each class, and the limit is reached when either the crank or rocker becomes collinear with the coupler.

TABLE 1.6

Limiting Angles of Non-Grashof Linkages

Class	Limiting Angles	Illustration
5	$\theta_2 = \pm\cos^{-1}\left(\dfrac{a^2+d^2-(b+c)^2}{2ad}\right)$	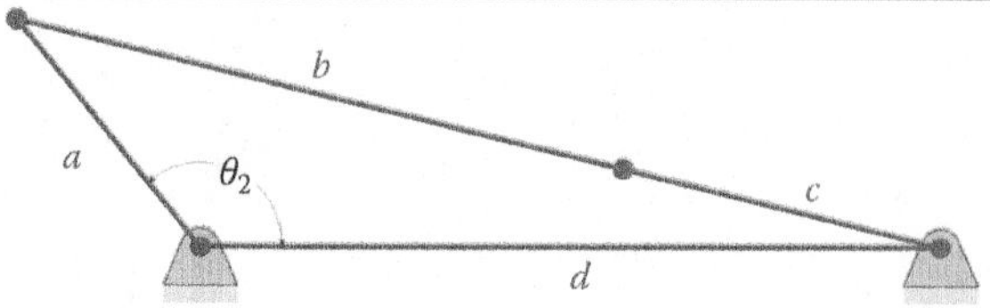
6	$\theta_4 = \pm\cos^{-1}\left(\dfrac{c^2+d^2-(a+b)^2}{2cd}\right)$	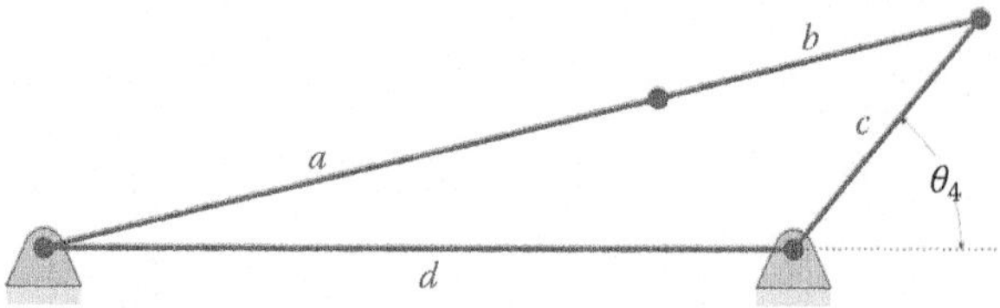
7	$\theta_4 = \pm\cos^{-1}\left(\dfrac{c^2+d^2-(a-b)^2}{2cd}\right)$	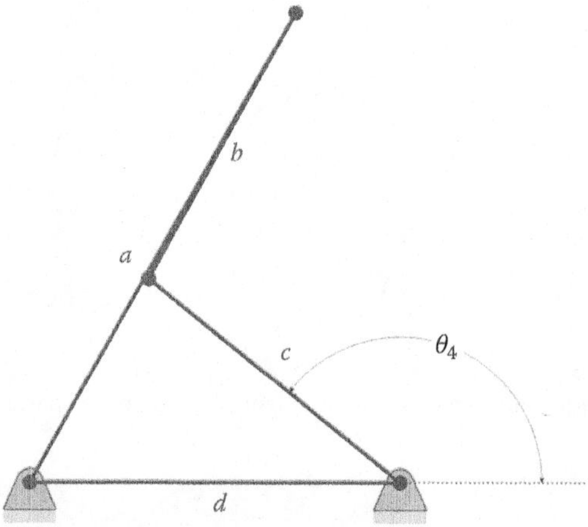
8	$\theta_2 = \pm\cos^{-1}\left(\dfrac{a^2+d^2-(b-c)^2}{2ad}\right)$	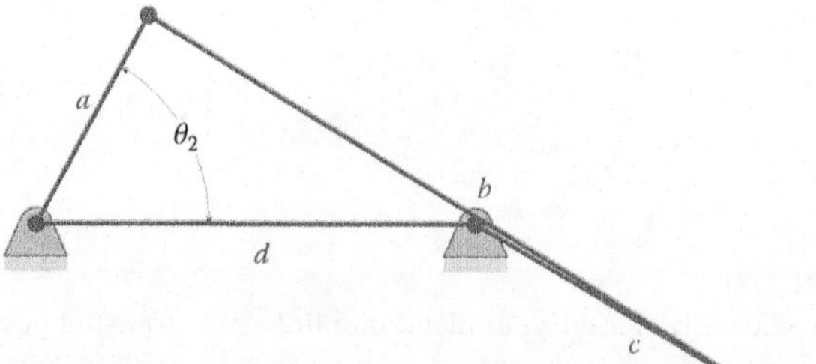
9	$\theta_4 = \pm\cos^{-1}\left(\dfrac{c^2+d^2-(b-a)^2}{2cd}\right)$	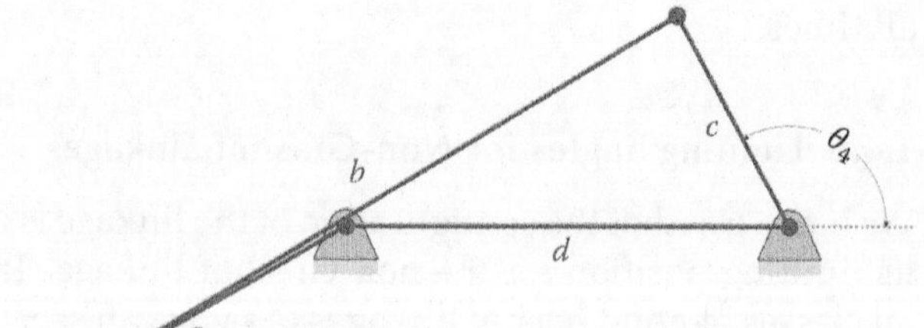

(Continued)

TABLE 1.6 (Continued)

Limiting Angles of Non-Grashof Linkages

Class	Limiting Angles	Illustration
10	$\theta_2 = \pm\cos^{-1}\left(\frac{a^2+d^2-(c-b)^2}{2ad}\right)$	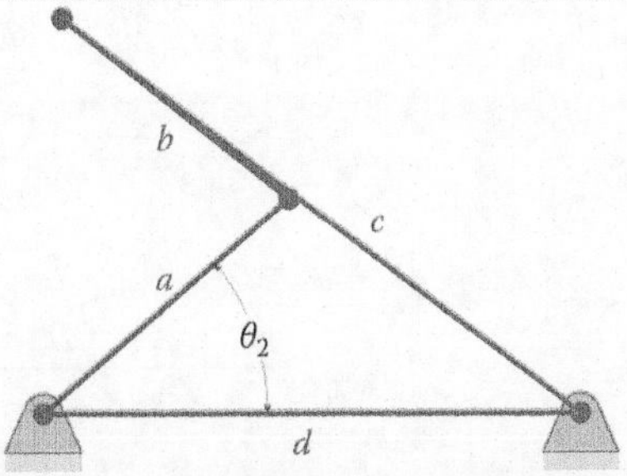

1.5 Practice Problems

Problem 1.1

What is the definition of a rigid body? Describe types of motion that a rigid body can experience.

Problem 1.2

Define the term *Degrees of Freedom*. What does it mean if a linkage has a negative *Degree of Freedom*?

Problem 1.3

How many degrees of freedom do the following joints on your body permit?

a. Your knee
b. Your ankle
c. Your shoulder
d. Your hip
e. A knuckle on one of your fingers

Problem 1.4

Figure 1.32 shows a simple roller bearing. How many DOF does the roller bearing have if the outer race is fixed to ground? Sketch the bearing, and indicate an appropriate set of coordinates that completely specify the configuration of the bearing.

Problem 1.5

How many degrees of freedom does the linkage in the Figure 1.33 have? Is it a mechanism, a structure, or a preloaded structure?

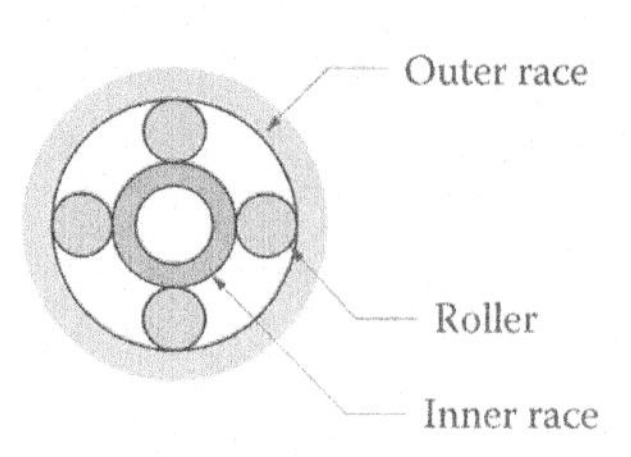

FIGURE 1.32
Problem 1.4.

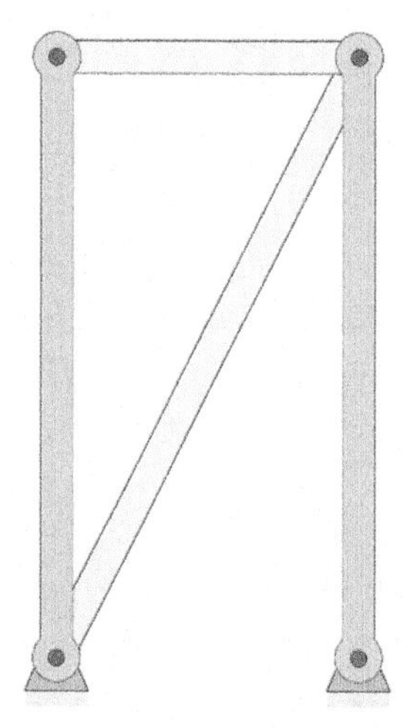

FIGURE 1.33
Problem 1.5.

Problem 1.6

How many degrees of freedom does the scissor-lift mechanism in Figure 1.34 have?

Problem 1.7

How many degrees of freedom does the cam-follower linkage in Figure 1.35 have? Is it a mechanism, a structure, or a preloaded structure? Sketch the figure and indicate an appropriate set of coordinates to completely specify the configuration of the linkage.

Problem 1.8

How many degrees of freedom does the linkage in Figure 1.36 have? Is it a mechanism, a structure, or a preloaded structure? Sketch the figure and indicate an appropriate set of coordinates to completely specify the configuration of the linkage.

Problem 1.9

How many degrees of freedom does the linkage in Figure 1.37 have? Is it a mechanism, a structure, or a preloaded structure? Sketch the figure and indicate an appropriate set of coordinates to completely specify the configuration of the linkage.

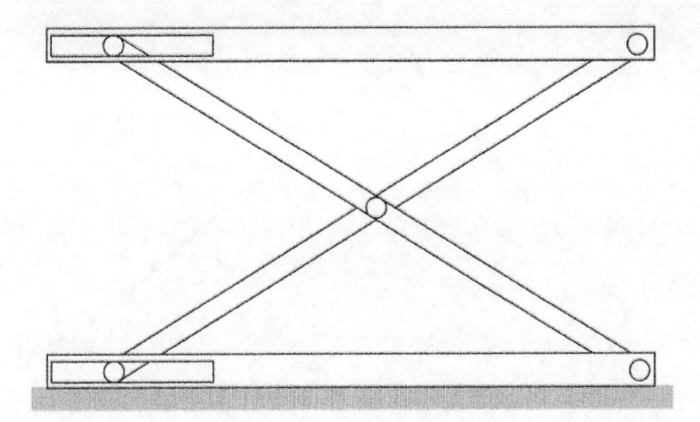

FIGURE 1.34
Problem 1.6.

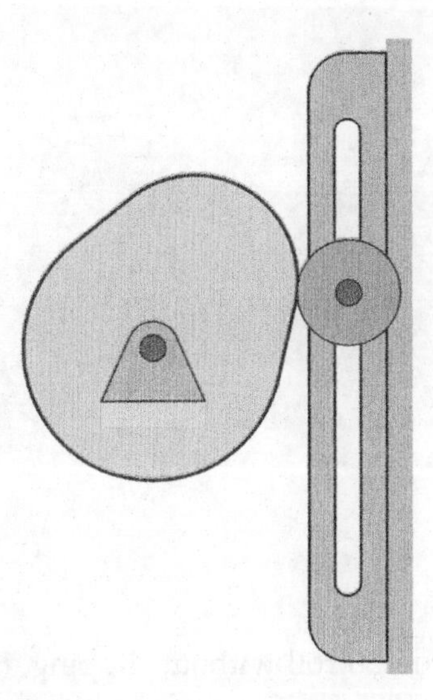

FIGURE 1.35
Problem 1.7.

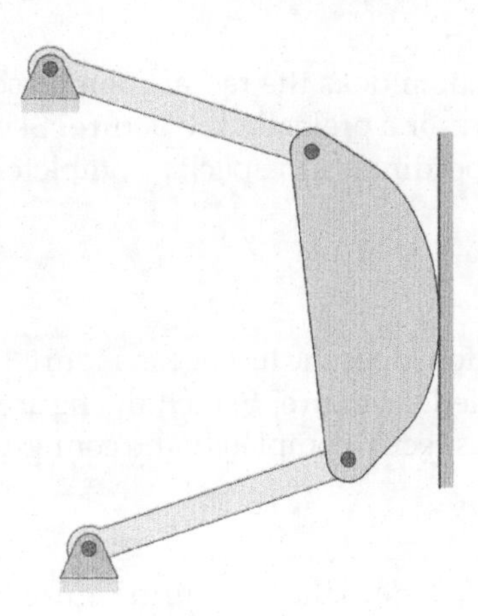

FIGURE 1.36
Problem 1.8.

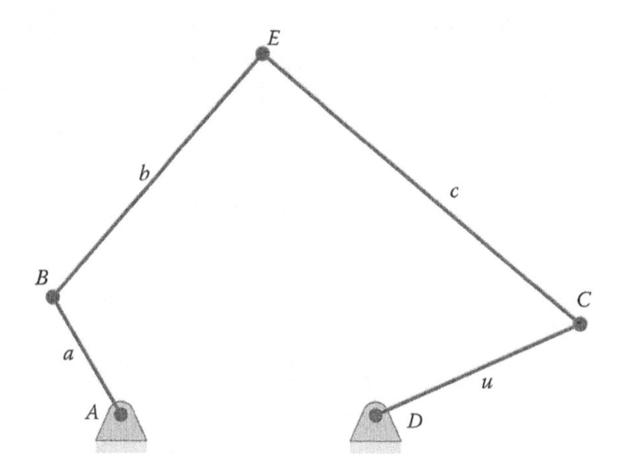

FIGURE 1.37
Problem 1.9.

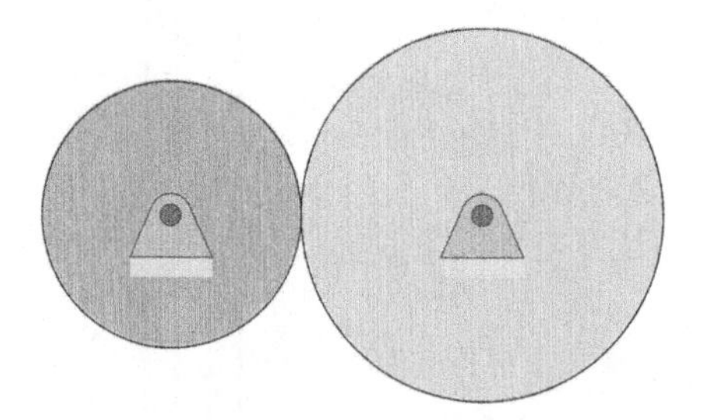

FIGURE 1.38
Problem 1.10.

Problem 1.10

The two cylinders in Figure 1.38 roll without slipping. How many degrees of freedom does the system have?

Problem 1.11

How many degrees of freedom does the radial compressor in Figure 1.39 have? Is it a mechanism, a structure, or a preloaded structure? Sketch the figure and indicate an appropriate set of coordinates to specify completely the configuration of the linkage.

Problem 1.12

How many degrees of freedom does the linkage in Figure 1.40 have? Is it a mechanism, a structure, or a preloaded structure? Sketch the figure and indicate an appropriate set of coordinates to specify completely the configuration of the linkage.

Problem 1.13

How many degrees of freedom does the linkage in Figure 1.41 have? Is it a mechanism, a structure, or a preloaded structure? Sketch the figure and indicate an appropriate set of coordinates to completely specify the configuration of the linkage.

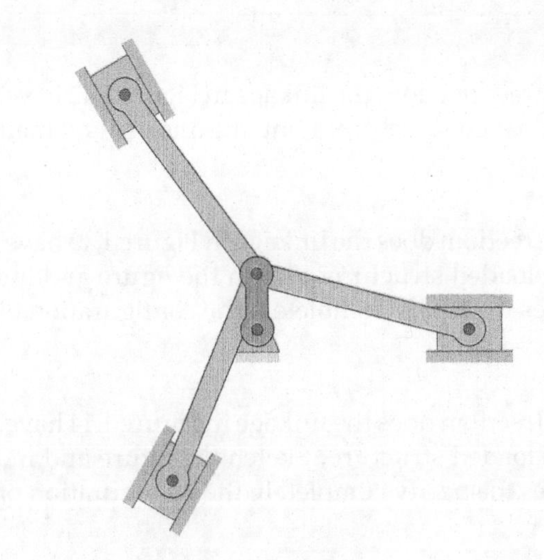

FIGURE 1.39
Problem 1.11.

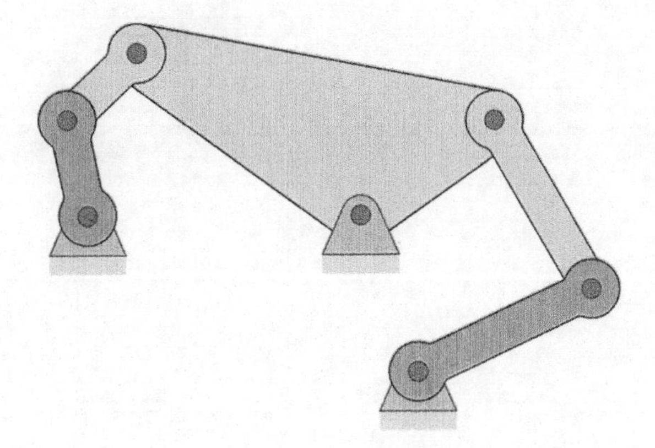

FIGURE 1.40
Problem 1.12.

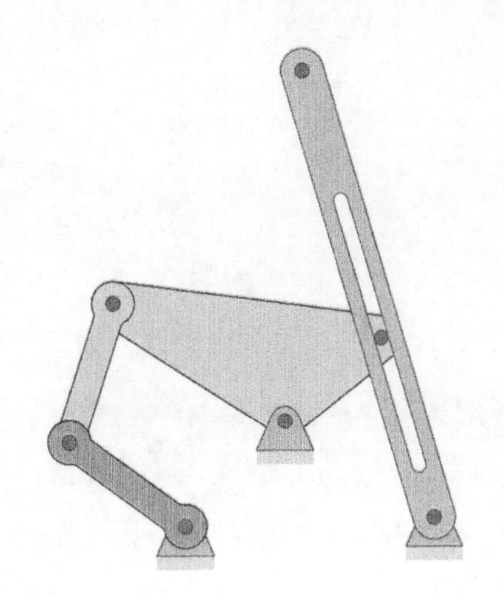

FIGURE 1.41
Problem 1.13.

Problem 1.14

How many degrees of freedom does the linkage in Figure 1.42 have? Is it a mechanism, a structure, or a preloaded structure? Hint: the answer is something of a paradox!

Problem 1.15

How many degrees of freedom does the linkage in Figure 1.43 have? Is it a mechanism, a structure, or a preloaded structure? Sketch the figure and indicate an appropriate set of coordinates to specify completely the configuration of the linkage.

Problem 1.16

How many degrees of freedom does the linkage in Figure 1.44 have? Is it a mechanism, a structure, or a preloaded structure? Sketch the figure and indicate an appropriate set of coordinates to specify completely the configuration of the linkage.

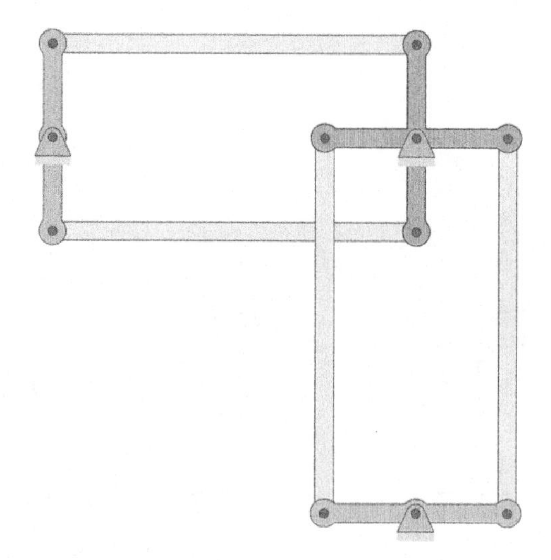

FIGURE 1.42
Problem 1.14.

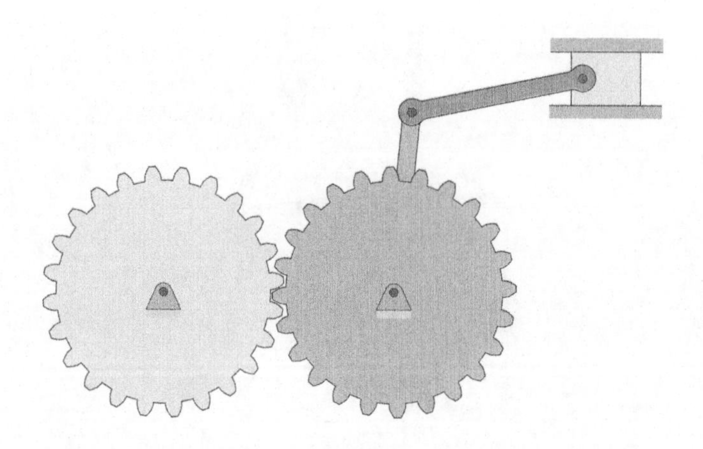

FIGURE 1.43
Problem 1.15.

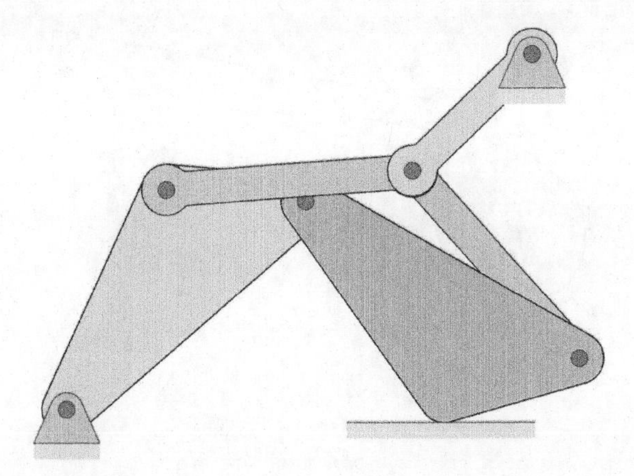

FIGURE 1.44
Problem 1.16.

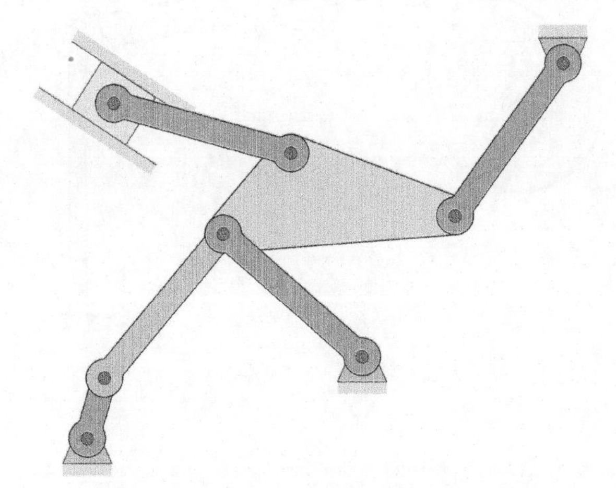

FIGURE 1.45
Problem 1.17.

Problem 1.17

How many degrees of freedom does the linkage in Figure 1.45 have? Is it a mechanism, a structure, or a preloaded structure? Sketch the figure and indicate an appropriate set of coordinates to specify completely the configuration of the linkage.

Problem 1.18

How many degrees of freedom does the linkage in Figure 1.46 have? Is it a mechanism, a structure, or a preloaded structure? Sketch the figure and indicate an appropriate set of coordinates to specify completely the configuration of the linkage.

Problem 1.19

How many degrees of freedom does the linkage in Figure 1.47 have? Is it a mechanism, a structure, or a preloaded structure? Sketch the figure and indicate an appropriate set of coordinates to specify completely the configuration of the linkage.

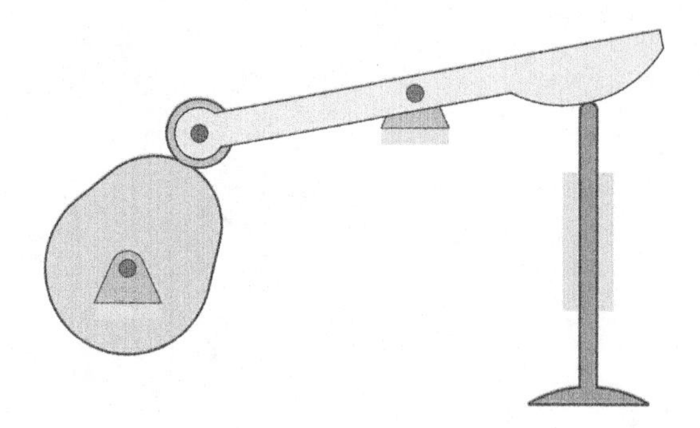

FIGURE 1.46
Problem 1.18.

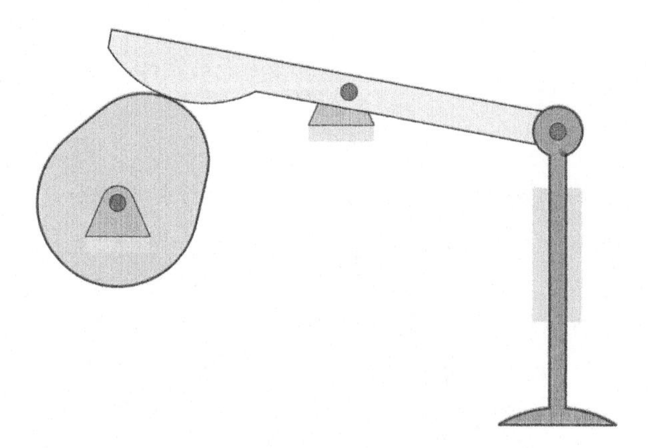

FIGURE 1.47
Problem 1.19.

Problem 1.20

How many degrees of freedom does the linkage in Figure 1.48 have? Is it a mechanism, a structure, or a preloaded structure? Sketch the figure and indicate an appropriate set of coordinates to specify completely the configuration of the linkage.

Problem 1.21

How many degrees of freedom does the linkage in Figure 1.49 have? Is it a mechanism, a structure, or a preloaded structure? Sketch the figure and indicate an appropriate coordinate that completely specifies the configuration of the linkage

Problem 1.22

Determine if the linkage in Figure 1.50 meets the Grashof condition. If it does, which link can make a full revolution?

Problem 1.23

Determine if the linkage shown in Figure 1.51 meets the Grashof condition.

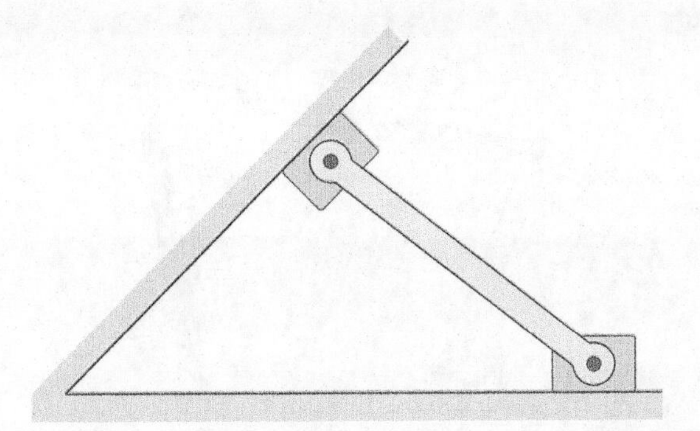

FIGURE 1.48
Problem 1.20.

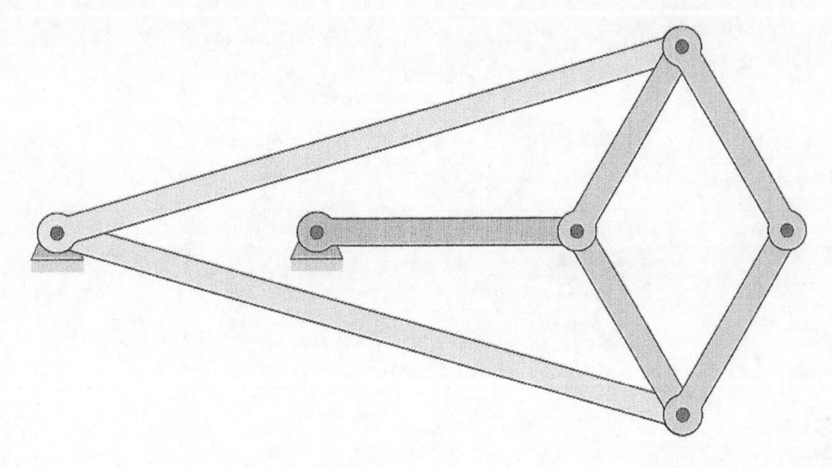

FIGURE 1.49
Problem 1.21.

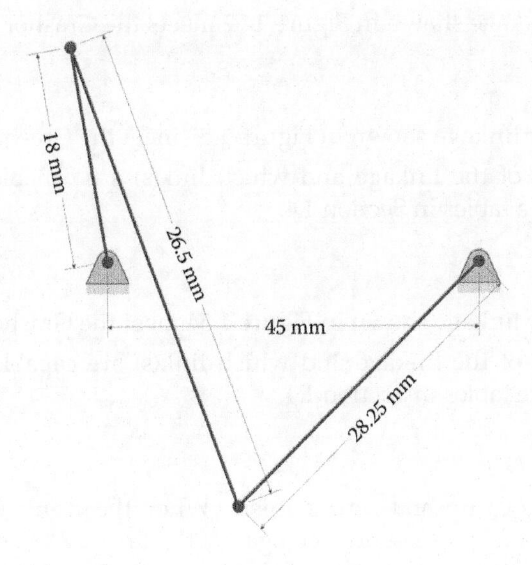

FIGURE 1.50
Problem 1.22.

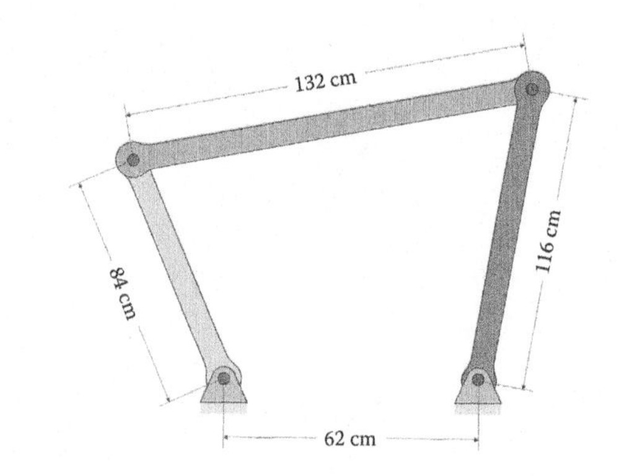

FIGURE 1.51
Problem 1.23.

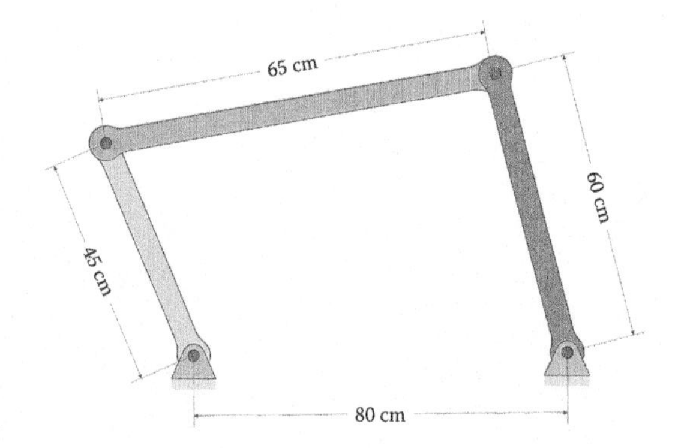

FIGURE 1.52
Problem 1.24.

Problem 1.24

Determine if the linkage shown in Figure 1.52 meets the Grashof condition.

Problem 1.25

a. Does the fourbar linkage shown in Figure 1.53 meet the Grashof condition?
b. Determine class of the linkage and which link(s) are capable of making a full rotation using the tables in Section 1.4.

Problem 1.26

a. Does the fourbar linkage shown in Figure 1.54 meet the Grashof condition?
b. Determine class of the linkage and which link(s) are capable of making a full rotation using the tables in Section 1.4.

Problem 1.27

Figure 1.55 shows a drum brake mechanism. When the crank AB rotates counterclockwise, the brake pads are pressed against the drum. The length AB is 100 mm, the length BC is 60 mm, then length CD is 156 mm and the distance between A and D is 110 mm. Is the linkage Grashof? Does it matter for this mechanism?

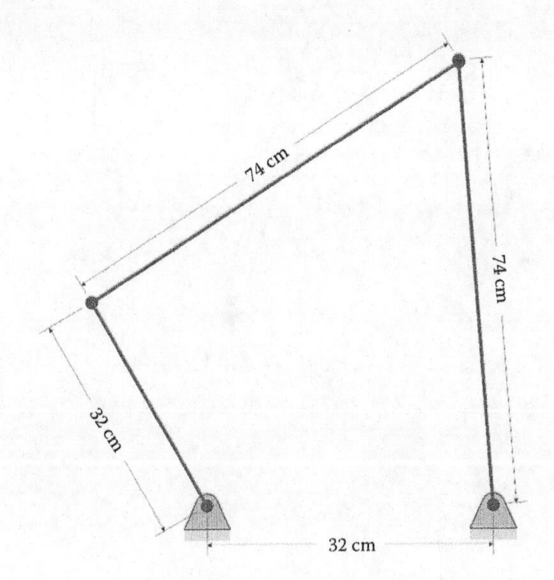

FIGURE 1.53
Problem 1.25.

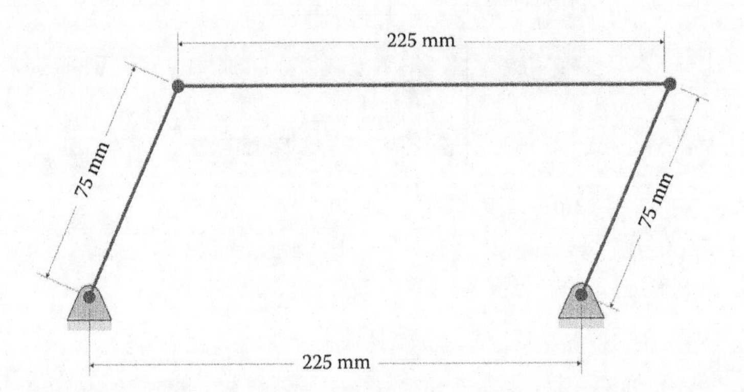

FIGURE 1.54
Problem 1.26.

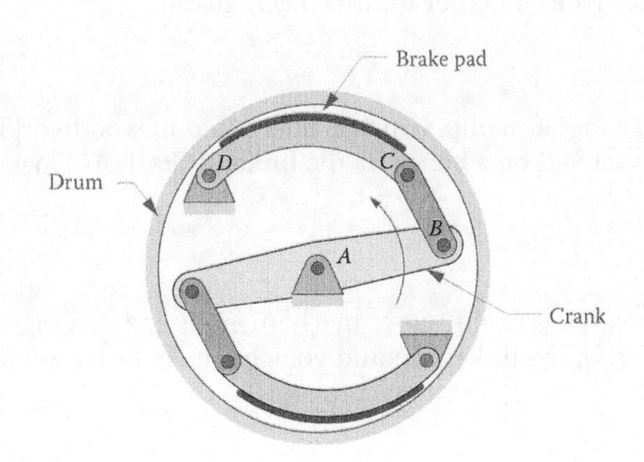

FIGURE 1.55
Problem 1.27.

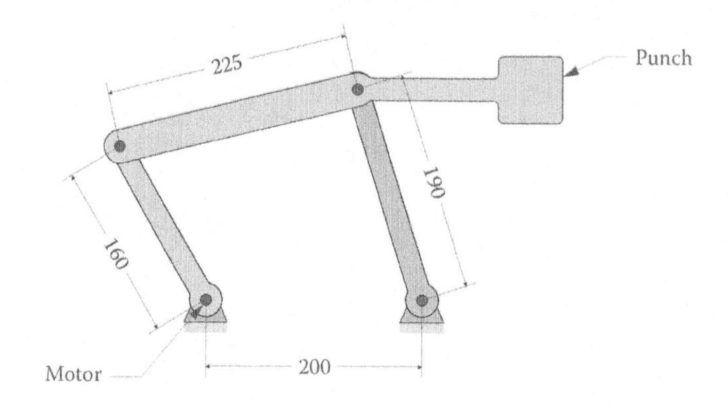

FIGURE 1.56
Problem 1.28.

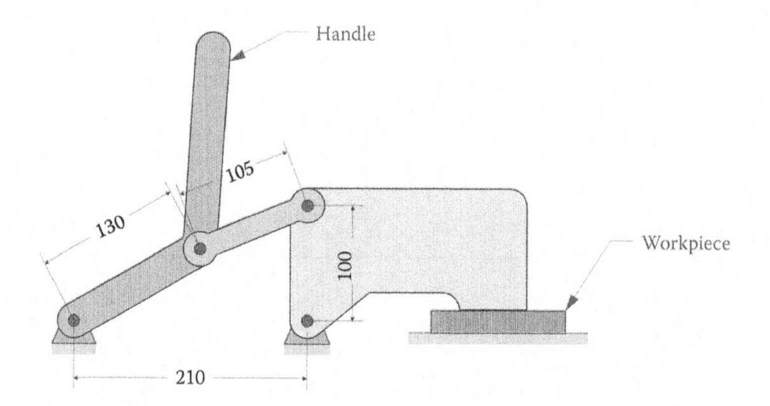

FIGURE 1.57
Problem 1.29.

Problem 1.28

The linkage in Figure 1.56 is used as part of a continuous stamping operation. It is intended that an AC motor drive the crank at a constant angular velocity. Is the linkage Grashof? Does it matter for this mechanism?

Problem 1.29

Figure 1.57 shows a toggle clamp, which is often used in woodworking operations to hold a workpiece fixed on a bench. Is the linkage Grashof? Does it matter for this mechanism?

Problem 1.30

Figure 1.58 shows a windshield wiper mechanism. Is the driving linkage Grashof? If not, which part of the linkage would you change in order to make it Grashof?

Problem 1.31

Figure 1.59 shows a trailing arm suspension design intended for use on an off-road vehicle. A third control arm has been added in order to regulate the "toe" of the

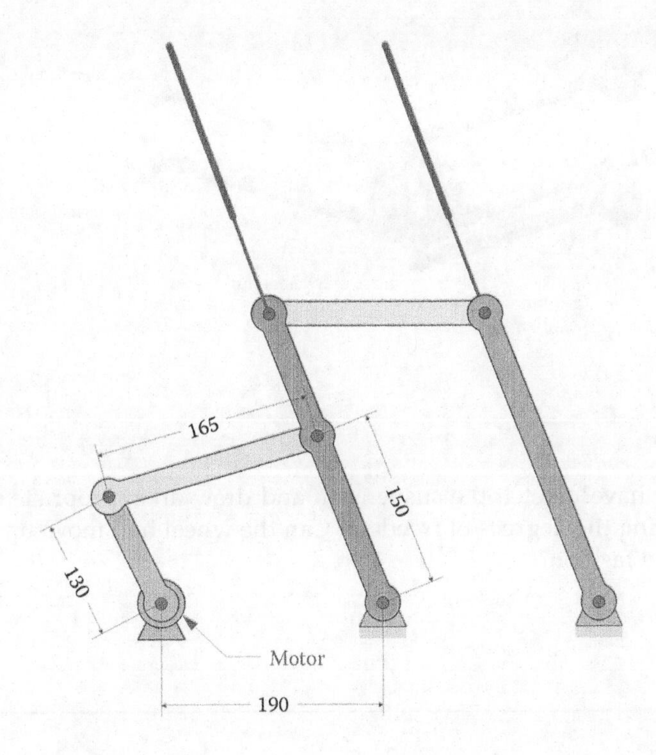

FIGURE 1.58
Problem 1.30.

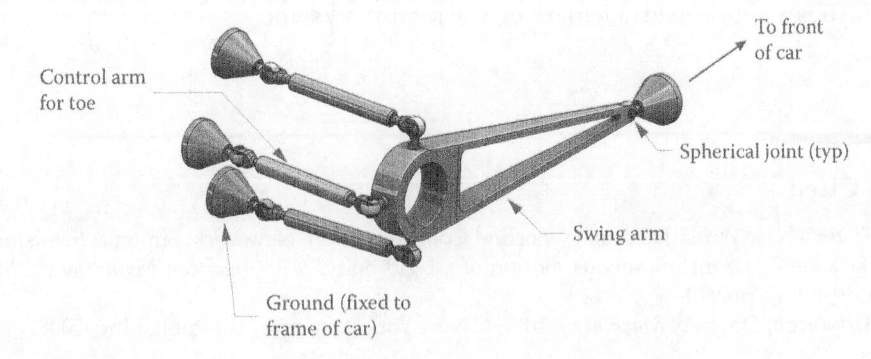

FIGURE 1.59
Problem 1.31.

wheel, which is the angle between the wheel centerline and the centerline of the vehicle. How many degrees of freedom does this suspension have? Can the wheel move up and down in the required fashion? Sketch the mechanism and draw a set of appropriate coordinates showing the available mobility.

Problem 1.32

Figure 1.60 shows a five-link suspension that is sometimes used in off-road vehicle design. Each angle of the wheel hub can be precisely tuned, although tuning is a finicky and time-consuming affair. How many degrees of freedom does this

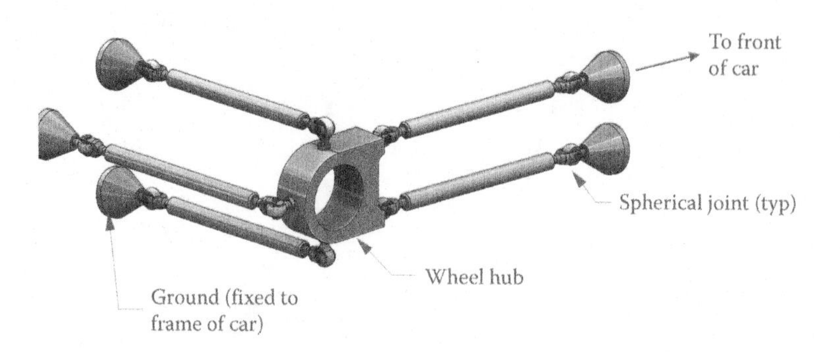

FIGURE 1.60
Problem 1.32.

suspension have? Sketch the suspension and draw an appropriate set of coordinates defining the degrees of freedom. Can the wheel hub move up and down in the required fashion?

Acknowledgments

SOLIDWORKS is a registered trademark of Dassault Systèmes SolidWorks Corporation.
MATLAB is a registered trademark of The MathWorks, Inc.

Works Cited

1. Webster's New World Dictionary, Second College Edition, New York: Simon & Schuster, 1982.
2. D. Jackson, "The instantaneous motion of a rigid body," *The American Mathematical Monthly*, vol. 49, no. 10, pp. 661–667, 1942.
3. R. L. Norton, *Design of Machinery*, 4th ed., New York: McGraw-Hill Publishing, 2008.

2

Graphical Linkage Synthesis Using SOLIDWORKS®

2.1 Introduction to Graphical Linkage Synthesis

Now that we have learned to classify the various types of fourbar linkage, we will turn our attention to linkage *design*, to accomplish a specified motion. Designing some types of linkages (e.g. the threebar and the slider-crank) is quite simple and will not be discussed here. Creating an appropriate fourbar linkage is more challenging, and that will be the focus of this chapter. The general problem statement is as follows: given a set of specified positions, design a fourbar linkage such that a portion of one of the links passes *exactly* through these positions as the crank makes a revolution. An example of such a scenario is shown in Figure 2.1. For this problem, we would need to find the lengths of each link such that the line *AB* passes through positions 1, 2, and 3, in order. In most cases, the line *AB* is attached to the coupler, since the coupler is capable of complex motion. But if the desired motion is a pure rotation then we may decide to attach line *AB* to the rocker instead.

It is possible to solve this problem using algebra and MATLAB®, but it is much simpler, and more intuitive, to use a CAD package such as SOLIDWORKS® for the design process. This is the technique that we will employ in this chapter. We will use the 2D drafting capabilities of SOLIDWORKS to create *layout sketches* for the linkages we design. Once the dimensions of each link have been finalized, we can revert to ordinary 3D environment in SOLIDWORKS to create each link as a separate part. Finally, we will mate each link together into an Assembly to check our work. In the old days, these problems would have been tackled using a straightedge, compass, and scale, but using SOLIDWORKS enables us to quickly and painlessly try several designs before settling on a final linkage. You can still use pencil and paper to employ the techniques in this chapter, but (with practice) you will be much faster and more efficient using CAD.

Our first example will be quite simple: design a fourbar linkage to sweep the rocker through a specified angle. We will gradually add complexity to our designs until we are able to design a fourbar to move the coupler through three specified positions. As we will see, it is not possible to meet more than three specified positions *exactly*, since it takes only three points to uniquely define a circle.

2.2 Two Specified Positions of the Rocker

We begin with the simplest possible case: designing a fourbar linkage such that the rocker sweeps out a specified angle, as shown in Figure 2.2. An example of this linkage in practical

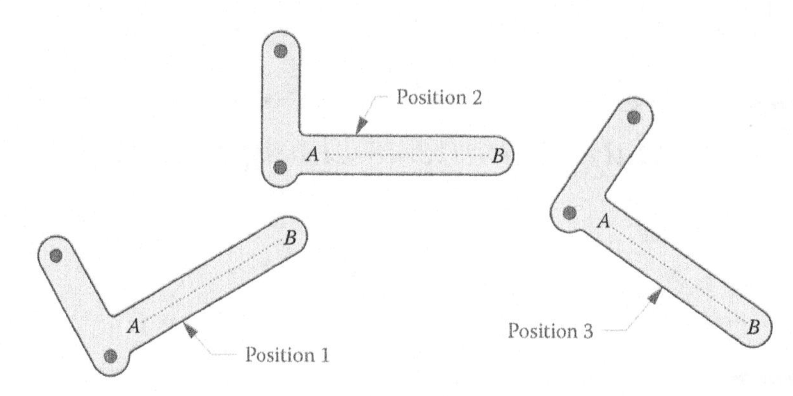

FIGURE 2.1
We wish to design a fourbar linkage to move the link through the specified positions.

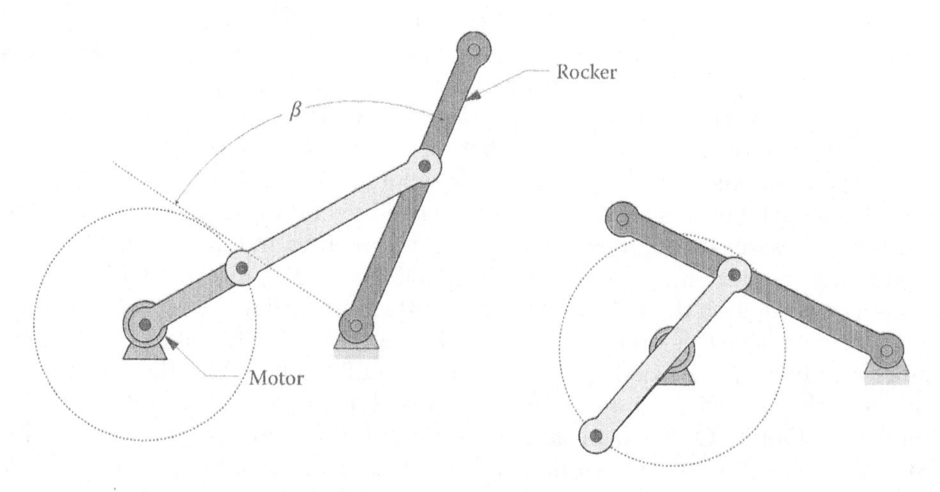

FIGURE 2.2
The rocker must sweep through the angle β as the crank makes a revolution.

use is a windshield wiper mechanism, where an inexpensive DC motor is attached to the crank. As the crank makes continuous revolutions, the windshield wiper attached to the end of the rocker sweeps back and forth across the windshield. Of course, this linkage must be Grashof to function properly, since the crank must be capable of making a full revolution.

Example 2.1: Sweep the Rocker through 80°

To begin the design process, open a new drawing in SOLIDWORKS. In this example, we will design a linkage that sweeps a 50 mm rocker through 80°. Create the drawing shown in Figure 2.3. The two positions of the rocker are shown as solid lines, and a *construction line* is used to define the ground. We have arbitrarily chosen an angle of 60° from the ground for the second position of the rocker, but any angle would work. The dots at the endpoints of the lines have been added to the figure to make them stand out; your drawing will not contain these dots (or the point labels C_1 and C_2). Add a *fixed* relation to the ground pin of the rocker to make the drawing fully defined. You should not proceed to the next step until your drawing is fully defined.

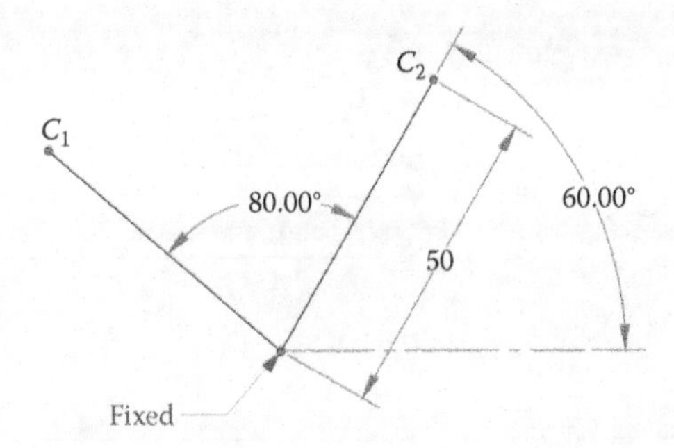

FIGURE 2.3
The rocker sweeps out an angle of 80°. Adding a fixed relation to the ground pivot of the rocker will make the drawing fully defined.

Our drawings will quickly become cluttered and hard to read when we begin adding the necessary construction lines, circles, etc. To keep the drawings neat, we will use a little-known toolbar in SOLIDWORKS: the *Layer* toolbar. Select the *Toolbars* from the *View* menu and click on *Layer*. The *Layer* toolbar enables us to place any set of objects in the drawing into its own, separate layer. We can turn the visibility of any of the layers on or off to hide objects that are not being used at the moment. Hiding a set of objects does not delete them, it merely makes them invisible for the moment. We can always show the objects later, if needed.

Once the Layer toolbar is visible, click on the icon that looks like a stack of folders to open the Layer dialog box. This is where you can set the properties of each layer, including color, line style, and visibility. Add the following five layers defined in Table 2.1 so that the dialog box looks like Figure 2.4.

Click OK to close the Layers dialog box once you have finished. There are two different methods for adding an object to a layer. In the first, you click on an object and select the desired layer from the Layers toolbar. We will need to do this for the objects we have already placed in the drawing (the lines and dimensions). In the second method, you select a layer from the Layers toolbar and begin drawing. All objects that you draw when a layer has been selected will be placed in that layer. If you find yourself drawing objects that disappear from view as soon as you finish them, it is likely that they are being placed on a hidden layer.

Select the two lines that define the positions of the rocker and place them in the Links layer. Select the dimensions and the horizontal construction line and place them in the HiddenDims layer. Once the two positions of the rocker have been defined, there is no need to have the angles cluttering up the drawing. In the Layers dialog box, click on the eyeball icon for the HiddenDims layer to make it hidden. Your drawing should appear to contain only the two lines showing the rocker positions.

Select the Construction layer from the Layers toolbar and draw a line connecting C_1 and C_2 as shown in Figure 2.5. Draw a circle centered at the midpoint of the construction

TABLE 2.1
Layers to Define in SOLIDWORKS Drawing

Name	Color	Line Style	Thickness (mm)
Links	Black	Solid	0.18
Dims	Blue	Solid	0.18
Construction	Gray	Dotted	0.18
NewLinks	Blue	Dashed	0.35
HiddenDims	Gray	Solid	0.18

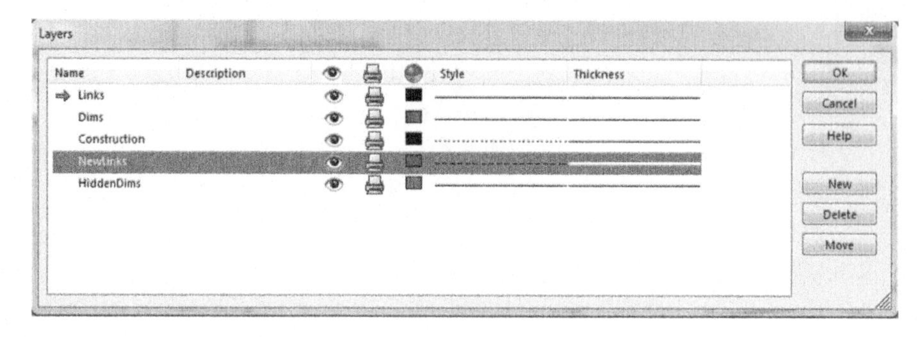

FIGURE 2.4
The Layers dialog box allows you to set the properties of each layer. Screenshot of SOLIDWORKS software.

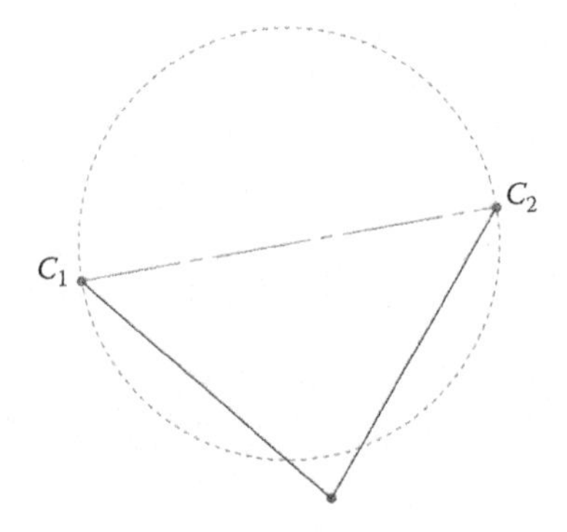

FIGURE 2.5
The circle is coincident with the rocker endpoints and centered at the midpoint of the construction line.

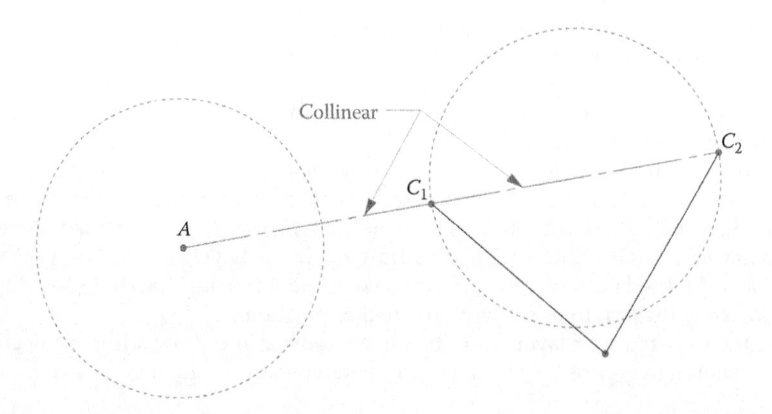

FIGURE 2.6
Draw a line collinear with the first construction line and place a circle equal to the first at the endpoint of the line.

line and coincident with C_2. Your drawing should still be fully defined. If it is not, make sure that the center of the circle has a Midpoint relation with the construction line and that the circle itself has a Coincident relation with C_2.

Next, draw a construction line starting at C_1 and going to the left. Add a Collinear relation to make both construction lines collinear, as shown in Figure 2.6. Create a new circle that is equal to the first at point A. After making sure that the new construction line is collinear with the first and that both circles have an Equal relation, your sketch should still be Under Defined. Can you see what is missing? Although both circles lie on the same line, we have not yet specified the distance between their centers.

Use Smart Dimension to make the distance between the centers of the two circles 90 mm, as shown in Figure 2.7. Your drawing should now be fully defined. Believe it or not, we have solved the problem! The point A is the ground pivot for the crank and the circle on the left traces out the path taken by the crank as it makes its revolution.

Change to the NewLinks layer and draw a line from A to B_2, and then from B_2 to C_2. The line AB_2 is the crank and B_2C_2 is the coupler. Use the Smart Dimension tool to find the lengths of each of these links (plus the ground link) as shown in Figure 2.8. Each time you add a dimension you will get a warning that the new dimensions are "driven." Since we are adding dimensions to a fully defined drawing, these dimensions cannot be changed, and are "driven" by the pre-existing dimensions in the drawing. In other words, when you specified the length of the rocker (50 mm) and the angle of sweep (80°) the crank length became fixed. The only free choice we had was the length of the coupler (90 mm) that specified the distance between the two circles. The crank is then 32.139 mm, the coupler is 90 mm, the rocker is 50 mm, and the distance between ground pins is 97.811 mm.

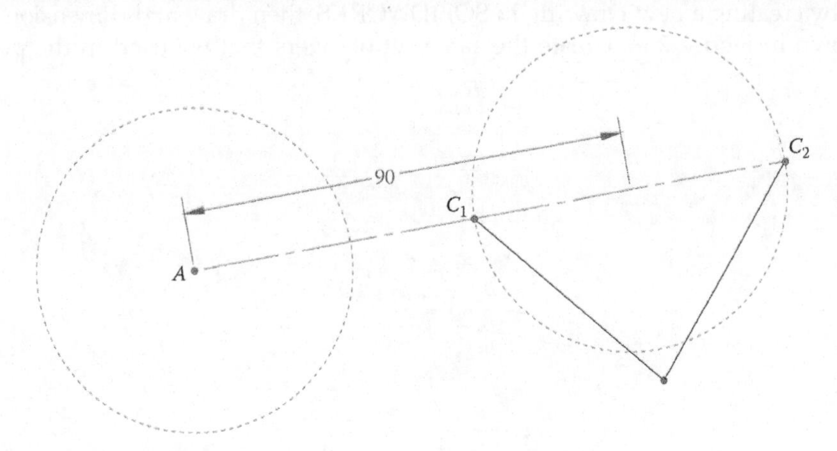

FIGURE 2.7
The distance between the centers of the two circles is 90 mm.

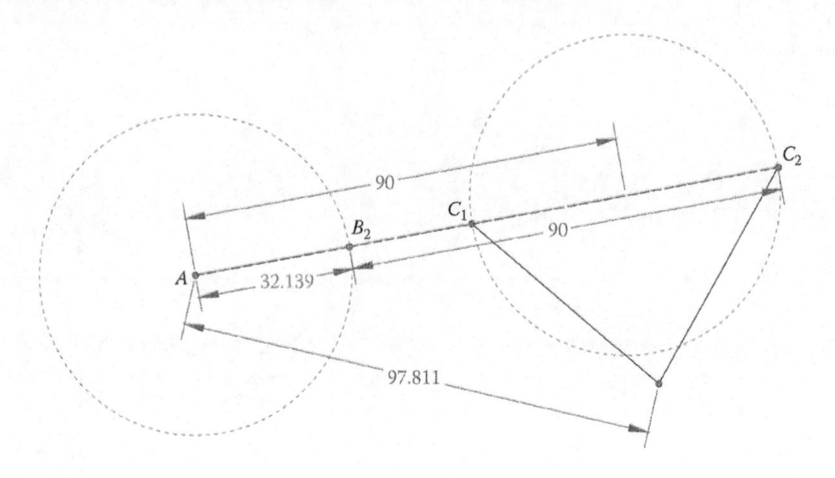

FIGURE 2.8
Dimensioning the links for the crank-rocker linkage. Each of the new dimensions is driven, since the drawing was fully defined before we added them.

Figures 2.9 and 2.10 show the linkage in its two specified positions. As a next step, we would use these dimensions to design a SOLIDWORKS part for each link, and place them together into an assembly. This very basic Grashof linkage is often added to a non-Grashof linkage to drive it between its two limiting angles. When used to drive a non-Grashof linkage it is referred to as a driver dyad. An example of how this may be added is shown in Figure 2.11 and will be used in later sections.

2.2.1 Two Positions of Rocker without Specified Ground Pin

In the preceding example we were given the desired angle of sweep of the rocker, which gave an implicit location for the ground pin at D. For the next example, we specify two positions of a *line on the rocker*, without specifying a ground pin – see Figure 2.12. Our first task will be to find the position of the ground pin at D such that the line on the rocker can reach the specified positions. Once we have the location of the ground pin, we can use the techniques in the preceding example to design the rest of the linkage (i.e., to find the crank and coupler lengths, and the distance between ground pins).

Begin by creating a new Drawing in SOLIDWORKS, then draw and dimension the two lines shown in Figure 2.13. Create the same set of layers that we used in the preceding

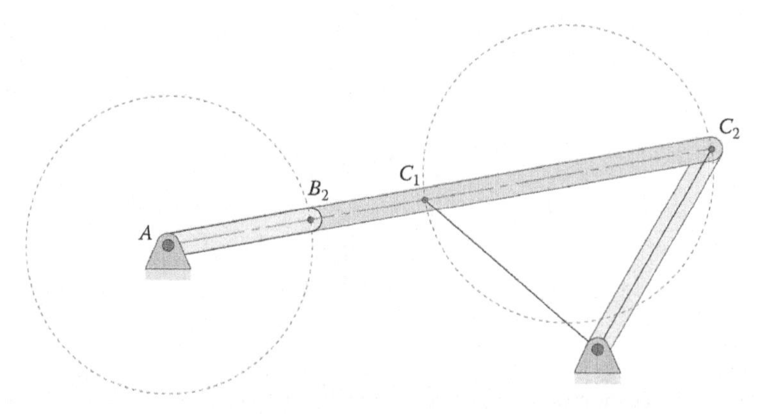

FIGURE 2.9
Position 1 for crank-rocker linkage.

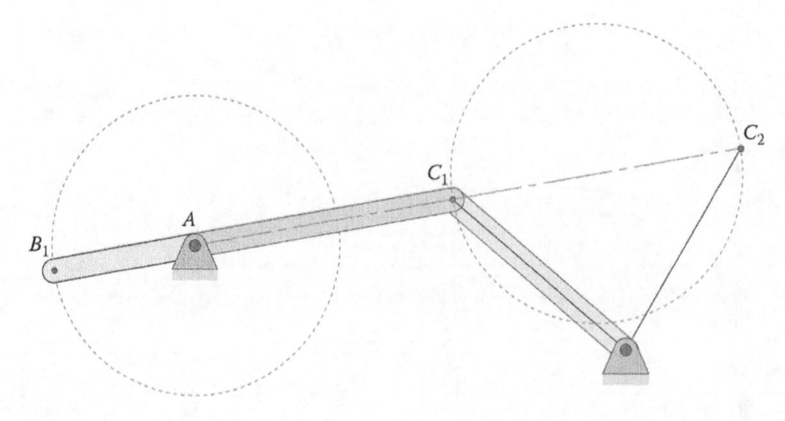

FIGURE 2.10
Position 2 for crank-rocker linkage.

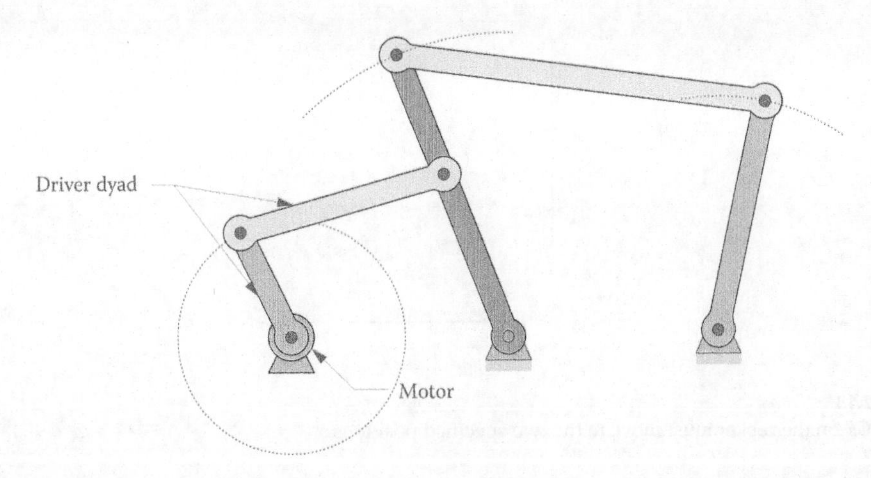

FIGURE 2.11
Simple two position rocker used as driver dyad for non-Grashof linkage.

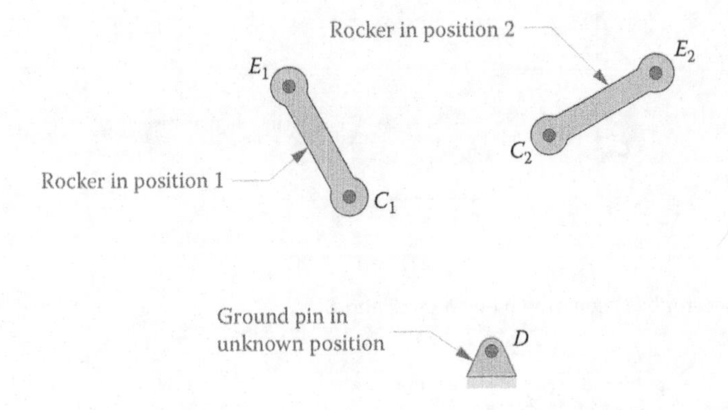

FIGURE 2.12
In this example, we are given two positions of a line on the rocker, but the ground pin location is unspecified.

example. The top line has a vertical relation added and the bottom line has a horizontal relation added. Of course, the two specified positions of the rocker need not be horizontal and vertical; we have done this for the sake of creating a simple example. The length of the vertical line is given an Equal relation with the horizontal line. Place a Fixed relation at point C_1 to fully define the drawing.

Now draw a line in the Construction layer between points E_1 and E_2. Create a *perpendicular bisector* on this line as shown in Figure 2.14. A perpendicular bisector starts at the midpoint of a given line (in this case the line E_1E_2) and is directed perpendicular to that line. The simplest way to do this in SOLIDWORKS is to hover over the center of the line E_1E_2 until the midpoint icon appears, then drag along the yellow perpendicular guide. The resulting line will be a perpendicular bisector of E_1E_2. Don't worry about the length of the bisector for now – we'll fix it in the next step.

Next, draw another line in the Construction layer between points C_1 and C_2. Create a perpendicular bisector for this line and extend it until it reaches the first perpendicular bisector. Use the Trim tool to make the two bisectors end at the same point, *D*, as shown in Figure 2.15. Your drawing should now be fully defined.

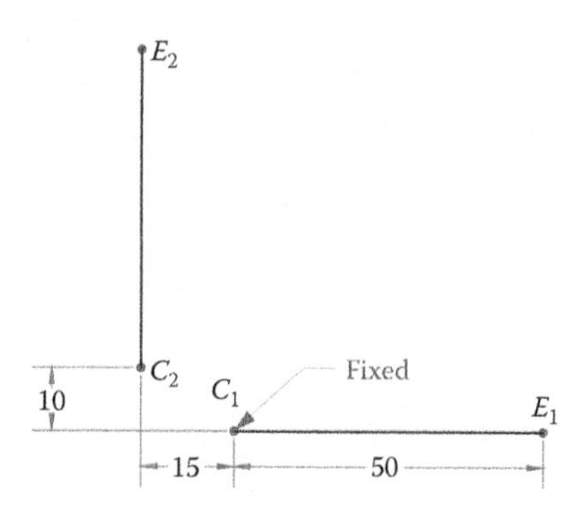

FIGURE 2.13
The line *CE* on the rocker must move to the two specified positions.

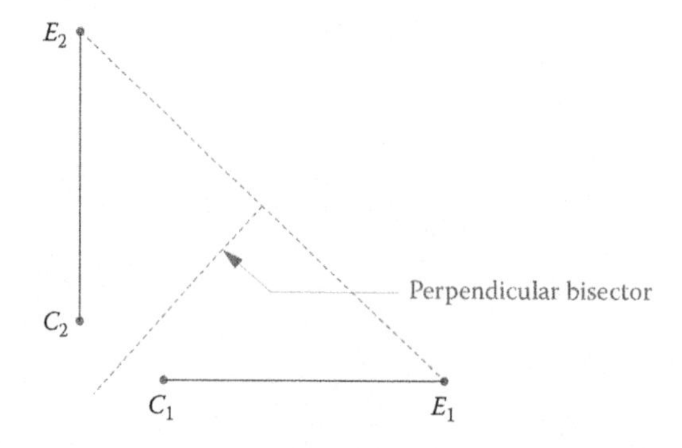

FIGURE 2.14
A perpendicular bisector has been drawn between E_1 and E_2.

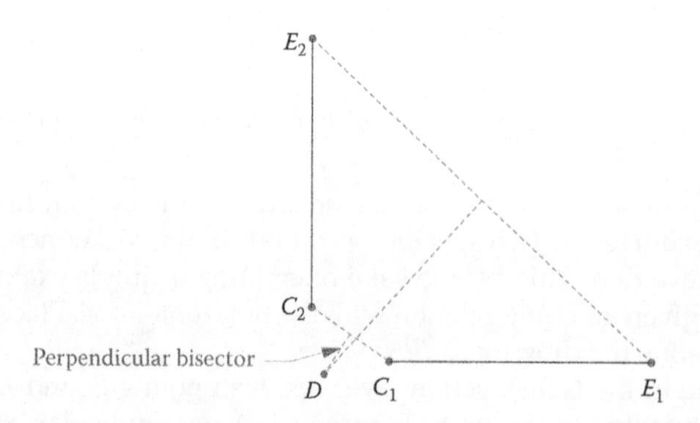

FIGURE 2.15
Another perpendicular bisector has been drawn for the line C_1C_2.

The point D is the missing ground pin! To see why, note that the distance from D to C_1 is the same as the distance from D to C_2, because D is on the perpendicular bisector of the line C_1C_2. This means that a circle centered at D that passes through C_1 will also pass through C_2. Similarly, a larger circle centered at D will pass through both E_1 and E_2. Since all points on the rocker sweep out circular arcs centered at D, we have found the necessary ground pin. This is easiest to visualize by drawing the complete rocker link as shown in

Figure 2.16. Instead of being a straight line, the rocker has a shallow "elbow" at point C that enables it to reach the ground pin at D.

We will now repeat the procedure used in the first example to design the remainder of the linkage. First, we choose the point C on the rocker to be the pin that connects it to the coupler. We are free to choose any point on the rocker that we wish for this purpose, but the point C is convenient for this example. In the Construction layer draw a circle centered at the midpoint of the line between C_1 and C_2 and passing through both C_1 and C_2. Draw a line starting at C_2 and running to the left, and use a relation to make the line Collinear with C_1C_2. Dimension the line as shown in Figure 2.17 and create an Equal circle at point A. Of course, point A is now the ground pin for the crank, and the crank length is equal to the radius of the circle.

As a final step, we should use Smart Dimension to find the length of the crank (the radius of the circle) and the distance between ground pins, as shown in Figure 2.18. To construct the rocker, we must also measure the distance between pin D and point C_1, and the angle between DC_1 and C_1E_1, but we have omitted the angular measurement to keep the diagram uncluttered.

The complete linkage for the second example is shown in Figure 2.19. The dotted arcs show the path that the rocker sweeps out when the crank makes a full revolution. The last step should be to check the Grashof condition of the linkage. The shortest link is the crank,

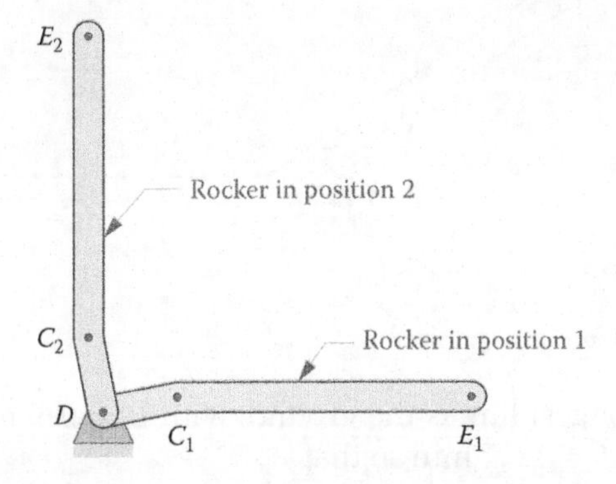

FIGURE 2.16
The rocker is formed by extending the line C_1E_1 to the ground pin D.

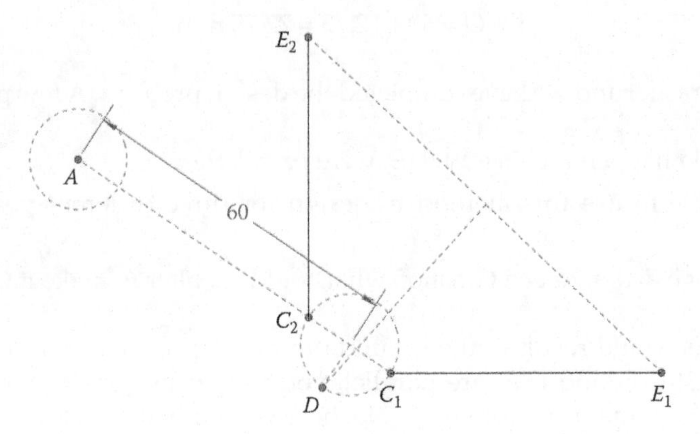

FIGURE 2.17
Draw a circle that centered at the midpoint between C_1 and C_2 that passes through C_1 and C_2.

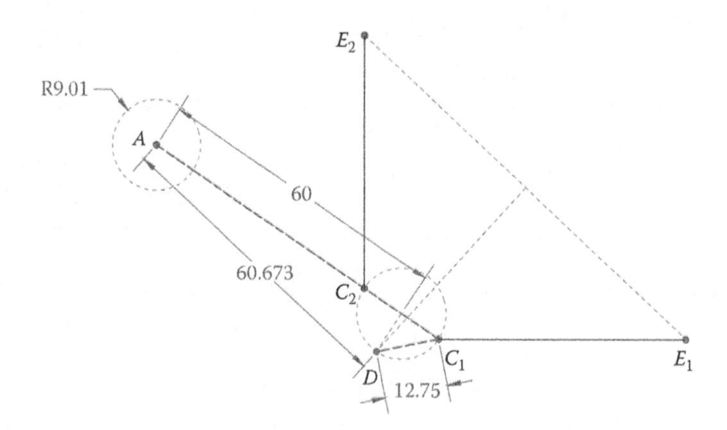

FIGURE 2.18
Finding the length of the crank and the distance between ground pins.

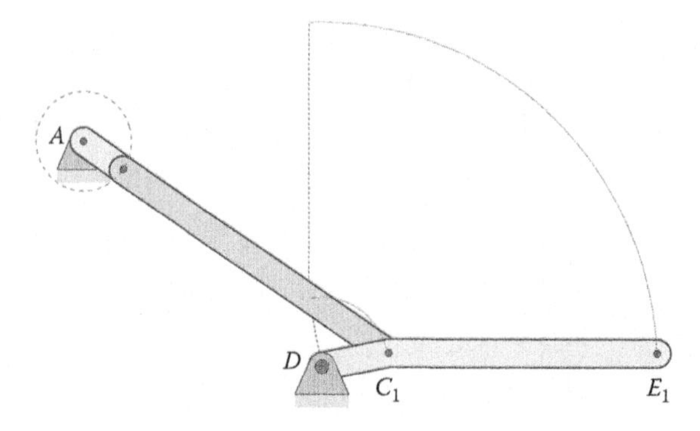

FIGURE 2.19
The completed linkage for the second example.

with $S = 9.01$ mm. The longest link is the ground, with $L = 60.67$ mm. The remaining two links are $P = 60$ mm and $Q = 12.75$ mm so that

$$S + L = 9.01 + 60.67 = 69.68 \text{ mm}$$

$$P + Q = 60 + 12.75 = 72.75 \text{ mm}$$

The linkage is Grashof and we have completed the design problem. A few points to ponder:

1. What would happen if C_1E_1 and C_2E_2 were parallel?
2. Is it possible to use this method to design for three different positions of the rocker?
3. If the linkage had not been Grashof, what could we alter to make it Grashof?

Answers: (1) This would result in the ground pin D residing at an infinite distance from the other links. If C_1E_1 and C_2E_2 are parallel, you should use one of the coupler design methods outlined in the next section. (2) No, because the point D is uniquely defined by the intersection of the perpendicular bisectors. If three positions are required, use one of the coupler design methods outlined in the next section. (3) There are two possible approaches: change the length of the coupler (i.e., the 60 mm dimension that we chose arbitrarily in this example) or modify the location where the coupler is pinned to the rocker (point C). Once we have found the proper shape for the rocker, we can attach the coupler at any point we wish.

2.2.2 Quick-Return Mechanisms

In the preceding example, we didn't worry about how long it took for the rocker to sweep out its angle. In the case of the windshield wiper mechanism, for example, it is probably best for the linkage to spend an equal time pushing the wipers to the left as to the right. But for some mechanisms, timing is critical. Consider the lifting mechanism shown in Figure 2.20. The purpose of this mechanism is to lift a fragile object from one conveyor belt to another. We wish to lift the object slowly and gently, so as not to damage it, but the return stroke (when the object is not on the platform) should be quick, so that the platform can be ready to receive another object as soon as possible.

To see how we might accomplish this, consider the *equal time* mechanism shown in Figure 2.21. As you can see, the rocker spends an equal amount of time moving forwards as backwards because the crank sweeps out 180° for each motion.

To change the timing, we need to change the angle that the *crank* sweeps through for the forward and return motions, as shown in Figure 2.22.

By simply lowering the fixed crank pivot, we change the portions of the crank's rotation that are spent moving forward or backward. As shown in Figure 2.23, the crank sweeps

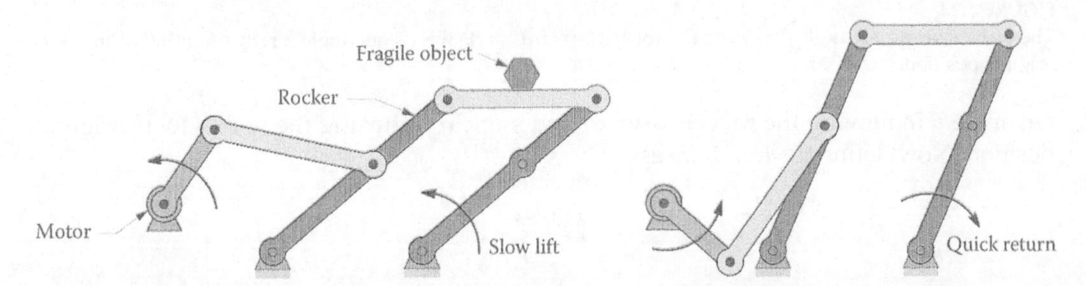

FIGURE 2.20
We wish to lift the fragile object up slowly, but the linkage should return to its bottom position quickly.

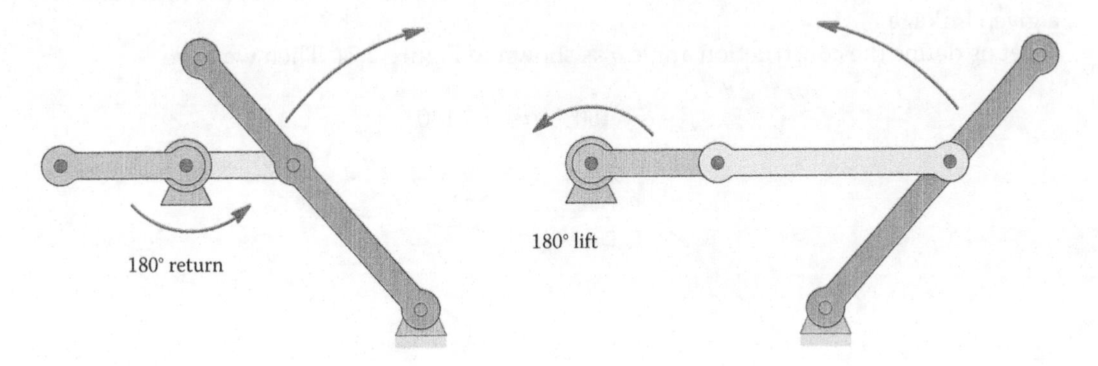

FIGURE 2.21
The equal time mechanism spends the same amount of time pushing the rocker forward as it does returning the rocker to its original position.

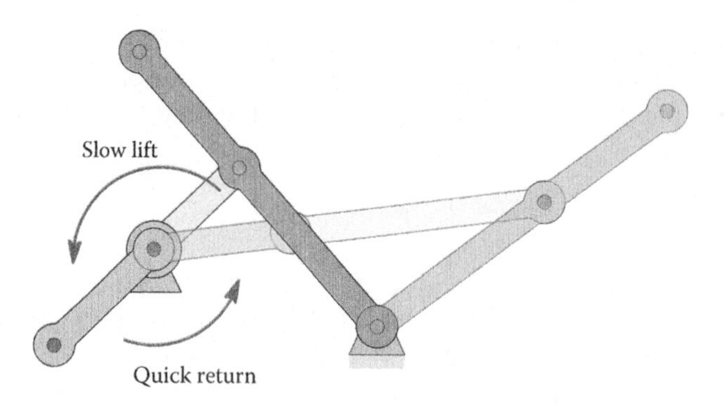

FIGURE 2.22
In this linkage, the crank sweeps out a larger angle in moving the rocker forward than it does in returning the rocker to its original position.

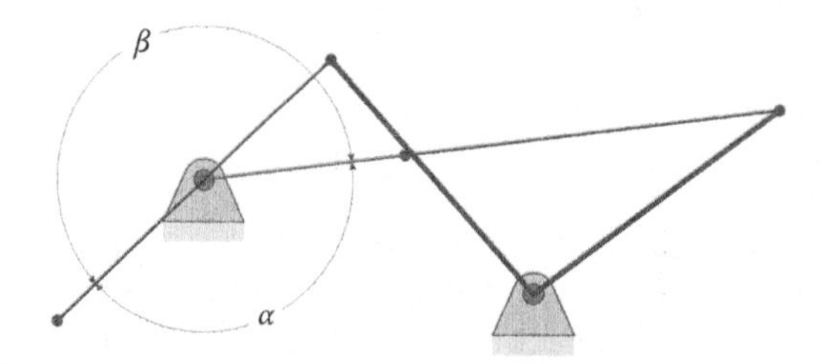

FIGURE 2.23
The crank sweeps out angle β moving the rocker forwards and sweeps out angle α returning the rocker to its original position.

out angle β in moving the rocker forward and angle α returning the rocker to its original position. Now define the *time ratio* as

$$T_R = \frac{\alpha}{\beta}$$

The time ratio gives the ratio of the return time to the forward time and is less than one for a quick-return mechanism. In general, we will design the linkage to meet a specific time ratio, although in some instances it is handy to be able to calculate the time ratio for a given linkage.

Let us define the construction angle δ as shown in Figure 2.24. Then we have

$$\beta - \delta = 180^\circ \quad \alpha + \delta = 180^\circ$$

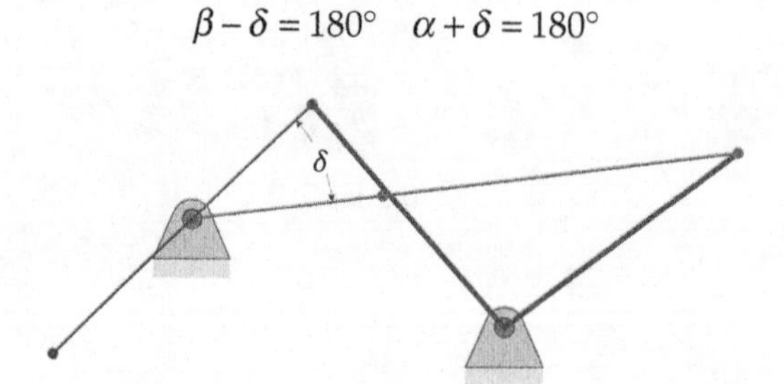

FIGURE 2.24
We use the construction angle δ to design the linkage.

Using these and the definition of the time ratio given in Equation (2.1) we can solve for the construction angle δ as

$$\delta = 180° \cdot \frac{1 - T_R}{1 + T_R} \tag{2.1}$$

The construction angle is useful for laying out a quick-return mechanism, as we will see in the next example.

Example 2.2: Quick-Return Mechanism

Design a fourbar linkage whose rocker sweeps out 40° with a time ratio of 1:1.25.

Solution

First, draw the rocker in its two positions as shown in Figure 2.25. The second position is drawn at an angle of 70° horizontally to make the two positions symmetric, but we could have chosen any other angle. Make sure to place a *Fixed* relation at the lowermost point so that the drawing is fully defined. Now solve for the time ratio as a decimal

$$T_R = \frac{1}{1.25} = 0.8$$

The construction angle is then

$$\delta = 180° \cdot \frac{1 - 0.8}{1 + 0.8} = 20°$$

Now draw intersecting lines from C_1 and C_2 as shown in Figure 2.26. Dimension the angle between the lines at δ, or 20°. The drawing should be underdefined at present, because we have not specified the length between C_1 and A.

Now examine the dimensions shown in Figure 2.27. In this figure, a is the crank length and b is the coupler length. In order for the rocker to take on the required positions, we must have

$$AC_1 = b - a$$

$$AC_2 = b + a$$

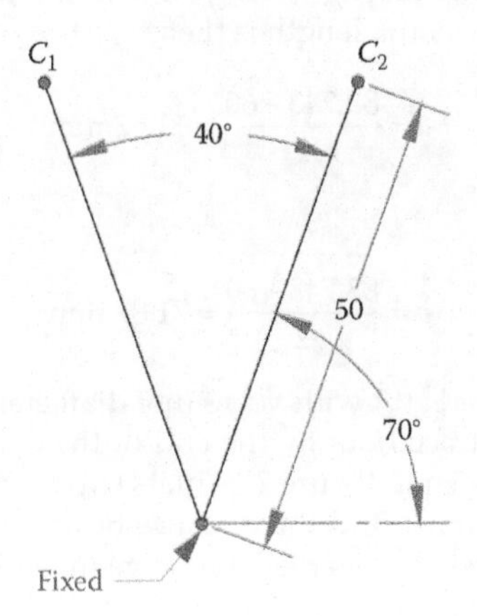

FIGURE 2.25
First draw the rocker in its two positions.

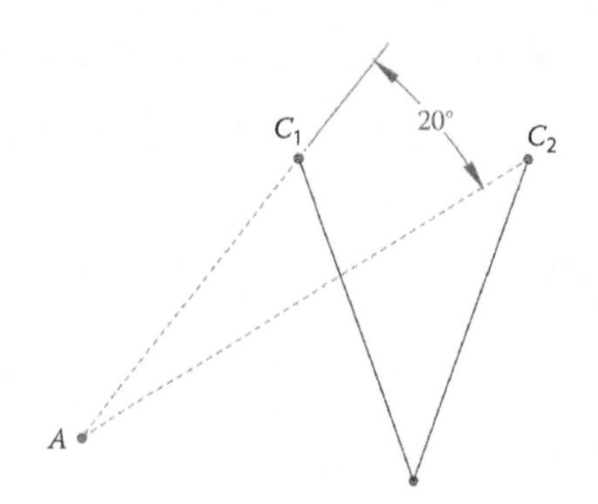

FIGURE 2.26
Draw intersecting lines from C_1 and C_2.

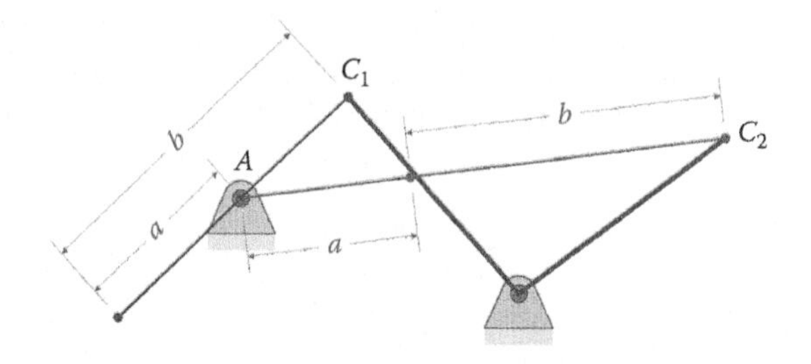

FIGURE 2.27
The crank and coupler lengths are determined by the distance between A and C_1.

or, solving for b and a, we have

$$a = \frac{AC_2 - AC_1}{2}$$

$$b = \frac{AC_2 + AC_1}{2}$$

Let us arbitrarily choose a length of 60 mm for AC_1 as shown in Figure 2.28. When you use Smart Dimension to find the length AC_2 you will find that it is a driven dimension that cannot be changed. The crank length is then

$$a = \frac{83.743 - 60}{2} = 11.87 \text{ mm}$$

and the coupler length is

$$b = \frac{83.743 + 60}{2} = 71.87 \text{ mm}$$

Finally, draw a circle centered at A with *radius* (not diameter) equal to the crank length. The circle shows the path traced out by the end of the crank. The coupler (shown as thick dashed blue lines) is seen in Figure 2.29 in its two extreme positions.

To finish the design, we must find the distance between ground pins, as shown in Figure 2.30. The figure shows the completed linkage in its right-most extreme position. The lengths of each link are then

Crank : 11.87 mm Coupler : 71.87 mm Rocker : 50 mm Ground: 50 mm

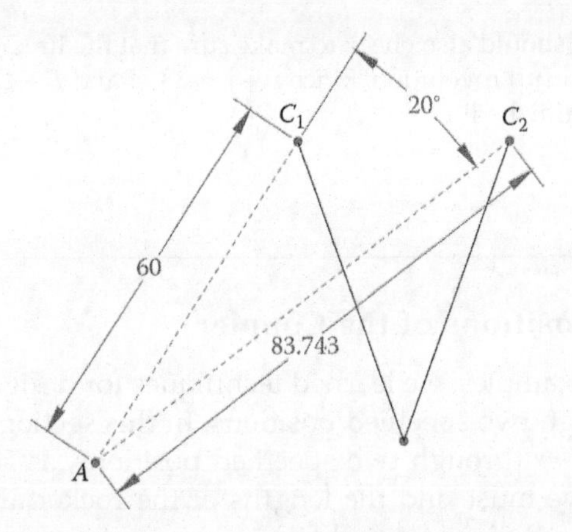

FIGURE 2.28
Choose 60 mm for the length AC_1. The length AC_2 is now driven and we cannot change it.

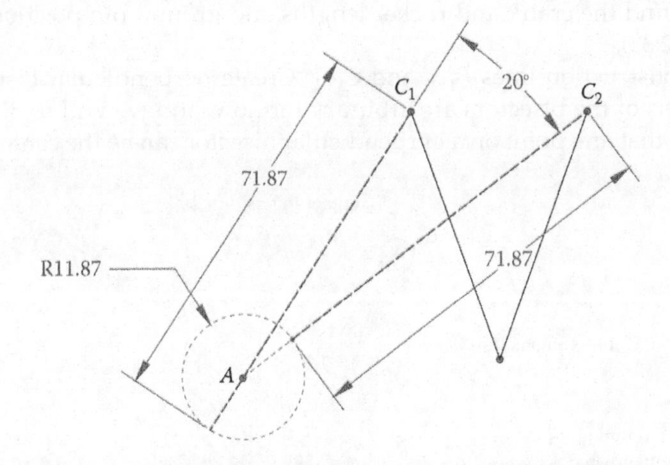

FIGURE 2.29
Draw a circle with the radius of the crank length. The coupler is shown in its two positions.

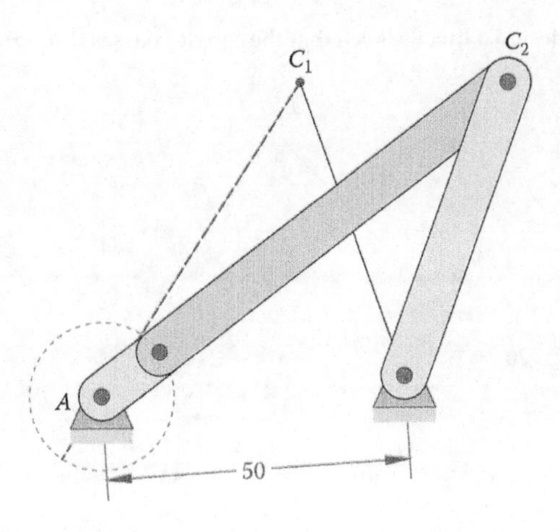

FIGURE 2.30
The last step is to find the distance between ground pins.

For completeness we should also check to make sure that the linkage is Grashof, so that the crank can make a full revolution. Since $S + L = 83.74$ and $P + Q = 100$ the linkage is Grashof, and we are finished!

2.3 Two Specified Positions of the Coupler

In the preceding three examples, we learned techniques for designing a linkage such that the rocker passed through two specified positions. In this section, we wish to find a linkage that passes the *coupler* through two specified positions, as shown in Figure 2.31. To complete the exercise, we must find the lengths of the rocker and crank, as well as the locations of the two ground pins at A and D.

Begin by opening a new Drawing in SOLIDWORKS and constructing the diagram shown in Figure 2.32. For this example, the coupler must move from B_1C_1 to B_2C_2, and we are required to find the crank and rocker lengths and ground pin positions to effect this motion (Figure 2.33).

Next, draw construction lines B_1B_2 and C_1C_2. Create perpendicular bisectors for these lines. The lengths of the bisectors are arbitrary for now, and we will fix them in the next step. Remember that any point on a perpendicular bisector can be the center of a circle that

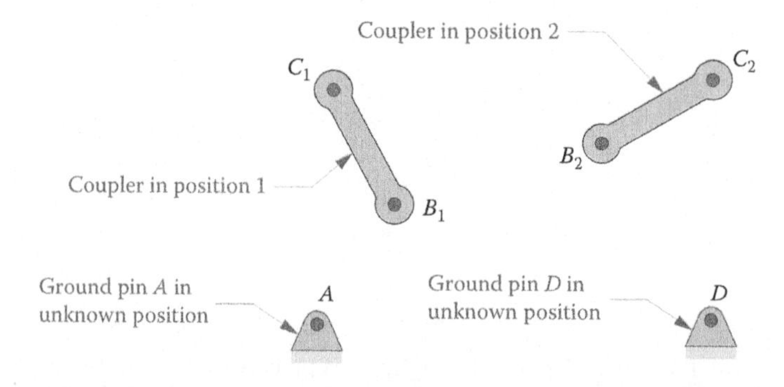

FIGURE 2.31
For this exercise we wish to design a linkage such that the coupler passes through the two specified positions.

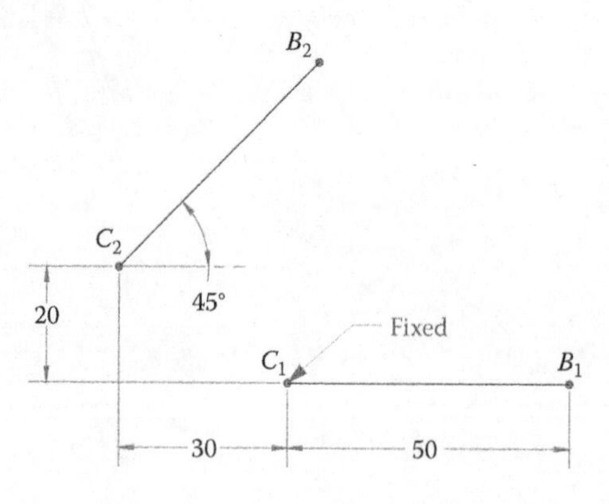

FIGURE 2.32
The coupler must move from B_1C_1 to B_2C_2.

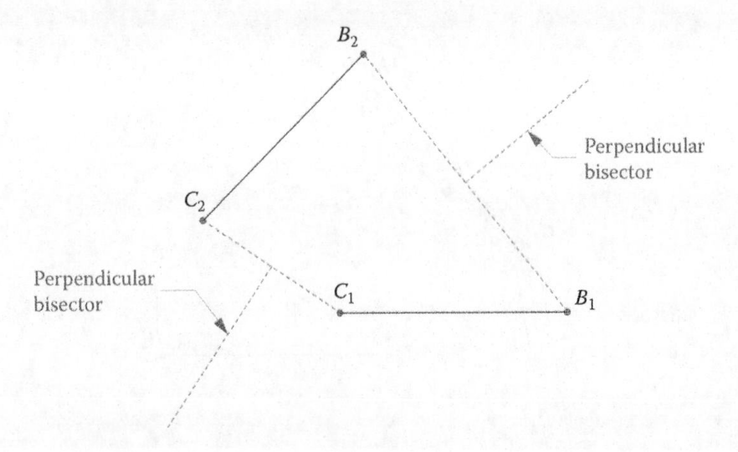

FIGURE 2.33
Draw perpendicular bisectors for the lines B_1B_2 and C_1C_2.

passes through both B_1 and B_2 (or C_1 and C_2), so any point on the perpendicular bisectors can serve as ground pins for the crank and rocker.

As a final step, we use the NewLinks layer to draw lines from the ends of the perpendicular bisectors to either end of the coupler, as shown in Figure 2.34. These new lines are the crank and rocker, and we may dimension them to any convenient length. Thus, we have two free choices in this problem: the length of the crank and the length of the rocker. The length between ground pins (118.81 mm in the figure) is determined once we have chosen the crank and rocker lengths.

The completed linkage, along with ground pins, is shown in Figure 2.35. You might be wondering why we have chosen the upper link to be the crank rather than the lower one. Try creating each link as a SOLIDWORKS Part, and then building the linkage in an

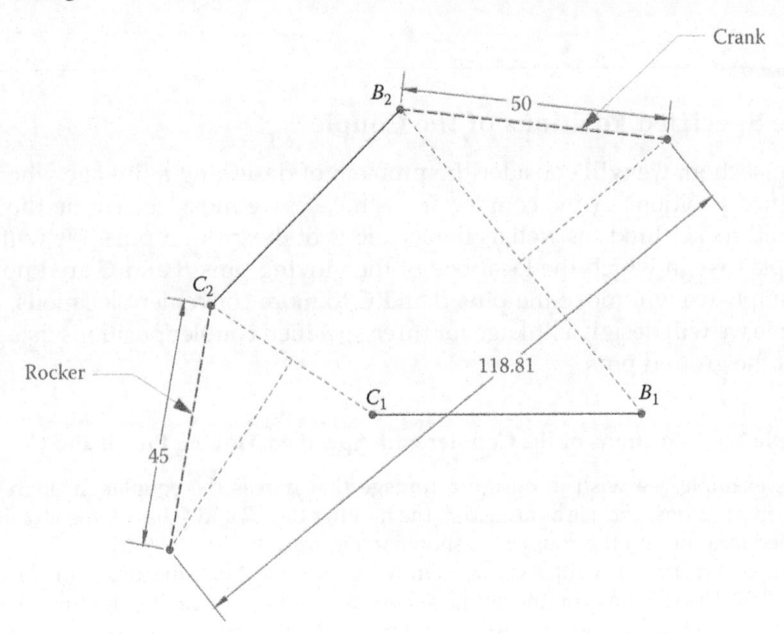

FIGURE 2.34
Draw the crank and rocker links from the ends of the coupler to the ends of the perpendicular bisectors.

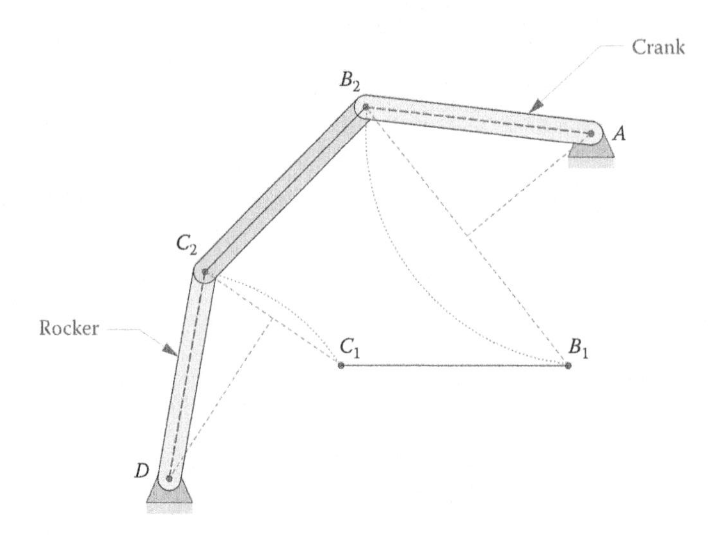

FIGURE 2.35
The completed linkage. The dotted blue arcs show the paths taken by points B and C.

Assembly. You will find that driving the linkage with the lower link results in "binding up" before B_1C_1 is reached. If we drive the linkage from the upper bar, on the other hand, the coupler can pass from B_2C_2 to B_1C_1 without binding. Thus, the logical choice for a driving link (the crank) is the upper bar. Design is an iterative process, and it is almost never possible to arrive at the best design on the first try.

As a final step, you may wish to construct a driver dyad linkage as was done in Example 1. This will permit the coupler to pass back and forth from B_1C_1 to B_2C_2 in a continuous motion.

2.4 Three Specified Positions of the Coupler

In this final section, we will consider the problem of designing a linkage when there are three specified positions of the coupler. In each case, we must determine the lengths of the crank and rocker links, as well as the locations of the ground pins. We will start with a very simple case in which the locations of the moving pins B and C are known. In the second example we will move the pins B and C to more convenient locations, and in the final example we will design a linkage for three specified coupler positions using specified locations of the ground pins.

Example 2.3: 3 Positions of the Coupler with Specified Moving Pins *B* and *C*

In this example, we wish to design a linkage that moves the coupler through three specified positions, and we assume that the moving pins B and C have been attached to specified locations on the coupler, as shown in Figure 2.36.

Begin by creating a SOLIDWORKS Drawing and construct the diagram shown in Figure 2.37. The two lines on the left have vertical relations, while the rightmost coupler position is inclined at an angle of 60°. Do not proceed until your drawing is Fully Defined.

Three points are required to fully define a circle. The three specified points that B passes through define a circle whose center is the ground pivot A, and the three

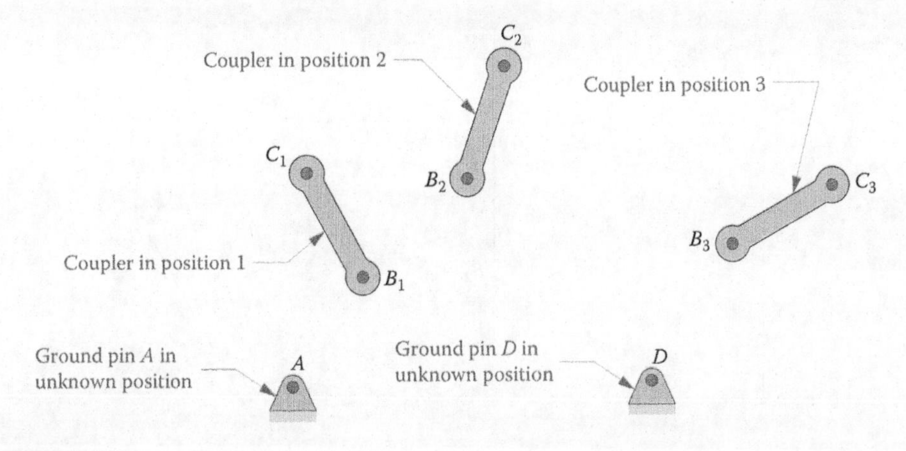

FIGURE 2.36
The coupler has three specified positions, and we must find the lengths of the rocker and crank, as well as the positions of the ground pins.

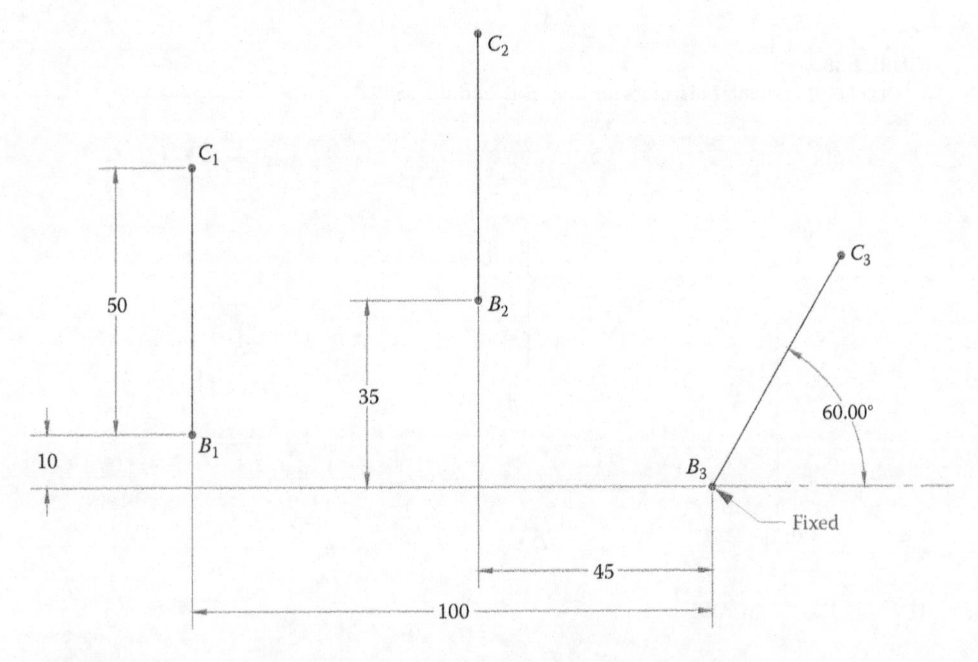

FIGURE 2.37
SOLIDWORKS Drawing showing the three specified positions of the coupler.

specified points for C define the ground pin D. Create lines B_1B_2 and B_2B_3 as shown in Figure 2.38, and construct perpendicular bisectors. The intersection of the bisectors defines the ground pin at A.

In a similar fashion, construct perpendicular bisectors for lines C_1C_2 and C_2C_3, as shown in Figure 2.39. The intersection of the bisectors defines the ground pin at D.

To conclude the exercise, draw the crank and rocker along lines AB and DC as shown in Figure 2.40. Since there were three specified positions for B and C, we have no free choices in this problem.

The complete linkage is shown in Figure 2.41. This was a relatively straightforward exercise since we had no free choices for link length or ground pin position.

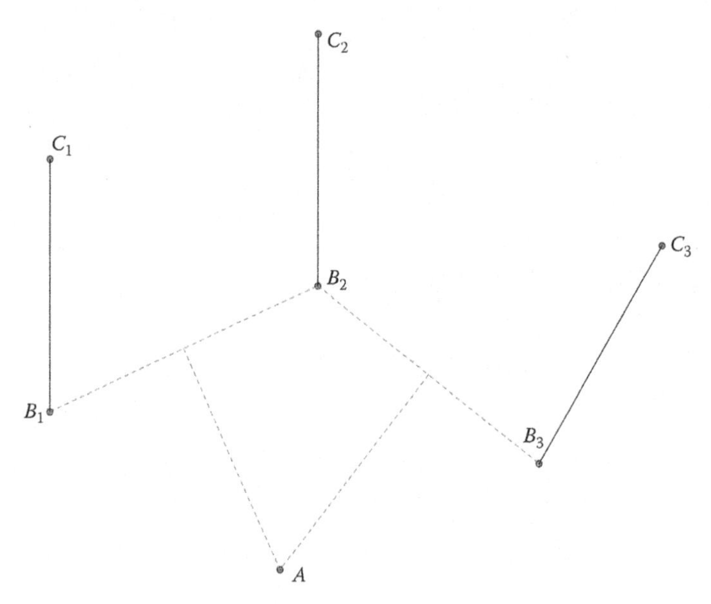

FIGURE 2.38
Construct perpendicular bisectors for lines B_1B_2 and B_2B_3.

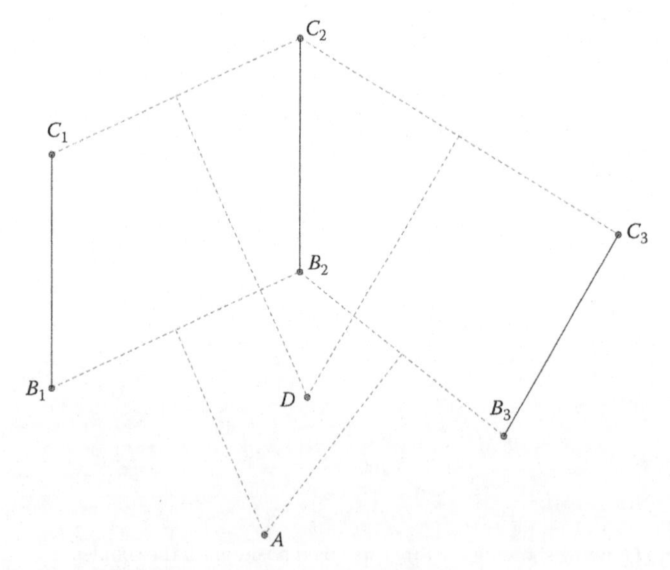

FIGURE 2.39
Construct perpendicular bisectors for lines C_1C_2 and C_2C_3.

Example 2.4: 3 Positions with Unspecified Moving Pivots

In the previous example, a line on the coupler moved through three specified positions and the moving pins *B* and *C* were located at either end of the line. It may happen that we need a line on the coupler to move through three positions, but the moving pins *B* and *C* must be located away from the line to prevent the crank and rocker from interfering with whatever action the coupler is intended to perform. For this situation, which is illustrated in Figure 2.42, we must find the lengths of the crank and rocker, as well as

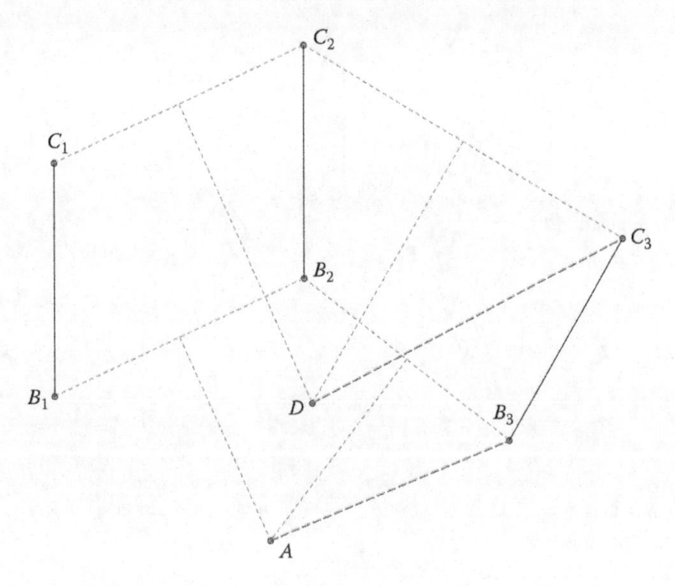

FIGURE 2.40
The crank and rocker are defined by the lines AB and DC, respectively.

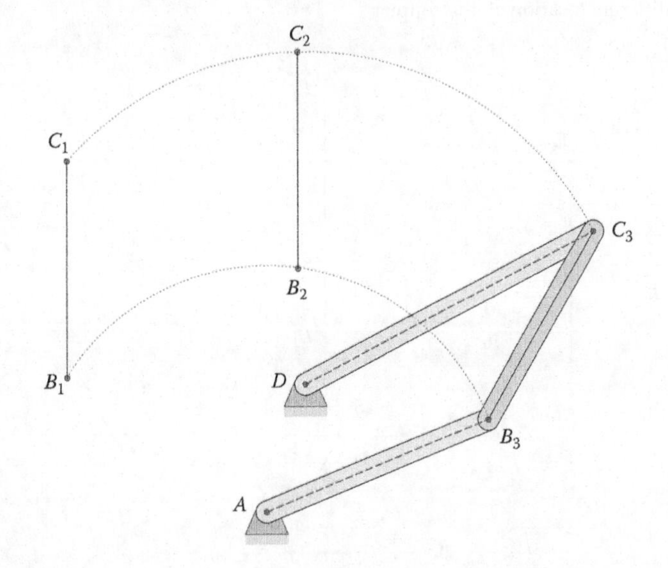

FIGURE 2.41
The completed linkage showing ground pins, crank, rocker, and coupler.

the location of the ground pins. We have free choice of location for the moving pins, but once we have chosen their location on the coupler the rest of the linkage is determined.

First, construct the drawing shown in Figure 2.43, and make sure that it is Fully Defined before proceeding. This drawing shows the three locations that a line on the coupler must pass through, but does not yet show the location of the moving pivots B and C.

The next step is to choose the location of the moving pivots on the coupler, as shown in Figure 2.44. These have been chosen arbitrarily for this example, but would be chosen to avoid interference in a practical setting. Place the lines B_1C_1 and C_1E_1 that define the moving pivots in your drawing.

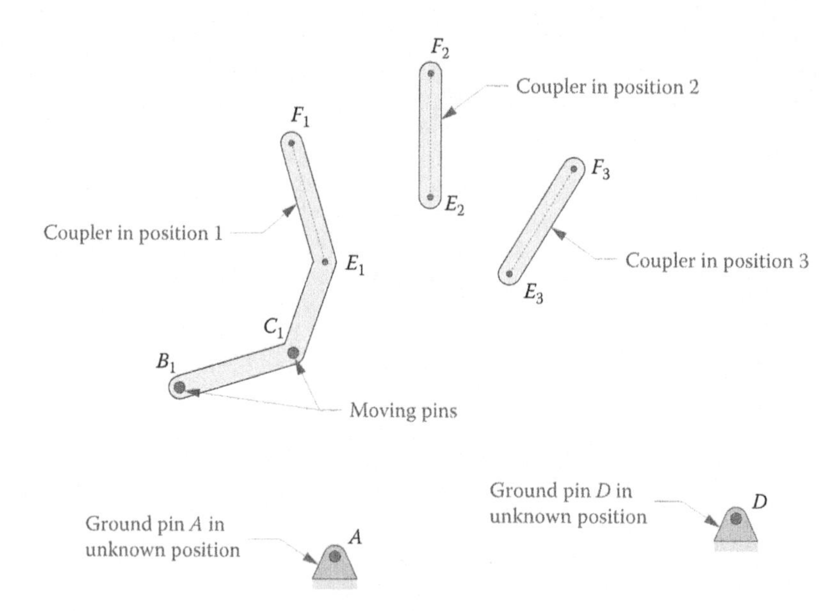

FIGURE 2.42
For this example, the line on the coupler moves through three specified positions, but the moving pins are at a different location on the coupler.

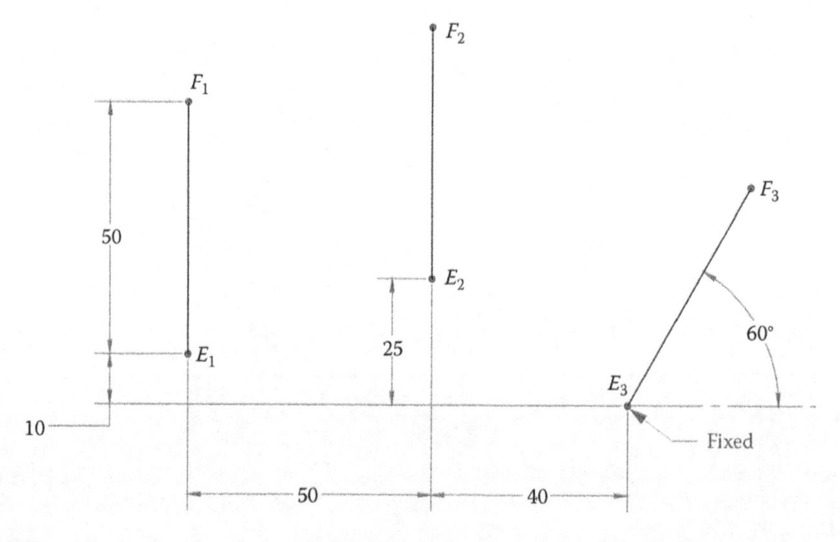

FIGURE 2.43
Construct the drawing that shows the locations of the three positions of a line on the coupler.

Now draw the locations of the moving pivots when the coupler has moved to positions 2 and 3, as shown in Figure 2.45. Make sure that the coupler maintains its shape as it moves from position to position. Your drawing should still be fully defined when you have completed this step.

Now that we know the locations of the moving pins in all three positions, we have reduced the problem to that of Example 2.3. We need only construct perpendicular bisectors for lines B_1B_2 and B_2B_3, as shown in Figure 2.46. The intersection of the bisectors defines the position of ground pin A.

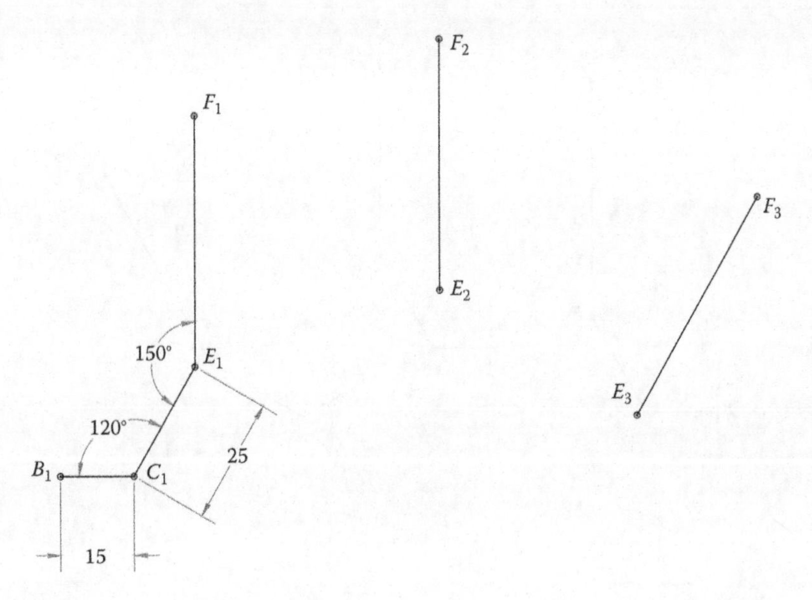

FIGURE 2.44
The locations for the moving pivots B and C have been chosen arbitrarily for this example.

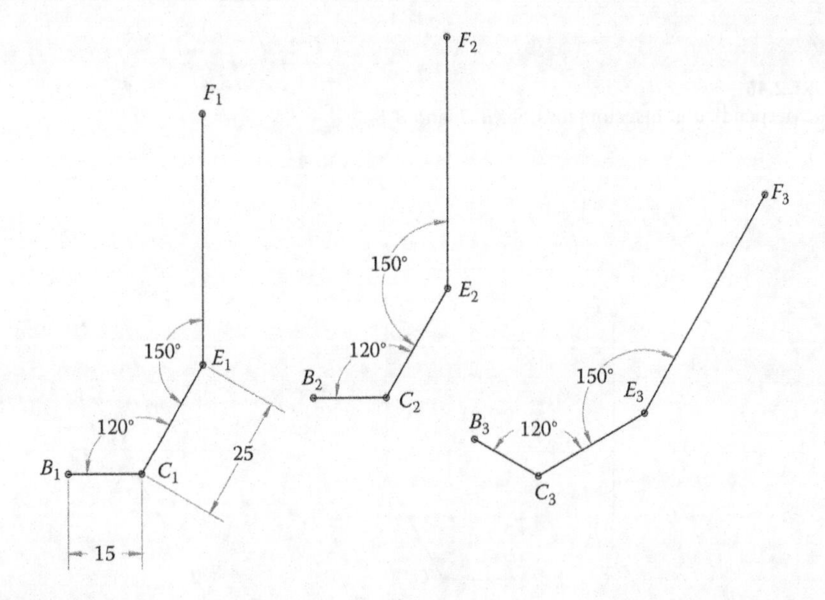

FIGURE 2.45
The moving pivots have been drawn in locations 2 and 3.

Similarly, find the location of ground pin D by constructing perpendicular bisectors for lines C_1C_2 and C_2C_3, as shown in Figure 2.47. All points on the linkage have now been defined.

To complete the exercise, draw lines for the crank and rocker along lines AB_1 and DC_1, as shown in Figure 2.48. All the link lengths and distance between ground pins can now be measured for the purpose of constructing SOLIDWORKS Parts and an Assembly.

The completed linkage is shown in Figure 2.49 with circular arcs showing the paths of moving pins B and C. Because we have moved pins B and C away from the specified line on the coupler, the coupler is now shaped like a badly drawn hockey stick.

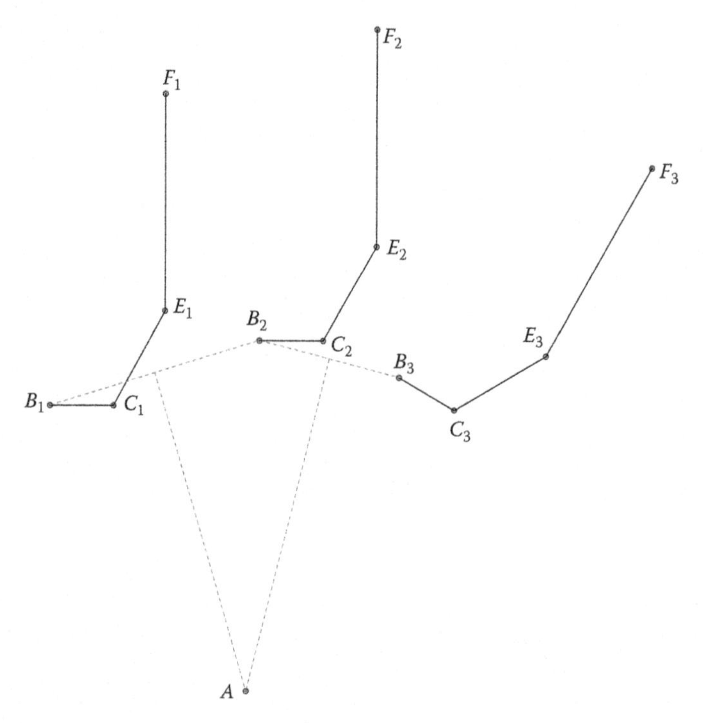

FIGURE 2.46
Create perpendicular bisectors for lines B_1B_2 and B_2B_3.

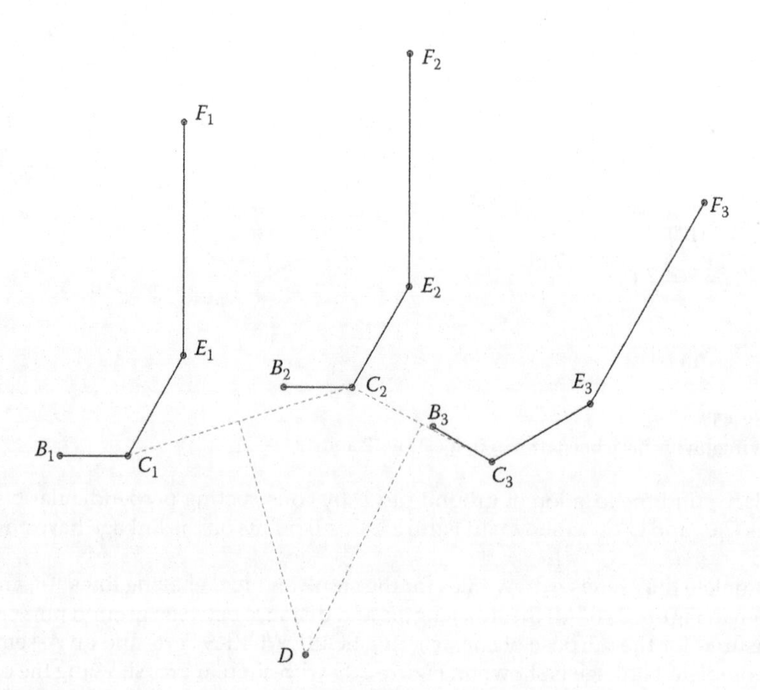

FIGURE 2.47
Perpendicular bisectors have been drawn for lines C_1C_2 and C_2C_3.

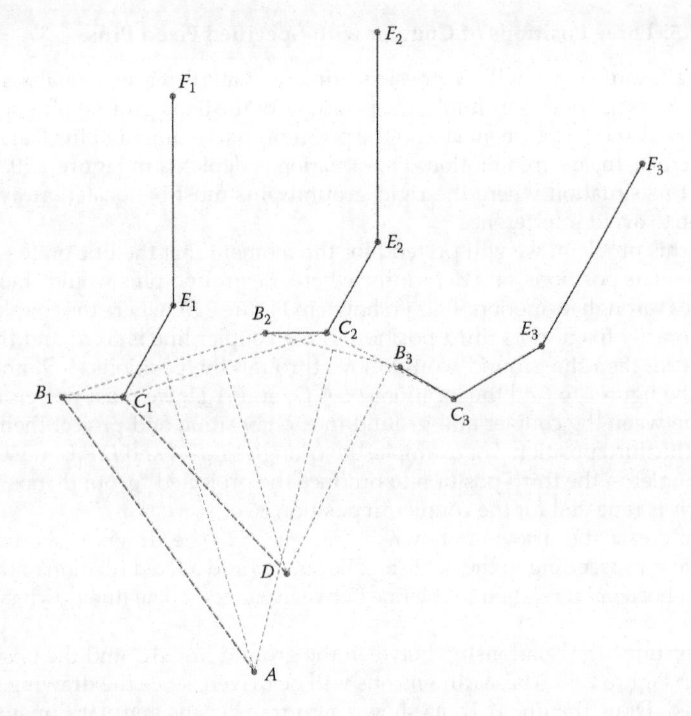

FIGURE 2.48
The crank lies along line AB_1 and the rocker lies along line DC_1.

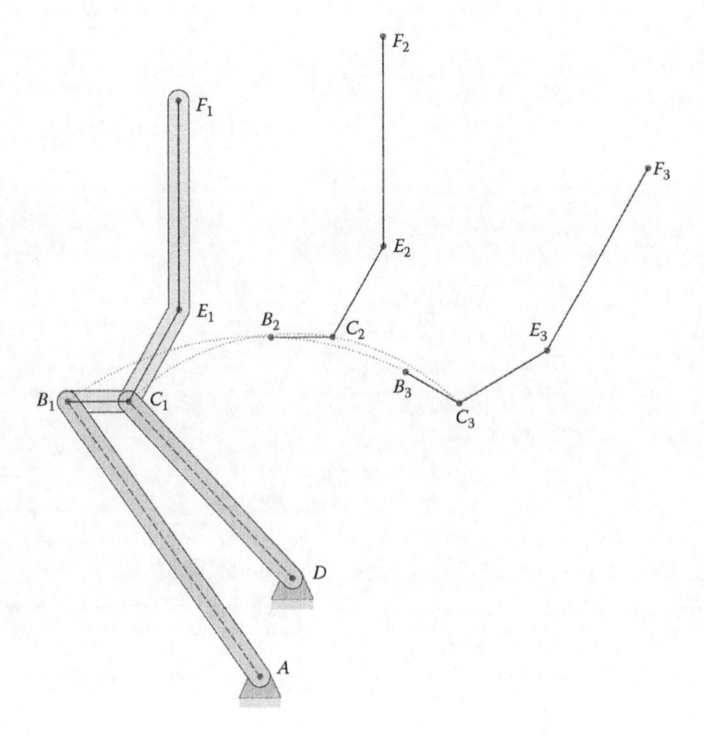

FIGURE 2.49
The completed linkage for Example 2.4.

Example 2.5: Three Positions of Coupler with Specified Fixed Pins

For our final example, we will synthesize a linkage with three specified positions of a line on the coupler. For this example, the positions of the fixed ground pins *A* and *D* are specified in advance, and we must find the positions of the moving pins *B* and *C* on the coupler to create the desired motion. The situation is depicted in Figure 2.50. We might encounter this situation where the rigid ground pins must be located away from the coupler line to avoid interference.

To solve this problem, we will pretend for the moment that the line on the coupler is fixed in one of its positions, and determine where the ground pins would "move" as the linkage goes through its motion. This is shown in Figure 2.51, where the coupler line has been temporarily fixed in its third position. If the coupler line is fixed and the ground pins are freed, then the ground would move through the positions A_1D_1 and A_2D_2 as shown in the figure. To find the locations of A_1D_1 and A_2D_2 we must preserve the relationships between the coupler and ground in each position and project them onto the coupler in the third position. For example, the triangles E_1F_1A and E_1F_1D must be copied onto the coupler in the third position to produce the projected "ground" position A_1D_1. This process is repeated for the coupler in position two.

To begin, create the drawing shown in Figure 2.52. The drawing should be fully defined before proceeding to the next step. Be sure to add a fixed relation to the ground pin *D* and a horizontal relation to the line between *A* and *D*. The line E_2F_2 has a vertical relation.

Next, determine the relationship between the ground line *AD* and the position E_2F_2, as shown in Figure 2.53. These dimensions will be driven, since the drawing is already fully defined. Draw the line A_2D_2 as shown, and transfer the angular dimensions. Use an Equal relation to make the line *AD* equal in length to A_2D_2 and the line E_2D equal to

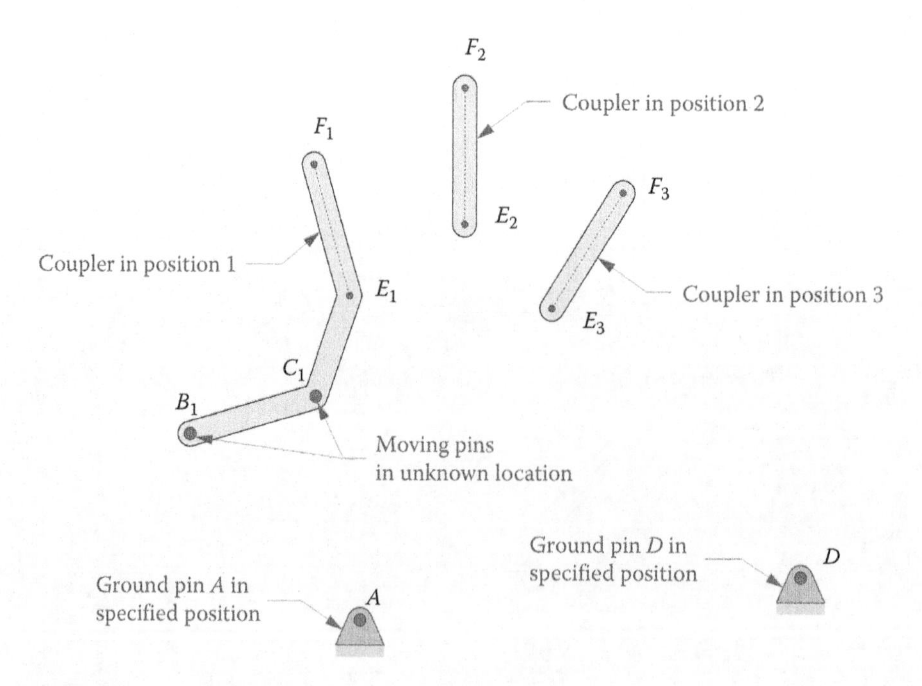

FIGURE 2.50
The line on the coupler moves through three specified positions and the ground pins are also specified.

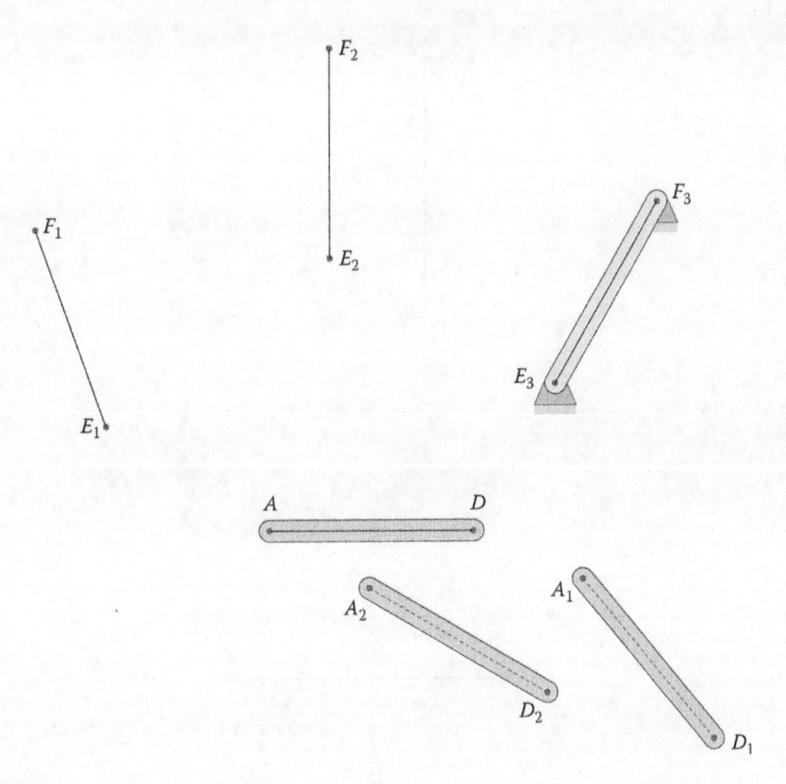

FIGURE 2.51
In the inversion problem we fix the line on the coupler at one of the positions and determine where the ground pins would "move."

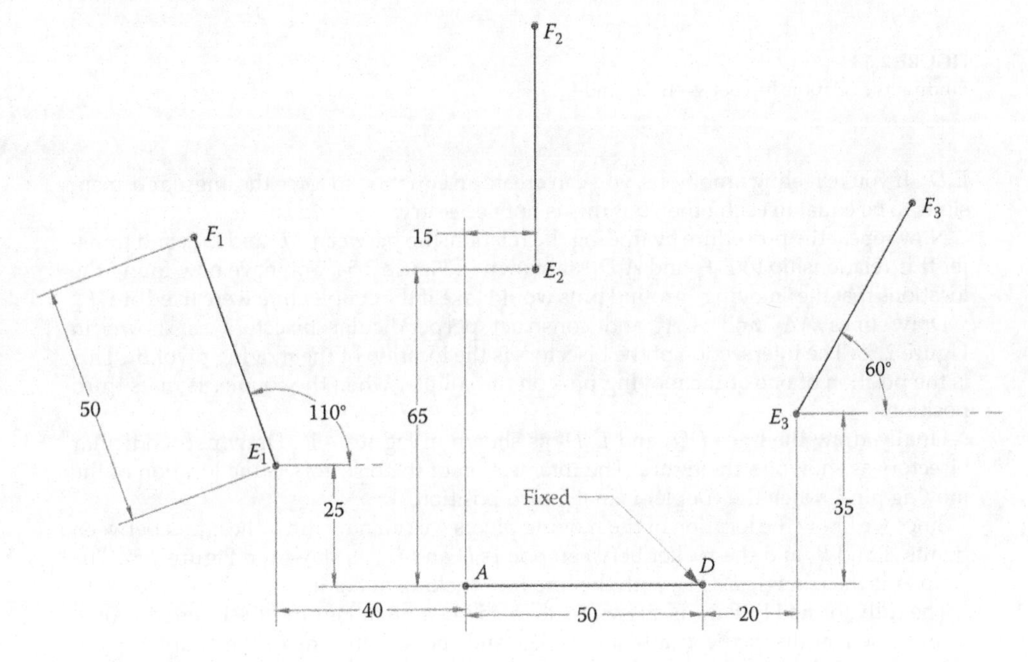

FIGURE 2.52
The three desired positions of a line on the coupler and the fixed ground pins A and D.

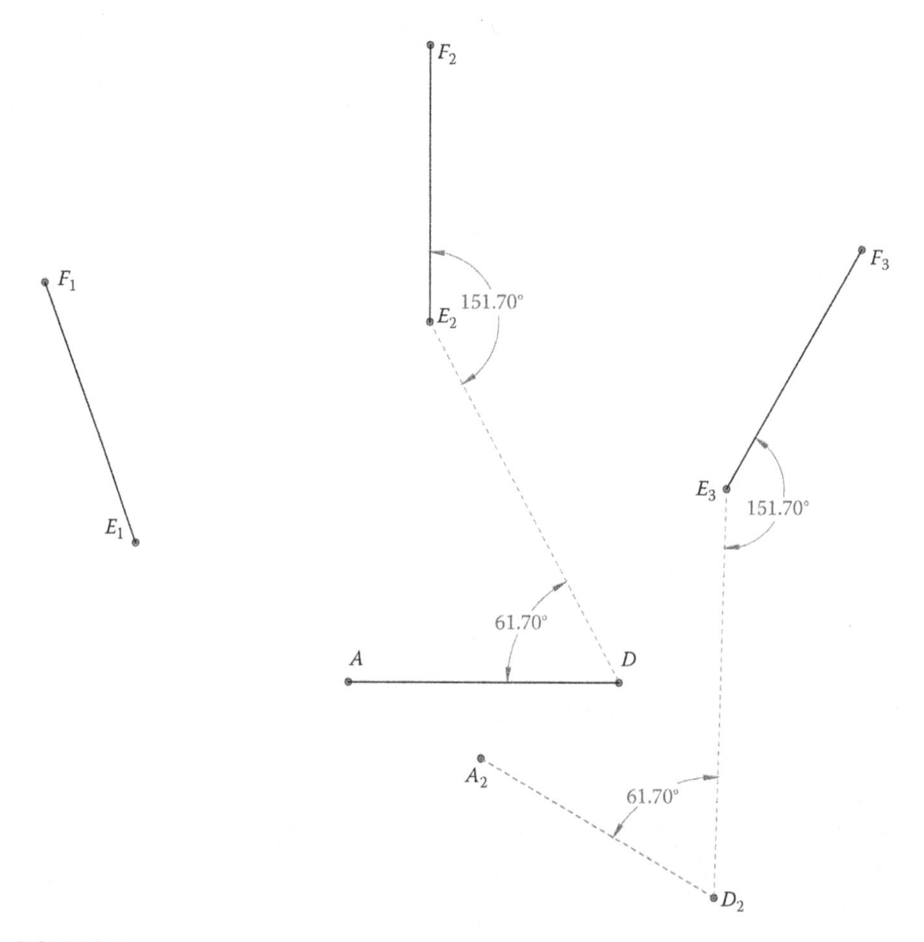

FIGURE 2.53
Finding the relationship between AD and E_2F_2.

E_3D_2. If you're feeling ambitious, you can create an equation to force the angular dimensions to be equal to each other, but this is not necessary.

Now repeat the procedure by finding the relationship between AD and E_1F_1 and transfer this relationship to E_3F_3 and A_1D_1, as shown in Figure 2.54. You have now found the locations that the "moving" ground pins would take if the coupler line were fixed at E_3F_3.

Draw lines AA_2 and A_1A_2 and construct perpendicular bisectors as shown in Figure 2.55. The intersection of the bisectors is the location of the moving pivot B_3. This is the position of one of the moving pins on the coupler when the coupler is in its third position.

Finally, draw the lines DD_2 and D_1D_2 as shown in Figure 2.56. Draw perpendicular bisectors as shown in the figure. The intersection of the bisectors is the location of the moving pin C when the coupler is in its third position.

Since we know the location of the moving pivots we can now draw the crank between points A and B_3 and the rocker between points D and C_3, as shown in Figure 2.57. The coupler is the triangle $B_3C_3E_3$ with the line E_3F_3 sticking out of it.

The fully formed linkage is shown in its third position in Figure 2.58. It remains only to find the lengths of the crank and rocker and the dimensions of the coupler. Since this linkage is likely to be non-Grashof, a driver dyad should be constructed using the techniques of Example 2.1 or 2.2 to drive the linkage through its desired range of motion.

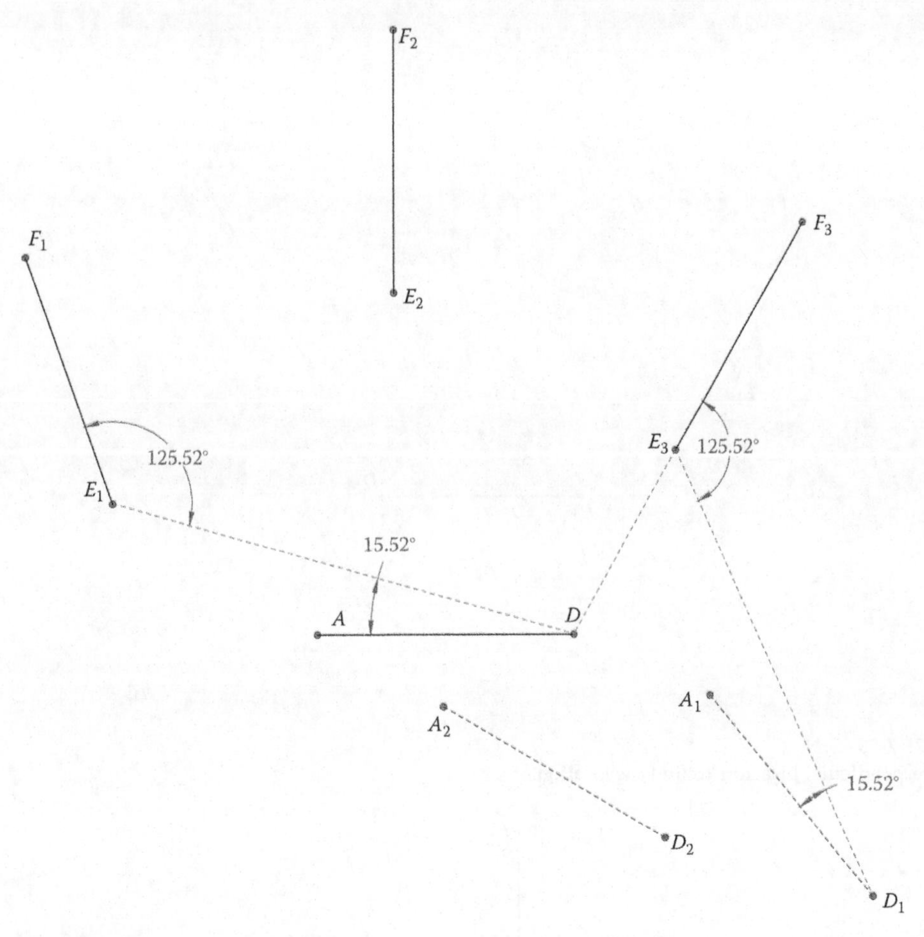

FIGURE 2.54
Finding the relationship between AD and E_1F_1.

Since it might be difficult to visualize the motion of this linkage it has been drawn in its three positions in Figure 2.59. Note that the coupler is a rigid body and maintains its shape throughout the motion of the linkage.

2.5 Summary

We have learned several techniques for designing fourbar linkages to create specified motions. The first few examples were quite simple, but very useful for creating driver dyad linkages for moving non-Grashof fourbars through their intended ranges of motion. The next few examples required three specified positions of the coupler and became progressively more complicated. In the simple examples we had several free choices (coupler length, location of moving coupler pins) that gave our designs a degree of flexibility. As the number of specified locations increased, however, our free choices were diminished until, as in the last example, the linkage was completely defined by our specified coupler positions and ground pins.

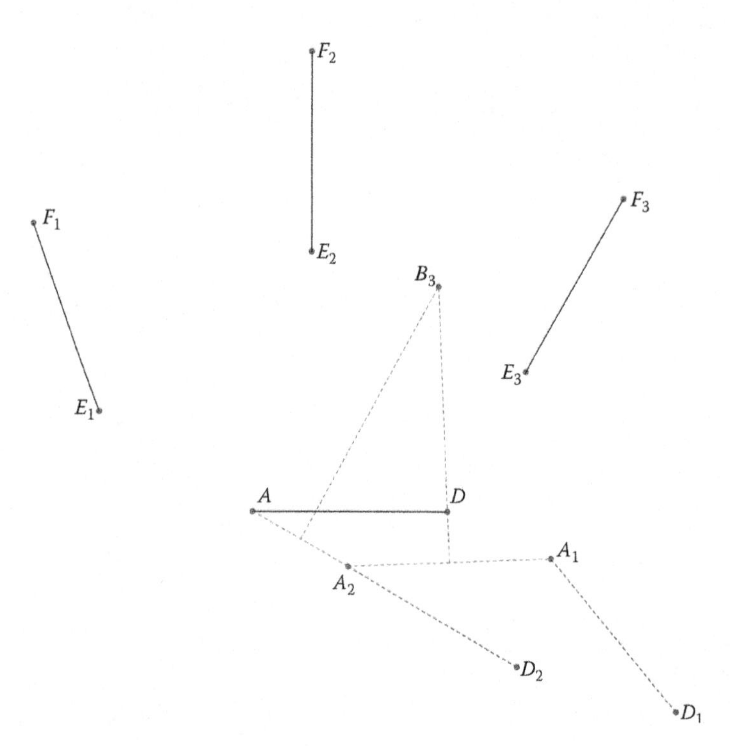

FIGURE 2.55
Use perpendicular bisectors to find the location of B_3.

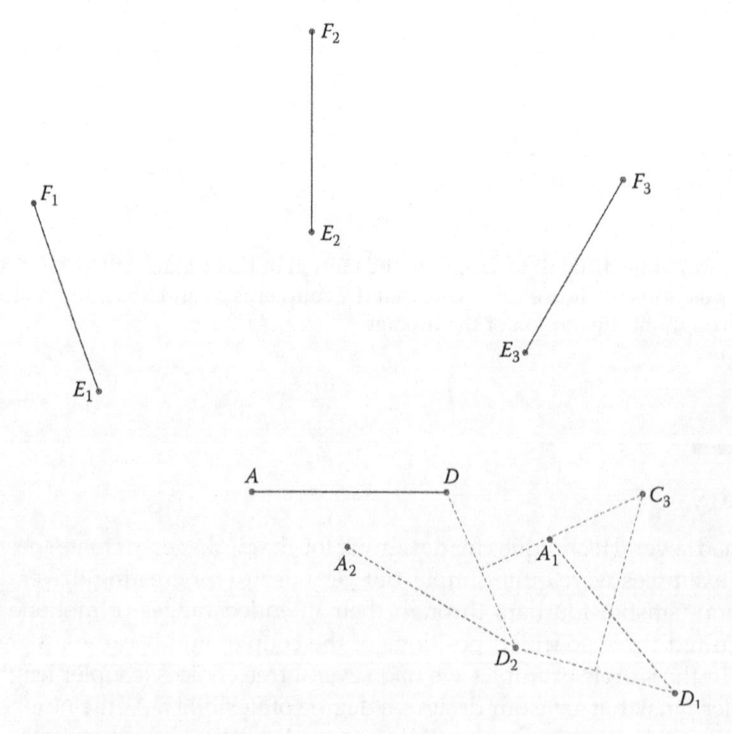

FIGURE 2.56
Use perpendicular bisectors to find the location of C_3.

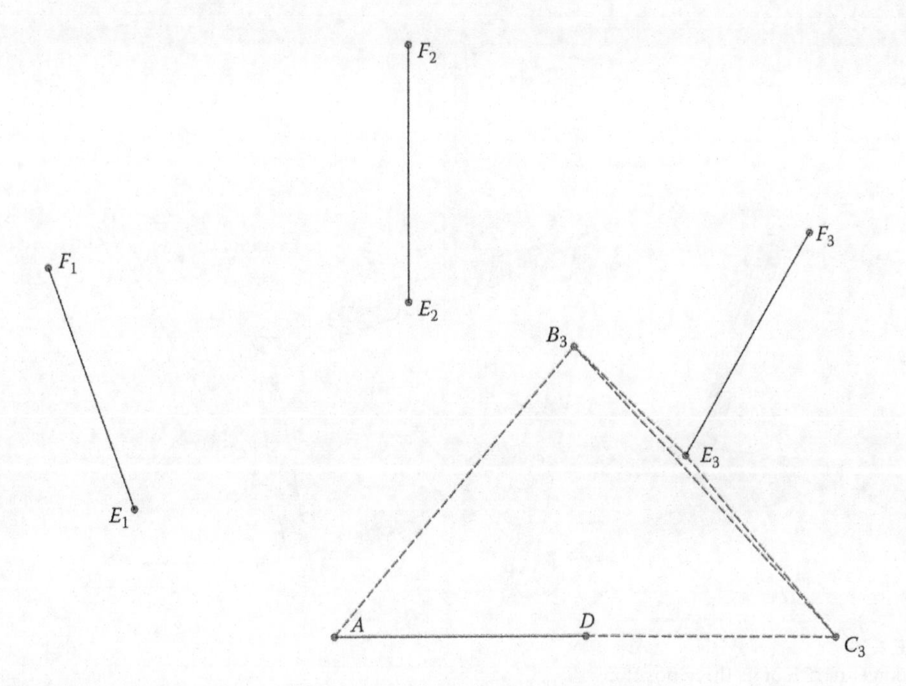

FIGURE 2.57
The crank is drawn between A and B_3 and the rocker is drawn between D and C_3.

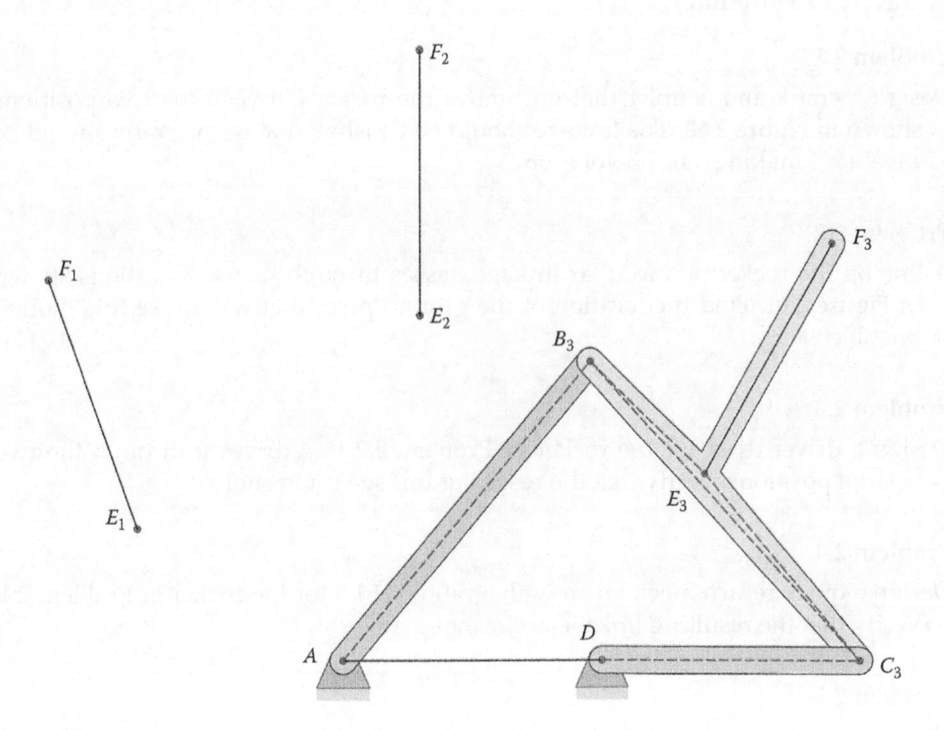

FIGURE 2.58
The fully formed linkage shown in position 3.

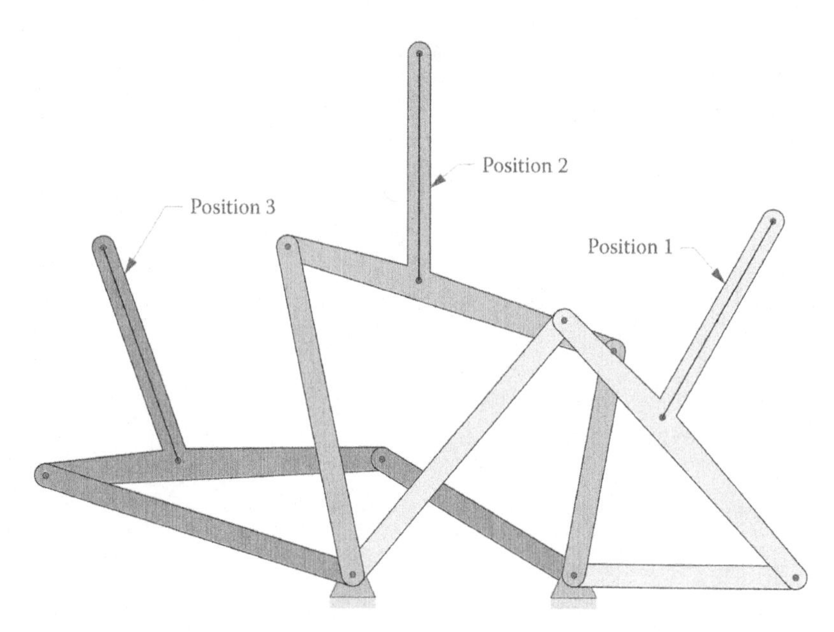

FIGURE 2.59
The linkage in each of its three positions.

2.6 Practice Problems

Problem 2.1

Design a crank and coupler that will move the rocker between the two positions shown in Figure 2.60. The linkage should be Grashof; that is, the crank should be capable of making a full revolution.

Problem 2.2

A line on the rocker of a fourbar linkage passes through the two locations shown in Figure 2.61. Find the location of the ground pivot that will make this motion possible.

Problem 2.3

Design a driver dyad for the rocker in Problem 2.2 that drives it through the two desired positions. Verify that the resulting linkage is Grashof.

Problem 2.4

Design a quick-return mechanism with a ratio of 1:1.5 for the rocker in Problem 2.1. Verify that the resulting linkage is Grashof.

Problem 2.5

Design a quick return mechanism with a ratio of 1:1.5 for the rocker in Problem 2.2. Verify that the resulting linkage is Grashof.

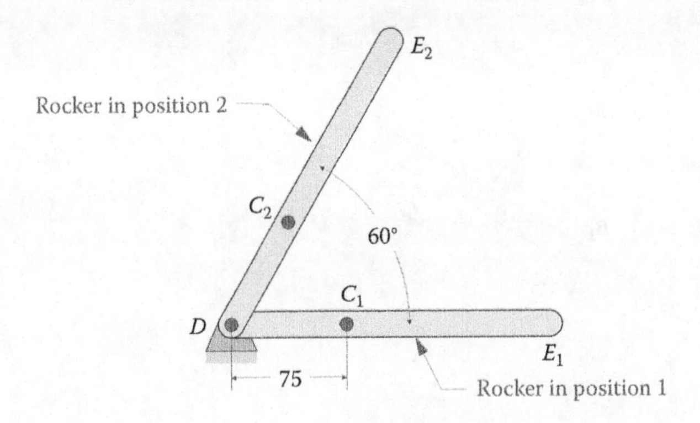

FIGURE 2.60
Problem 2.1.

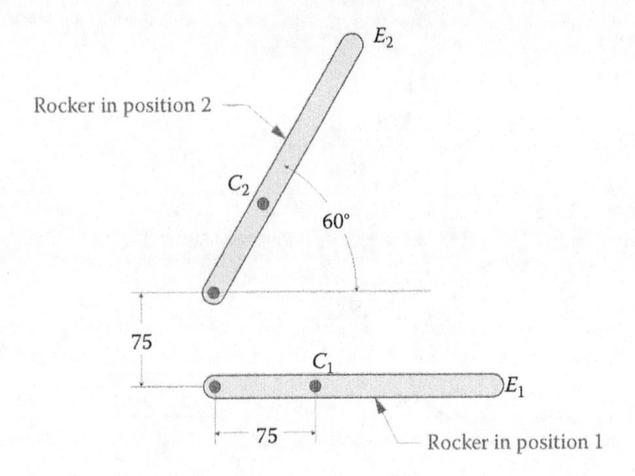

FIGURE 2.61
Problem 2.2.

Problem 2.6

Design a fourbar linkage that moves the coupler between the two positions specified in Figure 2.62. Next, design a driver dyad to drive the linkage. Verify that the driver dyad linkage is Grashof.

Problem 2.7

Design a fourbar linkage that moves the coupler between the three positions specified in Figure 2.63. Find the required positions of the ground pins.

Problem 2.8

Design a driver dyad linkage to drive the fourbar in Problem 2.7 through its specified positions. Verify that the driver dyad linkage is Grashof.

Problem 2.9

A "European Hinge" is found in many modern kitchens. While it is more complex than an ordinary hinge with a single pivot, it has the advantage of allowing the

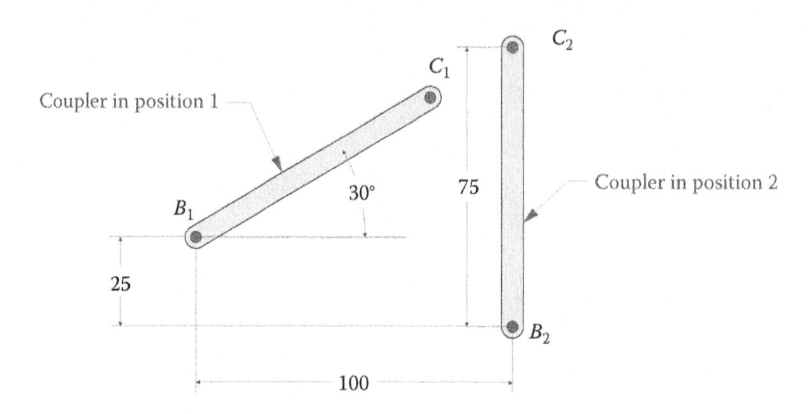

FIGURE 2.62
Problem 2.6.

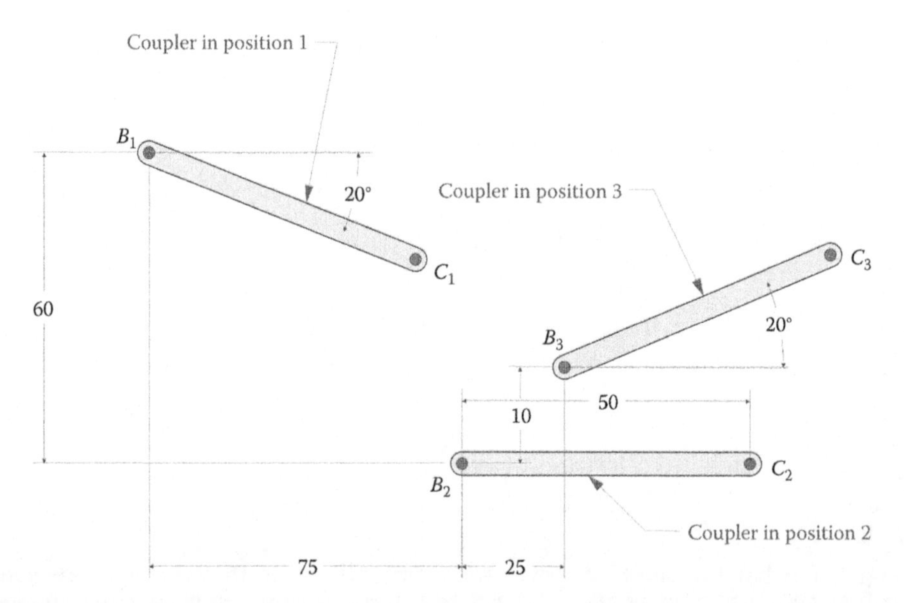

FIGURE 2.63
Problem 2.7.

door to swing fully open without interfering with neighboring cabinet doors. Figure 2.64 shows the three desired door positions, designated by line *BC*. Find suitable positions for the moving pivots on the hinge that will keep the fixed pivots roughly within the target area shown. This problem is more challenging than it looks, so do not be discouraged if your first few attempts do not succeed.

Problem 2.10

Repeat the European Hinge exercise of Problem 2.9, but this time use the fixed pivots shown in Figure 2.65. Find the locations of the moving pivots on the door that are necessary to reach the three required positions shown in the figure for Problem 2.9.

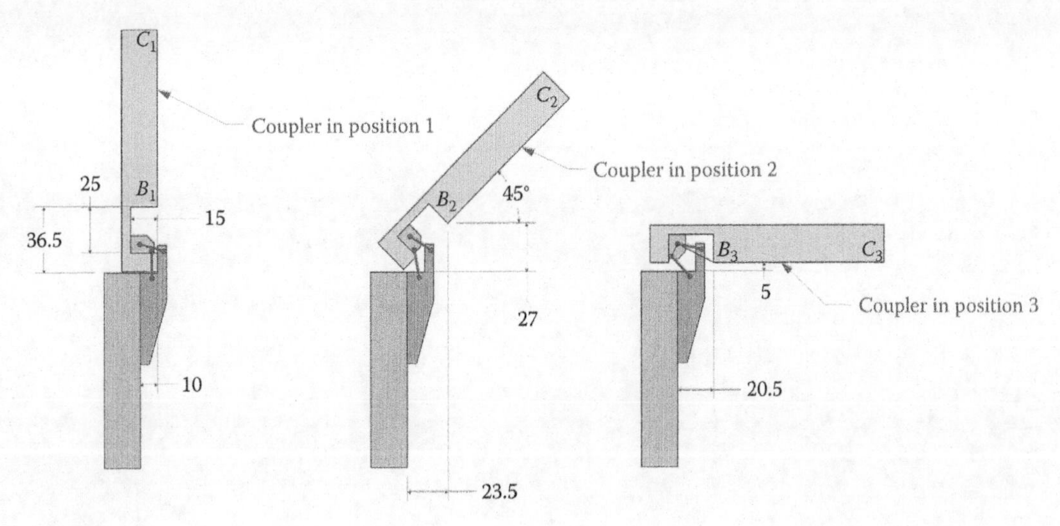

FIGURE 2.64
Problem 2.9.

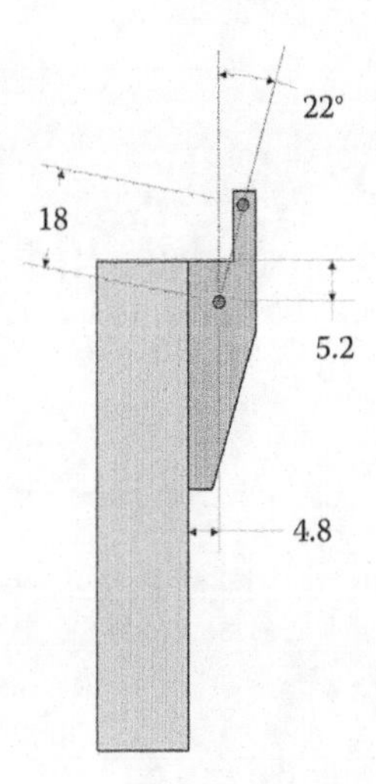

FIGURE 2.65
Problem 2.10.

Acknowledgments

Several images in this chapter were produced using SOLIDWORKS software. SOLIDWORKS is a registered trademark of Dassault Systèmes SolidWorks Corporation.

MATLAB is a registered trademark of The MathWorks, Inc.

3

Introduction to MATLAB®

3.1 Introduction

MATLAB (which stands for "MATrix LABoratory") is one of the most widely used pieces of engineering software in the world. The main reason for its wide adoption is that it makes many of the most common tasks in engineering (solving systems of equations, plotting) very easy. The plots generated by MATLAB are of sufficient quality to be "publication-ready" and it is easy to recognize MATLAB plots in much of the modern engineering literature. This book will emphasize the use of MATLAB in Mechanical Design, but you will find MATLAB to be useful in all your other engineering courses as well.

Despite the power and simplicity of MATLAB, many students find it very frustrating to use – especially at first. Since much of MATLAB was written during the 1980s, many of the error messages are cryptic, at best. The online help for MATLAB is thorough and well organized, but it is sometimes difficult for beginners to understand the "lingo" (e.g. it is difficult to understand how to use a *function handle* if you are not clear what a function handle is!).

This tutorial will introduce you to some of the basic concepts in MATLAB. We will focus on learning to do things that will be useful later in the textbook such as solving systems of equations and plotting. If you are already familiar with MATLAB you can safely skip this chapter and move on to Chapter 4.

3.2 Simple MATLAB® – The Command Window

When you first start up MATLAB you will see a window with a text prompt in it. This window is called the *Command Window*, and you can type commands at the prompt and receive an immediate response. Other common panels include the *Current Folder*, which contains a listing of the files in the current folder you are working in, and the *Workspace* panel, which lists all the MATLAB variables currently in memory. Since you just started, this window should be empty.

At its most basic level, you can use MATLAB as a glorified calculator. In the Command Window type

```
>> 2*2
```

and hit Enter. You'll get the response

```
ans =

     4
```

which is reassuring. The Workspace window now holds the variable ans, which stores the value of whatever you computed last. Instead of typing numbers directly, we can define variables to carry out the same operation.

```
>> a=2
a =
     2

>> b=2
b =
     2

>> a*b
ans =
     4
```

After typing this you should see the variables a and b appear in the Workspace window, along with their values. The Workspace window is a very handy feature that enables you to make sure that your variables are storing the values that you think they should be storing.

The main reason that so many engineers have adopted MATLAB is its ability to work easily with *vectors* and *matrices*. As we will see in the next chapter, a vector is simply a row (or column) of numbers and a matrix is a two-dimensional array of numbers. To enter a vector into MATLAB we use square brackets:

```
>> a=[2 4 6]
a =
     2     4     6
```

Because we used spaces between the numbers, MATLAB has stored a as a *row vector*. We would have achieved the same effect if we had used commas between the numbers. If you look in the Workspace window you'll see that a now has the value [2,4,6] and the previous value of 2 has been overwritten. It is sometimes confusing for beginners that a single letter (a) can store multiple values like this, but it is one of the things that make MATLAB so powerful. Another slightly tricky concept is that of the *index*. The index can be thought of as the *address* of a particular number in a vector. Since the number 6 is stored in the third position of a, its index is 3. If you want to access a particular entry in a, you would use the index enclosed in regular parentheses. In the Command Window type

```
>> a(3)
ans =
     6
```

Now type

```
>> a(2)
ans =
     4
```

In the first case, the *index* is 3 – we have accessed the third entry in a, and in the second case the index is 2 – the second entry in a has the value 4. Now let's try entering a *column vector*

```
>> b = [3;5;7]
b =
     3
     5
     7
```

Note that we have used semicolons to separate each row and that the semicolons have also appeared in the Workspace window. To emphasize the difference between a and b, type the following commands in the Command Window

```
>> size(a)
ans =
     1     3

>> size(b)
ans =
     3     1
```

The command `size()` is a built-in MATLAB function that gives the dimensions of a vector or matrix. The first dimension is the number of rows and the second dimension is the number of columns. Thus, the vector a has one row and three columns: it is a row vector. The vector b has three rows and one column: it is a column vector. Now try typing the following (slightly devious) command:

```
>> b(3)
ans =
     7
```

Can you see why this is devious? We used the command `a(3)` to access the third *column* in a and `b(3)` to access the third *row* in b, but the command worked just fine in both instances. This behavior is, unfortunately, the source of confusion for many beginning MATLAB programmers. To understand why this is, we'll have to look a little closer into vector operations. At the Command Window type

```
>> a*b
ans =
    68
```

Next, type

```
>> b*a

ans =

     6    12    18
    10    20    30
    14    28    42
```

Since we are used to the commutative property in multiplication, this result is very strange. Remember that in multiplying a and b, we are multiplying two *vectors*, not two numbers.

In the first instance, we have computed the *dot product*, which we will discuss more thoroughly in Chapter 4. The dot product is defined as

$$\mathbf{a} \cdot \mathbf{b} = a_1 b_1 + a_2 b_2 + a_3 b_3$$

and if you carry out these operations in your head you'll find that the answer is, indeed, 68. The other operation is known as an *outer product*

$$\mathbf{b} \otimes \mathbf{a} = \begin{bmatrix} a_1 b_1 & a_2 b_1 & a_3 b_1 \\ a_1 b_2 & a_2 b_2 & a_3 b_2 \\ a_1 b_3 & a_2 b_3 & a_3 b_3 \end{bmatrix}$$

Please do not memorize the formula for the outer product – this is the last time we'll see it! The important thing to observe is the size of the output in each instance. In the first instance, the size of `a*b` is 1×1, in other words a single number (a scalar). The size of `b*a`, on the other hand, is 3×3. In the first instance, we carried out the operation

$$(1 \times 3)^*(3 \times 1) \rightarrow (1 \times 1)$$

and in the second instance

$$(3 \times 1)^*(1 \times 3) \rightarrow (3 \times 3)$$

Thus, the result of a vector multiplication has the size of the *outer* dimensions of the two vectors. As we will see in a moment, the *inner* dimensions of the two vectors must be the same in order for the multiplication to work.

To convert a row vector into a column vector we can use the *transpose* operation. In mathematical terms this means that

$$\mathbf{a}^T = \begin{Bmatrix} 2 \\ 4 \\ 6 \end{Bmatrix}$$

$$\mathbf{b}^T = \begin{Bmatrix} 3 & 5 & 7 \end{Bmatrix}$$

In MATLAB we perform the transpose using the apostrophe. Thus, we can type

```
>> a'
ans =
     2
     4
     6
```

Now try using the transpose operator to multiply two column vectors together.

```
>> a'*b
Error using  *
Inner matrix dimensions must agree.
```

Congratulations – you've received your first cryptic MATLAB® error message! We received the message because we tried to perform a multiplication that is undefined:

$$(3\times 1)^*(3\times 1)\rightarrow \text{undefined}$$

Now that we know what the error message means, it should be a little less cryptic. If you receive this message when writing a complicated program, it almost always means that one of the variables that you thought was a row vector is actually a column vector, or vice versa. To avoid this type of problem in the future we will adopt the following convention.

1. If a variable has two dimensions – say the x and y coordinates of a point, we will store the two dimensions as a *column vector.* Thus, the coordinates of point C will be stored as

$$\mathbf{x}_C = \begin{Bmatrix} x_C \\ y_C \end{Bmatrix}$$

and *not*

$$x_C = \begin{Bmatrix} x_C & y_C \end{Bmatrix} \leftarrow \text{don't do this!}$$

2. When calculating a variable at several instants in time – say the coupler angle for multiple values of the crank angle, we will store the results in a *row vector.* Thus, the coupler angle found at three instants in time is

$$\theta_3 = \begin{Bmatrix} \theta_3(1) & \theta_3(2) & \theta_3(3) \end{Bmatrix}$$

3. Here's the tricky one: when we calculate a two-dimensional variable at multiple instants in time – say the position of point P for multiple crank angles, we will store each component in its own row, with each column representing a separate instant in time:

$$\mathbf{x}_P = \begin{bmatrix} x_P(1) & x_P(2) & x_P(3) \\ y_P(1) & y_P(2) & y_P(3) \end{bmatrix}$$

If we remain consistent with this convention, then we will avoid the error message shown above. The fact that you can access the third entry of a row *or* column vector by typing `a(3)` or `b(3)` can lead to confusion because MATLAB will choose the third entry of either one. It would have been better in the long run if MATLAB had required the user to type `a(1,3)` for the third entry of a row vector and `b(3,1)` for the third entry of a column vector, but this is what we are stuck with. The bottom line is this: if you receive an error message saying that inner matrix dimensions must be equal, the first thing to check is if a row vector has been entered as a column, or vice versa. The Workspace window is a good place to check this.

By the way, if you wish to enter a *matrix*, like the $\mathbf{x}_P$ array shown above, you would type

```
>> A =[2 4 6; 3 5 7]
A =
     2     4     6
     3     5     7
```

Notice that the rows have been separated by a semicolon, and the columns are separated by spaces. The A matrix has dimension 2×3. To access the first row, third column of A, type

```
>> A(1,3)
ans =
     6
```

And to access the second row, second column of A type

```
>> A(2,2)
ans =
     5
```

Now try entering the following

```
>> A(3,2)
Index exceeds matrix dimensions.
```

This is another commonly encountered error message. It means that we've asked to see a part of the A matrix that doesn't exist since it only has two rows and we asked for an entry in the *third* row. When you receive this message you should examine the definition of the variable in the Workspace to make sure it has the correct dimensions, and also check to make sure you haven't accidentally switched the row and column arguments within the parentheses. If you type

```
>> A(2,3)
ans =
     7
```

you get the expected result. To test your understanding of the above concepts, which of the following operations is defined (i.e., do the inner dimensions match)?

1. **A*b**	2. **A*a**
3. **b*A**	4. **a*A**
5. **b*A**^{T}	6. **a*A**^{T}

Answer: items 1 and 6 are defined, the rest are not.

By the way, we will adopt the convention of storing vectors as lowercase letters and matrices as uppercase letters. Both matrices and vectors are written in bold font.

3.3 Vector Notation in MATLAB®

Entering vectors one number at a time is fine for small vectors, but there will be times when we wish to define vectors with hundreds, or even thousands, of entries. Luckily,

MATLAB has an easy way to do this using the simple *colon* operator. Try typing in the Command Window

```
>> a=0:10
a =
     0     1     2     3     4     5     6     7     8     9    10
```

The colon tells MATLAB that we are specifying a vector that starts at zero and ends at ten, with a spacing of one between each entry. If we want a different spacing, we place the increment between the first and last numbers

```
>> a=0:2:10
a =
     0     2     4     6     8    10
```

gives the even numbers starting at zero and ending at ten. What if we wanted the list of odd numbers between 1 and 11? We could use

```
>> b=1:2:11
b =
     1     3     5     7     9    11
```

Or, even better, we could use the definition of a to calculate the odd numbers.

```
>> b = a + 1
b =
     1     3     5     7     9    11
```

The second method may be a little bit surprising. After all, a is the vector of even numbers from zero to ten, but "1" is a scalar. MATLAB interprets this as a command to add 1 to *each entry* in the vector a. Similarly, we could use the vector a to calculate the first six even powers of 2.

```
>> c=2^a
Error using  ^
One argument must be a square matrix and the other must be a scalar. Use
POWER (.^) for elementwise power.
```

Whoops! Here is another cryptic MATLAB error message. MATLAB uses the caret operator as a shorthand for multiplying a square matrix by itself, where a square matrix has the same number of rows and columns. If **A** is a square matrix, then **A**^2 = **A*****A**. What we wanted to do with `c = 2^a` is to raise the number 2 to the power of *each of the numbers* in a. Luckily, MATLAB has given us a hint on how to do this – the dot-carat notation.

```
>> c=2.^a
c =
           1           4          16          64         256        1024
```

Much better! Placing the dot before the carat (or a multiplication symbol) tells MATLAB to perform the operation on each element, rather than on the vector (or matrix) as a whole. It is not even clear what raising the number 2 to the power of an entire vector would mean!

3.4 A First Plot

We now have a vector c that contains the first six even powers of 2. It might be interesting to plot this, so as to observe its exponential increase. The command for plotting in MATLAB is very simple:

```
plot(c)
```

After you hit Enter, a new plot window should open that resembles Figure 3.1. It is an exponentially increasing curve, as we expected. The plot is incomplete as it stands because we haven't placed labels on the axes or a title. We'll learn to do this as part of a script in the next section.

3.5 Writing a Simple MATLAB® Script

In theory, we could do almost all our modeling in the Command Window by typing a long series of commands, one after the other. The power of MATLAB, however, lies in its ability to run *scripts*. A script is a series of MATLAB commands that we store in a file called an *m-file*. For example, we could write a script, called `SinePlot.m`, that would plot a sine wave for us and would contain all the commands necessary to make the plot ready to go in a report (axis labels, title, etc.) Storing the commands in a script enables us to make minor "tweaks" to the plot without having to retype everything. We will now leave the Command Window behind, because we'll be spending the rest of the tutorial in the Editor. On the Home tab in the Command Window click on the **New** button and choose **Script**

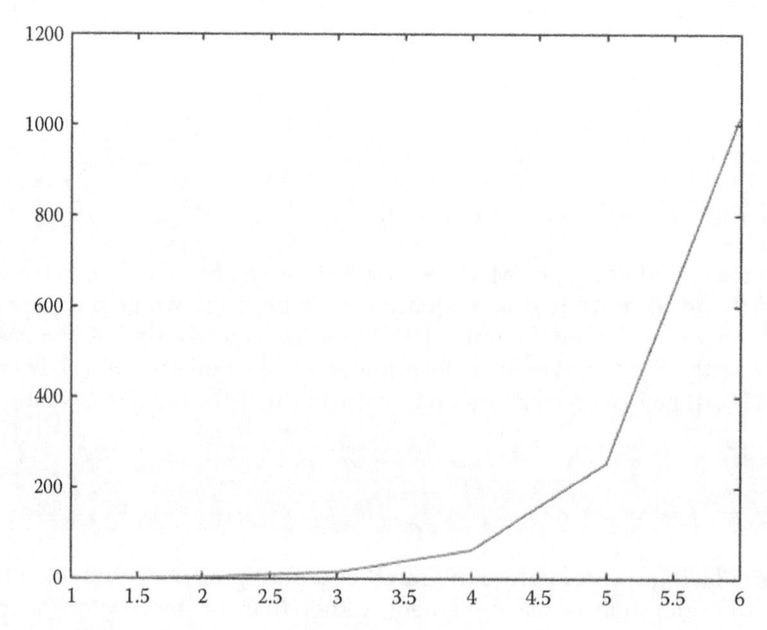

FIGURE 3.1
Our first MATLAB plot – the first six even powers of 2.

from the pulldown menu. A new window will open: the Editor window. We will use the Editor window to enter our scripts as text files. When we are done typing the script, we can hit the Run button (or the F5 key) to execute it. The result will be the same, for the most part, as if we had typed each command, one after the other, in the Command Window.

The first thing to do when entering a new script is to use **Save As** to name it. When the **Save As** dialog appears, browse to your desired folder (e.g. the Desktop) and save the file as `SinePlot.m`. Because MATLAB was originally written in the old days of DOS, it doesn't allow spaces or special characters in the file names. In other words, don't try to save it as `Sine Plot.m` – you'll get a very mysterious error message when you try to run the script later. MATLAB file names must begin with a letter and cannot contain spaces.

The first part of every script you write should contain a set of *comments* that describe the purpose of the program and its function. A comment is a part of the script that is not executed, but serves to give the reader important information about the script. Commenting your code is one of the most important things you will do as a programmer, because if you write anything useful, somebody else will want to use it. In addition, your future self will be grateful to your present self for commenting the code, since your future self will not remember the thought process you went through to create the program. Of course, it is possible to overdo it in commenting:

```
a = 2; % set a equal to 2
```

Your future self will know perfectly well that `a = 2` sets the variable a equal to two, so this particular comment can be left out. Anything that is nontrivial in your programs deserves a comment, though. At the top of the Editor window type

```
% SinePlot.m
% makes a nice plot of a sine wave from zero to 360 degrees
```

Use the percent sign to indicate a comment. Comments are displayed in green font by default. As the comments describe, the `SinePlot` program will plot a sine wave between 0° and 360°. The next line you should place in *all* your MATLAB scripts is the following:

```
clear variables; close all; clc
```

This clears all the variables that are in memory so that you can start with a "clean slate." Forgetting to do this can result in very mysterious errors if, for example, you forget to set a variable to its proper value and it still contains values from having run an earlier program. The command `close all` closes all open plot windows, and `clc` clears the Command Window. The semicolon tells MATLAB that you have finished entering a command (e.g `clear variables`) and that it should be ready for the next command (e.g. `close all`). We'll calculate the sine function in 1° increments, just to keep things simple.

```
x = 0:360;
```

We have defined `x` as a vector starting at 0 and counting up (by ones) to 360. How many elements does `x` contain? Since it starts at 0 and ends at 360, it contains 361 elements! This is somewhat counterintuitive, so consider that the first element, `x(1)` = 0. The second element is `x(2) = 1`. And the final element `x(361)` = 360. A general formula can be written `x(i) = i - 1`. Note that the first element in a MATLAB vector (or matrix) has the index 1, not zero as in many other languages.

The sine function in MATLAB requires that we give the angle in radians, not degrees. To convert from degrees to radians, simply multiply by $\pi/180$.

```
theta = x*pi/180;
```

This multiplies each value in the vector x by $\pi/180$. Note that pi is a built-in constant in MATLAB. Now that we have theta in radians, we can calculate the sine of theta using

```
y = sin(theta);
```

Remember that theta is a vector with 361 elements; calculating the sine of theta also results in a vector of 361 elements. Thus, y(1) = sin(0) and y(361) = sin(2π). Plotting the sine function is simplicity itself:

```
plot(x,y)
```

This is the general syntax for the plot command. The first argument is a vector of x values on the plot and the second argument is a vector of the corresponding y values. In this case, the x values give the angles (in degrees) and the y values give the sine of each angle. If we had wanted to have the angle in radians on the x axis we would have typed

```
plot(theta,y)
```

Let's test the program to see if it creates the plot that we expect. The simplest way to save and run the program is to hit the function key F5. If you hit F5, MATLAB will save all the currently open files in the Editor window and will execute the file that is on top (in this case, the SinePlot.m script). If you have typed everything correctly, you should obtain the plot shown in Figure 3.2. The plot is indeed a sine curve, but it is not very attractive.

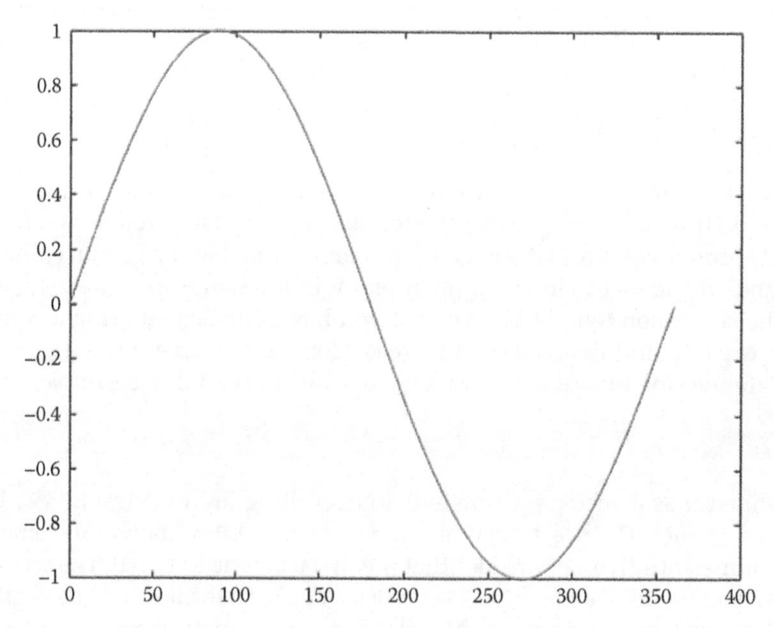

FIGURE 3.2
Our second plot – a sine function.

There are a few things we should add to make the plot more professional. The first thing might be to change the width of the line that was used to plot the sine wave. The default line width is pretty narrow in MATLAB, and a thicker line would be easier to see. Change the plot command to the following

```
plot(x,y,'LineWidth',2)
```

We have set the `LineWidth` parameter to 2. This is a very common way to modify a plotting command in MATLAB: we first tell MATLAB what to plot, and then we set the appropriate parameters for the plot. Since there are so many possible parameters to set, it is difficult to remember them all. Luckily, help is near at hand: type `help plot` in the Command Window and you'll get a full list of parameters you can set, along with some helpful examples.

The next things we might want to add to the plot are labels for the x and y axes. Below the plot command add the following lines:

```
xlabel('theta (degrees)'); xticks(0:60:360); xlim([0 360])
ylabel('sin(theta)')
```

There are quite a few interesting things going on here, so we'll take them one by one. The `xlabel` and `ylabel` commands add text to the x and y axes, respectively. The text must be enclosed in a single quote, otherwise MATLAB will think that you're trying to use a variable for the label. The x axis of the plot should run from 0° to 360°, but our initial plot went from 0° to 400°. To set the limits on the x axis, we use the `xlim` command, which requires a two-element vector specifying the lower and upper limits. Remember that vectors are enclosed in square brackets. Finally, we place a number on the x axis every 60° by using the `xticks` command. The argument in parentheses is simply another vector that starts at zero and proceeds in increments of 60° until it reaches 360°.

Finally, let us add a title and grid to the plot:

```
title('The sine function')
grid on
```

If you have typed everything correctly, you should obtain the plot shown in Figure 3.3. Congratulations – you have just created your first MATLAB script! The full script is shown below.

```
% SinePlot.m
% makes a nice plot of a sine wave from zero to 360 degrees

clear variables; close all; clc

x = 0:360;
theta = x*pi/180;

y = sin(theta);

plot(x,y,'LineWidth',2)
xlabel('theta (degrees)'); xticks(0:60:360); xlim([0 360])
ylabel('sin(theta)')
title('The sine function')
grid on
```

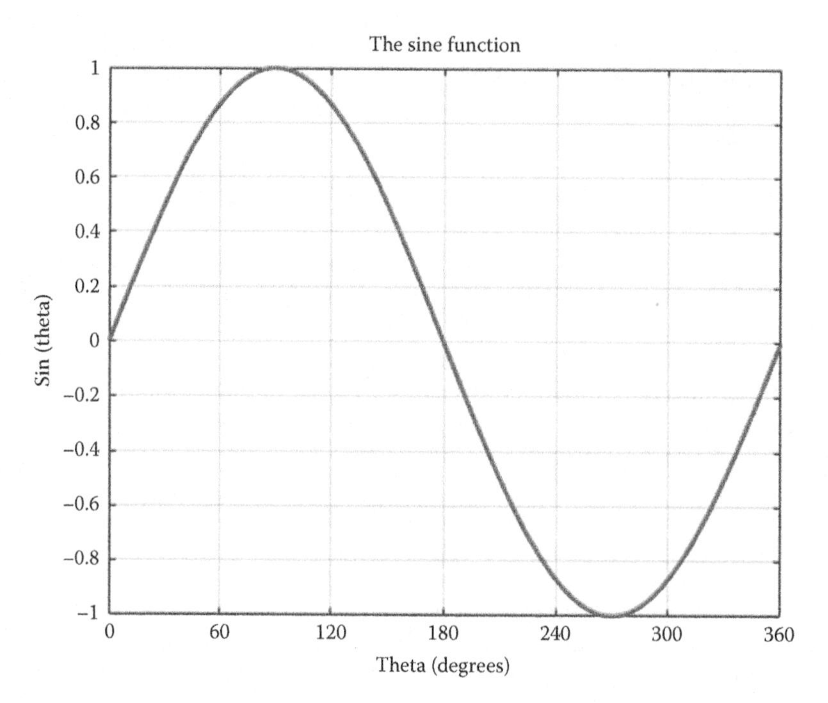

FIGURE 3.3
A much better-looking sine plot.

3.6 Plotting a Filled Square

The sine wave was a one-dimensional curve, so let us try plotting a filled 2D object: a square. Open a new MATLAB script and enter the following header:

```
% SquarePlot.m
% plots a filled square

clear variables; close all; clc
```

To create the square we will need to enter the coordinates of the vertices. To do this, we will create a matrix called `UnitSquare` that has two rows and four columns. The first row will contain all the x coordinates of the vertices and the second row will contain the y coordinates. It will be a *unit square* because the sides will have length 1, and it will be centered on the origin. If we create a table of coordinates for the vertices, it might look like the Table 3.1.

To enter this in MATLAB, type the following lines

```
UnitSquare = 0.5*[-1  1 1 -1;
                  -1 -1 1  1];
```

The 0.5 in front scales all the ones inside by one-half. The first line gives the x coordinates and the second line gives the y coordinates. Make sure you enter the semicolon after the first line so that MATLAB knows that there are two rows in this matrix. To plot the square, enter the following plotting commands

TABLE 3.1
Coordinates of Unit Square

	x	y
1	−0.5	−0.5
2	0.5	−0.5
3	0.5	0.5
4	−0.5	0.5

```
plot(UnitSquare(1,:),UnitSquare(2,:),'LineWidth',2)

xlabel('x (m)'); xlim([-2 2])
ylabel('y (m)'); ylim([-2 2])
grid on
```

The plot command may look a little strange at first. Remember that we enter the x coordinates first, followed by the y coordinates. The statement `UnitSquare(1,:)` tells MATLAB to use the first row of the `UnitSquare` matrix for the x coordinates and `UnitSquare(2,:)` gives the second row for the y coordinates. The colon in the column placeholder means that MATLAB should cycle through all the available columns in `UnitSquare` (in this case, columns 1–4). Finally, we have included the `LineWidth` parameter to make the lines stand out a little more from the grid lines.

When you save and execute the script, you should obtain the plot shown in Figure 3.4. This plot is unsatisfying for two reasons: first, the square is open on the left side and second, it appears to be squashed in the y direction. The plot command draws lines between the points that we specify, and we did not specify that the final point in the square should be the same as the starting point. Add one more column to the `UnitSquare` matrix as follows:

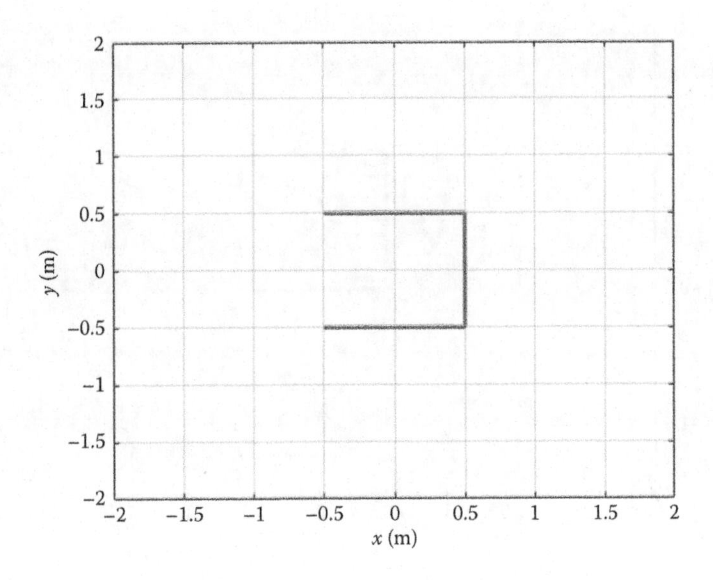

FIGURE 3.4
This is not a square!

```
UnitSquare = 0.5*[-1  1 1 -1 -1;
                  -1 -1 1  1 -1];
```

Note that the final column is the same as the first column. When you run the program again, you should obtain a closed rectangle (but not a square!) The squashing occurs because MATLAB tries to fill up as much of the plot window as it can with the requested plot. Since the x axis is much wider than the y axis, the plot ends up stretched in the x direction. To fix this, add the command

```
axis equal
```

at the end of the script.

You should now obtain the plot shown in Figure 3.5. It looks like a square, but it is not filled as required by the problem statement. To make a filled polygon we can use the `fill` command, which works much the same way as the `plot` command. Comment out the `plot` command and type the following `fill` command. You should obtain the plot shown in Figure 3.6.

```
fill(UnitSquare(1,:),UnitSquare(2,:),'b','LineWidth',2)
```

The fill command requires that you specify a color (in this case 'b' means blue) right after you give the x and y coordinates. We have now achieved the goal of plotting a filled unit square. What if we wanted to plot a square with side length 2 centered at the coordinates (1, 1)? We can simply define a new matrix, called `NewSquare`, as follows:

```
NewSquare = 2*UnitSquare + [1; 1];
```

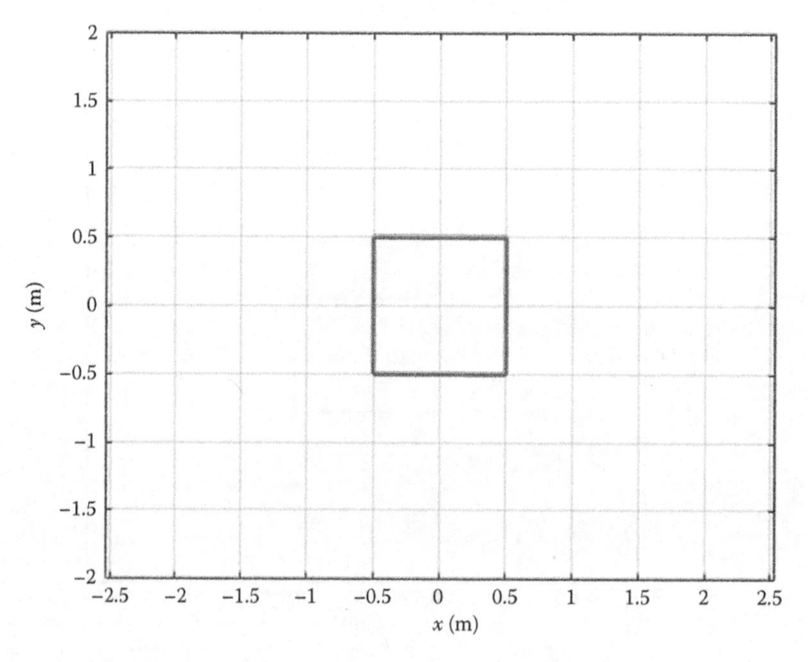

FIGURE 3.5
It's a square, but it's not filled.

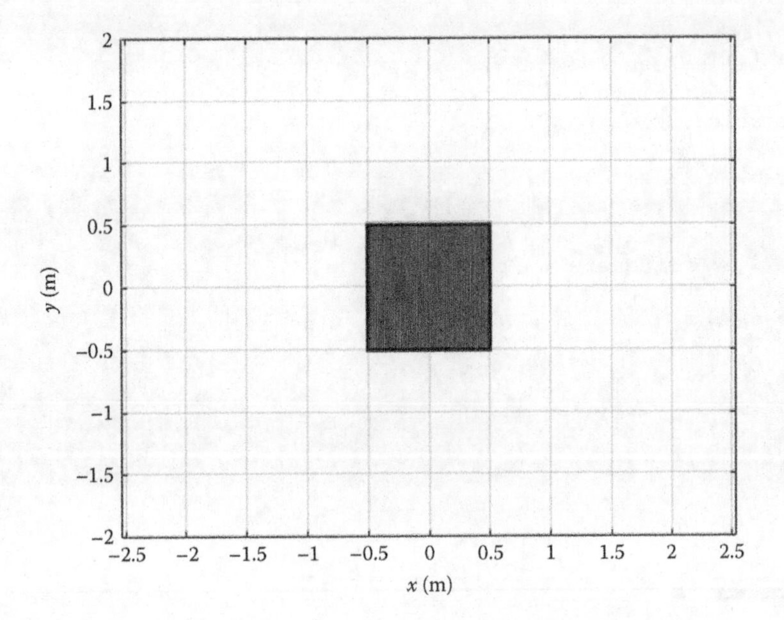

FIGURE 3.6
The filled square.

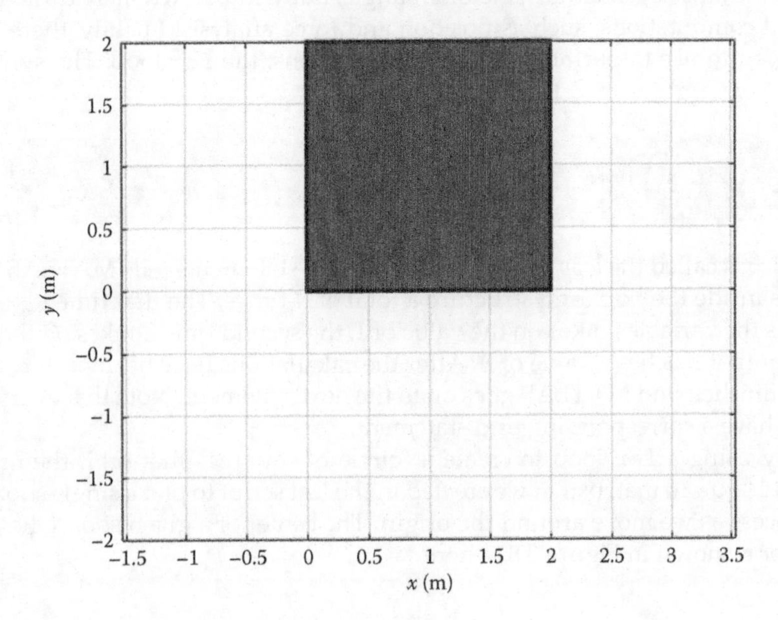

FIGURE 3.7
A larger square no longer centered at the origin.

This command multiplies each entry in the `UnitSquare` matrix by 2 (i.e., it scales the matrix by a factor of 2) and adds (1, 1) to each vertex. If you change the `fill` command to plot `NewSquare` instead of `UnitSquare`, you should obtain the plot shown in Figure 3.7. The full text of the `SquarePlot` script is shown below.

```
% SquarePlot.m
% plots a filled square

clear variables; close all; clc

UnitSquare = 0.5*[-1  1 1 -1 -1;
                  -1 -1 1  1 -1];

NewSquare = 2*UnitSquare + [1; 1];

fill(NewSquare(1,:),NewSquare(2,:),'b','LineWidth',2)

xlabel('x (m)'); xlim([-2 2])
ylabel('y (m)'); ylim([-2 2])
grid on
axis equal
```

3.7 Adding Some Structure – The for Loop

So far, we have used vector operations to perform repeated calculations. This works well for simple formulas, such as the sine of an angle, but will be extremely difficult for more complicated computations, such as position and force analysis. Luckily, there is a simple structure we can use to perform repeated calculations: the `for` loop. The syntax for this loop is

```
for i = 1:N
  do some stuff N times
end
```

The variable i is called the *loop counter*, and this set of statements tells MATLAB to perform whatever is inside the `for`...`end` structure a total of `N` times. The first time it performs the calculations the variable i takes on the value of 1, the second time it takes on the value of 2, and so on until it reaches a value of `N`. After the calculations have been performed `N` times the loop is finished and MATLAB goes on to the next statement. Note that every `for` statement must have a corresponding `end` statement.

Let us try using a `for` loop to create a "circle of squares" that orbit the origin. We'll use the `UnitSquare` matrix that we created in the last script to plot a single square, repeating the process as we move around the origin. The center of a given square lies at the end of the vector **r**, shown in Figure 3.8, where

$$\mathbf{r} = 2\begin{Bmatrix} \cos\theta \\ \sin\theta \end{Bmatrix}$$

Create a new script called `SquareCircle.m`. At the top of the file type the header:

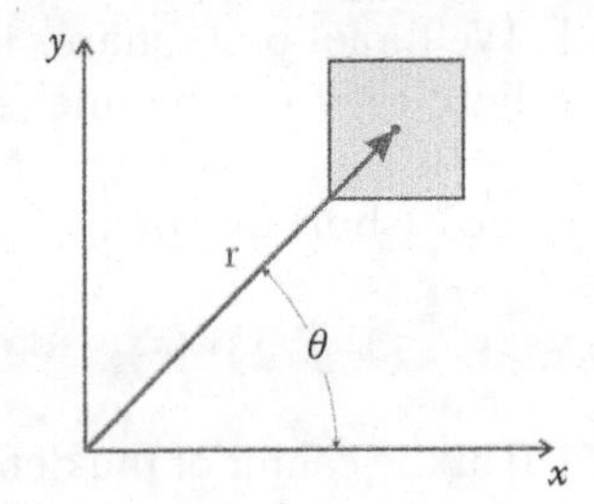

FIGURE 3.8
The origin of each square lies at the end of the vector **r**.

```
% SquareCircle.m
% makes a plot of eight squares arranged in a circle

clear variables; close all; clc

UnitSquare = 0.5*[-1  1 1 -1 -1;
                  -1 -1 1  1 -1];
```

We wish to plot eight squares, so type in the following for loop

```
for i = 1:8

end
```

The first thing we need to calculate inside the loop is the angle of the vector **r**, which we denote by `theta`. We want `theta` to start at zero and increment by 45° at each iteration of the loop. Since `i` starts at 1, we must subtract 1 if we wish to start at 0. The angle 45° is equal to $2\pi/8$ in radians. Thus, the first line inside the for loop should read

```
  theta = (i-1)*2*pi/8;              % angle at which to place square
```

It is customary to indent all the lines inside the for loop so that it is easy to see which commands are executed multiple times. The reader can easily verify that `theta` begins at 0 (when i = 1) and ends at $(7/8)\ 2\pi = 315°$ when i = 8. Next, we calculate the vector **r**, which gives the center of a given square

```
  r = 2*[cos(theta); sin(theta)]; % center of square
```

The vector **r** has been given a length of 2. Next, we'll create a new square centered at **r**

```
  NewSquare = UnitSquare + r;       % create new square
```

Since we will be creating filled squares, it would be nice if we could fill each square with a different color. MATLAB specifies colors as a three-element vector giving the red, green, and blue components, respectively. The values of `R`, `G`, and `B` must lie between zero and one with one giving full saturation of a given color. Thus, the triple [1 1 1] specifies the color white, [1 0 0] gives a red color, [0 0 0] gives black, and [0.5 0.5 0.5] gives the color gray halfway between white and black. We'll plot our squares in grayscale starting with black for the first square and ending in light gray for the final square. Since the color must be different for each square we must use the loop counter `i` to define the colors. On the next line in the loop, define the variable `col` (short for color).

```
col = (i-1)*[1 1 1]/8;                  % fill and edge color of square
```

Note the three-element vector [1 1 1] at the center of the definition. This triple is multiplied by `(i-1)/8`, which starts at 0 when `i` = 1 and ends at 7/8 when `i` = 8. Thus, the first color is defined as `[0 0 0]` (black) and the final color is defined as `[0.875 0.875 0.875]` (really light gray). We are now ready to plot the squares

```
fill(NewSquare(1,:),NewSquare(2,:),col,'EdgeColor',col,'LineWidth',2)
```

If you save and execute the program now you'll find that it seems to have plotted only the final square in the sequence. This is because the `fill` command automatically erases whatever was in the plot window before plotting the new square. Right after the `fill` command, type the command

```
hold on
```

This tells MATLAB to preserve everything in the plot window so that the next plotting command will simply add a new square to the existing plot. After the loop is finished, we should add the commands that make the plot look professional.

```
xlabel('x (m)'); xlim([-3 3])
ylabel('y (m)'); ylim([-3 3])
grid on
axis equal
```

If you save and execute the script (using F5) you will obtain the plot shown in Figure 3.9. The full listing of the script is shown below.

```
% SquareCircle.m
% makes a plot of eight squares arranged in a circle

clear variables; close all; clc

UnitSquare = 0.5*[-1  1 1 -1 -1;
                  -1 -1 1  1 -1];

for i = 1:8
  theta = (i-1)*2*pi/8;                 % angle at which to place square
  r = 2*[cos(theta); sin(theta)];       % center of square
  NewSquare = UnitSquare + r;           % create new square
  col = (i-1)*[1 1 1]/8;                % fill and edge color of square
  fill(NewSquare(1,:),NewSquare(2,:),col,'EdgeColor',col,'LineWidth',2)
  hold on
end
```

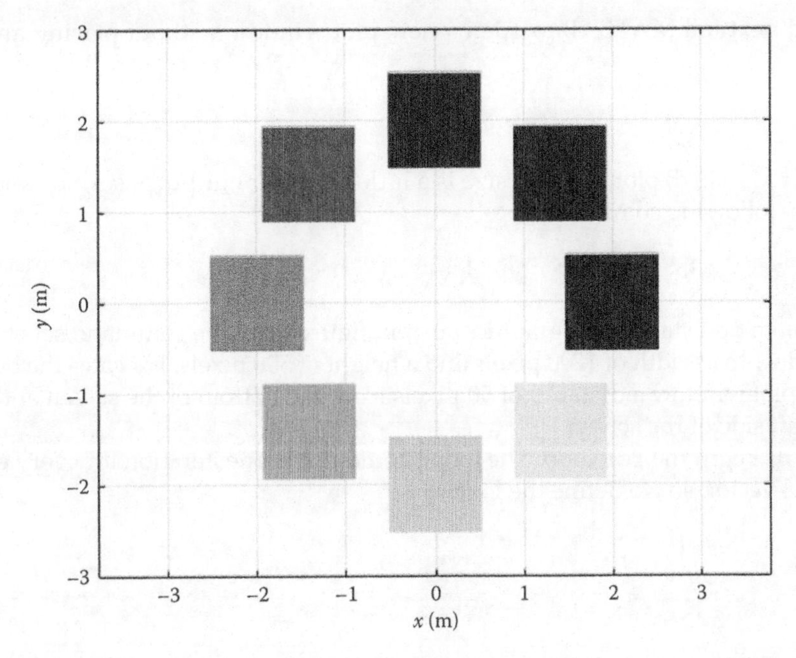

FIGURE 3.9
A circle of multi-colored squares.

```
xlabel('x (m)'); xlim([-3 3])
ylabel('y (m)'); ylim([-3 3])
grid on
axis equal
```

3.8 A Primitive Animation

Sometimes the best way to visualize the motion of a mechanism is to animate it. MATLAB has a full suite of features for making movie files, but we will concentrate here on animating a plot within a normal plot window. For this example, we will animate the motion of a projectile being thrown in the air at a 45° angle. The projectile has an initial velocity in both the x and y directions of 50 m/s. Open a new script called `Projectile.m` and type the following header:

```
% Projectile.m
% animates the path of a projectile

clear variables; close all; clc

t = 0:0.1:15;   % [s] vector of times
vx0 = 50;       % [m/s] initial velocity in the x direction
vy0 = 50;       % [m/s] initial velocity in the y direction
g = 9.81;       % [m/s^2] acceleration of gravity
```

We should next tell MATLAB to open a new plot window without placing anything in it yet.

```
figure
```

The default MATLAB plot window size is a little small for our purposes, so we will resize it using the following command

```
set(gcf,'Position',[50 50 1200 500])
```

The argument `gcf` stands for "graphics current figure" and the command sets the current `plot` window to a width of 1200 pixels and a height of 500 pixels. It locates the bottom corner of the plot window a distance of 50 pixels from the bottom of the screen and 50 pixels from the left side of the screen.

Now let us begin the `for` loop. The loop should make one iteration for every element in the `t` (time) vector, so we define the loop as

```
for i = 1:length(t)

end
```

The `length` statement gives the number of elements in the vector `t`. At the beginning of the loop, we should calculate the current position of the projectile. We do this by using some basic equations from introductory Physics:

$$x = v_{x0}t$$

$$y = v_{y0}t - \frac{1}{2}gt^2$$

Translating these equations into MATLAB gives

```
  x = vx0 * t(i);                    % x position
  y = vy0 * t(i) - 0.5*g*t(i)^2;    % y position
```

Remember that `t` is a vector, not a single number. The statement `t(i)` gives the *i*th number in the `t` vector. Since we have defined the loop such that `i` ranges from 1 to `length(t)` we will use every value of `t` as we iterate through the loop. The first iteration of the loop uses the first value within the `t` vector, the second iteration uses the second value in `t`, and so on.

We next plot the current position of the projectile.

```
  plot(x,y,'o','MarkerFaceColor','b')
```

Unlike our previous plots, the quantities x and y contain a single number each – the coordinates of the current position of the projectile. At a given time step, we are plotting the position of a single point, the current position of the projectile. For this reason, there is no line or curve to plot, since we don't have more than one point. The statement `'o'` tells MATLAB to plot the current point with a small circular marker, and the `'MarkerFaceColor'` command makes the circular marker a solid blue color. Next, enter the commands that make the plot presentable

```
axis equal
xlabel('x (m)'); xlim([0 600])
ylabel('y (m)'); ylim([0 150])
grid on
```

The axis limits were found through trial and error, and you will need to change them if your projectile has a different initial velocity. The final command in the loop is

```
drawnow
```

that tells MATLAB to place everything that is in the graphics buffer onto the plot window. This concludes the `for` loop, and you can save and execute the program now. When you execute the program, you should be rewarded with a beautiful animation of the projectile flying through the air and then landing (or falling below the x axis). The full listing of the program is shown below.

```
% Projectile.m
% animates the path of a projectile

clear variables; close all; clc

t = 0:0.1:15;   % vector of times
vx0 = 50;       % initial velocity in the x direction
vy0 = 50;       % initial velocity in the y direction
g = 9.81;       % acceleration of gravity

figure
set(gcf,'Position',[50 50 1200 500])

for i = 1:length(t)
  x = vx0 * t(i);                        % x position
  y = vy0 * t(i) - 0.5*g*t(i)^2; % y position

  plot(x,y,'o','MarkerFaceColor','b')

  axis equal
  xlabel('x (m)'); xlim([0 600])
  ylabel('y (m)'); ylim([0 150])
  grid on

  drawnow
end
```

3.9 Summary

We have now learned enough MATLAB to begin conducting position analysis on simple linkages. Some of what we learned here will be repeated in Chapter 4 when we learn to do position analysis on the threebar linkage. As you type and execute your programs, you will undoubtedly encounter many of the error messages described above. In most cases, the errors are the result of a vector or matrix not having the correct dimension (e.g. what you thought was a row vector was entered as a column vector, etc.) By using the definitions of the variables in the Workspace window, you can usually track down where the problem lies.

In other cases, the error is a result of using incorrect syntax for a particular command. For example, if we typed

```
fill(NewSquare(1,:),NewSquare(2,:))
```

in an effort to obtain a filled square, MATLAB would respond with

```
Error using fill
Not enough input arguments.

Error in SquareCircle (line 14)
fill(NewSquare(1,:),NewSquare(2,:))
```

When you encounter an error message of the type `error using` ... the best approach is to type `help fill` in the Command Window to see the correct syntax. As it happens, the `fill` command requires that we specify a fill color, in addition to the coordinates, which is what the error message was telling us in its own cryptic manner. Above all, don't become discouraged when you try to run your program for the first time and obtain a screen full of error messages. It is always possible to track down the source of the errors using a combination of deductive reasoning and the techniques described above. It is entirely normal to find MATLAB frustrating at first, in the same way that a bicycle or skateboard is frustrating until you learn to keep your balance. As in all things, practice makes perfect!

Acknowledgments

Several images in this chapter were produced using MATLAB software.

MATLAB is a registered trademark of The MathWorks, Inc.

4

Position Analysis of Linkages

4.1 Introduction to Position Analysis

In the previous chapter, we used graphical methods to design fourbar linkages to reach two or three specified positions. While these methods are handy for designing simple linkages, it is often the case that we require knowledge of the behavior of the linkage over its entire range of motion. Some important reasons include:

1. *Timing analysis*
   It may be important to predict the length of time a link requires to reach each specified position. Interfacing with other machinery, as done on assembly lines, often requires precise timing of each part.
2. *Prevention of interference*
   In most cases linkages must operate within a limited space (e.g. engine bay, assembly plant room, etc.). Knowledge of the "envelope" through which a linkage travels is therefore critical. A linkage with too large an envelope must be redesigned and reevaluated.
3. *Failure prevention*
   This may be divided into three categories:
   a. *Failure of the linkage*: If stresses within the links or at the joints of a linkage become too great then the linkage may fail. Example: failure of a crankshaft or connecting rod in a piston engine.
   b. *Failure of parts attached to a linkage*: Unless a linkage is properly balanced, its movement can create vibrations in the surrounding environment. This is always undesirable.
   c. *Failure of the object(s) being moved by the linkage*: All linkages are designed to move something. If the movement is done too quickly or with too much force, the objects being moved may fail. Examples: amusement park rides, overrevving an engine.

In Case (1), we need to predict the time it takes for a linkage to reach specified positions. For Case (2), we need to know the positions of the links at all times. For Case (3), we need the positions, velocities, and accelerations of each link to predict dynamic forces in and around the linkage.

As we learned in the design of the quick-return mechanism, we can predict the timing of a linkage using graphical methods fairly easily. In theory, we could also predict interference graphically, although this would quickly become tedious. Prediction and prevention of failure is very difficult and time-consuming using graphical means alone. Clearly, a better solution is needed.

In this chapter, we will develop methods for finding the overall configuration of some common linkages at any time. Since performing the calculations by hand will prove to be rather involved and time-consuming, our method should admit easy implementation in software, specifically MATLAB®. The approach we will use will be *vectorial* and *geometric.* For some of the linkages (e.g. inverted slider-crank, fourbar) we will solve for the position of each link using geometry alone. For others (e.g. the threebar and fourbar slider-crank) we will adopt the vector loop approach. Others (e.g. the geared fivebar) will use a combination of the two approaches.

To completely specify the configuration of a linkage, we must know at least one of the link angles in advance. For example, one of the links may be driven by a motor, as in the windshield wiper mechanism. In the fourbar linkage, we are given the *crank angle* (shown as θ_2 in Figure 4.1). The goal is then to find the angles of the other links (θ_3 and θ_4) as a function of the crank angle. Note that all angles are measured from the horizontal in the counterclockwise direction, and the ground link (link 1) is assumed to be horizontal. Because of this, θ_1 is always zero. If we wish to model a fourbar linkage that has a non-horizontal ground, we must employ a coordinate transformation, which will be discussed in a later section. We will begin our discussion with a review of vectors and matrices. Next, we will develop a method for calculating the positions of the simplest of linkages: the threebar slider-crank. From there we proceed to the fourbar linkage, the inverted slider-crank, the geared fivebar and finally the family of sixbar linkages.

4.2 Review of Vectors and Matrices

To begin our study of position analysis, a review of some basic vector operations is necessary. This will enable us to write our kinematic equations compactly and efficiently, and is also handy while using MATLAB. We begin with a short review of some properties of vectors and matrices, and conclude with coordinate transformations.

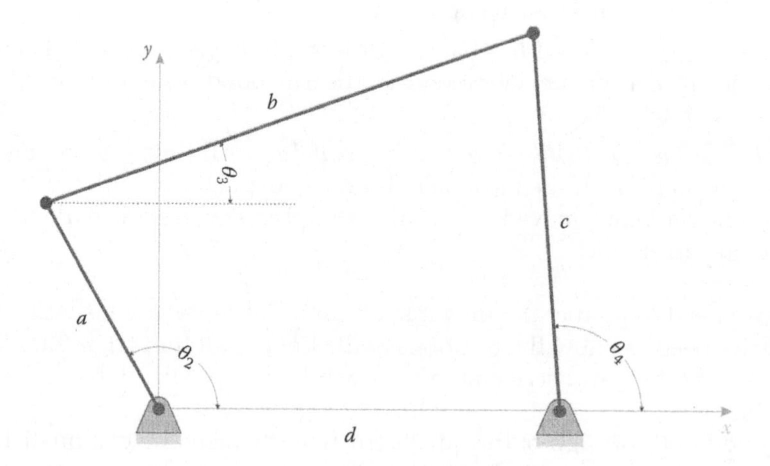

FIGURE 4.1
A typical fourbar linkage showing the link lengths and angles.

A vector is simply a row or column of numbers, as shown

$$\mathbf{v} = \begin{Bmatrix} v_1 \\ v_2 \\ \vdots \\ v_n \end{Bmatrix} \tag{4.1}$$

or

$$\mathbf{v} = \begin{Bmatrix} v_1 & v_2 & \dots & v_n \end{Bmatrix} \tag{4.2}$$

A matrix, on the other hand, is a set of numbers arranged in a rectangular grid.

$$\mathbf{A} = \begin{bmatrix} A_{11} & A_{12} & A_{13} \\ A_{21} & A_{22} & A_{23} \end{bmatrix} \tag{4.3}$$

Both vectors and matrices will be indicated in **bold** typeface. Lowercase letters will be used for vectors, and uppercase will be used for matrices. We specify the *dimension* of a vector or matrix by giving the number of rows first, followed by the number of columns. For example, the vector in Equation (4.1) has dimension $(n \times 1)$, the vector in Equation (4.2) is $(1 \times n)$ and the matrix in Equation (4.3) is (2×3). The vector in Equation (4.1) is often called a *column vector* because it inhabits a single column, while the vector in Equation (4.2) is called a *row vector* because it occupies a single row.

Often, we use vectors to represent directed line segments in space. In two dimensions (i.e. on a plane) vectors take the form:

$$\mathbf{r} = \begin{Bmatrix} r_x \\ r_y \end{Bmatrix} \tag{4.4}$$

as shown in Figure 4.2. The *length* or *magnitude* of a vector, denoted by vertical pipes on either side of a vector, is found using the Pythagorean theorem.

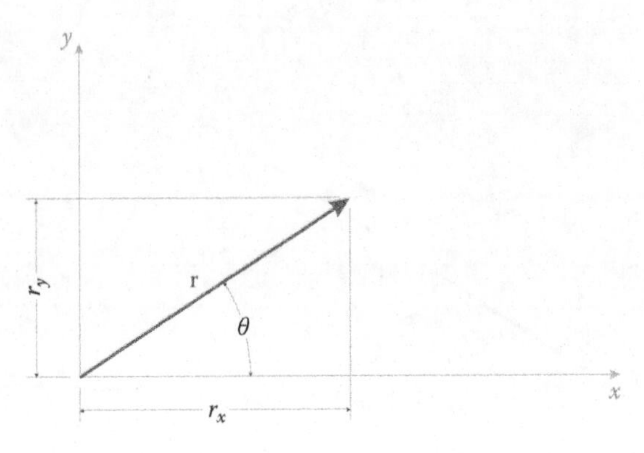

FIGURE 4.2
A two-dimensional vector has an x and y component.

$$|\mathbf{r}| = \sqrt{r_x^2 + r_y^2} \tag{4.5}$$

and the angle θ that a vector makes with the horizontal is

$$\tan\theta = \frac{r_y}{r_x} \tag{4.6}$$

It is frequently the case that we know the length of a vector and its angle, rather than its x and y components. This will occur, for example, when we perform position analysis on a linkage, since we know the lengths of each link in advance. The x and y components can be computed as

$$\begin{Bmatrix} r_x \\ r_y \end{Bmatrix} = \begin{Bmatrix} r\cos\theta \\ r\sin\theta \end{Bmatrix} \tag{4.7}$$

where $r = |\mathbf{r}|$ is the length of the vector (see Figure 4.3). Often, we will place the magnitude in front to more easily distinguish between magnitude and direction

$$\begin{Bmatrix} r_x \\ r_y \end{Bmatrix} = r\begin{Bmatrix} \cos\theta \\ \sin\theta \end{Bmatrix} \tag{4.8}$$

In this notation, the magnitude r is multiplied by both the cosine and sine terms. The quantity on the right in brackets is now a *unit vector*, which we will discuss in more detail below.

4.2.1 Vector Addition

Adding two vectors is equivalent to taking the sums of the individual components of each vector, as shown in Figure 4.4.

$$\mathbf{r} + \mathbf{s} = \begin{Bmatrix} r_x + s_x \\ r_y + s_y \end{Bmatrix} = \begin{Bmatrix} r\cos\theta_r + s\cos\theta_s \\ r\sin\theta_r + s\sin\theta_s \end{Bmatrix} \tag{4.9}$$

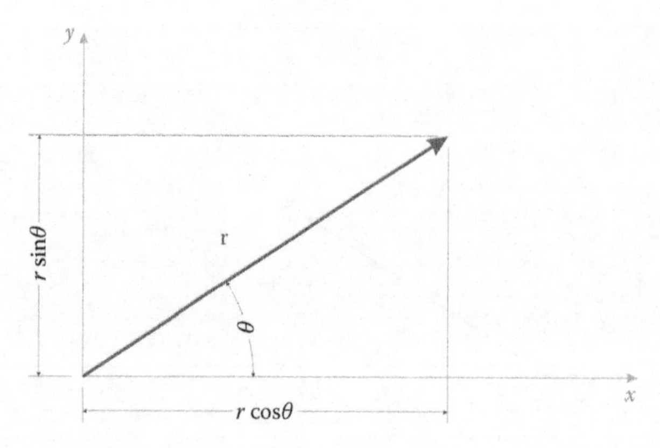

FIGURE 4.3
Horizontal and vertical components of a vector as found using trigonometry.

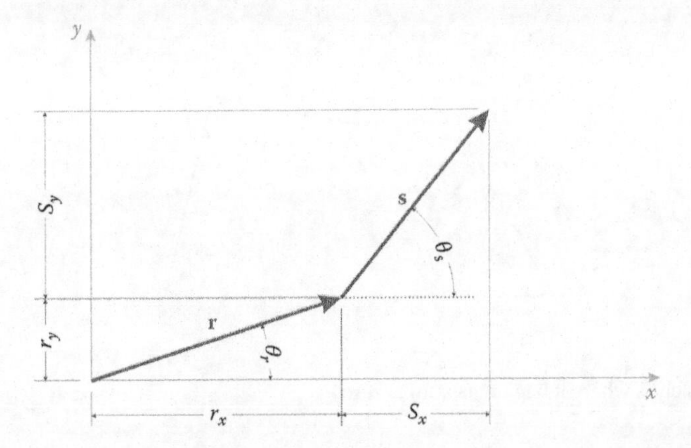

FIGURE 4.4
Geometric interpretation of adding two vectors.

The result of a vector sum is another vector.

It is often the case that we wish to find the coordinates of a point P, which is attached to a particular link. If we know the vectors **r** and **s** that are associated with links that form a "chain" to P, then we can write

$$\mathbf{p} = \mathbf{r} + \mathbf{s} \tag{4.10}$$

If the coordinates of point P relative to the origin are needed, then **r** must start at the origin (Figure 4.5).

4.2.2 The Vector Loop

Now consider the chain of vectors shown in Figure 4.6. We begin with $\mathbf{r}_2$, and proceed clockwise around the loop. Since the coordinates of the start of the loop and the end of the loop are the same, the sum of the vector loop is zero.

$$\mathbf{r}_2 + \mathbf{r}_3 - \mathbf{r}_4 - \mathbf{r}_1 = 0 \tag{4.11}$$

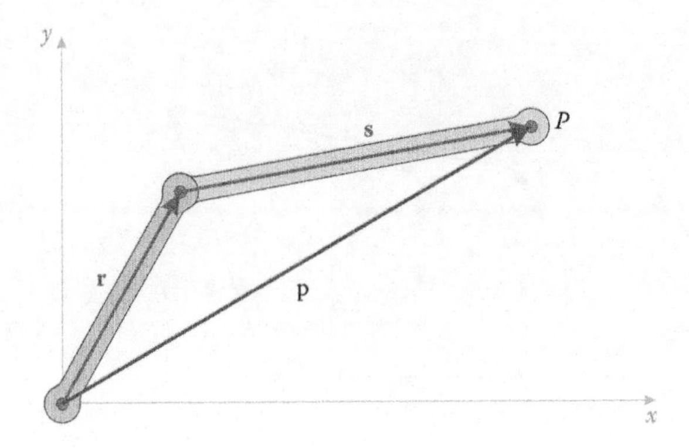

FIGURE 4.5
Finding the coordinates of point P by adding two vectors.

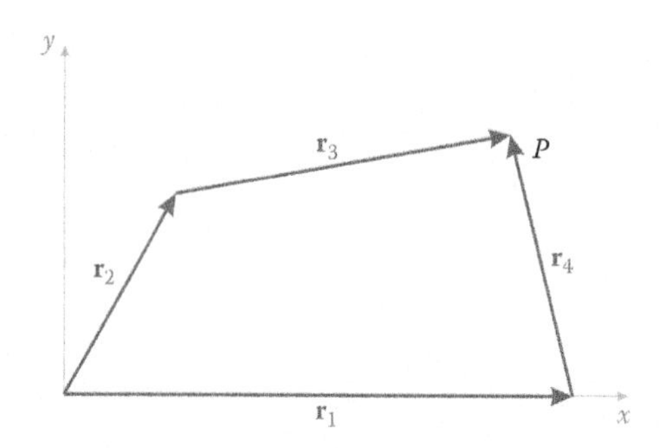

FIGURE 4.6
If a series of vectors ends where it began, its sum is zero.

Note in particular the negative signs associated with $\mathbf{r}_4$ and $\mathbf{r}_1$. In moving clockwise around the loop we move in the opposite sense of these two vectors (i.e., from the head to the tail), so we must subtract them instead of adding them. If we had chosen to move counterclockwise around the loop starting with $\mathbf{r}_1$ we would have

$$\mathbf{r}_1 + \mathbf{r}_4 - \mathbf{r}_3 - \mathbf{r}_2 = 0 \tag{4.12}$$

which is the same as the Equation (4.11), except multiplied by –1. It is important to remember that Equations (4.11) and (4.12) have two components each (an x and y component). While Equation (4.11) is written as a single equation in vector notation, it contains two separate equations, both of which must be satisfied. In other words

$$\begin{aligned} r_{2x} + r_{3x} - r_{4x} - r_{1x} &= 0 \\ r_{2y} + r_{3y} - r_{4y} - r_{1y} &= 0 \end{aligned} \tag{4.13}$$

You might be wondering how the direction for each vector was chosen when drawing the vector loop. In this text, we will strictly maintain the convention of measuring all angles from the positive x axis. This will make our vector computations much simpler and less prone to error. Another convention we will adopt is that of placing the tail of a vector on a ground pin, whenever possible. As shown in Figure 4.7, this convention enables us to

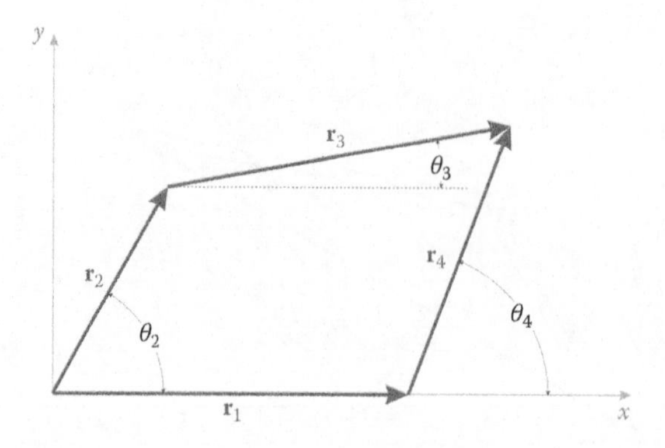

FIGURE 4.7
The direction of each vector is somewhat arbitrary, but we must remain consistent when we use the vector diagram to write the vector loop equation.

measure the angles of $\mathbf{r}_2$ and $\mathbf{r}_4$ directly from the fixed x axis. Unfortunately, the root of $\mathbf{r}_3$ is not fixed in space, so we must be content to measure its angle from the horizontal, as shown in the figure. It is important to understand that the choice of direction for the vector is arbitrary, but once a direction has been chosen the designer must remain consistent when developing the vector loop equations.

In other words, it would be perfectly valid to choose the directions for $\mathbf{r}_3$ and $\mathbf{r}_4$ shown in Figure 4.8, but the resulting vector loop equation would be changed to

$$\mathbf{r}_2 - \mathbf{r}_3 + \mathbf{r}_4 - \mathbf{r}_1 = 0$$

and the angles for each vector would need to be measured as shown in the figure. We have a great deal of flexibility in analyzing mechanisms, but consistency is critical.

4.2.3 The Dot Product

The dot product (or scalar product) of two vectors is defined as

$$\mathbf{r} \cdot \mathbf{s} = r_x s_x + r_y s_y = |\mathbf{r}||\mathbf{s}|\cos\theta \tag{4.14}$$

where θ is the angle between the two vectors. Note that if $\theta = 90°$ (i.e. if the vectors are orthogonal to each other) then $\cos\theta = 0$ and $\mathbf{r} \cdot \mathbf{s} = 0$. This is a good test as to whether two vectors are perpendicular to each other. We will often find it useful to compute a vector perpendicular to $\mathbf{r}$. There are two possibilities, given by

$$\mathbf{r}^{\perp} = \begin{Bmatrix} -r_y \\ r_x \end{Bmatrix} \qquad \mathbf{r}^{\perp} = \begin{Bmatrix} r_y \\ -r_x \end{Bmatrix} \tag{4.15}$$

Both of these are perpendicular to $\mathbf{r}$, but point in opposite directions. To prove that these are perpendicular to $\mathbf{r}$, we can use the dot product

$$\mathbf{r} \cdot \mathbf{r}^{\perp} = r_x(-r_y) + r_y r_x = 0 \qquad \mathbf{r} \cdot \mathbf{r}^{\perp} = r_x r_y + r_y(-r_x) = 0 \tag{4.16}$$

The perpendicular vector on the left in Equation (4.15) represents a rotation of the vector $\mathbf{r}$ *counterclockwise* by 90°, while the vector on the right represents a *clockwise* rotation of 90°.

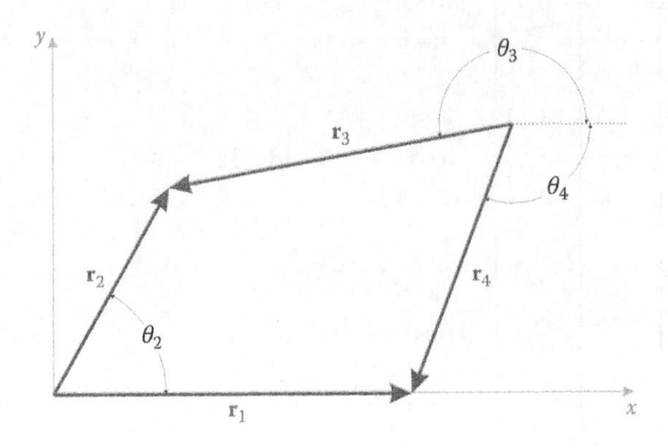

FIGURE 4.8
The direction of vectors $\mathbf{r}_3$ and $\mathbf{r}_4$ has been reversed in this diagram.

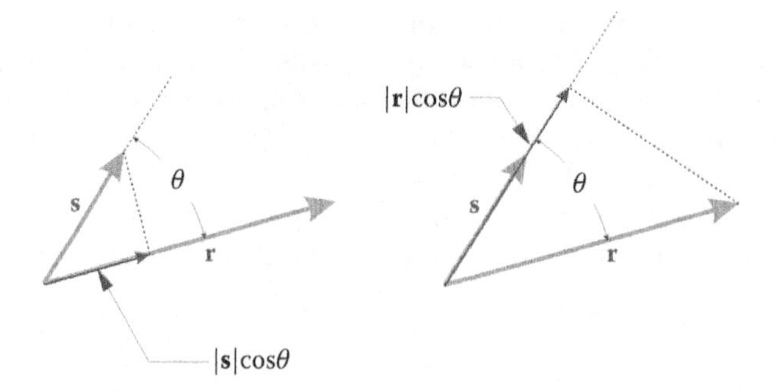

FIGURE 4.9
The dot product gives the projection of one vector onto another.

We will use the expression on the left more frequently, since a counterclockwise rotation is considered positive in a right-handed coordinate system.

The dot product may be interpreted geometrically as giving the projection of one vector onto another, as shown in Figure 4.9. If we denote the projection of **s** onto **r** as

$$s_r = |\mathbf{s}|\cos\theta$$

then the dot product **r** · **s** can be written

$$\mathbf{r}\cdot\mathbf{s} = |\mathbf{r}|\,s_r$$

and a similar expression can be written for r_s. If the angle between the two vectors is 90°, then the projection of one onto the other is zero, as is the dot product.

Example 4.1: Compute the Dot Product a · b for the Following Cases

1. $\mathbf{a} = \begin{Bmatrix} 5 \\ 2 \end{Bmatrix}$ $\mathbf{b} = \begin{Bmatrix} 1 \\ 6 \end{Bmatrix}$ $\mathbf{a}\cdot\mathbf{b} = 5 + 12 = 17$

2. $\mathbf{a} = \begin{Bmatrix} a_1 \\ a_2 \end{Bmatrix}$ $\mathbf{b} = \begin{Bmatrix} b_1 \\ b_2 \end{Bmatrix}$ $\mathbf{a}\cdot\mathbf{b} = a_1b_1 + a_2b_2$

3. $\mathbf{a} = \begin{Bmatrix} 2 \\ 4 \end{Bmatrix}$ $\mathbf{b} = \begin{Bmatrix} -4 \\ 2 \end{Bmatrix}$ $\mathbf{a}\cdot\mathbf{b} = -8 + 8 = 0$

4. $\mathbf{a} = \begin{Bmatrix} 1 \\ 2 \\ 3 \end{Bmatrix}$ $\mathbf{b} = \begin{Bmatrix} 4 \\ 5 \\ 6 \end{Bmatrix}$ $\mathbf{a}\cdot\mathbf{b} = 4 + 10 + 18 = 32$

5. $\mathbf{a} = \begin{Bmatrix} 1 \\ 2 \\ 3 \end{Bmatrix}$ $\mathbf{b} = \begin{Bmatrix} 0 \\ -3 \\ 2 \end{Bmatrix}$ $\mathbf{a}\cdot\mathbf{b} = 0 - 6 + 6 = 0$

6. $\mathbf{a} = \begin{Bmatrix} 1 \\ 2 \\ 3 \end{Bmatrix}$ $\mathbf{b} = \begin{Bmatrix} -2 \\ 1 \\ 0 \end{Bmatrix}$ $\mathbf{a}\cdot\mathbf{b} = -2 + 2 + 0 = 0$

7. $\mathbf{a} = \begin{Bmatrix} 1 \\ 1 \\ 1 \\ 1 \end{Bmatrix}$ $\mathbf{b} = \begin{Bmatrix} 1 \\ -1 \\ 1 \\ -1 \end{Bmatrix}$ $\mathbf{a} \cdot \mathbf{b} = 1 - 1 + 1 - 1 = 0$

8. $\mathbf{a} = \begin{Bmatrix} 1 \\ 1 \\ 1 \\ 1 \\ 1 \end{Bmatrix}$ $\mathbf{b} = \begin{Bmatrix} 1 \\ -1.5 \\ 1 \\ -1.5 \\ 1 \end{Bmatrix}$ $\mathbf{a} \cdot \mathbf{b} = 1 - 1.5 + 1 - 1.5 + 1 = 0$

As you can see by examples 5–8, orthogonal vectors can be found in 3, 4, or even 5 dimensions. Note that the result of a dot product is always a *scalar,* and not a vector. One common quantity computed with a dot product is work, which is defined as

$$dW = \mathbf{F} \cdot d\mathbf{r} \tag{4.17}$$

where **F** is a force and $d\mathbf{r}$ is the distance that the force has caused an object to move.

4.2.4 The Cross Product

Another useful vector tool is the cross product, which is usually denoted by the symbol "×". There are probably as many different methods for calculating cross products as there are math teachers, but one standard formula is

$$\mathbf{r} \times \mathbf{s} = \left(r_y s_z - r_z s_y\right)\hat{I} - \left(r_x s_z - r_z s_x\right)\hat{J} + \left(r_x s_y - r_y s_x\right)\hat{k} \tag{4.18}$$

Where $\hat{I}$, $\hat{J}$, $\hat{k}$ are the unit vectors (vectors of length 1) in the x, y and z directions, respectively. If both **r** and **s** are two-dimensional vectors (i.e. with zero z components) then the formula is quite simple:

$$\mathbf{r} \times \mathbf{s} = \left(r_x s_y - r_y s_x\right)\hat{k} \tag{4.19}$$

where $\hat{k}$ is the unit vector in the z direction. The cross product always produces a vector that is perpendicular to both of the vectors in the product—this is why we obtain a vector pointing in the z direction when we compute the cross product of two vectors confined to the xy plane. If the cross product of two nonzero vectors is zero, it indicates that the two vectors are parallel or antiparallel. One common quantity computed using a cross product is *torque,* where

$$\mathbf{T} = \mathbf{r} \times \mathbf{F} \tag{4.20}$$

where **F** is the applied force and **r** is the vector from the center of rotation to the point of application of the force. Note that if the force is directed inward to the center of rotation (i.e. it is parallel to **r**) then the resulting torque is zero. We will use the concept of torque quite often when we conduct force analysis on mechanisms in Chapter 7.

Example 4.2: Computing Cross Products

1. $\mathbf{a} = \begin{Bmatrix} 1 \\ 2 \end{Bmatrix} \quad \mathbf{b} = \begin{Bmatrix} 3 \\ 4 \end{Bmatrix} \quad \mathbf{a} \times \mathbf{b} = 4 - 6 = -2\hat{k}$

2. $\mathbf{a} = \begin{Bmatrix} 3 \\ 4 \end{Bmatrix} \quad \mathbf{b} = \begin{Bmatrix} 1 \\ 2 \end{Bmatrix} \quad \mathbf{a} \times \mathbf{b} = 6 - 4 = 2\hat{k} \quad (\mathbf{a} \times \mathbf{b} = -\mathbf{b} \times \mathbf{a})$

3. $\mathbf{a} = \begin{Bmatrix} 1 \\ 2 \end{Bmatrix} \quad \mathbf{b} = \begin{Bmatrix} 2 \\ 4 \end{Bmatrix} \quad \mathbf{a} \times \mathbf{b} = 4 - 4 = 0\hat{k}$ ($\mathbf{a}$ and $\mathbf{b}$ are parallel)

4. $\mathbf{a} = \begin{Bmatrix} 1 \\ 0 \\ 0 \end{Bmatrix} \quad \mathbf{b} = \begin{Bmatrix} 0 \\ 0 \\ 5 \end{Bmatrix} \quad \mathbf{a} \times \mathbf{b} = -5\hat{j}$ ($\mathbf{a}$ and $\mathbf{b}$ are both perpendicular to $\hat{j}$)

5. $\mathbf{a} = \begin{Bmatrix} -1 \\ 0 \\ 0 \end{Bmatrix} \quad \mathbf{b} = \begin{Bmatrix} 0 \\ 0 \\ 5 \end{Bmatrix} \quad \mathbf{a} \times \mathbf{b} = 5\hat{j}$ ($\mathbf{a}$ and $\mathbf{b}$ are both perpendicular to $\hat{j}$)

4.2.5 Unit Vectors

In later chapters on velocity, acceleration, and force analysis, we will find it convenient to work with *unit vectors*. Like any vector, a unit vector has a direction and magnitude, but the magnitude of a unit vector is always one. Figure 4.10 shows a situation where two points, P and Q, lie on the same link. The vector from the origin to point P is $\mathbf{r}$, and the vector from the origin to point Q is $\mathbf{s}$, where

$$\mathbf{r} = r\begin{Bmatrix} \cos\theta \\ \sin\theta \end{Bmatrix} \quad \mathbf{s} = s\begin{Bmatrix} \cos\theta \\ \sin\theta \end{Bmatrix} \tag{4.21}$$

Since both of these vectors have the same direction, but different magnitude, it is convenient to define the unit vector $\mathbf{e}$ as

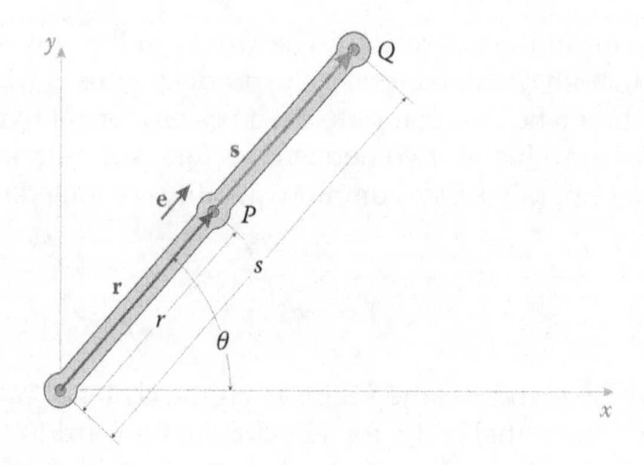

FIGURE 4.10
The unit vector **e** points in the same direction as **r** and **s**.

$$\mathbf{e} = \begin{Bmatrix} \cos\theta \\ \sin\theta \end{Bmatrix} \tag{4.22}$$

The reader may verify that the magnitude of this vector is in fact unity, since

$$\cos^2\theta + \sin^2\theta = 1 \tag{4.23}$$

Now **r** and **s** can be written

$$\mathbf{r} = r\mathbf{e} \quad \mathbf{s} = s\mathbf{e} \tag{4.24}$$

This is an altogether more compact notation, and will save us having to write out a seemingly endless stream of trigonometric functions. Since MATLAB was designed to handle vectors and matrices easily, we will find it quite simple to use unit vectors in carrying out our position, velocity, and acceleration analyses.

Any vector can be written as a product of the vector's length and a unit vector giving direction. As an example, consider the linkage chain shown in Figure 4.11. The vector from point A to point B can be written

$$\mathbf{r}_{AB} = a\mathbf{e}_2 \tag{4.25}$$

and the vector loop equation in terms of unit vectors is

$$a\mathbf{e}_2 + b\mathbf{e}_3 - c\mathbf{e}_4 - d\mathbf{e}_1 = 0 \tag{4.26}$$

Now recall the definition of the dot product given earlier

$$\mathbf{r}\cdot\mathbf{s} = |\mathbf{r}||\mathbf{s}|\cos\theta \tag{4.27}$$

If we write a dot product using only unit vectors, say $\mathbf{e}_2$ and $\mathbf{e}_3$ the result is

$$\mathbf{e}_2\cdot\mathbf{e}_3 = \cos\theta \tag{4.28}$$

since the magnitude of both vectors in the product is one. Thus, it is quite straightforward to find the angle between two unit vectors using the dot product.

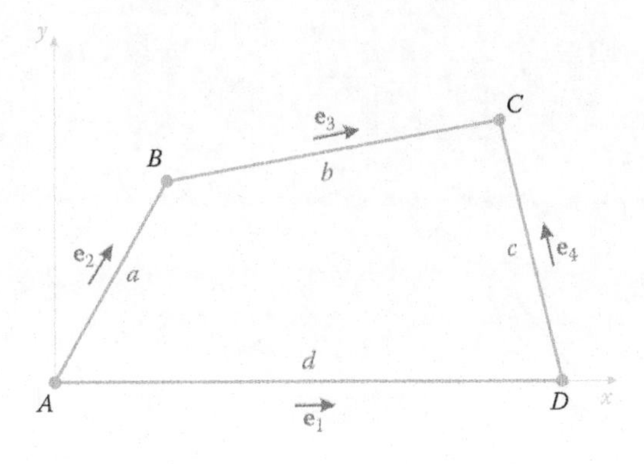

FIGURE 4.11
A vector loop using unit vectors to give direction.

Now let us define the *unit normal*, which is a vector perpendicular to **e**.

$$\mathbf{n} = \begin{Bmatrix} -\sin\theta \\ \cos\theta \end{Bmatrix} \tag{4.29}$$

We have chosen the vector that is rotated 90° counterclockwise from **e** as the definition of the unit normal, as shown in Figure 4.12, for reasons that will become apparent in the next section. The reader may also wish to verify that

$$\mathbf{e} \cdot \mathbf{n} = 0 \tag{4.30}$$

The unit normal is also a unit vector, but is defined as being perpendicular to **e**. It will come in handy when conducting velocity and acceleration analysis on linkages.

4.2.5.1 Time Derivatives of Unit Vectors

To conduct velocity and acceleration analysis of linkages, we will need to take time derivatives of unit vectors. Consider the unit vector **e**.

$$\mathbf{e} = \begin{Bmatrix} \cos\theta \\ \sin\theta \end{Bmatrix} \tag{4.31}$$

If we differentiate this with respect to time, we have

$$\frac{d\mathbf{e}}{dt} = \frac{d}{dt}\begin{Bmatrix} \cos\theta \\ \sin\theta \end{Bmatrix} \tag{4.32}$$

Taking the time derivative of a vector requires taking the derivative of each one of its components, in turn. Consider first the x component:

$$\frac{d\mathbf{e}_x}{dt} = \frac{d}{dt}(\cos\theta) \tag{4.33}$$

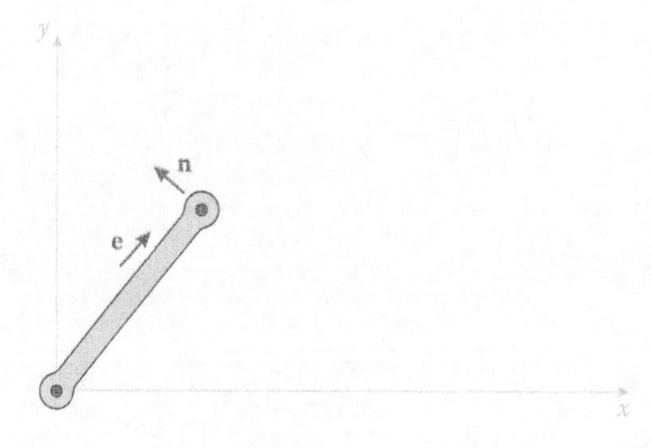

FIGURE 4.12
The unit normal is perpendicular to the unit vector, rotated 90° counterclockwise.

Since **e** is attached to a moving linkage, the angle θ is a function of time, and we must employ the chain rule of differentiation to take the derivative. If $u(x)$ is a function of x, and $v(u)$ is a function of u, then the chain rule states that

$$\frac{dv}{dx} = \frac{dv}{du} \times \frac{du}{dx} \tag{4.34}$$

In our example we have

$$u \to \theta \quad x \to t \quad v \to \cos\theta \tag{4.35}$$

so that

$$\frac{du}{dx} \to \frac{d\theta}{dt} \quad \frac{dv}{du} \to \frac{d}{d\theta}(\cos\theta) = -\sin\theta \tag{4.36}$$

Thus, the derivative of the x term of the unit vector is

$$\frac{d\mathbf{e}_x}{dt} = \frac{d}{dt}(\cos\theta) = -\frac{d\theta}{dt}\sin\theta \tag{4.37}$$

The term $d\theta/dt$ is the time rate of change of the angle θ, which we will define as the *angular velocity*, ω (Greek letter omega).

$$\omega \equiv \frac{d\theta}{dt} \tag{4.38}$$

Thus, the time derivative of the unit vector is

$$\frac{d\mathbf{e}}{dt} = \omega \begin{Bmatrix} -\sin\theta \\ \cos\theta \end{Bmatrix} \tag{4.39}$$

The alert reader may recognize the quantity in brackets as being the unit normal, and the derivative of the unit vector may be written

$$\frac{d\mathbf{e}}{dt} = \omega\mathbf{n} \tag{4.40}$$

Thus, the time derivative of the unit vector is the angular velocity multiplied by the unit normal.

What happens if we differentiate the unit normal with respect to time?

$$\frac{d\mathbf{n}}{dt} = \frac{d}{dt}\begin{Bmatrix} -\sin\theta \\ \cos\theta \end{Bmatrix} = \omega \begin{Bmatrix} -\cos\theta \\ -\sin\theta \end{Bmatrix} = -\omega\mathbf{e} \tag{4.41}$$

We have obtained unit vector again, multiplied (as before) by the angular velocity. Note that the minus sign in front of the result indicates that the direction is *opposite* the original unit vector direction, as shown in Figure 4.13. Therefore, taking the time derivative of the unit normal results in changing the direction by 90° counterclockwise and increasing the magnitude by a factor of ω. The reader may easily verify that

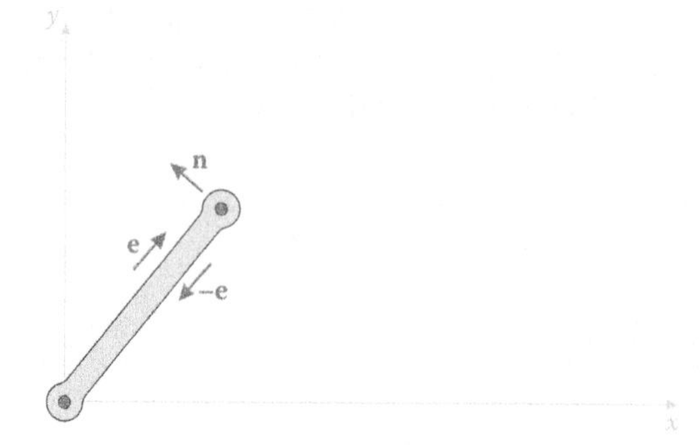

FIGURE 4.13
Differentiating the unit normal gives the unit vector again, but in the opposite direction.

$$\frac{d}{dt}(-\mathbf{e}) = -\omega\mathbf{n} \tag{4.42}$$

and that taking the time derivative of −**n** would bring us back where we started, pointing in the direction of **e**. Thus, time differentiation of the unit vector or unit normal results in a rotation of 90° counterclockwise accompanied by a multiplication by ω.

4.2.6 A Very Brief Introduction to Matrix Algebra

Matrices are used when we wish to solve sets of linear equations. In fact, MATLAB (which is short for "MATrix LABoratory") was originally developed for just this purpose. Suppose that we have the following two linear equations:

$$\begin{aligned} 1x_1 + 2x_2 &= 3 \\ 4x_1 + 5x_2 &= 6 \end{aligned} \tag{4.43}$$

For this simple example, we could solve the second equation for x_2, and substitute the result into the first equation to solve for x_1. This procedure works well for two or three equations, but quickly becomes tedious (and error-prone) for larger systems. Instead, we can arrange the equations in *matrix form*, and use software (such as MATLAB) to do the hard work for us. To do this, we take the coefficients of the variables and arrange them in a square grid, as shown below

$$\begin{bmatrix} 1 & 2 \\ 4 & 5 \end{bmatrix} \begin{Bmatrix} x_1 \\ x_2 \end{Bmatrix} = \begin{Bmatrix} 3 \\ 6 \end{Bmatrix} \tag{4.44}$$

Let us decompose this equation into its constituent parts

$$\begin{bmatrix} 1 & 2 \\ 4 & 5 \end{bmatrix} = \mathbf{A} = \text{matrix of coefficients}$$

$$\begin{Bmatrix} x_1 \\ x_2 \end{Bmatrix} = \mathbf{x} = \text{vector of unknowns}$$

$$\begin{Bmatrix} 3 \\ 6 \end{Bmatrix} = \mathbf{b} = \text{vector of knowns}$$

so that we can rewrite Equation (4.44) as

$$\mathbf{Ax} = \mathbf{b}$$

Note that everything on the right-hand side of the equation is *known*; that is, no variables appear on this side of the equation. The order that we arrange the numbers in the matrix (the square grid) is very important. Each row in the matrix corresponds to one of the equations, and each column corresponds to a variable we are solving for. If we rearrange the order of the vector of unknowns, we must also rearrange the *columns* of the matrix, so that

$$\begin{bmatrix} 2 & 1 \\ 5 & 4 \end{bmatrix} \begin{Bmatrix} x_2 \\ x_1 \end{Bmatrix} = \begin{Bmatrix} 3 \\ 6 \end{Bmatrix} \tag{4.45}$$

Note that the rows have not been affected by changing the order of the vector of unknowns.

Example 4.3: Arrange the following equations into matrix form

$5x_1 + 8x_2 = 13$ $8x_1 + 20x_2 = 25$	→	$\begin{bmatrix} 5 & 8 \\ 8 & 20 \end{bmatrix} \begin{Bmatrix} x_1 \\ x_2 \end{Bmatrix} = \begin{Bmatrix} 13 \\ 25 \end{Bmatrix}$
$8x_1 + 5x_2 - 13 = 0$ $20x_1 + 8x_2 - 25 = 0$	→	$\begin{bmatrix} 8 & 5 \\ 20 & 8 \end{bmatrix} \begin{Bmatrix} x_1 \\ x_2 \end{Bmatrix} = \begin{Bmatrix} 13 \\ 25 \end{Bmatrix}$
$6a + 8b = 10$ $8b + 20a = 30$	→	$\begin{bmatrix} 6 & 8 \\ 20 & 8 \end{bmatrix} \begin{Bmatrix} a \\ b \end{Bmatrix} = \begin{Bmatrix} 10 \\ 30 \end{Bmatrix}$
$1x + 2y + 3 = 4$ $4x + 3y + 2 = 1$	→	$\begin{bmatrix} 1 & 2 \\ 4 & 3 \end{bmatrix} \begin{Bmatrix} x \\ y \end{Bmatrix} = \begin{Bmatrix} 1 \\ -1 \end{Bmatrix}$
$x\cos\theta_2 + y\sin\theta_2 = x'$ $-x\sin\theta_2 + y\cos\theta_2 = y'$ (x and y are unknowns)	→	$\begin{bmatrix} \cos\theta_2 & \sin\theta_2 \\ -\sin\theta_2 & \cos\theta_2 \end{bmatrix} \begin{Bmatrix} x \\ y \end{Bmatrix} = \begin{Bmatrix} x' \\ y' \end{Bmatrix}$
$x_1 + x_2 = a$ $x_2 = b$ (x_1 and x_2 are unknowns)	→	$\begin{bmatrix} 1 & 1 \\ 0 & 1 \end{bmatrix} \begin{Bmatrix} x_1 \\ x_2 \end{Bmatrix} = \begin{Bmatrix} a \\ b \end{Bmatrix}$
$ax_1 + bx_2 + cx_3 + d = 0$ $g = ex_1 + fx_2$ $hx_2 + kx_3 = m$	→	$\begin{bmatrix} a & b & c \\ e & f & 0 \\ 0 & h & k \end{bmatrix} \begin{Bmatrix} x_1 \\ x_2 \\ x_3 \end{Bmatrix} = \begin{Bmatrix} -d \\ g \\ m \end{Bmatrix}$

A few other matrix definitions are in order. Consider the matrix

$$\mathbf{A} = \begin{bmatrix} 1 & 2 & 3 \\ 4 & 5 & 6 \end{bmatrix} \tag{4.46}$$

By exchanging columns and rows, we obtain the *transpose*

$$\mathbf{A}^T = \begin{bmatrix} 1 & 4 \\ 2 & 5 \\ 3 & 6 \end{bmatrix} \tag{4.47}$$

The *inverse* of a matrix is defined such that

$$\mathbf{A}^{-1}\mathbf{A} = \mathbf{U} \tag{4.48}$$

where **U** is the identity matrix.

$$\mathbf{U} = \begin{bmatrix} 1 & 0 & \cdots & 0 \\ 0 & 1 & \cdots & 0 \\ \vdots & \vdots & \ddots & \vdots \\ 0 & 0 & \dots & 1 \end{bmatrix} \tag{4.49}$$

Once we have the matrix equations written out, it is relatively easy to use MATLAB to solve them. To enter a matrix into MATLAB, we use the square bracket, with a semicolon separating lines. MATLAB will accept commas or spaces between columns, as shown.

```
>> A = [1 2;3 4]

A =

     1     2
     3     4
```

To access a particular entry in the matrix, we use normal parentheses in (*row, column*) format. For example

```
>> A(1,2)

ans =

     2
```

gives us the first row and second column of the matrix **A**. To enter a vector into MATLAB (e.g. the vector of knowns), use the same technique. A column (vertical) vector has one column and multiple rows, so we must use a semicolon between each entry

```
>> b = [5; 6]

b =
```

```
     5
     6
```

There are two methods for solving the matrix equation **Ax** = **b**, the *inverse* method and the *forward slash* method. Of these, MATLAB would prefer that you use the forward slash method, unless you really need the inverse of matrix **A** for some reason. To understand the reasoning behind the two methods, consider for a moment how you would solve the equation **Ax** = **b** if it were not a matrix equation:

$$Ax = b \tag{4.50}$$

The simplest technique would be to divide both sides by A, as shown

$$x = \frac{b}{A} \tag{4.51}$$

This is mathematically equivalent to multiplying both sides by the reciprocal (or inverse) of A.

$$x = A^{-1}b \tag{4.52}$$

In MATLAB, the forward-slash technique is the equivalent of Equation (4.51), while the inverse technique is the equivalent Equation (4.52). The first technique is computationally much faster, since calculating the inverse of a matrix can sometimes be rather involved.

```
>> x = A\b

x =

   -4.0000
    4.5000
```

or

```
>> x = inv(A)*b

x =

   -4.0000
    4.5000
```

When you execute both of these statements, the second takes noticeably longer. Let's check to see if MATLAB came up with the correct solution:

$$\begin{aligned} 1(-4)+2(4.5) &= -4+9=5 \\ 3(-4)+4(4.5) &= -12+18=6 \end{aligned} \tag{4.53}$$

It worked! We now have a convenient and powerful tool for solving sets of linear equations. We will make frequent use of this in conducting velocity, acceleration, and force analysis in later chapters.

It is important to observe that while

$$x = A^{-1}b = bA^{-1}$$

is perfectly valid for scalar operations, it will not work for matrix manipulations. In fact, the commutative property does not hold for matrix operations, and

$$\mathbf{A}^{-1}\mathbf{b} \neq \mathbf{b}\mathbf{A}^{-1}$$

To see why this is true, consider the dimensions of each term in the above equation. The inverse of the **A** matrix has the same dimension as the **A** matrix itself, so that $\mathbf{A}^{-1}$ has dimension (2 × 2) and **b** has dimension (2 × 1). In order for a matrix multiplication to be defined, the *inner dimensions* must be equal and the size of the result has the *outer dimensions.* That is, for the operation

$$\mathbf{A}^{-1}\mathbf{b} \rightarrow (2 \times 2) \times (2 \times 1)$$

the inner dimension is 2, and the outer dimensions are 2 and 1 so the result of this operation is a 2 × 1 vector. If we try

$$\mathbf{b}\mathbf{A}^{-1} \rightarrow (2 \times 1) \times (2 \times 2)$$

the inner dimensions are 1 and 2, which are unequal. Since they are unequal, the multiplication is undefined. To confirm this, try typing

```
>> b*inv(A)
```

The result is an error, as expected.

```
Error using  *
Inner matrix dimensions must agree.
```

This is MATLAB's way of telling us that the multiplication operation is undefined since the inner dimensions are unequal. Knowing the meaning of this rather cryptic error message will be very helpful in debugging the MATLAB programs we will write in later sections.

4.2.7 Transformation of Coordinates

In some situations, we will find it simplest to use an "auxiliary" or "local" coordinate system when modeling a linkage. Consider the slider-crank linkage shown in Figure 4.14. The slider is aligned with the x' axis, and we will find it relatively simple to perform our analysis in the local (x', y') system, to begin with. We require a method for transforming the results in the (x', y') system back into the "global" (x, y) system.

Consider the simplest case first, in which the moving (local) coordinate system shares the origin with the fixed (global) coordinate system but is rotated by an angle φ, as shown

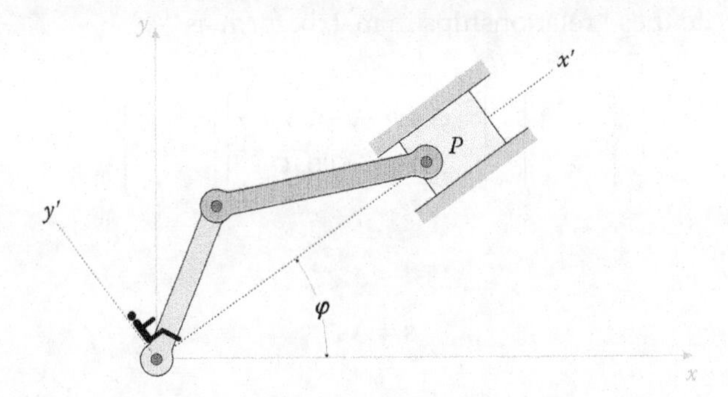

FIGURE 4.14
To the figure sitting at the origin and facing in the x' direction, the piston appears to be directly ahead.

in Figure 4.15. By breaking the vector **s** into its components in the moving coordinate system, we can derive the relationships below:

$$\begin{aligned} s_x &= s'_x \cos\varphi - s'_y \sin\varphi \\ s_y &= s'_x \sin\varphi + s'_y \cos\varphi \end{aligned} \tag{4.54}$$

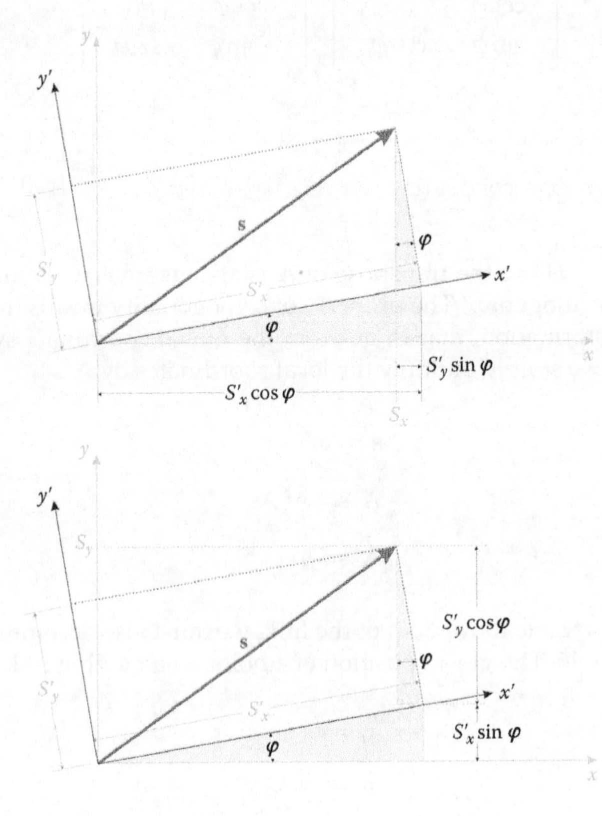

FIGURE 4.15
The moving (local) coordinate system is rotated by an angle φ from the fixed (global) coordinate system. The vector **s** is assumed known in the primed coordinate system.

It is common to write these relationships in matrix form as

$$\begin{Bmatrix} s_x \\ s_y \end{Bmatrix} = \begin{bmatrix} \cos\varphi & -\sin\varphi \\ \sin\varphi & \cos\varphi \end{bmatrix} \begin{Bmatrix} s'_x \\ s'_y \end{Bmatrix} \tag{4.55}$$

or, more compactly

$$\mathbf{s} = \mathbf{A}\mathbf{s}' \tag{4.56}$$

where

$$\mathbf{A} = \begin{bmatrix} \cos\varphi & -\sin\varphi \\ \sin\varphi & \cos\varphi \end{bmatrix} \tag{4.57}$$

is known as a *rotation matrix*. An interesting thing happens if we multiply the transformation matrix by its transpose:

$$\begin{bmatrix} \cos\varphi & -\sin\varphi \\ \sin\varphi & \cos\varphi \end{bmatrix} \times \begin{bmatrix} \cos\varphi & \sin\varphi \\ -\sin\varphi & \cos\varphi \end{bmatrix} =$$

$$\begin{bmatrix} \cos^2\varphi + \sin^2\varphi & \cos\varphi\sin\varphi - \sin\varphi\cos\varphi \\ \sin\varphi\cos\varphi - \cos\varphi\sin\varphi & \cos^2\varphi + \sin^2\varphi \end{bmatrix} = \begin{bmatrix} 1 & 0 \\ 0 & 1 \end{bmatrix} \tag{4.58}$$

In other words, $\mathbf{A}\mathbf{A}^T = \mathbf{U}$ or the transpose of $\mathbf{A}$ is also its inverse. A matrix that has this property is called "orthogonal." The property of orthogonality means that if we ever want to reverse the transformation, that is, go from the global coordinate system to the local coordinate system, we simply multiply the local coordinates by $\mathbf{A}^T$.

$$\begin{aligned} \mathbf{s} &= \mathbf{A}\mathbf{s}' \\ \mathbf{A}^T\mathbf{s} &= \mathbf{A}^T\mathbf{A}\mathbf{s}' \\ \mathbf{A}^T\mathbf{s} &= \mathbf{s}' \end{aligned} \tag{4.59}$$

To account for any possible movement of the link, we must also accommodate translation, as shown in Figure 4.16. The global position of a point on a moving link can be written

$$\begin{aligned} \mathbf{r}_P &= \mathbf{r} + \mathbf{s}_P \\ \mathbf{r}_P &= \mathbf{r} + \mathbf{A}\mathbf{s}'_P \end{aligned} \tag{4.60}$$

Where the angle φ is still measured between x and x' after the translation has occurred.

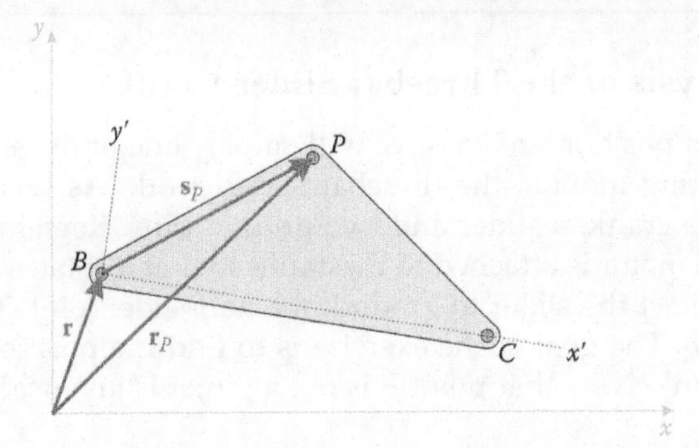

FIGURE 4.16
A link may be translating as well as rotating.

Example 4.4: A Simple Link

In Figure 4.17, the link has length r. The vector $\mathbf{r}$ gives the position of the point P at the end of the link. In the local system, we can write $\mathbf{r}'$ as

$$\mathbf{r}' = \begin{Bmatrix} r \\ 0 \end{Bmatrix} \tag{4.61}$$

The link is rotated away from the global x axis by an angle φ. To transform the coordinates of point P into the global system, we multiply by the rotation matrix $\mathbf{A}$

$$\mathbf{r} = \mathbf{A}\mathbf{r}'$$

$$\mathbf{r} = \begin{bmatrix} \cos\varphi & -\sin\varphi \\ \sin\varphi & \cos\varphi \end{bmatrix} \begin{Bmatrix} r \\ 0 \end{Bmatrix}$$

$$\mathbf{r} = \begin{Bmatrix} r\cos\varphi \\ r\sin\varphi \end{Bmatrix}$$

The reader will note that this is the same expression as Equation (4.7) with φ substituted for θ.

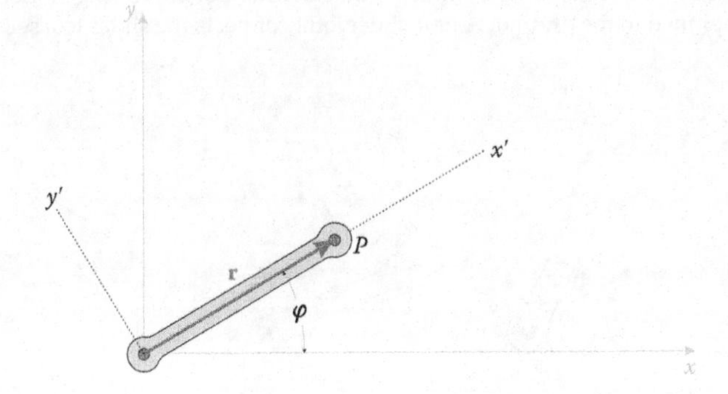

FIGURE 4.17
A moving coordinate system is attached to the link.

4.3 Position Analysis of the Threebar Slider-Crank

To begin our study of position analysis we will employ one of the simplest linkages that is capable of interesting motion: the threebar slider-crank. As seen in Figure 4.18, the threebar consists of a crank, a slider and two ground pins. Remember that the ground counts as one link. A motor is attached to the crank so that it rotates about ground pin A. Pin D is used to connect the slider to ground in a half-slider joint. The crank and slider are pinned at point B. The goal of the exercise is to find the position of point P for any orientation of the crank. Note that point P is not a pin; it is just used to define a point at the end of the slider.

There are only three fixed dimensions that are important for the position analysis, as shown in Figure 4.19. The length of the crank is given by a, the overall length of the slider is p and the distance between ground pins is d. The length b is defined as the distance from the crank pin A to the ground pin at D. This length will change as the crank rotates, and b is one of the variables we must solve for in our analysis.

We will now begin the position analysis for the threebar linkage. We assume at the outset that the crank length, a, the distance between ground pins d are known. Further, since we are driving the crank using a motor, we assume that the crank angle, θ_2, is also known. Begin by constructing a vector loop diagram on the linkage, as shown in Figure 4.20. The

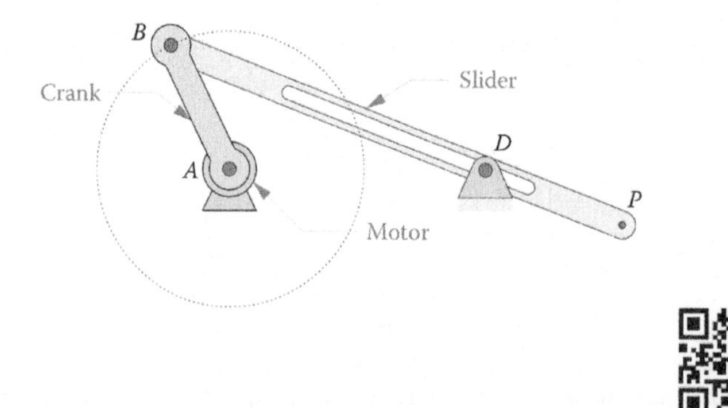

FIGURE 4.18
The threebar slider-crank is one of the simplest linkages that is capable of interesting motion. A motor is attached to the crank, which is pinned to the ground. A half-slider joint connects the slider to a second ground pin.

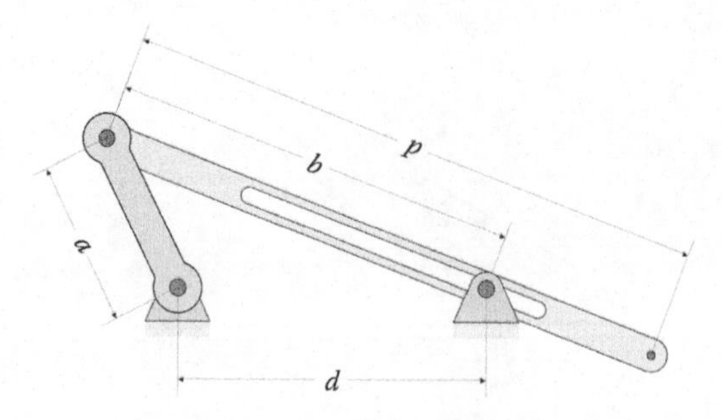

FIGURE 4.19
The dimensions of the threebar that are needed for position analysis.

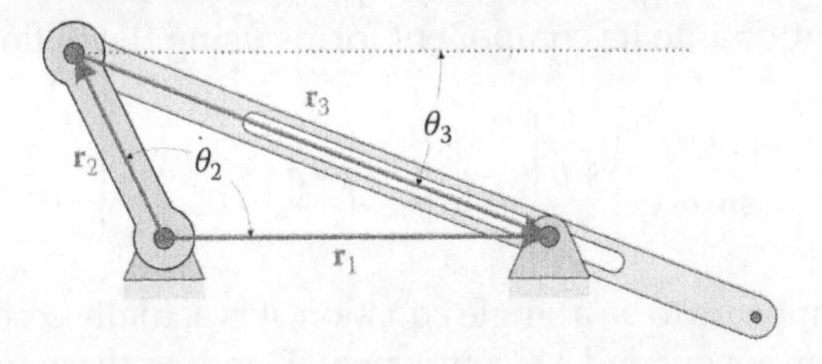

FIGURE 4.20
A vector loop diagram of the threebar linkage. The vector $\mathbf{r}_2$ is attached to the crank and $\mathbf{r}_3$ is attached to the slider.

vector $\mathbf{r}_2$ is attached to the crank, and has constant length a. Using unit vector notation, we can write $\mathbf{r}_2$ as

$$\mathbf{r}_2 = a\mathbf{e}_2 \tag{4.62}$$

where

$$\mathbf{e}_2 = \begin{Bmatrix} \cos\theta_2 \\ \sin\theta_2 \end{Bmatrix} \tag{4.63}$$

is the unit vector directed along the crank. The vector $\mathbf{r}_3$ is attached to the slider and connects the crank pin A to the ground pin D.

$$\mathbf{r}_3 = b\mathbf{e}_3 \tag{4.64}$$

The length b varies as the crank rotates, and the unit vector $\mathbf{e}_3$ is aligned with the slider.

$$\mathbf{e}_3 = \begin{Bmatrix} \cos\theta_3 \\ \sin\theta_3 \end{Bmatrix} \tag{4.65}$$

Finally, the vector $\mathbf{r}_1$ is

$$\mathbf{r}_1 = d\mathbf{e}_1 \tag{4.66}$$

Since $\mathbf{e}_1$ is aligned with the horizontal, it is defined as

$$\mathbf{e}_1 = \begin{Bmatrix} 1 \\ 0 \end{Bmatrix} \tag{4.67}$$

We will now write the *vector loop equation* for the threebar linkage. A vector loop is created by traveling around the linkage, one vector at a time, until we return to our starting point. Since we end at the same point as we began, the total distance traveled is zero. Beginning at point A, the vector loop can be written

$$\mathbf{r}_2 + \mathbf{r}_3 - \mathbf{r}_1 = 0 \tag{4.68}$$

Now expand this equation into its component forms using the definitions above

$$a\begin{Bmatrix} \cos\theta_2 \\ \sin\theta_2 \end{Bmatrix} + b\begin{Bmatrix} \cos\theta_3 \\ \sin\theta_3 \end{Bmatrix} - d\begin{Bmatrix} 1 \\ 0 \end{Bmatrix} = \begin{Bmatrix} 0 \\ 0 \end{Bmatrix} \tag{4.69}$$

While Equation (4.69) appears to be a single equation, it is actually composed of two separate equations with an x component and a y component. Dividing these two equations gives

$$\begin{aligned} x &: a\cos\theta_2 + b\cos\theta_3 - d = 0 \\ y &: a\sin\theta_2 + b\sin\theta_3 = 0 \end{aligned} \tag{4.70}$$

The crank length, a, and distance between ground pins, d, are known quantities, as is the crank angle θ_2. This leaves only the slider angle, θ_3, and the distance, b, as unknowns. The vector loop method has given us two equations, which means that the problem is solvable. A properly constructed vector loop diagram (or diagrams, for more complicated linkages) will always provide the same number of equations as unknowns.

To solve for the variables, first rearrange the x component of Equation (4.70) to solve for b.

$$b = \frac{d - a\cos\theta_2}{\cos\theta_3} \tag{4.71}$$

Insert this expression into the y component of Equation (4.70)

$$a\sin\theta_2 + \left(\frac{d - a\cos\theta_2}{\cos\theta_3}\right)\sin\theta_3 = 0 \tag{4.72}$$

Since

$$\frac{\sin\theta_3}{\cos\theta_3} = \tan\theta_3 \tag{4.73}$$

We may solve for θ_3 as

$$\tan\theta_3 = \frac{a\sin\theta_2}{a\cos\theta_2 - d} \tag{4.74}$$

Once θ_3 is known, we may use Equation (4.71) to solve for b. This concludes the hard part of the position analysis for the threebar. The problem statement, however, asks us to find the position of point P for any crank angle.

Let us define the vector $\mathbf{r}_{BP}$, which starts at point B and ends at point P. Since this vector has the same length as the overall length of the slider, p, and points in the same direction as the slider we may write

$$\mathbf{r}_{BP} = p\mathbf{e}_3 \tag{4.75}$$

A vector to point P can be found by adding the vectors $\mathbf{r}_2$ and $\mathbf{r}_{BP}$, as shown in Figure 4.21.

$$\mathbf{r}_P = \mathbf{r}_2 + \mathbf{r}_{BP} \tag{4.76}$$

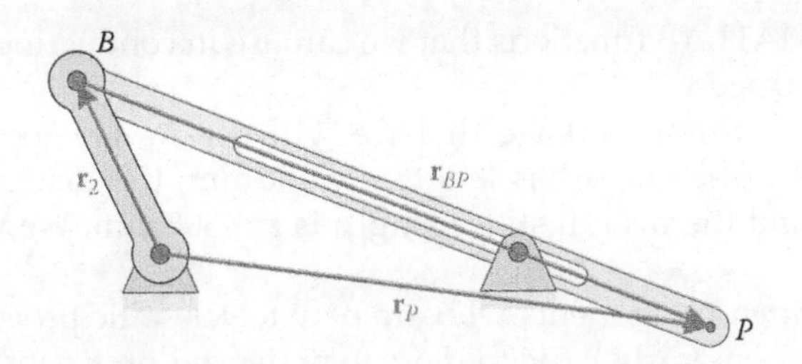

FIGURE 4.21
To find the point P, add the vectors $\mathbf{r}_2$ and $\mathbf{r}_{BP}$.

Or, using the unit vector notation, we have

$$\mathbf{r}_P = a\mathbf{e}_2 + p\mathbf{e}_3 \tag{4.77}$$

This formula is important enough to warrant special attention. It is often the case that we know the position of one point on a link, and desire to know the position of another point. Consider the link shown in Figure 4.22. If the position of point B is known, then we may use the *relative position formula* to find the position of point P.

$$\mathbf{r}_P = \mathbf{r}_B + \mathbf{r}_{BP} \tag{4.78}$$

We will use the relative position formula quite often in the sections that follow, and it is important enough that we will define a special MATLAB function to implement it. This formula can also be used to derive the *relative velocity* and *relative acceleration* formulas, as we will see.

4.4 Position Analysis of the Threebar Slider-Crank Using MATLAB®

Now that we have a set of formulas for calculating the angles and positions of the links on the threebar linkage, we will put our knowledge to work in writing a MATLAB program to perform the calculations for us. The goal of the program that we will write is to plot the position of point B and P as the crank makes a full revolution. Along the way, we will

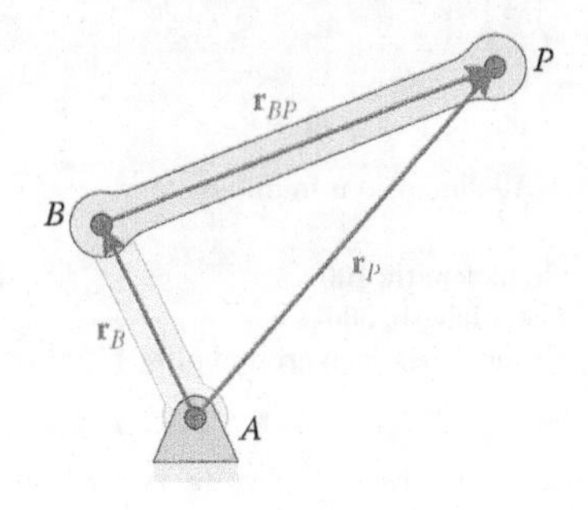

FIGURE 4.22
If the position of B is known, then the relative position formula can be used to solve for the position of point P.

create a few handy MATLAB functions that we can use in conducting position analysis of more complicated linkages.

A diagram of the threebar linkage that we will use in developing our program is shown in Figure 4.23. The crank has length $a = 100$ mm, the distance between ground pins is $d = 150$ mm and the overall slider length is $p = 300$ mm. We will place the origin at point A.

This section is written for students who are new to scientific programming, and some of the concepts will seem fairly basic to more experienced programmers. Many students seem to have difficulty in translating a set of formulas, as were derived in the previous section, to a program for evaluating these formulas. This section will demonstrate one method for writing a scientific program. Of course, each programmer has his or her own style, and you may feel free to tailor your own program as you see fit (assuming, of course, that the results are the same!). If you have never used MATLAB before, you should go through the simple MATLAB tutorial given in Chapter 3.

A program is a set of instructions that a programmer gives to a computer—like a recipe—with the goal of executing a particular task. In our case, we desire that the computer solve for the positions of the links of the threebar, and then produce plots of the paths of various points on the linkage. One of the first things to observe is that some of these tasks should be executed only once (e.g. defining the lengths of each link) and some are executed many times (e.g. solving for the slider length, b, and angle, θ_3, at a particular crank angle). We will place the tasks to be executed many times inside a *loop*, since we do not wish to repeatedly type in our set of formulas. Everything that should be executed only once (e.g. defining the link lengths and the plotting commands) will be placed outside the loop.

When writing a scientific program, the first thing to do is to make a list of its objectives. This is a good way to give an overall structure to the program; if you make a detailed enough list, the code will be relatively easy to write. In addition, the list of objectives can be copied and pasted into the program as comments that help to explain the purpose of each section of the program. The objectives of our linkage analysis program are to:

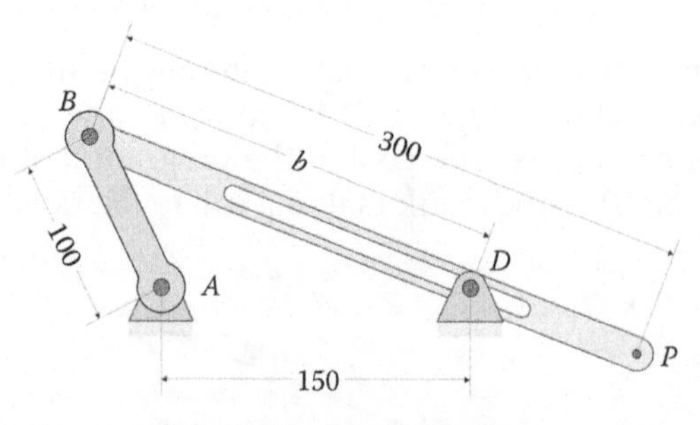

FIGURE 4.23
Dimensions of the threebar linkage used in the MATLAB code. We will use this linkage in later sections when we conduct velocity, acceleration, and force analysis.

1. Enter the linkage dimensions.
2. Conduct each of the following steps for every crank angle
   a. Calculate the angle of the slider, θ_3.
   b. Calculate the length b between point A and D.
   c. Use these values to calculate the positions of points B and P.
3. Once these calculations are complete, the program should generate a plot that shows the paths of points B and P as the crank makes a complete revolution.

Items (1) and (3) need only be executed once, while the tasks in item (2) are executed several times—once for each crank angle. Therefore, we will place the tasks in item (2) inside a loop, and all of the other tasks will be outside the loop.

At the top of every MATLAB program you should type a set of comments that describe the purpose of the program, its author (you!), and the date on which it was written. You might need to use this program for another class in a later semester, and it is very helpful to have a description of the program at the top so that you can remember what it does. At the top of a new MATLAB script, type:

```
% Threebar_Position_Analysis.m
% Conducts a position analysis on the threebar crank-slider linkage
% by Eric Constans, June 1, 2017
```

Note that the first line gives the name of the program: `Threebar_Position_Analysis.m`. This is not necessary (since it is just a comment) but it is good practice. Remember that MATLAB file names are not allowed to have spaces (or other special characters) in them, so We have used the underscore character instead. After you have typed the comments, save the script in a convenient location (e.g. the Desktop) using this file name. On the next two lines, type:

```
% Prepare Workspace
clear variables; close all; clc;
```

These lines should be typed at the top of all of your MATLAB scripts. Its purpose is to clear any variable definitions out of memory so that you start with a "clean slate." If you forget to do this, you will retain all of the variable definitions from the last time you executed the program, sometimes with very surprising and unexpected results. The `close all` command closes any plot windows that are open (as before, with the idea of starting with a clean slate) and the `clc` command clears the command window (`clc` stands for "command line clear").

Next, we should tell MATLAB the dimensions of the linkage, as shown:

```
% Linkage dimensions
a = 0.100;            % crank length (m)
d = 0.150;            % length between ground pins (m)
p = 0.300;            % slider length (m)
```

These lines specify the lengths of each link. Notice that we have specified the units for each dimension; this is important so that a reader of your program knows which system of units you are employing.

Next, we will enter the coordinates of the ground pins, since they do not change as the linkage moves. There are two ground pins, one at point A and one at point D.

```
% Ground pins
x0 = [0;0];          % point A (the origin)
xD = [d;0];          % point D
```

The square brackets indicate that MATLAB should define x0 and xD as vectors. Each vector has an x and y component. Separating the components by a semicolon defines them as column vectors with dimension 2×1. We will use point A as the origin in our calculations. Every variable that begins with the letter "x" will be used to store the coordinates of a point on the linkage. For example, xB will be used to store the position coordinates of the point B.

4.4.1 Data Structure for the Position Calculations

Defining the fixed ground pins was the final task that was to be executed a single time, other than the plotting commands, which must be done at the end of the program after all calculations are complete. We are now ready to begin framing the structure of the main loop, which will execute the set of position calculations for each angle of the crank. First, we must make an important decision: for how many different crank angles do we wish to calculate the positions of points B and P? If we tell MATLAB to perform the position calculations for very fine increments of the crank angle (e.g. every tenth, or hundredth of a degree) we will produce a very exact plot of the path of point P, but at the cost of a slow execution time. On the other hand, you can speed up execution of the program by only performing the position calculations for every ten degrees of rotation of the crank, but at the cost of a non-smooth, inaccurate position plot. This sort of tradeoff appears in programming quite often, and is part of the "art" of engineering.

For this example, we choose to perform the position calculations for every 1°of crank rotation. Since we will start with the crank-oriented horizontally (at 0°) and end once the crank is again oriented horizontally (at 360°) we will perform a total of 361 calculations. That is, if you count from 0 to 360 in increments of 1, you will have a total of 361 position calculations. Since we may wish to increase or decrease the number of calculations in the future, we will define a variable, N, to keep track of this number.

```
N = 361;   % number of times to perform position calculations
```

Next, we must determine how to store the results of our calculations. At each increment of crank rotation, we will calculate several variables: the slider angle and length (θ_3 and b), and the coordinates of points B and P on the linkage. When we have completed the main loop, we will have calculated (and stored) 361 values for θ_3 and b, as well as 361 x and y coordinates of the points B and P. It is most efficient to *preallocate* memory for all of these values so that MATLAB does not need to create new storage space at every iteration of the loop. After this is done, MATLAB can place newly calculated values into the preallocated space without having to find new space in memory every time it goes through the loop. The simplest way to preallocate space is to define theta2, theta3, and b as vectors of zeros.

```
theta2 = zeros(1,N);   % allocate space for crank angle
theta3 = zeros(1,N);   % allocate space for slider angle
b      = zeros(1,N);   % allocate space for slider length
```

These statements initialize theta2, theta3, and b as *row vectors* of 361 zeros each. As we work our way through the main loop, the zeros will all be overwritten by the calculated values for each variable. The statements above take three lines of code, and as we move to velocity and acceleration analysis these lines will expand until they consume an inordinate amount of space. A more space-conserving way to preallocate memory is to use the `deal` command, as shown:

```
[theta2,theta3,b] = deal(zeros(1,N)); % allocate space for link angles
```

The deal command "deals out" a row vector of 361 zeros to each one of the variables in the square brackets, as a card dealer would distribute cards in a game of poker. In this way, we can use a single line to preallocate memory for `theta2`, `theta3`, and `b`.

Defining the structure for the position variables is a little trickier. Each point (B and P) has an x and y coordinate, both of which must be stored for each crank angle. Thus, instead of using a single row vector as with the angles above, we require *two* rows for each position variable. Initialize the position variables using the following commands.

```
[xB,xP] = deal(zeros(2,N)); % allocate space for position of B,P
```

It is important that you understand the structure of the variables as defined above. Table 4.1 gives a graphical representation of the position coordinate `xB`. The first row gives the x coordinates while the second row gives the y coordinates. Each column represents the results of calculation for a single crank angle θ_2. Thus, the first column contains the coordinates of point B for $\theta_2 = 0°$, and the sixth column gives the coordinates of point B for $\theta_2 = 5°$, and so on.

If we wish to access the y coordinate of point B for the tenth crank angle ($\theta_2 = 9°$), we would type

```
>> xB(2,10)
```

at the command prompt. One of the most useful operators in MATLAB is the ordinary colon. If we wish to access *all* of the x coordinates of point B, we would type

```
>> xB(1,:)
```

at the command prompt. The result of this statement would be a row vector of length N containing the x coordinates of point B for each crank angle. Similarly, if we wish to find the x and y coordinates of point B for the twentieth crank angle we would enter

```
>> xB(:,20)
```

at the command prompt. The result of this statement would be a two-element column vector; the first element would be the x coordinate of point B at the twentieth crank angle

TABLE 4.1

The Structure of the Position Variable `xB`

	1	2	3	4	5		N
1	$xB_x(1)$	$xB_x(2)$	$xB_x(3)$	$xB_x(4)$	$xB_x(5)$	…	$xB_x(N)$
2	$xB_y(1)$	$xB_y(2)$	$xB_y(3)$	$xB_y(4)$	$xB_y(5)$	…	$xB_y(N)$

Each column corresponds to a single crank angle. The first row gives the x coordinate and the second row gives the y coordinate. The column and row labels represent the indices of the `xB` matrix; the row label gives the first index and the column label gives the second index.

and the second element would be the y coordinate. We will use the colon operator quite frequently in our MATLAB scripts, so it is important that you understand its syntax.

4.4.2 The Main Loop

We will use a `for` loop to perform the repeated position calculations. MATLAB purists may frown upon the use of `for` loops for such calculations, but sometimes readable, understandable code is more important than pure execution speed! If you are a MATLAB guru, you may try to vectorize the position calculations, but you will probably spend more time programming than you will save in execution speed. Type the following `for` loop into your script:

```
% Main Loop
for i = 1:N

end
```

Every `for` loop requires an `end` statement and it is a good idea to type it in now, so that you do not forget it later on. Every statement that we place between the `for` and the `end` will be executed `N` times. The variable `i` will take on the values 1, 2, 3 ... `N` depending upon which iteration we are currently executing.

The first thing to do inside the loop is to determine the crank angle, since all subsequent position calculations depend upon it. As stated earlier, we wish to perform the calculations at increments of 1°. Recall, however, that we wish to begin our calculations at 0°, and end at 360°. Our first guess at defining the crank angle, `theta2`, might look something like this

```
for i = 1:N
  theta2(i) = i-1;

end
```

This will make `theta2` take on the values 0, 1, 2, ... 360, as desired. However, these values are in *degrees*, and our calculations must be performed in *radians*. The conversion between degrees and radians is a multiplicative factor of $\pi/180$, so our next guess at defining `theta2` might be

```
for i = 1:N
  theta2(i) = (i-1)*pi/180;

end
```

where we have taken advantage of the fact that `pi` = π is a predefined constant within MATLAB. There is just one subtle difficulty with this formulation, however. Since `i` can only take on integer values, our crank rotation increments are limited to 1°. A better solution would be to have the crank angle increments be dependent upon the number of position calculations, `N`, so that the crank always makes a complete rotation. To do this, we must *map* the integers 1, 2, 3, ... `N` to angular values between 0 and 2π. One simple way to do this is shown below

```
for i = 1:N
  theta2(i) = (i-1)*(2*pi)/(N-1);


end
```

You should confirm that `theta2` takes on values between 0 and 2π by substituting `i = 0` and `i = N` into the formula above. Now the crank will make a full revolution, regardless of how many position calculations we perform. Note the presence of the `(i)` after `theta2`. We use this syntax to store each crank angle in the row vector `theta2`, overwriting the zeros that we initialized earlier. Thus, in the first iteration we will calculate `theta2(1)`, and in the twenty-fifth iteration we will calculate `theta2(25)`. After completing the calculations in the main loop, we can access the twentiethcrank angle (e.g.) by typing at the command prompt

```
>> theta2(20)

ans =

    0.3316
```

The variable `i` is known as the *index* of a particular value in `theta2` – think of it as the address of a particular number in the vector `theta2`.

4.4.3 Position Calculations

Now that we have the angle `theta2` defined, we can begin performing the position calculations. First solve for the angle `theta3`, using the formula

$$\tan\theta_3 = \frac{a\sin\theta_2}{a\cos\theta_2 - d} \tag{4.79}$$

Your first approach might be to divide the quantity a sinθ_2 by the quantity a cos$\theta_2 - d$ and take the inverse tangent. This will give the correct result if the angle θ_3 lies within the first or fourth quadrants. However, as shown in Figure 4.24, the inverse tangent function will give the same result if θ_3 lies in the third quadrant as it will if θ_3 lies in the first quadrant, since

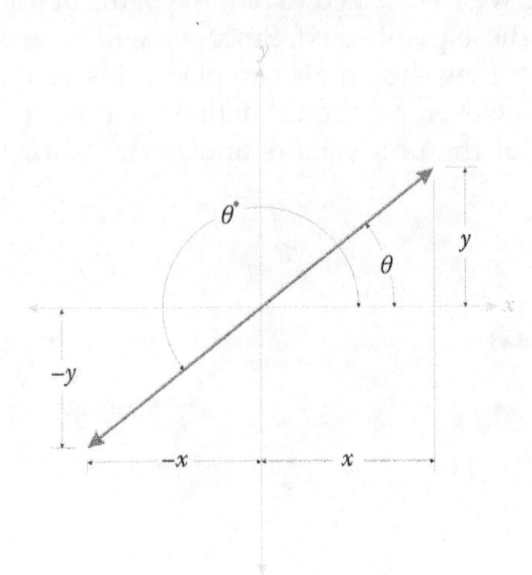

FIGURE 4.24
To the ordinary inverse tangent function the angles θ and θ^* look the same.

$$\frac{y}{x} = \frac{-y}{-x} \tag{4.80}$$

Luckily, MATLAB has a built-in function, `atan2`, which uses separate arguments for y and x thereby making all four quadrants distinct from one another.

As a further complication we observe that the angle θ_3 is drawn using our negative convention (moving CW from the horizontal) while θ_2 is positive (CCW from the horizontal), and vice versa (see Figure 4.25). Thus, instead of the formula given in Equation (4.79), we should use

$$\tan\theta_3 = \frac{-a\sin\theta_2}{d - a\cos\theta_2} \tag{4.81}$$

This is the same formula, but the numerator and denominator have both been multiplied by –1. After the calculating θ_3, we find the distance b easily using the Equation (4.71). In your script add the lines:

```
  theta3(i) = atan2(-a*sin(theta2(i)),d - a*cos(theta2(i)));
  b(i) = (d - a*cos(theta2(i)))/cos(theta3(i));
```

Make sure that you place a comma between the two arguments of the `atan2` function, and not a division symbol. Your complete for loop should now look like:

```
for i = 1:N
  theta2(i) = (i-1)*(2*pi)/(N-1);
  theta3(i) = atan2(-a*sin(theta2(i)),d - a*cos(theta2(i)));
  b(i) = (d - a*cos(theta2(i)))/cos(theta3(i));

end
```

If you execute the program now, you will be disappointed to find that nothing happens! we have told MATLAB to perform the position calculations, but not to plot anything. In the requirements listed above, we were asked to plot the paths of points B and P, so we should calculate the positions of these points next. Since we will be executing these calculations once for every crank angle, they should also be placed inside the loop. Before calculating the positions of B and P, however, we should define the unit vector (and normal) for each link. We will not use all of the unit vectors and normals for the position calculations,

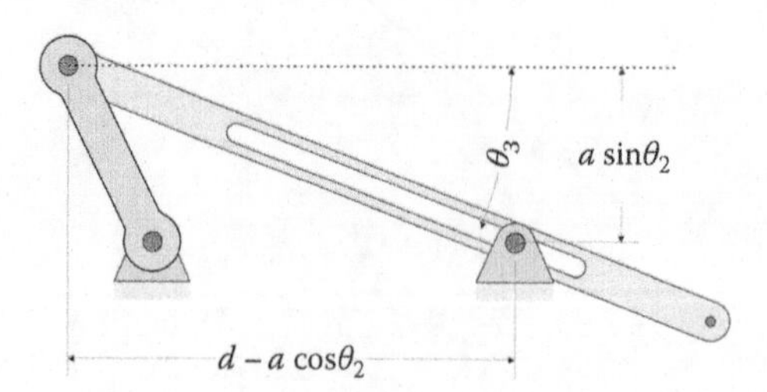

FIGURE 4.25
The angle θ_3 is negative in this figure, since it is measured from the horizontal.

but they will all come in handy later when we conduct velocity and acceleration analysis. Recall that the formulas for a general unit vector and normal are given by

$$\mathbf{e} = \begin{Bmatrix} \cos\theta \\ \sin\theta \end{Bmatrix} \quad \mathbf{n} = \begin{Bmatrix} -\sin\theta \\ \cos\theta \end{Bmatrix} \tag{4.82}$$

Since we will be calculating several unit vectors in our analysis, it is worthwhile to develop a general piece of code that we can use repeatedly for any linkage. The most common way to create reusable code in MATLAB is to create a *function*, which is a separate file that is called by the main program as needed. One common example of a function is the humble sine function

```
h = c*sin(delta);
```

which is used to evaluate the sine of an angle. The function `sin` is a piece of code that resides deep in the bowels of MATLAB. Luckily, we need never be concerned with the internal workings of the `sin` function, we simply supply it with an *argument* (`delta` in this case), and it returns an answer: the sine of the angle `delta`.

While many languages allow you to define functions within the main program file, MATLAB would prefer that you define the function in a separate file in the same folder as the main program. The syntax for a MATLAB function is

```
function [a, b, c, ...] = functionName(A, B, C, ...)
```

The variables `A`, `B`, `C`, etc., are values that we pass *to* the function in order for it to do its calculations. Once the calculations are complete, the function *returns* the variables `a`, `b`, `c`, etc. We must give the function a name (shown as `functionName`) that follows the normal MATLAB file naming conventions (no spaces, must start with a letter, etc.). We will define a function called `UnitVector` that will calculate the unit vector and unit normal, given an input angle `theta`. Create a new script in MATLAB and type in the following function:

```
% UnitVector.m
% Calculates the unit vector and unit normal for a given angle
%
% theta = angle of unit vector
% e     = unit vector in the direction of theta
% n     = unit normal to the vector e

function [e,n] = UnitVector(theta)

e = [ cos(theta); sin(theta)];
n = [-sin(theta); cos(theta)];
```

This function returns a unit vector `e` and a unit normal `n`, given the angle `theta`. As expected, each of these vectors has both an x and y component, and each is a column vector of dimension 2×1. Of course this is a very, very simple function, and most MATLAB functions are more complicated – consider the `atan2` function, for example. Save the function as `UnitVector.m` in the same folder as the `Threebar_Position_Analysis.m` script. One very important fact about functions is that it does not matter what you call the variables inside the function, the variables inside the function are erased as soon as the function has finished executing. To calculate the unit vector for the crank, we would type in the main program

```
[e2,n2]    = UnitVector(theta2(i));
```

It doesn't matter that the unit vector for the crank is called `e2` in the main program, whereas it is called `e` in the function. All that matters is that the order of the arguments matches between the main program and the function. For example, if we were to mistakenly type

```
[n2,e2]    = UnitVector(theta2(i));
```

we would end up with the unit vector being stored in `n2`, and the unit normal stored in `e2`. As our functions get more complicated, be sure to pay attention to the order of the arguments. Having entered and saved the `UnitVector.m` function, enter the following in the main program, right after the calculation for `b(i)`.

```
% calculate unit vectors
  [e2,n2]    = UnitVector(theta2(i));
  [e3,n3]    = UnitVector(theta3(i));
```

We can now use these unit vectors to calculate the coordinates of points B and P. The point B is at the end of the crank, so its position is found by

$$\mathbf{x}_B = a\mathbf{e}_2 \tag{4.83}$$

where $\mathbf{x}_B$ is a two-dimensional vector containing the x and y coordinates of point B. The position of point P is

$$\mathbf{x}_P = \mathbf{x}_B + p\mathbf{e}_3 \tag{4.84}$$

where we have used the relative position formula described earlier. Since we will have much occasion to use the relative position formula, it makes sense to define a separate function for it. Open a new MATLAB script and enter the following:

```
% Function FindPos.m
% calculates the position of a point on a link using the
% relative position formula
%
% x0 = position of first point on the link
% L  = length of vector between first and second points
% e  = unit vector between first and second points
% x  = position of second point on the link

function x = FindPos(x0, L, e)

x = x0 + L * e;
```

Again, this is a very simple function. It returns the position of a point on a link as `x` given the position of another point on the link, `x0`, as well as the length and unit vector associated with the link (`L` and `e`). For the point B on the crank, `x0` would simply be the origin. Save the function `FindPos.m` in the same folder as the main program, and enter the following in the main program after the unit vector calculations

```
% solve for positions of points B and P on the linkage
  xB(:,i) = FindPos(      x0,a,e2);
  xP(:,i) = FindPos(xB(:,i),p,e3);
```

The syntax in these two statements might be a little confusing at first. Remember that we are calculating the position of points B and P for every crank angle: 361 calculations in all. Since each calculation for the position of point B results in two values (the x and y coordinates) we must use the colon operator for the first index of `xB`. The quantity `xB(:,i)` refers to both the x and y coordinates of the ith calculation for the position of point B. In other words, the colon tells MATLAB to cycle through all possible values of this first index (in this case, 1 and 2). At any iteration in the loop, i takes on a single value; thus, the quantity `xB(:,i)` refers to a single 2×1 vector: the coordinates of point B at the current crank angle. The complete main loop should now be

```
for i = 1:N
  theta2(i) = (i-1)*(2*pi)/(N-1);
  theta3(i) = atan2(-a*sin(theta2(i)),d - a*cos(theta2(i)));
  b(i) = (d - a*cos(theta2(i)))/cos(theta3(i));

% calculate unit vectors
  [e2,n2]   = UnitVector(theta2(i));
  [e3,n3]   = UnitVector(theta3(i));

% solve for positions of points B and P on the linkage
  xB(:,i) = FindPos(      x0,a,e2);
  xP(:,i) = FindPos(xB(:,i),p,e3);
end
```

We are now ready to plot the position of the point B. Immediately after the loop, type the command

```
plot(xB(1,:),xB(2,:))
```

Note that the colon is now in the second index in `xB`. The syntax for the plot command is

```
plot(vector of x coordinates, vector of y coordinates)
```

The quantity `xB(1,:)` gives a vector of the x coordinates for point B, while `xB(2,:)` gives all of the y coordinates. When you run the program you are rewarded by the plot of what appears to be an ellipse. Since we are plotting the path of the point B, which is fixed to the end of the crank, the expected motion is a circle. Recall that all motion on a link with one pin grounded is confined to circular arcs. What is happening is that MATLAB is scaling the x and y axes of the plot to best fit the plot window, and the x axis inevitably ends up stretched out a little. We can remedy the situation by typing

```
axis equal
```

immediately after the `plot` command. Try this, and your plot should become a circle. Since we probably want to measure the coordinates of points on the plot, a grid would also be helpful. Type

```
grid on
```

after the `axis` command to see the grid. To plot the position of point *P* on the same figure, modify your plot statement as

```
plot(xB(1,:),xB(2,:),xP(1,:),xP(2,:))
```

4.4.4 Making a Fancy Plot and Verifying your Code

When you execute the program described above, you should obtain a set of two curves as shown in Figure 4.26. We have solved the problem as it was given in the problem statement, but there are still a few things we should add to make the plot more professional. First, we should add a title and axis labels, as follows:

```
title('Paths of points B and P on the Threebar Linkage')
xlabel('x-position [m]')
ylabel('y-position [m]')
```

We should also add a legend to the plot so that a viewer can distinguish between the two curves.

```
legend('Point B', 'Point P','Location','SouthEast')
```

Here we have used the `Location` property to place the legend at the lower right corner of the plot. We must be sure to place type the legend titles in the same order as our plot command. If we had typed `'Point P'` as the first argument, the colors in the legend would

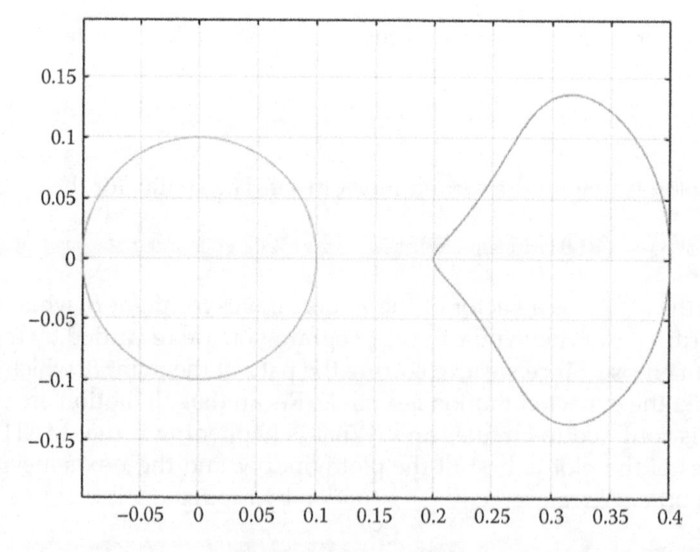

FIGURE 4.26
Paths of the points *B* and *P*. This is the plot you should obtain by running the code described above. Your colors will be different from the plot above, but the shapes of the traces should be the same.

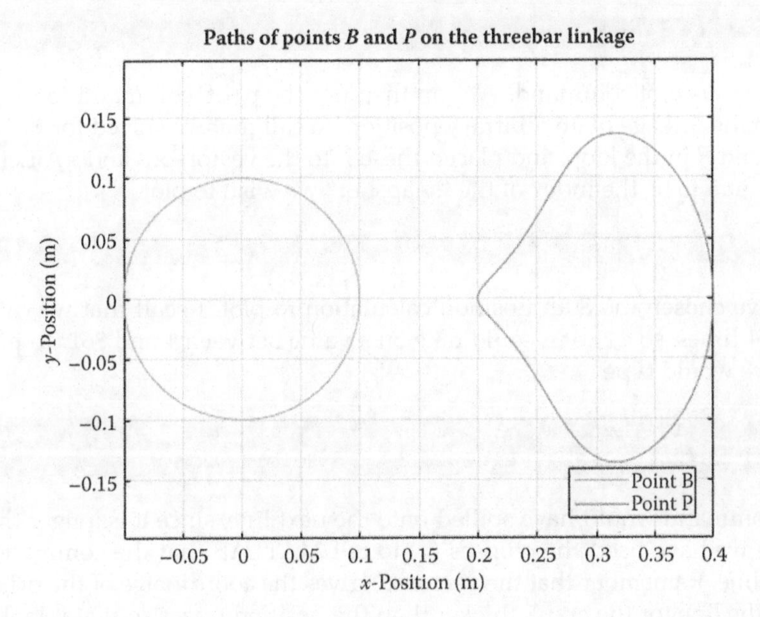

FIGURE 4.27
Threebar position plot with title, legend, and axis labels.

not properly match the plot. Once the legend is in place, the plot is complete, as shown in Figure 4.27.

4.4.5 Verifying Your Calculations

But we are not quite finished! As you have been typing in the code, part of your brain should have been asking, "how do I know that this code accurately models a threebar linkage? How can I be sure that I haven't made a typo somewhere that would produce a valid-seeming, but inaccurate plot?" Checking and verifying calculations is one of the most important roles you will play as a professional engineer, and should be taken very seriously. There are a number of methods we could use to verify the code, but the two methods used by the authors for the code in this chapter are:

1. Draw a sketch of the linkage in SOLIDWORKS®. Use the SmartDimension tool to measure the slider angle at a few different crank angles. Compare these dimensions with the ones calculated in the code.
2. Use MATLAB to draw a "snapshot" of the linkage overlaid on the plot produced by the code given above. If the linkage and the curves line up, there is a good chance that the code is producing correct results. This method is described in more detail below.

4.4.6 Drawing the Linkage in MATLAB®

To overlay a plot of the links on our traces, we must first tell MATLAB to keep plotting in the same window; otherwise any new `plot` command will open up a new plot window. To do this, simply type

```
hold on
```

after the previous plot command. We can then use the `plot` command to draw a line for each link on the linkage in an arbitrary position. Recall that we solved for the coordinates of points *B* and *P* in the loop, and placed these into the vectors `xB` and `xP`. Let us define a variable `iTheta` to be the index of the "snapshot" we wish to plot.

```
iTheta = 80;
```

Here we have chosen the 80th position calculation to plot. Recall that we calculated the positions 361 times, so `iTheta` could take on a value between 1 and 361. To plot a line for the crank, we would type

```
plot([x0(1) xB(1,iTheta)],...
    [x0(2) xB(2,iTheta)],'Linewidth',2,'Color','k');
```

The `plot` command would have spilled onto the next line, since it is longer than 80 characters. Here we have used the ellipses (...) to tell MATLAB that the command continues on the next line. Remember that the vector `x0` gives the coordinates of the origin (0,0). We have made the line for the crank thicker than the position traces so that it looks more like a solid link. The color is black, which MATLAB abbreviates 'k' (to distinguish from blue, 'b'). To plot the slider, type

```
plot([xB(1,iTheta) xP(1,iTheta)],...
    [xB(2,iTheta) xP(2,iTheta)],'Linewidth',2,'Color','k');
```

If you execute the code, you should see the plot in Figure 4.28. To make the plot even fancier, we might wish to add "pins" to each of the points *A*, *B*, *D*, and *P*.

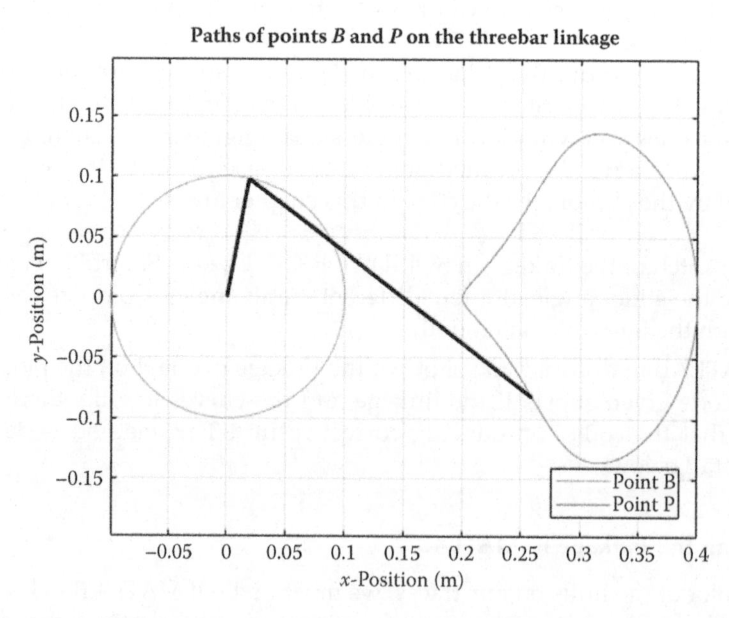

FIGURE 4.28
Plot of the paths of points *B* and *P* with the links overlaid.

```
plot([x0(1) xD(1) xB(1,iTheta) xP(1,iTheta)],...
     [x0(2) xD(2) xB(2,iTheta) xP(2,iTheta)],...
     'o','MarkerSize',5,'MarkerFaceColor','k','Color','k');
```

The `'o'` argument specifies that only small circles are to be used without lines connecting them, and the other arguments specify the dimension and color of the circles. As a final "tweak" we will label each of the points whose paths we are plotting. Use the following `text` commands to place text on your plot:

```
% plot the labels of each pin
text(        x0(1),          x0(2),'A','HorizontalAlignment','center');
text(xB(1,iTheta),xB(2,iTheta),'B','HorizontalAlignment','center');
text(        xD(1),          xD(2),'D','HorizontalAlignment','center');
text(xP(1,iTheta),xP(2,iTheta),'P','HorizontalAlignment','center');
```

The first two arguments give the x and y coordinates of the text. The third argument is the text itself, and the final arguments specify the alignment of the text relative to the coordinates you have provided. Your final plot should look like Figure 4.29.

To prevent the text from overlapping the pins, thus making it more readable, we can add small offsets to their position; the author has used a value of 0.015. A complete listing of the threebar position analysis code is given below. Make sure that your code produces the same plot as shown in Figure 4.29, as we will use this as a basis for all of the programs in future chapters.

```
% Threebar_Position_Analysis.m
% Conducts a position analysis on the threebar crank-slider linkage
% by Eric Constans, June 1, 2017
```

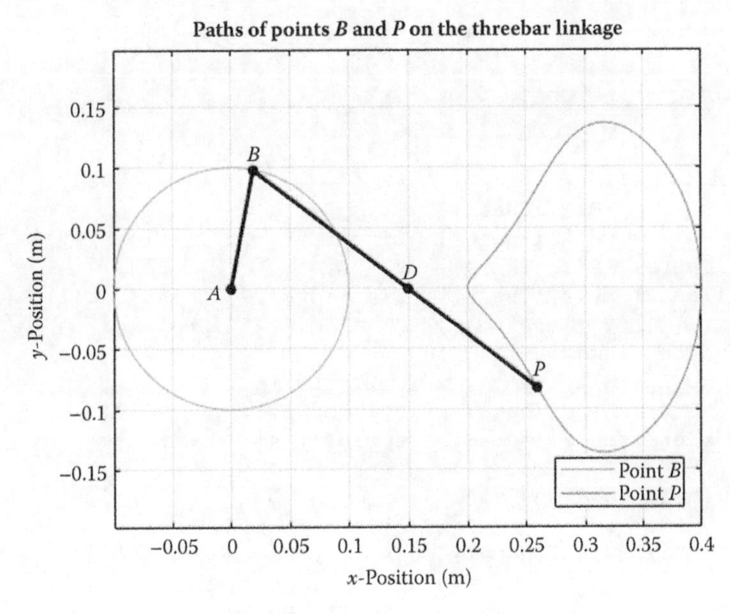

FIGURE 4.29
The complete plot with path traces, overlaid linkage, point markers, and labels.

```
% Prepare Workspace
clear variables; close all; clc;

% Linkage dimensions
a = 0.100;          % crank length (m)
d = 0.150;          % length between ground pins (m)
p = 0.300;          % slider length (m)

% Ground pins
x0 = [0;0];         % point A (the origin)
xD = [d;0];         % point D

N = 361;   % number of times to perform position calculations
[xB,xP]            = deal(zeros(2,N)); % allocate space for position of B,P
[theta2,theta3,b] = deal(zeros(1,N)); % allocate space for link angles

for i = 1:N
  theta2(i) = (i-1)*(2*pi)/(N-1);
  theta3(i) = atan2(-a*sin(theta2(i)),d - a*cos(theta2(i)));
  b(i) = (d - a*cos(theta2(i)))/cos(theta3(i));

% calculate unit vectors
  [e2,n2]   = UnitVector(theta2(i));
  [e3,n3]   = UnitVector(theta3(i));

% solve for positions of points B and P on the linkage
  xB(:,i) = FindPos(     x0,a,e2);
  xP(:,i) = FindPos(xB(:,i),p,e3);
end

plot(xB(1,:),xB(2,:),'Color',[153/255 153/255 153/255])
hold on
plot(xP(1,:),xP(2,:),'Color',[0 110/255 199/255])

% specify angle at which to plot linkage
iTheta = 80;

% plot crank and slider
plot([x0(1)         xB(1,iTheta)],...
     [x0(2)         xB(2,iTheta)],'Linewidth',2,'Color','k');
plot([xB(1,iTheta) xP(1,iTheta)],...
     [xB(2,iTheta) xP(2,iTheta)],'Linewidth',2,'Color','k');

% plot joints on linkage
plot([x0(1) xD(1) xB(1,iTheta) xP(1,iTheta)],...
     [x0(2) xD(2) xB(2,iTheta) xP(2,iTheta)],...
     'o','MarkerSize',5,'MarkerFaceColor','k','Color','k');

% plot the labels of each joint
text( x0(1)-0.015,                  x0(2),'A','HorizontalAlignment','center');
text(xB(1,iTheta),xB(2,iTheta)+0.015,'B','HorizontalAlignment','center');
text(       xD(1),        xD(2)+0.015,'D','HorizontalAlignment','center');
text(xP(1,iTheta),xP(2,iTheta)+0.015,'P','HorizontalAlignment','center');
```

```
title('Paths of points B and P on the Threebar Linkage')
xlabel('x-position [m]')
ylabel('y-position [m]')
legend('Point B', 'Point P','Location','SouthEast')
axis equal
grid on
```

4.5 Position Analysis of the Slider-Crank

We will now turn our attention to finding the position of any point on another simple linkage – the slider-crank. A typical slider-crank mechanism (from a one-cylinder internal combustion engine) is shown in Figure 4.30. In most cases of interest, we wish to find the position of the piston in the cylinder as a function of crank angle. We may also be interested in the angle between the connecting rod and the cylinder, since an excessive connecting rod angle will create undue friction between the piston and cylinder.

Without loss of generality, we may rotate the cylinder so that it is oriented horizontally, and place the crank pin at the origin as shown in Figure 4.31. If the cylinder is not horizontal, we may use a coordinate transformation described in Section 4.2.7 to rotate the results of our calculations as needed. The cylinder is located a vertical distance c from the crank pin. In an engine, the distance c would be zero (i.e. the crank pin would be aligned with the axis of the cylinder.) The crank length is a and the connecting rod length is b. The horizontal position of the slider is d. For a given slider-crank mechanism the dimensions a, b, and c are fixed, and assumed known. The horizontal position of the slider, d, is time varying, and is one of the quantities we must solve for.

To solve for the positions of the links in the slider-crank, we first construct a vector loop diagram, as shown in Figure 4.32. In this diagram, the vector $\mathbf{r}_2$ is attached to the crank and $\mathbf{r}_3$ is attached to the connecting rod. The vector $\mathbf{r}_1$ is horizontal, and connects the ground pin with the point below the piston pin. The vector $\mathbf{r}_4$ is vertical, and slides back and forth with the piston. The length of each vector is constant, except for $\mathbf{r}_1$, which changes as the

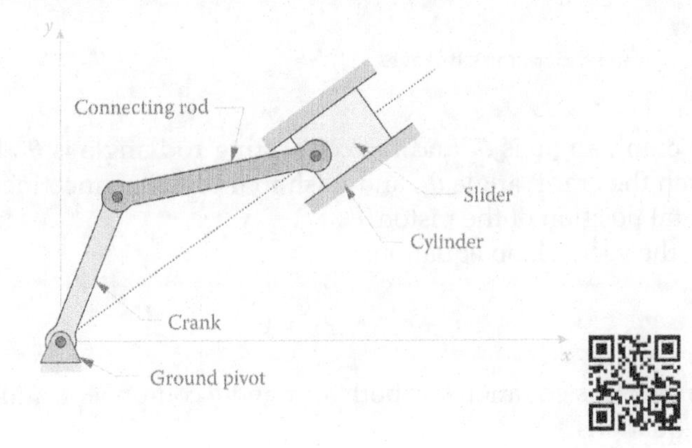

FIGURE 4.30
The slider-crank mechanism consists of a crank, slider, and connecting rod. The crank is attached to the ground at one end, and the cylinder is grounded as well.

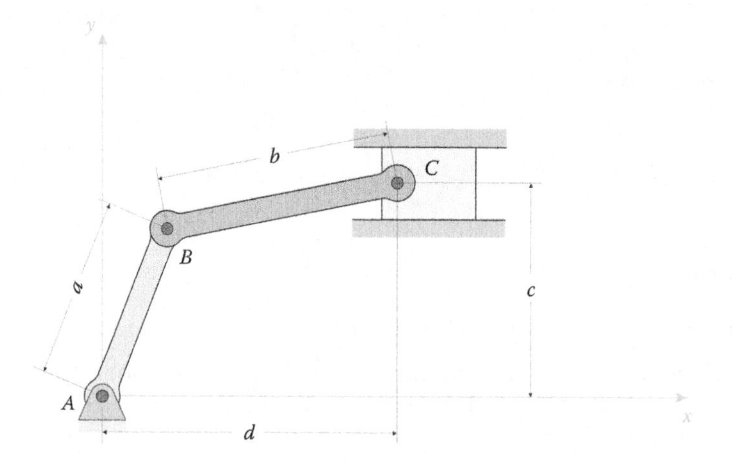

FIGURE 4.31
Dimensions of the slider-crank linkage. The crank length is a, the connecting rod length is b, the vertical distance to the slider is c and the horizontal distance to the slider is d. The crank is grounded at pin A, pin B attaches the crank to the connecting rod, and pin C attaches the connecting rod to the piston.

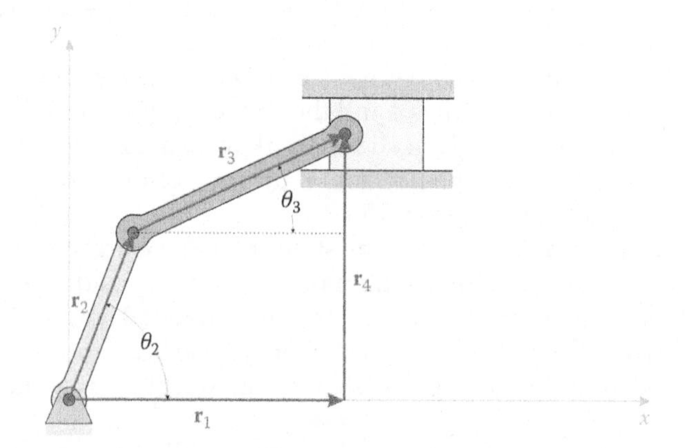

FIGURE 4.32
The vector loop diagram for the slider-crank linkage.

piston moves. The crank angle is θ_2 and the connecting rod angle is θ_3. In most cases of interest we are given the crank angle θ_2, and wish to find the connecting rod angle θ_3, as well as the horizontal position of the piston, d.

First, let us write the vector loop equation:

$$\mathbf{r}_2 + \mathbf{r}_3 - \mathbf{r}_4 - \mathbf{r}_1 = 0 \tag{4.85}$$

As noted in Section 4.2, this equation has both an x and y component, and may be divided into two separate equations.

$$\begin{Bmatrix} a\cos\theta_2 \\ a\sin\theta_2 \end{Bmatrix} + \begin{Bmatrix} b\cos\theta_3 \\ b\sin\theta_3 \end{Bmatrix} - \begin{Bmatrix} 0 \\ c \end{Bmatrix} - \begin{Bmatrix} d \\ 0 \end{Bmatrix} = \begin{Bmatrix} 0 \\ 0 \end{Bmatrix} \tag{4.86}$$

Or, more simply

$$a\cos\theta_2 + b\cos\theta_3 - d = 0$$
$$a\sin\theta_2 + b\sin\theta_3 - c = 0 \tag{4.87}$$

We can use the second equation to solve for θ_3

$$\theta_3 = \sin^{-1}\left(\frac{c - a\sin\theta_2}{b}\right) \tag{4.88}$$

Once we have solved for θ_3, we can use the first equation in (4.87) to solve for the position of the piston.

$$d = a\cos\theta_2 + b\cos\theta_3 \tag{4.89}$$

4.5.1 Extreme Positions of the Slider-Crank

In some cases (as in the design of an internal combustion engine), it is necessary to know the two extreme positions of the slider as the crank makes its revolution. As seen in Figure 4.33 and Figure 4.34, both extremes occur when the crank and connecting rod are in alignment. In this configuration the linkage forms a right triangle, such that

$$d_{max} = \sqrt{(a+b)^2 - c^2}$$
$$d_{min} = \sqrt{(b-a)^2 - c^2} \tag{4.90}$$

Of course, if $c = 0$ then the extreme positions are given by $b \pm a$. The reader may have noticed that the second formula in Equation (4.90) gives an imaginary result if c is greater than $b - a$. This will be discussed in the section that follows. If c is greater than $b + a$ then the linkage cannot be assembled!

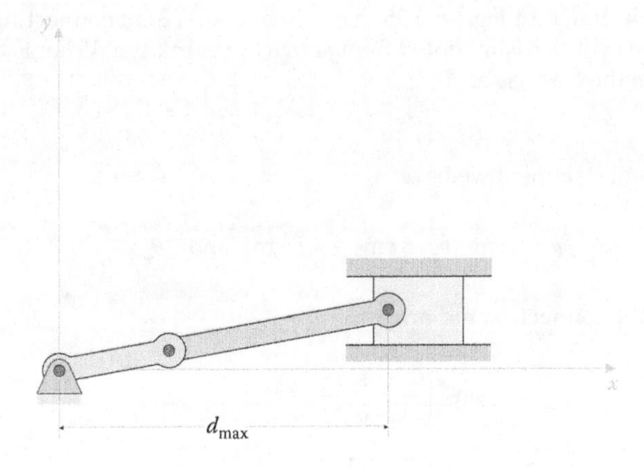

FIGURE 4.33
The piston has its maximum displacement when the crank and connecting rod are aligned.

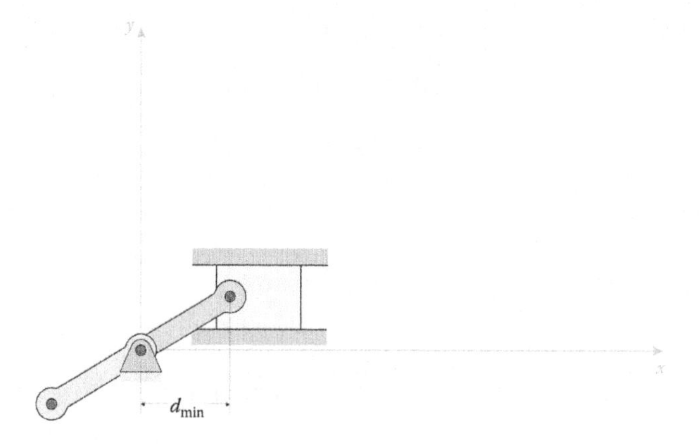

FIGURE 4.34
The piston has its minimum displacement when the crank and connecting rod are anti-aligned.

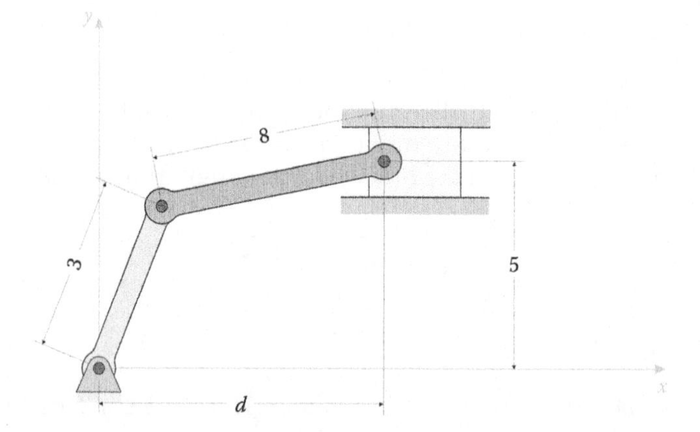

FIGURE 4.35
Slider-crank used in Example 4.3. All dimensions are in cm.

Example 4.5

The slider-crank shown in Figure 4.35 has a 3 cm crank, 8 cm connecting rod, and the centerline of the cylinder is mounted 5 cm above the crank pin. What is the position of the piston when the crank is at 90°?

Solution

From the problem statement we have

$$a = 3\text{ cm} \quad b = 8\text{ cm} \quad c = 5\text{ cm} \quad \text{and} \quad \theta_2 = \frac{\pi}{2}$$

First, solve for the connecting rod angle

$$\theta_3 = \sin^{-1}\left(\frac{c - a\sin\theta_2}{b}\right)$$

$$\theta_3 = \sin^{-1}\left(\frac{5 - 3(1)}{8}\right) = 0.253\text{ rad} = 14.5°$$

Then the piston position is

$$d = a\cos\theta_2 + b\cos\theta_3$$

$$d = 3(0) + 8\cos(0.253) = 7.75\ \text{cm}$$

Example 4.6

Repeat Example 4.3 with crank length 4 cm, connecting rod length 8 cm, slider offset 5 cm, and crank angle 270°.

Solution
From the problem statement we have

$$a = 4\ \text{cm} \quad b = 8\ \text{cm} \quad c = 5\ \text{cm} \quad \text{and} \quad \theta_2 = \frac{3}{2}\pi$$

First, solve for the connecting rod angle

$$\theta_3 = \sin^{-1}\left(\frac{c - a\sin\theta_2}{b}\right)$$

$$\theta_3 = \sin^{-1}\left(\frac{5-4(-1)}{8}\right) = \text{error}$$

It appears as though something has gone wrong with our formula for the connecting rod angle! Examining the argument in the arcsine function, we see that

$$\left(\frac{5-4(-1)}{8}\right) = 1.125$$

Since the arcsine cannot accept arguments with magnitude greater than 1, the solution fails.

For a physical interpretation, see Figure 4.36. The crank has two limiting angles, $\theta_{2\min}$ and $\theta_{2\max}$, beyond which it cannot travel without the linkage "binding up." In each case, the connecting rod is trying to push/pull the piston through the cylinder wall; that is, θ_3 is 90°. Substituting this into Equation (4.88) gives

$$1 = \frac{c - a\sin\theta_2}{b} \tag{4.91}$$

or

$$\sin\theta_2 = \frac{c-b}{a} \tag{4.92}$$

The arcsine function has two solutions:

$$\theta_{2\min} = \sin^{-1}\frac{c-b}{a}$$

$$\theta_{2\max} = \pi - \sin^{-1}\frac{c-b}{a} \tag{4.93}$$

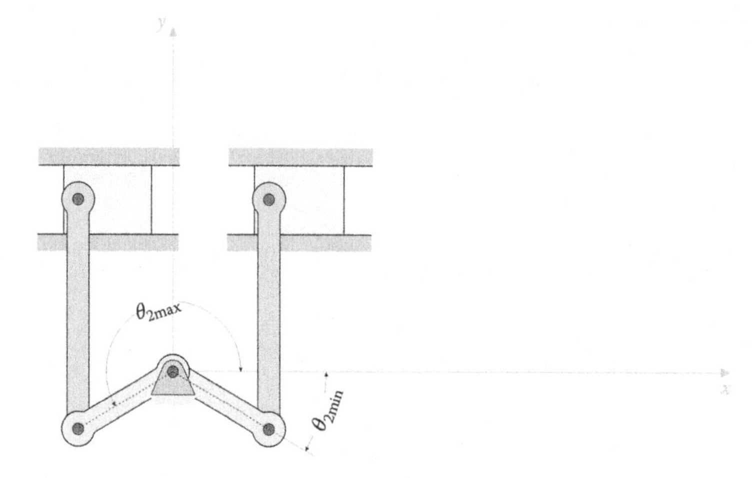

FIGURE 4.36
The two limiting positions of the slider-crank in Example 4.4.

Then, for the current example

$$\theta_{2\min} = -48.6^\circ$$
$$\theta_{2\max} = 228.6^\circ$$

The crank is only allowed to range between these two values, and thus the crank angle of 270° is invalid. This situation is analogous to the Grashof condition for the fourbar linkage in that a full rotation of the crank is only permitted for certain values of a, b and c. In particular, we must have

$$\begin{aligned} &a + |c| \leq b \text{ if } |c| > 0 \\ &a \leq b \text{ if } c = 0 \end{aligned} \tag{4.94}$$

if the crank is to be capable of making a full rotation.

Example 4.7

A slider-crank linkage has crank length 8 cm, connecting rod length 16 cm and slider offset −5 cm. Is the crank capable of making a full revolution? If not, what is its range of motion? What are the minimum and maximum positions of the slider?

Solution
Since c is less than zero, we note that $a + |c| = 13$ cm, and $b = 16$ cm. Thus, the crank can make a full revolution. Using Equation (4.90), we have

$$\begin{aligned} d_{\max} &= 23.5 \text{ cm} \\ d_{\min} &= 6.2 \text{ cm} \end{aligned} \tag{4.95}$$

We have developed a set of formulas for analyzing the position of all links on the slider-crank. Our next step will be to implement these into a MATLAB code so that we can conduct a full position analysis at any crank angle.

4.6 Position Analysis of the Slider-Crank Using MATLAB®

We will now translate the formulas we derived into a MATLAB code that will conduct a position analysis of the slider-crank for all crank angles. The goal of the exercise is to predict the position of the piston within the cylinder as a function of crank angle. Figure 4.37 shows the mechanism that we wish to analyze: a simple one-cylinder compressor. For this mechanism the crank takes the form of a flywheel, and the slider is the piston. We will call the pin at A the *ground pin*, the pin at B the *crank pin*, and the pin at C the *wrist pin*. The shape of the flywheel will not affect our position analysis; only the distance between the ground pin and crank pin matters. There is no vertical offset for the compressor; thus the distance $c = 0$.

Figure 4.38 shows the dimensions of the example slider-crank. We will use the same dimensions when we conduct velocity, acceleration, and force analysis in later chapters. The connecting rod is three times the length of the crank, which is a good "rule of thumb" length to avoid excessive side loads on the cylinder walls.

Open a new MATLAB script and save it as `SliderCrank_Position_Analysis.m` (no spaces!). You should save it in the same folder as the previous example so that we can use the MATLAB functions we defined earlier. At the top of the file, type

```
% SliderCrank_Position_Analysis.m
% performs a position analysis on the slider-crank linkage and
% plots the piston position as a function of crank angle
% by Eric Constans, June 2, 2017
```

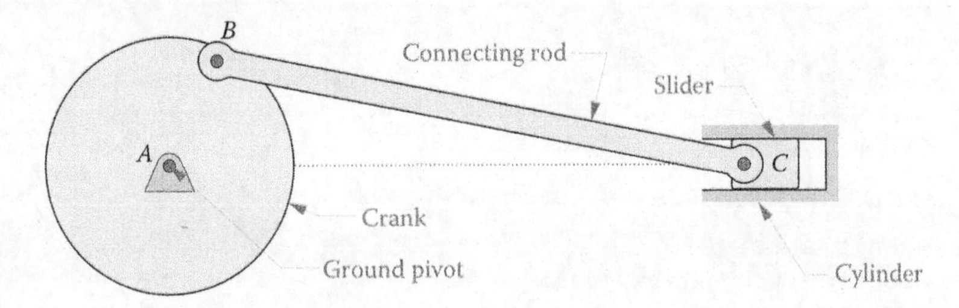

FIGURE 4.37
The slider-crank mechanism used in the MATLAB example code. Note that the axis of the cylinder is aligned with the ground pin – there is no vertical offset.

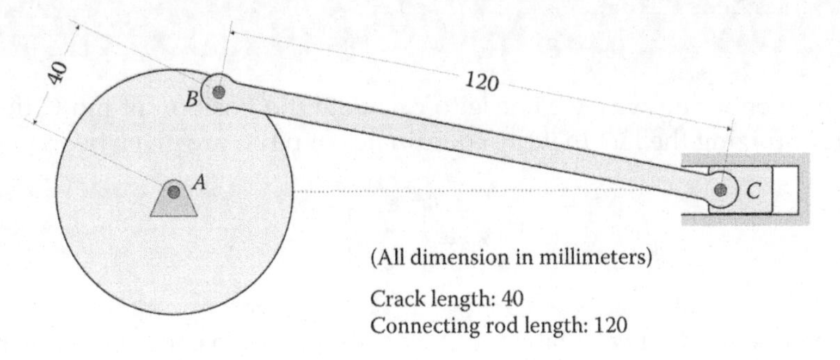

FIGURE 4.38
Dimensions of the slider-crank mechanism for the MATLAB example code.

```
% Prepare Workspace
clear variables; close all; clc;
```

We next enter the dimensions of the linkage. Even though this mechanism has no vertical offset, we will keep the code general so that we may use it with any slider-crank mechanism.

```
% Linkage dimensions
a = 0.040;              % crank length (m)
b = 0.120;              % connecting rod length (m)
c = 0.0;                % vertical slider offset (m)

% Ground pins
x0 = [0;0];             % ground pin at A (origin)
```

Next, allocate space in memory for the position variables that we will calculate. There are only two pins that move, B and C. We will be solving for the connecting rod angle, `theta3`, and the piston position, `d`.

```
N = 361;    % number of times to perform position calculations
[xB,xC] = deal(zeros(2,N));              % allocate space for pins B and C
[theta2,theta3,d] = deal(zeros(1,N));% allocate space for link angles
```

Now, we are ready for the main loop. Since the position calculations are simple, the loop will be very short.

```
% Main loop
for i = 1:N
  theta2(i) = (i-1)*(2*pi)/(N-1);
  theta3(i) = asin((c - a*sin(theta2(i)))/b);
  d(i) = a*cos(theta2(i)) + b*cos(theta3(i));

% calculate unit vectors
  [e1,n1] = UnitVector(0);
  [e2,n2] = UnitVector(theta2(i));
  [e3,n3] = UnitVector(theta3(i));

% solve for position of point B on the linkage
  xB(:,i) = FindPos(      x0,    a,    e2);
  xC(:,i) = FindPos(xB(:,i),    b,    e3);
end
```

You might be wondering why we chose to calculate the position of pin C the hard way, instead of recognizing the fact that the coordinates of pin C are given by

$$\mathbf{x}_C = \begin{Bmatrix} d \\ c \end{Bmatrix} \tag{4.96}$$

This is indeed a redundant calculation, but we will use it as a check on our solution when we are finished. Finally, enter the plot commands to plot the position of the piston versus crank angle.

```
% plot the piston position
plot(theta2*180/pi,d*1000,'Color',[0 110/255 199/255])
title('Piston Position versus Crank Angle for Slider-Crank')
xlabel('Crank angle (degrees)')
ylabel('Position (mm)')
grid on
set(gca,'xtick',0:60:360)
xlim([0 360])
```

Note that the plotting units have been changed to millimeters by multiplying d by 1000 in the `plot` command.

Since the piston moves horizontally, its y coordinate remains constant. For this reason we did not plot the y coordinate of the piston versus the x coordinate – we would have ended up with a straight line! Instead, it makes more sense to plot the piston position versus crank angle for the compressor.

Execute the code and admire the resulting plot. If you have entered everything correctly, you should obtain a plot resembling Figure 4.39. As we expect, the piston reaches its maximum displacement when the crank angle is 0° (or 360°) and a minimum at a crank angle of 180°.

4.6.1 Verifying the Code

The first thing to check is whether the maximum and minimum positions agree with the formulas we derived in Section 4.5. As we found earlier, the extreme positions of the piston are given by

$$\begin{aligned} d_{max} &= b + a \\ d_{min} &= b - a \end{aligned} \tag{4.97}$$

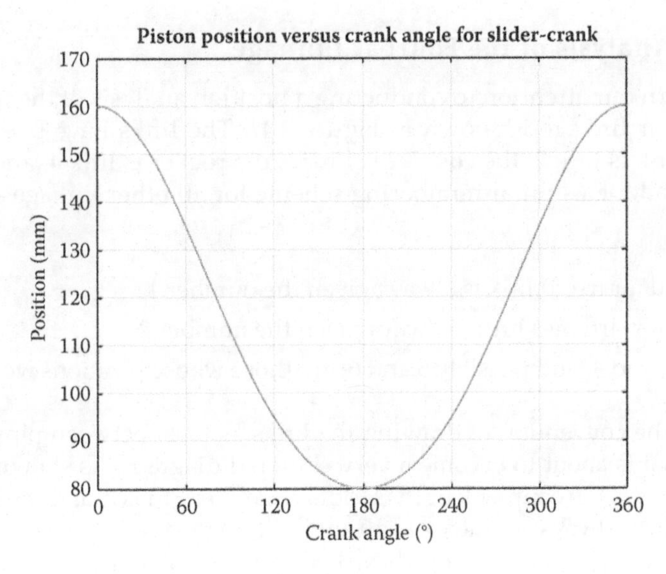

FIGURE 4.39
Position of the piston versus crank angle for the example mechanism.

when the vertical offset is zero. For our example mechanism, this would give

$$d_{max} = 120\text{ mm} + 40\text{ mm} = 160\text{ mm}$$
$$d_{min} = 120\text{ mm} - 40\text{ mm} = 80\text{ mm} \tag{4.98}$$

These numbers agree with the plot, so we have passed our first check.

As a second check, let us plot the piston position d versus the x coordinate of pin C. As discussed above, these should be identical. Change your plotting code to the following, and execute the program.

```
% plot the piston position
plot(theta2*180/pi,d*1000,'o','Color',[153/255 153/255 153/255])
hold on
plot(theta2*180/pi,xC(1,:)*1000,'Color',[0 110/255 199/255])
title('Piston Position versus Crank Angle for Slider-Crank')
xlabel('Crank angle (degrees)')
ylabel('Position (mm)')
legend('d','xC')
grid on
```

As seen in Figure 4.40, the two traces match exactly, and we have a second verification of the code. Make sure that your code produces identical results before moving on to the homework exercises.

As a final note, you might wish to incorporate a check at the beginning of your code to ensure that the linkage can be assembled. Look in Section 4.5 for the discussion on the extreme positions of the piston and limiting crank angles for the appropriate formulas.

4.7 Position Analysis of the Fourbar Linkage

We will now turn our attention to conducting a position analysis of the fourbar linkage. A typical fourbar linkage is shown in Figure 4.41. The links have been numbered as follows: the crank is link 2, the coupler is link 3, the rocker is link 4, and the ground is link 1. We will adopt a similar numbering scheme for all other linkages in the sections that follow.

- The ground (or fixed) link is always given the number 1.
- The crank (or driving) link is always given the number 2.
- The links 3 and 4 (and possibly 5 and 6) are those whose positions we wish to find.

We now adopt the convention of drawing the links as lines between pins; this will help to de-clutter what is about to become a very cluttered diagram. As shown in Figure 4.42, the crank length is a, the coupler length is b, the rocker length is c, and the length between ground pins is d. As always, we measure all angles from the positive x axis. We first assume that all of the link lengths are given (or have been measured). Further, we assume that link

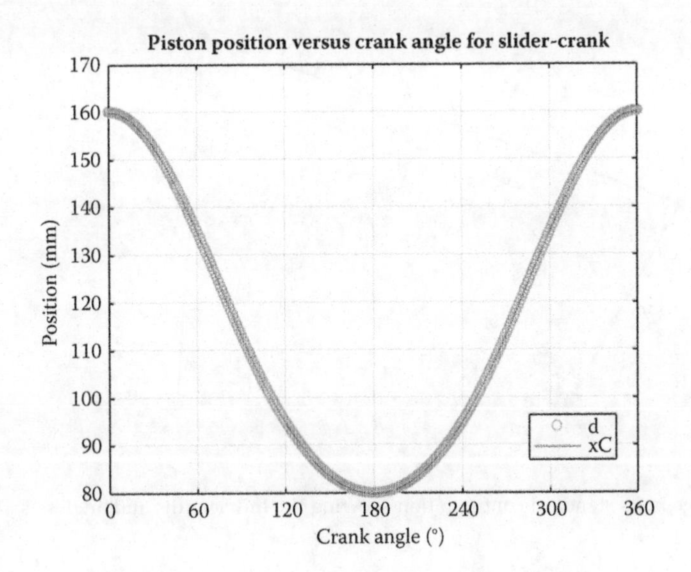

FIGURE 4.40
Piston position d plotted against the x-coordinate of pin C. The two traces overlay each other.

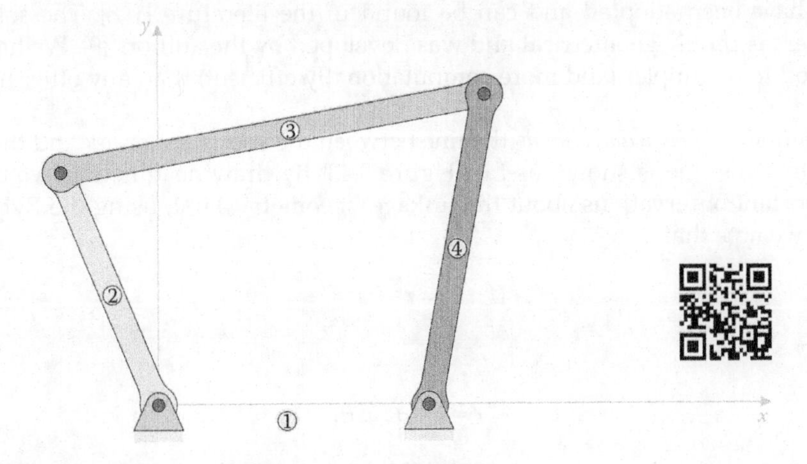

FIGURE 4.41
A typical fourbar linkage with the origin at the left ground pin.

2 (the crank) is driving the linkage, and that its angle, θ_2, is known. Thus, our list of known quantities is

$$\text{known: } a, b, c, d, \theta_2$$

Our goal, then, is to find a method for calculating the coupler and rocker angles: θ_3 and θ_4.

$$\text{unknown: } \theta_3, \theta_4$$

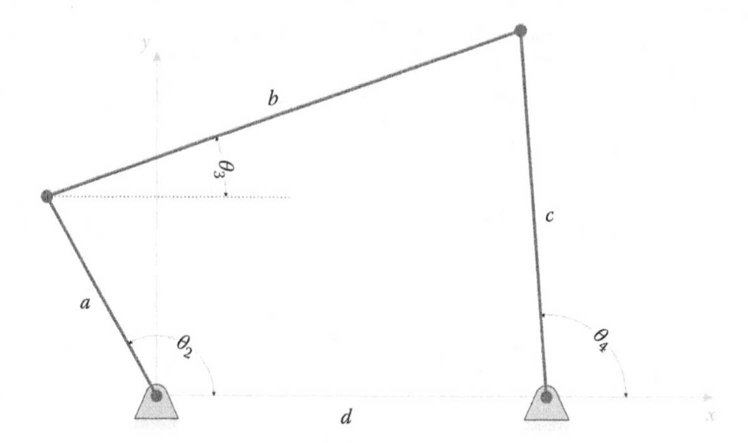

FIGURE 4.42
The fourbar linkage in its standard configuration showing the link lengths and angles as measured from the horizontal.

Once we have found the coupler and rocker angles, it will prove to be a simple matter to find the position of any point on the linkage using vector addition. The problem of solving for the coupler and rocker angles is surprisingly challenging, and a wide variety of solutions have been adopted and can be found in the literature [1–8]. The solution we provide here is purely geometrical and was developed by the authors [9]. We have found this method to be simpler (and more computationally efficient) than any other method in the literature.

Let us define the *prime diagonal* as the line between the end of the crank and the rocker's ground pin. This line is shown as f in Figure 4.43. By drawing this line, we can make some important observations about the linkage's geometry. First, using the Pythagorean Theorem, we note that

$$f^2 = r^2 + s^2 \tag{4.99}$$

where

$$\begin{aligned} r &= d - a\cos\theta_2 \\ s &= a\sin\theta_2 \end{aligned} \tag{4.100}$$

Thus,

$$f^2 = a^2 + d^2 - 2ad\cos\theta_2 \tag{4.101}$$

Since we are given θ_2 and the link lengths, f is simple to calculate. You might recognize the expression above as a restatement of the Law of Cosines. Now define the angle opposite θ_2 as δ. We can also use the Law of Cosines to write

$$f^2 = b^2 + c^2 - 2bc\cos\delta \tag{4.102}$$

or, solving for δ, we have

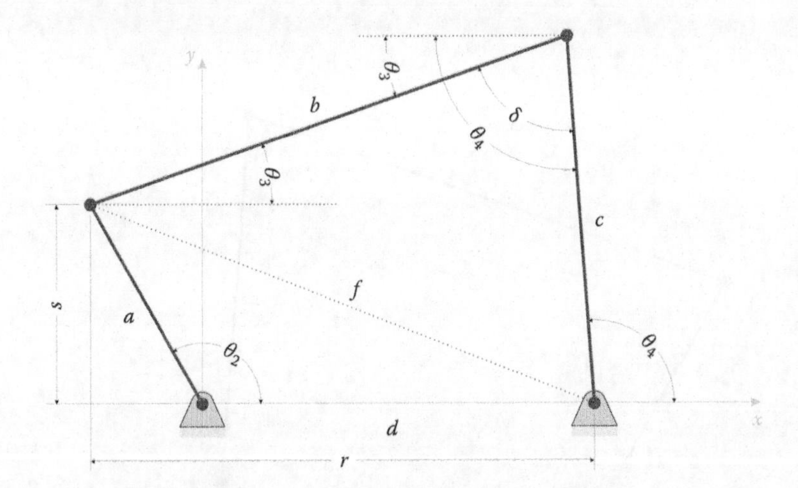

FIGURE 4.43
The prime diagonal, *f*, extends from the crank pin to the opposite ground pin.

$$\cos\delta = \frac{b^2 + c^2 - f^2}{2bc} \tag{4.103}$$

In Figure 4.43, we can also see that

$$\delta = \theta_4 - \theta_3 \tag{4.104}$$

which means that we need only solve for θ_3, since Equation (4.104) can be used to find θ_4.

Now that we know the angle δ, we can use it to calculate a few more interesting quantities. Project a perpendicular line from the coupler to the rocker pin, as shown in Figure 4.44. Define the new lengths

$$\begin{aligned} g &= b - c\cos\delta \\ h &= c\sin\delta \end{aligned} \tag{4.105}$$

Next, going back to the variables r and s defined earlier, we can write

$$\begin{aligned} r &= g\cos\theta_3 + h\sin\theta_3 \\ s &= h\cos\theta_3 - g\sin\theta_3 \end{aligned} \tag{4.106}$$

as shown in Figures 4.45 and 4.46. We now have two equations with one unknown, θ_3. Each equation is transcendental, and difficult to solve on its own. Therefore, we will employ a few tricks to isolate θ_3. First, divide both equations by cos θ_3

$$\begin{aligned} \frac{r}{\cos\theta_3} &= g + h\tan\theta_3 \\ \frac{s}{\cos\theta_3} &= h - g\tan\theta_3 \end{aligned} \tag{4.107}$$

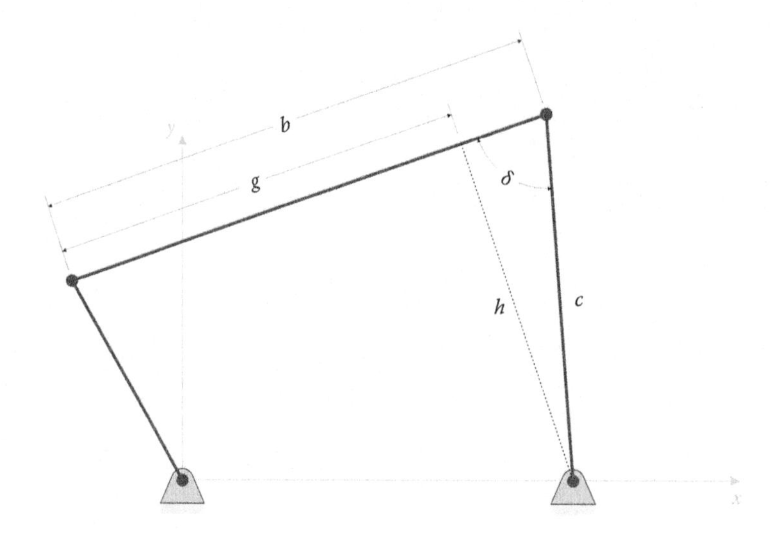

FIGURE 4.44
The lengths g and h can be found once the angle δ is known.

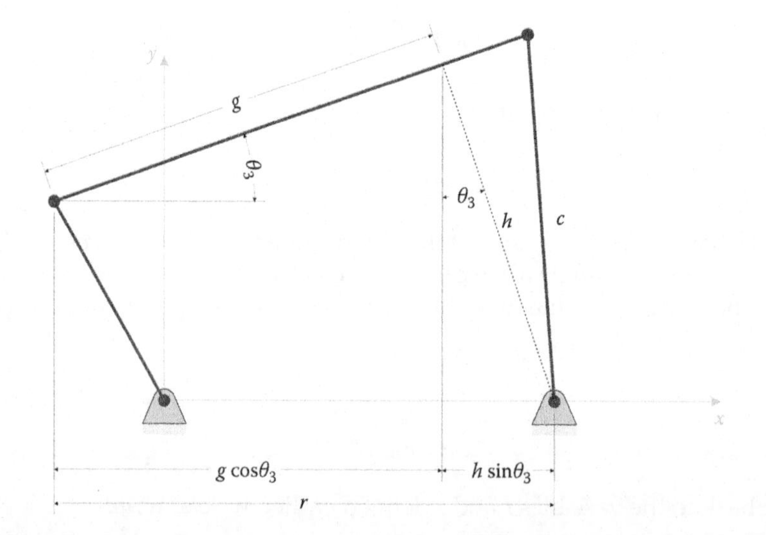

FIGURE 4.45
The dimensions g and h can be related to r through the angle θ_3.

Then, solve both for cos θ_3

$$\cos\theta_3 = \frac{r}{g + h\tan\theta_3}$$
$$\cos\theta_3 = \frac{s}{h - g\tan\theta_3} \tag{4.108}$$

Set the two equations equal to each other

$$\frac{r}{g + h\tan\theta_3} = \frac{s}{h - g\tan\theta_3} \tag{4.109}$$

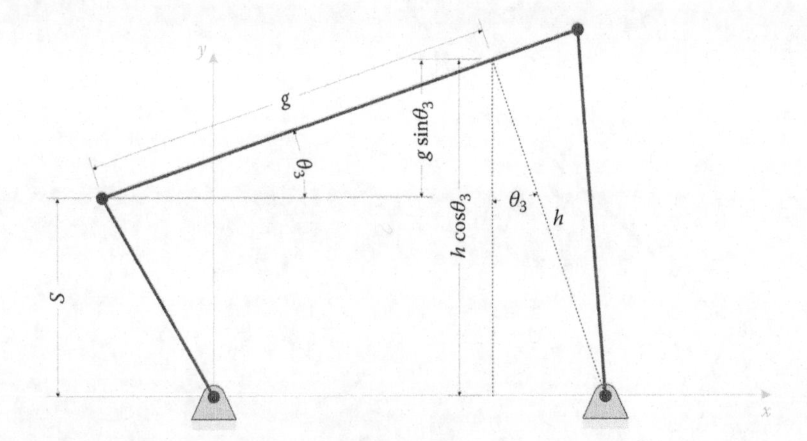

FIGURE 4.46
The dimensions g and h can be related to s through the angle θ_3.

And finally, solve for tan θ_3.

$$hr - gr\tan\theta_3 = gs + hs\tan\theta_3 \tag{4.110}$$

$$(gr + hs)\tan\theta_3 = hr - gs \tag{4.111}$$

$$\tan\theta_3 = \frac{hr - gs}{gr + hs} \tag{4.112}$$

Once we have calculated θ_3, we can use (4.104) to calculate θ_4. Thus, we have achieved our goal of finding the two unknown angles of the fourbar linkage. This method has the added feature of employing the tangent function (as opposed to sine or cosine). When we solve these equations using MATLAB or Excel, we can use the **atan2** function to solve for θ_3 in any quadrant. Similar four-quadrant functions do not exist for sine or cosine.

4.7.1 Finding the Position of Any Point on the Linkage

We are often required to trace the path of a point on the coupler that is not at one of the pins. Such a point is shown as P in Figure 4.47, where the coupler is represented as a triangle with internal angle γ. A simple vector sum will do the trick, as shown

$$\mathbf{r}_P = \mathbf{r}_2 + \mathbf{r}_{BP} \tag{4.113}$$

$$\mathbf{r}_P = a\mathbf{e}_2 + p\mathbf{e}_{BP} \tag{4.114}$$

where

$$\mathbf{e}_{BP} = \begin{Bmatrix} \cos(\theta_3 + \gamma) \\ \sin(\theta_3 + \gamma) \end{Bmatrix} \tag{4.115}$$

is the unit vector pointing from B to P.

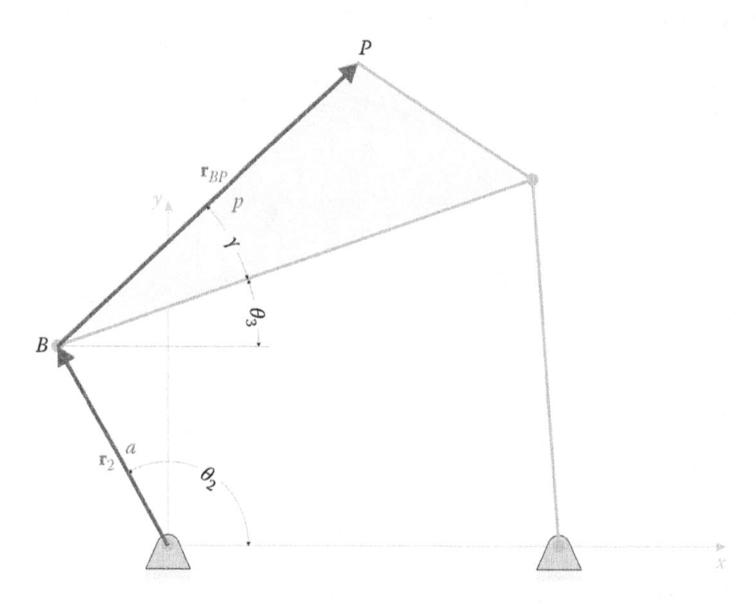

FIGURE 4.47
The point P moves with the coupler, which is triangular in this linkage. A vector sum can be used to travel from the origin to the point P.

Example 4.8: Find the Position of P for One Crank Angle

Let us now conduct a simple example problem for one position of the linkage. For this linkage the crank length is 2 cm, the coupler is 3.2 cm, the rocker is 3 cm, and the distance between ground pins is 1.5 cm. The length BP is 2 cm and the angle γ is 20°. Find the position of point P if the crank angle is 30°.

Solution
First, write out the information given in the problem statement.

$$a = 2\text{ cm} \quad b = 3.2\text{ cm} \quad c = 3\text{ cm} \quad d = 1.5\text{ cm} \quad p = 2\text{ cm} \quad \gamma = 20°$$

Also, the crank angle is specified as

$$\theta_2 = 30°$$

We begin by calculating r and s

$$r = d - a\cos\theta_2 = -0.2321\text{ cm}$$

$$s = a\sin\theta_2 = 1.0\text{ cm}$$

Next, calculate f.

$$f = \sqrt{r^2 + s^2} = 1.0266\text{ cm}$$

And, solving for the angle δ we find

$$\delta = \cos^{-1}\left(\frac{b^2 + c^2 - f^2}{2bc}\right) = 18.7027°$$

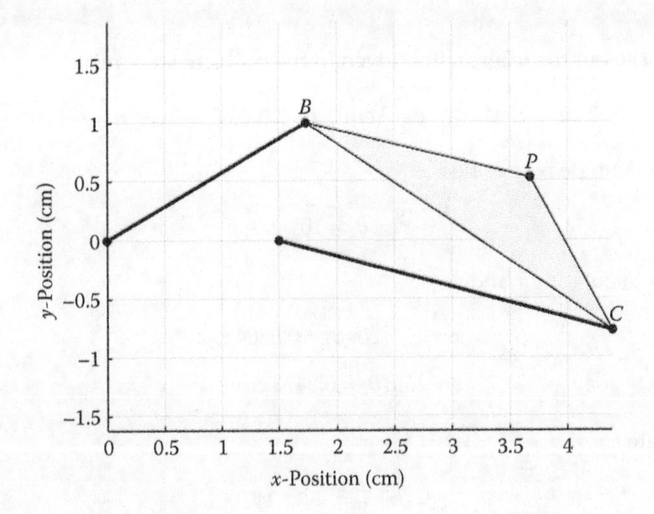

FIGURE 4.48
Configuration of fourbar linkage for Example 4.3.

We can now solve for g and h

$$g = b - c\cos\delta = 0.3584 \text{ cm}$$
$$h = c\sin\delta = 0.9620 \text{ cm}$$

Now solve for the coupler angle, θ_3

$$\tan\theta_3 = \frac{hr - gs}{gr + hs}$$

$$\theta_3 = -33.4988°$$

The rocker angle is found through Equation (4.104)

$$\theta_4 = \theta_3 + \delta = -14.7962°$$

Finally, the coordinates of point P are found through Equation (4.114)

$$\mathbf{r}_P = a\begin{Bmatrix} \cos\theta_2 \\ \sin\theta_2 \end{Bmatrix} + p\begin{Bmatrix} \cos(\theta_3 + \gamma) \\ \sin(\theta_3 + \gamma) \end{Bmatrix} = \begin{Bmatrix} 3.6768 \\ 0.5331 \end{Bmatrix} \text{cm}$$

Figure 4.48 shows the configuration of the linkage for the present example. As you can see, the coordinates of point P do appear to lie at (3.68, 0.53) cm.

Example 4.9: Coordinates of Point P for a Slightly Different Linkage

Now consider a slightly modified fourbar linkage, with crank length 2cm, the coupler length 4cm, rocker length 3cm, and distance between ground pins 2.5cm. The length AP is 2cm and the angle γ is 20°. Find the position of point P if the crank angle is 10°.

Solution

As before, write out the information given in the problem statement.

$$a = 2\text{ cm} \quad b = 4\text{ cm} \quad c = 3\text{ cm} \quad d = 2.5\text{ cm} \quad p = 2\text{ cm} \quad \gamma = 20°$$

Also, the crank angle is specified as

$$\theta_2 = 10°$$

We begin by calculating r and s

$$r = d - a\cos\theta_2 = 0.5304\text{ cm}$$

$$s = a\sin\theta_2 = 0.3473\text{ cm}$$

Next, calculate f.

$$f = \sqrt{r^2 + s^2} = 0.6340\text{ cm}$$

And, solving for the angle δ we find

$$\delta = \cos^{-1}\left(\frac{b^2 + c^2 - f^2}{2bc}\right) = \text{ERROR!}$$

Oh no! We haven't gotten very far, and already we have encountered an error! If we examine the argument in the arccosine function, we can see why

$$\frac{b^2 + c^2 - f^2}{2bc} = 1.0249$$

Since cosine can never return a value greater than one, this will result in an error. To see what happened, let us take a step back and examine the linkage. First, conduct a Grashof analysis.

$$S + L = 6.0\text{ cm}$$

$$P + Q = 5.5\text{ cm}$$

The linkage is not Grashof, so the crank can't make a full revolution. Instead, the crank binds up at a certain minimum angle, and can't go beyond this. To see what the minimum and maximum angles of revolution are, we should sketch the linkage as shown in Figure 4.49.

As seen in the sketch, the linkage "binds up" when the coupler and rocker become collinear. When this occurs, the linkage forms a triangle with sides (a, b–c, d). We can again employ the Law of Cosines to find the maximum and minimum crank angles

$$\cos\theta_{2\min} = \frac{a^2 + d^2 - (b - c)^2}{2ad} = 22.33°$$

$$\cos\theta_{2\max} = 360° - 22.33° = 337.67°$$

Thus, the crank can swing between 22.33° and 337.67°. To continue this example, let us find the position of point P when the crank angle is 30°.

$$r = d - a\cos\theta_2 = 0.7679\text{ cm}$$

$$s = a\sin\theta_2 = 1.0\text{ cm}$$

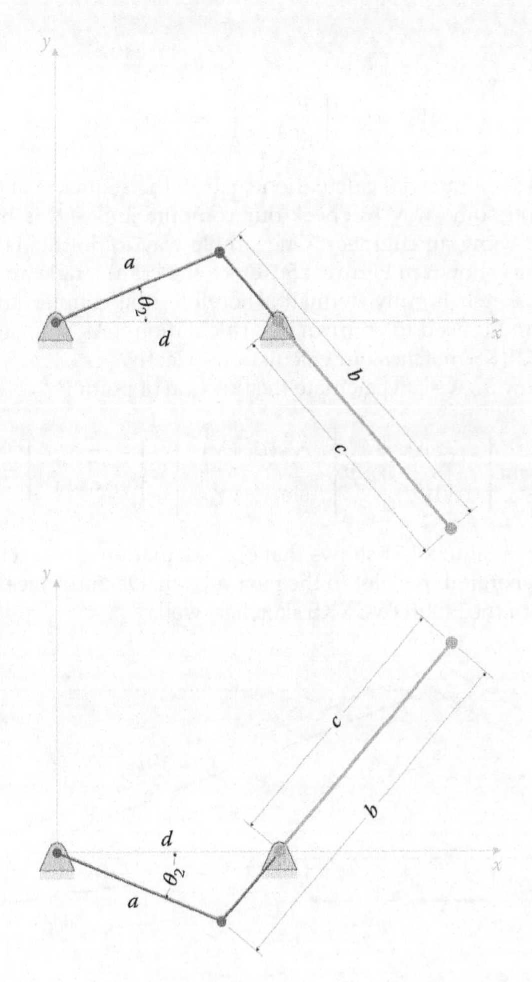

FIGURE 4.49
The two extreme positions of the non-Grashof linkage in the example occur when the rocker and coupler are collinear.

Next, calculate f.

$$f = \sqrt{r^2 + s^2} = 1.2609 \text{ cm}$$

And the angle δ

$$\delta = \cos^{-1}\left(\frac{b^2 + c^2 - f^2}{2bc}\right) = 12.7279°$$

Next, calculate g and h

$$g = b - c\cos\delta = 1.0737 \text{ cm}$$

$$h = c\sin\delta = 0.6610 \text{ cm}$$

Finally, calculate θ_3

$$\theta_3 = \tan^{-1}\left(\frac{hr - gs}{gr + hs}\right) = -20.8617°$$

Whenever we make a nontrivial calculation such as the sequence above, it is our duty as engineers to find some way to check our computations – this is one of the most important parts of being an engineer! One simple way to do this is to make a sketch in SOLIDWORKS, as shown in Figure 4.50. Once the crank angle and link lengths are dimensioned, the sketch is fully defined. Therefore, the coupler angle is a "driven" dimension, and can be used to confirm our calculations. As you can see, the answer given by SOLIDWORKS matches our calculations exactly.

Now that we know θ_3, we can calculate the position of point P

$$r_P = a\begin{Bmatrix} \cos\theta_2 \\ \sin\theta_2 \end{Bmatrix} + p\begin{Bmatrix} \cos(\theta_3 + \gamma) \\ \sin(\theta_3 + \gamma) \end{Bmatrix} = \begin{Bmatrix} 3.7318 \\ 0.9699 \end{Bmatrix} \text{cm}$$

The MATLAB plot in Figure 4.51 shows that our calculation is correct, and we will create the code that generated this plot in the next section. Of course, we can double-check this calculation with the SOLIDWORKS sketch as well.

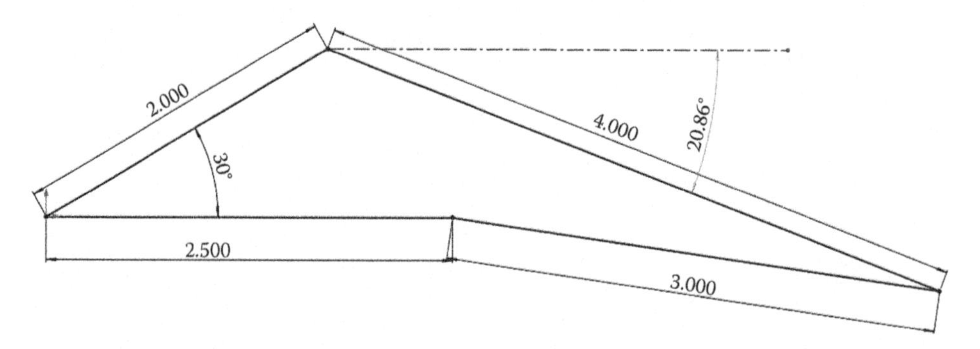

FIGURE 4.50
SOLIDWORKS sketch of the linkage in the example. The angle θ_3 is a driven dimension, and it confirms our calculations.

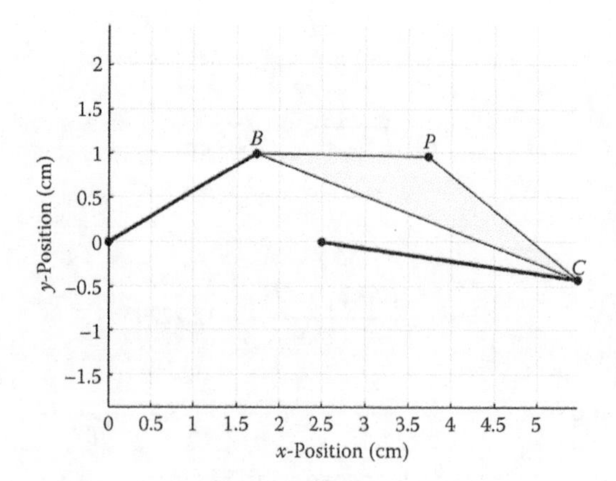

FIGURE 4.51
MATLAB plot of the non-Grashof linkage in Example 4.4.

4.7.2 A Digression into Trigonometric Identities

Let us approach the tangent formula given in Equation (4.112) from a different angle, as it were. If we examine a table of trigonometric identities, we will usually find a tangent sum formula

$$\tan(u \pm v) = \frac{\tan u \pm \tan v}{1 \mp \tan u \tan v} \tag{4.116}$$

Examining Figure 4.52, we see that

$$\theta_3 = \beta - \alpha \tag{4.117}$$

so that

$$\tan\theta_3 = \frac{\tan\beta - \tan\alpha}{1 + \tan\beta\tan\alpha} \tag{4.118}$$

where

$$\tan\alpha = \frac{s}{r} \quad \tan\beta = \frac{h}{g} \tag{4.119}$$

Substituting these into Equation (4.118) gives

$$\tan\theta_3 = \frac{\frac{h}{g} - \frac{s}{r}}{1 + \frac{h}{g}\cdot\frac{s}{r}} = \frac{hr - gs}{gr + hs} \tag{4.120}$$

as before. There is more than one way to arrive at our position formula!

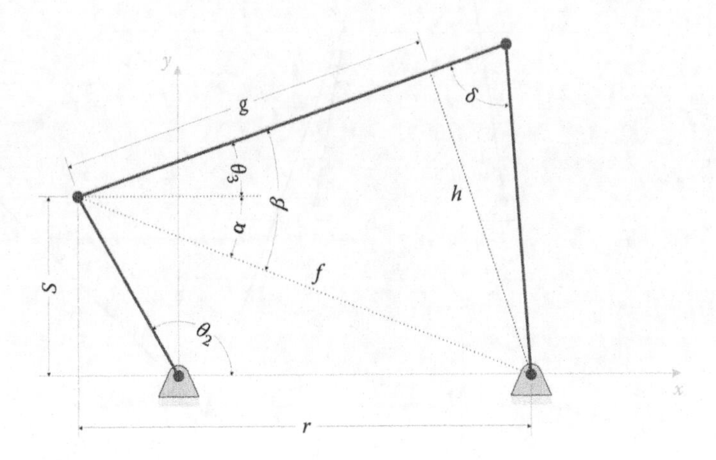

FIGURE 4.52
The fourbar linkage in the open configuration

4.7.3 Open and Crossed Configurations of the Fourbar

Figure 4.53 shows a typical fourbar linkage in its "open" and "crossed" configurations. We have used the open configuration to define the sense of the angles in our formulas. For example, the angle δ was defined as the angle *from* the coupler *to* the rocker, as shown in Figure 4.53 at left. Since the direction of this angle is counterclockwise, we consider it to have a positive value. In the crossed configuration, shown at right, the angle from coupler to rocker sweeps in the clockwise direction, and is therefore negative. Thus, to switch between the open and crossed configurations in our calculations, we can simply change the sign of δ.

$$\delta = \cos^{-1}\left(\frac{b^2 + c^2 - f^2}{2bc}\right) \text{ for open}$$
$$\delta = -\cos^{-1}\left(\frac{b^2 + c^2 - f^2}{2bc}\right) \text{ for crossed} \tag{4.121}$$

This operation is mathematically valid because the cosine function gives the same result for positive and negative angles

$$\cos(\delta) = \cos(-\delta) \tag{4.122}$$

All of the remaining formulas for θ_3 and θ_4 are the same as before.

4.7.4 Summary

We have developed a simple, yet robust method for finding the coupler and rocker angles on the fourbar linkage for a specified crank angle. Minimum and maximum

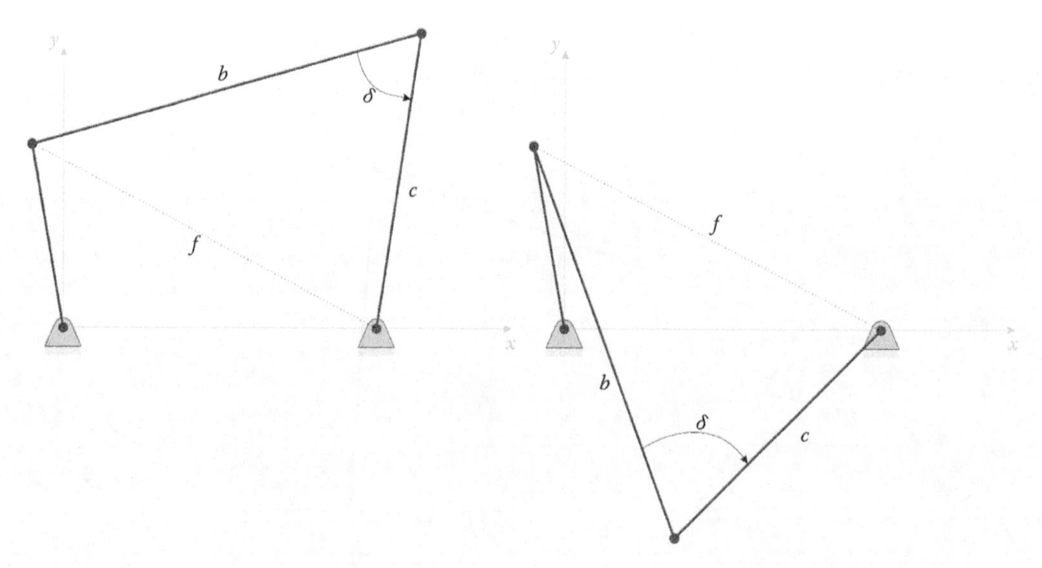

FIGURE 4.53
Open and crossed configurations of a fourbar linkage.

crank angles for non-Grashof linkages were calculated and switching between the open and crossed configurations was found to be as simple as changing the sign of the angle δ. Using a handheld calculator to solve the equations was found to be straightforward, if tedious. In the next section, we will use MATLAB to quickly and efficiently perform the calculations described in this section. This will enable us to create plots of the trajectories of various points on the linkage to ensure that the linkage is behaving as desired.

4.8 Position Analysis of the Fourbar Linkage Using MATLAB®

Now that we have a set of formulas for calculating the angles and positions of the various links on the fourbar linkage we will put our knowledge to work in writing a MATLAB program to perform the calculations for us. For now, we will limit ourselves to solving the Grashof Class 2 linkage, where the crank is the shortest link and can make a full revolution. In a later section, we will extend this to the non-Grashof Classes 5, 8, and 10, which are also driven by the crank, but have limiting angles.

A diagram of the fourbar linkage that we will use in developing our program is shown in Figure 4.54. The linkage has link lengths a, b, c, and d, and we will place the origin at point A. Our goal is to plot the paths of the points B, C, and P on the linkage as the crank makes a full revolution.

We begin the modeling process by making an outline of the program structure. The objectives of our fourbar linkage analysis program are to:

1. Determine whether the linkage is Grashof.
2. If it is not Grashof, the program should inform the user and then terminate.

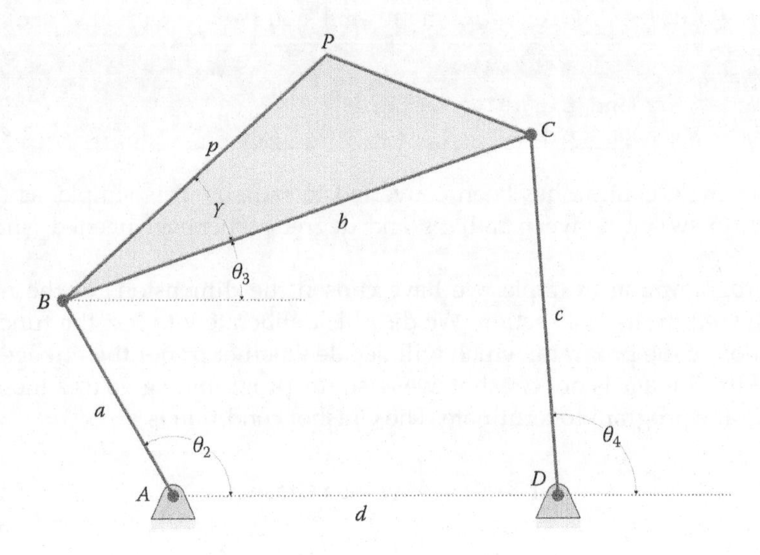

FIGURE 4.54
Critical dimensions of the fourbar linkage. The coupler has been drawn as a triangle with the point P at its top.

3. If it is Grashof, then the program should conduct each of the following steps for every crank angle
   a. Calculate the internal angle δ.
   b. Calculate θ_3 and θ_4, the coupler and rocker angles, respectively.
   c. Use these angles to calculate the positions of points B, C, and P.
4. Once these calculations are complete, the program should generate a plot that shows the paths of points B, C, and P as the crank makes a complete revolution.

Items 1, 2, and 4 need to be executed only once, while the tasks in item 3 are executed several times – once for each crank angle. Therefore, we will place the tasks in item 3 inside a loop, and all of the other tasks will be outside the loop. At the top of a new MATLAB script, type:

```
% Fourbar_Position_Analysis.m
% conducts a position analysis of the fourbar linkage and plots the
% positions of points B, C and P.
% by Eric Constans, June 2, 2017

% Prepare Workspace
clear variables; close all; clc;
```

After you have typed the comments, save the script in a convenient location using this file name. Next, we should tell MATLAB the dimensions of the linkage, as shown:

```
% Linkage dimensions
a = 0.2;                % crank length (m)
b = 0.4;                % coupler length (m)
c = 0.3;                % rocker length (m)
d = 0.25;               % length between ground pins (m)
p = 0.2;                % length from B to P (m)
gamma = 20*pi/180; % angle between BP and coupler (converted to rad)

% ground pins
x0 = [0;0];    % ground pin at A (origin)
xD = [d;0];    % ground pin at D
```

Note that the angle gamma has been converted to radians. It is simple, as the formula above shows, to switch between radians and degrees whenever needed (such as when plotting).

For this programming example, we have chosen the dimensions of the non-Grashof linkage discussed in the last section. We did this deliberately to test the functionality of the next section of the program, which will decide whether or not the linkage is Grashof. Recall that if the linkage is not Grashof, we wish the program to give us a message to this effect, and for the program to terminate. The Grashof condition is

$$\text{if } S+L<P+Q \text{ then Grashof} \tag{4.123}$$

so we must first determine which are the shortest and longest links. Luckily, MATLAB has built-in "minimum" and "maximum" functions, which find the minimum and maximum numbers in a vector, respectively. Add a comment denoting a new section of the code for checking the Grashof Condition and use the `min` function to find `S`.

```
% Grashof Check
S = min([a b c d]);   % length of shortest link
```

This statement will search through the four link lengths (*a*, *b*, *c* and *d*) and assign the minimum value to the variable *S*. Note that we have enclosed the link lengths in square brackets to convert them into a single (1×4) vector. Similarly, the `max` function is written

```
L = max([a b c d]);   % length of longest link
```

It is probably not immediately obvious how to find *P* and *Q*, since these are neither the minimum nor the maximum links. Instead, we will employ a trick to arrive at *the sum* of *P* and *Q*, since this is what we really require for the Grashof condition. Let

```
T = sum([a b c d]);   % total of all link lengths
```

be the total of all the link lengths. The `sum` function is another built-in MATLAB function that calculates the sum of all the elements in a vector. Then

```
PQ = T - S - L;   % length of P plus length of Q
```

will give the *sum* of *P* and *Q*. Make sure that you call the variable `PQ`, and not `P+Q`. Now we are ready to test the Grashof condition. Type the following "if-else" statement into your script.

```
if (S+L < PQ)   % Grashof condition
  disp('Linkage is Grashof.')
else   % if not Grashof, terminate program
  disp('Linkage is not Grashof')
  return
end
```

If the Grashof condition is met, then MATLAB will write a confirmation message to the command window and the script will continue executing. If it is not met, MATLAB will display the appropriate message and the `return` statement will cause control to be "returned" to the command window; in other words, the program will terminate. Test your script now, to make sure that it detects the non-Grashof linkage correctly.

Once you have gotten this part of the script to function correctly, change the link lengths to a Grashof Class 2 linkage as shown below. The dimensions are also shown in Figure 4.55.

```
% Linkage dimensions
a = 0.130;             % crank length (m)
b = 0.200;             % coupler length (m)
c = 0.170;             % rocker length (m)
d = 0.220;             % length between ground pins (m)
p = 0.150;             % length from B to P (m)
gamma = 20*pi/180;  % angle between BP and coupler (converted to rad)
```

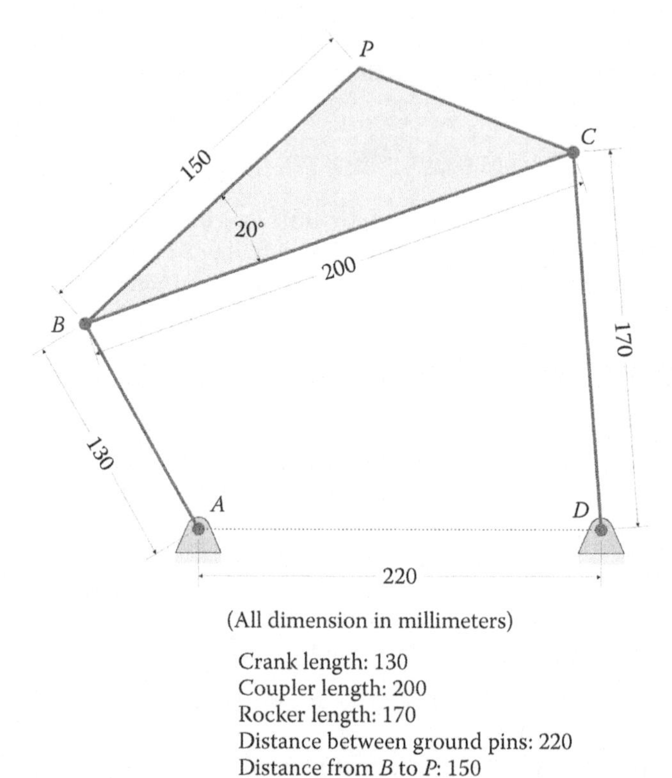

FIGURE 4.55
Dimensions of the example linkage used in the MATLAB code.

If you run the script again, the command window should inform you that the linkage is Grashof. Do not continue until this part of your program functions correctly!

4.8.1 Data Structure for the Position Calculations

The Grashof test was the final task that was to be executed a single time, other than the plotting commands, which must be done at the end of the program after all calculations are complete. As before, we choose to perform the position calculations for every 1° of crank rotation.

```
N = 361;    % number of times to perform position calculations
[xB,xC,xP] = deal(zeros(2,N));                % allocate space for positions
[theta2,theta3,theta4] = deal(zeros(1,N)); % allocate space for angles
```

4.8.2 The Main Loop

As before, we will use a `for` loop to perform the repeated position calculations. Type the following `for` loop into your script:

```
for i = 1:N
  theta2(i) = (i-1)*(2*pi)/(N-1);


end
```

Note the presence of (i) after theta2. The other variables within the loop (e.g. r, s, delta, etc.) will not be given an index; therefore, they are *scalars*, and are overwritten every time the loop executes anew. Another way of saying this is that we use the same place in memory to store the values of r and s every time we conduct a position calculation, so the old value of r is overwritten by the newest value of r when it is calculated. In contrast, we find a new place in memory to store the latest value of theta2, so that the previous values are preserved.

4.8.3 Position Calculations

Now that we have the angle theta2 defined, we can begin performing the position calculations. Start by calculating r and s, which we will use to determine the angle delta.

```
  r = d - a*cos(theta2(i));
  s = a*sin(theta2(i));
```

We could now calculate the variable f, as given in

$$f = \sqrt{r^2 + s^2} \tag{4.124}$$

but to save a square root operation, we make use of the fact that only the *square* of f is used in subsequent calculations. We, therefore, use

```
  f2 = r^2 + s^2;  % f squared
```

The only difference between the open and crossed configurations was the sign of delta. We choose the positive sign here (open), but can easily change it if we desire to calculate the crossed configuration later.

```
  delta = acos((b^2 + c^2 - f2)/(2*b*c));   % open configuration
```

The acos function in MATLAB gives the inverse (or arc) cosine. Now that we have delta, we can calculate g and h

```
  g = b - c*cos(delta);
  h = c*sin(delta);
```

We are now ready to solve for the angle theta3, using the formula

$$\tan\theta_3 = \frac{hr - gs}{gr + hs} \tag{4.125}$$

As we did with the threebar, we use the atan2 function to calculate the inverse tangent, since it is valid in all four quadrants. In your script add the lines:

```
  theta3(i) = atan2((h*r - g*s),(g*r + h*s));
  theta4(i) = theta3(i) + delta;
```

Make sure that you place a comma between the two arguments of the `atan2` function, and not a division symbol. Your complete for loop should now look like:

```
for i = 1:N
  theta2(i) = (i-1)*(2*pi)/(N-1);   % crank angle

% conduct position analysis to solve for theta3 and theta4
  r = d - a*cos(theta2(i));
  s = a*sin(theta2(i));
  f2 = r^2 + s^2;                        % f squared
  delta = acos((b^2+c^2-f2)/(2*b*c)); % angle between coupler and rocker

  g = b - c*cos(delta);
  h = c*sin(delta);

  theta3(i) = atan2((h*r - g*s),(g*r + h*s));
  theta4(i) = theta3(i) + delta;
 end
```

Since we have not calculated the positions of points *B*, *C* or *P*, we still cannot plot anything. The next step is to ask MATLAB to calculate the appropriate unit vectors (before the `end` statement):

```
% calculate unit vectors
  [e1,n1] = UnitVector(0);
  [e2,n2] = UnitVector(theta2(i));
  [e3,n3] = UnitVector(theta3(i));
  [e4,n4] = UnitVector(theta4(i));
  [eBP,nBP] = UnitVector(theta3(i) + gamma);
```

In addition to the usual unit vectors that run along each link, we have also defined the unit vector for the line that extends from point *B* to point *P* on the coupler, since

$$\mathbf{x}_P = a\mathbf{e}_2 + p\mathbf{e}_{BP} \tag{4.126}$$

To get from point *A* to point *C*, we may choose between two different paths, as shown in Figure 4.56.

$$\mathbf{x}_C = a\mathbf{e}_2 + b\mathbf{e}_3 \tag{4.127}$$

or

$$\mathbf{x}_C = d\mathbf{e}_1 + c\mathbf{e}_4 \tag{4.128}$$

We (arbitrarily) choose the second of these expressions, and we can easily calculate the coordinates of points *B*, *C*, and *P* with the `FindPos` function.

```
% solve for positions of points B, C and P on the linkage
  xB(:,i) = FindPos(      x0,     a,    e2);
  xC(:,i) = FindPos(      xD,     c,    e4);
  xP(:,i) = FindPos(xB(:,i),     p,   eBP);
```

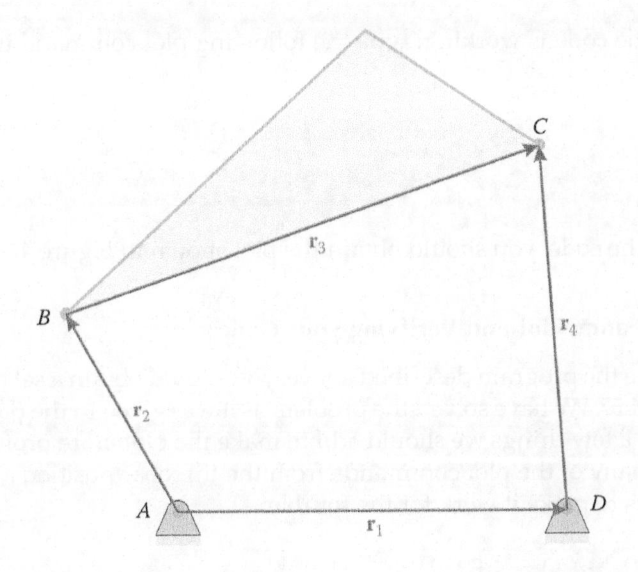

FIGURE 4.56
To get to point *C*, we can choose the path *ABC* or *ADC*.

The complete loop should now be:

```
for i = 1:N
  theta2(i) = (i-1)*(2*pi)/(N-1);  % crank angle

% conduct position analysis to solve for theta3 and theta4
  r = d - a*cos(theta2(i));
  s = a*sin(theta2(i));
  f2 = r^2 + s^2;                        % f squared
  delta = acos((b^2+c^2-f2)/(2*b*c)); % angle between coupler and rocker

  g = b - c*cos(delta);
  h = c*sin(delta);

  theta3(i) = atan2((h*r - g*s),(g*r + h*s));
  theta4(i) = theta3(i) + delta;

% calculate unit vectors
  [e1,n1] = UnitVector(0);
  [e2,n2] = UnitVector(theta2(i));
  [e3,n3] = UnitVector(theta3(i));
  [e4,n4] = UnitVector(theta4(i));
  [eBP,nBP] = UnitVector(theta3(i) + gamma);

% solve for positions of points B, C and P on the linkage
  xB(:,i) = FindPos(      x0,    a,    e2);
  xC(:,i) = FindPos(      xD,    c,    e4);
  xP(:,i) = FindPos(xB(:,i),    p,   eBP);
end
```

To see whether the code is working, type the following plot command immediately after the loop.

```
plot(xB(1,:),xB(2,:),xC(1,:),xC(2,:),xP(1,:),xP(2,:))
axis equal
grid on
```

Upon executing the code, you should obtain the plot shown in Figure 4.57.

4.8.4 Making a Fancy Plot and Verifying your Code

When you execute the program described above, you should obtain a set of three curves as shown in Figure 4.57. We have solved the problem as it was given in the problem statement, but there are still a few things we should add to make the plot more professional. You can copy and paste many of the plot commands from the threebar position analysis program, with a few tweaks to make it work for the fourbar.

```
% specify angle at which to plot linkage
hold on
iTheta = 120;

% plot the coupler as a triangular patch
patch([xB(1,iTheta) xC(1,iTheta) xP(1,iTheta)],...
      [xB(2,iTheta) xC(2,iTheta) xP(2,iTheta)],[229/255 240/255
249/255]);

% plot crank and rocker
plot([x0(1) xB(1,iTheta)],[x0(2) xB(2,iTheta)],'Linewidth',2,'Color'
,'k');
```

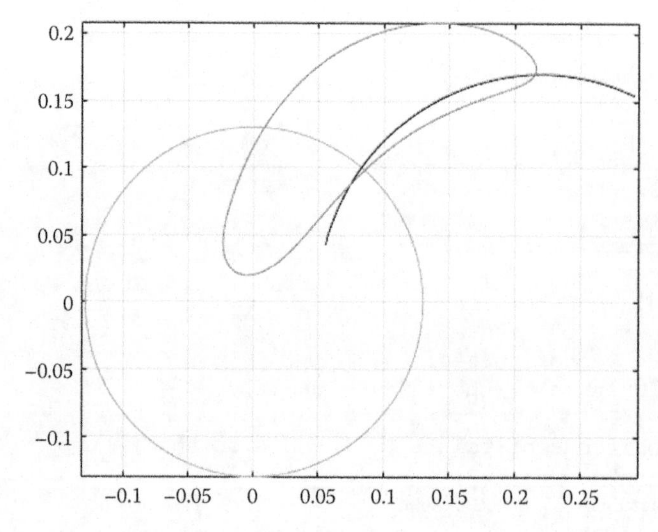

FIGURE 4.57
Paths of the points *B*, *C*, and *P*. This is the plot you should obtain by running the code described above. Your colors will be different from the plot above, but the shapes of the traces should be the same.

```
plot([xD(1) xC(1,iTheta)],[xD(2) xC(2,iTheta)],'Linewidth',2,'Color'
,'k');

% plot joints on linkage
plot([x0(1) xD(1) xB(1,iTheta) xC(1,iTheta) xP(1,iTheta)],...
     [x0(1) xD(2) xB(2,iTheta) xC(2,iTheta) xP(2,iTheta)],...
     'o','MarkerSize',5,'MarkerFaceColor','k','Color','k');

% plot the labels of each joint
text(xB(1,iTheta),xB(2,iTheta)+.015,'B','HorizontalAlignment','center');
text(xC(1,iTheta),xC(2,iTheta)+.015,'C','HorizontalAlignment','center');
text(xP(1,iTheta),xP(2,iTheta)+.015,'P','HorizontalAlignment','center');

axis equal
grid on

title('Paths of points B, C and P on Fourbar Linkage')
xlabel('x-position [m]')
ylabel('y-position [m]')
legend('Point B', 'Point C', 'Point P','Location','SouthEast')
```

The purpose of most of this code, recall, is to make a "snapshot" of the linkage at a single crank angle (in this case, 120°) so that we have an initial verification that the position analysis code is working correctly. Figure 4.58 shows the final position plot of the fourbar linkage used in the example.

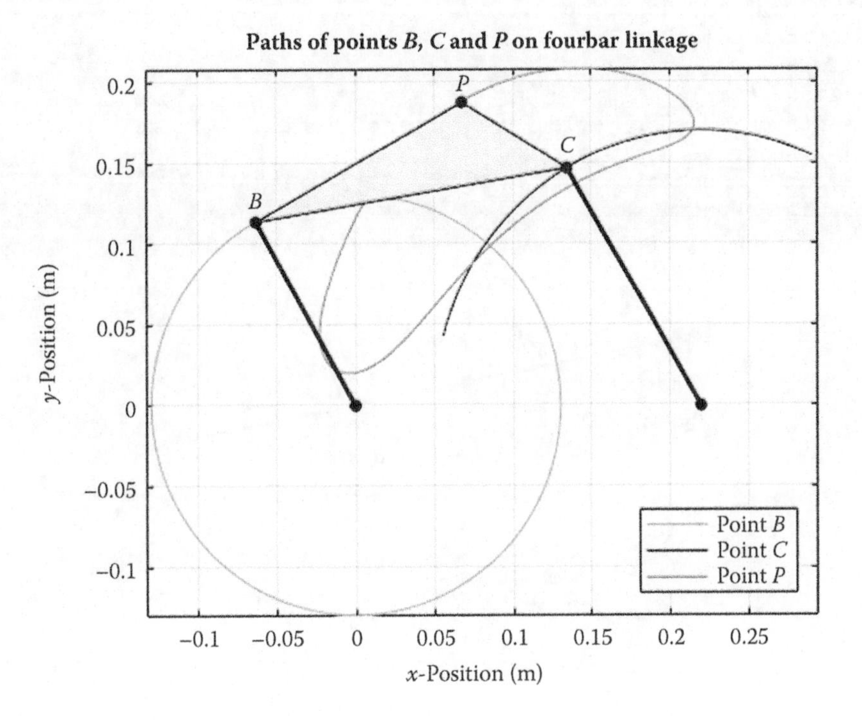

FIGURE 4.58
Fourbar position plot with title, legend, and axis labels and snapshot.

4.8.5 Plotting the Non-Grashof Linkage

The code given above will generate a plot for Grashof linkages only. Now consider the case where the crank length is 150 mm, the coupler is 200 mm, the rocker is 170 mm, and the distance between ground pivots is 250 m. Here we have

$$S + L = 400 \text{ mm}$$

$$P + Q = 370 \text{ mm}$$

so that the linkage is clearly non-Grashof. If we use these lengths in our code it will inform us that the linkage is non-Grashof and will terminate without producing a plot. It is likely, however, that we would prefer that the code produce position plots for the linkage in its possible range of motion, even if the crank cannot undergo a full revolution.

As seen in Figure 4.59, the linkage binds up when the coupler and rocker are aligned. We can use the Law of Cosines to determine the angles θ_{2min} and θ_{2max}, which will define the range of motion for the crank.

$$\cos\theta_{2\,max} = \frac{a^2 + d^2 - (b+c)^2}{2ad} \tag{4.129}$$

Of course, θ_{2min} is the negative of θ_{2max}. If the linkage is Grashof, we would like the code to plot the motion of the linkage for a full revolution of the crank. If it is not Grashof, we wish to plot the resulting motion when the crank swings between θ_{2min} and θ_{2max}. Therefore, we must first modify our Grashof condition checking logic as follows:

```
if (S+L < PQ)  % Grashof condition
  disp('Linkage is Grashof.')
  theta2min = 0;
  theta2max = 2*pi;
else  % if not Grashof, calculate range of motion
  disp('Linkage is not Grashof')
  theta2max = acos((a^2 + d^2 - (b + c)^2)/(2*a*d));
  theta2min = -theta2max;
end
```

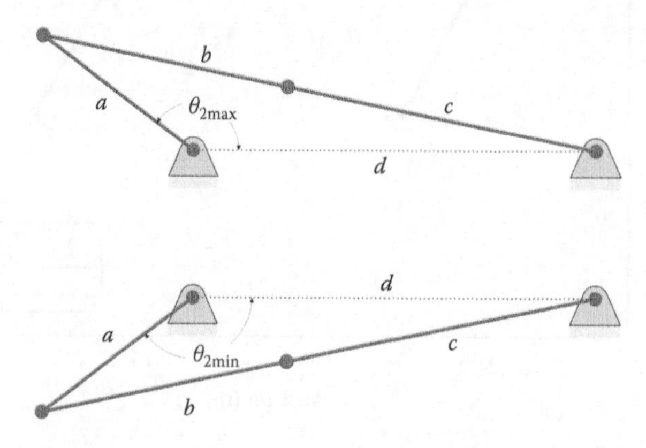

FIGURE 4.59
Limits of motion for the non-Grashof linkage.

Next, we must modify our calculation of θ_2 in the main loop. For either case, we wish θ_2 to range between θ_{2min} and θ_{2max}. We can use a linear interpolation formula to effect this:

$$\theta_2(i) = \left(\frac{\theta_{2\max} - \theta_{2\min}}{N-1}\right)(i-1) + \theta_{2\min} \tag{4.130}$$

The reader should verify that the formula produces `theta2 = theta2min` when `i` = 1 and `theta2 = theta2max` when `i = N`. Modify your `theta2` calculation in the main loop as

```
theta2 = (i-1)*(theta2max - theta2min)/(N-1) + theta2min;
```

One disadvantage of this method is that the index, `i`, no longer corresponds directly to the crank angle, `theta2`, so you must use Equation (4.130) to calculate the crank angle for a given index.

If you execute the code with `iTheta` set to 200 (crank angle 14.12°) you should obtain the plot shown in Figure 4.60. Thus, we now have a MATLAB script that can plot the full range of motion for both Grashof and (some) non-Grashof linkages.

The linkage shown in Figure 4.59 has the ground as the longest link, and our code will plot the motion of this type of non-Grashof linkage. In Figure 4.61 we see a non-Grashof linkage with the coupler as the longest link, and whose limiting positions are quite different. It is left as an exercise for the reader to modify the MATLAB code to account for this type of non-Grashof linkage (hint: use an if statement to check which link is the longest, and calculate θ_{2min} and θ_{2max} accordingly.)

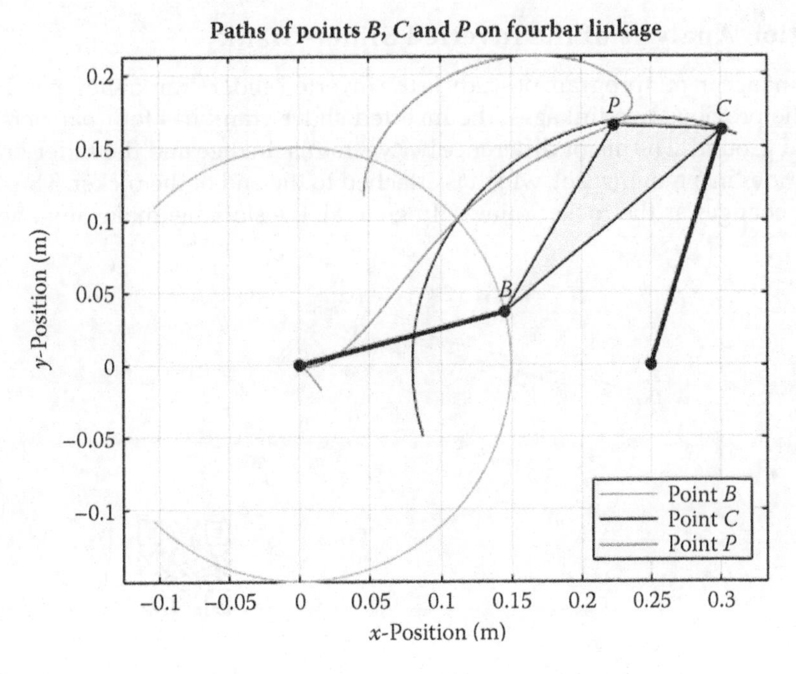

FIGURE 4.60
Plot of the non-Grashof linkage at `iTheta` = 200.

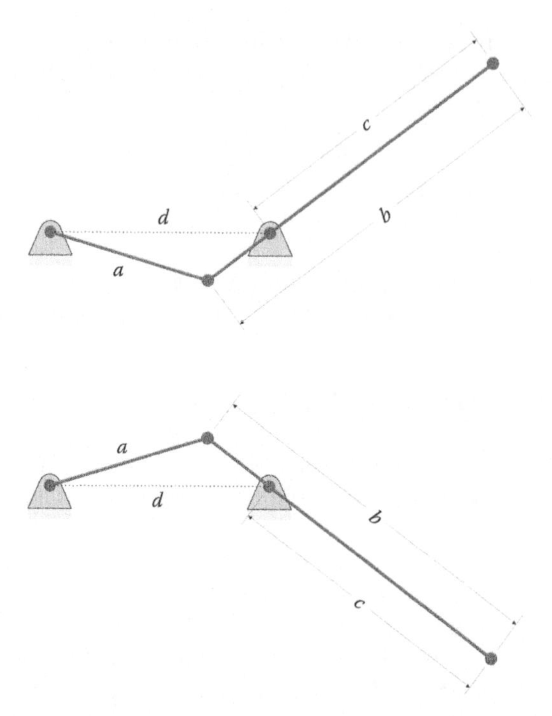

FIGURE 4.61
Another type of non-Grashof linkage with the coupler as the longest link.

4.9 Position Analysis of the Inverted Slider-Crank

The next linkage in our program of study is the inverted slider-crank, shown in Figure 4.62. As with the previous two linkages, the inverted slider-crank has four bars: crank, slider, rocker, and ground. The major difference between this linkage and the slider-crank is that the slider rides in a moving slot, which is attached to the end of the rocker. Thus, the angle of the slot changes as the rocker rotates. In general, the slot may make any angle, δ, with

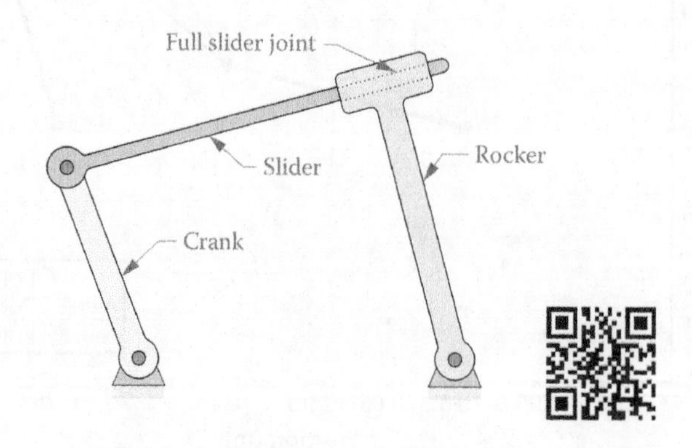

FIGURE 4.62
The inverted slider-crank consists of a crank, slider, rocker, and ground. The rocker and slider are connected with a full-slider joint.

the rocker, but this angle must remain constant. Despite the obvious differences between the fourbar linkage and the inverted slider-crank, position analysis of the two linkages will prove to be remarkably similar.

Figure 4.63 shows the dimensions of the inverted slider-crank. The crank length is a, the rocker length is c, and the distance between ground pins is d. The point C is defined as the intersection of the slider and the rocker. This point will travel up and down the slider as the linkage moves, which will cause the length b to change with time. The angle between slider and rocker is defined as δ. Recall that the fourbar linkage had four fixed lengths: a, b, c, and d, and the angle between coupler and rocker was variable. In contrast, the inverted slider-crank as fixed lengths a, c, and d, and a fixed angle δ, while the length b is variable. Following a similar line of reasoning as with the fourbar, we can write:

$$\theta_4 = \theta_3 + \delta \tag{4.131}$$

Thus, we can easily find θ_4 once we have found θ_3 since δ is known. For the inverted slider-crank, then, the list of known quantities is

$$\text{known}: a,\ c,\ d,\ \delta,\ \theta_2$$

and the list of unknowns is

$$\text{unknown}: b,\ \theta_3$$

We begin by drawing the prime diagonal, f as shown in Figure 4.64. We also define the quantities r and s

$$\begin{aligned} r &= d - a\cos\theta_2 \\ s &= a\sin\theta_2 \end{aligned} \tag{4.132}$$

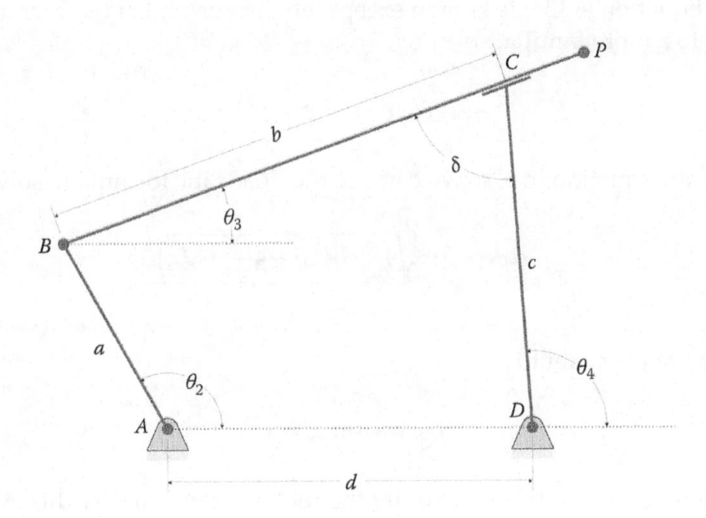

FIGURE 4.63
Dimensions of the inverted slider-crank linkage. The distance between points B and C is defined as b, which changes as the linkage moves.

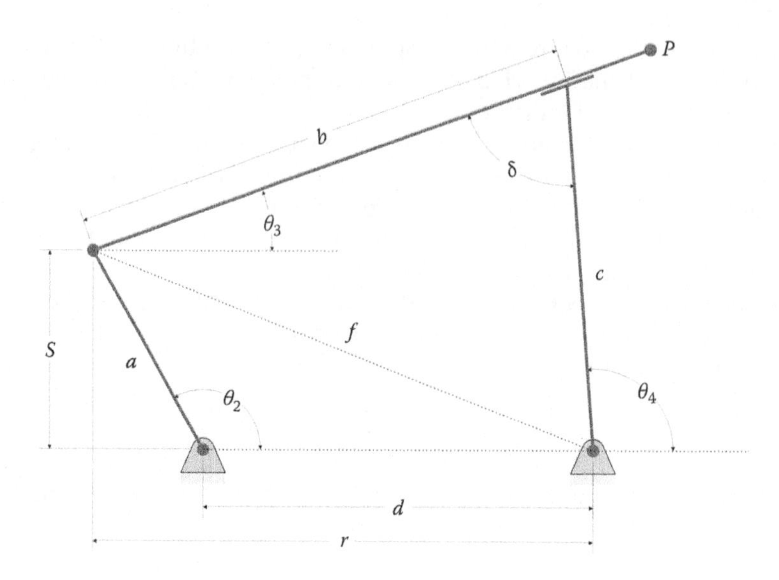

FIGURE 4.64
The prime diagonal stretches from the crank pin to the ground pin of the rocker, as with the fourbar.

Using the Pythagorean Theorem, we find again that:

$$f^2 = a^2 + d^2 - 2ad\cos\theta_2 \tag{4.133}$$

We can make use of the Law of Cosines with the angle δ

$$f^2 = b^2 + c^2 - 2bc\cos\delta \tag{4.134}$$

Everything in Equation (4.134) is known except the distance b. Let us rearrange this equation to put it into a more familiar form

$$b^2 - (2c\cos\delta)b + (c^2 - f^2) = 0 \tag{4.135}$$

This is a quadratic equation in b, so we can use the quadratic formula to solve it

$$b = \frac{2c\cos\delta \pm \sqrt{4c^2\cos^2\delta - 4(c^2 - f^2)}}{2} \tag{4.136}$$

Using the trigonometric identity

$$\cos^2\delta + \sin^2\delta = 1 \tag{4.137}$$

and factoring like terms out from the under the root, we can simplify this expression to

$$b = c\cos\delta \pm \sqrt{f^2 - c^2\sin^2\delta} \tag{4.138}$$

We also define the quantities g and h, by drawing a perpendicular from the slider through pin D (Figure 4.65).

$$g = b - c\cos\delta$$
$$h = c\sin\delta \tag{4.139}$$

The formula for b then simplifies to

$$b = c\cos\delta \pm \sqrt{f^2 - h^2} \tag{4.140}$$

and once we have solved for b, we can calculate g. Finally, we can calculate the angle θ_3 just as we did for the fourbar

$$\tan\theta_3 = \frac{hr - gs}{gr + hs} \tag{4.141}$$

You might be wondering how to interpret the ± symbol in Equation (4.140). It comes about because there are two different ways to assemble the inverted slider-crank, as shown in Figure 4.66. The assembly on the left is known as the the *open* configuration, since none of the lines cross. The other configuration is called *crossed*. In the crossed configuration, the value for b is negative, which means that the slider points in the direction opposite from the way its angle is defined. Both are valid configurations for the linkage, but we usually are more interested in the open configuration.

4.9.1 Limiting Positions for the Inverted Slider-Crank

Owing to the fact that the slider and rocker must meet at a specified angle, δ, it is more challenging to assemble a working inverted slider-crank than it is for the fourbar or

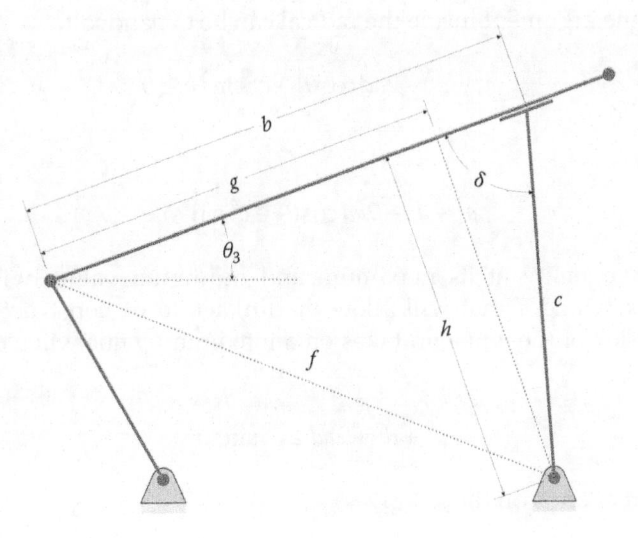

FIGURE 4.65
The lengths g and h are defined by dropping a perpendicular line from the slider through pin D.

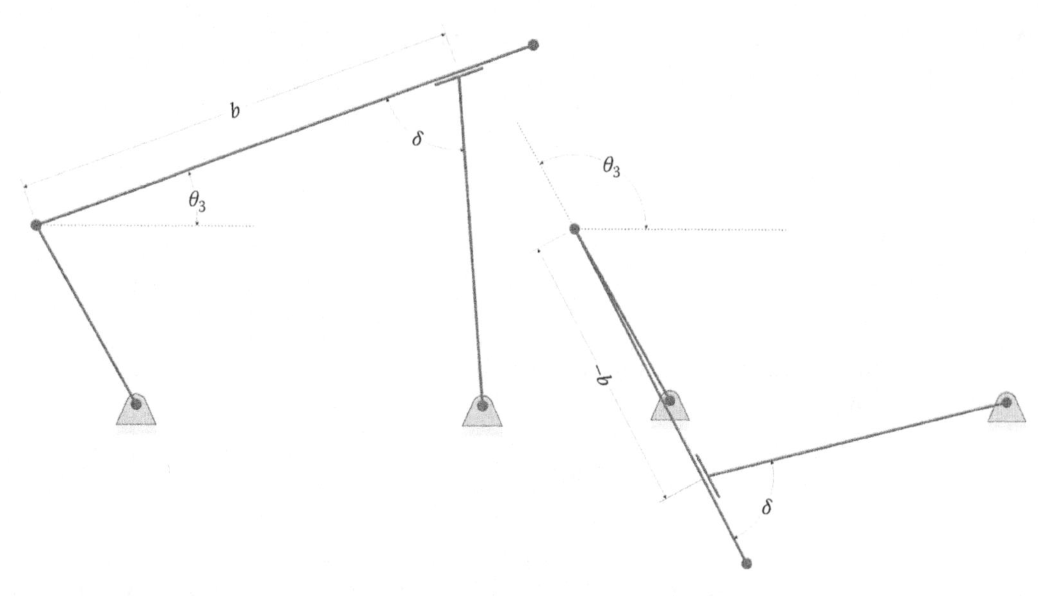

FIGURE 4.66
The open and crossed configurations of the inverted slider-crank are found by taking the two different solutions for the length b.

slider-crank. To determine the viability of a given set of link lengths, let us first examine Equation (4.138), which determines the length of the slider between the crank pin and the rocker.

$$b = c\cos\delta \pm \sqrt{f^2 - c^2\sin^2\delta}$$

To obtain a valid solution, the argument inside the radical must be positive (or zero). Using Equation (4.133), the argument inside the radical can be expanded to

$$a^2 + d^2 - 2ad\cos\theta_2 - c^2\sin^2\delta \geq 0 \tag{4.142}$$

or

$$a^2 + d^2 - 2ad\cos\theta_2 \geq c^2\sin^2\delta \tag{4.143}$$

Evaluating this inequality at its maximum and minimum values will determine the lengths of links a, c, and d that will allow the linkage to be constructed. The quantity on the left-hand side of the equation takes on a maximum value when $\theta_2 = 180°$, and cos $\theta_2 = -1$, so that

$$a^2 + d^2 + 2ad \geq c^2\sin^2\delta \tag{4.144}$$

Collecting the three terms on the left gives

$$(a + d)^2 \geq c^2\sin^2\delta \tag{4.145}$$

And taking the square root

$$|a+d| \geq c\sin\delta \tag{4.146}$$

If this condition is not met, then the linkage cannot be assembled. In this situation, the rocker is too long so that there is no configuration that the coupler can pass through the slider at the end of the rocker (Figure 4.67).

Even if the linkage can be assembled, it is still possible for it to be "non-Grashof"; that is, the crank may not be able to make a full revolution. Evaluating the minimum value of Equation (4.143) when $\theta_2 = 0$, and $\cos\theta_2 = 1$:

$$a^2 + d^2 - 2ad\cos\theta_2 \geq c^2\sin^2\delta$$

Factoring and simplifying as before, we find that

$$|a-d| \geq c\sin\delta \tag{4.147}$$

If this condition is not met, then the crank cannot make a full revolution.

If we barely meet the condition by setting the two sides of the equation equal to each other

$$|a-d| = c\sin\delta \tag{4.148}$$

then b can equal zero if $\delta = 90$. This is also an undesirable situation since it means that the crank will collide with the rocker when $\theta_2 = 0$ (see Figure 4.68).

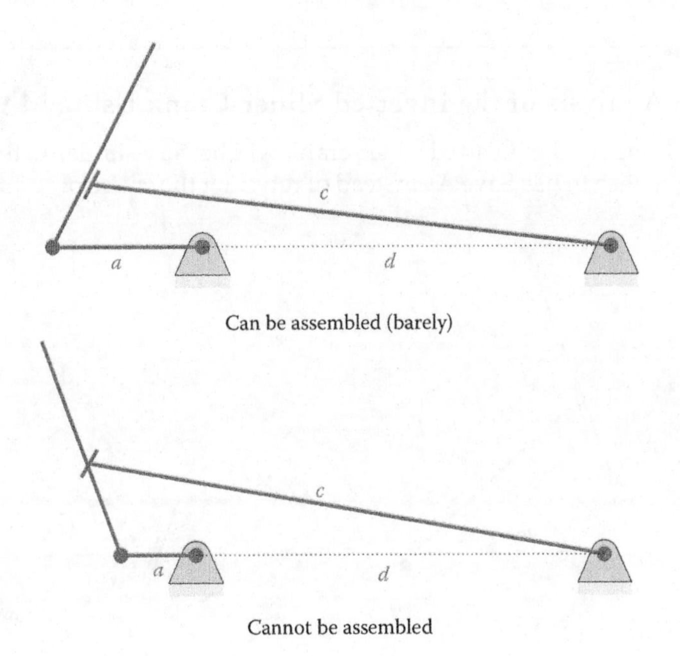

FIGURE 4.67
In the top figure the combined length of a and d are greater than $c\sin\delta$, so the linkage can be assembled. In the bottom figure $a + d < c\sin\delta$ so the linkage cannot be assembled.

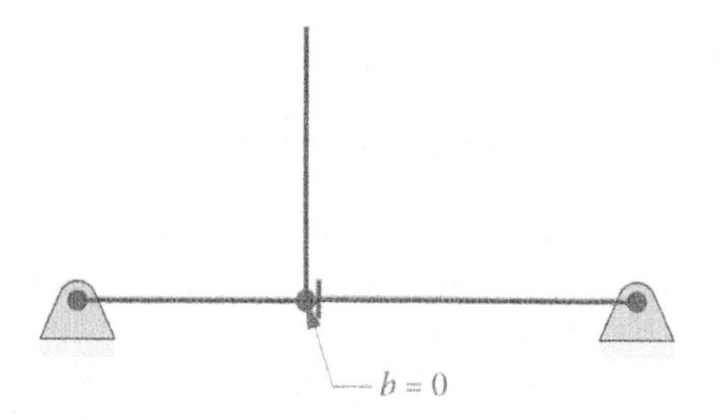

FIGURE 4.68
If $(a - d) = c \sin\delta$ then the length b will be zero when $\theta_2 = 0$, causing a collision between the rocker and crank if $\delta = 0$.

The limiting positions for the crank are found by setting the argument under the radical equal to zero in Equation (4.138).

$$\cos\theta_2 = \frac{a^2 + d^2 - c^2 \sin^2 \delta}{2ad} \tag{4.149}$$

The limiting angles are shown in Figure 4.69. It is interesting to note that we have once again arrived at the Law of Cosines. When the linkage is at its limiting positions a triangle can be constructed with sides a, d, and $c \sin\delta$. As can be seen in Figure 4.69, the length b takes on the value $c \cos\delta$ in the limiting position, which is confirmed by Equation (4.138). It is a good idea to check for the "Grashof" condition of the linkage in your code before starting the position calculations.

4.10 Position Analysis of the Inverted Slider-Crank Using MATLAB®

The code we develop for the inverted slider-crank will be very similar to the fourbar code. In fact, you might wish to use **Save As** instead of retyping the entire program. The example

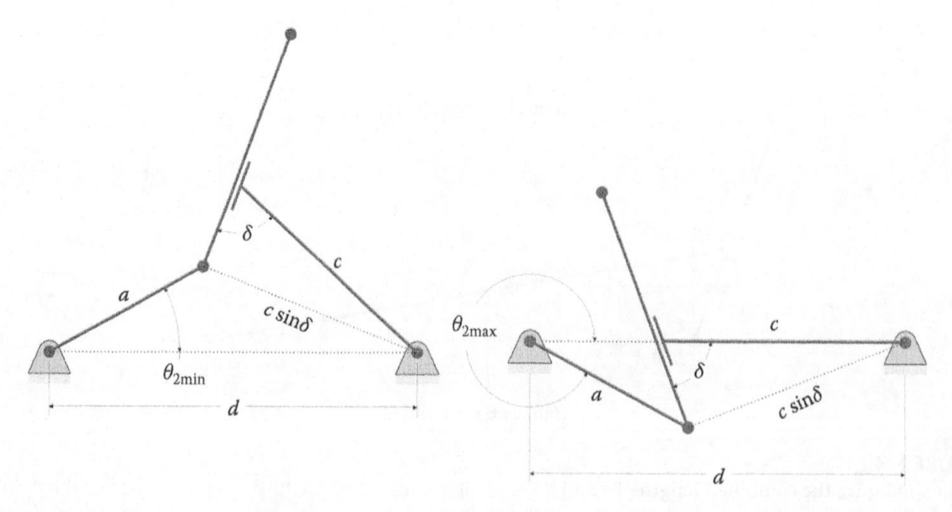

FIGURE 4.69
Limiting positions for the "non-Grashof" inverted slider-crank.

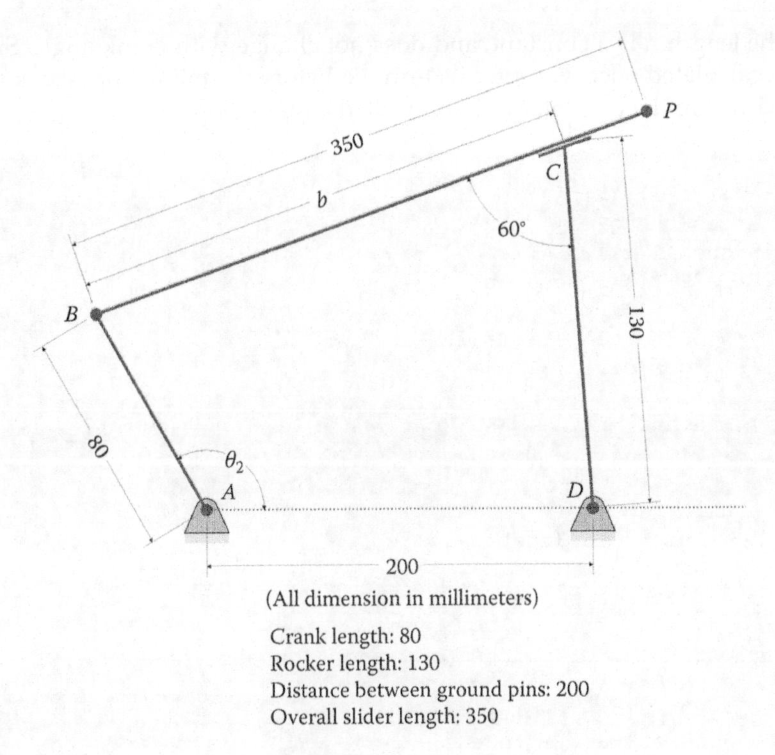

FIGURE 4.70
Example inverted slider-crank linkage used in the MATLAB code.

linkage used here is shown in Figure 4.70. This linkage is "Grashof"; that is, its crank can make a full revolution. We will check the Grashof condition to be able to plot non-Grashof inverted slider-cranks at the end of this section.

Enter the dimensions and allocate space for the variables as below

```
% InvSlider_Position_Analysis.m
% Conducts a position analysis of the inverted slider-crank linkage and
% plots the positions of points B, C and P
% by Eric Constans, June 5, 2017

% Prepare Workspace
clear variables; close all; clc;

% Linkage dimensions
a = 0.080;              % crank length (m)
c = 0.130;              % rocker length (m)
d = 0.200;              % length between ground pins (m)
p = 0.350;              % slider length (m)
delta = 60*pi/180;      % angle between slider and rocker (converted to rad)
h = c*sin(delta);       % h is a constant, only calculate it once

% ground pins
x0 = [0;0];    % ground pin at A (origin)
xD = [d;0];    % ground pin at D
```

Note that the length h is a constant, and does not change with crank angle. Since it only needs to be calculated once, we enter its formula before the main loop. We now enter the formulas learned in the previous section inside the main loop.

```
for i = 1:N
  theta2(i) = (i-1)*(2*pi)/(N-1);
  r = d - a*cos(theta2(i));
  s = a*sin(theta2(i));
  f2 = r^2 + s^2;  % f squared

  b(i) = c * cos(delta) + sqrt(f2 - h^2);
  g = b(i) - c*cos(delta);

  theta3(i) = atan2((h*r - g*s),(g*r + h*s));
  theta4(i) = theta3(i) + delta;

% calculate unit vectors
  [e2,n2]   = UnitVector(theta2(i));
  [e3,n3]   = UnitVector(theta3(i));
  [e4,n4]   = UnitVector(theta4(i));

% solve for positions of points B, C and P on the linkage
  xB(:,i) = FindPos(      x0,a,e2);
  xC(:,i) = FindPos(      xD,c,e4);
  xP(:,i) = FindPos(xB(:,i),p,e3);
end
```

Most of this code, with the exception of the b calculation, is familiar from the fourbar code. Finally, modify the plotting commands to plot the links, and point P, of the inverted slider-crank.

```
plot(xB(1,:),xB(2,:),xC(1,:),xC(2,:),xP(1,:),xP(2,:))
hold on

% specify angle at which to plot linkage
iTheta = 120;

% plot crank, slider and rocker
plot([         x0(1) xB(1,iTheta)],[         x0(2) xB(2,iTheta)],...
       'Linewidth',2,'Color','k');
plot([xB(1,iTheta) xP(1,iTheta)],[xB(2,iTheta) xP(2,iTheta)],...
       'Linewidth',2,'Color','k');
plot([xD(1)        xC(1,iTheta)],[xD(2)        xC(2,iTheta)],...
       'Linewidth',2,'Color','k');

% plot joints on linkage
 plot([x0(1) xD(1) xB(1,iTheta) xC(1,iTheta) xP(1,iTheta)],...
       [x0(2) xD(2) xB(2,iTheta) xC(2,iTheta) xP(2,iTheta)],...
       'o','MarkerSize',5,'MarkerFaceColor','k','Color','k');

% plot the labels of each joint
text(xB(1,iTheta),xB(2,iTheta)+.015,'B','HorizontalAlignment','center');
text(xC(1,iTheta),xC(2,iTheta)+.015,'C','HorizontalAlignment','center');
```

```
text(xP(1,iTheta),xP(2,iTheta)+.015,'P','HorizontalAlignment','center');

axis equal
grid on

title('Paths of points B, C and P on Inverted Slider-Crank')
xlabel('x-position [m]')
ylabel('y-position [m]')
legend('Point B', 'Point C', 'Point P','Location','SouthEast')
```

The plot in Figure 4.71 shows the paths of important points on an inverted slider-crank as the crank makes a full revolution. Use this plot to check the results of your own code before attempting the homework problems.

4.10.1 Position Analysis of the Non-Grashof Linkage

If you change the crank length to 100 mm, you will have created a "non-Grashof" inverted slider-crank, since

$$|a-d| = 100 \text{ mm}$$
$$c\sin\delta = 113 \text{ mm} \tag{4.150}$$

Since the condition

$$|a-d| \geq c\sin\delta \tag{4.151}$$

is not met, the crank cannot make a full revolution. As we did with the fourbar, let us check the condition of the linkage at the beginning of the code, right after defining

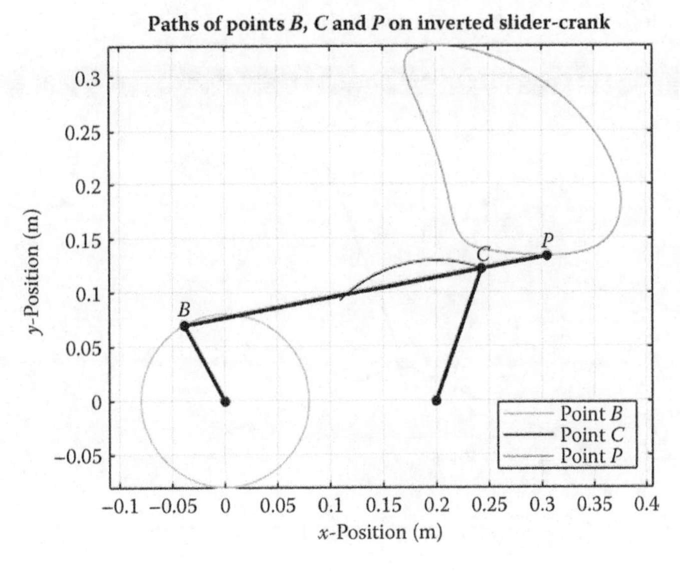

FIGURE 4.71
Position plot for the example inverted slider-crank linkage.

the link lengths. We should first check if the linkage could be assembled at all, using the condition

$$a + d \geq c \sin \delta \tag{4.152}$$

If this condition is not met, we should terminate the program and return control to the command window. If it is met, we should then check whether or not the linkage is Grashof. Enter the following lines of code after the definitions of the link lengths

```
if ((a + d) < c*sin(delta))
  disp('Linkage cannot be assembled')
  return
else
  if (abs(a - d) >= c*sin(delta))
    disp('Linkage is Grashof')
    theta2min = 0;
    theta2max = 2*pi;
  else
    disp('Linkage is not Grashof')
    theta2min = acos((a^2 + d^2 - (c*sin(delta))^2)/(2*a*d));
    theta2max = 2*pi - theta2min;
  end
end
```

We use the check on the Grashof condition to define minimum and maximum values for `theta2`. You have probably noticed that the definition of `theta2min` and `theta2max` are slightly different than they were for the fourbar (Figure 4.72). This is because the crank starts its motion at an angle `theta2min` above the horizontal and sweeps counterclockwise until it reaches `theta2max`. To ensure that `theta2max` is greater than `theta2min`, we have used

$$\theta_{2\max} = 2\pi - \theta_{2\min} \tag{4.153}$$

Be sure that the definition of `theta2` inside the loop is

```
  theta2(i) = (i-1)*(theta2max - theta2min)/(N-1) + theta2min;
```

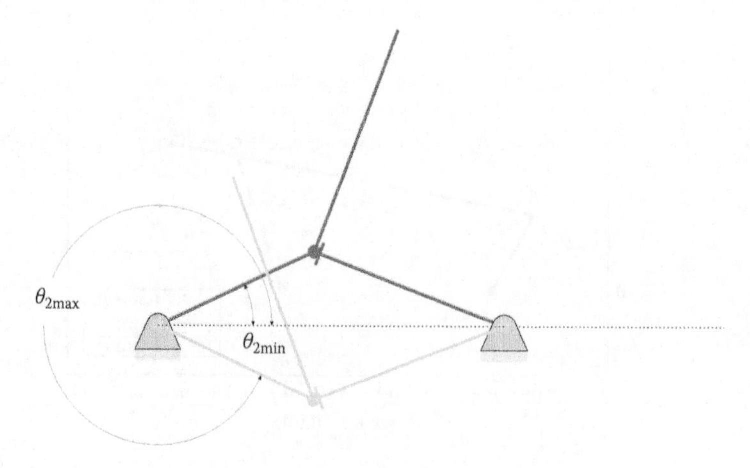

FIGURE 4.72
Lower and upper limits of the crank angle, θ_2, for the non-Grashof inverted slider-crank

After entering these lines of code, and changing a to a length of 100 mm try running the code. The classic MATLAB "ping" lets you know that you have encountered an error. When you switch to the command line window you are greeted with the message:

```
Linkage is not Grashof
Error using atan2
Inputs must be real.

Error in InvSlider_Position_Analysis (line 50)
  theta3(i) = atan2((h*r - g*s),(g*r + h*s));
```

It appears that one of the arguments in the `atan2` function is imaginary. To find out which one is the culprit, type each argument (followed by the enter key) at the command line.

```
>> h
h =
     0.1126
>> r
r =
     0.1067
>> g
g =
    0.0000e+00 + 1.8626e-09i
>> s
s =
     0.0360
```

Everything looks as expected except for the variable g. The formula for g is

$$g = b - c\sin\delta \tag{4.154}$$

Since c and $\sin\delta$ are both real, we know that the problem must lie with b. Type `b(1)` at the command line to look at a single calculated value within the b vector

```
>> b(1)
ans =
    0.0650 + 0.0000i
```

It appears as though b has a small (but finite) imaginary component. Since we calculated b using the square root function,

$$b = c\cos\delta + \sqrt{f^2 - h^2} \tag{4.155}$$

a slight roundoff error must have made h^2 greater than f^2. It is relatively straightforward to prove that the quantity $f^2 - h^2$ should be zero at θ_{2min} such that $b = c\cos\delta$. Instead, a slight numerical error in calculating θ_{2min} has caused b to have an imaginary component, which results in an error in implementing the `atan2` function. If MATLAB's calculated value for θ_{2min} is less than the true value by even a tiny amount, b will end up with an imaginary component. A quick and easy solution to this problem is to add a very small amount to θ_{2min} (and subtract the same amount from θ_{2max}) to ensure that θ_2 stays within bounds. Immediately after defining the link lengths, add the statement

```
eps = 0.000001;       % tiny number added to theta2 to keep it in bounds
```

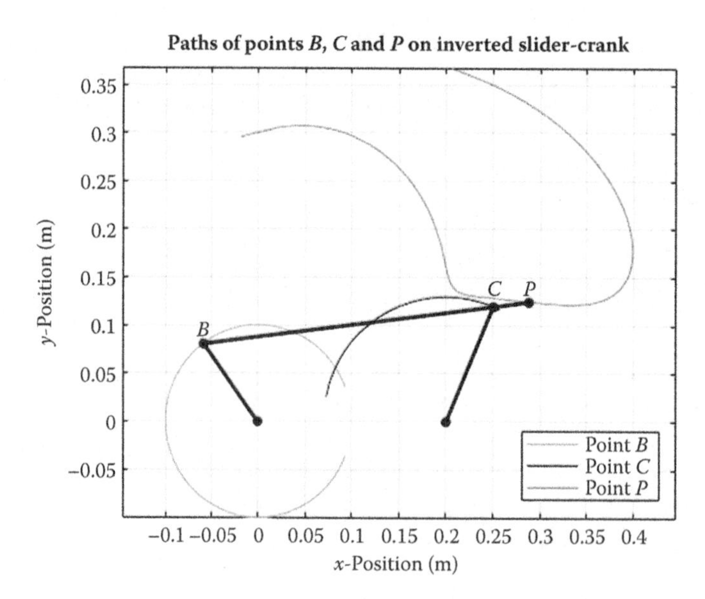

FIGURE 4.73
Position plot for the non-Grashof inverted slider-crank linkage in the example.

Then change the formula for `theta2min` to the following

```
theta2min = acos((a^2 + d^2 - (c*sin(delta))^2)/(2*a*d)) + eps;
```

When you run the code again, the error should be gone, and you will be rewarded with the position plot shown in Figure 4.73. We now have a general purpose program to calculate the positions of points on a Grashof and non-Grashof inverted slider-crank linkage. Be sure that your plot matches Figure 4.73 before moving on.

4.11 Position Analysis of the Geared Fivebar Linkage

We now turn our attention to a slightly more complicated linkage, the geared fivebar. As shown in Figure 4.74, the geared fivebar consists of two gears, two couplers, a crank, and a rocker. The crank is rigidly attached to the first gear and must rotate with it; the rocker is attached to the second gear and must rotate with it as well. The two gears are in mesh so that the rotation of the second gear is proportional to that of the first. We assume that the input to the linkage is the rotation of the first gear. A quick degree of freedom (DOF) analysis will reassure us that the geared fivebar has only one DOF, as required. The gear constraint removes one DOF (rotation) from the second gear and there are five pin joints. Using the modified Gruebler equation we find that

$$DOF = 3(5-1) - 2(5) - 1 = 1 \tag{4.156}$$

As before, we wish to find the location of any point on the linkage for a given input crank angle.

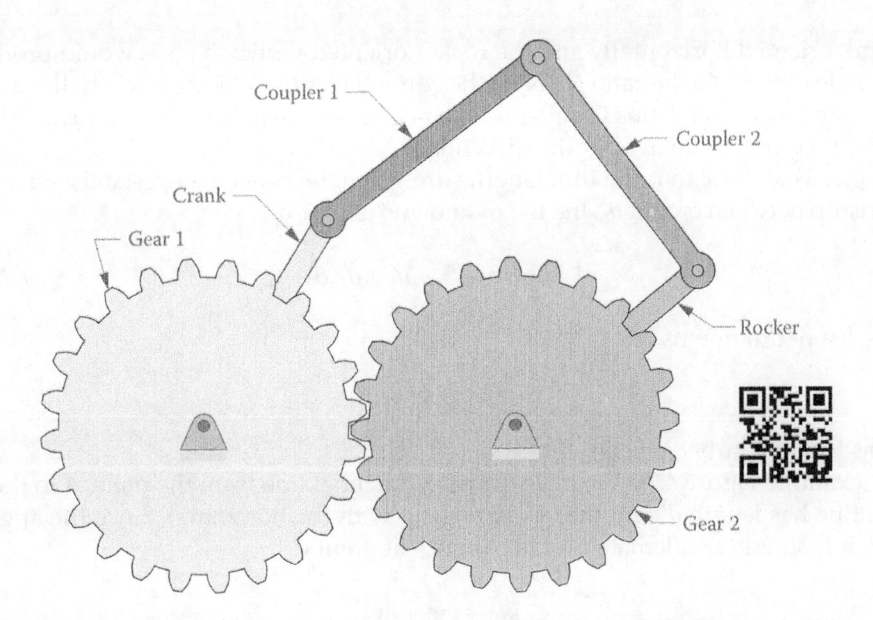

FIGURE 4.74
The geared fivebar linkage has two gears, two couplers, a crank, and a rocker. The crank is attached to the first gear and the rocker is attached to the second gear.

Figure 4.75 shows the important dimensions of the geared fivebar linkage. The crank has length a and the first coupler has length b. The second coupler has length c, while the rocker has length u. The distance between ground pins is d. Suppose that the first gear has N_1 teeth and the second gear has N_2 teeth. Then the relationship between θ_5 and θ_2 is

$$\theta_5 = -\frac{N_1}{N_2}\theta_2 + \varphi \tag{4.157}$$

where φ is a constant offset angle that arises because we may place the two gears into mesh in any manner we choose. For example, we may decide to assemble the linkage with

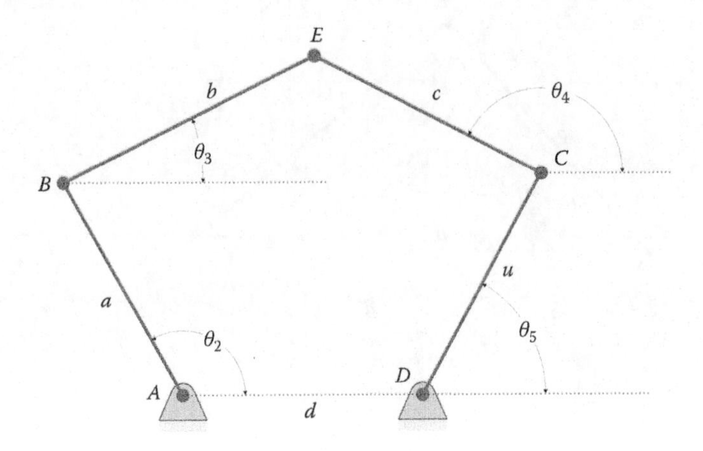

FIGURE 4.75
Dimensions of the geared fivebar linkage.

the crank oriented horizontally and the rocker oriented vertically: this would produce an offset angle of $\varphi = 90°$. The ratio N_1/N_2 is the *gear ratio* for the linkage – we shall have much more to say about gear ratios Chapter 8. The negative sign arises from the fact that meshing external gears rotate in opposite directions.

Assume, as always, that the link lengths are given. Since we have established a simple relationship between θ_2 and θ_5, the list of knowns is

$$\text{known: } a, b, c, d, u, \theta_2, \theta_5 \tag{4.158}$$

and the list of unknowns is

$$\text{unknown: } \theta_3, \theta_4 \tag{4.159}$$

as it was for the fourbar linkage.

Now examine Figure 4.76. We have drawn a diagonal line from the point A to the point C. This line has length d' and makes an angle β with the horizontal. Since the angle θ_5 is known, it is simple to calculate the coordinates of point C.

$$\mathbf{x}_C = d\mathbf{e}_1 + u\mathbf{e}_5 \tag{4.160}$$

The length d' can be found using the Pythagorean theorem:

$$d' = \sqrt{(d + u\cos\theta_5)^2 + (u\sin\theta_5)^2} \tag{4.161}$$

Or

$$d' = |\mathbf{x}_C| = \sqrt{\mathbf{x}_{Cx}^2 + \mathbf{x}_{Cy}^2} \tag{4.162}$$

and the angle β is seen to be

$$\tan\beta = \frac{\mathbf{x}_{Cy}}{\mathbf{x}_{Cx}} \tag{4.163}$$

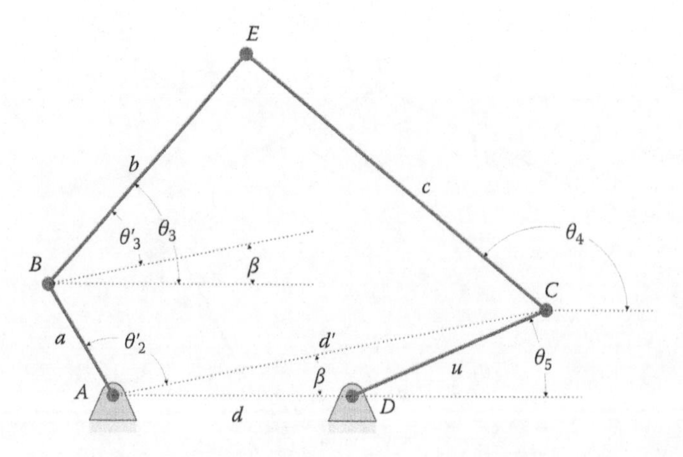

FIGURE 4.76
A line has been drawn from point A to point C. The links a, b, c, and d' form a fourbar linkage.

If we rotate our frame of reference by the angle β, the links a, b, c, and d' form a simple fourbar linkage whose solution is known from the previous section. Note that we must use a slightly modified crank angle, θ'_2 as the input to the solution procedure, to account for the fact that we have rotated an angle β from the horizontal

$$\theta_2' = \theta_2 - \beta \tag{4.164}$$

We can use the fourbar formulas from earlier to solve for a modified coupler angle, θ'_3, and then rotate this back into the original coordinate frame by noting that

$$\theta_3 = \theta_3' + \beta \tag{4.165}$$

In other words, you may reuse almost all of the MATLAB code developed in Section 4.8 to conduct the position analysis for the geared fivebar linkage! Of course, the length d' and the angle β change with every crank angle so that these calculations must be placed inside the main loop.

4.12 Position Analysis of the Geared Fivebar Using MATLAB®

We will now implement the formulas we have learned into a MATLAB code for plotting the positions of all points on the geared fivebar linkage. Figure 4.77 shows the example linkage we will consider. Begin, as usual, with the initial comments and definitions of link lengths:

```
% Fivebar_Position_Analysis.m
% conducts a position analysis on the geared fivebar linkage and
% plots the positions of points P and Q
% by Eric Constans, June 6, 2017

% Prepare Workspace
clear variables; close all; clc;

% Linkage dimensions
a = 0.120;                % crank length (m)
b = 0.250;                % coupler 1 length (m)
c = 0.250;                % coupler 2 length (m)
d = 0.180;                % distance between ground pins (m)
u = 0.120;                % length of link on gear 2 (m)
N1 = 24;                  % number of teeth on gear 1
N2 = 24;                  % number of teeth on gear 2
rho = N1/N2;              % gear ratio
phi = 0;                  % offset angle between gears
gamma3 =  20*pi/180;      % angle to point P on coupler 1
gamma4 = -20*pi/180;      % angle to point Q on coupler 2
p = 0.200;                % distance to point P on coupler 1
q = 0.200;                % distance to point Q on coupler 2
```

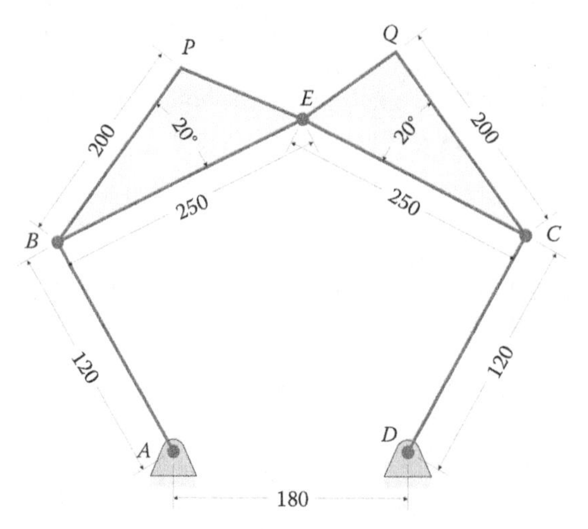

(All dimensions in millimeters)

Length of link on gear 1: 120
Coupler 1 length: 250
Coupler 2 length: 250
Length of link on gear 2: 120
Distance between ground pins: 180

Length from B to P: 200
Length from C to Q: 200
Teeth on gear 1: 48
Teeth on gear 2: 24

FIGURE 4.77
The geared fivebar linkage used in the first example.

```
% ground pins
x0 = [0;0];  % ground pin at A (origin)
xD = [d;0];  % ground pin at D

% Main Loop
N = 361;    % number of times to perform position calculations
[xB,xC,xE,xP,xQ]           = deal(zeros(2,N));  % allocate space for
positions
[theta2,theta3,theta4] = deal(zeros(1,N));  % allocate space for angles
```

To keep things simple at first, we will let the numbers of teeth in the two gears be equal at 24. Once we have developed and verified a working code, we will change the number of teeth in the first gear to 48. Note that we have not allocated space for `theta5`, since

$$\theta_5 = -\rho\theta_2 \tag{4.166}$$

it can be easily calculated at any time. The internal angle of the coupler on the left is set to positive 20° since we rotate counterclockwise from the line BE to BP. The second coupler has a negative angle since we rotate –20° from the line CE to CQ.

```
for i = 1:N
  theta2(i) = (i-1)*(2*pi)/(N-1);   % crank angle
  theta5 = -rho*theta2(i) + phi;    % angle of second gear
  [e2,n2] = UnitVector(theta2(i)); % unit vector for crank
  [e5,n5] = UnitVector(theta5);     % unit vector for second gear
```

```
    xC(:,i) = FindPos(xD, u, e5);      % coords of pin C

    dprime = sqrt(xC(1,i)^2 + xC(2,i)^2);   % distance to pin C
    beta = atan2(xC(2,i),xC(1,i));          % angle to pin C
```

Next, begin typing the main loop, as shown above. To conduct the position analysis, we must first determine dprime and beta, which are calculated using the coordinates of point C. Since we need the unit vector for link 5 in the calculation of xC, we calculate e2 and e5 as soon as the angles theta2 and theta5 are known. The remainder of the main loop is very similar to that of the fourbar position analysis, except that the angles must be modified by beta.

```
    r = dprime - a*cos(theta2(i) - beta);
    s = a*sin(theta2(i) - beta);
    f2 = r^2 + s^2;                        % f squared

    delta = acos((b^2+c^2-f2)/(2*b*c));
    g = b - c*cos(delta);
    h = c*sin(delta);

    theta3(i) = atan2((h*r - g*s),(g*r + h*s)) + beta;
    theta4(i) = theta3(i) + delta;

    [e3,n3] = UnitVector(theta3(i));   % unit vector for first coupler
    [e4,n4] = UnitVector(theta4(i));   % unit vector for second coupler

    [eBP,nBP] = UnitVector(theta3(i) + gamma3); % unit vec from B to P
    [eCQ,nCQ] = UnitVector(theta4(i) + gamma4); % unit vec from C to Q

    xB(:,i) = FindPos(      x0,   a,   e2);
    xE(:,i) = FindPos(xB(:,i),   b,   e3);
    xP(:,i) = FindPos(xB(:,i),   p,  eBP);
    xQ(:,i) = FindPos(xC(:,i),   q,  eCQ);
end
```

4.12.1 Verifying Your Code

Of course, no simulation is complete without some form of code verification. As discussed earlier, quick way to check your results is to make a sketch of the linkage in SOLIDWORKS (without the gears) and see if the coordinates produced by your code match the ones in the sketch. A second method of verification is to note that there are two vector paths to the point E, as shown in Figure 4.78.

$$\mathbf{r}_E = a\mathbf{e}_2 + b\mathbf{e}_3$$
$$\mathbf{r}_E = d\mathbf{e}_1 + u\mathbf{e}_5 + c\mathbf{e}_4 \tag{4.167}$$

As a first check on the accuracy of your code, create a plot of the trajectory of point E calculated in both of the ways given in Equation (4.167). If both traces are exactly alike, then you can have some confidence that your code is producing accurate results.

As seen in Figure 4.79 both methods for reaching point E produce the same result. The resulting curve resembles the symbol for infinity or a sideways figure eight, and is known as a *lemniscate*.

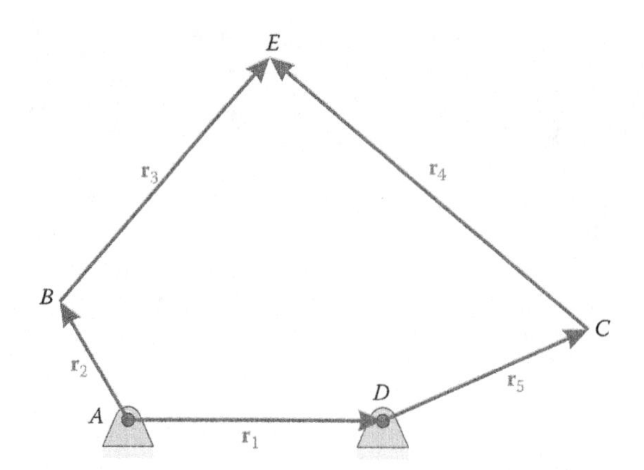

FIGURE 4.78
There are two paths to point C, which can be used as a means of verifying the fivebar code.

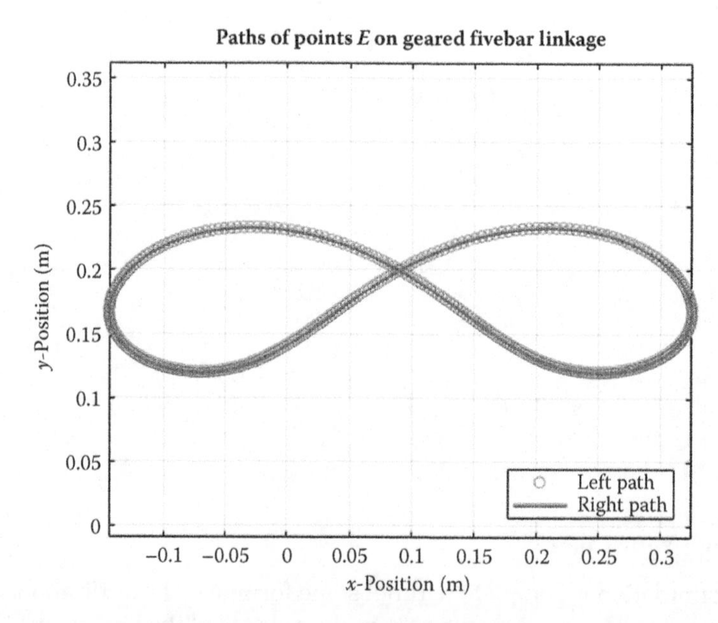

FIGURE 4.79
Trajectory of point E for the example fivebar with gear ratio = 1. Both methods for reaching point E produce the same trace.

4.12.2 Position of Any Point on the Linkage

As with the fourbar linkage, we may be interested in the trajectory of a point on the fivebar that is not located at a pin joint. In Figure 4.80, we have a diagram showing vector paths to two points, P and Q. Both of the couplers are represented as triangles in this diagram, to show that they may have arbitrary shape. The vector path to points P and Q are

$$\begin{aligned} \mathbf{x}_P &= \mathbf{r}_2 + \mathbf{r}_{BP} \\ \mathbf{x}_Q &= \mathbf{r}_1 + \mathbf{r}_5 + \mathbf{r}_{CQ} \end{aligned} \tag{4.168}$$

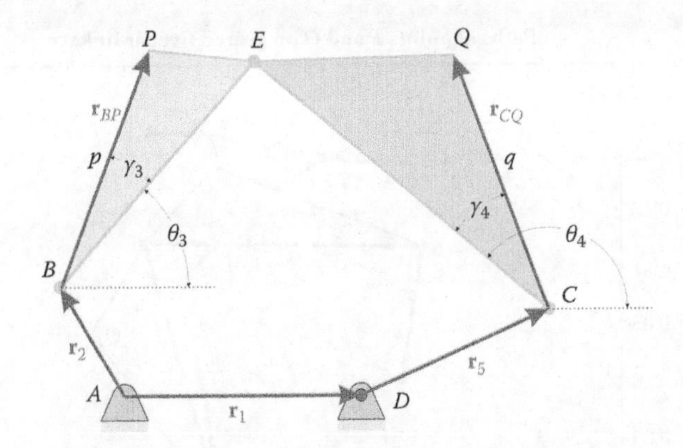

FIGURE 4.80
Finding the coordinates of the points P and Q on the geared fivebar.

or, using unit vector notation

$$\mathbf{x}_P = a\mathbf{e}_2 + p\mathbf{e}_{BP} \tag{4.169}$$

$$\mathbf{x}_Q = d\mathbf{e}_1 + u\mathbf{e}_5 + q\mathbf{e}_{CQ} \tag{4.170}$$

The following section contains a few plots of sample geared fivebar linkages. Try to get your code to match these results before tackling any of the geared fivebar homework problems.

Example 4.10: Gears with Equal Numbers of Teeth

Figure 4.81 shows the trajectory of points P and Q for the example linkage with equal numbers of teeth. It is perhaps not surprising that the curves are symmetric, since the linkage is symmetric and both gears are identical.

Example 4.11: Gears with Unequal Numbers of Teeth

Figure 4.82 shows the trajectory of points P and Q for the example fivebar with gear 1 having 48 teeth and gear 2 having 24 teeth. Note the presence of several cusps in the trajectories. A cusp (or "tooth") is a location where a point on the linkage changes direction suddenly – this will provide some interesting challenges when we conduct velocity and acceleration analysis. Even a very simple linkage can generate complicated paths!

4.13 Position Analysis of the Sixbar Linkage

As the final mechanism in our study of position analysis we will examine the one DOF sixbar linkage. There are five types of 1 DOF sixbar linkage, as shown in Figures 4.83 through 4.87. Each linkage consists of four two-pin links and two three-pin links. In the Stephenson linkages the three-pin links are separated by the two-pin links (two single links and one two-link pair), while in the Watt linkages the three-pin links are directly

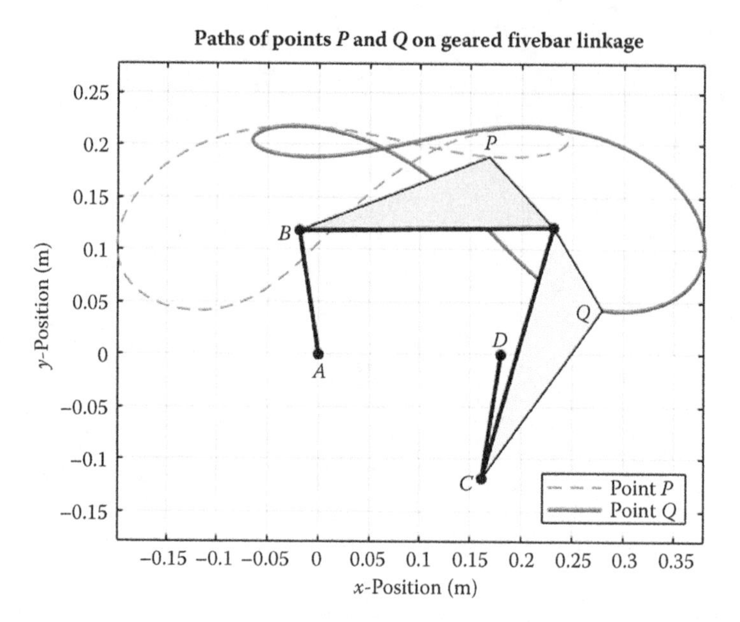

FIGURE 4.81
Paths of points P and Q for the example linkage with gear ratio = 1.

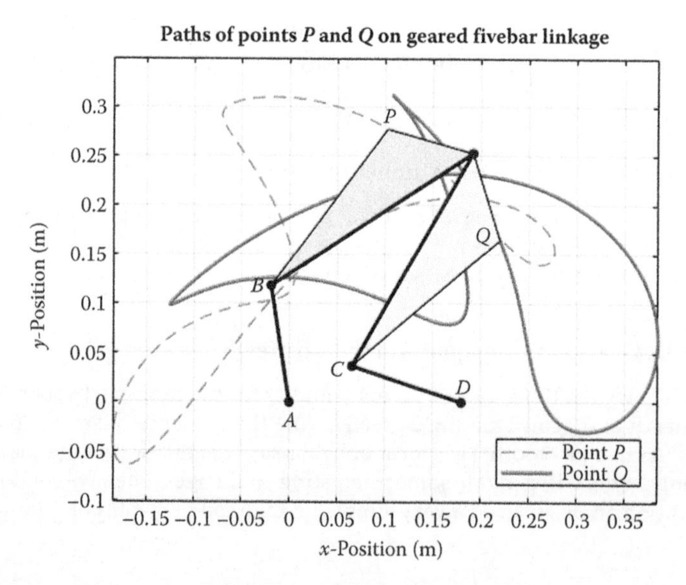

FIGURE 4.82
Paths of points P and Q with gear 1 = 48 teeth and gear 2 = 24 teeth.

connected, and the two-pin links are connected in pairs. The types are distinguished by which link is grounded.

Four of the five sixbar linkages can easily be analyzed by first solving the fourbar problem for the lower set of links and then finding the angles of the remaining two links using another fourbar position analysis. Only the Stephenson Type II sixbar resists solution using this method, since its lower linkage chain consists of five bars.

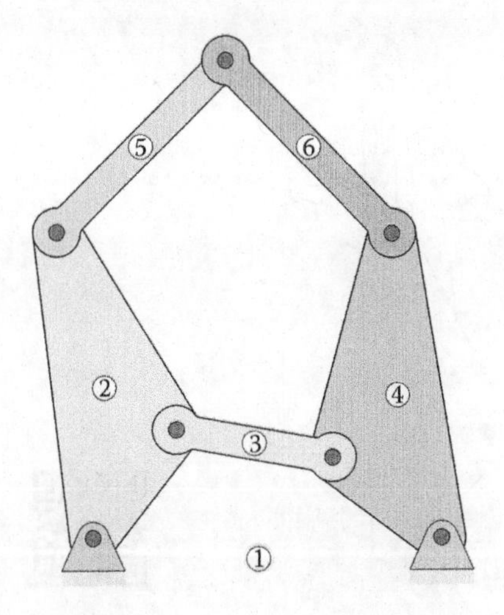

FIGURE 4.83
Stephenson Type I sixbar linkage.

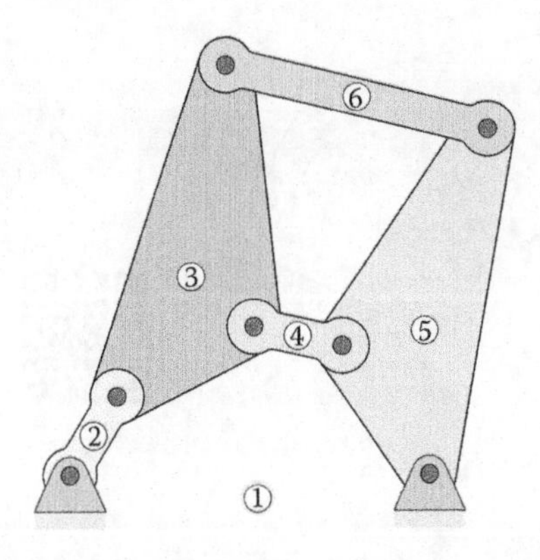

FIGURE 4.84
Stephenson Type II sixbar linkage

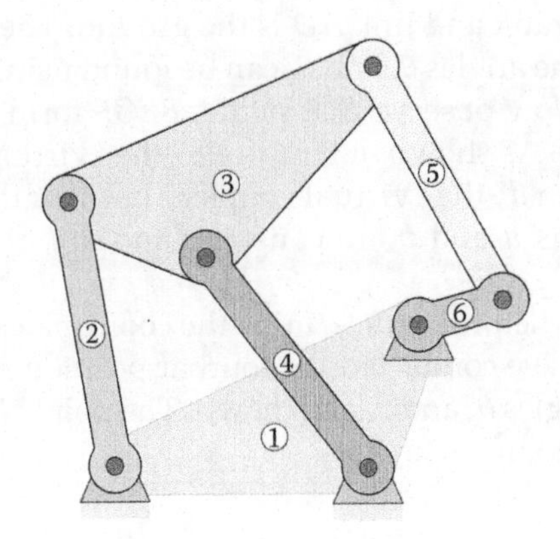

FIGURE 4.85
Stephenson Type III sixbar linkage.

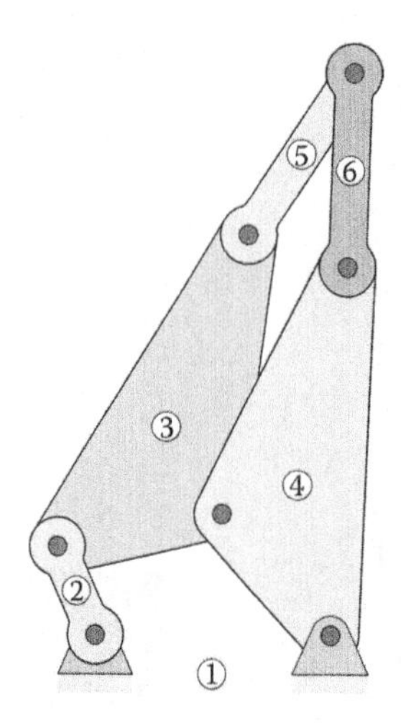

FIGURE 4.86
Watt Type I sixbar linkage.

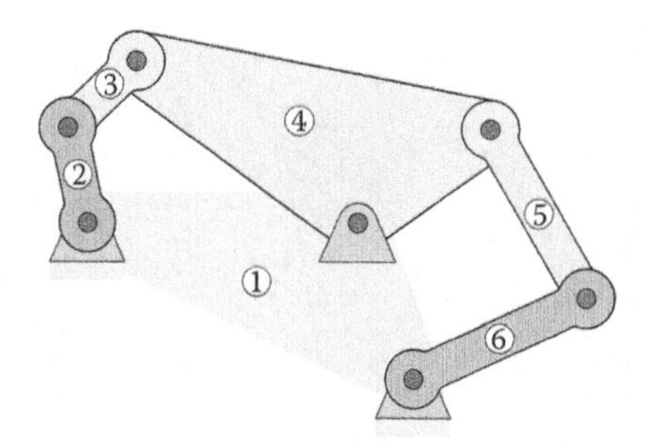

FIGURE 4.87
Watt Type II sixbar linkage.

4.13.1 Stephenson Type I Sixbar Linkage

Figure 4.88 shows the critical dimensions of the Stephenson Type I sixbar linkage. Here we assume that link ABE is the crank and link AD is the ground. The links of dimension a, b, c, and d form a fourbar, and the angles θ_3 and θ_4 can be found using the ordinary fourbar solution developed earlier. Now observe that points $BEGF$ form a second, "virtual" fourbar linkage on top of the first. As shown in Figure 4.89 the "virtual crank" has length a', the "virtual ground" has length d', the "virtual coupler" has length u and the "virtual rocker" has length v. The lengths u and v are constant, and are given in the problem statement.

To find the lengths a' and d', it is helpful to determine the coordinates of a few key points on the linkage. Assume that we have conducted the fourbar position analysis for the linkage given by $ABCD$, so that the angles θ_3 and θ_4 are known. The points B and E are attached to the three-pin crank, and their coordinates are

$$\mathbf{x}_B = a\mathbf{e}_2 \tag{4.171}$$

$$\mathbf{x}_E = p\mathbf{e}_{AE} \tag{4.172}$$

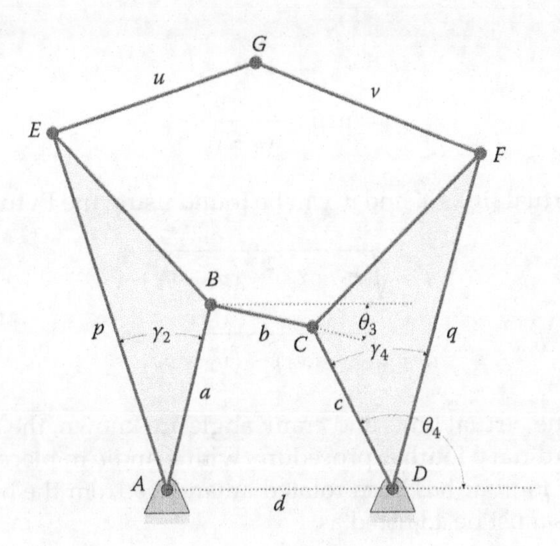

FIGURE 4.88
Dimensions of the Stephenson Type I sixbar linkage. The links a, b, c, and d form a fourbar linkage.

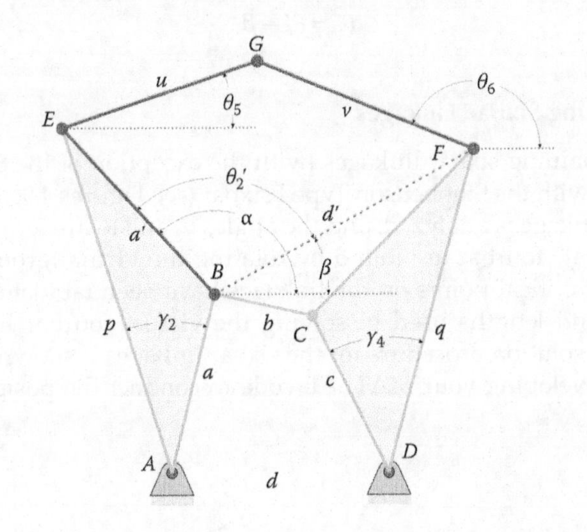

FIGURE 4.89
The upper "virtual" fourbar linkage is formed by a', u, v, and d'.

The point F is attached to the three-pin rocker, and its coordinates are

$$\mathbf{x}_F = d\mathbf{e}_1 + q\mathbf{e}_{DF} \tag{4.173}$$

Since the virtual link d' serves as the ground for the upper fourbar, the "virtual crank angle" θ_2' is given by

$$\theta_2' = \alpha - \beta \tag{4.174}$$

where

$$\tan\alpha = \frac{y_E - y_B}{x_E - x_B} \tag{4.175}$$

and

$$\tan \beta = \frac{y_F - y_B}{x_F - x_B} \tag{4.176}$$

The lengths of the virtual links a' and d' can be found using the Pythagorean theorem

$$a' = \sqrt{(x_E - x_B)^2 + (y_E - y_B)^2} \tag{4.177}$$

$$d' = \sqrt{(x_F - x_B)^2 + (y_F - y_B)^2} \tag{4.178}$$

Once the lengths of the virtual links and crank angle are known, the angles θ_5' and θ_6' can be found using the ordinary fourbar procedure, with u and v in place of b and c. Since the ground on the upper fourbar has been rotated an angle β from the horizontal, the angles of the upper two bars must be adjusted as

$$\begin{aligned} \theta_5 &= \theta_5' + \beta \\ \theta_6 &= \theta_6' + \beta \end{aligned} \tag{4.179}$$

4.13.2 The Remaining Sixbar Linkages

Analysis of the remaining sixbar linkages (with the exception of the Stephenson Type II sixbar) proceeds as with the Stephenson Type I sixbar (see Figures 4.90 through 4.92). First, the lower fourbar linkage is analyzed, and the angles θ_3 and θ_4 are found. Then the angles in the upper "virtual" fourbar are found by rotating the virtual ground by the angle β. The coordinates for critical points on the linkages have been tabulated in Table 4.2. The important angles and lengths used in solving the virtual fourbar linkage are seen in Table 4.3. A general solution procedure for the sixbar linkages is shown below. Follow this procedure when developing your MATLAB code to conduct the position analysis of the sixbar linkages.

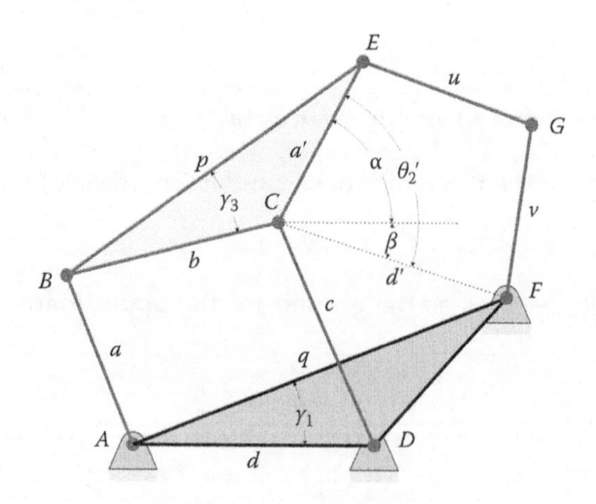

FIGURE 4.90
Critical dimensions of the Stephenson Type III sixbar linkage.

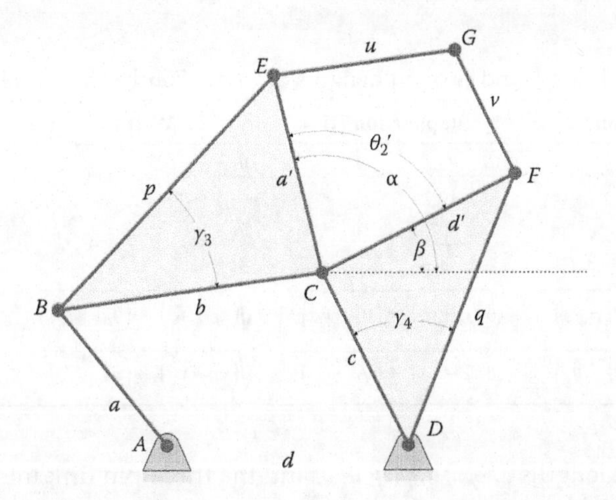

FIGURE 4.91
Critical dimensions of the Watt Type I sixbar linkage.

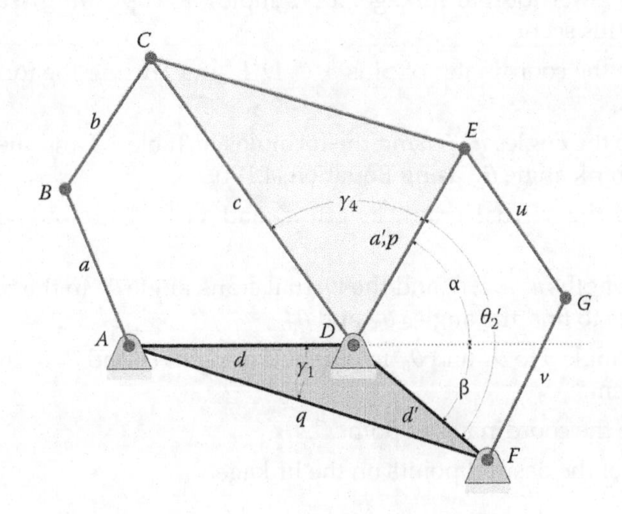

FIGURE 4.92
Critical dimensions of the Watt Type II sixbar linkage.

TABLE 4.2

Coordinates of Critical Points on Sixbar Linkages

	Stephenson I	Stephenson III	Watt I	Watt II
$\mathbf{x}_B$	$a\mathbf{e}_2$	$a\mathbf{e}_2$	$a\mathbf{e}_2$	$a\mathbf{e}_2$
$\mathbf{x}_C$	$\mathbf{x}_B + b\mathbf{e}_3$	$\mathbf{x}_B + b\mathbf{e}_3$	$\mathbf{x}_B + b\mathbf{e}_3$	$\mathbf{x}_B + b\mathbf{e}_3$
$\mathbf{x}_D$	$d\mathbf{e}_1$	$d\mathbf{e}_1$	$d\mathbf{e}_1$	$d\mathbf{e}_1$
$\mathbf{x}_E$	$p\mathbf{e}_{AE}$	$\mathbf{x}_B + p\mathbf{e}_{BE}$	$\mathbf{x}_B + p\mathbf{e}_{BE}$	$\mathbf{x}_D + p\mathbf{e}_{DE}$
$\mathbf{x}_F$	$\mathbf{x}_D + q\mathbf{e}_{DF}$	$q\mathbf{e}_{AF}$	$\mathbf{x}_D + q\mathbf{e}_{DF}$	$q\mathbf{e}_{AF}$
$\mathbf{x}_G$	$\mathbf{x}_E + u\mathbf{e}_5$	$\mathbf{x}_E + u\mathbf{e}_5$	$\mathbf{x}_E + u\mathbf{e}_5$	$\mathbf{x}_E + u\mathbf{e}_5$

TABLE 4.3
Important Angles and Crank and Ground Lengths of Upper Fourbar for the Sixbar Linkages

	Stephenson I	Stephenson III	Watt I	Watt II
$\tan\beta$	$\frac{y_F - y_B}{x_F - x_B}$	$\frac{y_F - y_C}{x_F - x_C}$	$\frac{y_F - y_C}{x_F - x_C}$	$\frac{y_F - y_D}{x_F - x_D}$
$\tan\alpha$	$\frac{y_E - y_B}{x_E - x_B}$	$\frac{y_E - y_C}{x_E - x_C}$	$\frac{y_E - y_C}{x_E - x_C}$	$\frac{y_E - y_D}{x_E - x_D}$
a'	$\sqrt{(x_E - x_B)^2 + (y_E - y_B)^2}$	$\sqrt{(x_E - x_C)^2 + (y_E - y_C)^2}$	$\sqrt{(x_E - x_C)^2 + (y_E - C_y)^2}$	$\sqrt{(x_E - x_D)^2 + (y_E - y_D)^2}$
d'	$\sqrt{(x_F - x_B)^2 + (y_F - y_B)^2}$	$\sqrt{(x_F - x_C)^2 + (y_F - y_C)^2}$	$\sqrt{(x_F - x_C)^2 + (y_F - C_y)^2}$	$\sqrt{(x_F - x_D)^2 + (y_F - y_D)^2}$

1. Enter the link lengths a, b, c, d, p, q, u, v, and the three-pin link internal angles γ as given in the problem statement.
2. Main Loop
   a. Use the lengths a, b, c, and d, and the crank angle θ_2 to conduct a position analysis of the lower fourbar linkage. It is simplest to copy and paste your fourbar code for this section.
   b. Calculate the coordinates of pins B, C, D, E, and F using the formulas given in Table 4.2.
   c. Calculate the angles α, β using the formulas in Table 4.3 and then compute the virtual crank angle θ_2' using Equation (4.174).
   d. Calculate the virtual crank length a' and virtual ground length d' using Table 4.3.
   e. Use the lengths a', u, v, d', and the virtual crank angle θ_2' in the fourbar solution procedure to find the angles θ_5' and θ_6'.
   f. Add the angle β to θ_5' and θ_6' to obtain the angles θ_5 and θ_6 in the fixed coordinate system.
   g. Compute the coordinates of point G.
3. Create plots of the desired points on the linkage.

In Section 4.14, we will use MATLAB to carry out the procedure described above. We will develop a code for only one of the linkages (the Stephenson Type I sixbar) since the remaining three are so similar. Only the Stephenson Type II sixbar cannot be analyzed using this method.

4.13.3 The Stephenson Type II Sixbar Linkage

As seen in Figure 4.84, the Stephenson Type II sixbar linkage does not contain a grounded fourbar. Instead, the links 1, 2, 3, 4, and 5 form a grounded *fivebar* linkage, which cannot be solved using the algebraic methods discussed thus far. Figure 4.93 shows a vector loop diagram of the linkage. We may write two loop equations

$$\begin{aligned} \mathbf{r}_2 + \mathbf{r}_3 + \mathbf{r}_5 - \mathbf{r}_4 - \mathbf{r}_1 &= 0 \\ \mathbf{r}_2 + \mathbf{r}_{3E} + \mathbf{r}_6 - \mathbf{r}_{4F} - \mathbf{r}_1 &= 0 \end{aligned} \tag{4.180}$$

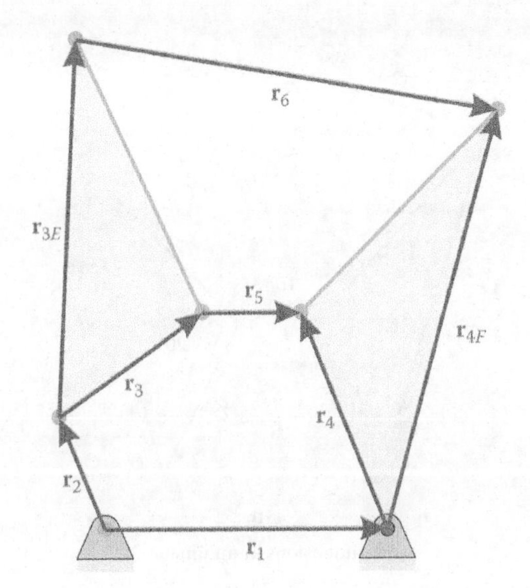

FIGURE 4.93
Vector loop diagram of the Stephenson Type II sixbar linkage. This linkage cannot be analyzed by algebraic means alone.

and the result will be four nonlinear, transcendental equations with four unknown angles, θ_3, θ_4, θ_5, and θ_6. To solve these, we may employ one of the nonlinear equation solvers in MATLAB, or write a simple Newton–Raphson solver. The Newton–Raphson solver is discussed in the final section of this chapter, and this method has been used to construct the animations of the Stephenson Type II sixbar linkages on the website.

4.13.4 Summary

We have presented solution techniques for four of the five types of one DOF sixbar linkages. Each of these is based upon performing the fourbar linkage position analysis twice: once for a grounded lower linkage and again for a "virtual" upper fourbar. MATLAB code is given in the next section for the Stephenson Type I linkage, but you should try programming the remaining linkages (and checking your answers with the plots in the next section) on your own. Only the Stephenson Type II linkage defies solution using algebraic techniques.

4.14 Position Analysis of the Sixbar Linkage Using MATLAB®

We will now use the procedure developed in the preceding section to create a MATLAB script to perform a position analysis on the Stephenson Type I sixbar linkage. This will be the most complicated script we have developed so far, with more than 100 lines of code! It is worthwhile to go through the development in some detail, so that writing the programs for the other sixbar linkages will be relatively straightforward.

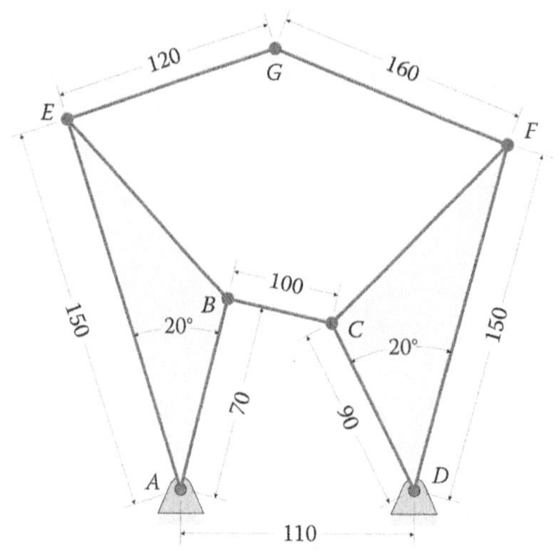

(All dimensions in millimeters)

Crank length: 70
Length AE on crank: 150
Internal angle of crank: 20°
Coupler length: 100
Distance between ground pins: 110

Rocker length: 90
Length DF on rocker: 150
Internal angle of rocker: –20°
Length of link 5: 120
Length of link 6: 160

FIGURE 4.94
The Stephenson Type I sixbar linkage used in the example.

Figure 4.94 gives the dimensions of the sixbar linkage we will analyze. We will closely follow the procedure outlined in the previous section. The top portion of the code is similar to the programs we have already written; first define the link lengths and angles, then allocate space for the position variables as shown in Step 1 of the procedure.

```
% Sixbar_S1_Position_Analysis.m
% conducts a position analysis of the Stephenson Type I sixbar linkage
% and plots the positions of points E, F and G.
% by Eric Constans, June 2, 2017

% Prepare Workspace
clear variables; close all; clc;

% Linkage dimensions
a = 0.070;              % crank length (m)
b = 0.100;              % coupler length (m)
c = 0.090;              % rocker length (m)
d = 0.110;              % length between ground pins (m)
p = 0.150;              % length to third pin on crank triangle (m)
q = 0.150;              % length to third pin on rocker triangle (m)
u = 0.120;              % length of link 5 (m)
v = 0.160;              % length of link 6 (m)
gamma2 = 20*pi/180;     % internal angle of crank triangle
gamma4 = -20*pi/180;    % internal angle of rocker triangle
```

```
% Ground pins
x0 = [ 0; 0];     % ground pin at A (origin)
xD = [ d; 0];     % ground pin at D

% allocate space for variables
N = 361;    % number of times to perform position calculations
[xB,xC,xE,xF,xG]                           = deal(zeros(2,N)); % positions
[theta2,theta3,theta4,theta5,theta6] = deal(zeros(1,N)); % angles
```

Since the lower fourbar linkage is composed of links with length *a, b, c,* and *d,* we can copy the fourbar position analysis code directly, as given in Step 2*a*.

```
% Main Loop
for i = 1:N

% solve lower fourbar linkage
  theta2(i) = (i-1)*(2*pi)/(N-1); % crank angle
  r = d - a*cos(theta2(i));
  s = a*sin(theta2(i));
  f2 = r^2 + s^2;
  delta = acos((b^2+c^2-f2)/(2*b*c));
  g = b - c*cos(delta);
  h = c*sin(delta);

  theta3(i) = atan2((h*r - g*s),(g*r + h*s)); % coupler angle
  theta4(i) = theta3(i) + delta;              % rocker angle
```

We then use the formulas in Table 4.2 to calculate the coordinates of pins *B, C, E,* and *F,* as suggested by Step 2*b*. The location of pin *D* is fixed, and was calculated before the main loop.

```
% calculate unit vectors
  [e2,n2] = UnitVector(theta2(i));
  [e3,n3] = UnitVector(theta3(i));
  [e4,n4] = UnitVector(theta4(i));
  [eAE,nAE] = UnitVector(theta2(i) + gamma2);
  [eDF,nDF] = UnitVector(theta4(i) + gamma4);

% solve for positions of points B, C, E, F
  xB(:,i) = FindPos(x0,a, e2);
  xC(:,i) = FindPos(xD,c, e4);
  xE(:,i) = FindPos(x0,p,eAE);
  xF(:,i) = FindPos(xD,q,eDF);
```

Next, we must calculate the angles α and β using the formulas given in Table 4.3, as well as the virtual crank and ground lengths, a' and d'. Since these formulas require the repeated use of differences in coordinates, it will save some computational effort (and make the code neater) to define the temporary variables `xFB`, `yFB`, `xEB`, and `yEB`.

```
% solve upper fourbar linkage
  xFB = xF(1,i) - xB(1,i);   yFB = xF(2,i) - xB(2,i);
  xEB = xE(1,i) - xB(1,i);   yEB = xE(2,i) - xB(2,i);
  beta =  atan2(yFB, xFB);
```

```
  alpha = atan2(yEB, xEB);
  aPrime = sqrt(xEB^2 + yEB^2); % virtual crank length on upper fourbar
  dPrime = sqrt(xFB^2 + yFB^2); % virtual ground length on upper fourbar
  theta2Prime = alpha - beta;   % virtual crank angle on upper fourbar
```

As given in Step 2e, we now compute the angles in the upper fourbar linkage

```
  r = dPrime - aPrime*cos(theta2Prime);
  s = aPrime*sin(theta2Prime);
  f2 = r^2 + s^2;
  delta = acos((u^2+v^2-f2)/(2*u*v));
  g = u - v*cos(delta);
  h = v*sin(delta);

  theta5Prime = atan2((h*r - g*s),(g*r + h*s)); % coupler and rocker
  theta6Prime = theta5Prime + delta;            % angles on upper fourbar
```

Finally, use the angle β to rotate back to the original coordinate system and solve for the coordinates of point G.

```
  theta5(i) = theta5Prime + beta;               % return angles to fixed
  theta6(i) = theta6Prime + beta;               % fixed CS

% calculate remaining unit vectors
  [e5,n5]   = UnitVector(theta5(i));
  [e6,n6]   = UnitVector(theta6(i));

% calculate position of point G
  xG(:,i) = FindPos(xE(:,i),    u,    e5);
end
```

4.14.1 Making the Sixbar Plot

Because of the large number of links, the sixbar position plots will be somewhat crowded. To make them more readable, we will introduce a few new plotting tricks here. First, it will be handy to make use of more than the default eight MATLAB colors.

There are multiple methods for handling colors in MATLAB (see, e.g. `colormap`, etc.), but the method presented here is simple to implement. It is common in computer systems to specify a color using an "RGB triple" in the form [R G B] where R, G, and B are values between 0 and 255 that give the red, green, and blue content of a color, respectively. For example, pure red would be given by the triple [255 0 0], white is [255 255 255] and medium gray is [128 128 128]. In our system, we will define a *base color*, and use a MATLAB function to calculate the RGB triples for ten shades of that color. The first shade will be the base color itself, and the remaining shades will be progressively lighter until the final shade is almost white.

We will use a linear interpolation formula to calculate the RGB value for each shade. If the base value for a given color is S_0, then the formula for shade i is

$$S_i = \frac{255 - S_0}{10}(i-1) + S_0 \tag{4.181}$$

The reader will confirm that the formula gives $S_i = S_0$ when $i = 0$ and $S_i = 255$ when $i = 11$. Let us create a function, `DefineColor.m`, to create our own custom palette of colors. Open up a new script, and type the following function.

```
% DefineColor.m
% makes a palette of custom colors for plotting in MATLAB
% the first color in the palette gives the full base color, and
% the rest fade gradually to white
%
% colorBase = 1x3 input array giving RGB values (between 0 and 255)
% C         = 10x3 output array giving RGB values (between 0 and 1)

function C = DefineColor(colorBase)

C = zeros(11,3);
for i = 1:11
  for j = 1:3
    C(i,j) = (i-1)*(255 - colorBase(j))/10 + colorBase(j);
  end
end

C = C/255;
```

The final line in the function divides the entire array by 255, since MATLAB specifies its RGB values in the range between 0 and 1. In the main program immediately after the end of the main loop type

```
cBlu = DefineColor([  0 110 199]); % Pantone 300C
cBlk = DefineColor([  0   0   0]); % grayscale
```

These lines define two color palettes: `cBlu` and `cBlk` that give shades of blue and grayscales, respectively. Coincidentally, these are the palettes used for the illustrations in this textbook! Figure 4.95 shows the colors created by the DefineColor function.

To plot the trajectories of points *E*, *F*, and *G*, type

```
figure; hold on
plot(xE(1,:),xE(2,:),'Color',cBlu(1,:))
plot(xF(1,:),xF(2,:),'Color',cBlu(7,:))
plot(xG(1,:),xG(2,:),'Color',cBlk(5,:))
```

Instead of using the default MATLAB colors, we have used our newly defined palettes to customize the traces. Next, plot the two triangular links, the crank and rocker:

```
iTheta = 120;  % specify angle at which to plot linkage
% plot the two three-pin links as triangular patches
patch([x0(1) xB(1,iTheta) xE(1,iTheta)],...
      [x0(2) xB(2,iTheta) xE(2,iTheta)],cBlu(9,:),...
      'EdgeColor',cBlk(6,:),'LineWidth',2,'FaceAlpha',0.5);
patch([xD(1) xF(1,iTheta) xC(1,iTheta)],...
      [xD(2) xF(2,iTheta) xC(2,iTheta)],cBlu(10,:),...
      'EdgeColor',cBlk(6,:),'LineWidth',2,'FaceAlpha',0.5);
```

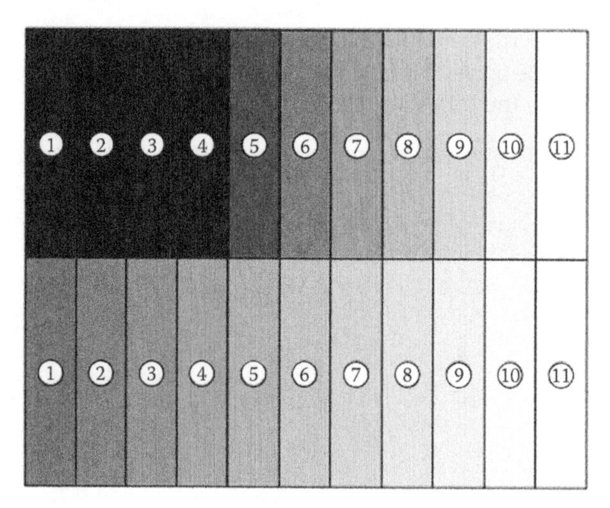

FIGURE 4.95
Color palette defined by the function `DefineColor.m`.

Since the triangular links are plotted after the traces, it is likely that some of the traces have been obscured. To avoid this, we have used the `FaceAlpha` parameter to make the links semitransparent. We have also made the edges a medium gray so that the traces will stand out.

```
% plot the two-pin links
plot([xB(1,iTheta) xC(1,iTheta)],[xB(2,iTheta) xC(2,iTheta)],...
     'Linewidth',2,'Color',cBlk(5,:));
plot([xE(1,iTheta) xG(1,iTheta)],[xE(2,iTheta) xG(2,iTheta)],...
     'Linewidth',2,'Color',cBlk(5,:));
plot([xF(1,iTheta) xG(1,iTheta)],[xF(2,iTheta) xG(2,iTheta)],...
     'Linewidth',2,'Color',cBlk(5,:));

% plot fixed pins
 plot([x0(1) xD(1)],[x0(2) xD(2)],'o','MarkerSize',5,...
      'MarkerFaceColor',cBlk(1,:),'Color',cBlk(1,:));

% plot joints on linkage
 plot([xB(1,iTheta) xC(1,iTheta) xE(1,iTheta) xF(1,iTheta) xG(1,iTheta)],...
      [xB(2,iTheta) xC(2,iTheta) xE(2,iTheta) xF(2,iTheta) xG(2,iTheta)],...
      'o','MarkerSize',5,'MarkerFaceColor',cBlk(6,:),'Color',cBlk(6,:));
```

Plotting the two pin links and pin joints is the same as before, although we have changed the color of the moving pins to medium gray.

```
% plot the labels of each joint
text(x0(1), x0(2) - 0.012, 'A','HorizontalAlignment','center');
text(xD(1), xD(2) - 0.012, 'D','HorizontalAlignment','center');
text(xB(1,iTheta),xB(2,iTheta)+0.012,'B','HorizontalAlignment','center');
text(xC(1,iTheta)+0.012,xC(2,iTheta),'C','HorizontalAlignment','center');
text(xE(1,iTheta)-0.012,xE(2,iTheta),'E','HorizontalAlignment','center');
```

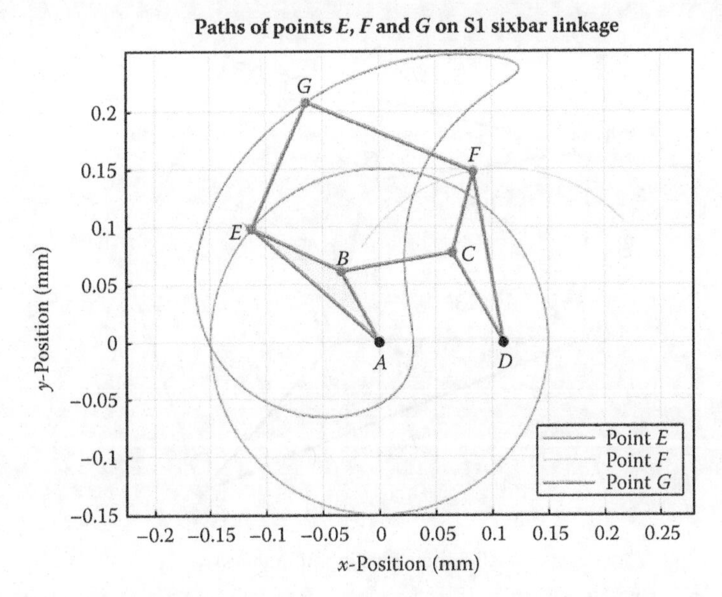

FIGURE 4.96
Paths of *E*, *F*, and *G* for the example Stephenson Type I sixbar linkage.

```
text(xF(1,iTheta),xF(2,iTheta)+0.012,'F','HorizontalAlignment','center');
text(xG(1,iTheta),xG(2,iTheta)+0.012,'G','HorizontalAlignment','center');

axis equal
grid on
title('Paths of points E, F and G on S1 Sixbar Linkage')
xlabel('x-position [mm]')
ylabel('y-position [mm]')
legend('Point E', 'Point F', 'Point G','Location','SouthEast')
```

Finally, the commands for labeling the points are the same as we used for the previous linkages. If you have entered everything correctly, you should obtain the position plot shown in Figure 4.96.

4.14.2 The Remaining Sixbar Linkages

A MATLAB script that solves the Stephenson Type I linkage has been developed above. Programs for the remaining sixbar linkages (with the exception of the Stephenson Type II) are left as an exercise for the reader. Figures 4.97 through 4.102 show a set of example plots of the four types of sixbar linkages that we have discussed. Ensure that your plots match these before beginning the homework exercises.

4.15 Advanced Topic: The Newton–Raphson Method

We have seen in the preceding sections that there are some linkages whose positions cannot be analyzed using algebraic methods, most notably the Stephenson Type II sixbar linkage.

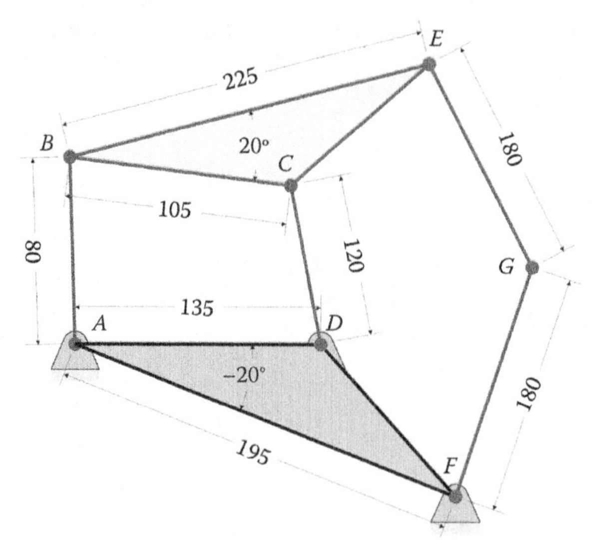

(All dimensions in millimeters)

Crank length: 80
Coupler length: 105
Internal angle of coupler: 20°
Length *BE* on coupler: 225
Distance between ground pins: 135

Internal angle of ground: −20°
Length *AF* on ground: 195
Rocker length: 120
Length of link 5: 180
Length of link 6: 180

FIGURE 4.97
Dimensions of example Stephenson Type III linkage. The internal angle of the ground triangle is negative since it sweeps clockwise from the line between ground pivots.

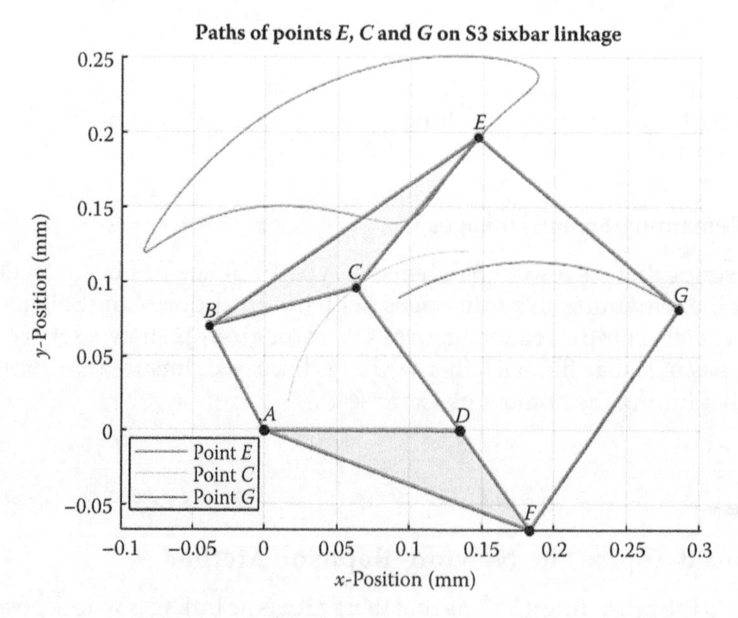

FIGURE 4.98
Position solution for the example Stephenson Type III linkage.

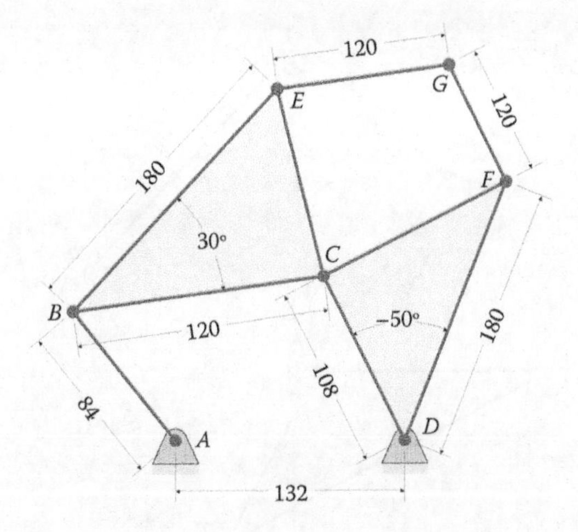

(All dimensions in millimeters)

Crank length: 84	Rocker length: 108
Coupler length: 120	Internal angle of rocker: −50°
Internal angle of coupler: 30°	Length DF on rocker: 180
Length BE on coupler: 180	Length of link 5: 120
Distance between ground pins: 132	Length of link 6: 120

FIGURE 4.99
Dimensions of the example Watt Type I sixbar linkage.

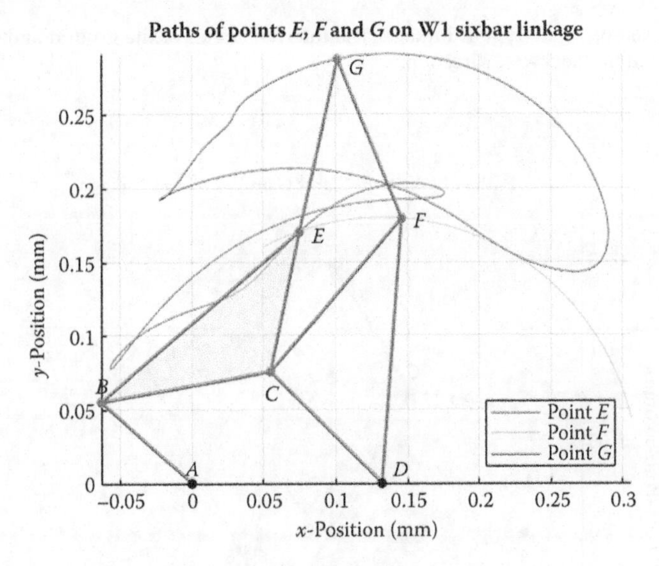

FIGURE 4.100
Position solution for the Watt Type I linkage.

This section presents a method for conducting the position analysis for *any* one DOF linkage: the Newton–Raphson algorithm. In contrast with the algebraic methods presented so far, the Newton–Raphson algorithm is *iterative*, which means that we use a "guess and check" approach to finding the positions. This means that the Newton–Raphson algorithm will generally be slower than the algebraic approach, if it is available. As we will see, our

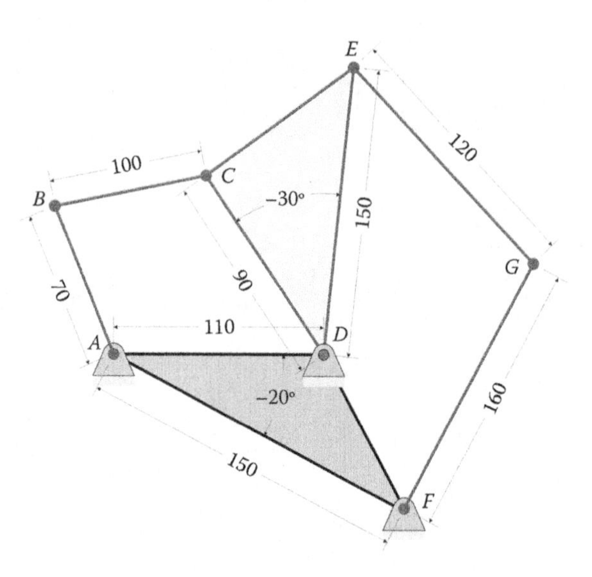

(All dimensions in millimeters)

Crank length: 70	Rocker length: 90
Coupler length: 100	Internal angle of rocker: −30°
Distance between ground pins: 110	Length *DF* on rocker: 150
Internal angle of ground: −20°	Length of link 5: 120
Length *AE* on ground: 150	Length of link 6: 160

FIGURE 4.101
Dimensions of the example Watt Type II linkage. The internal angles of the ground and rocker are negative since they sweep out in the clockwise direction.

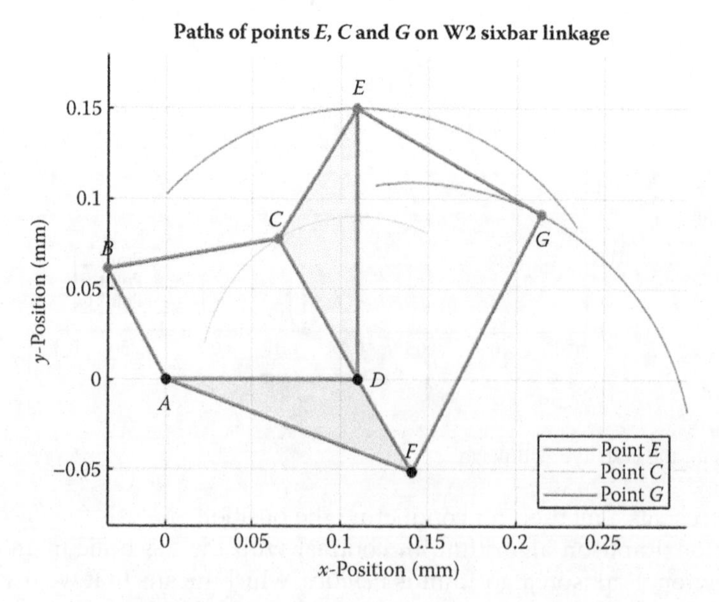

FIGURE 4.102
Position solution for the example Watt Type II linkage.

guesses will not be completely random, and the algorithm usually converges on the correct solution after only a few tries. And for some linkages (e.g. the Stephenson Type II sixbar) we have no choice but to use an iterative approach.

As a first example, we will conduct a position analysis of the fourbar linkage, since its equations are simpler than the sixbar. Consider the vector loop diagram shown in Figure 4.103. Adding the vectors around the loop brings us back where we started, so that the vector sum is zero.

$$\mathbf{r}_2 + \mathbf{r}_3 - \mathbf{r}_4 - \mathbf{r}_1 = 0 \tag{4.182}$$

Expanding this equation into its x and y components gives

$$\begin{aligned} a\cos\theta_2 + b\cos\theta_3 - c\cos\theta_4 - d &= 0 \\ a\sin\theta_2 + b\sin\theta_3 - c\sin\theta_4 &= 0 \end{aligned} \tag{4.183}$$

These equations are sometimes called *constraint equations,* since they must both be satisfied in order for the linkage to be properly assembled. Both equations have the form

$$f(\theta_2, \theta_3, \theta_4) = 0 \tag{4.184}$$

In other words, the left-hand sides of both equations are functions of θ_2, θ_3, and θ_4, and both must equal zero at all times. Since we normally specify the crank angle in advance, it is more correct to write Equation (4.184) as

$$f(\theta_3, \theta_4) = 0 \tag{4.185}$$

Let us denote the vector of unknowns in this equation as $\mathbf{q}$

$$\mathbf{q} = \begin{Bmatrix} \theta_3 \\ \theta_4 \end{Bmatrix} \tag{4.186}$$

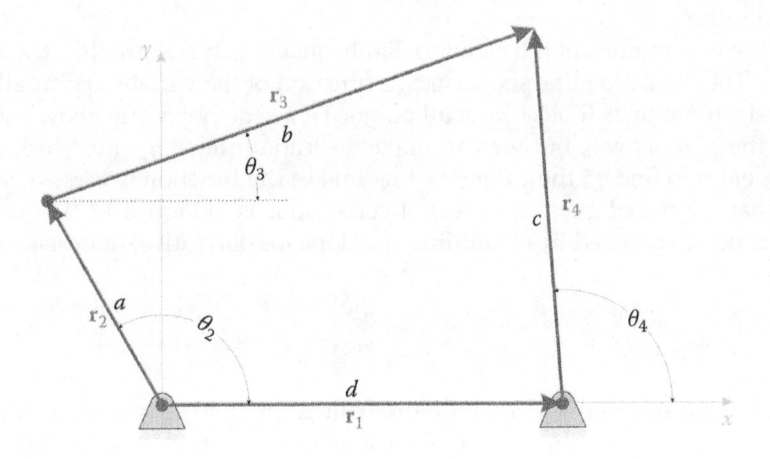

FIGURE 4.103
Vector loop diagram for the fourbar linkage.

Since we have two equations of constraint, which both must equal zero, we can also write this as a vector in the form

$$\Phi(\mathbf{q}) = 0 \tag{4.187}$$

where

$$\begin{aligned} \Phi_1 &= a\cos\theta_2 + b\cos\theta_3 - c\cos\theta_4 - d \\ \Phi_2 &= a\sin\theta_2 + b\sin\theta_3 - c\sin\theta_4 \end{aligned} \tag{4.188}$$

The Newton–Raphson algorithm is a method for solving systems of nonlinear equations of the type in Equation (4.187). We will first discuss how the method works with a single equation and single unknown, and then generalize it for multiple equations and multiple unknowns.

4.15.1 The One-Dimensional Newton-Raphson Algorithm

Consider a single constraint equation of the form

$$\Phi(q) = 0 \tag{4.189}$$

where Φ can be any function of q, for example

$$\begin{aligned} &\sin q = 0 \\ &q^2 = 0 \\ &5q + 3 = 0 \end{aligned} \tag{4.190}$$

In each case, we would like to solve for q such that the left-hand side of the equation equals zero. In other words, we are interested in finding the *root* of each equation. For the expressions above, finding the roots is simple, but for more complicated equations, an analytical solution is not always possible. For this type of equation, we need to develop an automatic procedure that we can implement in MATLAB. The procedure we will use is the Newton–Raphson algorithm.

The simplest way to present the Newton–Raphson algorithm is graphically, as shown in Figure 4.104. The thick blue line shows $\Phi(q)$, a function of the variable q. Initially, we don't know what the function Φ looks like at all points (i.e. we do not really know enough about Φ to make the plot above!), but we can make an initial guess, q_1, and work from there. Again, our goal is to find q^*, the value of q that makes the function Φ zero.

Assume that our initial guess, q_1, was not very good, as in Figure 4.104. We can deduce the best direction to proceed by examining the slope (or derivative) of Φ at q_1

$$\text{slope} = \left.\frac{d\Phi}{dq}\right|_{q_1} \tag{4.191}$$

In Figure 4.105, the slope tells us that a more fruitful place to search for q^* would be at a lower value of q. In fact, we can make a pretty good guess as to the location of q^* by noting that the points $\Phi(q_1)$, q^*, and q_1 form a triangle, whose base is $(q_1 - q^*)$, height is $\Phi(q_1)$, and slope is $d\Phi/dq$. Using this triangle, we can rewrite the slope at q_1 as

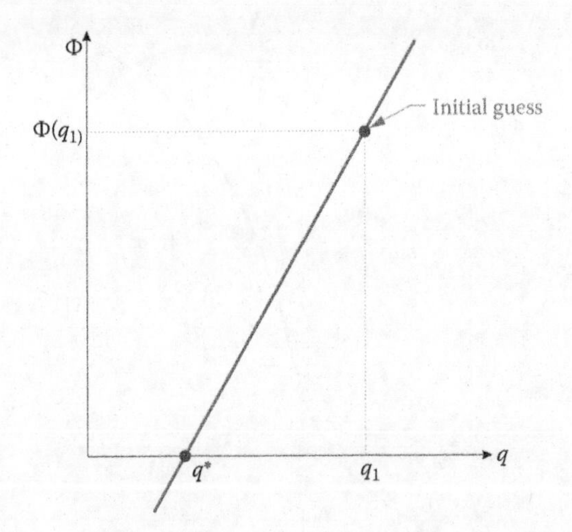

FIGURE 4.104
The initial guess in the Newton–Raphson algorithm.

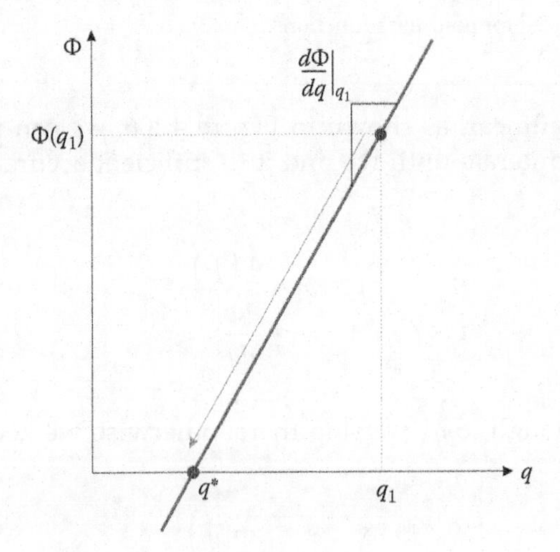

FIGURE 4.105
Direction to move for the second guess.

$$\left.\frac{d\Phi}{dq}\right|_{q_1} \approx \frac{\Phi(q_1) - 0}{q_1 - q^*}$$

Solving this equation for q^*, we have

$$q^* = q_1 - \frac{\Phi(q_1)}{\left.\dfrac{d\Phi}{dq}\right|_{q_1}} \tag{4.192}$$

This is the estimating procedure for the Newton–Raphson algorithm. If the constraint equation is linear, we can find the root in one step! Of course, our linkage constraint equations are not linear, and it will take multiple steps to find the roots.

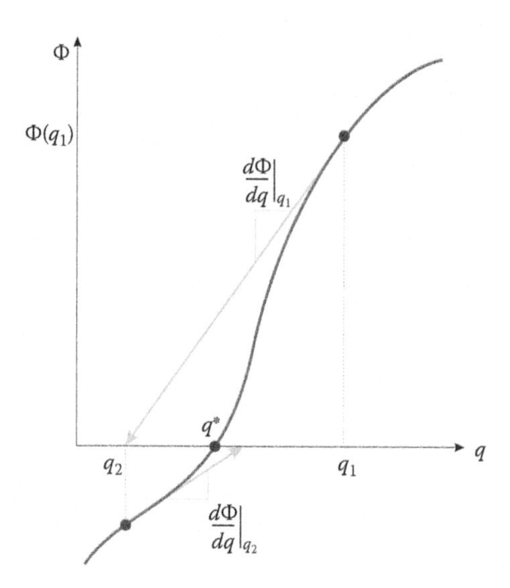

FIGURE 4.106
More than one move is required for nonlinear functions.

If the equation is nonlinear, as shown in Figure 4.106, we can use Equation (4.192) to make a guess at q^*, and iterate until we find it to sufficient accuracy. Starting at point q_1, we find q_2

$$q_2 = q_1 - \frac{\Phi(q_1)}{\left.\frac{d\Phi}{dq}\right|_{q_1}} \tag{4.193}$$

If this point is close enough to q^*, we stop there; otherwise we keep estimating until we have converged upon q^*.

$$q_{i+1} = q_i - \frac{\Phi(q_i)}{\left.\frac{d\Phi}{dq}\right|_{q_i}} \tag{4.194}$$

4.15.2 One Dimensional Examples

Let us try a few examples to become familiar with the algorithm. We'll start simple, and work our way to more complicated functions. First, consider the following linear function:

$$\Phi = 2q + 3 \tag{4.195}$$

We wish to find the value of q that makes the function Φ equal to zero. To start the algorithm, we make an initial (random) guess.

$$\text{guess}: q_1 = 5 \tag{4.196}$$

If we substitute this into the function Φ, we have

$$\Phi(5) = 13 \tag{4.197}$$

This isn't anywhere near zero, so we need to update our guess using the Newton–Raphson formula. To use the formula, we need to know the derivative of the function

$$\frac{d\Phi}{dq} = \frac{d}{dq}(2q+3) = 2 \tag{4.198}$$

The next guess is then

$$q_2 = q_1 - \frac{\Phi(q_1)}{\left.\frac{d\Phi}{dq}\right|_{q_1}} = 5 - \frac{13}{2} = -\frac{3}{2} \tag{4.199}$$

Substituting this into our function, we see that

$$\Phi\left(-\frac{3}{2}\right) = 2\cdot\left(-\frac{3}{2}\right) + 3 = 0 \tag{4.200}$$

Thus, with one guess, we have found the value of q that makes Φ zero. This is always the case for linear functions.

4.15.3 A More Complicated Function

Let us try something a little more complicated: a cubic function that has been plotted in Figure 4.107.

$$\Phi = q^3 - q^2 - 10q \tag{4.201}$$

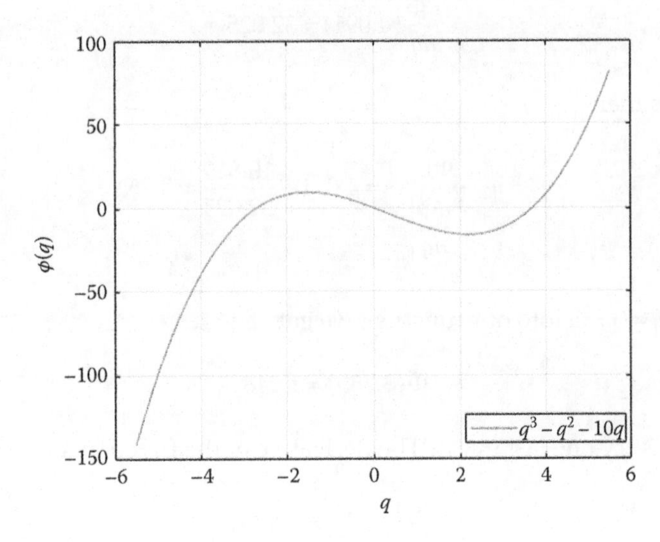

FIGURE 4.107
The cubic function used in the example. This function crosses the x axis at three locations.

The derivative of this function is

$$\frac{d\Phi}{dq} = 3q^2 - 2q - 10 \tag{4.202}$$

As an initial guess, let's try

$$q_1 = 5 \tag{4.203}$$

as before. Substituting this into the function, we have

$$\Phi(5) = 50 \tag{4.204}$$

Nowhere near zero! The derivative of the function at $q = 5$ is

$$\frac{d\Phi}{dq}(5) = 55 \tag{4.205}$$

The next guess is then

$$q_2 = q_1 - \frac{\Phi(q_1)}{\left.\frac{d\Phi}{dq}\right|_{q_1}} = 5 - \frac{50}{55} = 4.091 \tag{4.206}$$

If we substitute $q = 4.091$ into the function, we get

$$\Phi(4.091) = 10.819 \tag{4.207}$$

We are getting closer! The derivative of the function at $q = 4.091$ is

$$\frac{d\Phi}{dq}(4.091) = 32.025 \tag{4.208}$$

The next guess is then

$$q_3 = q_2 - \frac{\Phi(q_2)}{\left.\frac{d\Phi}{dq}\right|_{q_2}} = 4.091 - \frac{10.819}{32.025} = 3.753 \tag{4.209}$$

If we substitute $q = 3.753$ into our function, we get

$$\Phi(3.753) = 1.248 \tag{4.210}$$

Much closer! Let's try one last guess. The derivative of the function at $q = 3.753$ is

$$\frac{d\Phi}{dq}(3.753) = 24.751 \tag{4.211}$$

The next guess is then

$$q_4 = q_3 - \frac{\Phi(q_3)}{\left.\dfrac{d\Phi}{dq}\right|_{q_3}} = 3.753 - \frac{1.248}{24.751} = 3.703 \tag{4.212}$$

The value of the function here is

$$\Phi(3.703) = 0.026 \tag{4.213}$$

This is close enough to zero! Although this function was considerably more complicated than the first one, it took us only four iterations to find the root. The root is $q^* = 3.703$. The tables below give examples of what happens when we try different initial guesses. Table 4.4 shows the initial guess we just tried, $q_1 = 5$.

Next, we'll try a smaller initial guess, $q_1 = 1.5$. As you can see in Table 4.5, we have arrived at different root, $q^* = 0$. Since our function is cubic, we can expect there to be three roots.

Next, we try an initial guess $q_1 = -1.5$. As seen in Table 4.6, the algorithm has some difficulty in converging to an answer. The reason for this is that the slope at $q_1 = -1.5$ is very shallow (−0.25). The updated guess is therefore very far away from the initial guess ($q_2 = 36$). That is, because of the shallow slope, we move very far away from the location of the roots. This is a danger in using the Newton–Raphson algorithm – if our initial guess happens to lie at a point of zero slope, the algorithm may have trouble converging.

In our final example, shown in Table 4.7, we converge to the root −2.702. Convergence occurs very quickly, in four or five iterations.

TABLE 4.4

Initial Guess = 5.0

i	q_i	Φ	$d\Phi/dq$
1	5.000	50.000	55.000
2	4.091	10.819	32.025
3	3.753	1.248	24.751
4	3.703	0.026	23.724
5	3.702	0.0000121	

TABLE 4.5

Initial Guess = 1.5

i	q_i	Φ	$d\Phi/dq$
1	1.500	−13.875	−6.250
2	−0.72	6.308	−7.005
3	0.181	−1.832	−10.263
4	0.00203	−0.02	−10.004
5	4.101×10^{-7}	4.101×10^{-6}	−10.000

TABLE 4.6

Initial Guess = –1.5

i	q_i	Φ	$d\Phi/dq$
1	–1.500	9.375	–0.25
2	36.000	45,000	3,806
3	24.177	13,305	1,695
4	16.328	3,923	757
5	11.146	1149	340
6	7.771	331.145	155.614
7	5.643	91.403	75.238
8	4.412	22.280	39.562
9	3.848	3.702	26.734
10	3.710	0.1995	23.871

TABLE 4.7

Initial Guess = –4.0

i	q_i	Φ	$d\Phi/dq$
1	–4.000	–40	46
2	–3.13	–9.172	25.66
3	–2.773	–1.282	18.614
4	–2.704	–0.0439	17.345
5	–2.702	–0.000058	17.298

4.15.4 Newton–Raphson in Multidimensional Space

We can use the same type of iterative procedure to solve multiple nonlinear equations simultaneously. Our task now is to create a version of Equation (4.194) for multiple constraint equations.

$$q_{i+1} = q_i - \frac{\Phi(q_i)}{\left.\frac{d\Phi}{dq}\right|_{q_i}} \tag{4.214}$$

We will need to do some rearranging to get this into a multidimensional form that MATLAB can work with. First, multiply both sides by the derivative term and rearrange to obtain

$$\left.\frac{d\Phi}{dq}\right|_{q_i} \cdot (q_{i+1} - q_i) = -\Phi(q_i) \tag{4.215}$$

Define

$$\Delta q_i = q_{i+1} - q_i \tag{4.216}$$

Then

$$\left.\frac{d\Phi}{dq}\right|_{q_i} \cdot \Delta q_i = -\Phi(q_i) \tag{4.217}$$

We can generalize this to multidimensional systems as

$$\left.\frac{d\boldsymbol{\Phi}}{d\mathbf{q}}\right|_{q_i} \cdot \Delta\mathbf{q}_i = -\boldsymbol{\Phi}(\mathbf{q}_i) \tag{4.218}$$

Remember that $\boldsymbol{\Phi}$ is now a vector of functions, not a single function. You may well wonder what it means to take the derivative of a vector of functions – with respect to another vector! This is actually simpler than it might seem. As an example, let

$$\mathbf{g} = \begin{Bmatrix} g_1(\mathbf{q}) \\ g_2(\mathbf{q}) \\ g_3(\mathbf{q}) \\ g_4(\mathbf{q}) \end{Bmatrix} \quad \mathbf{q} = \begin{Bmatrix} x \\ y \\ \theta \end{Bmatrix} \tag{4.219}$$

Here, g_1, g_2, etc. are functions of the variables in the vector **q**. The derivative of **g** with respect to **q** is

$$\frac{d\mathbf{g}}{d\mathbf{q}} = \begin{bmatrix} \frac{\partial g_1}{\partial x} & \frac{\partial g_1}{\partial y} & \frac{\partial g_1}{\partial \theta} \\ \frac{\partial g_2}{\partial x} & \frac{\partial g_2}{\partial y} & \frac{\partial g_2}{\partial \theta} \\ \frac{\partial g_3}{\partial x} & \frac{\partial g_3}{\partial y} & \frac{\partial g_3}{\partial \theta} \\ \frac{\partial g_4}{\partial x} & \frac{\partial g_4}{\partial y} & \frac{\partial g_4}{\partial \theta} \end{bmatrix} \tag{4.220}$$

Note that if **g** has n elements and **q** has k elements, then $dg/d\mathbf{q}$ is a $n \times k$ *matrix*. To reiterate, the derivative of a *vector of functions* with respect to a *vector of variables* is a *matrix*. This matrix of first derivatives has a special name – it is known as the *Jacobian* matrix. In this text, we will denote the Jacobian matrix with a bold **J**. Thus, at the ith iteration we have

$$\mathbf{J}_i = \left.\frac{d\boldsymbol{\Phi}}{d\mathbf{q}}\right|_{\mathbf{q}_i} \tag{4.221}$$

Example 4.12: The Fourbar Linkage

Earlier, we found the constraint equations for the fourbar linkage

$$\boldsymbol{\Phi} = \begin{Bmatrix} a\cos\theta_2 + b\cos\theta_3 - c\cos\theta_4 - d \\ a\sin\theta_2 + b\sin\theta_3 - c\sin\theta_4 \end{Bmatrix} \quad \mathbf{q} = \begin{Bmatrix} \theta_3 \\ \theta_4 \end{Bmatrix} \tag{4.222}$$

The Jacobian matrix for this is

$$\mathbf{J} = \begin{bmatrix} -b\sin\theta_3 & c\sin\theta_4 \\ b\cos\theta_3 & -c\cos\theta_4 \end{bmatrix} \tag{4.223}$$

Example 4.13: The Slider-Crank Linkage

The vector loop equations for the slider-crank linkage are

$$\Phi = \left\{ \begin{array}{c} a\cos\theta_2 + b\cos\theta_3 - d \\ a\sin\theta_2 + b\sin\theta_3 - c \end{array} \right\} \quad \mathbf{q} = \left\{ \begin{array}{c} \theta_3 \\ d \end{array} \right\} \tag{4.224}$$

Thus, the Jacobian matrix for the slider-crank is

$$\mathbf{J} = \begin{bmatrix} -b\sin\theta_3 & -1 \\ b\cos\theta_3 & 0 \end{bmatrix} \tag{4.225}$$

4.15.5 The Newton–Raphson Algorithm in MATLAB®

Using the procedure outlined above, we can develop a flowchart for the Newton–Raphson algorithm as shown in Figure 4.108. As an example, we will conduct the position analysis on the example fourbar linkage discussed in Section 4.8. Begin by preparing the workspace and initializing the linkage dimensions, as shown below.

```
% Fourbar_Position_Analysis.m
% Solves for the positions of the links on a fourbar linkage and
% plots the paths of points on the linkage using the Newton-Raphson
% algorithm.
% by Eric Constans, June 9, 2017

% Prepare workspace
clear variables; close all; clc
```

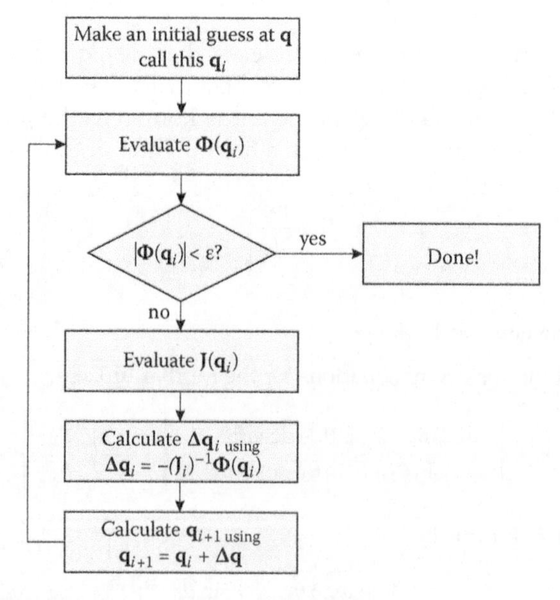

FIGURE 4.108
Flowchart for the Newton–Raphson algorithm.

```
% Linkage dimensions
a = 0.130;           % crank length (m)
b = 0.200;           % coupler length (m)
c = 0.170;           % rocker length (m)
d = 0.220;           % length between ground pins (m)
p = 0.150;           % length from B to P (m)
gamma = 20*pi/180; % angle between BP and coupler (converted to rad)

% ground pins
x0 = [0;0];    % ground pin at A (origin)
xD = [d;0];    % ground pin at D

N = 361;    % number of times to perform position calculations
[xB,xC,xP]                 = deal(zeros(2,N));  % allocate for pos of B, C, P
[theta2,theta3,theta4] = deal(zeros(1,N));  % allocate space for angles
```

Before beginning the main loop, we must first give our initial guesses for the unknowns. These initial guesses will have a significant impact on whether the Newton–Raphson algorithm can find a solution to the position equations. The rest of main loop will be the same as before, but we will replace the position calculations at each time step with a Newton–Raphson calculation.

```
t3 = 0; t4 = 0;  % initial guesses for Newton-Raphson algorithm
for i = 1:N
  theta2(i) = (i-1)*(2*pi)/(N-1);

% Newton-Raphson Calculations go here

% calculate unit vectors
  [e2,n2] = UnitVector(theta2(i));
  [e3,n3] = UnitVector(theta3(i));
  [e4,n4] = UnitVector(theta4(i));
  [eBP,nBP] = UnitVector(theta3(i) + gamma);

% solve for positions of points B, C and P on the linkage
  xB(:,i) = FindPos(      x0,    a,    e2);
  xC(:,i) = FindPos(      xD,    c,    e4);
  xP(:,i) = FindPos(xB(:,i),   p,   eBP);
end
```

Inside the main loop, and after the calculation of `theta2` begin typing in the Newton–Raphson algorithm. Since it is an iterative procedure, we will begin with a second `for` loop. We will limit the `for` loop to 20 iterations, because the algorithm normally converges very quickly. If we do not reach convergence after 20 iterations, we will probably not reach convergence at all!

```
% Newton-Raphson Calculations
  for j = 1:5
    phi(1,1) = a*cos(theta2(i)) + b*cos(t3) - c*cos(t4) - d;
    phi(2,1) = a*sin(theta2(i)) + b*sin(t3) - c*sin(t4);

% If constraint equations are satisfied, then terminate
    if (norm(phi) < 0.000001)
```

```
        theta3(i) = t3;
        theta4(i) = t4;
        break
    end
```

The first statements inside the `for` loop calculate the constraint equations. We have included two indices in `phi` to ensure that it is formed as a column vector, instead of a row vector. Next, the *norm* of phi is calculated. Here "norm" is another word for magnitude; that is the square root of the sum of squares of the vector. This provides an indication as to whether the algorithm has converged upon a valid solution. If we have reached a valid solution, then we substitute the values stored in the temporary variables `t3` and `t4` into `theta3` and `theta4`, and break out of the inner `for` loop. If the convergence criterion is not met, we continue with the algorithm.

```
% calculate Jacobian matrix
    J = [-b*sin(t3)  c*sin(t4);
          b*cos(t3) -c*cos(t4)];

% update variables using Newton-Raphson equation
    dq = -J\phi;
    t3 = t3 + dq(1);
    t4 = t4 + dq(2);
  end
```

Next, the Jacobian matrix is calculated, and we update the variables `t3` and `t4` according to the Newton–Raphson procedure. The end statement indicates the end of the inner `for` loop. The remainder of the program, including the plotting routines, is the same as for the fourbar position analysis in Section 4.8. Try running the program now to see what happens.

The results are not encouraging! Instead of a beautiful plot, we obtain the following error message:

```
> In Fourbar_NewtonRaphson (line 47)
Warning: Matrix is singular to working precision.
```

This is MATLAB's way of telling you that it cannot invert the Jacobian matrix, and in trying to do so it was forced to divide by zero. Unless there is a typographical error somewhere, this is usually caused by our initial guesses being too far away from the actual solution. In many cases, it is difficult to obtain initial estimates by merely guessing, so a common approach is to draw a sketch of the linkage in SOLIDWORKS with crank angle $\theta_2 = 0$ and then to use the Smart Dimension tool to find the initial values for the other angles. Try changing the initial guesses to

```
t3 = pi/4; t4 = pi/2;  % initial guesses for Newton-Raphson algorithm
```

The result should be the plot shown in Figure 4.109. A full listing of the code is given below. It is left as an exercise for the reader to perform the position analysis on the Stephenson Type II sixbar linkage using the Newton–Raphson algorithm . The dimensions of a sample Stephenson Type II linkage are shown in Figure 4.110, and the position plot is shown in Figure 4.111.

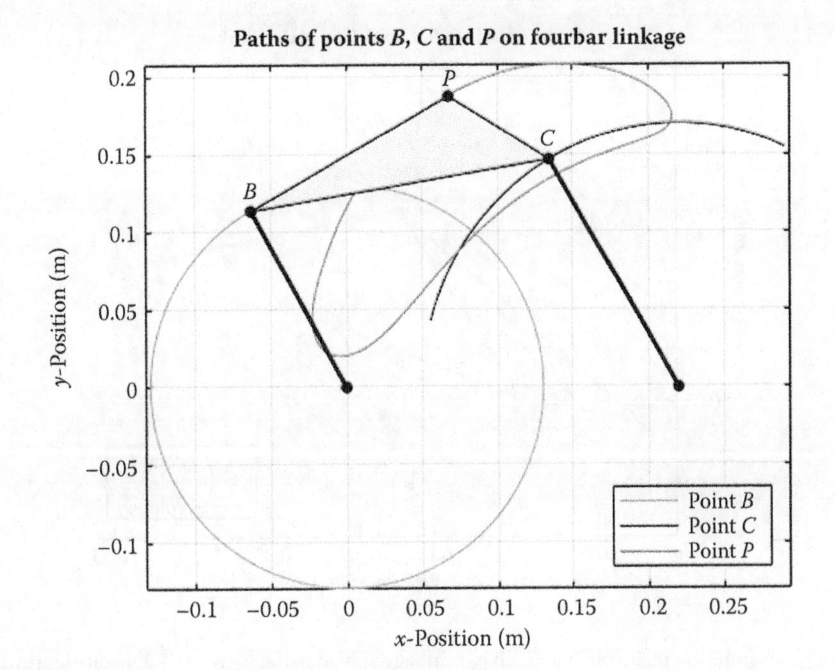

FIGURE 4.109
Fourbar position plot obtained using the Newton–Raphson algorithm.

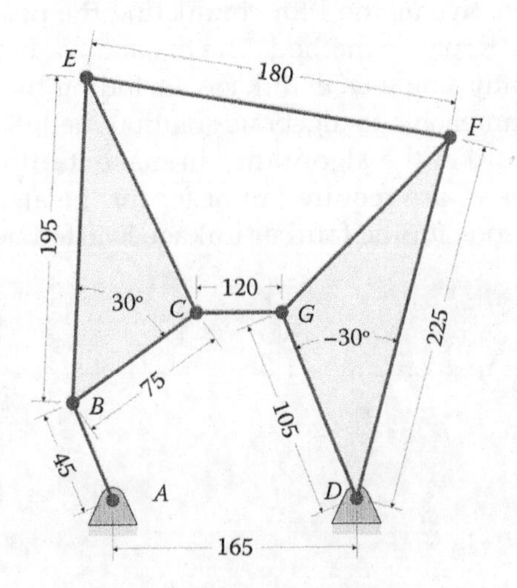

(All dimensions in millimeters)

Crank length: 45
Coupler length: 75
Length *BE* on coupler: 195
Internal angle of coupler: 30°
Distance between ground pins: 165

Rocker length: 105
Length *DF* on rocker: 225
Internal angle of rocker: −30°
Length of link 5: 120
Length of link 6: 180

FIGURE 4.110
Dimensions of example Stephenson Type II sixbar linkage.

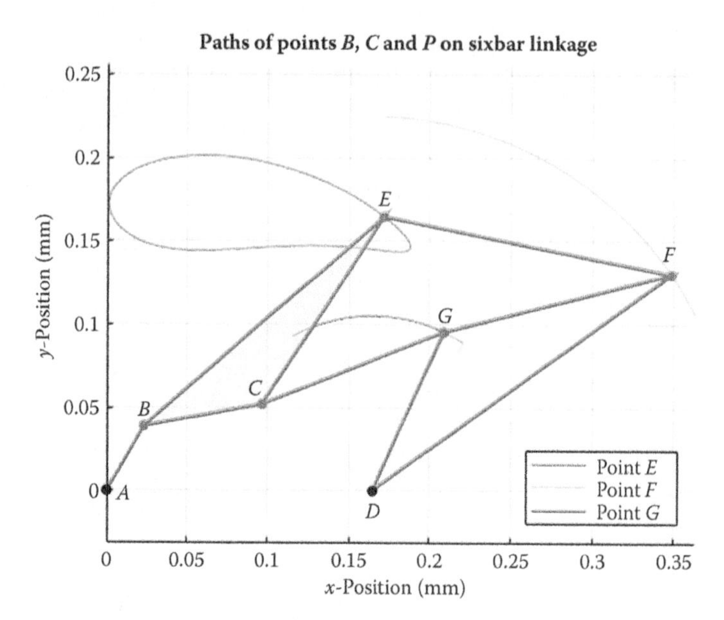

FIGURE 4.111
Position plot for example Stephenson Type II sixbar linkage solved using Newton–Raphson algorithm.

4.15.6 Summary

We have presented an alternative method for conducting the position analysis on single DOF linkages – the Newton–Raphson method. The major advantage to the method is that it can be made to work for any single DOF linkage, including those (like the Stephenson Type II sixbar) that are not amenable to algebraic solution methods. The major disadvantage is that the abstract nature of the algorithm can make it difficult to debug, and that reasonably good initial guesses are required in order for the algorithm to converge. An example Newton–Raphson code for the fourbar linkage is listed below.

```
% Fourbar_Position_Analysis.m
% Solves for the positions of the links on a fourbar linkage and
% plots the paths of points on the linkage using the Newton-Raphson
% algorithm.
% by Eric Constans, June 9, 2017

% Prepare workspace
clear variables; close all; clc

% Linkage dimensions
a = 0.130;           % crank length (m)
b = 0.200;           % coupler length (m)
c = 0.170;           % rocker length (m)
d = 0.220;           % length between ground pins (m)
p = 0.150;           % length from B to P (m)
gamma = 20*pi/180; % angle between BP and coupler (converted to rad)

% ground pins
x0 = [0;0];    % ground pin at A (origin)
```

```
xD = [d;0];    % ground pin at D

N = 361;    % number of times to perform position calculations
[xB,xC,xP]                 = deal(zeros(2,N));  % allocate for pos of B, C, P
[theta2,theta3,theta4] = deal(zeros(1,N));  % allocate space for angles

t3 = pi/4; t4 = pi/2;  % initial guesses for Newton-Raphson algorithm

for i = 1:N
  theta2(i) = (i-1)*(2*pi)/(N-1);

% Newton-Raphson Calculations
  for j = 1:5
    phi(1,1) = a*cos(theta2(i)) + b*cos(t3) - c*cos(t4) - d;
    phi(2,1) = a*sin(theta2(i)) + b*sin(t3) - c*sin(t4);
% If constraint equations are satisfied, then terminate
    if (norm(phi) < 0.000001)
      theta3(i) = t3;
      theta4(i) = t4;
      break
    end

% calculate Jacobian matrix
    J = [-b*sin(t3)  c*sin(t4);
          b*cos(t3) -c*cos(t4)];

% update variables using Newton-Raphson equation
    dq = -J\phi;
    t3 = t3 + dq(1);
    t4 = t4 + dq(2);
  end

% calculate unit vectors
  [e2,n2] = UnitVector(theta2(i));
  [e3,n3] = UnitVector(theta3(i));
  [e4,n4] = UnitVector(theta4(i));
  [eBP,nBP] = UnitVector(theta3(i) + gamma);

% solve for positions of points B, C and P on the linkage
  xB(:,i) = FindPos(      x0,    a,    e2);
  xC(:,i) = FindPos(      xD,    c,    e4);
  xP(:,i) = FindPos(xB(:,i),    p,   eBP);
end

plot(xB(1,:),xB(2,:),'Color',[153/255 153/255 153/255])
hold on
plot(xC(1,:),xC(2,:),'Color',[25/255 25/255 25/255])
plot(xP(1,:),xP(2,:),'Color',[0 110/255 199/255])

% specify angle at which to plot linkage
iTheta = 120;

% plot the coupler as a triangular patch
patch([xB(1,iTheta) xC(1,iTheta) xP(1,iTheta)],...
```

```
    [xB(2,iTheta) xC(2,iTheta) xP(2,iTheta)],[229/255 240/255
249/255]);

% plot crank and rocker
plot([x0(1) xB(1,iTheta)],[x0(2) xB(2,iTheta)],'Linewidth',2,'Color'
,'k');
plot([xD(1) xC(1,iTheta)],[xD(2) xC(2,iTheta)],'Linewidth',2,'Color'
,'k');

% plot joints on linkage
plot([x0(1) xD(1) xB(1,iTheta) xC(1,iTheta) xP(1,iTheta)],...
     [x0(1) xD(2) xB(2,iTheta) xC(2,iTheta) xP(2,iTheta)],...
     'o','MarkerSize',5,'MarkerFaceColor','k','Color','k');

% plot the labels of each joint
text(xB(1,iTheta),xB(2,iTheta)+.015,'B','HorizontalAlignment','center');
text(xC(1,iTheta),xC(2,iTheta)+.015,'C','HorizontalAlignment','center');
text(xP(1,iTheta),xP(2,iTheta)+.015,'P','HorizontalAlignment','center');

axis equal
grid on

title('Paths of points B, C and P on Fourbar Linkage')
xlabel('x-position [m]')
ylabel('y-position [m]')
legend('Point B', 'Point C', 'Point P','Location','SouthEast')
```

4.16 Practice Problems

Problem 4.1

Explain the difference between a Grashof and non-Grashof fourbar linkage.

Problem 4.2

The positions of the links on the slider-crank linkage in Figure 4.112 can be solved using simple hand calculations, and can be checked using SOLIDWORKS. Solve for the slider position, *d*, and the connecting rod angle θ_3 for the following two situations:

a. The crank length is 1.4 cm, the coupler length is 4 cm, the vertical offset is 1 cm, and the crank angle is 45°.
b. The crank length is 7 cm, the coupler length is 25 cm, the vertical offset is 10 cm, and the crank angle is 330°.

Problem 4.3

Conduct the position analysis of the inverted slider-crank in Figure 4.113 using hand calculations and verify your solution using SOLIDWORKS. Solve for θ_3 and θ_4 using the following data:

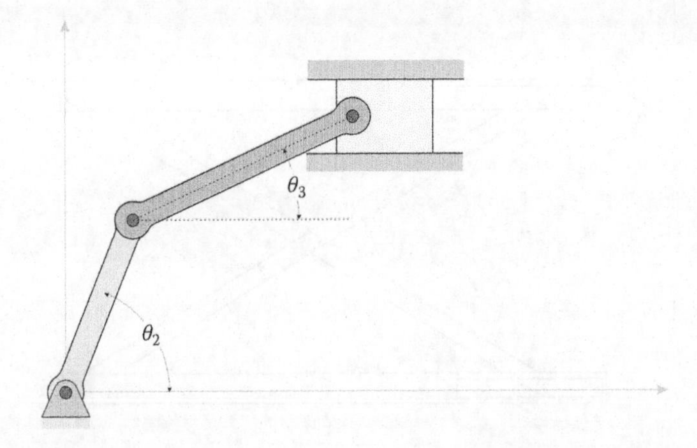

FIGURE 4.112
Problem 4.2.

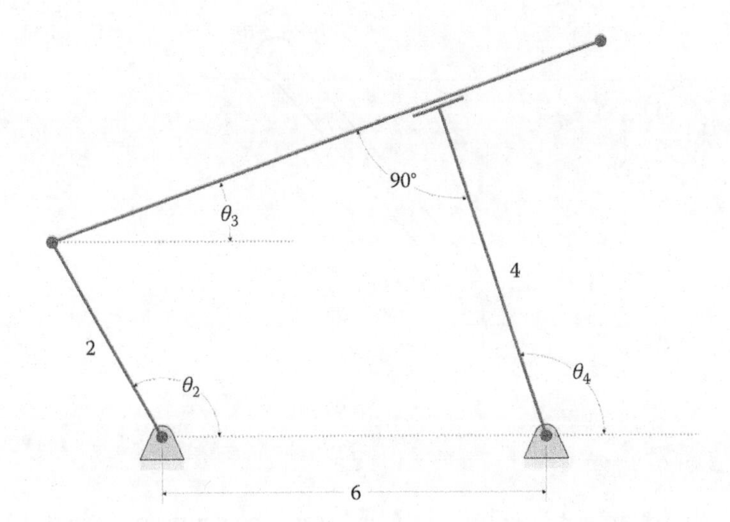

FIGURE 4.113
Problem 4.3.

a. Solve the problem using the values shown in the figure. The crank angle is 30°.
b. Using the same figure, alter the values such that the crank length is 10 cm, the length of the rocker is 6 cm, and the distance between ground pins is 3 cm. The crank angle is 45° and δ is 45°.

Problem 4.4

Figure 4.114 shows a common "scissor-lift" linkage that is used to lift heavy objects. A hydraulic cylinder controls the length d, which raises or lowers the load. Create the linkage in SOLIDWORKS, then record the height h when d takes on the values 2.5, 2.25, and 2 m.

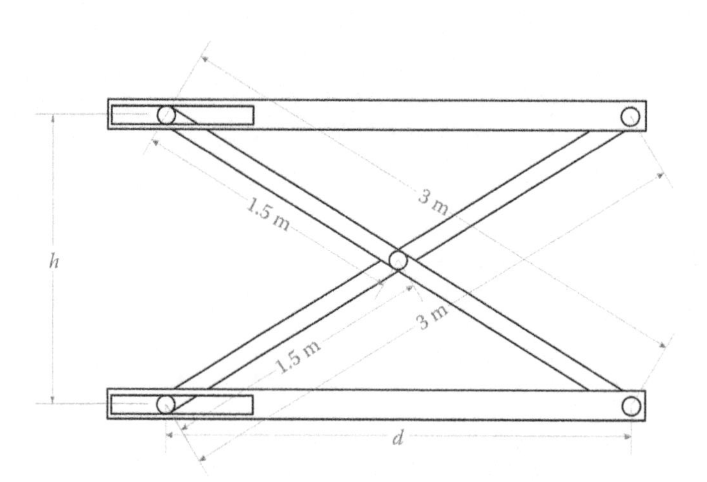

FIGURE 4.114
Problem 4.4.

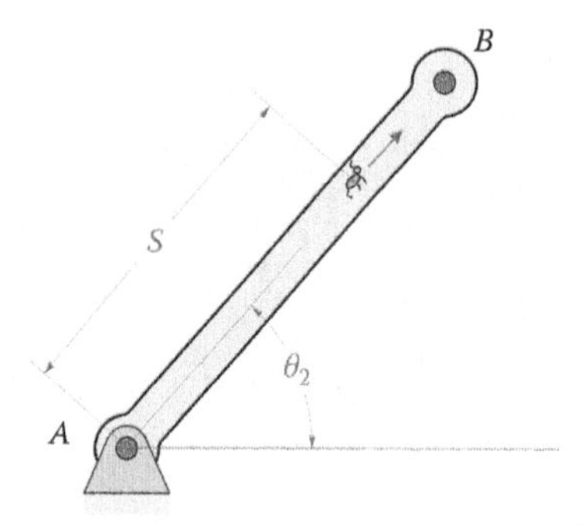

FIGURE 4.115
Problem 4.5.

Problem 4.5

Figure 4.115 shows an ant crawling outward from the center of a rotating link such that its distance from point A is given by the function $s = 10t$ (in millimeters). The crank angle is also a function of time given by $\theta_2 = 4t$ (in radians). Plot the trajectory of the ant for 10 s.

Problem 4.6

Figure 4.116 shows the ant from Problem 4.5 is now crawling from point C to point B such that the distance s is given by the function $s = 200 - 10t$ (in millimeters). The length AB and BC are both 200 mm. The angle θ_2 is given by the function $\theta_2 = 2t$ (in radians) and the angle $\theta_3 = -4t$ (in radians). Plot the trajectory of the ant for 10 s. Hint: this is an excellent application for the `FindPos` function.

Problem 4.7

Use MATLAB to plot the trajectory of point B at the end of the crank and point P on the threebar linkage shown in Figure 4.117 for one revolution of the crank.

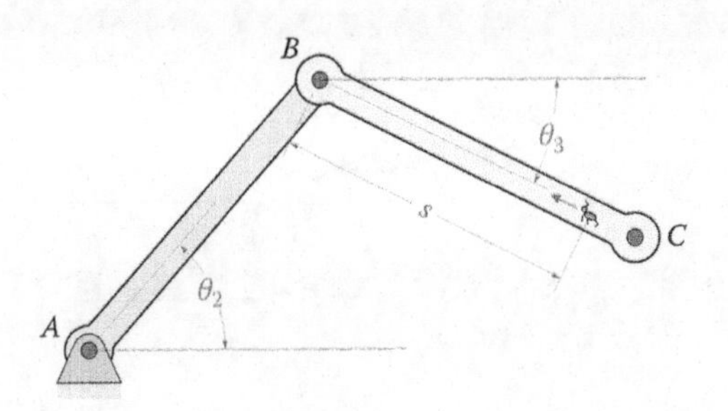

FIGURE 4.116
Problem 4.6.

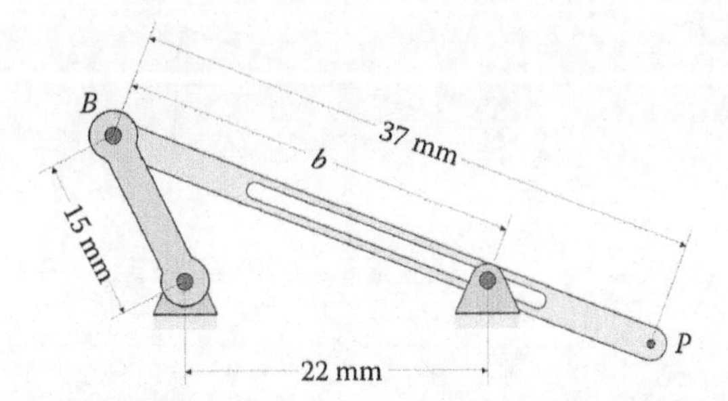

FIGURE 4.117
Problem 4.7.

Problem 4.8

Plot the trajectory of point P on the threebar linkage shown in Figure 4.118 as the crank makes a single revolution. All dimensions are in meters. Hint: the ground pins are not horizontally aligned, so you will need to perform a new vector loop analysis on the linkage.

Problem 4.9

A practical example of a slider-crank linkage is the piston/connecting rod/crank assembly in an engine. For the linkage shown in Figure 4.119, use MATLAB to plot the position of the piston as a function of the crank angle.

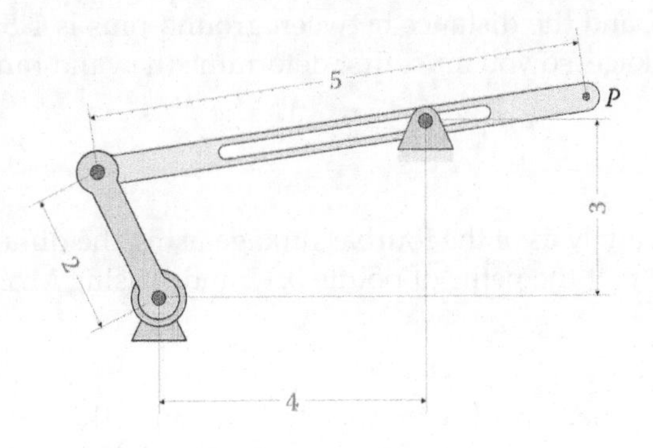

FIGURE 4.118
Problem 4.8.

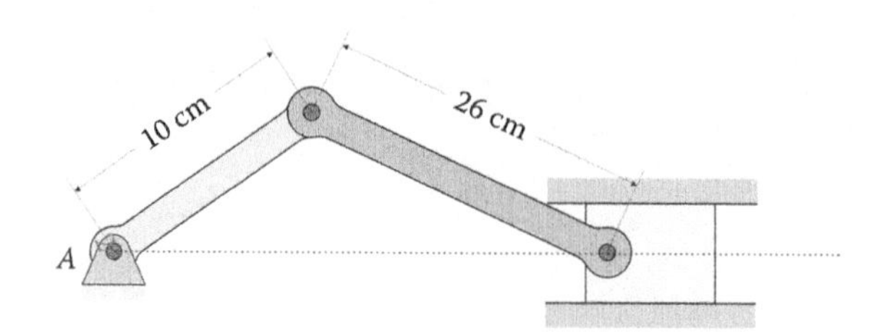

FIGURE 4.119
Problem 4.9.

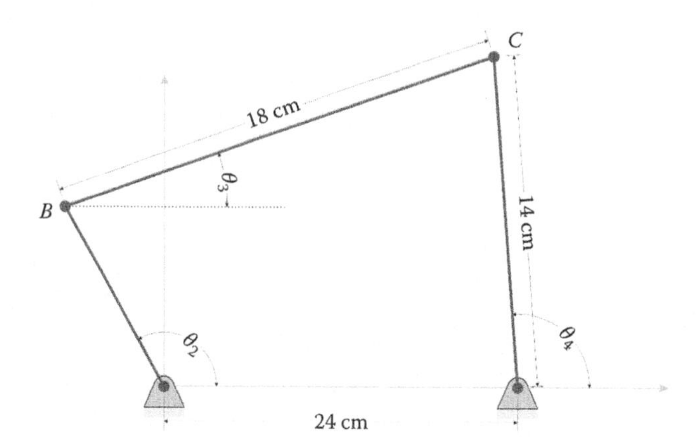

FIGURE 4.120
Problem 4.10.

Problem 4.10

Use MATLAB to plot the paths of points *B* and *C* on the fourbar linkage shown in Figure 4.120 for one revolution of the crank.

Problem 4.11

Plot the paths of points *B* and *C* on the fourbar linkage shown in the figure in Problem 4.10 if the crank length is 50 mm, the coupler length is 75 mm, the rocker length is 85 mm, and the distance between ground pins is 115 mm. Hint: this is a non-Grashof linkage, so you must first determine the valid range of motion of the crank.

Problem 4.12

Conduct a position analysis of the fourbar linkage using the dimensions provided in Figure 4.121 and plot the paths of points *B*, *C*, and *P* using MATLAB.

Problem 4.13

Conduct a position analysis of a fourbar linkage with the following links: crank length = 75 mm, coupler length = 100 mm, rocker length = 85 mm, length between ground pins = 125 mm, distance between points *B* and *P* = 75 mm, and internal

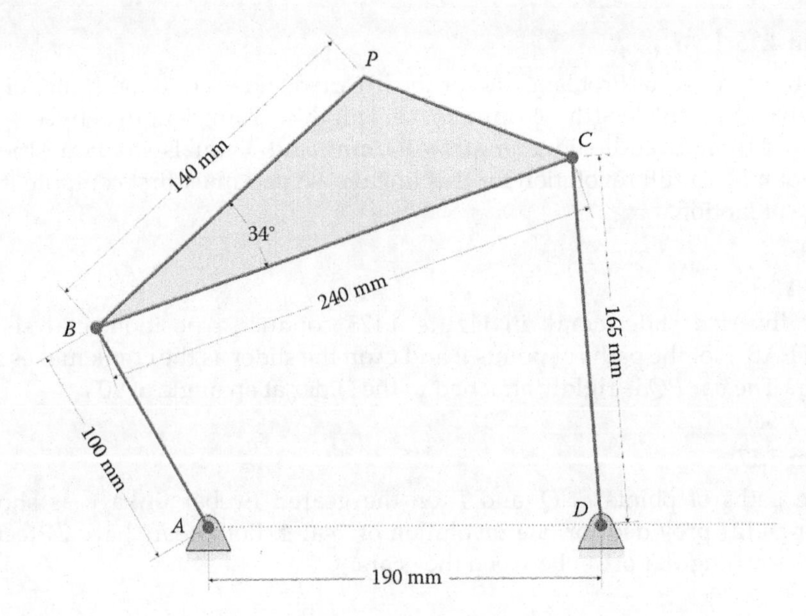

FIGURE 4.121
Problem 4.12.

coupler angle = 15°. Plot the paths of points B, C, and P using MATLAB and refer to the figure for Problem 4.12. Hint: this is a non-Grashof linkage, so you must first find the valid range of motion for the crank.

Problem 4.14

For the inverted slider-crank in Figure 4.122 conduct a position analysis using MATLAB. Plot the paths of points C on the rocker and P at the end of the slider, using the dimensions provided.

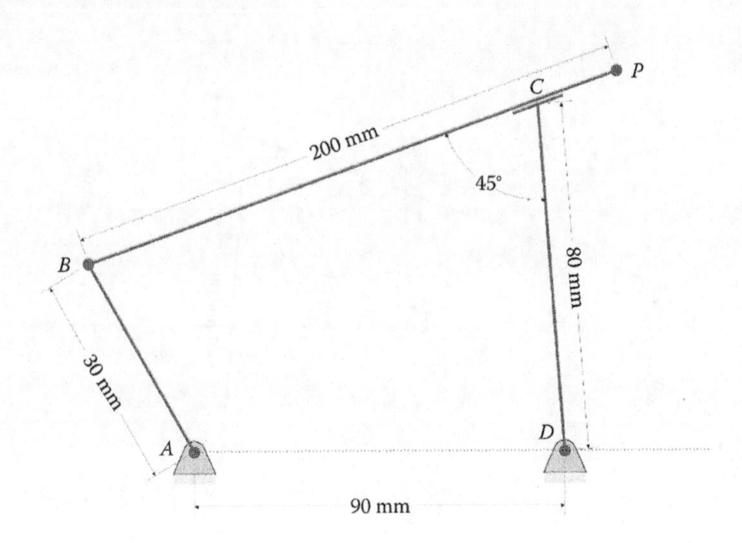

FIGURE 4.122
Problem 4.14.

Problem 4.15

Repeat the analysis of Problem 4.14 for the inverted slider-crank with the following dimensions: crank length = 45 mm, rocker length = 95 mm, length between ground pins = 80 mm, overall slider length = 160 mm, and δ equals 45°. Hint: the crank cannot make a full revolution for this linkage, so you must first compute its valid range of motion.

Problem 4.16

For the inverted slider-crank in Figure 4.123 conduct a position analysis using MATLAB. Plot the paths of points P and Q on the slider as the crank makes a revolution. The bar PQ is rigidly attached to the slider at an angle of 90°.

Problem 4.17

Plot the paths of points E, Q, and P on the geared fivebar linkage as shown in Figure 4.124 provided for one revolution of gear 1. Both gears have 24 teeth and there is no angular offset between the gears.

Problem 4.18

Plot the paths of points E, Q, and P on the geared fivebar linkage from Problem 4.17 where the crank length is 10 cm, both coupler lengths are 23 cm, the distance between ground pins is 17 cm, the length of the link on gear 2 is 11 cm, gear 1 has 50 teeth, gear 2 has 25 teeth, there is no angle offset between the gears. The internal coupler angles are $\gamma_3 = 25°$ and $\gamma_4 = -25°$ and the distances BP and CQ are 18 cm.

Problem 4.19

Use MATLAB to conduct a position analysis on the geared fivebar linkage shown in Figure 4.125. The gear at point A has 50 teeth and the gear at point D has 25 teeth. Gear D is rotated 180° when gear A is at 0°. Plot the trajectories of points E, P, and Q.

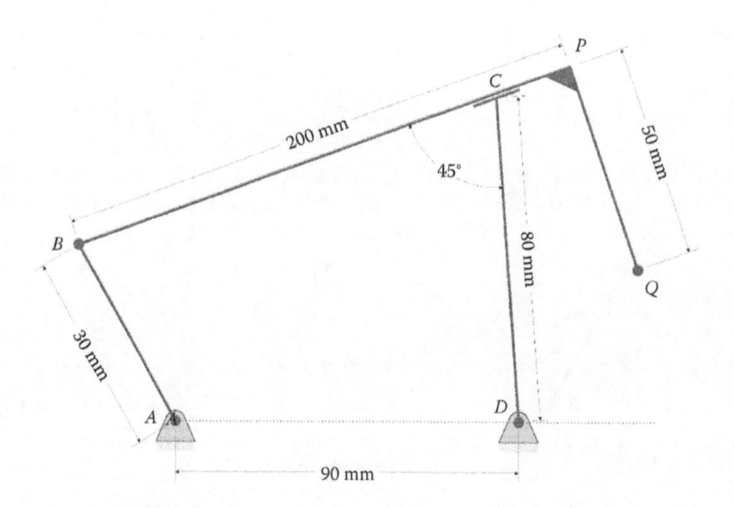

FIGURE 4.123
Problem 4.16.

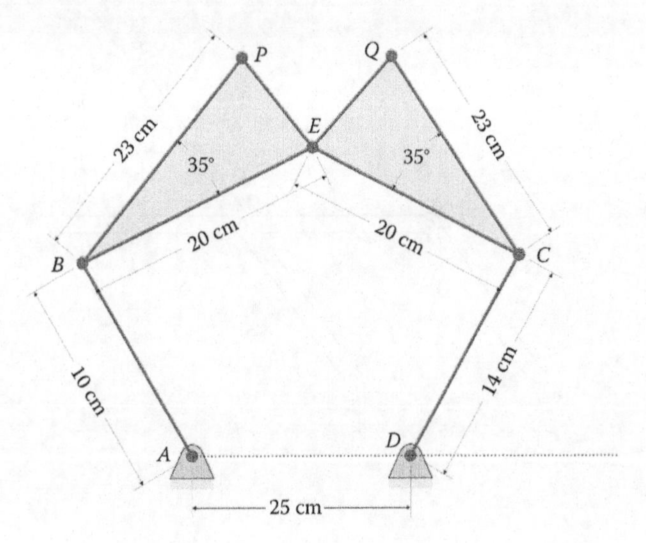

FIGURE 4.124
Problem 4.17.

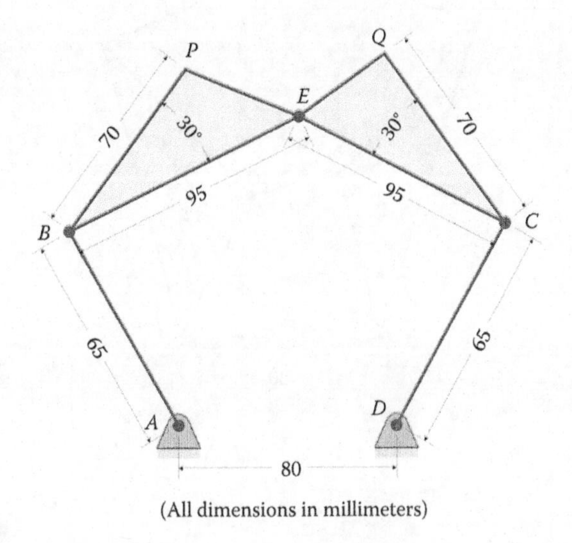

FIGURE 4.125
Problem 4.19.

Problem 4.20

The linkage shown in Figure 4.126 is a claw mechanism. Conduct a position analysis to plot the paths of the points *B*, *C*, *P*, and *Q* for one revolution of the crank. All dimensions are in centimeters.

Problem 4.21

Plot the paths of points *E*, *F*, and *G* of the sixbar linkage with the dimensions in Figure 4.127. All dimensions are in millimeters.

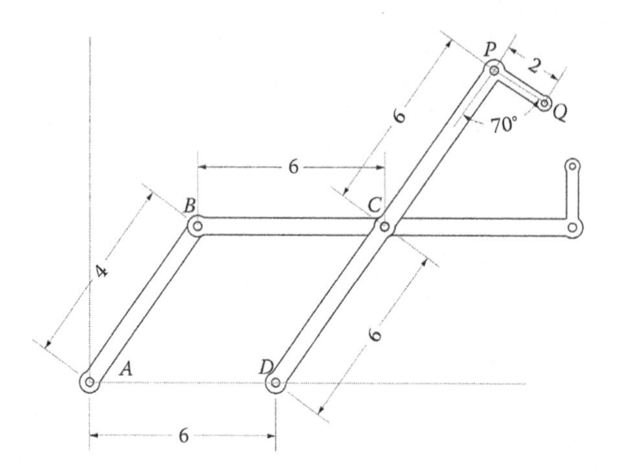

FIGURE 4.126
Problem 4.20.

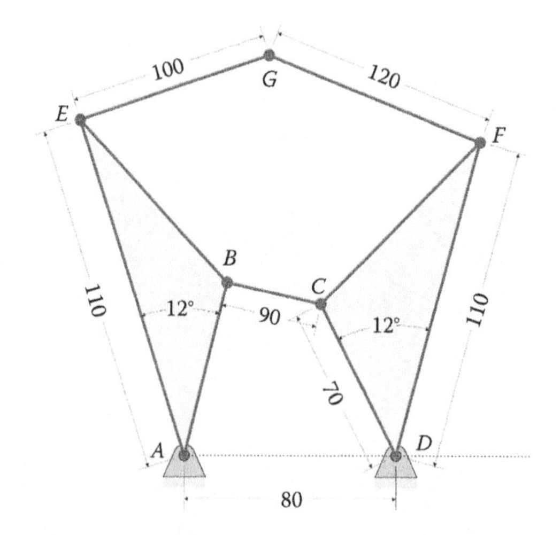

FIGURE 4.127
Problem 4.21.

Problem 4.22

Use MATLAB to conduct a position analysis on the sixbar linkage shown in Figure 4.128. Plot the trajectory of points *E*, *F*, *G*, and *P* for one revolution of the crank.

Problem 4.23

Plot the paths of points *E*, *C*, and *G* of the sixbar with the dimensions in Figure 4.129. All dimensions are in millimeters.

Problem 4.24

Plot the paths of points *E*, *F*, and *G* on the sixbar linkage shown in Figure 4.130, using the given dimensions. All dimensions are in millimeters.

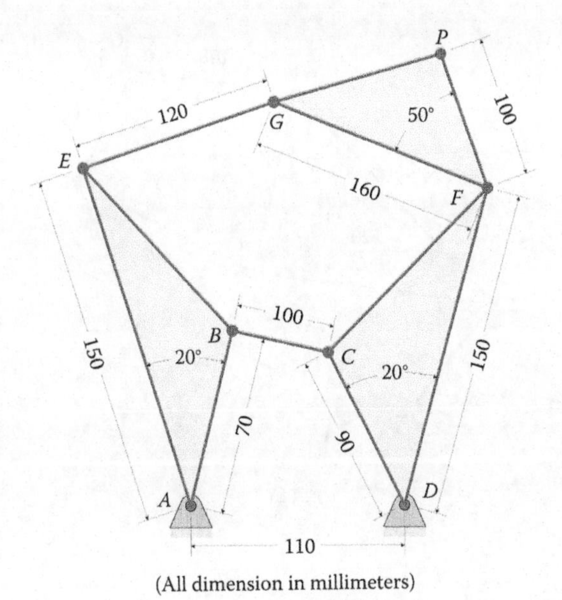

FIGURE 4.128
Problem 4.22.

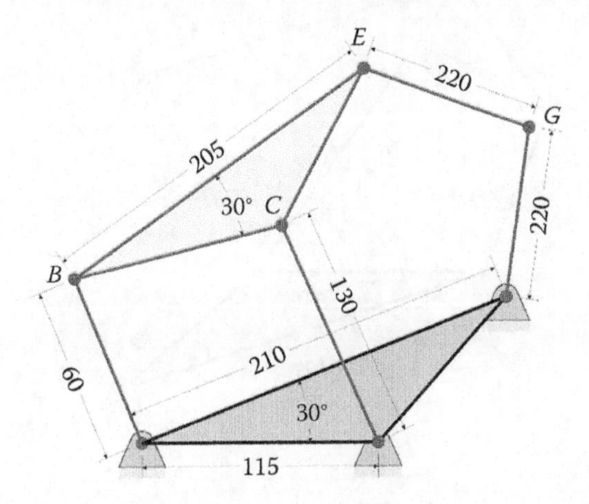

FIGURE 4.129
Problem 4.23.

Problem 4.25

Plot the paths of Points E, C, and G on the sixbar linkage in Figure 4.131, use the dimensions provided. All dimensions are in millimeters.

Problem 4.26

Use MATLAB to conduct a position analysis on the sixbar linkage shown in Figure 4.132. Plot the trajectory of points C, E, G, and P for one revolution of the crank.

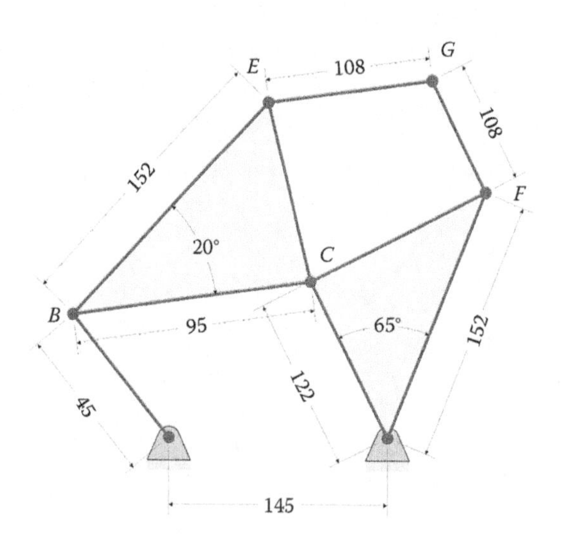

FIGURE 4.130
Problem 4.24.

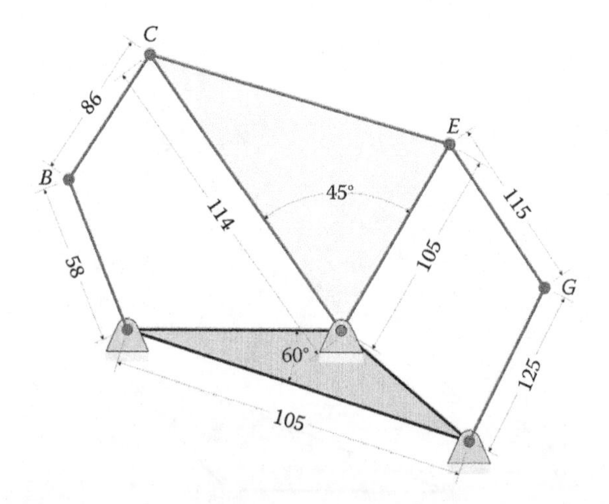

FIGURE 4.131
Problem 4.25.

Problem 4.27: Design Problem

Design a claw mechanism for the purpose of picking up an object of arbitrary shape. Your design must include a fourbar linkage, and you must provide a brief description of how the mechanism works.

Problem 4.28

Sixbar linkages are sometimes used in walking mechanisms, as with the Klann linkage. A modified version of the Klann linkage is shown in Figure 4.133. The link lengths are as follows:

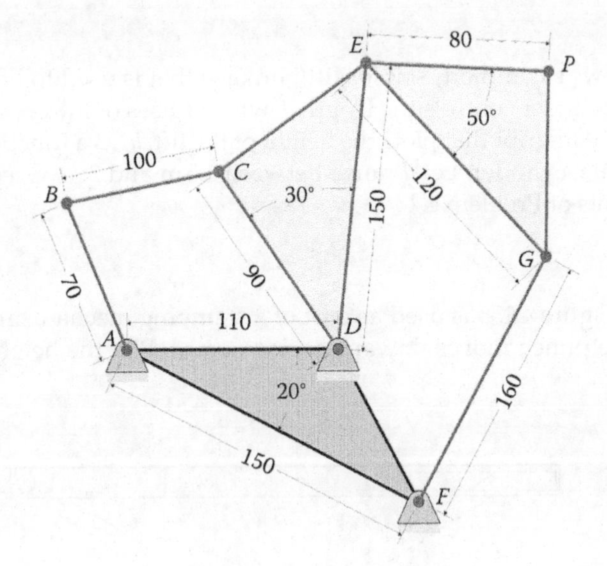

FIGURE 4.132
Problem 4.26.

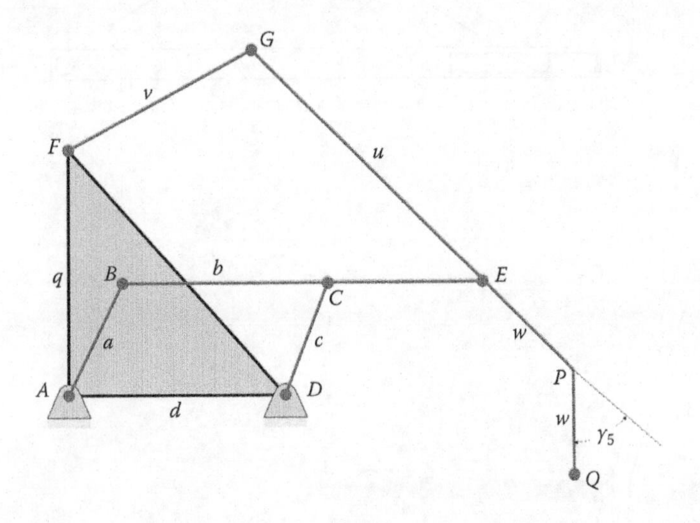

FIGURE 4.133
Problem 4.28.

$a = 30$ mm $\quad b = 50$ mm $\quad c = 30$ mm $\quad d = 50$ mm

$q = 40$ mm $\quad u = 60$ mm $\quad v = 40$ mm $\quad w = 30$ mm

$BE = 60$ mm

The angle $DAF = 90°$ and the angle $\gamma_5 = 40°$. Plot the trajectory of the "foot" at point Q for one revolution of the crank, AB.

Problem 4.29

Figure 4.134 shows a common "scissor-lift" linkage that is used to lift heavy objects. A hydraulic cylinder controls the length *d*, which raises or lowers the load. Write a simple MATLAB script that plots the height of the lift, *h*, as a function of the length of the hydraulic cylinder. Let *d* range between 1.5 m and 3 m. Check your answer with the results of Problem 4.4.

Problem 4.30

The linkage in Figure 4.135 is used as part of a stamping mechanism. As the handle is rotated, the punch moves upward or downward. Plot the height of the punch,

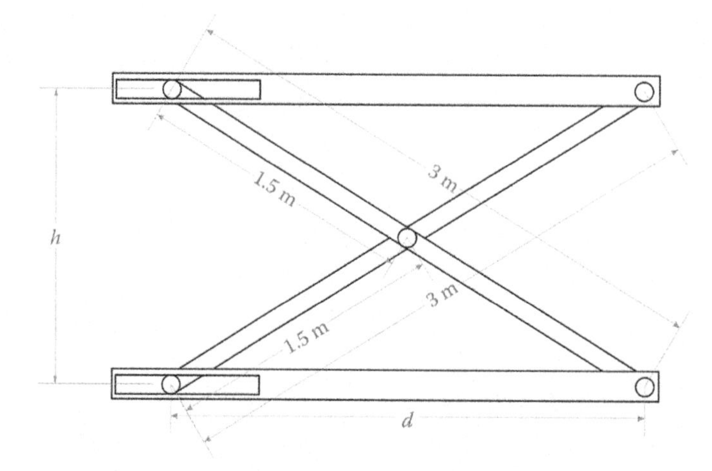

FIGURE 4.134
Problem 4.29.

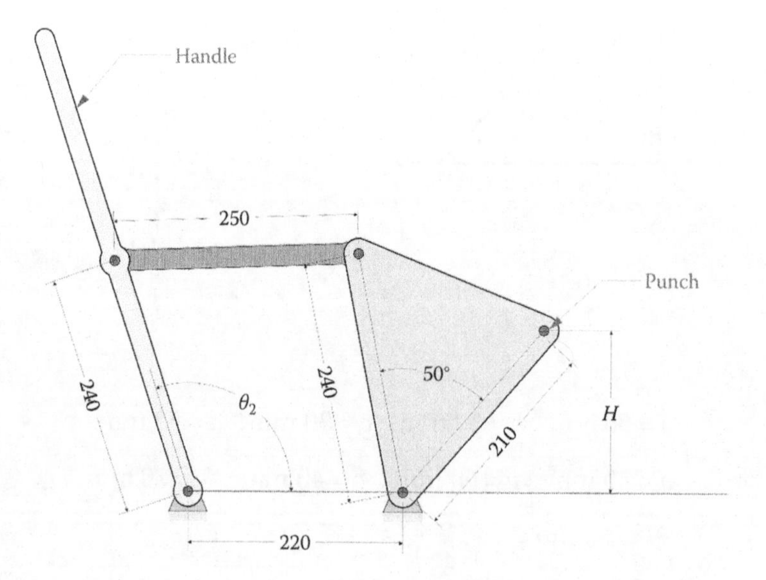

FIGURE 4.135
Problem 4.30.

H, as a function of the handle angle, θ_2. The handle is only capable of moving between 45° and 135°.

Acknowledgments

Several images in this chapter were produced using SOLIDWORKS software. SOLIDWORKS is a registered trademark of Dassault Systèmes SolidWorks Corporation.

Several images in this chapter were produced using MATLAB software.

MATLAB is a registered trademark of The MathWorks, Inc.

Excel is a registered trademark of Microsoft Corporation.

Works Cited

1. A. Cayley, "On three-bar motion," *Proceedings of the London Mathematical Society*, vol. VII, pp. 136–166, 1876.
2. S. Roberts, "On three-bar motion in plane space," *Proceedings of the London Mathematical Society*, vol. VII, pp. 14–23, 1875.
3. C. W. Wampler, "Solving the kinematics of planar mechanisms," *Journal of Mechanical Design*, vol. 121, no. 3, pp. 387–391, 1999.
4. R. L. Norton, *Design of Machinery*, New York: McGraw-Hill, 2012.
5. K. Waldron and G. Kinzel, *Kinematics, Dynamics and Design of Machinery*, Hoboken, NJ: Wiley, 2004.
6. G. Martin, *Kinematics and Dynamics of Machines*, Prospect Heights, IL: Waveland Press, 1982.
7. D. Myszka, *Machines and Mechanisms*, Upper Saddle River, NJ: Prentice Hall, 1999.
8. C. Wilson and J. Sadler, *Kinematics and Dynamics of Machinery*, Upper Saddle River, NJ: Prentice Hall, 2003.
9. E. Constans, T.R. Chandrupatla, and H. Zhang, "An efficient position solution for the fourbar linkage," *International Journal of Mechanisms and Robotic Systems*, vol. 2, no 3/4, pp. 365–373, 2015.

5

Velocity Analysis of Linkages

5.1 Introduction to Velocity Analysis

We have now developed a powerful toolkit for finding the positions of moving links on several mechanisms. It is time to set the linkages in motion! In this section, we will develop methods for finding the velocities at various points on the linkage using two different approaches:

1. Graphical Approach (pencil and paper or SOLIDWORKS®)
2. Analytical Approach (MATLAB®)

We will begin with a few general observations about velocity. First, velocity is a *vector*, which means that it has a *magnitude* and a *direction*. It is common (but inaccurate) to say that a car moves with a velocity of 100 km/h. To completely specify the velocity of the car, we must also tell which direction it is moving in. Thus, to state that a car is moving at a *speed* of 100 km/h in a *northward* direction would completely specify its velocity. In all of the analysis that follows, we will find both speed and direction for a point on a linkage.

As we observed with Chasle's Theorem, there are three general types of motion (see Figure 5.1). In the case of pure translation, all points on the body move with equal speed, and in the same direction. Thus, as shown in Figure 5.2, the points A and B have equal velocity.

$$\mathbf{v}_B = \mathbf{v}_A \tag{5.1}$$

This fact will be important when we discuss relative velocities below.

5.1.1 Pure Rotation

Figure 5.3 shows a rigid body pinned to the ground at point A. Because the body is pinned to the ground, it can only experience pure rotation, and the point B traces a circular arc centered at point A. The velocity vector, $\mathbf{v}_B$, is tangential to the circle, which means that it is perpendicular to the vector $\mathbf{r}_B$. On the right side of Figure 5.3 the magnitude and direction of the position vector $\mathbf{r}_B$ are shown.

$$\mathbf{r}_B = a\mathbf{e} = a\begin{Bmatrix} \cos\theta \\ \sin\theta \end{Bmatrix} \tag{5.2}$$

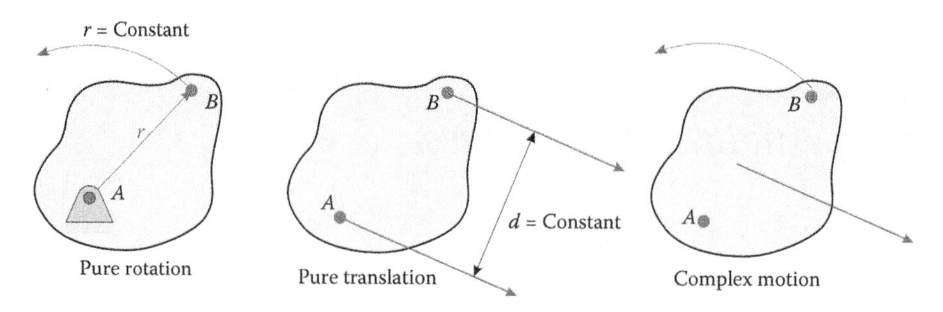

FIGURE 5.1
The three types of motion – left: pure rotation, center: pure translation, right: complex motion.

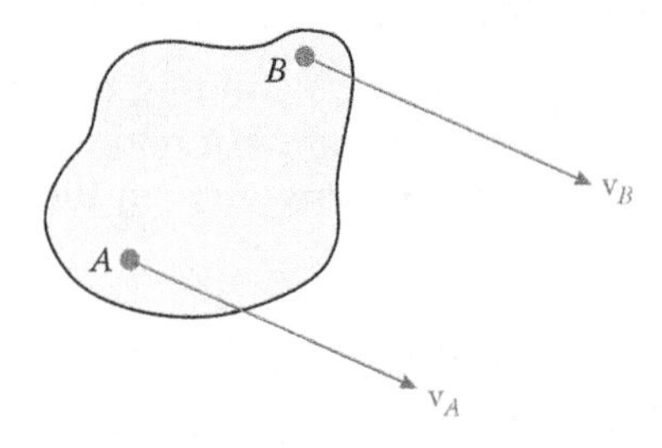

FIGURE 5.2
In pure translation, the velocities of all points on the body are equal.

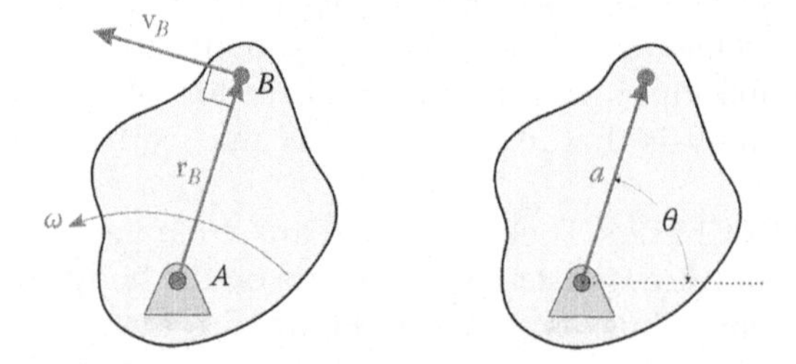

FIGURE 5.3
In pure rotation, the point B describes a circular arc about the center of rotation at point A.

To find the velocity of point B, we must take the time derivative of $\mathbf{r}_B$.

$$\mathbf{v}_B = \frac{d\mathbf{r}_B}{dt} = \frac{d}{dt}a\mathbf{e} \tag{5.3}$$

The length a is constant for this vector and, as discussed in Section 4.2.5 the derivative of the unit vector is the angular velocity multiplied by the unit normal:

$$\mathbf{v}_B = \frac{d\mathbf{r}_B}{dt} = a\frac{d\mathbf{e}}{dt} = a\omega\mathbf{n} \tag{5.4}$$

The magnitude of the velocity is then

$$|\mathbf{v}_B| = a\omega \tag{5.5}$$

and the direction is

$$\mathbf{v}_B \perp \mathbf{r}_B \tag{5.6}$$

In other words, the velocity at point B is rotated 90° counterclockwise from the position vector out to point B. The magnitude of the velocity is the product of the distance from the center of rotation and the angular velocity as given in Equation (5.5). As we will encounter this formula quite often in the sections that follow, it is worth committing to memory.

5.1.2 Complex Motion

Complex motion is, as you might expect, more complex. As Chasles' theorem states, any motion can be described as a combination of rotational and translational movements. Consider the motion of the rigid body in Figure 5.4. The point A moves with velocity $\mathbf{v}_A$. Since the body is rigid, the point B must also share this velocity. The rotation of the body also causes the point B to move *relative* to A in the same manner as for pure rotation.

$$|\mathbf{v}_{BA}| = a\omega \tag{5.7}$$

$$\mathbf{v}_{BA} \perp \mathbf{r}_B$$

The total velocity at point B is then a combination of the translational and rotational components, written as a vector sum:

$$\mathbf{v}_B = \mathbf{v}_A + \mathbf{v}_{BA} \tag{5.8}$$

5.1.3 Velocity of a Point Moving on a Rotating Link

Before we begin graphical methods, let us examine one more configuration. In Figure 5.5, we see a block mounted in a slider on a rotating link. The block is free to slide up and down in the slot, and the rotating link is pinned to ground. The distance from point A to point B is still a, but since the block slides, the quantity a is no longer a constant. Thus, the position vector looks the same as before:

$$\mathbf{r}_B = a\mathbf{e} \tag{5.9}$$

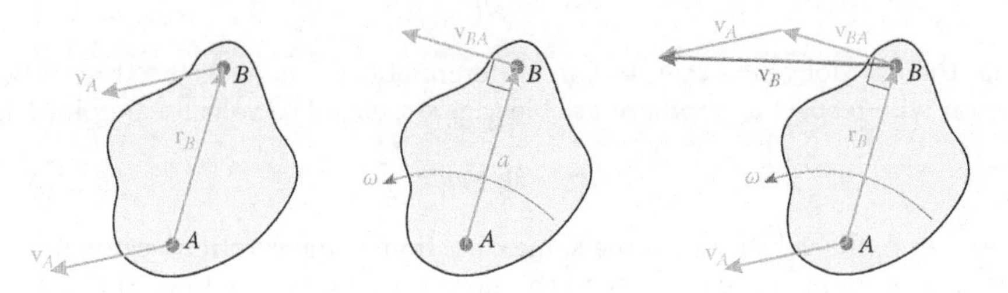

FIGURE 5.4
The velocity in complex motion is a combination of pure rotation and pure translation.

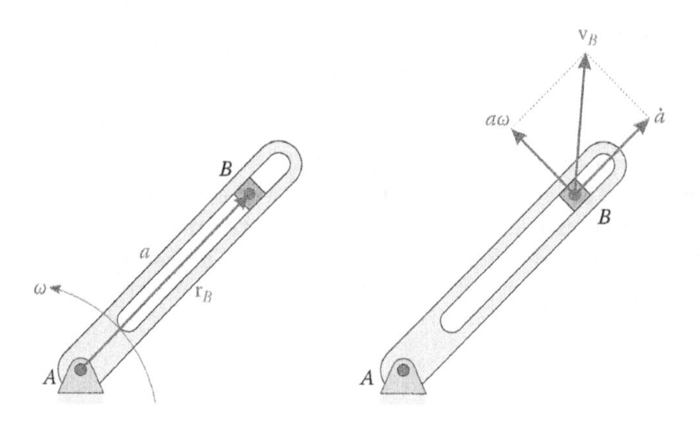

FIGURE 5.5
The square block slides within the slot at the same time as the link rotates.

In taking the time derivative, we note that the position vector is the *product* of two varying quantities; we must, therefore, employ the *product rule* from calculus. Given two functions, $u(x)$ and $v(x)$, the derivative with respect to x is

$$\frac{d}{dx}(u \times v) = \frac{du}{dx}v + \frac{dv}{dx}u \tag{5.10}$$

Thus, the velocity of point B is

$$\frac{d\mathbf{r}_B}{dt} = \frac{d}{dt}(a\mathbf{e}) = \frac{da}{dt}\mathbf{e} + a\frac{d}{dt}\mathbf{e} \tag{5.11}$$

The second term is the same derivative as before, and we solve it using the chain rule. In the first derivative, the term da/dt gives the rate of change of a; that is, it gives the speed of the block within the slot. To simplify our expressions, we will introduce *overdot notation*. A single dot over a quantity indicates differentiation with respect to time. Thus,

$$\dot{a} = \frac{da}{dt} \tag{5.12}$$

If we were to take the time derivative of a twice (e.g. to find acceleration) we would simply write

$$\ddot{a} = \frac{d^2a}{dt^2} \tag{5.13}$$

Note that the overdot refers only to *time* differentiation. It is never used to indicate differentiation with respect to another variable (e.g. x, θ, etc.). The velocity at point B is then

$$\mathbf{v}_B = \dot{a}\mathbf{e} + a\omega\mathbf{n} \tag{5.14}$$

In the first term, the velocity is in the same direction as the position vector. The velocity in the second term is perpendicular to the position vector, as before. The total velocity of B is the vector sum of these two as shown in the right-hand side of Figure 5.5. We will encounter a similar expression when we find the velocity of the threebar slider-crank later in this chapter. To summarize, we have found two expressions for the velocity at the end of a rotating link that is pinned to ground:

$$\begin{aligned} \mathbf{v}_B &= a\omega\mathbf{n} \text{ (length } a \text{ is constant)} \\ \mathbf{v}_B &= \dot{a}\mathbf{e} + a\omega\mathbf{n} \text{ (length } a \text{ is changing with time)} \end{aligned} \tag{5.15}$$

5.2 The Method of Instant Centers

We now turn our attention to a different graphical method of finding velocities: the method of instant centers. This method is based upon the following theorem:

Theorem 1:

There exists a point, I, common to two bodies in motion, which has the same instantaneous velocity.

As a simple example, consider the two links joined by a pin shown in Figure 5.6. It is obvious that the two links have the same velocity at point *B*, since that is where they are pinned together. The point *B* is called the *instant center* between the two links. A pin joint is *always* the instant center between two bodies pinned together, since the pin enforces identical motion at the joint.

This seems rather obvious, but interestingly, the theorem above applies to *any* two bodies on the plane – these may be connected or unconnected. In the case of unconnected bodies, the instant center may not lie directly on one of the bodies, as in the case of a pin joint, but may lie off in space somewhere. For some cases, e.g. the slider joint, an instant center may lie at infinity! More on this later.

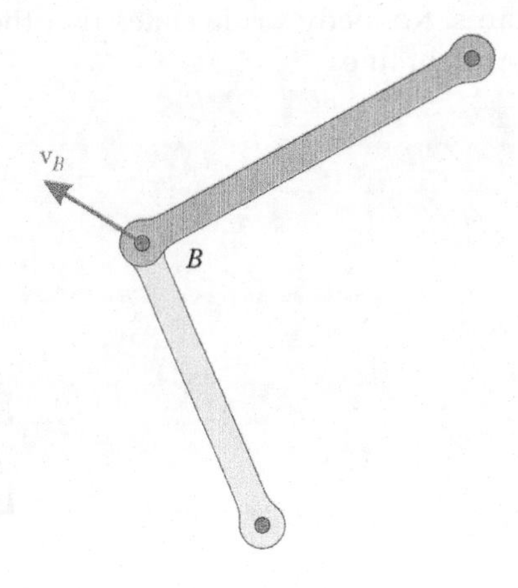

FIGURE 5.6
The two links have the same velocity at the pin joint.

Since there exists one instant center (IC) per pair of bodies, we may calculate the total number of ICs with the following formula

$$C = \frac{n(n-1)}{2} \tag{5.16}$$

where C is the number of instant centers and n is the number of bodies (or links). Thus, a fourbar linkage has six instant centers, and a sixbar linkage has 15. The reasoning behind this formula is also relatively straightforward. If we have a collection of n bodies, each body must connect with n–1 links (since it does not connect with itself). We divide by two because instant center, I_{ij}, is the same point as instant center, I_{ji}, where the subscripts indicate the two-paired bodies but order is irrelevant.

The concept of an instant center would be a relatively useless mathematical exercise if it were not for the following theorem, known as Kennedy's Rule:

Theorem 2:

Any three bodies in a plane will have exactly three instant centers and all three will lie in a straight line.

The first part of the theorem is obvious: if we apply $n = 3$ in Equation (5.16) we get three instant centers. The second part is not obvious at all, but is quite useful in finding instant centers that do not lie on pin joints. To see this, let us find the instant centers of a fourbar linkage.

5.2.1 Instant Centers of the Fourbar Linkage

The instant centers at the pin joints are easy to find, as shown in Figure 5.7. According to Equation (5.16) we may expect to find six instant centers for the fourbar linkage. The pin joints give us four and we can use Kennedy's rule to find the other two. Kennedy's rule applies to groups of three links each, so let us make a table of all the possible permutations of three links on the fourbar. Below each group of three, we will list all the possible permutations of two links in the group since an instant center occurs between two bodies (links). This is shown in Table 5.1.

All of the entries in bold are the instant centers that we have not found yet; that is, the ones that do not lie at pin joints. Kennedy's rule states that the three instant centers in a given column must lie on a straight line.

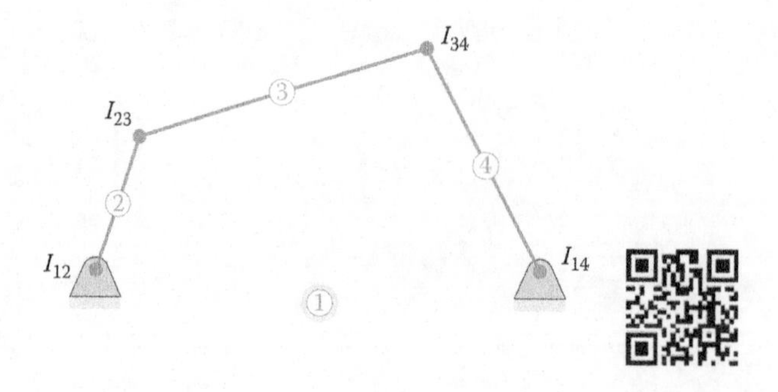

FIGURE 5.7
The pins on the fourbar linkage give four of its six instant centers. The instant center between bodies 1 and 2 is labeled I_{12}, etc.

TABLE 5.1

All Instant Centers of the Fourbar Linkage

123	124	134	234
I_{12}	I_{12}	$\mathbf{I_{13}}$	I_{23}
$\mathbf{I_{13}}$	I_{14}	I_{14}	$\mathbf{I_{24}}$
I_{23}	$\mathbf{I_{24}}$	I_{34}	I_{34}

The entries in bold are the ones that are not at pin joints.

For example, the instant centers for links 2,3,4 must lie on a straight line; as must the instant centers for 1,2,4. In each case, we know two of the instant centers, but by extending the lines between these, we can find the third. This is shown in Figure 5.8, where we have found the location of I_{24}.

The same procedure can be used to find I_{13}, the instant center between ground and link 3 (see Figure 5.9). This instant center is in fact the most important one for our purposes. Since I_{13} is the instant center between the coupler and ground, this means that (at this instant

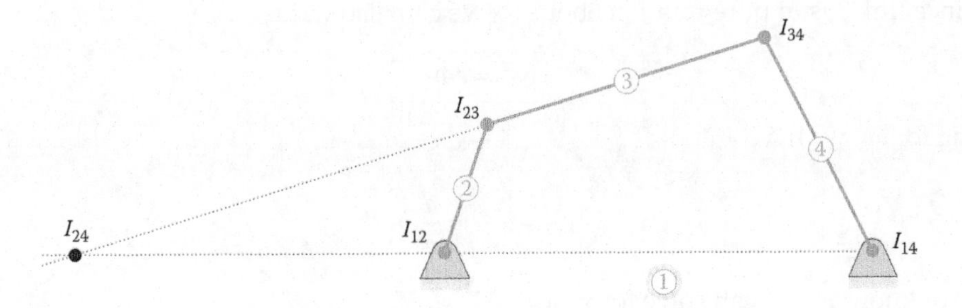

FIGURE 5.8
The instant center I_{24} lies at the intersection of the two lines shown.

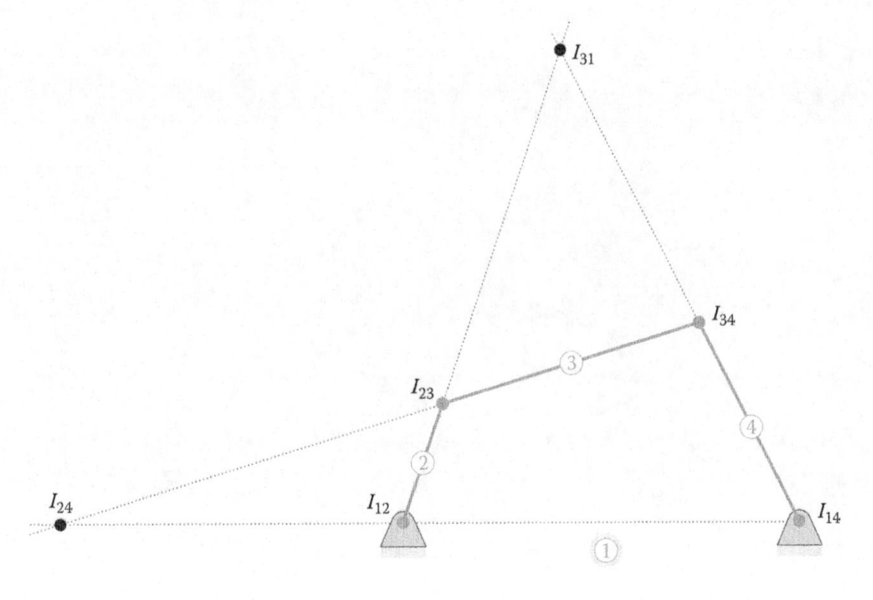

FIGURE 5.9
The location of I_{13} can be found by extending links 2 and 4.

in time) link 3 and ground have the same velocity at this point (which is zero, since the ground doesn't move). The only way for this to be possible is that link 3 is in *pure rotation* about I_{13} at this instant in time. We can use this fact to easily find the instantaneous angular velocity of link 3, since it is easy to move between translational velocities and angular velocities in the case of pure rotation.

Another way to state this is that *at this instant in time* we could pin link 3 to ground, and link 3 would be in pure rotation about this ground pivot. At the next instant in time the location of the instant center, I_{13}, will be different, and we could pin link 3 to ground at this new location. The same goes for the other instant centers: at the current instant in time, the instant center I_{24} gives the location where we could temporarily pin links 2 and 4 together; that is, link 2 is in pure rotation with respect to link 4 at I_{24}, but only at this instant in time. At the next instant in time, link 2 will be in pure rotation with respect to link 4 at a different location.

From Figure 5.10, we see that

$$|\mathbf{v}_B| = a\omega_2 \tag{5.17}$$

But, since link 3 is in pure rotation about I_{13}, we can also write

$$|\mathbf{v}_B| = e\omega_3 \tag{5.18}$$

Solving for ω_3, we have

$$\omega_3 = \frac{|\mathbf{v}_B|}{e} = \frac{a}{e}\omega_2 \tag{5.19}$$

Since we know ω_3, we can solve for $\mathbf{v}_C$ as

$$|\mathbf{v}_C| = f\omega_3 \tag{5.20}$$

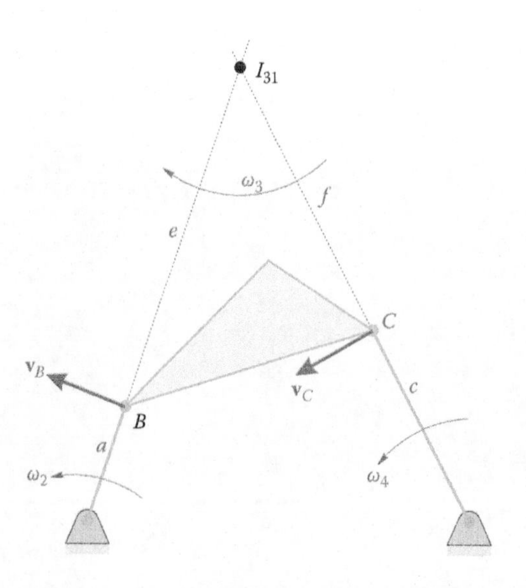

FIGURE 5.10
The instant center I_{13} can be used to find the remaining velocities on the fourbar, if the crank speed is known.

And finally,

$$\omega_4 = \frac{|\mathbf{v}_C|}{c} = \frac{f}{c}\omega_3 \tag{5.21}$$

Thus, the instant center method gives us a quick, graphical method for finding velocities on a linkage.

5.2.2 SOLIDWORKS® Tutorial – Velocity Analysis of the Fourbar Linkage

We will now present a simple method for using SOLIDWORKS to find the velocities of important points on the fourbar linkage using the method of instant centers. Figure 5.11 shows a fourbar linkage with crank length 1.5 m, coupler length 2.0 m, rocker length 2.25 m, and ground length 3.0 m. The crank is inclined 60° from the horizontal at this moment in time, and has angular velocity $\omega_2 = 10$ rad/s.

First, draw the linkage as shown in Figure 5.11. It is simplest to use a SOLIDWORKS Drawing for this purpose, although a sketch in a Part may be used as well. We have used the Layer toolbar to define four separate layers for the links: the linkage dimensions, the construction lines used to find the instant centers and the instant center dimensions. We have also placed a **Fix** relationship on the lower left ground pin to make the drawing fully defined.

Now draw construction lines (centerlines) in the direction of the crank and rocker lengths until they intersect at I_{13}. Control-click the crank and left construction line, and add a **Collinear** relationship. Do the same with the rocker and right construction line. Use an **Annotation** to label the instant center I_{13} (Figure 5.12).

Finally, add Smart Dimensions to the construction lines you just drew, as shown in Figure 5.13. We now have all the information needed to complete the velocity analysis. First, the velocity at point B is found

$$|\mathbf{v}_B| = \omega_2 a = (10\text{ rad/s})(1.5\text{ m}) = 15\text{ m/s}$$

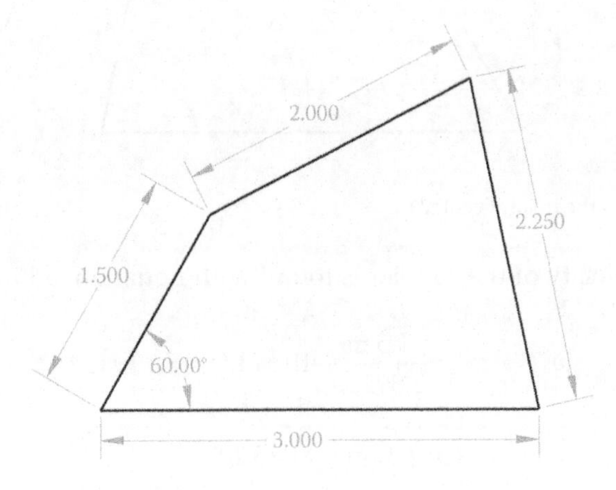

FIGURE 5.11
The fourbar linkage used in the tutorial.

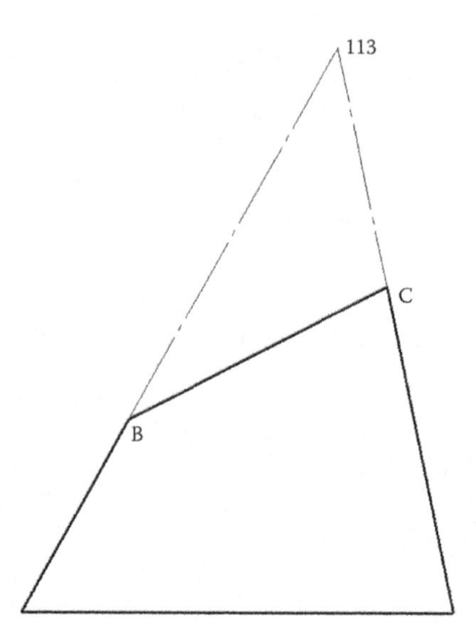

FIGURE 5.12
Construction lines used to find instant center I_{13}.

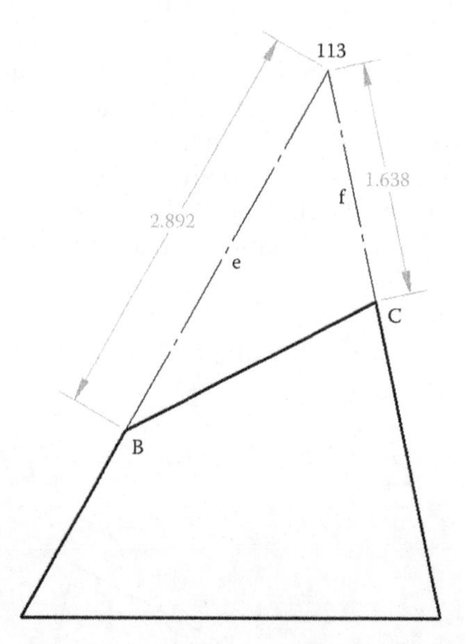

FIGURE 5.13
Finding the dimensions to the instant center I_{13}.

Next, the angular velocity of the coupler is found with Equation (5.19)

$$\omega_3 = \frac{a}{e}\omega_2 = \frac{1.5\text{ m}}{2.892\text{ m}} \cdot 10\text{ rad/s} = 5.19\text{ rad/s}$$

The velocity at point C is found using Equation (5.20)

$$|\mathbf{v}_C| = f\omega_3 = 1.638\text{ m} \cdot 5.19\text{ rad/s} = 8.50\text{ m/s}$$

And finally, the angular velocity of the rocker is found with Equation (5.21)

$$\omega_4 = \frac{f}{c}\omega_3 = \frac{1.638 \text{ m}}{2.25 \text{ m}} \cdot 5.19 \text{ rad} = 3.78 \text{ rad/s}$$

As you can see, the method of instant centers gives a quick and straightforward way of finding velocities on a linkage at a given instant in time.

5.2.3 Instant Centers of the Slider-Crank Linkage

Figure 5.14 shows a slider-crank linkage, with a vertically offset slider. The instant centers that lie at pin joints have been marked. The locations of the other instant centers are a little more challenging to find.

First, consider the piston inside the full-slider joint. An alternative way of looking at the joint is to consider it to be at the end of a rocker, whose pivot is an infinite distance above or below the slider joint. Recall that a link pinned to ground must trace out a circular arc as it moves. If the radius of the circular arc is sufficiently large (or infinite) then it approximates a straight line. This is shown in Figure 5.15, where the slider is connected to ground through a "virtual link" of infinite length (shown in gray). The instant center I_{14} is on a line perpendicular to the slider at infinity.

Consulting Table 5.1, we see that the instant center I_{13} must lie on the line connecting I_{14}–I_{34} and also on the line connecting I_{12}–I_{23}. The intersection of these two lines gives the location of I_{13}, as shown in Figure 5.16.

To find the last instant center, draw another line perpendicular to the slider through I_{12}. This line passes through the instant centers I_{14} (at infinity), I_{12} and I_{24}. Draw another line through I_{23} and I_{34}. The intersection of the two lines is I_{24}. It may strike you as odd that we shifted the perpendicular line to the left to find I_{24}. As it happens, we are making use of a little-known definition of parallel lines. It is common to define parallel lines as lines that never intersect, no matter how long we extend them. An alternative definition is that parallel lines *do* in fact intersect, but only at infinity. Thus, shifting the location of I_{14} to the right or left has no effect, since the length of the virtual link is infinite. We must take care, however, that the *direction* of the virtual link remains perpendicular to the direction of travel of the slider. If we march off to infinity in the wrong direction, we will not find the instant center I_{14}!

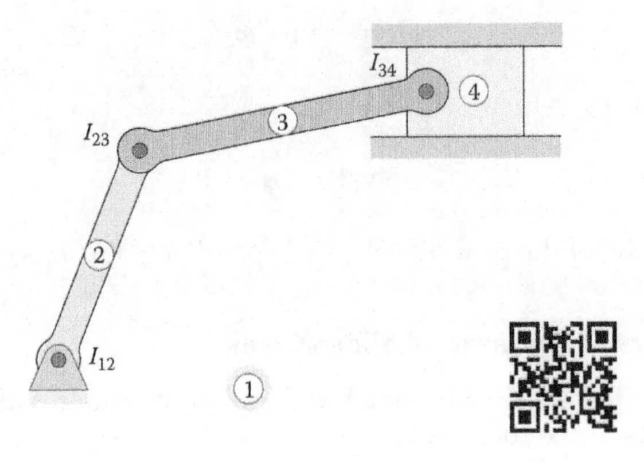

FIGURE 5.14
The pin joints on the slider-crank give three of the six instant centers.

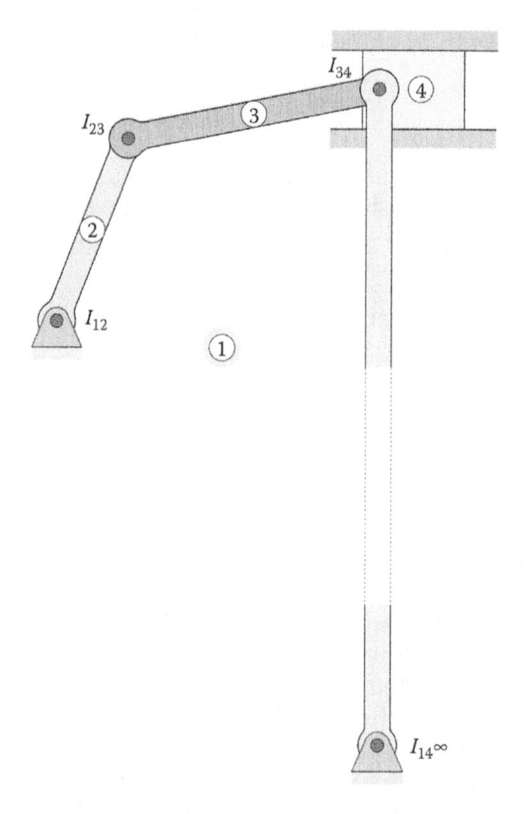

FIGURE 5.15
The slider is attached to a virtual link of infinite length. The virtual link is perpendicular to the slider.

We may now use the locations of the instant centers to find the velocities of the links on the slider-crank. As shown in Figure 5.18, the velocity at the end of the crank is

$$|\mathbf{v}_B| = a\omega_2 \tag{5.22}$$

Since the connecting rod is pinned to ground (at this instant) at I_{13}, we may adopt the same procedure as with the fourbar, and write

$$\omega_3 = \frac{|\mathbf{v}_B|}{e} = \frac{a}{e}\omega_2 \tag{5.23}$$

The velocity of the piston is then

$$|\mathbf{v}_C| = f\omega_3 \tag{5.24}$$

Of course, the direction of the piston's velocity is parallel to the axis of the cylinder.

5.2.4 Instant Centers of the Inverted Slider-Crank

The inverted slider-crank linkage is shown in Figure 5.19, along with the instant centers at the three pin joints. Since the links 3 and 4 are connected with a full-slider joint, they move in pure translation relative to each other. Because of this, there is no obvious location where the two links have the same velocity.

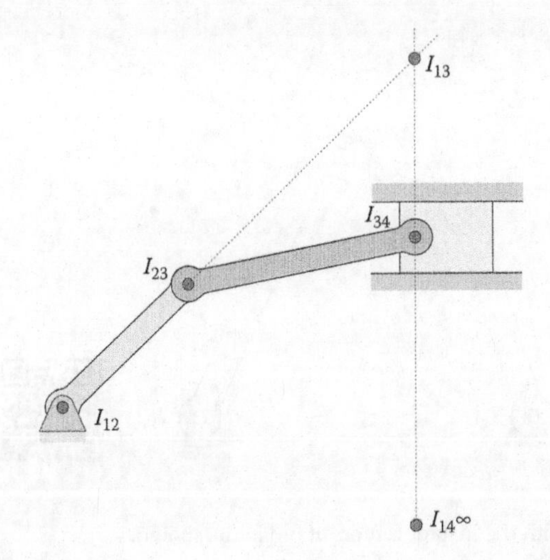

FIGURE 5.16
The instant center I_{13} is found at the intersection of the perpendicular line and link 2.

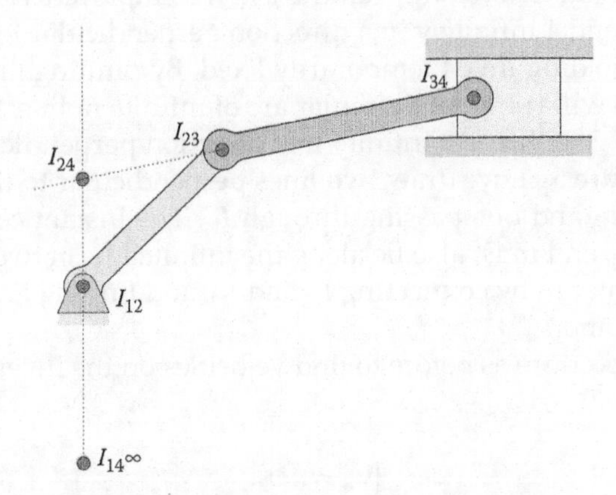

FIGURE 5.17
The instant center I_{24} lies at the intersection of the perpendicular line and link 3.

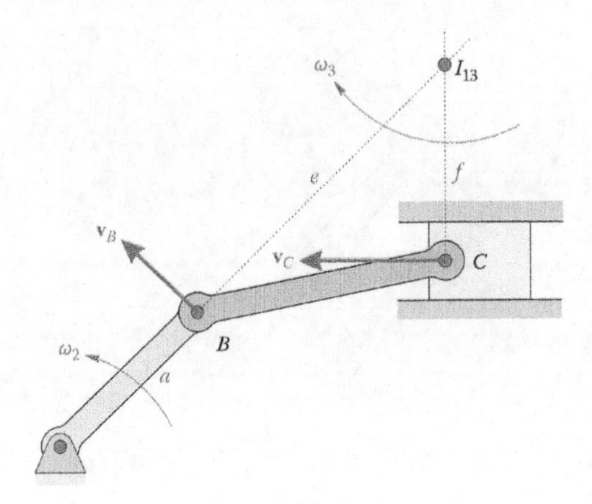

FIGURE 5.18
Finding the velocity of the slider using the instant center method.

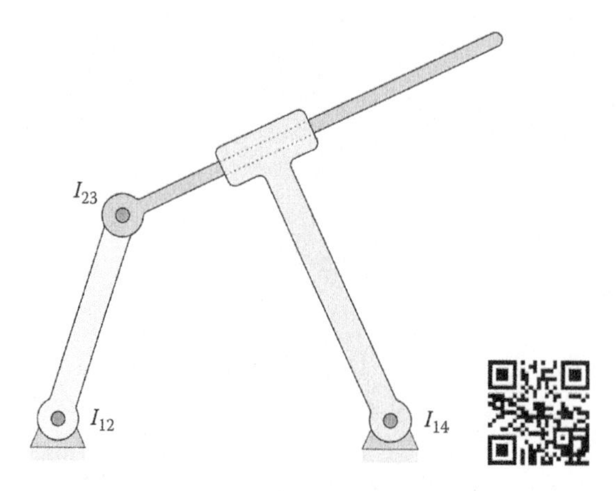

FIGURE 5.19
The inverted slider-crank, with the instant centers at pin joints shown.

However, we can make use of the same trick we employed for the slider-crank by extending the links 3 and 4 infinitely in a direction perpendicular to the slot, as shown in Figure 5.20. Imagine holding link 4 temporarily fixed. By pinning link 3 to link 4 at infinity any point on link 3 will trace out a circular arc of infinite radius; that is, a straight line. Thus, we may pin link 3 to link 4 at infinity in a direction perpendicular to the slot.

Thus, as seen in Figure 5.21, we draw two lines perpendicular to the slot out to infinity: one passing through I_{23} and one passing through I_{14}. The instant center I_{24} must lie on a line between I_{12} and I_{14}, and must also lie along the infinite line between I_{23} and I_{34}. Finally, we find the instant center I_{13} by connecting I_{12} and I_{23}, and finding its intersection with the line going between I_{14} and I_{34}

We use the same procedure as before to find velocities on the inverted slider-crank. The velocity at point B is

$$|\mathbf{v}_B| = a\omega_2. \tag{5.25}$$

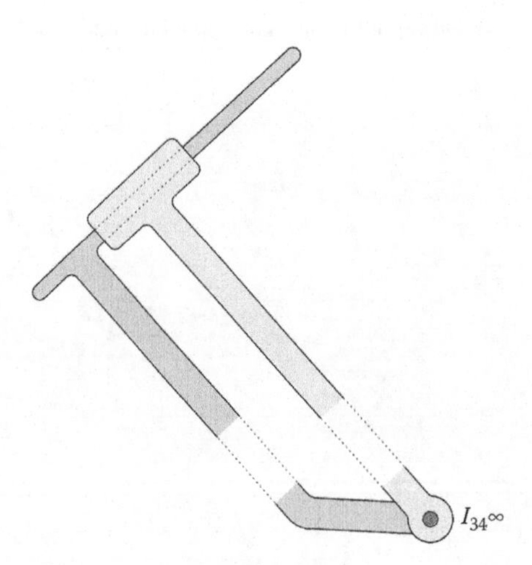

FIGURE 5.20
By extending links 3 and 4 infinitely in a direction perpendicular to the slot, we can pin them together.

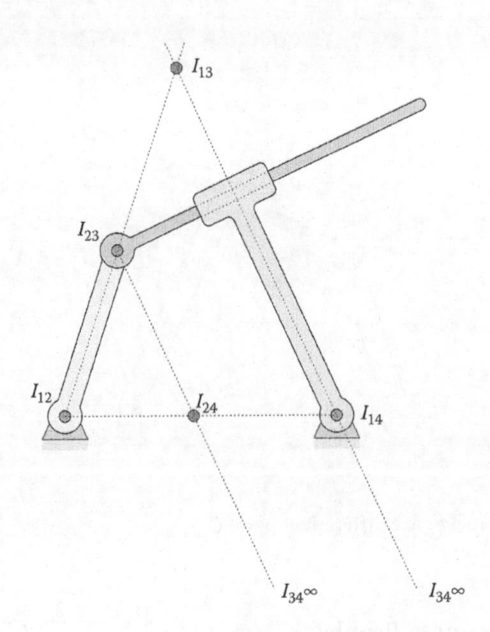

FIGURE 5.21
The remaining two instant centers are found in a similar fashion as the slider-crank.

and the angular velocity of the slider is

$$\omega_3 = \frac{|\mathbf{v}_B|}{e} = \frac{a}{e}\omega_2 \tag{5.26}$$

Because the slider and rocker are connected with a full-slider joint, they cannot rotate relative to one another. Therefore, the angular velocity of the rocker is the same as that for the slider.

$$\omega_4 = \omega_3 \tag{5.27}$$

Despite having the same *angular* velocity, the *translational* velocity of the rocker and slider are different at point C; that is, they slide relative to one another (see Figure 5.22). The velocity of the slider at point C is

$$|\mathbf{v}_{C3}| = \omega_3 f \tag{5.28}$$

but the velocity of the rocker at point C is

$$|\mathbf{v}_{C4}| = \omega_4 c \tag{5.29}$$

We are sometimes interested in the relative velocity between the two links at the point C, known as the *velocity of slip*:

$$|\mathbf{v}_{slip}| = \big||\mathbf{v}_{C4}| \pm |\mathbf{v}_{C3}|\big| \tag{5.30}$$

The velocities are added if they point in opposite directions, and subtracted if they point in the same direction. The direction of each velocity must be determined through examination of the direction of the angular velocities in the diagram.

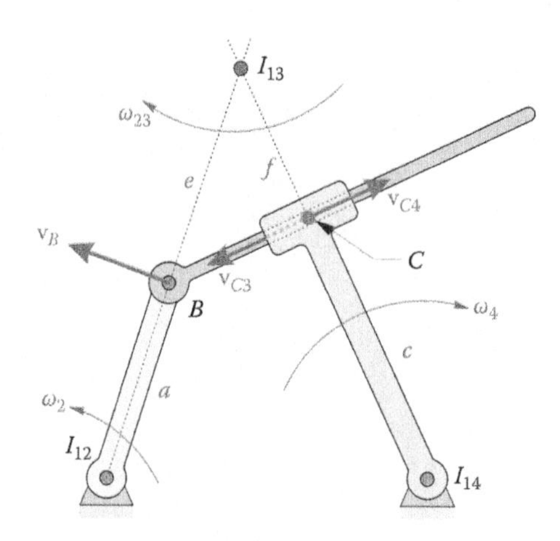

FIGURE 5.22
The velocity of the slider and rocker are different at point *C*.

5.2.5 Instant Center Example Problems

Example 5.1: Gears in Mesh

Figure 5.23 shows a pair of gears in mesh with one another. Since there are three bodies (gear 2, gear 3, and ground) we expect to find three instant centers. The easiest to locate, as always, are at the two pin joints. The third lies at the point of contact of the two gears, since both gears have the same velocity at this point. Note that all three lie on a straight line, as dictated by Kennedy's rule.

Example 5.2: Cam-Follower Mechanism

Figure 5.24 shows a typical cam-follower mechanism. Since there are four bodies in the mechanism, we must find six instant centers. The pin joints form three instant centers. The point of contact between the roller and the cam is another instant center, I_{34}, if we assume that there is no slip between the roller and cam (see next example problem). Using Kennedy's rule enables us to find the remaining two instant centers.

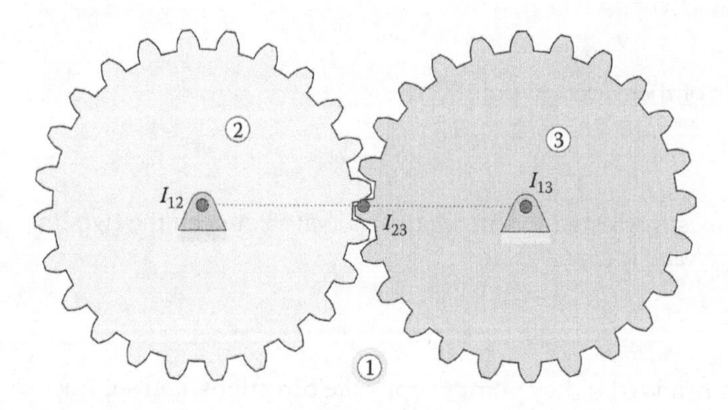

FIGURE 5.23
Two gears in mesh have three instant centers.

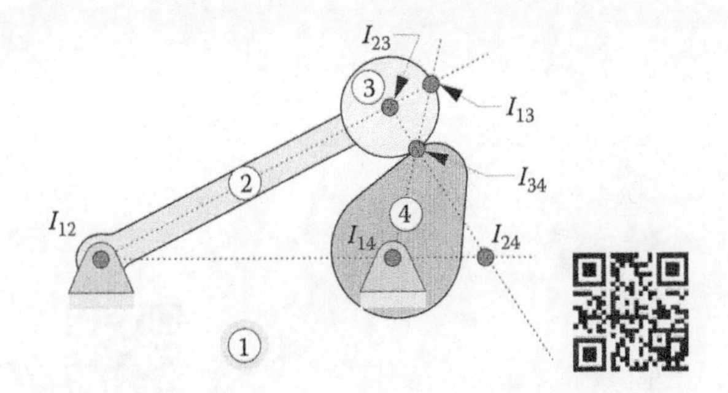

FIGURE 5.24
The cam-follower mechanism has six instant centers.

Example 5.3: Cam-Follower Mechanism, Part 2

Figure 5.25 shows a cam-follower mechanism in which the follower slides along the surface of the cam, instead of rolling without slipping. Since there are only three bodies, we must find three instant centers. Your first thought might have been to place an instant center at the point of contact between the cam and follower, as in the previous example. This would be incorrect, for two reasons:

1. The cam slides on the follower as it rotates, which means that the velocities of the two bodies are different at the point of contact.
2. Placing an instant center at the point of contact would violate Kennedy's rule, since the three instant centers would not be in a straight line.

As with the slider-crank and inverted slider-crank mechanisms, we must draw a line perpendicular to the line of contact, and instant center I_{23} lies at the intersection of this line and the ground line.

Example 5.4: The Threebar Linkage

Figure 5.26 shows the threebar linkage that consists of a crank, slider, and ground. Since there are three links, we expect to find three instant centers. The first two are simple to find, since the crank has two pins. The third, unfortunately, is not so simple. Your first thought might be to place the third instant center at the pin joint between the slider and ground, but since the slider is free to slide along this pin, it does not have the same (zero) velocity as the ground at this point. Kennedy's rule states that the three instant centers must lie on a straight line, so we should find the third instant center somewhere on the dashed line shown in Figure 5.26.

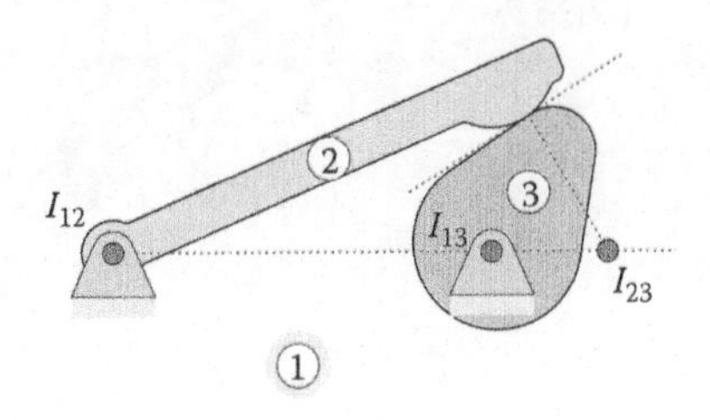

FIGURE 5.25
This cam mechanism has a follower that does not roll.

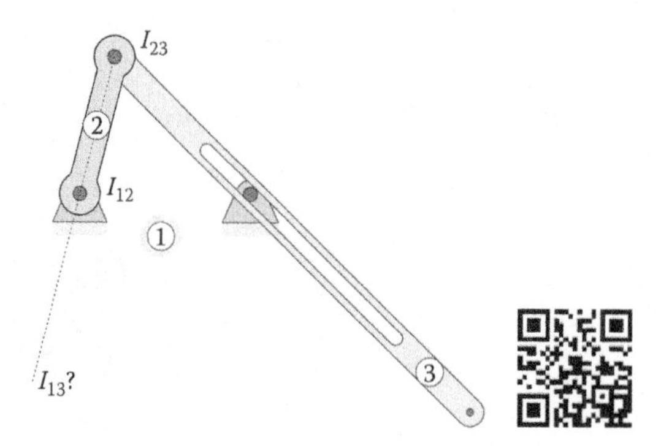

FIGURE 5.26
The threebar linkage has three instant centers, two of which are at the pin joints on the crank.

As we have seen in the earlier linkages, the instant center I_{13} represents the point that the link 3 (the slider) may be considered pinned to ground at this instant in time – that is, the slider is in pure rotation about I_{13} at this moment. We also know, from the analysis of the inverted slider-crank, that the instant center I_{13} must lie on a line perpendicular to the slot. It is not immediately obvious, however, where in the slot we should begin the perpendicular line.

As a first guess, let us draw a line perpendicular to the slot starting at the arbitrary point P, as shown in Figure 5.27. Since body 3 is pinned to ground at I_{13} at this instant in time, the velocities on body 3 must be perpendicular to radial lines extending from I_{13}. Thus, the velocity at point P is parallel to the slot, but the velocity at point D (where the pin is located) has a component normal to the slot. Because the slot is pinned to ground at point D, there can be no velocity component normal to the slot, and so we have reached a contradiction.

If we construct a virtual link CD that pins the slider to ground at I_{13} (see Figure 5.28), we see that the velocity of the slider at D must be perpendicular to CD since the slider is in pure rotation about point C at this instant in time. Since CD is perpendicular to the slot, the velocity $\mathbf{v}_D$ is parallel to the slot, as required. Thus, the instant center I_{13} lies at the intersection of the crank line and a line perpendicular to the slot that passes through ground pin D. While Kennedy's rule may seem straightforward at first glance, the *reasoning* behind it may sometimes be used to locate "tricky" instant centers.

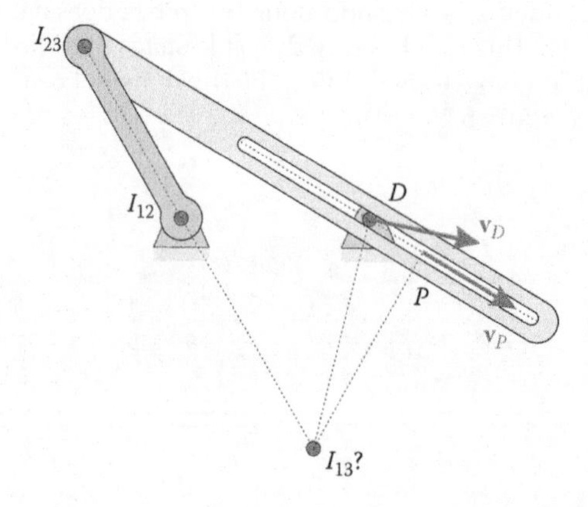

FIGURE 5.27
The instant center I_{13} has been placed (incorrectly) on an arbitrary line perpendicular to the slot.

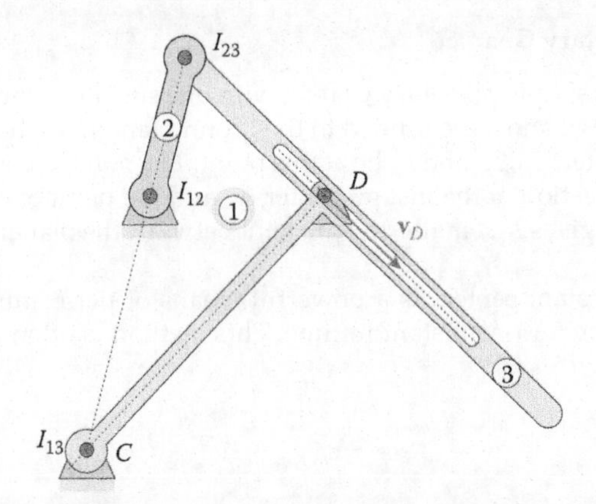

FIGURE 5.28
Drawing a line perpendicular to the slot and intersecting the ground pin D lets us find the instant center I_{13}.

Example 5.5: The Geared Fivebar Linkage

A geared fivebar linkage is shown in Figure 5.29. Since there are five bodies, we must find 10 instant centers. The five pins provide five instant centers, and the point of contact between the two gears provides a sixth. We may use bodies 235 and 345 to find the location of I_{35}, and then use bodies 234 and 245 to find I_{24}.

The remaining two instant centers are found as shown in Figure 5.30. Since these are the instant centers with ground, they would most likely be the "important" ones used for velocity analysis.

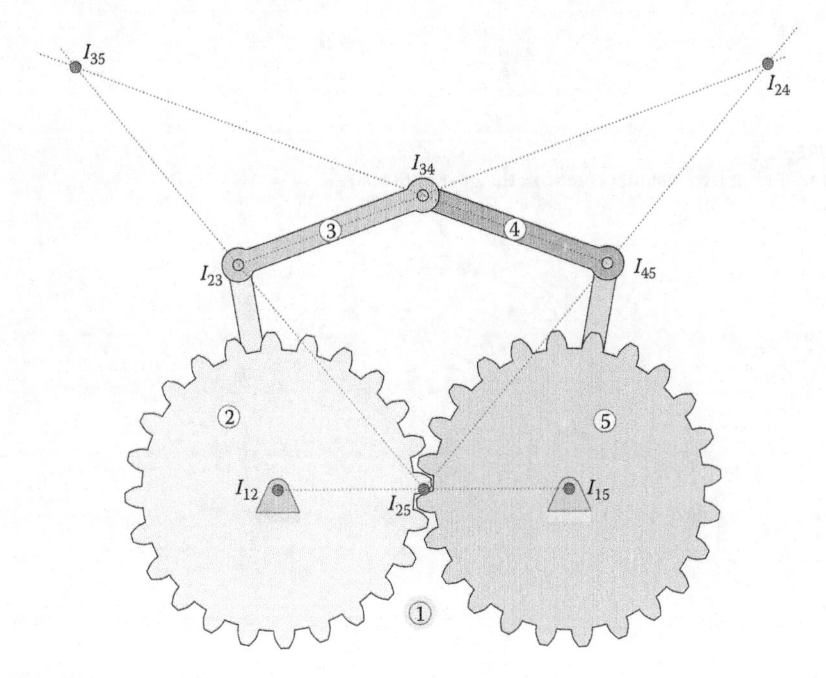

FIGURE 5.29
The geared fivebar linkage has ten instant centers.

Example 5.6: Planetary Gearset

Figure 5.31 shows a simple planetary gearset with one sun, one planet, and a fixed ring gear. Because bodies 2 and 4 are pinned to the ground (and to each other) at the central axis, the instant centers I_{12}, I_{14}, and I_{24} lie at this point. The point of contact between gears 2 and 3 gives the location of the instant center I_{23}, and the point of contact between the planet and the ring gives I_{13}. Finally, the pin joint between the planet and bar 4 gives the location of I_{34}.

The method of instant centers is a powerful means of performing a quick velocity analysis on a linkage for one instant in time. This method is often used in automotive

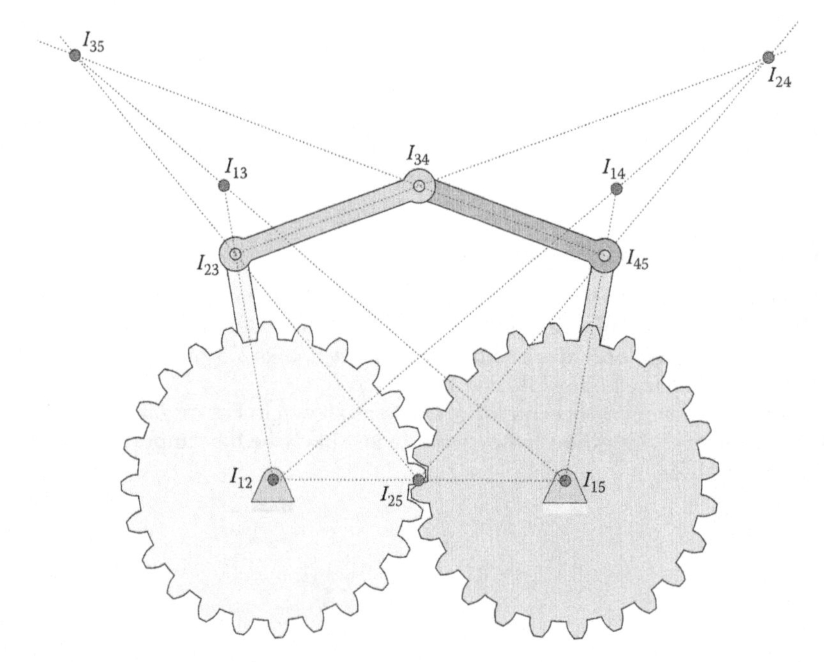

FIGURE 5.30
The remaining two instant centers on the geared fivebar.

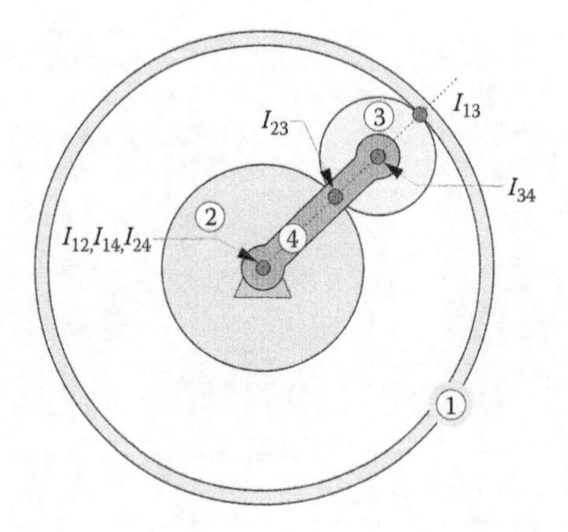

FIGURE 5.31
This simple planetary gearset has six instant centers, three of which are located at the central axis.

suspension design, and can also be used as a "reality check" when conducting the analytical velocity calculations that we will develop in later sections. Some instant centers (e.g. pin joints) are simple to find, but others, particularly where sliding is involved, require more careful consideration.

5.2.6 Velocity Ratios

Now is a good time to pause and reconsider Kennedy's rule in a slightly different way.

> Any three bodies in a plane will have exactly three instant centers and all three will lie in a straight line.

In the previous section, we made use of the idea that the instant center I_{13} gives the point about which the coupler (link 3) is in pure rotation. We were able to use this to find the velocities at both moving pins, and the angular velocities of links 3 and 4. We will now develop a "shortcut" to finding the angular velocity of the rocker (link 4) without first calculating ω_3. We will use this technique in the next section to find the *mechanical advantage* of the fourbar linkage.

Consider the fourbar linkage shown in Figure 5.32. In the previous section, we learned that the instant center I_{24} lies at the intersection of lines drawn between I_{23}–I_{34} and I_{12}–I_{14}. The point I_{24} is the point where links 2 and 4 have the *same velocity* at this instant in time. In order for two velocities to be equal, their magnitudes and *directions* must be the same.

Both links 2 and 4 are in pure rotation about their respective ground pins. Thus, the velocity at any point must be perpendicular to a line drawn radially from the center of rotation. A simple example is shown in Figure 5.33: since the links extend radially outward from the ground pins, the velocities $\mathbf{v}_B$ and $\mathbf{v}_C$ are perpendicular to the links.

The next example is a little more complicated – we have expanded the crank and rocker into triangular shapes without changing the critical dimensions (*a*, *b*, *c*, and *d*) of the fourbar linkage. The same principle holds; however, the velocities point in a direction perpendicular to lines drawn radially from the center of rotation. In Figure 5.34, the velocities at points *E* and *F* are clearly in different directions, and cannot be equal. The problem is then to find a locus of points where the velocity vectors point in the same direction.

Part of the solution to the puzzle is shown in Figure 5.35. Since the radial lines are collinear with the line between the two ground pins, both velocity vectors must point perpendicular to this line, and thus in the same direction. In fact, the line between ground

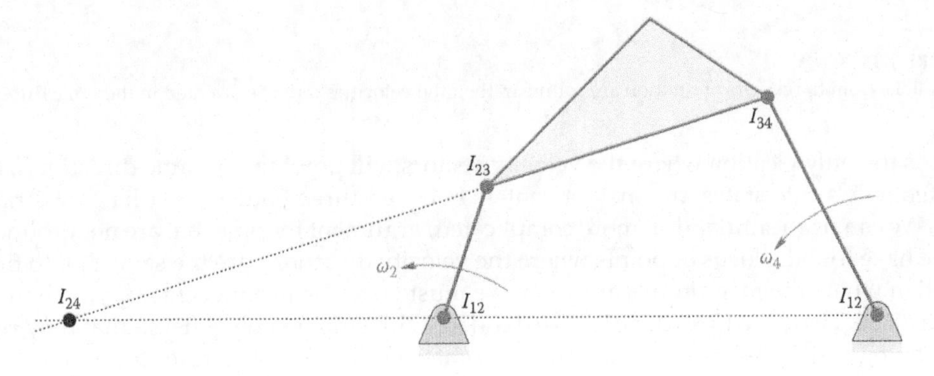

FIGURE 5.32
The instant center I_{24} is found at the intersection of lines I_{23}–I_{34} and I_{12}–I_{14}.

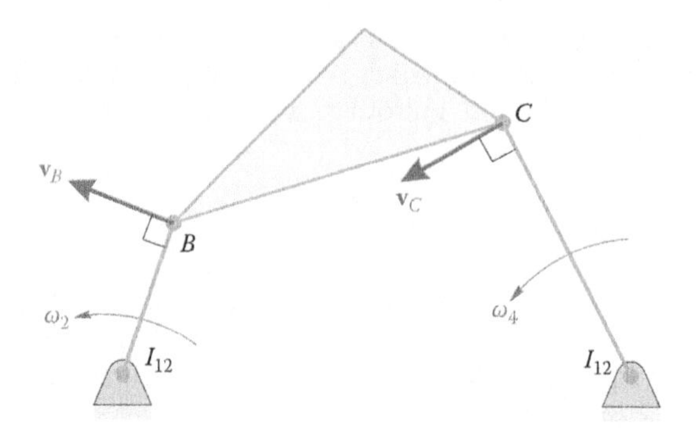

FIGURE 5.33
The velocities $\mathbf{v}_B$ and $\mathbf{v}_C$ are perpendicular to the lines drawn to their respective centers of rotation.

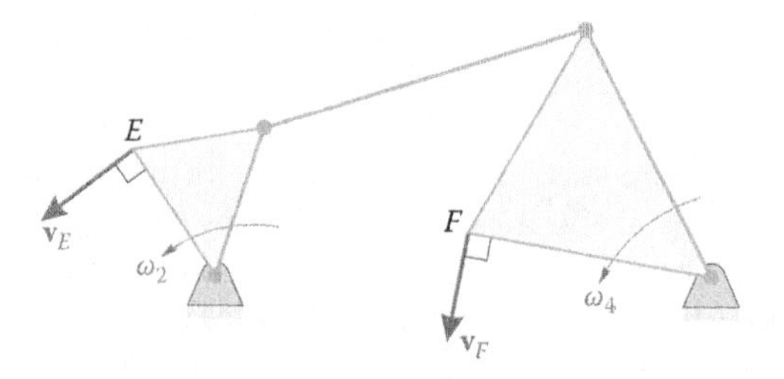

FIGURE 5.34
The crank and rocker have become triangles, but the velocities are still normal to the lines to the centers of rotation.

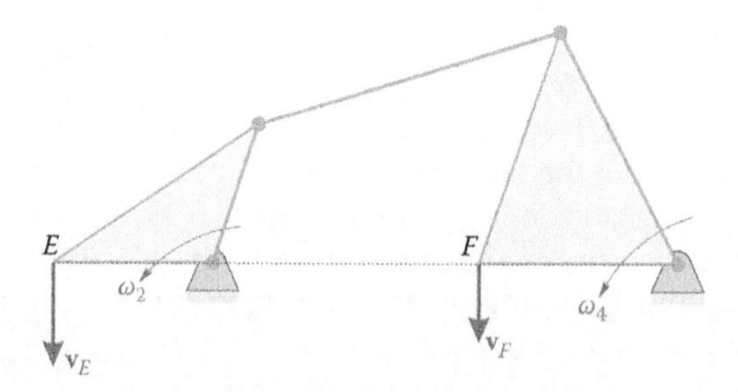

FIGURE 5.35
If the lines from the centers of rotation are collinear, then the velocities must be oriented in the same direction.

pins is the only location where the velocity vectors will point in the same direction. Thus, as Kennedy's rule states, the instant centers between three bodies must lie on a straight line. We can use a similar (but more complicated) argument for pins that are not grounded.

We have found a locus of points where the velocity directions are the same, but to find a location where the magnitudes are equal we must go to the instant center I_{24}, as shown in Figure 5.36. Here, the links have been extended (with strange polygons) so that they reach I_{24}. Owing to the definition of the instant center (and Kennedy's rule) the translational velocity of both links must be equal at this point. For simplicity, let us denote the instant center I_{24} as point E. Then

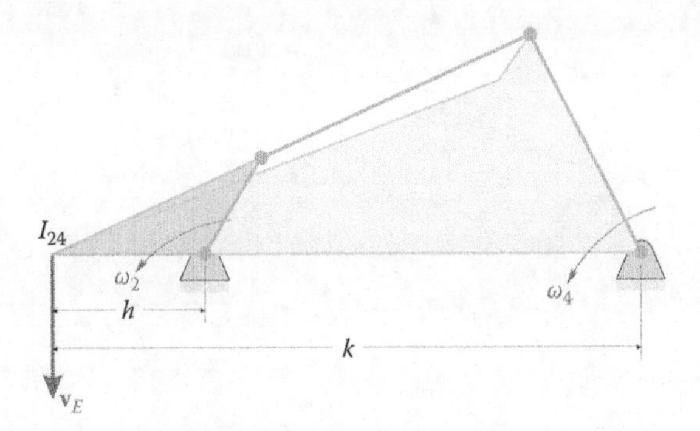

FIGURE 5.36
Extend the crank and rocker until they meet at I_{24}.

$$\mathbf{v}_{2E} = \mathbf{v}_{4E} = \mathbf{v}_E$$

Since each link is in pure rotation about its ground pin, we can easily calculate the velocity as

$$|\mathbf{v}_E| = \omega_2 h = \omega_4 k$$

Thus, if we know the location of instant center I_{24}, we can easily calculate the angular velocity of the rocker

$$\omega_4 = \frac{h}{k}\omega_2 \tag{5.31}$$

In this way, we have found a "shortcut" to calculating the angular velocity of the rocker, without the intermediate step of finding the angular velocity of the coupler. While this may not seem like such an important achievement, we will soon find a use for it in calculating mechanical advantage in the next section.

5.2.7 Mechanical Advantage

Another use for the method of instant centers is in calculating the mechanical advantage produced by a linkage. In many cases, we wish to increase the force delivered by the linkage to a load, as in the rock crusher linkage shown in Figure 5.37. The purpose of the rock crusher is to impart sufficient force to a rock to crush it. The input force to the linkage is applied at the crank pin, and the output force is taken in the middle of the rocker. In Figure 5.37 we have two different configurations of the same linkage. On the left, the input link (the crank) is nearly perpendicular to the coupler, while in the linkage at right, the input link is nearly parallel with the coupler. In the analysis to follow, we assume that the input and output forces are perpendicular to the input and output links. The distances from the ground pivots to the points of application of the forces are r_{in} and r_{out}, so that the input and output torques are:

$$T_{in} = F_{in} r_{in} \quad T_{out} = F_{out} r_{out} \tag{5.32}$$

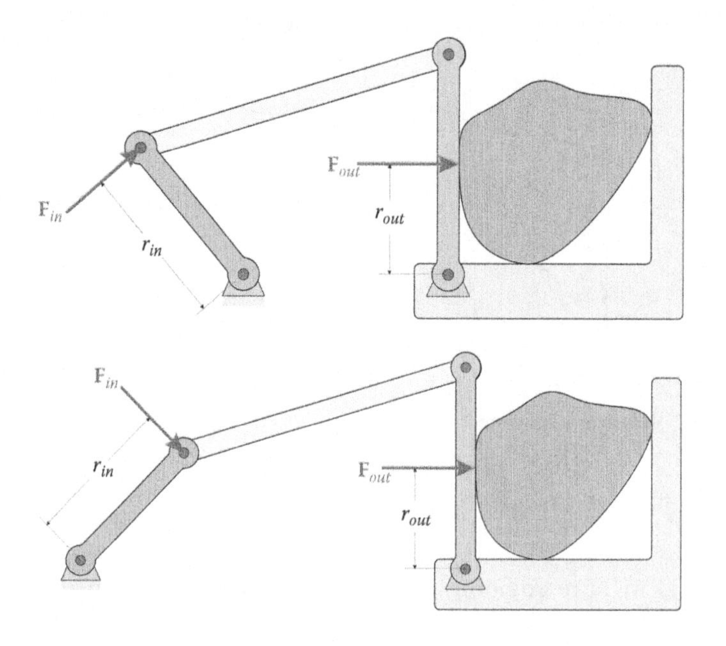

FIGURE 5.37
Two possible designs for a "rock crusher". In the top figure, the crank and coupler are nearly perpendicular. In the bottom figure they are nearly parallel.

The *power* transmitted by a linkage can be found in two ways. First, we may multiply output force by output velocity:

$$P_{out} = F_{out} V_{out} \tag{5.33}$$

Alternatively, we may multiply output torque by output angular velocity

$$P_{out} = T_{out}\omega_{out} = T_{out}\omega_4 \tag{5.34}$$

Losses in linkages are typically quite small; here we assume that all input power is transmitted to the output

$$P_{out} = P_{in} \tag{5.35}$$

so that the input and output powers are equal

$$T_{out}\omega_4 = T_{in}\omega_2 \tag{5.36}$$

Substituting the expressions for torque given in Equation (5.32), we have

$$F_{out} r_{out}\omega_4 = F_{in} r_{in}\omega_2 \tag{5.37}$$

Let us define the *mechanical advantage* of the linkage as the ratio of output force to input force. In the case of the rock crusher, this would provide a measure of the usefulness of the linkage, since we wish to achieve the highest possible output force (for crushing rocks) with the smallest possible input force.

$$\frac{F_{out}}{F_{in}} = \frac{r_{in}}{r_{out}}\left(\frac{\omega_2}{\omega_4}\right). \tag{5.38}$$

This is the same velocity ratio that we derived in the previous section. Using Equation (5.31), we find that

$$\frac{F_{out}}{F_{in}} = \frac{r_{in}}{r_{out}}\left(\frac{k}{h}\right) \tag{5.39}$$

We can use Equation (5.39) to design linkages with considerable mechanical advantage. Consider the rock crusher linkages shown in Figure 5.38, where the ratio of k/h is greater than one. This indicates that output force will be greater than input force (assuming that the ratio r_{in}/r_{out} is greater than or equal to one). In the linkage on the right, however, the value of h is very small, which will make the ratio k/h very large. This linkage has considerable force amplification and would make crushing the rock that much easier.

Now consider what happens when the crank and coupler are aligned, as shown in Figure 5.39. Here, the instant center I_{24} is coincident with the fixed pivot O_2. The force multiplication for this linkage is infinite! Of course, deformation of links and "slop" in the pins will prevent infinite forces from occurring, but the force multiplication is quite significant.

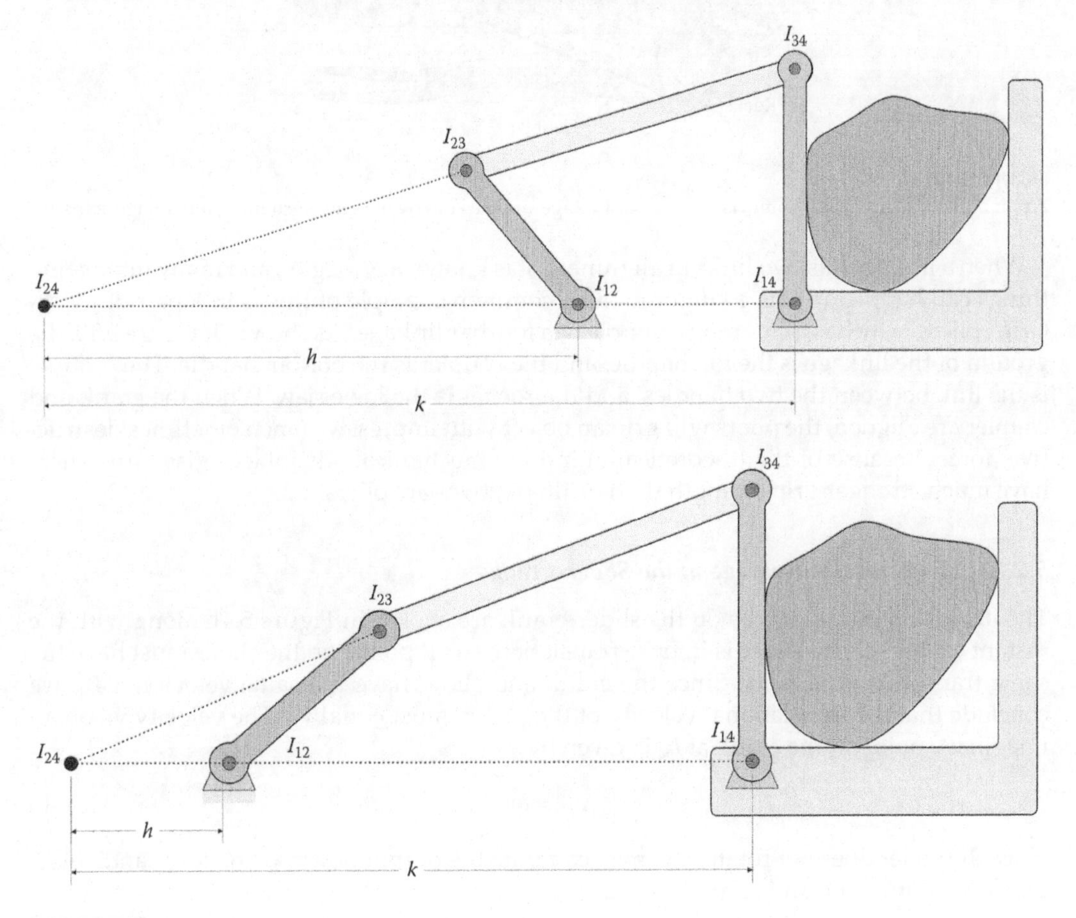

FIGURE 5.38
The ratio k/h is quite different for the two rock crushers shown.

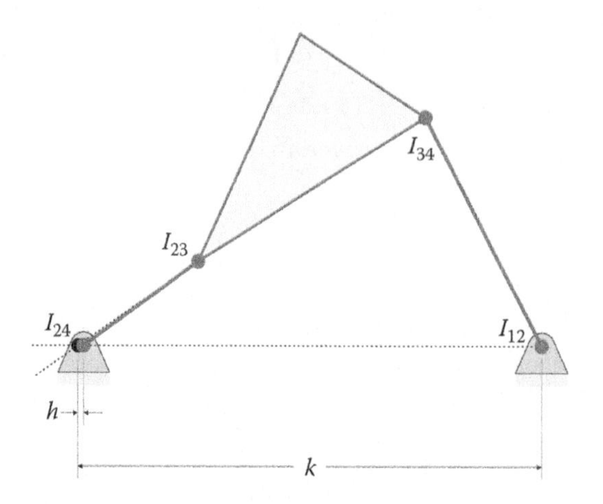

FIGURE 5.39
When the crank and coupler are collinear, the mechanical advantage is (theoretically) infinite.

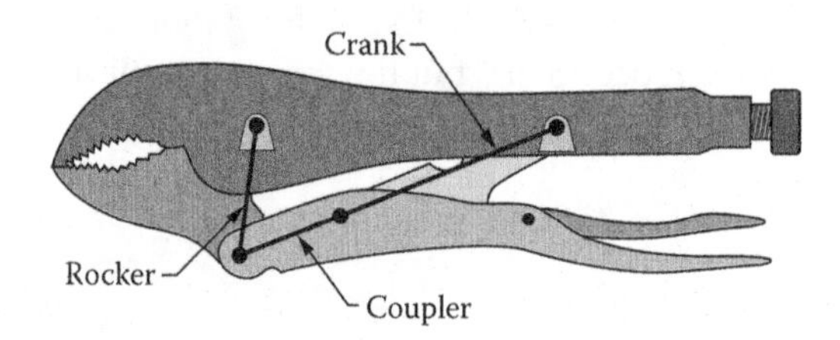

FIGURE 5.40
A pair of "Vise-Grip" pliers comprises a fourbar linkage, and has nearly infinite mechanical advantage at toggle.

When a linkage has two links in alignment, it is known as "toggle", and many interesting things can happen in this configuration. A common example of this is in a pair of "Vise-Grip" pliers, which is a common example of a fourbar linkage. As shown in Figure 5.40, the ground of the linkage is the top handle, and the coupler is the bottom handle. The "crank" is the link between the two handles, and the rocker is the lower jaw. When the crank and coupler are aligned, the pliers will grip an object with impressive (and sometimes destructive) force. Because of the (theoretically) infinite mechanical advantage, Vise-Grip pliers have much stronger grip strength than ordinary, two-jaw pliers.

5.2.7.1 Mechanical Advantage in the Slider-Crank

The input and output forces on the slider-crank are shown in Figure 5.41, along with the instant center I_{24}. The slider is in pure translation, so all points on the slider must have the same translational velocity. Since the crank and slider have the same velocity at I_{24}, we conclude that the translational velocity of the piston must equal $\mathbf{v}_D$. The velocity $\mathbf{v}_D$ on an imaginary point on the crank at I_{24} is given by

$$|\mathbf{v}_D| = \omega_2 h \tag{5.40}$$

Since the slider does not rotate, we must compute the output power using force and velocity, instead of torque and angular velocity.

$$P_{out} = F_{out}|\mathbf{v}_D| \tag{5.41}$$

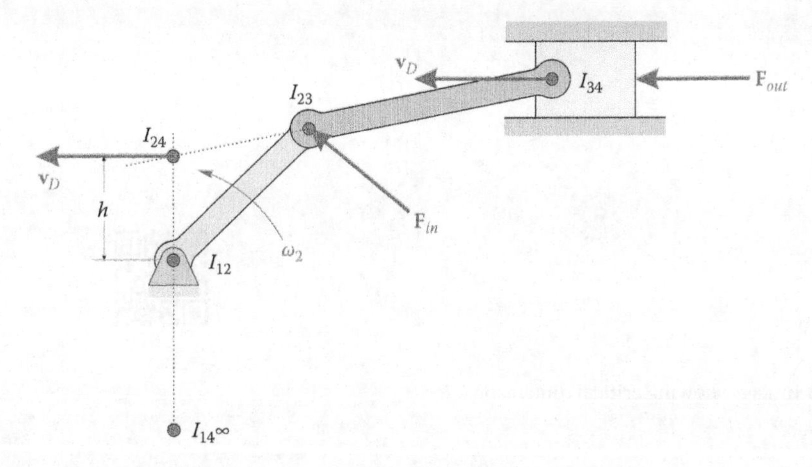

FIGURE 5.41
The slider-crank linkage showing input and output forces.

If we apply the input force at the end of the crank as shown in the figure, then the input power is

$$P_{in} = F_{in}\omega_2 a \tag{5.42}$$

Thus, the mechanical advantage of the slider-crank is

$$\frac{F_{out}}{F_{in}} = \frac{a}{h} \tag{5.43}$$

As before, if the crank and connecting rod are collinear, the mechanical advantage becomes very large or infinite.

5.3 Velocity Analysis of the Threebar Slider-Crank

While the instant-center method shown earlier can be used to easily find the velocities on a linkage at a single point in time, it is tedious to use if the velocities over the entire range of motion are needed. For this case, it is much more efficient to use a vectorial/algebraic method. Here we will take advantage of the ability of MATLAB to solve matrix equations with a minimum of effort on the part of the user. In addition, MATLAB's plotting capabilities will allow us to easily visualize the velocities during the entire motion of the linkage. The tradeoff, of course, is that some physical intuition is lost during this process, since the method is purely algebraic.

Figure 5.42 shows the threebar linkage with its critical dimensions. Figure 5.43 shows a vector loop diagram of the threebar linkage, similar to the one we constructed in Chapter 4. As before, we simply add the vectors going around the loop. Since we end up where we started, the vector sum is zero.

$$\mathbf{r}_2 + \mathbf{r}_3 - \mathbf{r}_1 = 0 \tag{5.44}$$

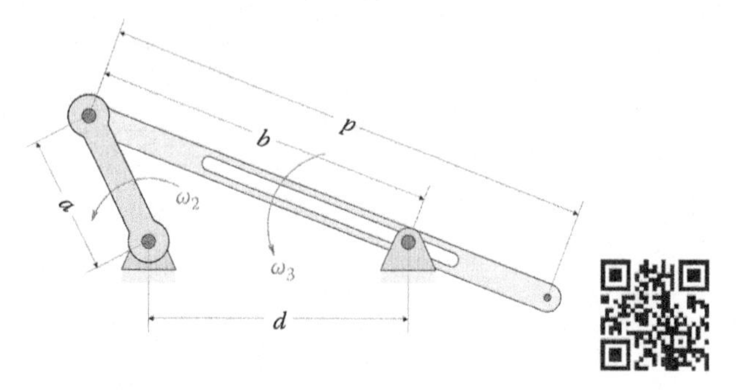

FIGURE 5.42
The threebar linkage showing critical dimensions.

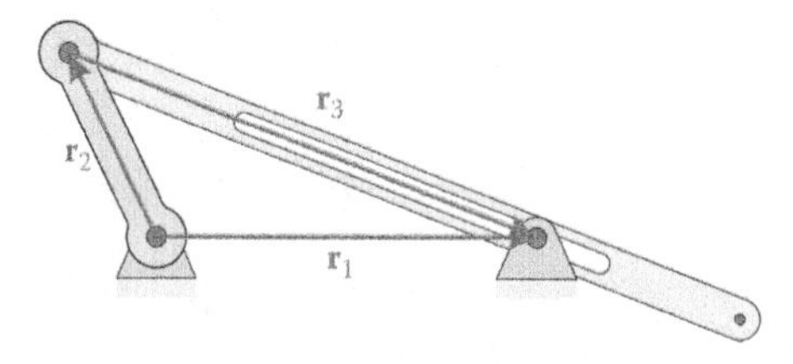

FIGURE 5.43
Vector loop diagram of the threebar linkage.

If we expand this into unit vector form, we have

$$a\mathbf{e}_2 + b\mathbf{e}_3 - d\mathbf{e}_1 = 0 \tag{5.45}$$

Before we begin the velocity analysis, we must first have conducted a position analysis as was done in Chapter 4. In the sections that follow, we assume that the position angle θ_3 and slider length b are known for every crank angle θ_2. To begin the velocity analysis, we take the time derivative of Equation (5.2). As discussed in Section 4.2.5, we must employ the chain rule of differentiation to take the derivative.

$$\frac{d}{dt}(a\mathbf{e}_2) = a\omega_2\mathbf{n}_2 \tag{5.46}$$

The derivative of the second term is slightly more complicated, since both b and $\mathbf{e}_3$ are functions of time. We must, therefore, use the *product rule* to differentiate it, since it is a product of two functions of time:

$$\frac{d}{dt}(u{\cdot}v) = \frac{du}{dt}v + u\frac{dv}{dt} \tag{5.46}$$

Applying this to the second term gives

$$\frac{d}{dt}(b\mathbf{e}_3) = \frac{db}{dt}\mathbf{e}_3 + b\omega_3\mathbf{n}_3 \tag{5.47}$$

Let us define

$$\dot{b} = \frac{db}{dt} \tag{5.48}$$

so that

$$\frac{d}{dt}(b\mathbf{e}_3) = \dot{b}\mathbf{e}_3 + b\omega_3\mathbf{n}_3 \tag{5.49}$$

The term $\dot{b}$ represents how quickly the length between points B and D is changing, or how quickly the slider is moving past the fixed pin at D. Accordingly, we will denote this quantity the *velocity of slip*. Differentiating the final term in Equation (5.45) gives

$$\frac{d}{dt}(d\mathbf{e}_1) = 0 \tag{5.50}$$

since both d and $\mathbf{e}_1$ are constant. Thus, the final result of differentiating Equation (5.45) is

$$a\omega_2\mathbf{n}_2 + \dot{b}\mathbf{e}_3 + b\omega_3\mathbf{n}_3 = 0 \tag{5.51}$$

At this point in the analysis it is useful to remember which variables above are known, and which are unknown. As stated earlier, we have used the position analysis formulas to solve for the angle θ_3 and the length b, thus $\mathbf{e}_3$ and $\mathbf{n}_3$ are known. The lengths of each link are known, and we assume that the crank angle, θ_2, is given. Further, let us assume that the crank is connected to a motor whose speed, ω_2, is also known. Thus, the list of known quantities is

$$\text{Known}: a, b, d, \theta_2, \theta_3, \omega_2$$

The only unknowns, then, are

$$\text{Unknown}: \omega_3, \dot{b}$$

Since Equation (5.51) is a vector equation with two components, we could solve for ω_3 in the y equation, substitute this into the x equation and solve for $\dot{b}$. While this approach is perfectly feasible and valid, it is also somewhat cumbersome and error-prone. A simpler approach will be used here. First, put the terms with unknown quantities on one side of the equation, with the fully known terms on the other side.

$$\dot{b}\mathbf{e}_3 + b\omega_3\mathbf{n}_3 = -\omega_2 a\mathbf{n}_2 \tag{5.52}$$

Next, put the above equations into matrix form

$$\left[b\mathbf{n}_3 \,\vdots\, \mathbf{e}_3\right]\begin{Bmatrix}\omega_3 \\ \dot{b}\end{Bmatrix} = -\omega_2 a\mathbf{n}_2 \tag{5.53}$$

Further, if we define

$$\mathbf{A} = \left[b\mathbf{n}_3 \,\vdots\, \mathbf{e}_3\right] \qquad \boldsymbol{\omega} = \begin{Bmatrix}\omega_3 \\ \dot{b}\end{Bmatrix} \qquad \mathbf{b} = -\omega_2 a\mathbf{n}_2 \tag{5.54}$$

Then we can write the matrix equation more compactly as

$$\mathbf{A}\boldsymbol{\omega} = \mathbf{b} \tag{5.55}$$

It is now quite simple to enter the above matrix equation into MATLAB, and have MATLAB solve and plot the velocities of the linkage throughout the range of motion. In fact, this is what MATLAB was designed to do! Since the matrix/vector notation in Equation (5.54) may be unfamiliar (and cryptic) to you, you might wish to verify that

$$\mathbf{A} = \begin{bmatrix} -b\sin\theta_3 & \cos\theta_3 \\ b\cos\theta_3 & \sin\theta_3 \end{bmatrix} \quad \mathbf{b} = \begin{Bmatrix} \omega_2 a\sin\theta_2 \\ -\omega_2 a\cos\theta_2 \end{Bmatrix} \tag{5.56}$$

when written out in full.

5.3.1 Velocity of Any Point on the Linkage

We will use an expression similar to the one we used in the instant-center method to find the velocities at points on the linkage as shown in Figure 5.44. Let us start with a simple problem – finding the velocity at point B. First, recall that the position vector of point B is given by

$$\mathbf{r}_B = \mathbf{r}_2 = a\mathbf{e}_2 \tag{5.57}$$

To find velocity, differentiate the vector $\mathbf{r}_B$ with respect to time.

$$\mathbf{v}_B = \frac{d}{dt}(a\mathbf{e}_2) \tag{5.58}$$

$$\mathbf{v}_B = a\omega_2\mathbf{n}_2 \tag{5.59}$$

The position vector of point P is

$$\mathbf{r}_P = \mathbf{r}_B + \mathbf{r}_{BP} \tag{5.60}$$

$$\mathbf{r}_P = a\mathbf{e}_2 + p\mathbf{e}_3 \tag{5.61}$$

Differentiating with respect to time gives

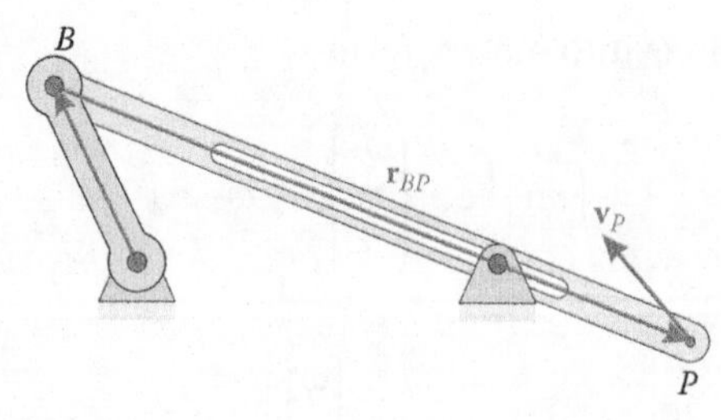

FIGURE 5.44
Finding the velocity of point P on the threebar linkage.

$$\mathbf{v}_P = \frac{d}{dt}\left(a\mathbf{e}_2 + p\mathbf{e}_3\right) \tag{5.62}$$

$$\mathbf{v}_P = a\omega_2\mathbf{n}_2 + p\omega_3\mathbf{n}_3 \tag{5.63}$$

$$\mathbf{v}_P = \mathbf{v}_B + p\omega_3\mathbf{n}_3 \tag{5.64}$$

You may have noticed that this is identical to the relative velocity formula

$$\mathbf{v}_P = \mathbf{v}_B + \mathbf{v}_{PB} \tag{5.65}$$

with $\mathbf{v}_B$ given in Equation (5.59) and $\mathbf{v}_{PB}$ the velocity of P relative to B. We now have a simple and powerful method for finding the velocity at any point on the linkage. In fact, this formula is reminiscent of the relative position formula, and we will define a MATLAB function to compute it, just as we did for position.

5.3.2 Velocity Analysis of the Threebar Slider-Crank Using MATLAB®

At the end of the last section, we obtained a matrix equation whose unknowns were the velocities, ω_3 and $\dot{b}$.

$$\mathbf{A}\boldsymbol{\omega} = \mathbf{b} \tag{5.66}$$

where

$$\mathbf{A} = \left[b\mathbf{n}_3 \ \vdots \ \mathbf{e}_3\right] \qquad \boldsymbol{\omega} = \begin{Bmatrix} \omega_3 \\ \dot{b} \end{Bmatrix} \qquad \mathbf{b} = \{-\omega_2 a\mathbf{n}_2\} \tag{5.67}$$

We will now develop some MATLAB code to solve the velocity problem using the dimensions shown in Figure 5.45. You should use the threebar position analysis code that you developed earlier as a starting point. To conduct the velocity analysis we must first know

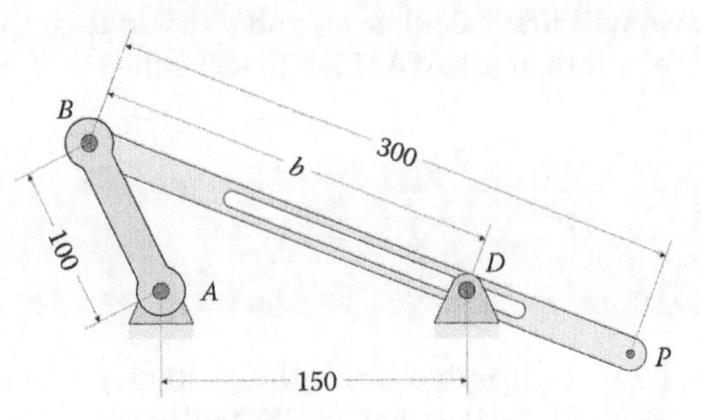

(All dimensions in millimeters)

Crank length: 100
Slider length: 300
Distance between ground pins: 150
Crank angular velocity: 10 rad/s

FIGURE 5.45
The threebar linkage used in the example MATLAB code. This is the same threebar that was used in the position analysis example.

the angular velocity of the crank, ω_2. Enter a value for omega2 at the top of the program, immediately after the statements that provide link lengths. Also, make sure to allocate some space in memory for omega3 and bdot, as you did for theta3 and b. The portion of the code before the main loop should now read

```
% Threebar_Velocity_Analysis.m
% Conducts a velocity analysis on the threebar crank-slider linkage
% and plots the velocity of point P
% by Eric Constans, June 1, 2017

% Prepare Workspace
clear variables; close all; clc;

% Linkage dimensions
a = 0.100;            % crank length (m)
d = 0.150;            % length between ground pins (m)
p = 0.300;            % slider length (m)

% Ground pins
x0 = [0;0];           % point A (the origin)
xD = [d;0];           % point D
v0 = [0;0];           % velocity of pin A (zero)

% Angular velocity of crank
omega2 = 10;          % angular velocity of crank (rad/sec)

N = 361;    % number of times to perform position calculations
[xB,xP] = deal(zeros(2,N)); % allocate space for position of B,P
[vB,vP] = deal(zeros(2,N)); % allocate space for velocity of B,P

[theta2,theta3,b] = deal(zeros(1,N)); % allocate space for link angles
[omega3,bdot]     = deal(zeros(1,N)); % allocate space for velocities
```

Note that we have defined a variable v0 to specify that the velocity at pin A is zero. To solve the velocity problem, we will first calculate ω_3 and $\dot{b}$. To do this, we need to define the **A** matrix and **b** vector. We will then ask MATLAB to solve the set of matrix equations using the "\" operator.

```
% conduct velocity analysis to solve for omega3 and bdot
  A_Mat = [b(i)*n3     e3];
  b_Vec = -a*omega2*n2;
  omega_Vec = A_Mat\b_Vec;   % solve for velocities
```

This set of statements should be inserted inside the main loop immediately before the end statement. It is not necessary to use the _Mat or _Vec suffixes shown in the code above, but we have included them to prevent confusion between these and other variables (especially b, the slider length). The solution vector, omega _ Vec, includes ω_3 as its first component and $\dot{b}$ as its second component. Thus, after solving the matrix equation, we decompose the vector into its omega3 and bdot components.

```
  omega3(i) = omega_Vec(1);     % decompose omega_Vec into
  bdot(i)   = omega_Vec(2);     % individual components
```

To solve for the translational velocities of points B and P, we will use the relative velocity formula. Because we will employ this formula so often, it is helpful to create a new MATLAB function for it. Open up a new MATLAB script and enter the following function:

```
% Function FindVel.m
% calculates the translational velocity at a point on the linkage
% using the relative velocity formula
%
% v0    = velocity of first point
% L     = length of vector to second point on the link
% omega = angular velocity of link
% n     = unit normal to vector btw first and second points
% v     = velocity of second point

function v = FindVel(v0, L, omega, n)

v = v0 + omega * L * n;
```

The form of this function is strikingly similar to the FindPos.m function created earlier. We can now use this function in the main program to calculate the translational velocities of the points B and P.

```
% calculate velocity at important points on linkage
  vB(:,i) = FindVel(       v0,    a,     omega2,   n2);
  vP(:,i) = FindVel(vB(:,i),    p, omega3(i),   n3);
```

The complete velocity analysis code is shown at the end of this section. We have eliminated the position plotting portion of the code for brevity, but it is a good idea to keep it in your code to ensure it still works.

5.3.2.1 Verifying the Code

Of course, we are not finished yet! We must next verify the code, to make sure we haven't made any typographical (or mathematical) errors along the way. We will use the example threebar linkage in Figure 5.45 as a check on our calculations. We will check our calculations at a crank angle of $\theta_2 = 120°$.

Draw the linkage in a SOLIDWORKS drawing. Next, find the location of the instant center I_{13} and the critical dimensions e and f, as shown in Figure 5.46. The angular velocity of the slider is found through Equation (5.19).

$$\omega_3 = \frac{a}{e}\omega_2 = \frac{100\text{ mm}}{271.43\text{ m}} \cdot 10\text{ rad/s} = 3.68\text{ rad/s}$$

At the command prompt in MATLAB, type

```
>> omega3(121)

ans =

   3.6842
```

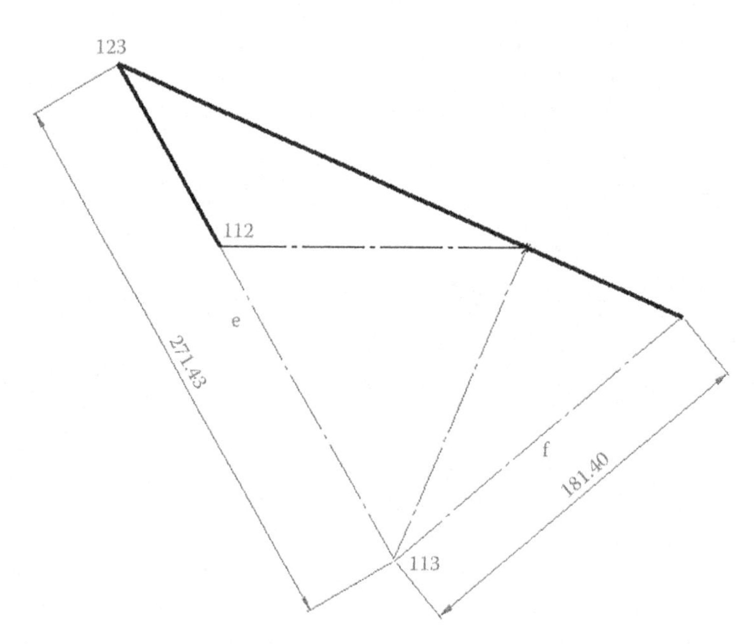

FIGURE 5.46
Instant center I_{13} of the threebar linkage is used to verify our velocity analysis code.

Recall that the index 121 corresponds to a crank angle of $\theta_2 = 120°$ because MATLAB requires that we start our index numbering at 1, which corresponded to a crank angle of $\theta_2 = 0°$. The MATLAB and SOLIDWORKS values agree with each other with only a small error, most likely due to roundoff error in the SOLIDWORKS dimensions.

The velocity at point P can be calculated from the instant center diagram as

$$|\mathbf{v}_P| = f\omega_3 = 181.40\ \text{mm} \cdot 3.68\ \text{rad/s} = 668.3\ \text{mm/s}$$

But we have asked MATLAB to calculate the x and y components of $\mathbf{v}_P$, so we must take the magnitude to calculate the total velocity (remember we have specified our code units in meters so we will need mentally account for the factor of 1000).

```
>> sqrt(vP(1,121)^2 + vPy(2,121)^2)

ans =

    0.6683
```

Thus, the MATLAB prediction is in agreement with our instant center analysis, and we can have confidence that our code is giving correct results.

5.3.2.2 Verifying the Code – An Alternative Approach

When we conduct acceleration analysis in the next chapter, we will be unable to use the instant center method to verify the code since it can only be used to calculate velocity. A second method of verifying the code uses the definition of velocity

$$\omega_3 = \frac{d\theta_3}{dt} \quad \dot{b} = \frac{db}{dt} \tag{5.68}$$

We can use MATLAB to make a numerical approximation to the derivative as:

$$\omega_3(i) \approx \frac{\theta_3(i+1) - \theta_3(i)}{\Delta t} \tag{5.69}$$

for $i = 1{:}N - 1$. In other words, dividing the difference between the current and previous slider angles by the change in time between calculations gives an approximation to the angular velocity at the current time step. Note that this is a *forward* estimation – we could also use the *backward* estimation formula

$$\omega_3(i) \approx \frac{\theta_3(i) - \theta_3(i-1)}{\Delta t} \tag{5.70}$$

but for sufficiently small time steps, it won't make an appreciable difference. Finding the time increment between calculations is a little tricky, but recall that the increment in crank angle between calculations was defined to be

$$\Delta\theta_2 = \frac{2\pi}{N-1} \tag{5.71}$$

We can find the time increment, Δt, by using the definition of angular velocity

$$\omega_2 = \frac{\Delta\theta_2}{\Delta t} \tag{5.72}$$

where we assume that the crank has constant angular velocity ω_2. The time increment is then

$$\Delta t = \frac{2\pi}{(N-1)\omega_2} \tag{5.73}$$

Since we will be verifying several codes by using the derivative estimation technique, it is worthwhile to develop a derivative-estimating function. The goal of the function will be to estimate the derivative of a variable, and plot it alongside the calculated derivative. The function should be general enough that we can use it for verifying velocities, accelerations, or whatever other derivative we may desire.

Create a new MATLAB script and type the following at the top of the file:

```
% Function Derivative_Plot.m
% plots an approximation to the derivative of a function along with
% the function itself
%
% crankAngle = used for the x axis of the derivative plots
% theta      = the function whose derivative is to be estimated
% omega      = the calculated derivative
% dt         = time step
```

Use `Save As` to save the file as `Derivative_Plot.m` (no spaces!). Make sure you save this file in the same folder as your velocity analysis program, otherwise MATLAB won't know where to find it. The first line of code in the function should give the function name, along with any arguments we need

```
function Derivative_Plot(crankAngle, theta, omega, dt)
```

The first argument (`crankAngle`) provides the vector of crank angles as it sweeps out its rotation. We will use this as the x axis when we plot the derivatives. The second argument, `theta`, gives the vector of angular positions that we will differentiate. When we calculate accelerations in a later chapter, `theta` will consist of velocities; this doesn't matter as long as we remember that `theta` is the vector of values that we wish to differentiate.

The third argument gives the calculated values of the derivatives. For the present situation, this is the `omega3` or `bdot` vector that we are trying to verify. Finally, the `dt` argument gives the time increment so that we can calculate our estimate of the derivative.

Below the function declaration, initialize the following variables:

```
N = length(theta);          % length of position and velocity vectors
omegaStar = zeros(N,1);     % estimate of derivative
```

The estimated angular velocities will be stored in the vector `omegaStar`. Since we wish our function to be as general as possible, we will not assume that the length of each vector is 361. For non-Grashof linkages, the vector may be shorter. And if we wish to perform our calculations on a finer time scale the vector may be longer. Now, begin a `for` loop to calculate the derivative estimates

```
% estimate derivative
for i = 1:N-1
  omegaStar(i) = (theta(i+1) - theta(i))/dt;
end
```

The `for` loop must terminate at N–1 because we have used `theta(i+1)` in the calculation and `theta(N+1)` is undefined. Let us assume that the final derivative estimate is the same as the first estimate:

```
% assume final derivative is the same as the first
omegaStar(N) = omegaStar(1);
```

This will not be correct in the case of non-Grashof linkages, and our function will give a spurious datapoint at the end of the plot in such a situation. However, it will provide correct plots for most situations we will encounter in this text where the crank runs through a complete cycle. If you wish, you can make a few simple modifications to the function to force it to plot only N–1 points; this will make it perfectly general. Finally, add the series of `plot` commands shown below.

```
% plot estimated and analytical derivatives
figure
plot(180*crankAngle/pi, omegaStar,'Color', [0 110/255 199/255])
hold on
plot(180*crankAngle/pi, omega, '.','Color', [51/255 51/255 51/255])
legend('Estimated','Analytical')
title('Comparison of Calculated and Analytical Derivatives')
xlabel('Crank angle (degrees)')
ylabel('Derivative')
grid on
```

Note that `crankAngle` has been converted to degrees for plotting purposes.

That completes our first nontrivial function! Make sure you save the file again, and then go back to the threebar-velocity analysis script. Immediately after the main loop, type the following two statements:

```
dt = 2*pi/((N-1)*omega2);   % time increment between calculations
Derivative_Plot(theta2, theta3, omega3, dt) % verify derivatives
```

The first statement calculates the value of `dt`, and the second calls our newly-created function. Here we are comparing the calculated values of `omega3` against the estimates provided by the position angle `theta3`.

When you run the code, you should obtain the plot shown in Figure 5.47. The analytical and estimated values are almost identical, and we conclude that the code is producing accurate results.

Now change the function call statement in the main program to the following

```
Derivative_Plot(theta2, b, bdot, dt) % verify derivatives
```

This will estimate the derivative of the length b for estimating $\dot{b}$. Figure 5.48 shows the comparison for the velocity of slip. So far, the code seems to be producing accurate results. As a final check, let us verify the x component of the velocity at point P.

```
Derivative_Plot(theta2, xP(1,:), vP(1,:), dt) % verify derivatives
```

The comparison is shown in Figure 5.49. Since the code seems to be producing accurate results, you may comment out the `Derivative_Plot` function call and modify the plotting commands to trace out the x and y velocities of point P, as shown below.

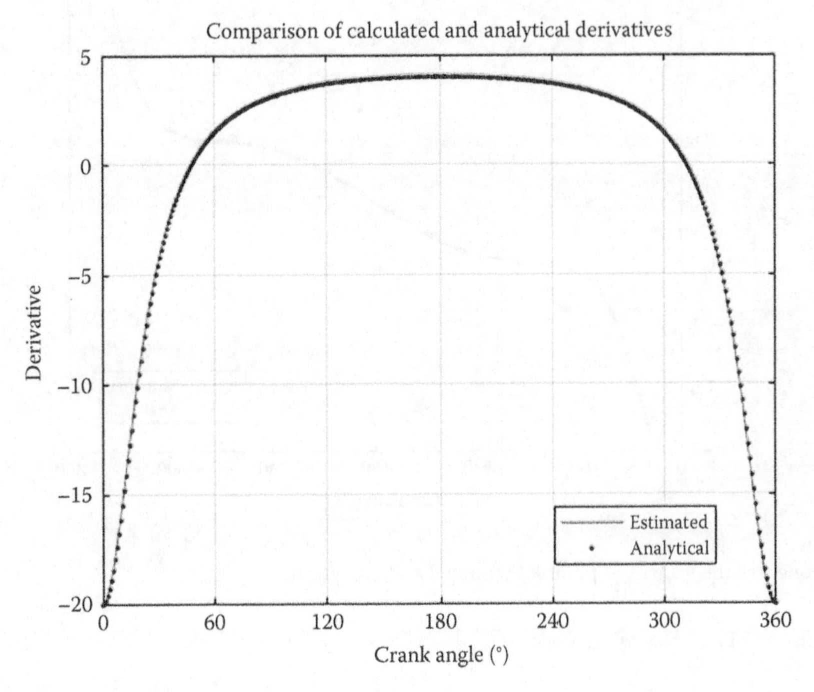

FIGURE 5.47
A comparison of the analytical and estimated values for ω_3, the angular velocity of the slider.

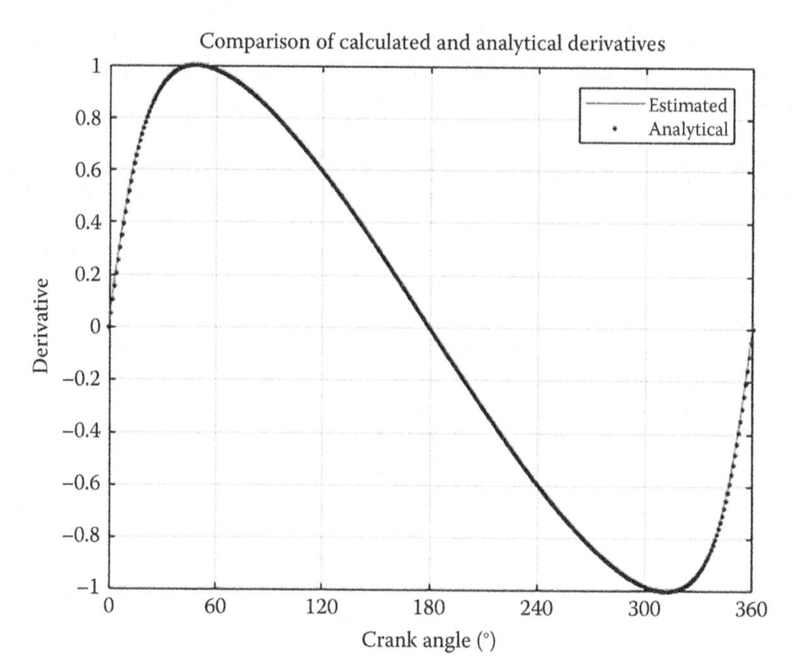

FIGURE 5.48
A comparison of the analytical and estimated values for the velocity of slip, $\dot{b}$.

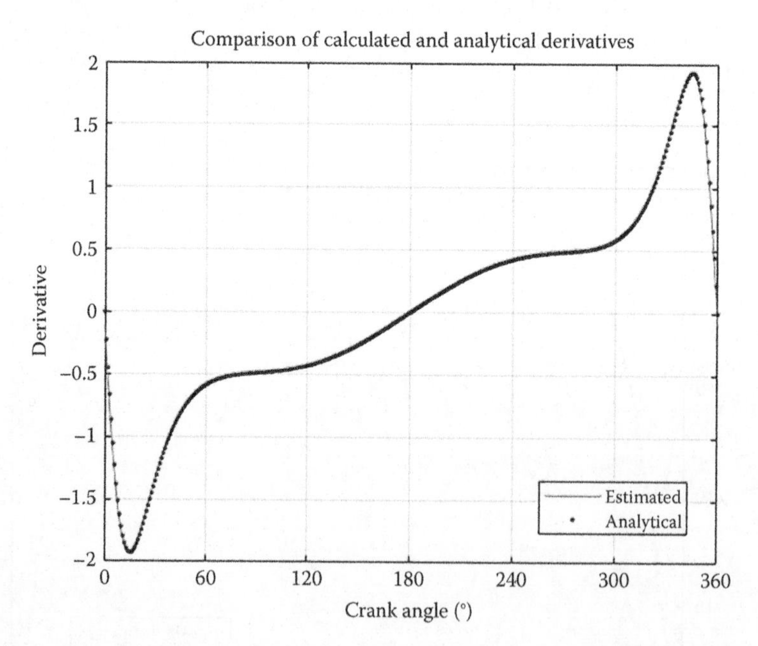

FIGURE 5.49
The x component of the velocity at point P: estimated and analytical.

```
% plot the velocity of point P
plot(theta2*180/pi,vP(1,:),'Color',[153/255 153/255 153/255])
hold on
plot(theta2*180/pi,vP(2,:),'Color',[0 110/255 199/255])
```

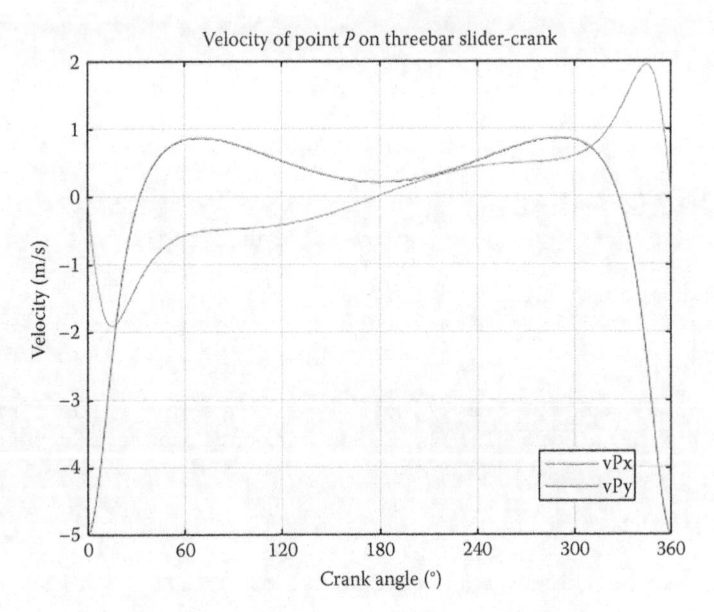

FIGURE 5.50
The velocity of point P in the example problem.

```
legend('vPx','vPy')
title('Velocity of point P on Threebar Slider-Crank')
xlabel('Crank angle (degrees)')
ylabel('Velocity (m/s)')
grid on
set(gca,'xtick',0:60:360)
xlim([0 360])
```

The plot in Figure 5.50 gives the velocity of point P as the crank makes a full revolution. Make sure that your own code gives the same results as the example problem before attempting the homework problems.

To summarize, we have developed a simple code for calculating the velocity of any point on the threebar linkage. More importantly, we have also developed a robust, general purpose function for verify our derivative estimates. Remember, verifying your simulation is one of the most important roles you will take on as an engineer. In the next few sections, we will use the same procedure to find the velocities of the more complicated linkages we studied in Chapter 4.

```
% Threebar_Velocity_Analysis.m
% Conducts a velocity analysis on the threebar crank-slider linkage
% and plots the velocity of point P
% by Eric Constans, June 1, 2017

% Prepare Workspace
clear variables; close all; clc;

% Linkage dimensions
a = 0.100;          % crank length (m)
d = 0.150;          % length between ground pins (m)
p = 0.300;          % slider length (m)
```

```
% Ground pins
x0 = [0;0];           % point A (the origin)
xD = [d;0];           % point D
v0 = [0;0];           % velocity of pin A (zero)

% Angular velocity of crank
omega2 = 10;          % angular velocity of crank (rad/sec)

N = 361;    % number of times to perform position calculations
[xB,xP] = deal(zeros(2,N)); % allocate space for position of B,P
[vB,vP] = deal(zeros(2,N)); % allocate space for velocity of B,P

[theta2,theta3,b] = deal(zeros(1,N)); % allocate space for link angles
[omega3,bdot] = deal(zeros(1,N));      % allocate space for velocities

for i = 1:N
  theta2(i) = (i-1)*(2*pi)/(N-1);
  theta3(i) = atan2(-a*sin(theta2(i)),d - a*cos(theta2(i)));
  b(i) = (d - a*cos(theta2(i)))/cos(theta3(i));
```

```
% calculate unit vector
  [e1,n1]   = UnitVector(0);
  [e2,n2]   = UnitVector(theta2(i));
  [e3,n3]   = UnitVector(theta3(i));

% solve for positions of points B, C and P on the linkage
  xB(:,i) = FindPos(  [0;0],a,e2);
  xP(:,i) = FindPos(xB(:,i),p,e3);

% conduct velocity analysis to solve for omega3 and bdot
  A_Mat = [b(i)*n3 e3];
  b_Vec = -a*omega2*n2;
  omega_Vec = A_Mat\b_Vec;  % solve for velocities

  omega3(i) = omega_Vec(1);    % decompose omega_Vec into
  bdot(i)   = omega_Vec(2);    % individual components

% calculate velocity at important points on linkage
  vB(:,i) = FindVel(      v0,    a,    omega2,  n2);
  vP(:,i) = FindVel(vB(:,i),    p, omega3(i),  n3);
end

% plot the velocity of point P
plot(theta2*180/pi,vP(1,:),'Color',[153/255 153/255 153/255])
hold on
plot(theta2*180/pi,vP(2,:),'Color',[0 110/255 199/255])
legend('vPx','vPy')
title('Velocity of point P on Threebar Slider-Crank')
xlabel('Crank angle (degrees)')
ylabel('Velocity (m/s)')
grid on
set(gca,'xtick',0:60:360)
xlim([0 360])
```

5.4 Velocity Analysis of the Slider-Crank

The vector loop for the slider-crank mechanism is shown in Figure 5.51. In the position analysis for the slider-crank we found the vector loop equation to be

$$\mathbf{r}_2 + \mathbf{r}_3 - \mathbf{r}_4 - \mathbf{r}_1 = 0 \tag{5.74}$$

or, in unit vector form

$$a\mathbf{e}_2 + b\mathbf{e}_3 - c\mathbf{e}_4 - d\mathbf{e}_1 = 0 \tag{5.75}$$

Let us differentiate each term individually with respect to time. The first two are familiar from previous examples:

$$\frac{d}{dt}(a\mathbf{e}_2) = \omega_2 a\mathbf{n}_2 \tag{5.76}$$

$$\frac{d}{dt}(b\mathbf{e}_3) = \omega_3 b\mathbf{n}_3 \tag{5.77}$$

The third term differentiates to zero, since c is a constant and $\mathbf{e}_4$ is always vertical (i.e., the vertical distance from the cylinder to the crank pin is fixed).

$$\frac{d}{dt}(c\mathbf{e}_4) = 0 \tag{5.78}$$

The derivative of the fourth term, however, is not zero, since the horizontal position of the piston varies with time. We still must use the product rule, but the time derivative of $\mathbf{e}_1$ is zero, since it is always horizontal. Thus, we have:

$$\frac{d}{dt}(d\mathbf{e}_1) = \dot{d}\mathbf{e}_1 \tag{5.79}$$

where we have used the overdot notation to indicate differentiation with respect to time. The term $\dot{d}$ represents the horizontal velocity of the piston. We could equivalently write this as

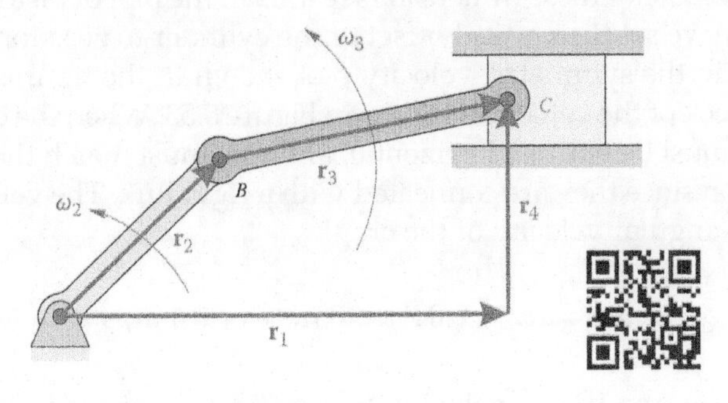

FIGURE 5.51
The vector loop for the slider-crank linkage.

$$\dot{d} = v_{Dx} \tag{5.80}$$

but the overdot notation is simpler. The differentiated vector loop equation is then

$$\omega_2 a\mathbf{n}_2 + \omega_3 b\mathbf{n}_3 - \dot{d}\mathbf{e}_1 = 0 \tag{5.81}$$

It is useful again, at this point, to take stock of what variables we know, and which are unknown. As before, we assume that a full position analysis has been performed, so that we know

$$\text{Known} : a, b, c, d, \theta_2, \theta_3, \omega_2$$

We also assume that the crank is driven by a motor with known speed, ω_2. The only unknown quantities in the equation above are

$$\text{Unknown} : \omega_3, \dot{d}$$

As before, let us rearrange the equations into matrix form, so that we can use MATLAB to solve them.

$$\left[b\mathbf{n}_3 \vdots -\mathbf{e}_1 \right] \begin{Bmatrix} \omega_3 \\ \dot{d} \end{Bmatrix} = \{ -\omega_2 a\mathbf{n}_2 \} \tag{5.82}$$

If we define

$$\mathbf{A} = \left[b\mathbf{n}_3 \vdots -\mathbf{e}_1 \right] \qquad \boldsymbol{\omega} = \begin{Bmatrix} \omega_3 \\ \dot{d} \end{Bmatrix} \qquad \mathbf{b} = \{ -\omega_2 a\mathbf{n}_2 \} \tag{5.83}$$

then we may use MATLAB to solve for the velocities as we did with the fourbar linkage. Because the code is so similar to that of the threebar, its development is left as an exercise for the reader.

5.4.1 Example Slider-Crank

The dimensions and crank angular velocity for the example slider-crank linkage are shown in Figure 5.52. Using these dimensions results in the plot of piston velocity, $\dot{d}$ shown in Figure 5.53. We have set the vertical offset of the cylinder to zero for the example problem, which results in the symmetric velocity plot shown in the figure. This enables us to perform a quick check of the velocities shown in Figure 5.53. When the crank is vertical the velocity at point B must be entirely horizontal, and this must match the purely horizontal velocity of the piston since they are connected with a rigid link. The velocity at point B can be found using the angular velocity of the crank

$$|\mathbf{v}_B| = a\omega_2 = 0.04\ \text{m} \cdot 10\ \text{rad/s} = 0.4\ \text{m/s}$$

As seen in the figure, the piston velocity crosses 0.4 m/s at a crank angle of 90°. It is interesting to note that the maximum piston velocity is a little higher than 0.4 m/s, and

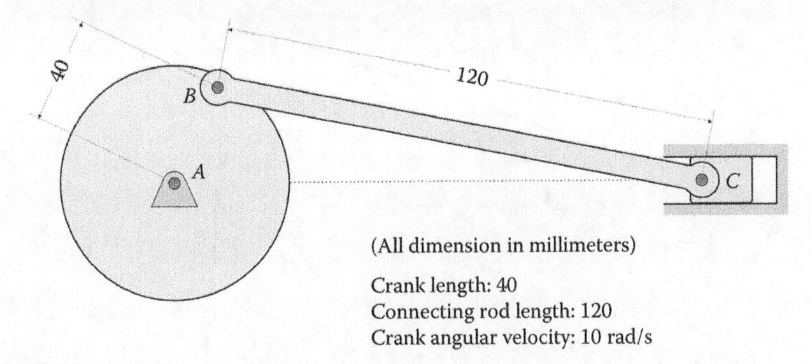

FIGURE 5.52
Dimensions of the example slider-crank linkage. Note that the vertical offset of the cylinder has been set to zero for the example.

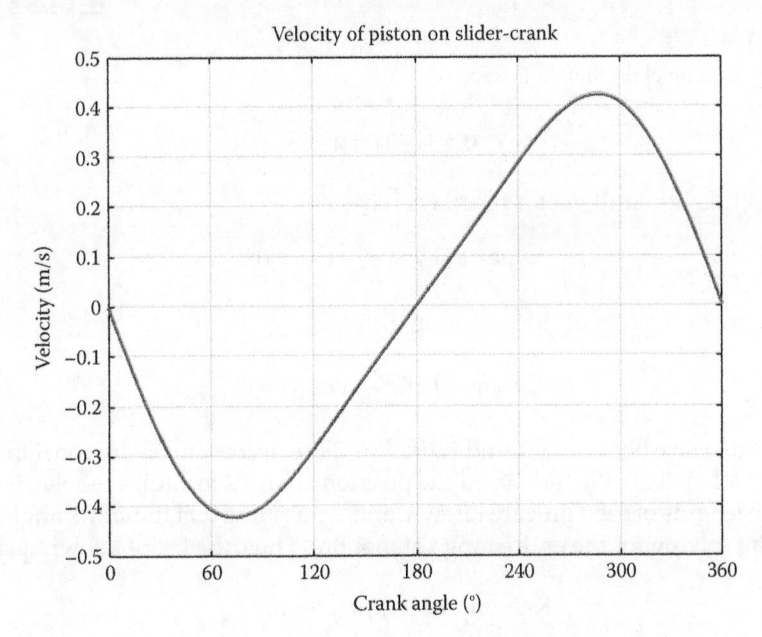

FIGURE 5.53
Piston velocity for the example slider-crank linkage. The plot is symmetric about the x axis because the vertical offset has been set to zero.

occurs at a crank angle of 73°, instead of at 90° as might have been expected. As we will see in the next chapter, the acceleration of the piston is zero at this crank angle.

5.5 Velocity Analysis of the Fourbar Linkage

Figure 5.54 shows a vector loop diagram of the fourbar linkage, similar to the one we constructed for the previous two sections. As before, we simply add the vectors going around the loop. Since we end up where we started, the vector sum is zero.

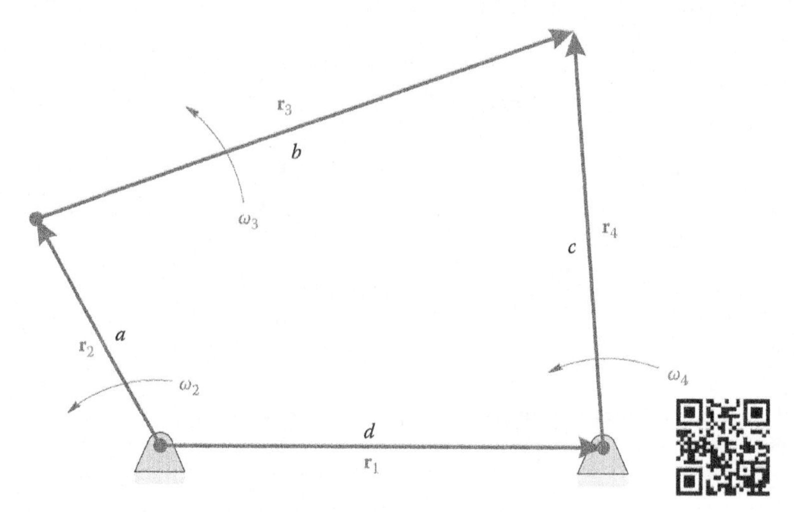

FIGURE 5.54
The vector loop diagram of the fourbar linkage.

$$\mathbf{r}_2 + \mathbf{r}_3 - \mathbf{r}_4 - \mathbf{r}_1 = 0 \tag{5.84}$$

If we expand this into unit vector form, we have

$$a\mathbf{e}_2 + b\mathbf{e}_3 - c\mathbf{e}_4 - d\mathbf{e}_1 = 0 \tag{5.85}$$

Taking the time derivative of this equation gives

$$a\omega_2\mathbf{n}_2 + b\omega_3\mathbf{n}_3 - c\omega_4\mathbf{n}_4 = 0 \tag{5.86}$$

Note that the distance between ground pins, d, and the unit vector $\mathbf{e}_1$ are constant and do not appear in Equation (5.51). We have used the position analysis formulas to solve for the angles θ_3 and θ_4. The lengths of each link are known, and we assume that the crank angle, θ_2, is given (or that we are solving for the entire range of motion). Thus, the list of known quantities is

$$\text{Known}: a, b, c, d, \theta_2, \theta_3, \theta_4, \omega_2$$

The only unknowns, then, are

$$\text{Unknown}: \omega_3, \omega_4$$

We first put the terms with unknown quantities on one side of the equation, with the fully known terms on the other side.

$$\omega_3 b\mathbf{n}_3 - \omega_4 c\mathbf{n}_4 = -\omega_2 a\mathbf{n}_2 \tag{5.87}$$

Next, put the above equations into matrix form

$$\left[b\mathbf{n}_3 \vdots - c\mathbf{n}_4\right]\begin{Bmatrix}\omega_3 \\ \omega_4\end{Bmatrix} = -\omega_2 a\mathbf{n}_2 \tag{5.88}$$

Further, if we define

$$\mathbf{A} = \left[b\mathbf{n}_3 \,\vdots\, -c\mathbf{n}_4\right] \qquad \boldsymbol{\omega} = \begin{Bmatrix} \omega_3 \\ \omega_4 \end{Bmatrix} \qquad \mathbf{b} = -\omega_2 a\mathbf{n}_2 \tag{5.89}$$

Then we can write the matrix equation more compactly as

$$\mathbf{A}\boldsymbol{\omega} = \mathbf{b} \tag{5.90}$$

We may use this matrix equation to solve for the unknown angular velocities on the four-bar linkage.

5.5.1 Velocity of Any Point on the Linkage

As with the slider-crank, the velocity of pin B on the crank is given by (see Figure 5.55)

$$\mathbf{v}_B = \omega_2 a\mathbf{n}_2 \tag{5.91}$$

The position vector of point C is

$$\mathbf{r}_C = \mathbf{r}_1 + \mathbf{r}_4 \tag{5.92}$$

$$\mathbf{r}_C = d\mathbf{e}_1 + c\mathbf{e}_4 \tag{5.93}$$

Differentiating with respect to time gives

$$\mathbf{v}_C = \omega_4 c\mathbf{n}_4 \tag{5.94}$$

Thus, it is relatively straightforward to find the velocity at any joint on the linkage. But what happens if we need to find the velocity at a point not located at a pin? In Figure 5.56, the point P is attached to the coupler, which has been drawn as a triangle. The distance

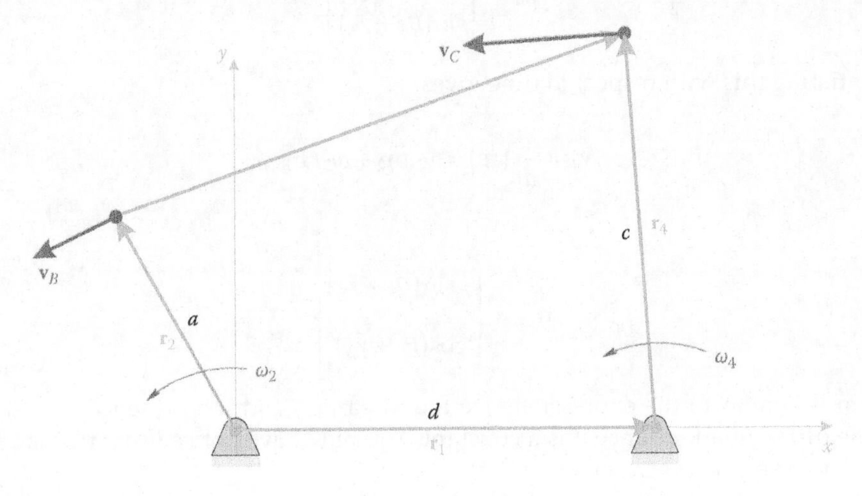

FIGURE 5.55
Finding the velocities of points B and C on the fourbar linkage.

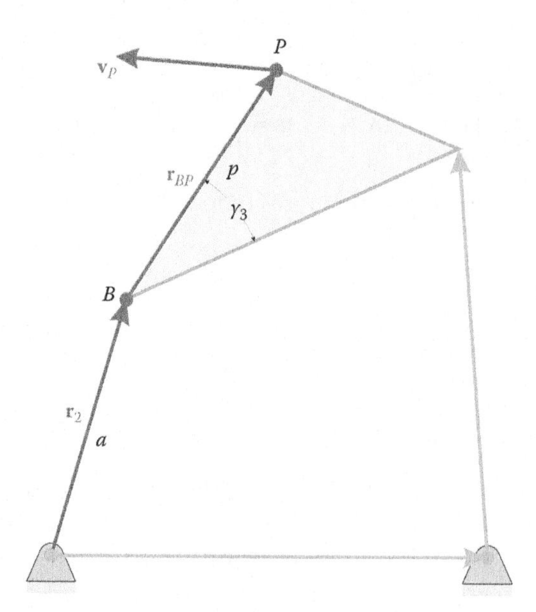

FIGURE 5.56
Finding the velocity of point P on the coupler of the fourbar linkage.

from point B to P is a constant, p, and the angle between the coupler and the line BP is also a constant, γ_3. The position vector to point P is

$$\mathbf{r}_P = \mathbf{r}_2 + \mathbf{r}_{BP} \tag{5.95}$$

$$\mathbf{r}_P = a\mathbf{e}_2 + p\mathbf{e}_{BP} \tag{5.96}$$

where

$$\mathbf{e}_{BP} = \begin{Bmatrix} \cos(\theta_3 + \gamma_3) \\ \sin(\theta_3 + \gamma_3) \end{Bmatrix} \tag{5.97}$$

Differentiating this with respect to time gives

$$\mathbf{v}_P = \frac{d}{dt}(\mathbf{r}_P) = \omega_2 a\mathbf{n}_2 + \omega_3 p\mathbf{n}_{BP} \tag{5.98}$$

where

$$\mathbf{n}_{BP} = \begin{Bmatrix} -\sin(\theta_3 + \gamma_3) \\ \cos(\theta_3 + \gamma_3) \end{Bmatrix} \tag{5.99}$$

The form is similar to the expressions we found earlier, and the presence of γ_3 does not affect the differentiation, since it is a constant. You may have noticed that this is identical to the relative velocity formula

$$\mathbf{v}_P = \mathbf{v}_B + \mathbf{v}_{PB} \tag{5.100}$$

with $\mathbf{v}_B$ given in Equation (5.59) and $\mathbf{v}_{PB}$ the velocity of P relative to B. We now have a simple and powerful method for finding the velocity at any point on the linkage.

5.5.2 Fourbar Velocity Analysis Using MATLAB®

We will now develop some MATLAB code to solve the velocity problem for the linkage shown in Figure 5.57. You should use the fourbar position analysis code that you developed earlier as a starting point. To conduct the velocity analysis we must first know the angular velocity of the crank, ω_2. Enter a value for `omega2` at the top of the program, immediately after the statements that provide link lengths. Also, make sure to allocate some space in memory for `omega3` and `omega4`, as you did for `theta3` and `theta4`.

```
% Fourbar_Velocity_Analysis.m
% Conducts a velocity analysis on the fourbar linkage
% and plots the velocity of point P
% by Eric Constans, June 1, 2017

% Prepare Workspace
clear variables; close all; clc;

% Linkage dimensions
```

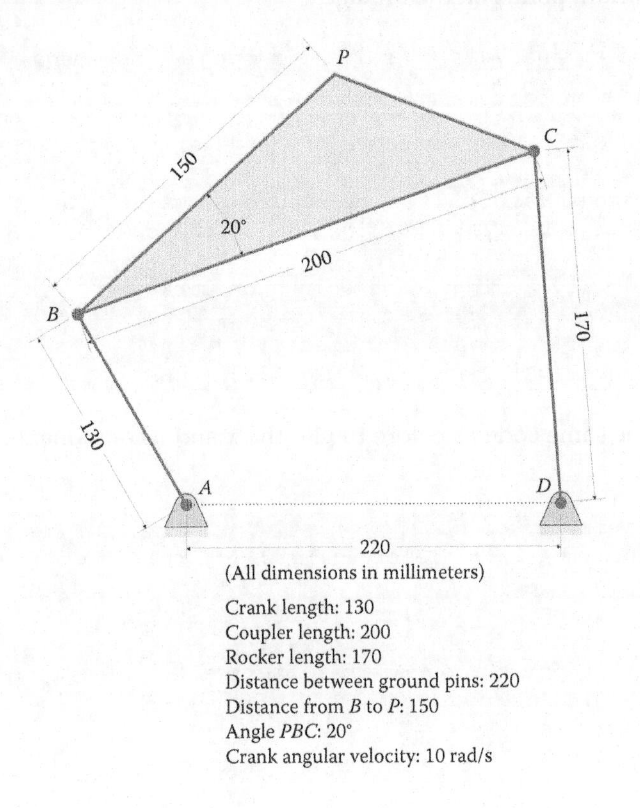

FIGURE 5.57
Example linkage used in the fourbar velocity analysis. This is the same linkage as was used for position analysis in Chapter 4.

```
a = 0.130;              % crank length (m)
b = 0.200;              % coupler length (m)
c = 0.170;              % rocker length (m)
d = 0.220;              % length between ground pins (m)
p = 0.150;              % length from B to P (m)
gamma3 = 20*pi/180; % angle between BP and coupler (converted to rad)

% Ground pins
x0 = [0;0];             % point A (the origin)
xD = [d;0];             % point D
v0 = [0;0];             % velocity of pin A (zero)

% Angular velocity and acceleration of crank
omega2 = 10;          % angular velocity of crank (rad/s)

N = 361;    % number of times to perform position calculations
[xB,xC,xP] = deal(zeros(2,N)); % allocate space for pos of B, C, P
[vB,vC,vP] = deal(zeros(2,N)); % allocate space for vel of B, C, P
[theta2,theta3,theta4] = deal(zeros(1,N));  % allocate for link angles
[omega3,omega4] = deal(zeros(1,N));  % allocate for angular velocities
```

After the position analysis portion of the main loop, enter the code for performing the velocity analysis. Again, we employ the `FindVel` function to find the translational velocities at important points on the linkage.

```
% conduct velocity analysis to solve for omega3 and omega4
  A_Mat = [b*n3    -c*n4];
  b_Vec = -a*omega2*n2;
  omega_Vec = A_Mat\b_Vec;  % solve for angular velocities

  omega3(i) = omega_Vec(1);     % decompose omega_Vec into
  omega4(i) = omega_Vec(2);     % individual components

% calculate velocity at important points on linkage
  vB(:,i) = FindVel(      v0,    a,     omega2,    n2);
  vC(:,i) = FindVel(      v0,    c, omega4(i),    n4);
  vP(:,i) = FindVel(vB(:,i),    p, omega3(i),   nBP);
```

Finally, we use the same code as before to plot the x and y components of the velocity at point P.

```
% plot the velocity of point P
plot(theta2*180/pi,vP(1,:),'Color',[153/255 153/255 153/255])
hold on
plot(theta2*180/pi,vP(2,:),'Color',[0 110/255 199/255])
legend('vPx','vPy','Location','Southeast')
title('Velocity of point P on Fourbar Linkage')
xlabel('Crank angle (degrees)')
ylabel('Velocity (m/s)')
grid on
set(gca,'xtick',0:60:360)
xlim([0 360])
```

5.5.3 Verifying the Code

Of course, we are not finished yet! We should perform a few quick validations to ensure that our code is producing accurate results. As a first check, let us conduct an instant center analysis using SOLIDWORKS. The example fourbar linkage has been recreated in a SOLIDWORKS drawing and is shown in Figure 5.58. We will check our calculations at a crank angle of $\theta_2 = 30°$.

Find the location of the instant center I_{13} and the critical dimensions e, f, and q, as shown in Figure 5.59. The angular velocity of the coupler is found through Equation (5.19)

$$\omega_3 = \frac{a}{e}\omega_2 = \frac{130 \text{ mm}}{217.082 \text{ mm}} \cdot 10 \text{ rad/s} = 5.989 \text{ rad/s}$$

At the command prompt in MATLAB, type

```
>> omega3(31)

ans =

  -5.9885
```

Recall that the index 31 corresponds to a crank angle of $\theta_2 = 30°$. The MATLAB and SOLIDWORKS values agree with each other with only a small error, most likely due to roundoff error in the SOLIDWORKS dimensions. The negative sign in the MATLAB answer indicates that the coupler is rotating clockwise, which we can easily see in the SOLIDWORKS drawing.

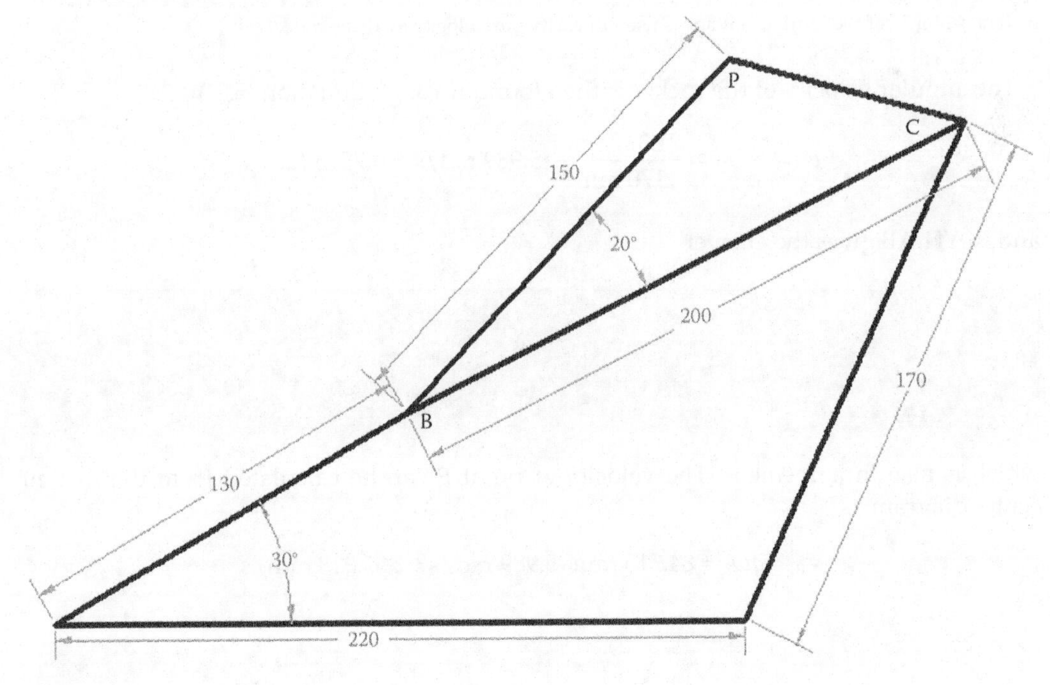

FIGURE 5.58
Fourbar linkage drawn in SOLIDWORKS that is used as a check on the MATLAB calculations.

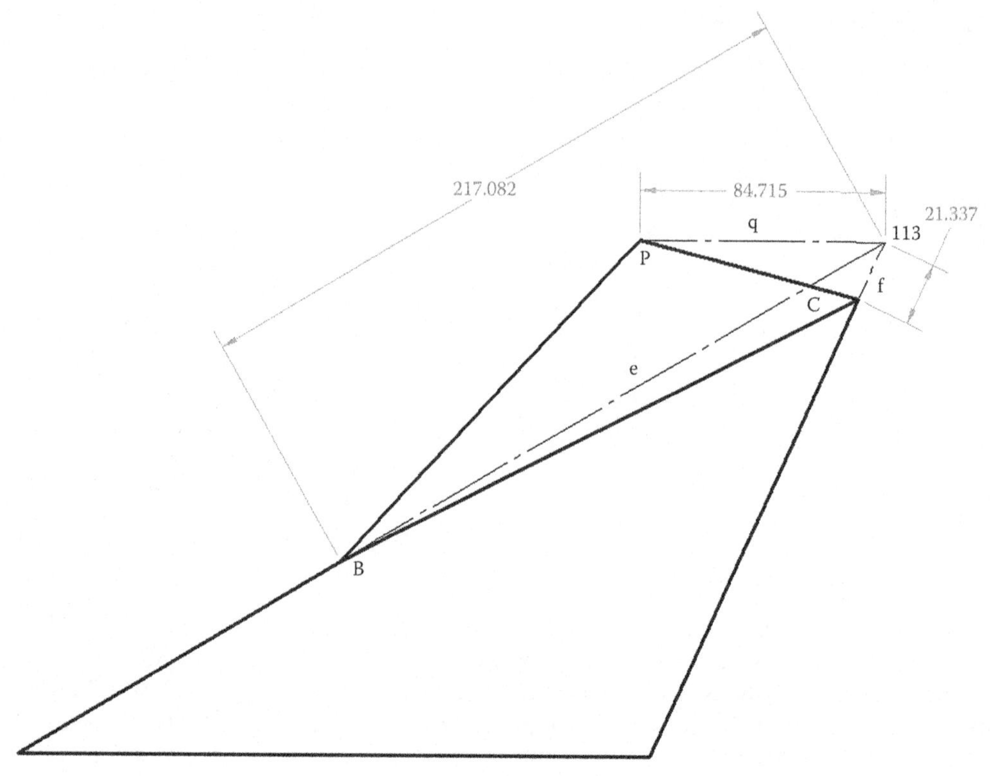

FIGURE 5.59
Instant center I_{13} of the fourbar linkage is used to verify our velocity analysis code.

The angular velocity of the rocker is then found through Equation (5.21)

$$\omega_4 = \frac{f}{c}\omega_3 = \frac{21.337 \text{ mm}}{170 \text{ mm}} \cdot 5.989 \text{ rad/s} = 0.7516 \text{ rad/s}$$

and MATLAB gives the answer

```
>> omega4(31)

ans =

    0.7516
```

which is also in agreement. The velocity at point P can be calculated from the instant center diagram as

$$|\mathbf{v}_P| = q\omega_3 = 84.715 \text{ mm} \cdot 5.989 \text{ rad/s} = 507.318 \text{ mm/s}$$

```
>> sqrt(vP(1,31)^2 + vP(2,31)^2)

ans =

  0.5073
```

Remember that the MATLAB code uses meters, not millimeters, so the answer is different by a factor of 1000. Thus, the MATLAB prediction is in agreement with our instant center analysis, and we can have confidence that our code is giving correct results.

Next, we will use the `Derivative _ Plot` function to verify that the angular velocities are being computed correctly for the full revolution of the crank. Figures 5.60 and 5.61 show the calculated and analytical values for the angular velocities of the coupler and rocker, respectively. The calculated and estimated values are almost identical, and we conclude that the code is producing accurate results.

As a final check, let us verify the x component of the velocity at point P.

```
Derivative_Plot(theta2, xP(1,:), vP(1,:), dt) % verify derivatives
```

The comparison is shown in Figure 5.62. Since the code seems to be producing accurate results, you may comment out the function call and modify the plotting commands to trace out the x and y velocities of point P, as shown below.

```
% plot the velocity of point P
plot(theta2*180/pi,vP(1,:),'Color',[153/255 153/255 153/255])
hold on
plot(theta2*180/pi,vP(2,:),'Color',[0 110/255 199/255])
legend('vPx','vPy','Location','Southeast')
title('Velocity of point P on Fourbar Linkage')
xlabel('Crank angle (degrees)')
ylabel('Velocity (m/s)')
grid on
set(gca,'xtick',0:60:360)
xlim([0 360])
```

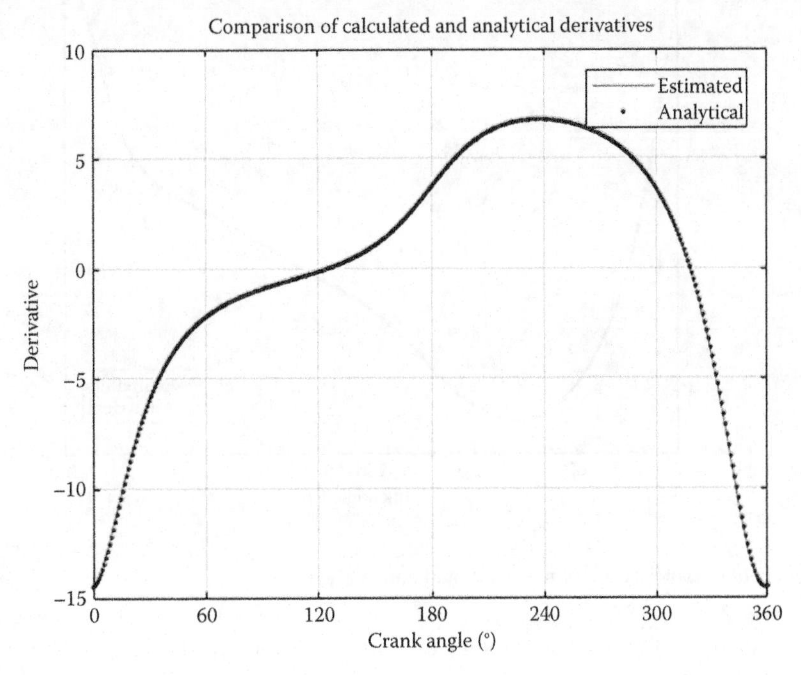

FIGURE 5.60
A comparison of the estimated and analytical values for ω_3, the angular velocity of the coupler.

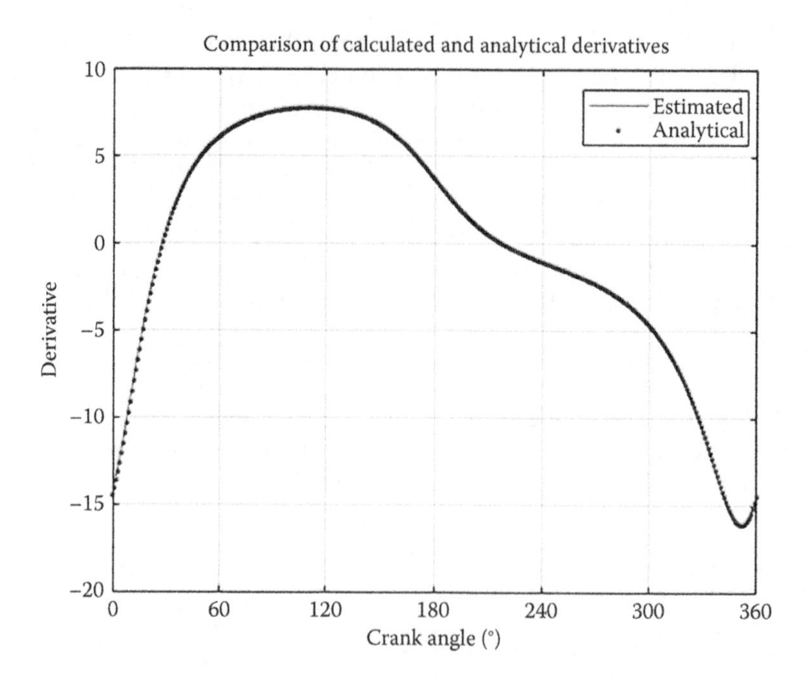

FIGURE 5.61
A comparison of the estimated and analytical values for the angular velocity of the rocker, ω_4.

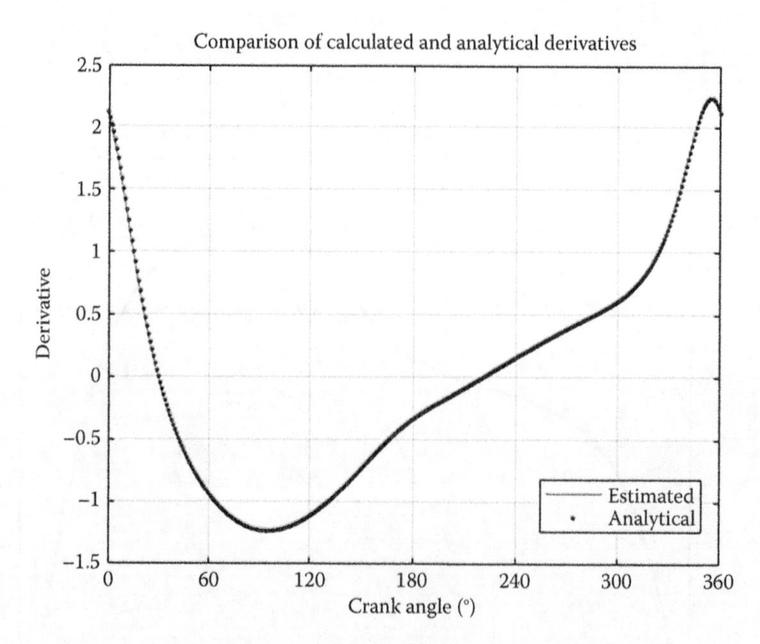

FIGURE 5.62
The x component of the velocity at point P: estimated and analytical.

The plot in Figure 5.63 gives the velocity of point P as the crank makes a full revolution. Make sure that your own code gives the same results as the example problem before attempting the homework problems.

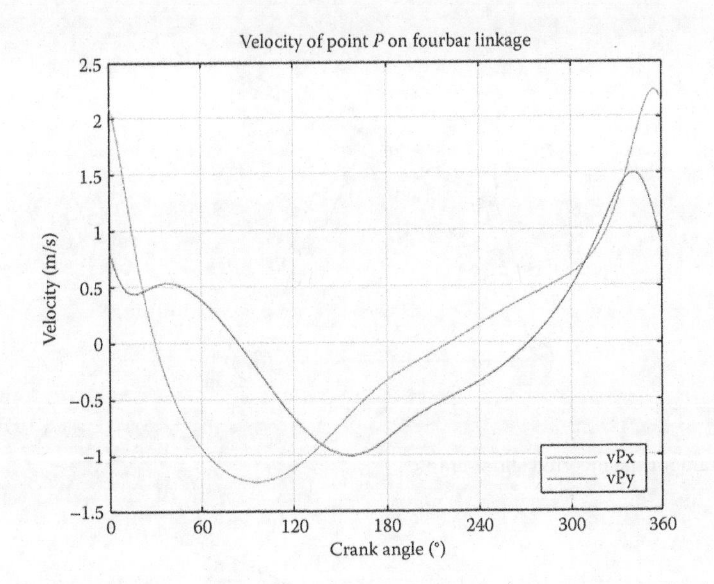

FIGURE 5.63
The velocity of point P in the example problem.

5.6 Velocity Analysis of the Inverted Slider-Crank

Figure 5.64 shows the vector loop diagram for the inverted slider-crank linkage, while Figure 5.65 shows the relevant translational and angular velocities. Following the loop gives

$$\mathbf{r}_2 + \mathbf{r}_3 - \mathbf{r}_4 - \mathbf{r}_1 = 0 \tag{5.101}$$

or, in unit vector form:

$$a\mathbf{e}_2 + b\mathbf{e}_3 - c\mathbf{e}_4 - d\mathbf{e}_1 = 0 \tag{5.102}$$

But θ_3 and θ_4 are not independent of one another, and are related by the constant δ.

$$\theta_4 = \theta_3 + \delta \tag{5.103}$$

If we differentiate Equation (5.103) with respect to time, we find that

$$\frac{d\theta_4}{dt} = \frac{d\theta_3}{dt} + \frac{d\delta}{dt} \tag{5.104}$$

Since δ is a constant, we have

$$\omega_4 = \omega_3 \tag{5.105}$$

This makes sense, since the slider and the rocker must rotate together at the same speed. We will now differentiate each term in the vector loop equation individually, as we did earlier. The first term is the same as before

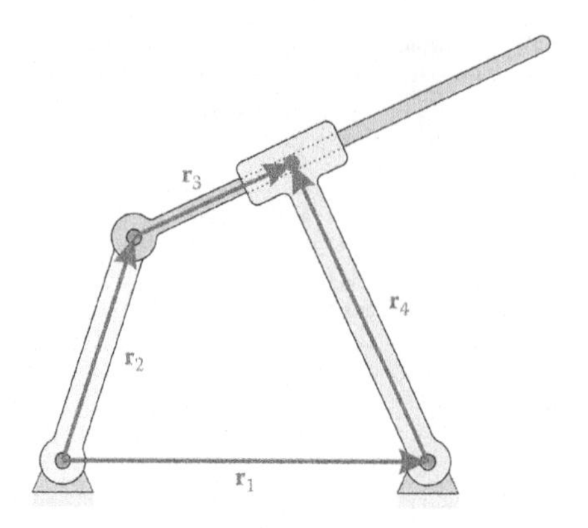

FIGURE 5.64
Vector loop diagram for the inverted slider-crank.

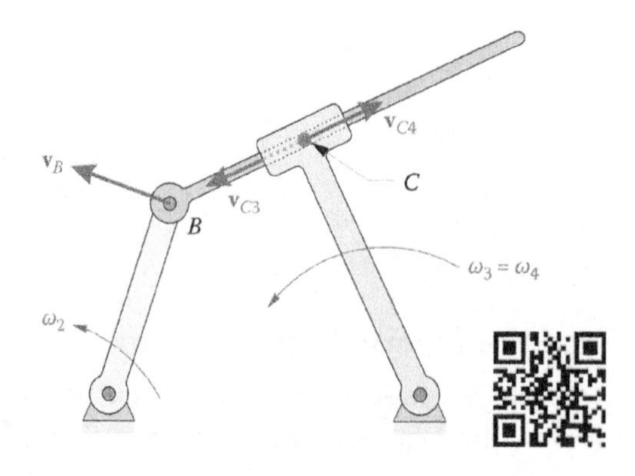

FIGURE 5.65
The inverted slider-crank linkage showing translational and angular velocities.

$$\frac{d}{dt}(a\mathbf{e}_2) = \omega_2 a\mathbf{n}_2 \tag{5.106}$$

The second term, however, is a little trickier. For this linkage, the length b is not a constant but varies with the motion of the linkage. Neither is the angle θ_3 a constant. Thus, we must use the product rule to differentiate this term:

$$\frac{d}{dt}(b\mathbf{e}_3) = \frac{db}{dt}\mathbf{e}_3 + b\omega_3\mathbf{n}_3 \tag{5.107}$$

As before, let us define

$$\dot{b} = \frac{db}{dt} \tag{5.108}$$

Then

$$\frac{d}{dt}(b\mathbf{e}_3) = \dot{b}\mathbf{e}_3 + b\omega_3\mathbf{n}_3 \tag{5.109}$$

The third term differentiates in the familiar manner, since c is a constant

$$\frac{d}{dt}(c\mathbf{e}_4) = c\omega_3\mathbf{n}_4 \tag{5.110}$$

where we have used $\omega_4 = \omega_3$, as shown above. The fourth term vanishes, since d and $\mathbf{e}_1$ are constant. Thus, the differentiated vector loop is

$$a\omega_2\mathbf{n}_2 + \dot{b}\mathbf{e}_3 + b\omega_3\mathbf{n}_3 - c\omega_3\mathbf{n}_4 = 0 \tag{5.111}$$

As before, let us separate knowns from unknowns

$$\text{Known}: a,\, b,\, c,\, d,\, \theta_2,\, \theta_3,\, \theta_4, \omega_2$$

$$\text{Unknown}: \omega_3,\, \dot{b}$$

And finally, putting this into matrix form we have:

$$\left[b\mathbf{n}_3 - c\mathbf{n}_4 \vdots \mathbf{e}_3\right]\begin{Bmatrix} \omega_3 \\ \dot{b} \end{Bmatrix} = \{-a\omega_2\mathbf{n}_2\} \tag{5.112}$$

The velocity analysis code for the inverted slider-crank is sufficiently similar to the previous linkages that it is left as an exercise for the reader. Figure 5.66 shows the example of inverted slider-crank that was used for the position analysis in Chapter 4.

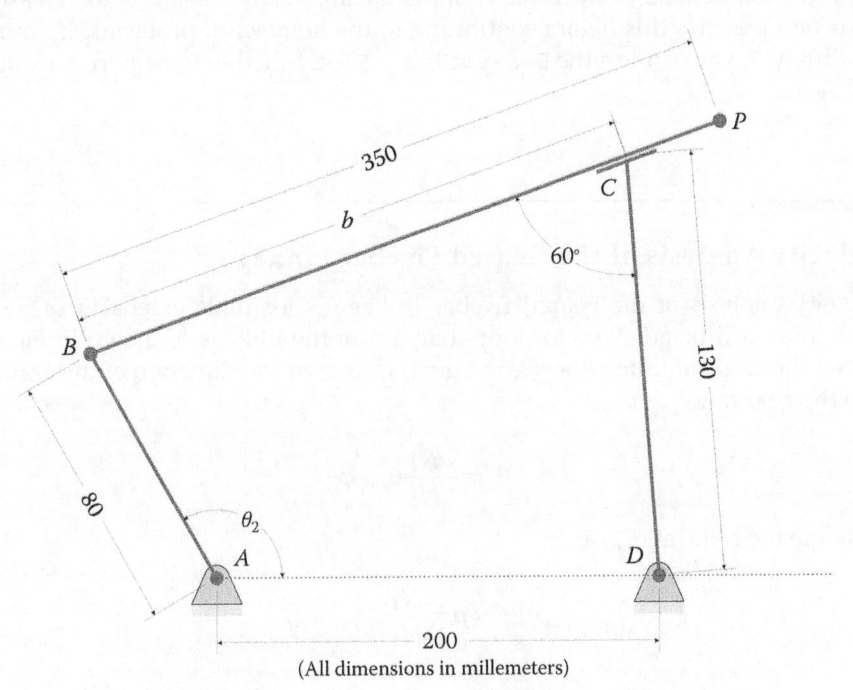

FIGURE 5.66
Dimensions of the example inverted slider-crank linkage.

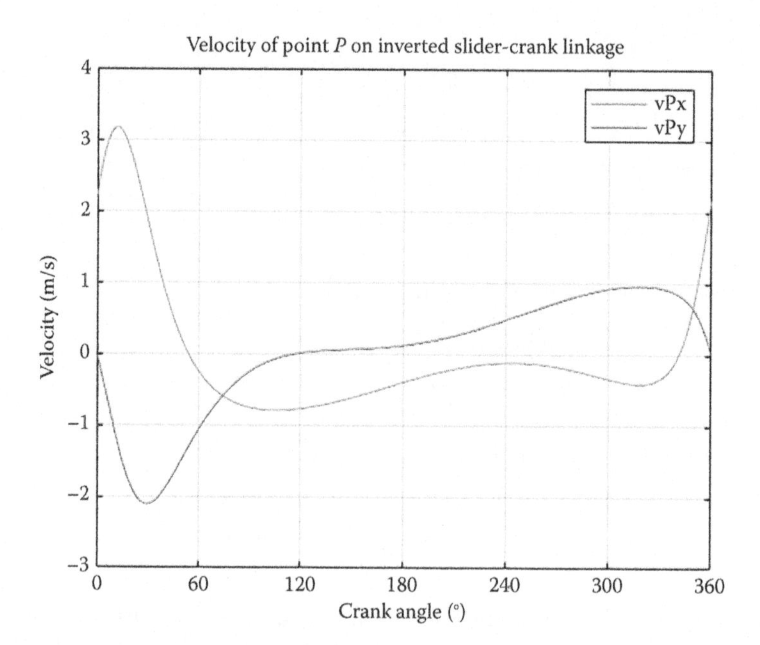

FIGURE 5.67
x and y velocity of point P on the example inverted slider-crank linkage.

The x and y components of the velocity at point P are shown in Figure 5.67. Please ensure that your plot matches this before continuing to the homework problems. If your velocities are different, you can use the `Derivative_Plot` function to help to track down the source of error.

5.7 Velocity Analysis of the Geared Fivebar Linkage

The velocity analysis of the geared fivebar linkage is a simple extension of the analysis of the fourbar linkage. A vector loop diagram of the linkage is shown in Figure 5.68. Recall that the angle of link 5 (the second gear) is known as a function of the crank angle, through the gear ratio

$$\theta_5 = -\frac{N_1}{N_2}\theta_2 + \varphi \tag{5.113}$$

Let us define the gear ratio, ρ, as

$$\rho = \frac{N_1}{N_2} \tag{5.114}$$

Then

$$\theta_5 = -\rho\theta_2 + \varphi \tag{5.115}$$

If we differentiate this expression with respect to time, we arrive at the speed ratio between the two gears

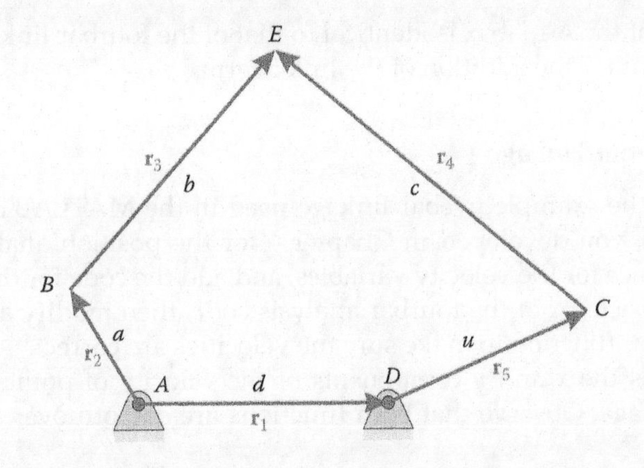

FIGURE 5.68
Vector loop diagram of the geared fivebar linkage.

$$\frac{d\theta_5}{dt} = -\rho\frac{d\theta_2}{dt} \tag{5.116}$$

or

$$\omega_5 = -\rho\omega_2 \tag{5.117}$$

Writing the vector loop equation gives

$$\mathbf{r}_2 + \mathbf{r}_3 - \mathbf{r}_4 - \mathbf{r}_5 - \mathbf{r}_1 = 0 \tag{5.118}$$

and writing in unit vector form

$$a\mathbf{e}_2 + b\mathbf{e}_3 - c\mathbf{e}_4 - u\mathbf{e}_5 - d\mathbf{e}_1 = 0 \tag{5.119}$$

Differentiating this with respect to time gives

$$a\omega_2\mathbf{n}_2 + b\omega_3\mathbf{n}_3 - c\omega_4\mathbf{n}_4 - u\omega_5\mathbf{n}_5 = 0 \tag{5.120}$$

Collecting all known terms on the right-hand side of the equation and rearranging into matrix form results in a familiar formula

$$\mathbf{A}\boldsymbol{\omega} = \mathbf{b} \tag{5.121}$$

with

$$\mathbf{A} = \left[b\mathbf{n}_3 \,\vdots\, -c\mathbf{n}_4\right] \qquad \mathbf{b} = \{-a\omega_2\mathbf{n}_2 + u\omega_5\mathbf{n}_5\} \tag{5.122}$$

and

$$\boldsymbol{\omega} = \begin{Bmatrix} \omega_3 \\ \omega_4 \end{Bmatrix} \tag{5.123}$$

It is noteworthy that the **A** matrix is identical to that of the fourbar linkage, and the **b** vector is unchanged but for the addition of the link 5 term.

5.7.1 Example Fivebar Linkage

Figure 5.69 shows the example fivebar linkage used in the MATLAB calculations below. Start with the code you developed in Chapter 4 for the position analysis of the geared fivebar. Allocate space for the velocity variables, and add the code for the velocity analysis (you may cut and paste from the fourbar analysis code, then modify as needed). Use the `Derivative_Plot` function to make sure the velocities are correct.

Figure 5.70 shows the x and y components of the velocity of point P for the example geared fivebar linkage. Observe that both functions are smooth over the entire range of motion.

Something interesting happens if we change the link lengths slightly such that a = 150 mm, u = 150 mm and d = 200 mm. The resulting velocity of point P is shown in Figure 5.71. Observe the discontinuity in velocity at $\theta_2 = 180°$.

To see what is happening at this point, it is helpful to plot the x and y coordinates of point P as shown in Figure 5.72. At $\theta_2 = 180°$ the plots exhibit sharp corners caused by a sudden change in direction.

Figure 5.73 shows the trajectory plot for the same linkage. The bold black circle highlights the sudden change in direction of the point P as the crank makes its revolution. Since velocity consists of both magnitude and direction, a sudden change in direction will result in a discontinuity in velocity, as we have seen. Viewing the plot in Figure 5.71, it appears as

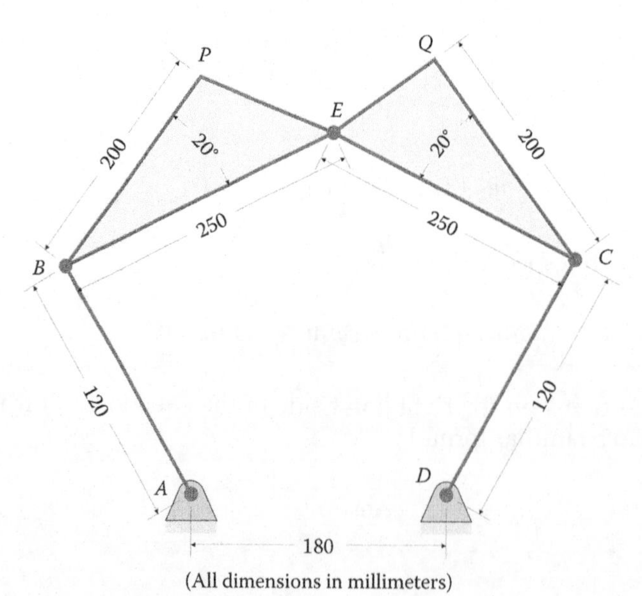

FIGURE 5.69
Example geared fivebar linkage.

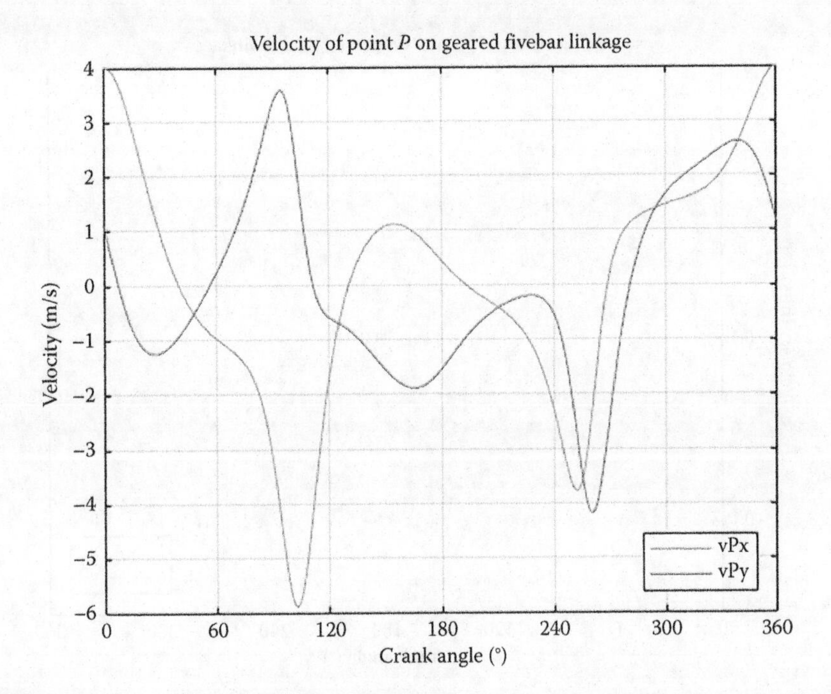

FIGURE 5.70
Velocity of point P on the example geared fivebar linkage.

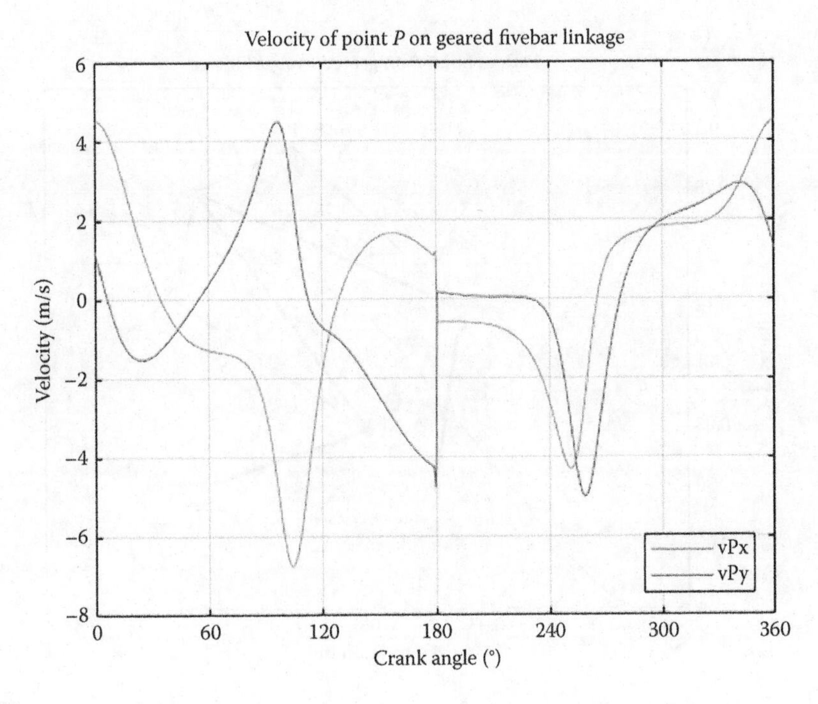

FIGURE 5.71
By changing the link lengths slightly we obtain discontinuities in the velocity of point P.

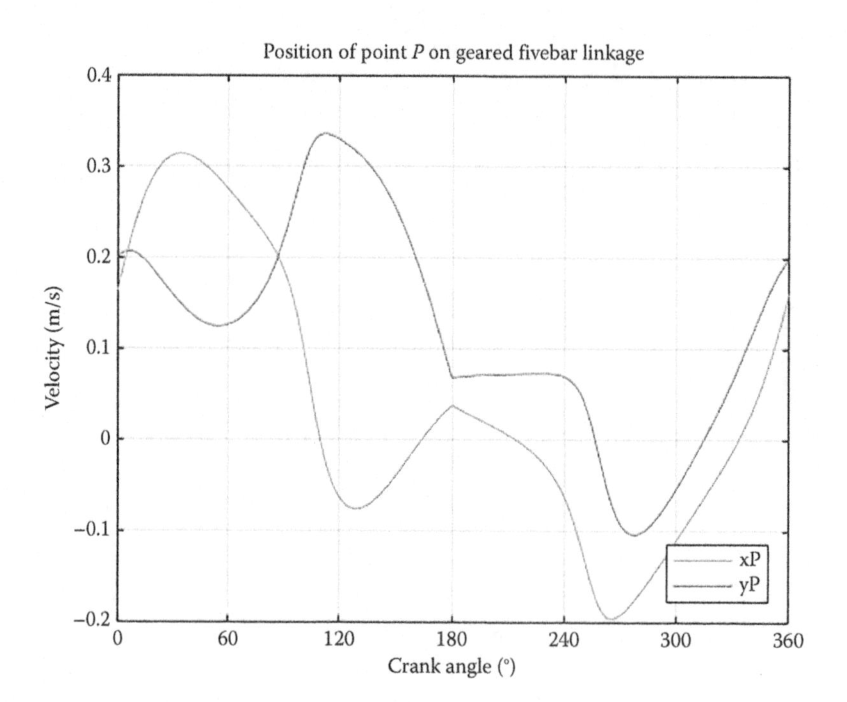

FIGURE 5.72
x and y coordinates of point P for modified fivebar linkage.

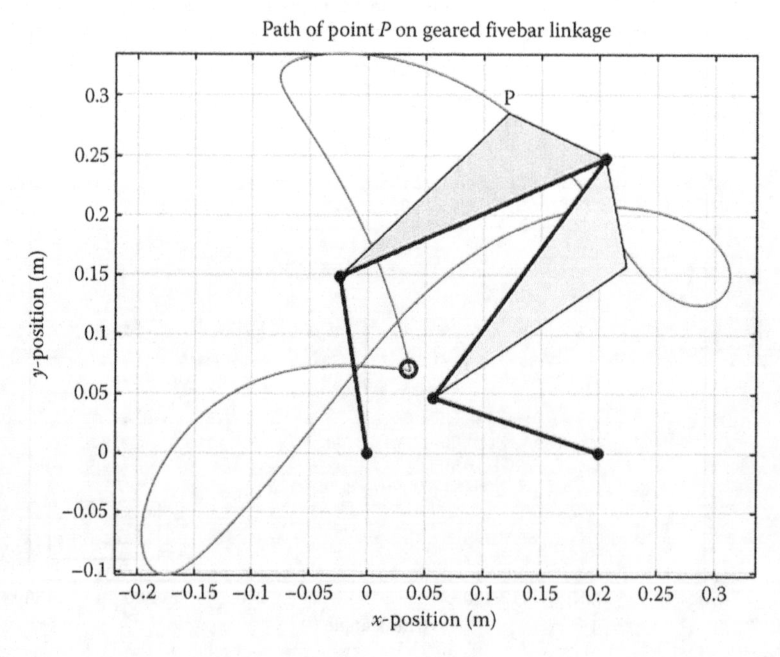

FIGURE 5.73
Trajectory plot for modified fivebar linkage.

though the velocity at point P changes value almost instantly when $\theta_2 = 180°$. This type of behavior will cause difficulties when we perform acceleration analysis, since

$$a = \frac{dv}{dt} \tag{5.124}$$

If dv is finite, but dt is zero, then we have infinite acceleration at the discontinuity. Since force is proportional to acceleration, it would appear that a discontinuity would cause infinite forces (and stresses) to occur somewhere in the linkage. Luckily, imprecisions in the fit of the pins will prevent infinite forces, but high forces will occur nonetheless, which may lead to excessive vibration and premature failure of the parts of the linkage. This type of discontinuity should be avoided, if possible.

5.8 Velocity Analysis of the Sixbar Linkage

We now turn our attention to conducting the velocity analysis of the sixbar linkage. We will use the vector loop method as before, with the caveat that the sixbar requires *two* vector loop equations, instead of just one. The vector loop diagram of the Stephenson Type I sixbar linkage is shown in Figure 5.74. The lower fourbar vector loop is

$$\mathbf{r}_2 + \mathbf{r}_3 - \mathbf{r}_4 - \mathbf{r}_1 = 0 \tag{5.125}$$

And the upper loop is

$$\mathbf{r}_{AE} + \mathbf{r}_5 - \mathbf{r}_6 - \mathbf{r}_{DF} - \mathbf{r}_1 = 0 \tag{5.126}$$

Writing each equation in unit vector form gives

$$a\mathbf{e}_2 + b\mathbf{e}_3 - c\mathbf{e}_4 - d\mathbf{e}_1 = 0 \tag{5.127}$$

$$p\mathbf{e}_{AE} + u\mathbf{e}_5 - v\mathbf{e}_6 - q\mathbf{e}_{DF} - d\mathbf{e}_1 = 0 \tag{5.128}$$

We then differentiate with respect to time, again employing the chain rule of differentiation:

$$a\omega_2\mathbf{n}_2 + b\omega_3\mathbf{n}_3 - c\omega_4\mathbf{n}_4 = 0 \tag{5.129}$$

$$p\omega_2\mathbf{n}_{AE} + u\omega_5\mathbf{n}_5 - v\omega_6\mathbf{n}_6 - q\omega_4\mathbf{n}_{DF} = 0 \tag{5.130}$$

As before, the angles γ_2 and γ_4 do not play a role in the differentiation, since they are constant. Now collect all unknown terms on the right side of the equations, and the known terms on the left.

$$b\omega_3\mathbf{n}_3 - c\omega_4\mathbf{n}_4 = -a\omega_2\mathbf{n}_2 \tag{5.131}$$

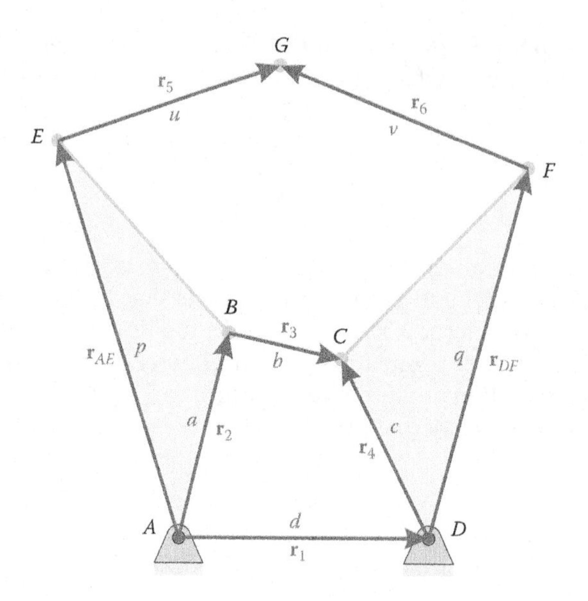

FIGURE 5.74
Vector loop diagram of the Stephenson Type I sixbar linkage.

$$u\omega_5\mathbf{n}_5 - v\omega_6\mathbf{n}_6 - q\omega_4\mathbf{n}_{DF} = -p\omega_2\mathbf{n}_{AE} \tag{5.132}$$

We now have four equations, with four unknowns

$$\text{unknown}: \omega_3,\ \omega_4,\ \omega_5, \omega_6 \tag{5.133}$$

and we can arrange the equations into matrix form as before

$$\mathbf{A\omega} = \mathbf{b} \tag{5.134}$$

where

$$\mathbf{A} = \begin{bmatrix} b\mathbf{n}_3 \vdots -c\mathbf{n}_4 & \vdots\ 0_{21} \vdots 0_{21} \\ 0_{21} \ \vdots -q\mathbf{n}_{DF} & \vdots\ u\mathbf{n}_5 \vdots -v\mathbf{n}_6 \end{bmatrix} \tag{5.135}$$

$$\mathbf{b} = \begin{Bmatrix} -a\omega_2\mathbf{n}_2 \\ -p\omega_2\mathbf{n}_{AE} \end{Bmatrix} \qquad \mathbf{\omega} = \begin{Bmatrix} \omega_3 \\ \omega_4 \\ \omega_5 \\ \omega_6 \end{Bmatrix} \tag{5.136}$$

Of course, this set of equations is only valid for the Stephenson Type I sixbar linkage. The MATLAB syntax for entering the **A** matrix is a little tricky for the sixbar linkages. Since there are four unknown velocities, the **A** matrix will have dimension (4 × 4). It is important to remember that each "row" in Equation (5.135) is actually two rows in the matrix corresponding to the x and y components of each vector. Each zero entry in the matrix is therefore a column vector of two zeros

$$0_{21} = \begin{Bmatrix} 0 \\ 0 \end{Bmatrix} \tag{5.137}$$

Near the top of the MATLAB script, after defining the link lengths and ground pivots, define the vector

```
Z21 = zeros(2,1);
```

Then to define the **A** matrix we would enter

```
A_Mat = [b*n3    -c*n4    Z21    Z21;
         Z21   -q*nDF   u*n5  -v*n6];
```

Each instance of Z21 produces a column vector of two zeros, as required.

A similar analysis can be carried out for the other sixbar linkages, whose vector loop diagrams are shown in Figures 5.75–5.78. It is interesting to note that no special measures are needed for velocity analysis of the Stephenson Type II sixbar. Once the position analysis has been performed using the Newton–Raphson approach, the velocity analysis proceeds as with the other sixbar linkages. Table 5.2 shows the resulting **A** matrices and **b** vectors for performing velocity analysis on the family of sixbar linkages.

5.8.1 Some Example Solutions for the Sixbar Linkage

Figures 5.79–5.83 show dimensions for the example sixbar linkages discussed in Chapter 4. In each case, the angular velocity of the crank is 10 rad/s. Figures 5.84–5.88 shows the x and y components of velocity for an interesting point on each linkage. Be sure to check your answers against these plots before attempting the homework problems.

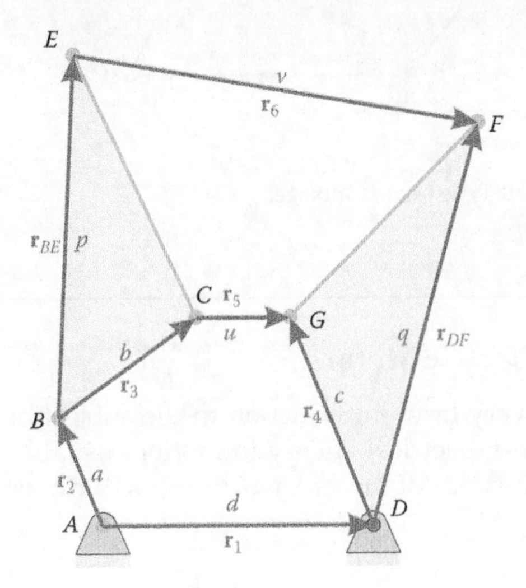

FIGURE 5.75
Vector loop diagram for the Stephenson Type II sixbar linkage.

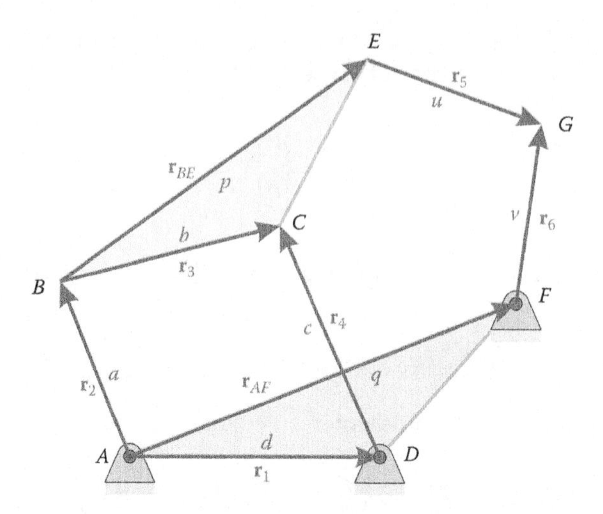

FIGURE 5.76
Vector loop diagram for the Stephenson Type III sixbar linkage.

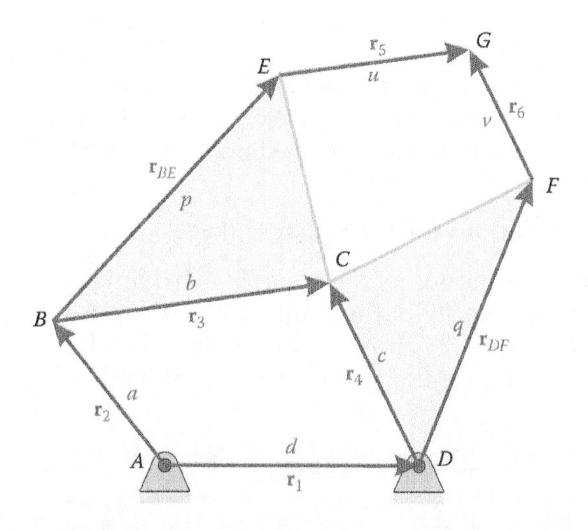

FIGURE 5.77
Vector loop diagram for the Watt Type I sixbar linkage.

5.9 Introduction to Electric Motors

This section will give a very brief introduction to the subject of electric motors. Electric motors are used to convert electrical energy to motion, usually of the rotational variety. There are many different types of motors, which we will divide here into the following categories:

- AC Motors
- DC Motors

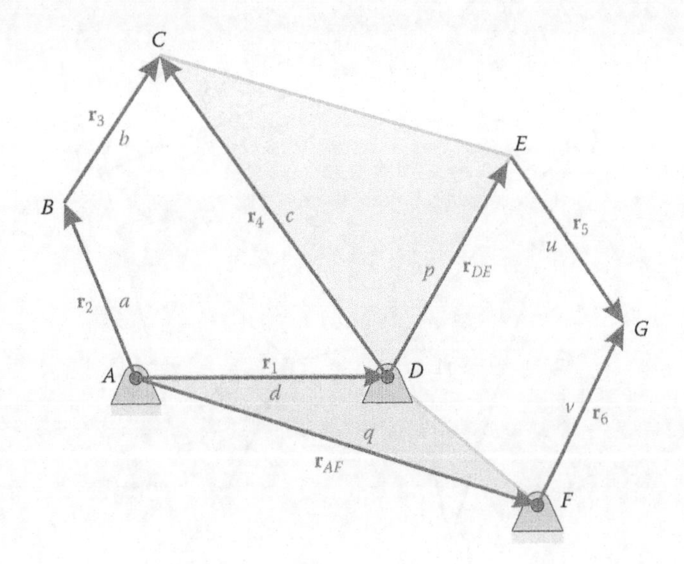

FIGURE 5.78
Vector loop diagram for the Watt Type II sixbar linkage.

TABLE 5.2
A Matrices and **b** Vectors for Velocity Analysis of the Sixbar Linkages

Linkage	**A**	**b**
Stephenson I	$\begin{bmatrix} b\mathbf{n}_3 \vdots -c\mathbf{n}_4 & \vdots 0_{21} & \vdots 0_{21} \\ 0_{21} \vdots -q\mathbf{n}_{DF} & \vdots u\mathbf{n}_5 & \vdots -v\mathbf{n}_6 \end{bmatrix}$	$\begin{Bmatrix} -a\omega_2\mathbf{n}_2 \\ -p\omega_2\mathbf{n}_{AE} \end{Bmatrix}$
Stephenson II	$\begin{bmatrix} b\mathbf{n}_3 \vdots -c\mathbf{n}_4 & \vdots u\mathbf{n}_5 & \vdots 0_{21} \\ p\mathbf{n}_{BE} \vdots -q\mathbf{n}_{DF} & \vdots 0_{21} & \vdots v\mathbf{n}_6 \end{bmatrix}$	$\begin{Bmatrix} -a\omega_2\mathbf{n}_2 \\ -a\omega_2\mathbf{n}_2 \end{Bmatrix}$
Stephenson III	$\begin{bmatrix} b\mathbf{n}_3 \vdots -c\mathbf{n}_4 & \vdots 0_{21} & \vdots 0_{21} \\ p\mathbf{n}_{BE} \vdots 0_{21} & \vdots u\mathbf{n}_5 & \vdots -v\mathbf{n}_6 \end{bmatrix}$	$\begin{Bmatrix} -a\omega_2\mathbf{n}_2 \\ -a\omega_2\mathbf{n}_2 \end{Bmatrix}$
Watt I	$\begin{bmatrix} b\mathbf{n}_3 \vdots -c\mathbf{n}_4 & \vdots 0_{21} & \vdots 0_{21} \\ p\mathbf{n}_{BE} \vdots -q\mathbf{n}_{DF} & \vdots u\mathbf{n}_5 & \vdots -v\mathbf{n}_6 \end{bmatrix}$	$\begin{Bmatrix} -a\omega_2\mathbf{n}_2 \\ -a\omega_2\mathbf{n}_2 \end{Bmatrix}$
Watt II	$\begin{bmatrix} b\mathbf{n}_3 \vdots -c\mathbf{n}_4 & \vdots 0_{21} & \vdots 0_{21} \\ 0_{21} \vdots p\mathbf{n}_{DE} & \vdots u\mathbf{n}_5 & \vdots -v\mathbf{n}_6 \end{bmatrix}$	$\begin{Bmatrix} -a\omega_2\mathbf{n}_2 \\ 0 \end{Bmatrix}$

- Servomotors (hobby and industrial)
- Stepper motors

AC motors run using line current (from an outlet or electrical panel). DC motors are powered using batteries or a "wall wart" power supply. Servomotors and stepper motors are usually powered with DC motors, but require sophisticated controllers to function. Each type of motor is described in more detail below. Note that entire books are devoted to the

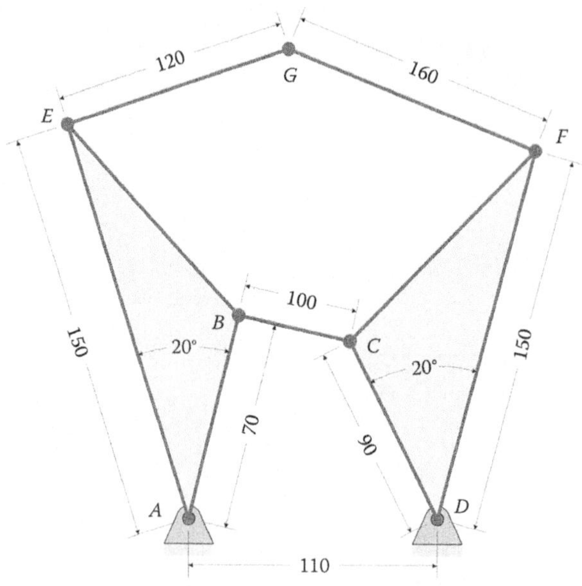

(All dimensions in millimeters)

Crank length: 70	Rocker length: 90
Length *AE* on crank: 150	Length DF on crank: 150
Internal angle of crank: 20°	Internal angle of rocker: –20°
Coupler length: 100	Length of link 5: 120
Distance between ground pins: 110	Length of link 6: 160
Crank angular velocity: 10 rad/s	

FIGURE 5.79
Dimensions of example Stephenson Type I sixbar linkage.

subject of *each type* of motor described here – this section can provide only the briefest of introductions. There are many excellent online resources to choose from once you have decided upon a particular type of motor for your application.

5.9.1 AC Motors

An AC motor is powered by plugging it into the wall, or wiring it up to an electrical panel. A typical AC motor is shown in Figure 5.89. AC motors can be divided into two major types: single phase and polyphase.

The abbreviation AC stands for "alternating current." The voltage present at the "hot" lead in a typical wall outlet is shown in Figure 5.90. The voltage signal takes the form of a sine wave at 60 Hz (or 50 Hz in Europe). The amplitude of the voltage is normally specified by its root-mean-square value, or rms, which is calculated as

$$V_{pp} = \sqrt{2} \times V_{\text{rms}}$$

The rms value gives an "average" voltage over the period of one sine wave, and is helpful for calculating the power consumed during a full cycle. In the United States, the rms voltage present at a wall outlet is nominally 120 V, but this can vary as much as ±5 V depending upon how much electricity is being consumed at a given moment, and how far the outlet is from the power source.

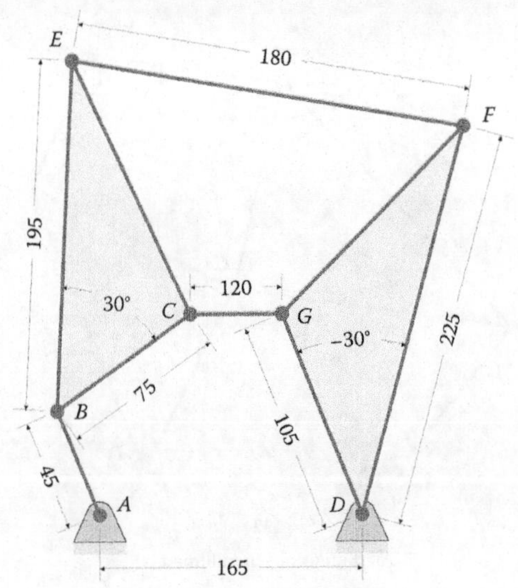

(All dimensions in millimeters)

Crank length: 45
Coupler length: 75
Length *BE* on coupler: 195
Internal angle of coupler: 30°
Distance between ground pins: 165
Crank angular velocity: 10 rad/s

Rocker length: 105
Length *DF* on rocker: 225
Internal angle of rocker: −30°
Length of link 5: 120
Length of link 6: 180

FIGURE 5.80
Dimensions of example Stephenson Type II sixbar linkage.

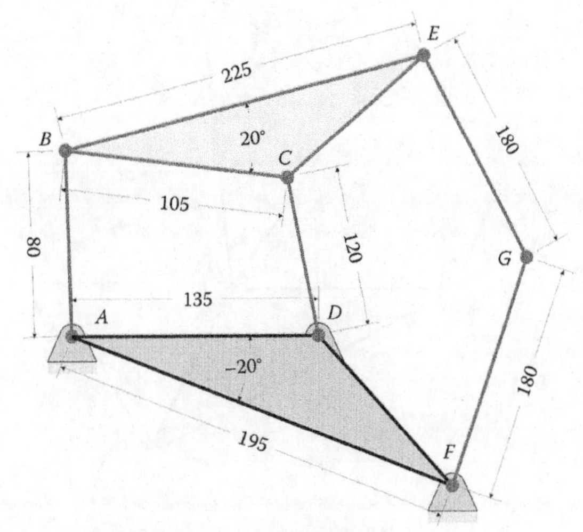

(All dimensions in millimeters)

Crank length: 80
Coupler length: 105
Internal angle of coupler: 20°
Length *BE* on coupler: 225
Distance between ground pins: 135
Crank angular velocity: 10 rad/s

Internal angle of ground: −20°
Length *AF* on ground: 195
Rocker length: 120
Length of link 5: 180
Length of link 6: 180

FIGURE 5.81
Dimensions of example Stephenson Type III sixbar linkage.

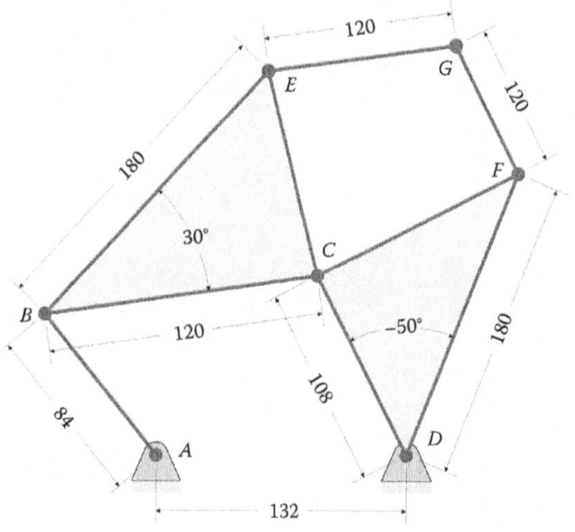

(All dimensions in millimeters)

Crank length: 84
Coupler length: 120
Internal angle of coupler: 30°
Length *BE* on coupler: 180
Distance between ground pins: 132
Crank angular velocity: 10 rad/s

Rocker length: 108
Internal angle of rocker: −50°
Length *DF* on rocker: 180
Length of link 5: 120
Length of link 6: 120

FIGURE 5.82
Dimensions of example Watt Type I sixbar linkage.

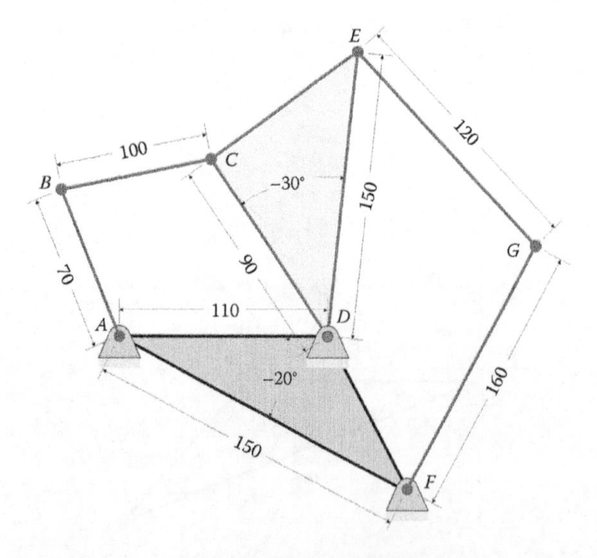

(All dimensions in millimeters)

Crank length: 70
Coupler length: 100
Distance between ground pins: 110
Internal angle of ground: −20°
Length *AF* on ground: 150
Crank angular velocity: 10 rad/s

Rocker length: 90
Internal angle of rocker: −30°
Length *DE* on rocker: 150
Length of link 5: 120
Length of link 6: 160

FIGURE 5.83
Dimensions of example Watt Type II sixbar linkage.

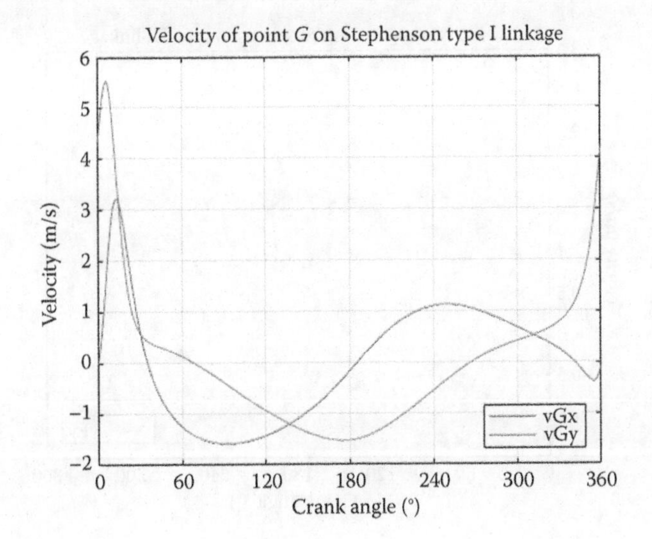

FIGURE 5.84
Velocity of point *G* on Stephenson Type I sixbar linkage.

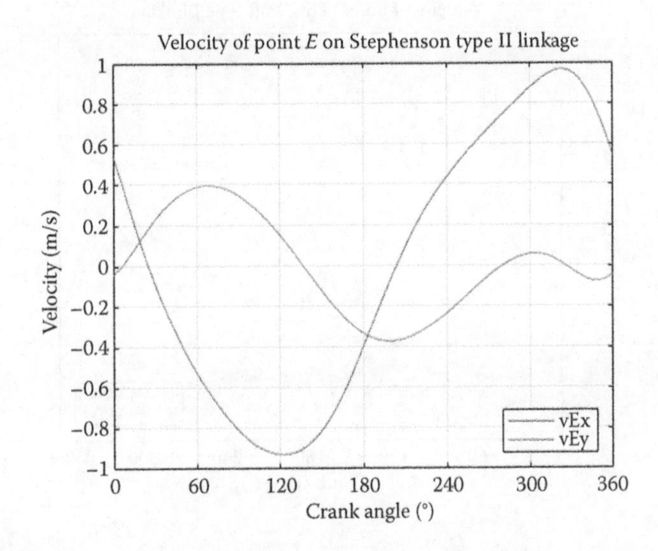

FIGURE 5.85
Velocity of point *E* on Stephenson Type II sixbar linkage.

The operation of a typical AC motor is shown in Figure 5.91. The changes in current direction create changes in magnetic field polarity, which causes the magnet in the center to rotate. Either the magnet in the center is an electromagnet, or has a magnetic field induced by the alternating current in the coils outside. A single-phase AC motor is easily identified by the fact that there are three wires to hook up: "hot," neutral, and ground.

Three-phase power has three "hot" leads, each of which is 120° out of phase with the other leads. The voltage versus time signal for a low-voltage three-phase power supply is shown in Figure 5.92. Three-phase power is only available in industrial or commercial settings, so you are very unlikely to encounter a three-phase outlet at your house or

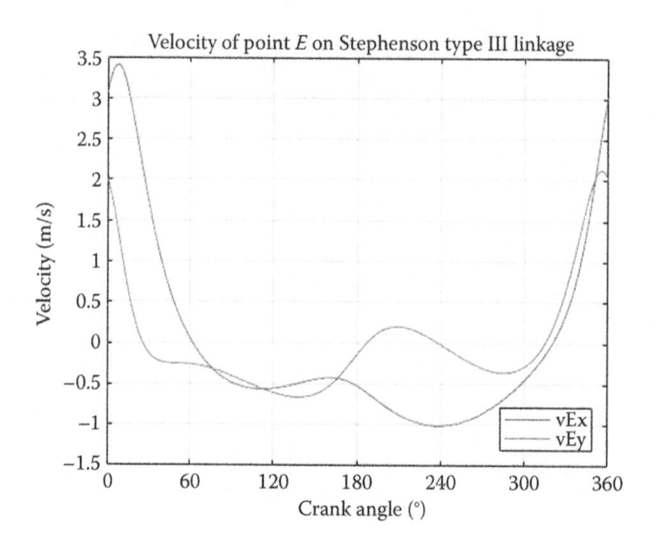

FIGURE 5.86
Velocity of point E on Stephenson Type III sixbar linkage.

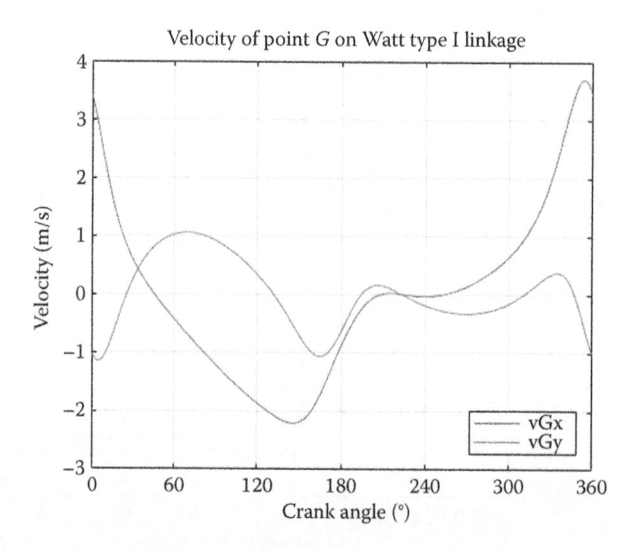

FIGURE 5.87
Velocity of point G on Watt Type I sixbar linkage.

apartment. A three-phase plug is easily identified by its four prongs – three are the "hot" leads and one is ground.

Three-phase motors have six coils that are powered in sequence, as shown in Figure 5.93. Since each coil is spaced 60° apart, the three-phase motor has much smoother and efficient operation than the single-phase motor. In fact, three-phase motors are comparatively much smaller than their single-phase brethren for a given power. The main disadvantage of three-phase motors is the rarity of three-phase power outlets. For this reason, they are mainly used in industrial settings. Large industrial motors (e.g. for milling machines, lathes, HVAC blowers, etc.) are three-phase, while small, handheld power tools (e.g. sanders, grinders, etc.) are single phase.

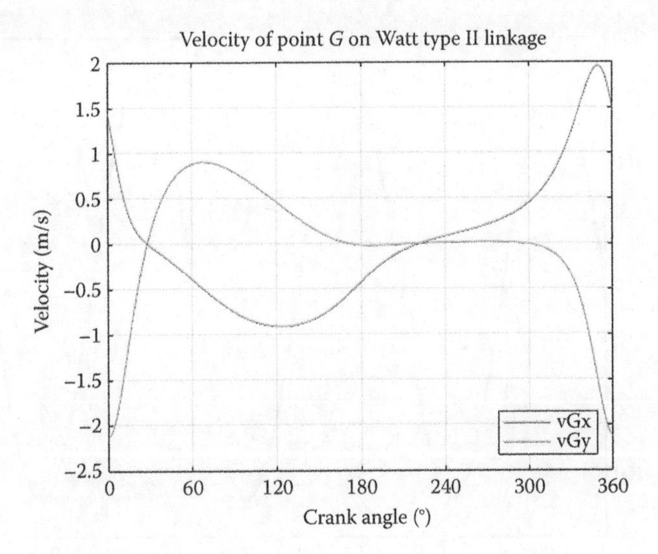

FIGURE 5.88
Velocity of point G on Watt Type II sixbar linkage.

FIGURE 5.89
A typical single-phase AC motor.

The load-speed curve for a typical AC motor is shown in Figure 5.94. This curve shows the speed of the motor for a given load; as the load on the motor is increased, the speed decreases. Because the speed of an AC motor is governed by the switching of current direction, the AC motor speed will remain relatively constant under varying load. If an AC motor is overloaded, it will usually stall suddenly, rather than slow down gradually.

Until recently, the speed of an AC motor was governed strictly by the frequency of the power supply voltage and the number of coils. In most cases, there were two AC motor speeds: 1745 rpm and 3450 rpm, which are very close to multiples of 60 Hz (1800 and 3600). This is still the case for most inexpensive, consumer-grade motors. The advent of economical high-power switching transistors has made "variable-frequency drives" feasible in industrial settings. These controllers vary the speed of an AC motor by changing the

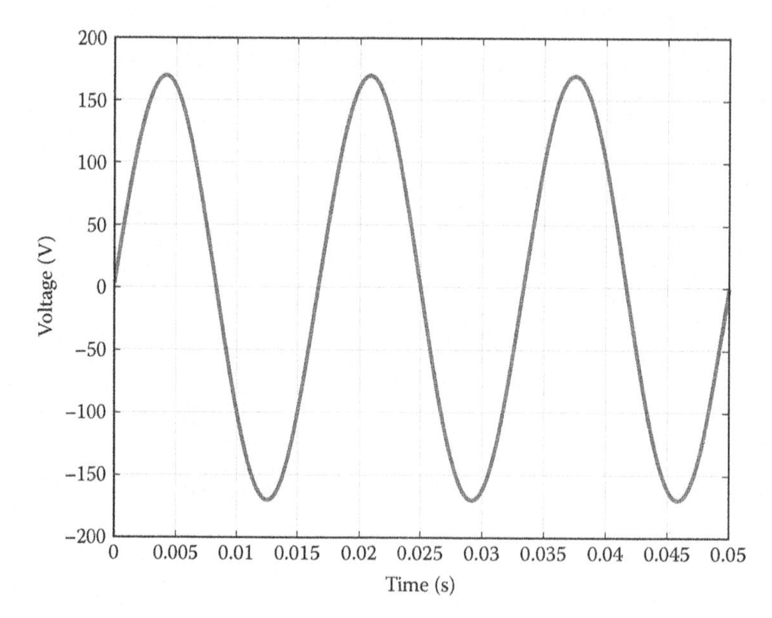

FIGURE 5.90
Voltage versus time for single-phase AC power at an ordinary outlet.

frequency of the AC pulses. The controllers are too expensive for use outside industrial settings, however.

5.9.2 DC Motors

DC motors, such as the one shown in Figure 5.95, are ubiquitous today. They are simple, cheap, and reliable. In fact, it is common to find DC motors for sale for less than one dollar! The most common DC motors are powered by batteries, and can be very small. A cell phone vibrator, for instance, is a DC motor with an unbalanced weight on its shaft.

A cutaway view of a typical permanent-magnet DC motor is shown in Figure 5.96. The magnets are stationary, and are placed outside the coil. The coil is attached to the rotor, and it rotates with the shaft. Current passes from the battery through the *brushes* to the coil. The current passing through the coil creates a magnetic field – the attraction between this field and that of the permanent magnets creates a torque on the coil. Once the field of the coil has begun to align itself with the field from the magnets, the current in the coil reverses direction because of the *commutator.* In this way, the rotor keeps spinning, with the coil current reversing direction every half revolution.

A typical load-speed curve for a DC motor is shown in Figure 5.97. The greater the load applied to the motor, the slower it will go, until it stalls. In fact, the load-speed curve for a DC motor is almost linear, which is quite different from that of an AC motor, which rotates at an almost constant speed until the stall torque is reached.

The speed of a DC motor increases as we increase the voltage applied to the leads, as shown in Figure 5.98. Thus, a typical DC motor will spin faster with a 9 V battery than with 2 AA batteries (3 V). Of course, if too high a voltage is applied to a DC motor the resulting high current will melt the insulation in the motor coils and cause it to "short out."

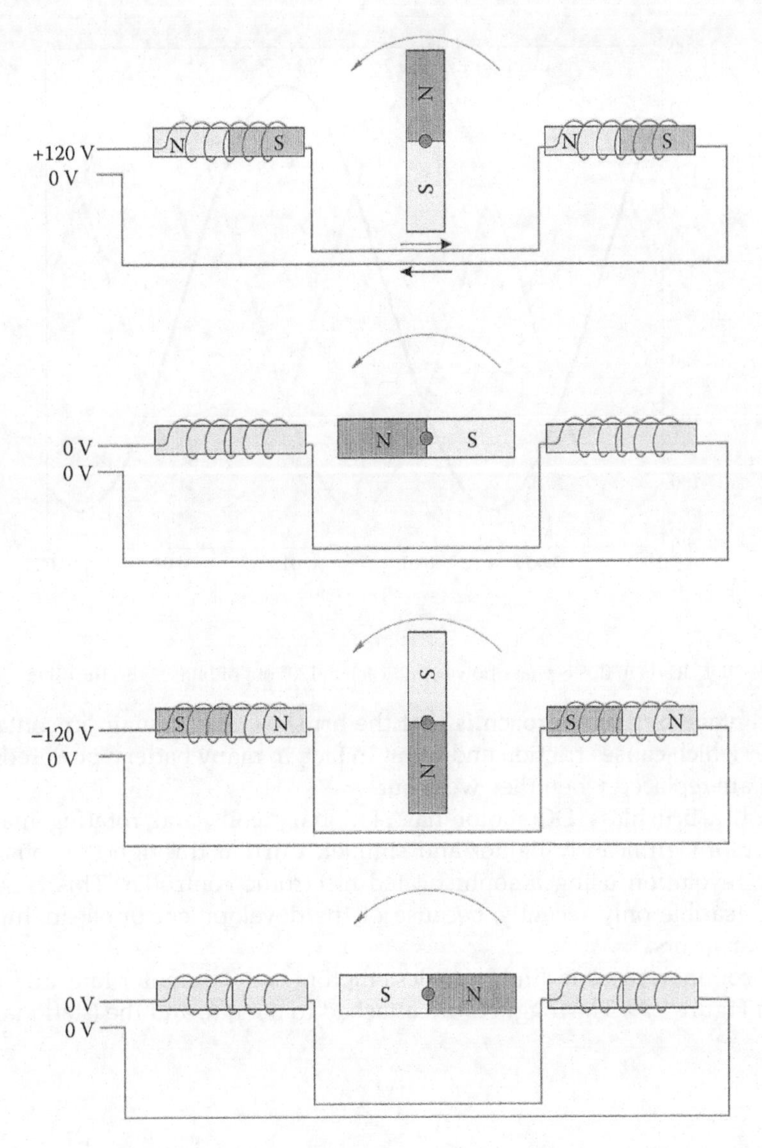

FIGURE 5.91
One rotation cycle for an AC motor.

The simplest, cheapest DC motor controllers work this way – speed is increased by increasing voltage and vice versa. Unfortunately, this type of controller does not account for the load placed on the motor – the motor will still slow down if a load is applied. More sophisticated controllers use *feedback* to measure the instantaneous speed of the motor and adjust the applied voltage accordingly.

5.9.3 Brushless Motors

An interesting newer motor technology is found in *brushless DC motors*. Recall that the current must switch direction every half revolution in order for the motor to keep spinning. This is accomplished by the brushes/commutator in a traditional DC motor.

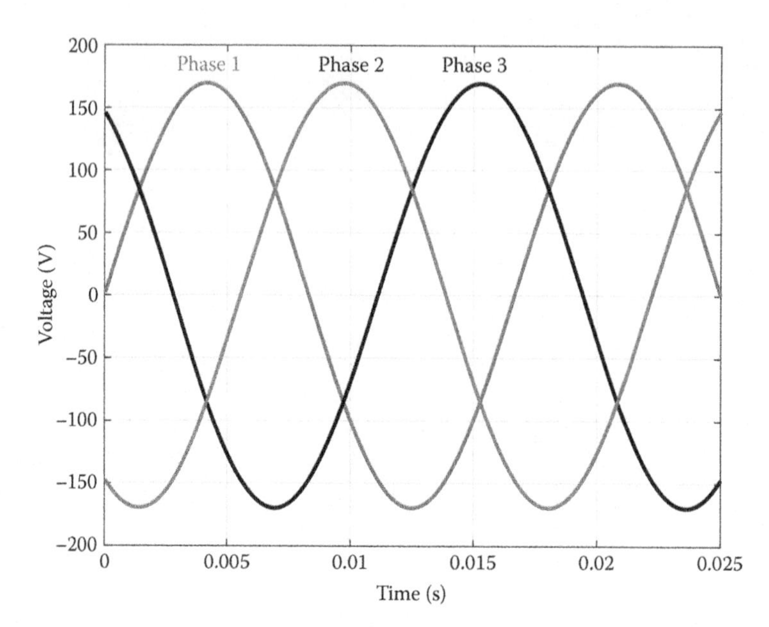

FIGURE 5.92
There are three "hot" leads in three-phase power, and each is 120° out of phase with the other.

One disadvantage of this approach is that the brushes must remain in contact with the commutator, which causes friction and wear. In fact, in many battery-powered hand tools, the brushes are replaced when they wear out.

In contrast, a brushless DC motor has stationary coils and rotating magnets. This makes the rotor significantly lighter and simpler. Current through the coils is reversed twice every revolution using a sophisticated electronic controller. This type of control has become feasible only recently, because of the development of cheap, high-powered switching transistors.

The most common place to find brushless motors is in computer fans and disk drives, as shown in Figure 5.99. The magnets are attached to the fan, and the (stationary) coils are

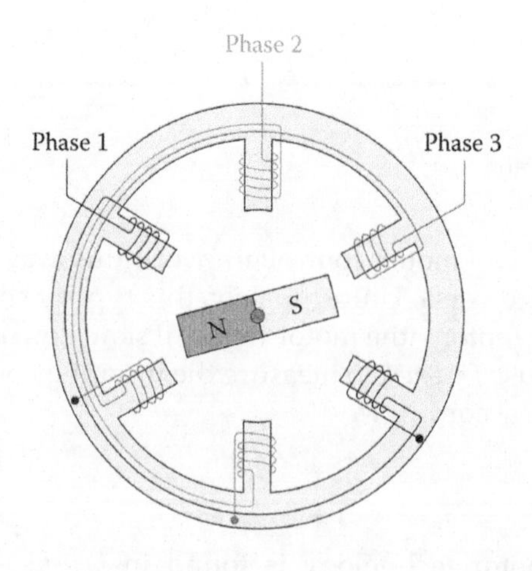

FIGURE 5.93
Wiring of the coils in a typical high-voltage three-phase motor. Note that opposing coils are wired in series. For a low-voltage three-phase motor, these would be wired in parallel.

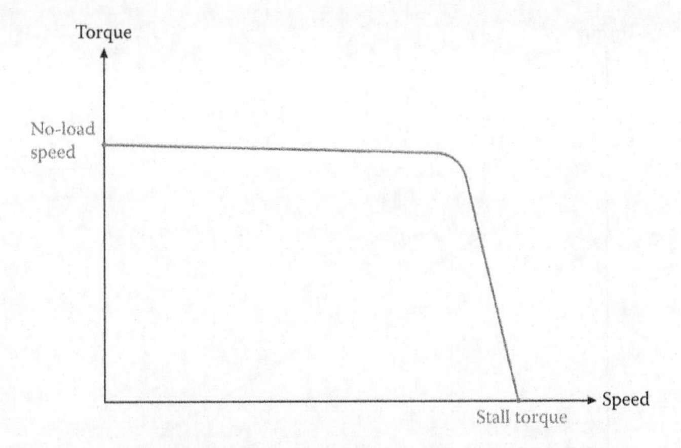

FIGURE 5.94
The load-speed curve for a typical single-phase AC motor.

FIGURE 5.95
A typical small, battery-powered DC motor.

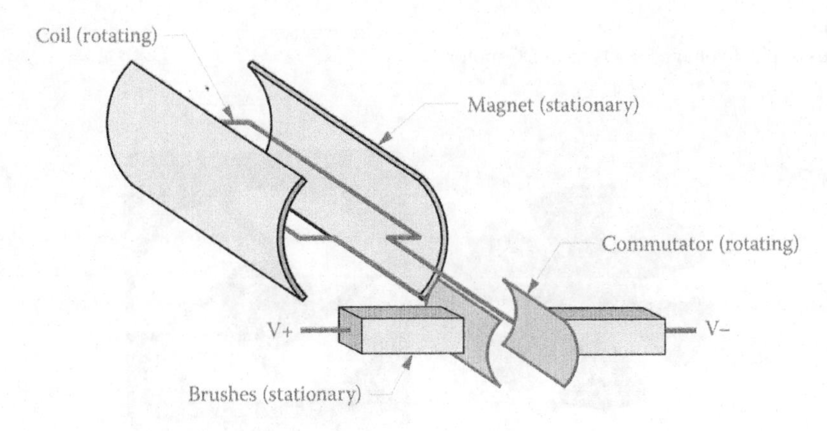

FIGURE 5.96
Cutaway view of a typical DC motor. The commutator rotates with the motor shaft and changes the direction of current within the coil every half revolution.

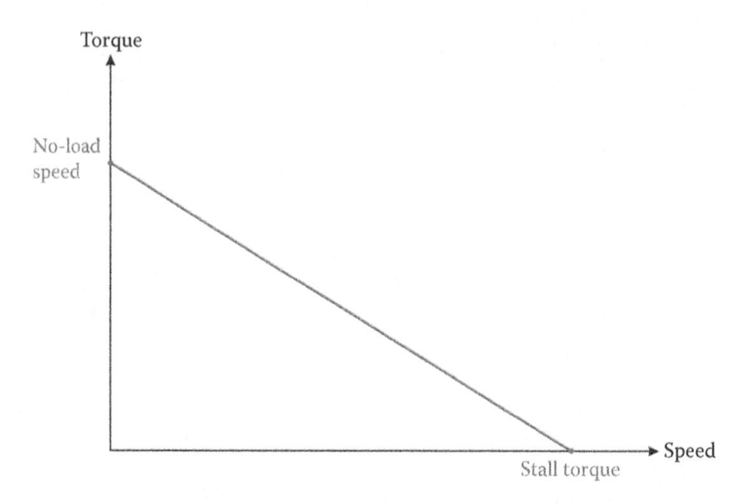

FIGURE 5.97
Load-speed curve for a typical DC motor.

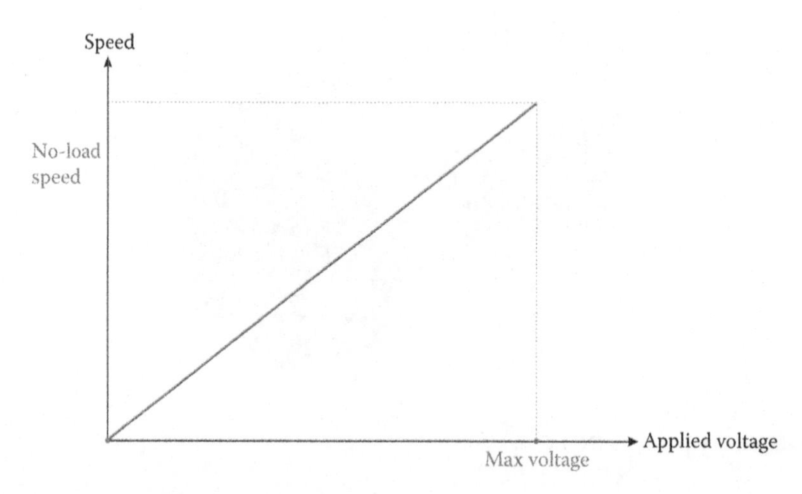

FIGURE 5.98
Speed versus applied voltage for a typical DC motor.

FIGURE 5.99
Brushless motors are often used for computer fans, since they generate very little electrical noise.

inside the hub. The advantage of a brushless motor in this application is the lack of electrical noise caused by sparking between brush and commutator.

5.9.4 Servo Motors

All of the motors discussed so far will rotate continuously when power is applied. A servo motor, in contrast, will rotate to a specified position, and hold that position until requested to move. Figure 5.100 shows a small hobby servo, used in building RC airplanes, cars or in robotics. There are three wires leading to the servo: two are for DC power, and one wire signals the requested position. Servomotors are usually DC motors, and come in all sizes from hobby to industrial. Hobby servomotors are very inexpensive, but suffer from low available torque and speed. High torque/speed industrial servomotors, which are commonly used for assembly-line robots, can be very expensive.

FIGURE 5.100
A common hobby-type servomotor.

FIGURE 5.101
A common stepper motor.

TABLE 5.3
Comparison of the Most Common Motor Types

	AC motor	DC motor	Brushless DC	Servo	Stepper
Cost	Cheap	Cheap	Less cheap	Expensive	Expensive
Controller needed?	No	No	Yes	Yes	Yes
Variable speed?	No	Yes	Yes	Yes	Yes
Drive torque	High	High	High	High	Low
Holding torque	Low	Moderate	Moderate	High	High

5.9.5 Stepper Motors

A stepper motor is similar to a brushless motor, in that the coils are stationary and magnets are fixed to the rotor. The difference is that stepper motors are designed to rotate in steps, not continuously. Because of this, they can be used for precise position control, as needed in robotics and similar applications. These motors can hold a position with reasonably high holding torque, but have relatively low torque to move between positions. A stepper motor, like a servomotor, requires sophisticated control hardware and software to operate. Stepper motors are most commonly found in printers and copiers, where they can regulate the position of the paper with precision and speed. A common stepper motor is shown in Figure 5.101.

Table 5.3 summarizes the properties of each motor discussed in this section. You may use this table as a first-pass selection guide for your particular application.

5.10 Practice Problems

Problem 5.1

In Figure 5.102, an ant is crawling at a speed of 10 mm/s outward on a link that is rotating at 2 rad/s in the counterclockwise direction. At the time of interest, the ant is located at a distance of 50 mm from the center of rotation. What is the total velocity of the ant?

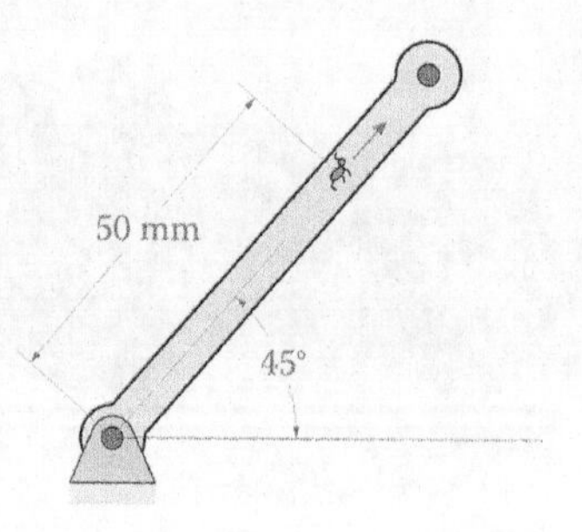

FIGURE 5.102
Problem 5.1.

Problem 5.2

In the linkage shown in Figure 5.103 the length *AB* and *BC* are both 200 mm. Link *AB* rotates at 100 rpm clockwise and link *BC* rotates at 200 rpm counterclockwise. Find the total velocity at point *C*.

Problem 5.3

The ant from Problem 5.1 is now crawling at a speed of 10 mm/s from point *C* to point *B* as shown in Figure 5.104. The length *AB* and *BC* are both 200 mm. Link *AB* rotates at 10 rpm clockwise and link *BC* rotates at 20 rpm counterclockwise. At the time of interest, the ant is 150 mm from point *B*. Find the total velocity of the ant.

Problem 5.4

The rider of an amusement park ride sits at point *D* at the end of arm *CD* as shown in Figure 5.105. To enhance his enjoyment of the ride, he is attempting to bring a sandwich toward his mouth at 20 mm/s. The length *AB* is 10 m, the length *BC* is 5 m, and the length *CD* is 2 m. What is the overall velocity of the sandwich?

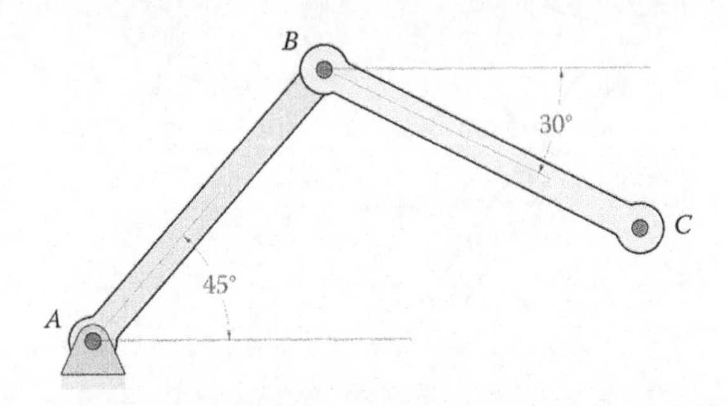

FIGURE 5.103
Problem 5.2.

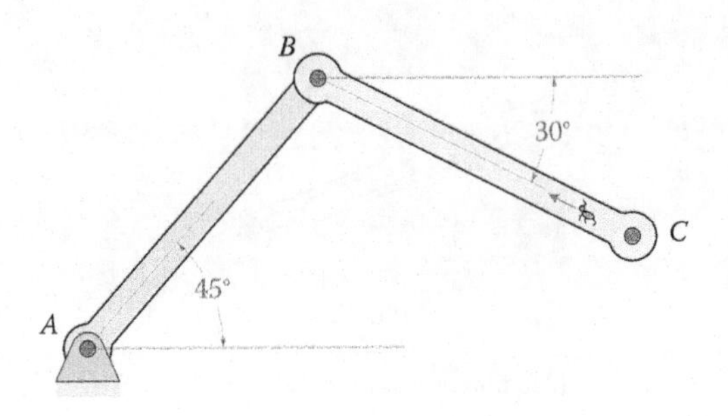

FIGURE 5.104
Problem 5.3.

Problem 5.5

Figure 5.106 shows a threebar linkage whose crank rotates at 10 rad/s. Use SOLIDWORKS and the method of instant centers to find the velocity at point P at the end of the slider. The crank is at 120° from the horizontal. What is the angular velocity of the slider?

Problem 5.6

Figure 5.107 shows a slider linkage whose crank rotates at 10 rad/s. Use SOLIDWORKS and the method of instant centers to find the velocity at point C on the piston. The crank is at 60° from the horizontal. What is the angular velocity of the connecting rod?

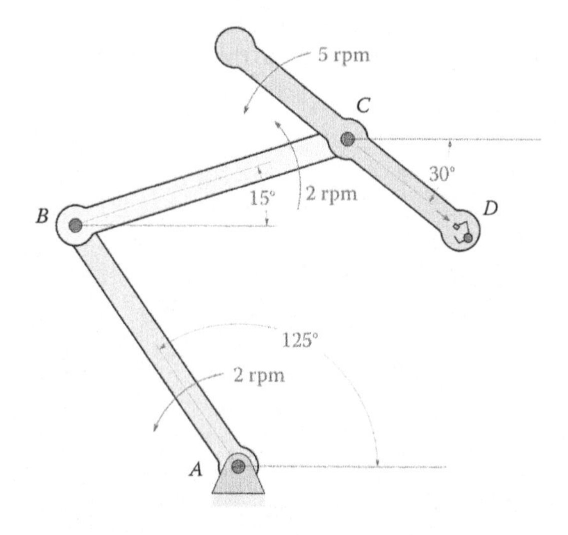

FIGURE 5.105
Problem 5.4.

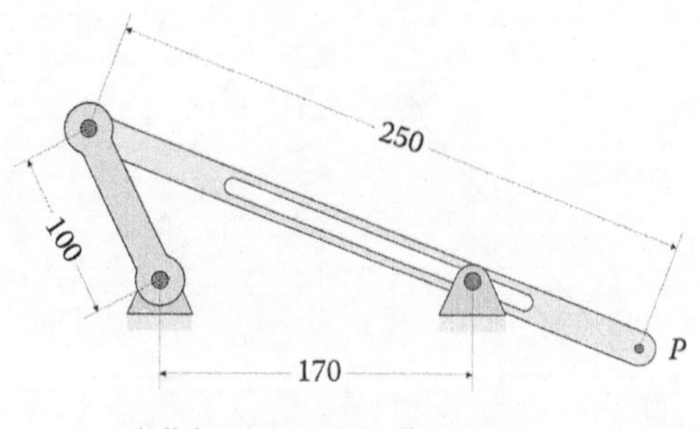

FIGURE 5.106
Problem 5.5.

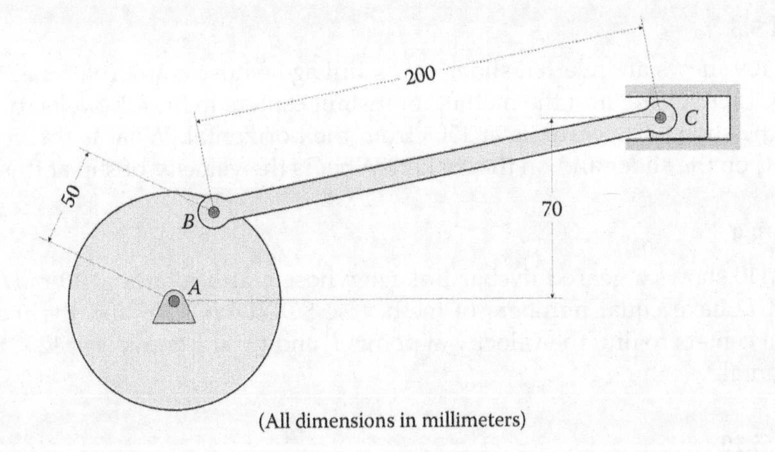

FIGURE 5.107
Problem 5.6.

Problem 5.7

Figure 5.108 shows a fourbar linkage whose crank rotates at 10 rad/s. Use SOLIDWORKS and the method of instant centers to find the velocity at point P on the coupler. The crank is at 120° from the horizontal. What is the angular velocity of the coupler and rocker?

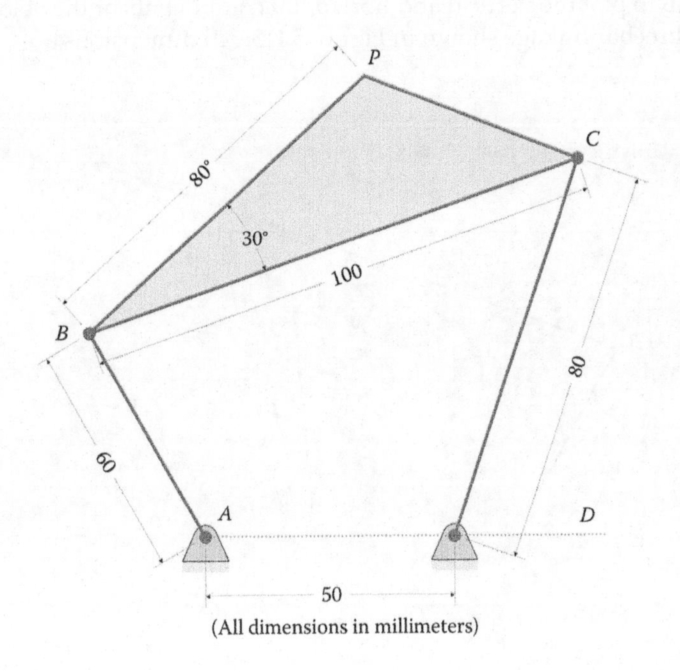

FIGURE 5.108
Problem 5.7.

Problem 5.8

Figure 5.109 shows an inverted slider-crank linkage whose crank rotates at 10 rad/s. Use SOLIDWORKS and the method of instant centers to find the velocity at point P on the slider. The crank is at 150° from the horizontal. What is the velocity at point C on the slider and on the rocker? What is the velocity of slip at this point?

Problem 5.9

Figure 5.110 shows a geared fivebar linkage whose crank rotates at 10 rad/s. Gears A and D have equal numbers of teeth. Use SOLIDWORKS and the method of instant centers to find the velocity at points P and Q. The crank is at 120° from the horizontal.

Problem 5.10

Figure 5.111 shows a sixbar linkage whose crank (link AB) rotates at 10 rad/s. Use SOLIDWORKS and the method of instant centers to find the velocity at point G. The crank is at 120° from the horizontal.

Problem 5.11

Use MATLAB to plot the x and y components of the velocity of point P of the threebar linkage shown in Figure 5.112. The crank is 0.5 m, the slider is 2 m, and the distance between ground pins is 1 m. The crank has a constant angular velocity of 8 rad/s. What is the amplitude of the slider's maximum angular velocity?

Problem 5.12

Use MATLAB to plot the vertical and horizontal components of the velocity of point P on the threebar linkage shown in Figure 5.113. All dimensions are in meters and

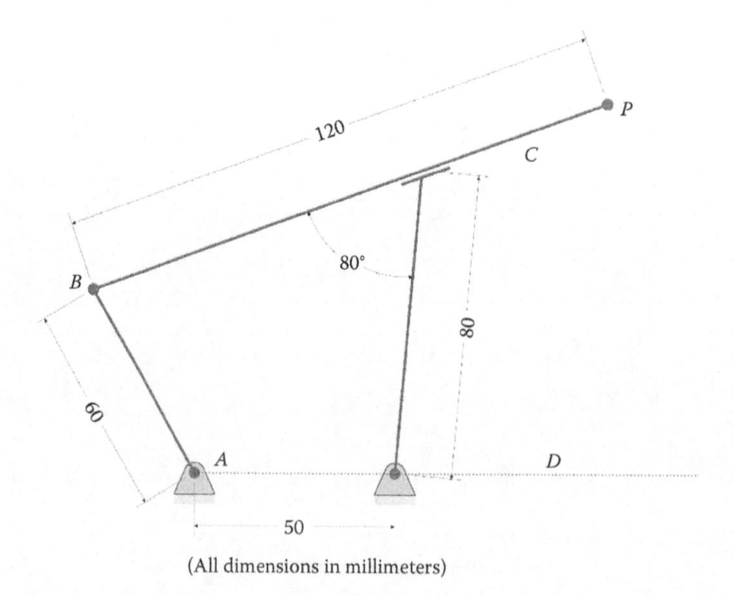

FIGURE 5.109
Problem 5.8.

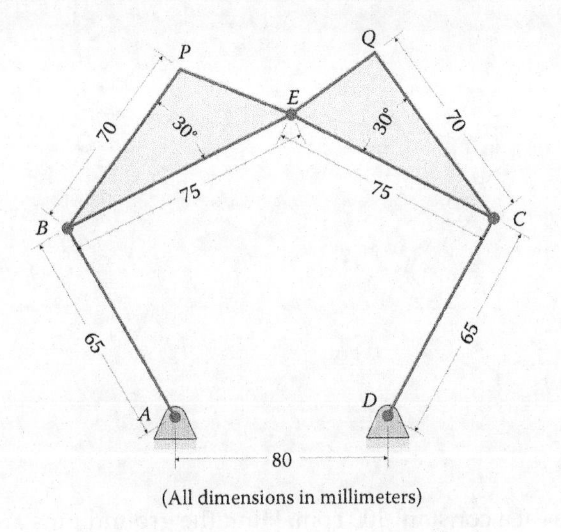

FIGURE 5.110
Problem 5.9.

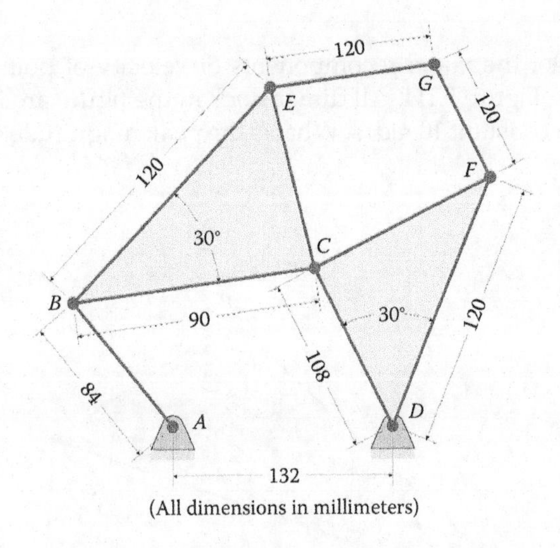

FIGURE 5.111
Problem 5.10.

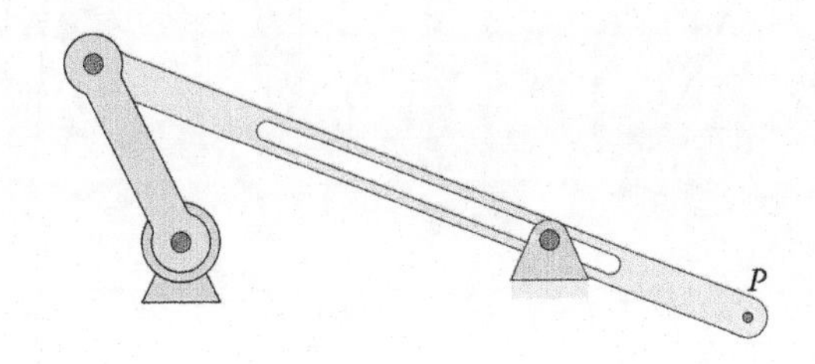

FIGURE 5.112
Problem 5.11.

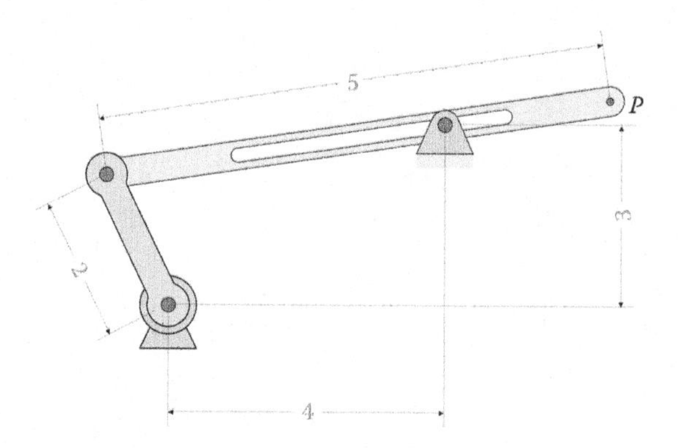

FIGURE 5.113
Problem 5.12.

the crank rotates at a constant 100 rpm. Hint: the ground pins are not horizontally aligned, so you will need to perform a new vector loop analysis on the linkage, and differentiate to find velocity.

Problem 5.13

Use MATLAB to plot the x and y components of velocity of point P for the fourbar linkage shown in Figure 5.114. All dimensions in the figure are in centimeters. The crank rotates at a constant 10 rad/s. What is the peak magnitude of the velocity at P?

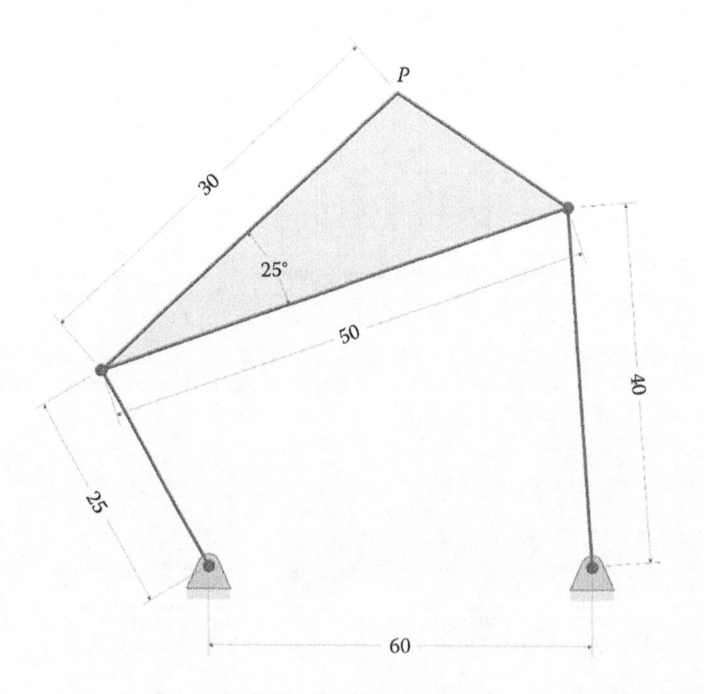

FIGURE 5.114
Problem 5.13.

Problem 5.14

The crank of the compressor shown in Figure 5.115 spins at a constant 1,700 rpm. Calculate and plot the velocity of the piston. What is the maximum velocity of the piston? All dimensions are given in millimeters.

Problem 5.15

Use MATLAB to conduct a velocity analysis on the inverted slider-crank linkage shown in Figure 5.116 if the crank rotates at a constant 10 rad/s. Plot the x and y components of the velocity at point P. Also, plot the velocity of slip ($\dot{b}$) between the slider and rocker.

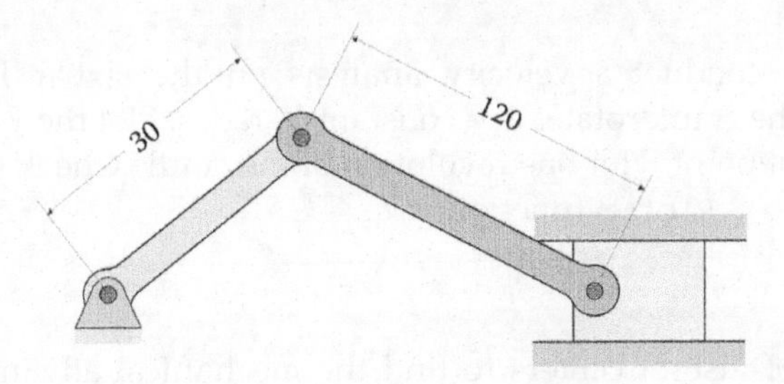

FIGURE 5.115
Problem 5.14.

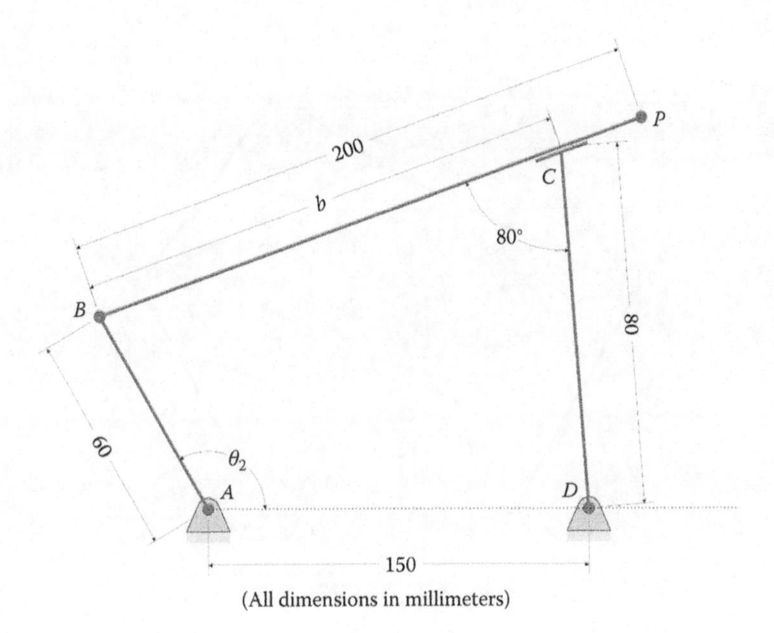

FIGURE 5.116
Problem 5.15.

Problem 5.16

Use MATLAB to conduct a velocity analysis on the geared fivebar linkage shown in Figure 5.117 if the crank rotates at a constant 10 rad/s. The gear at point A has 50 teeth and the gear at point D has 25 teeth. Gear D is rotated 180° when gear A is at 0°. Plot the x and y components of the velocity at point P. Check your answer using the `Derivative _ Plot.m` function.

Problem 5.17

Use MATLAB to conduct a velocity analysis on the sixbar linkage shown in Figure 5.118 if the crank rotates at a constant 10 rad/s. Plot the x and y components of the velocity at point P for one revolution of the crank. Check your answer using the `Derivative _ Plot.m` function.

Problem 5.18

Use MATLAB to conduct a velocity analysis on the sixbar linkage shown in Figure 5.119 if the crank rotates at a constant 10 rad/s. Plot the x and y components of the velocity at point P for one revolution of the crank. Check your answer using the `Derivative _ Plot.m` function.

Problem 5.19

Use the method of instant centers to find the mechanical advantage of the slider-crank in Figure 5.120. The mechanical advantage is defined as F_{out}/F_{in}.

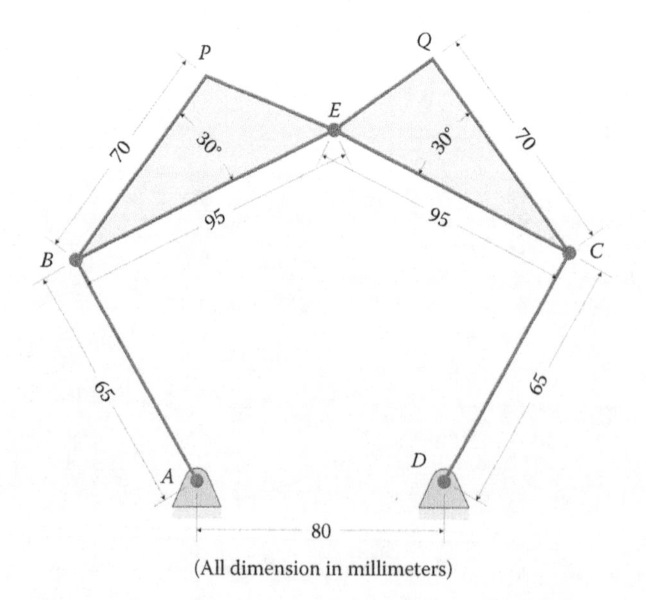

FIGURE 5.117
Problem 5.16.

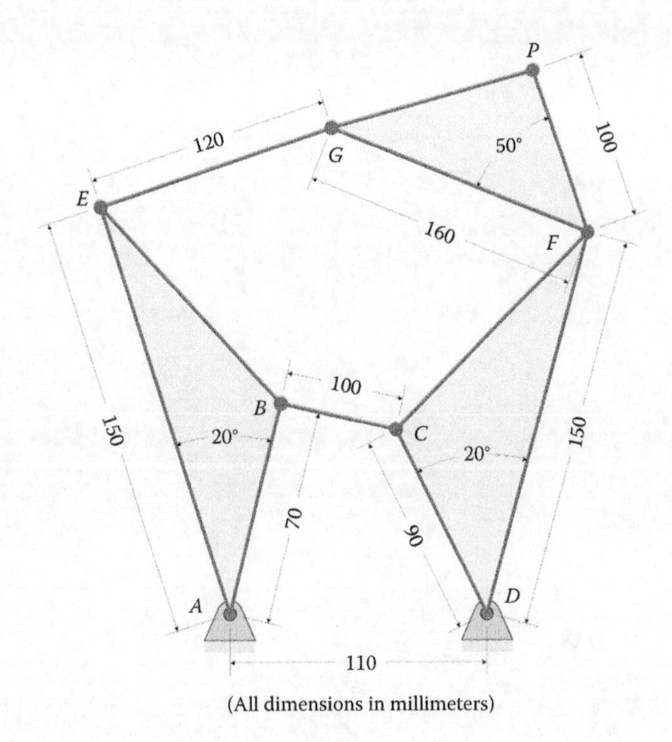

FIGURE 5.118
Problem 5.17.

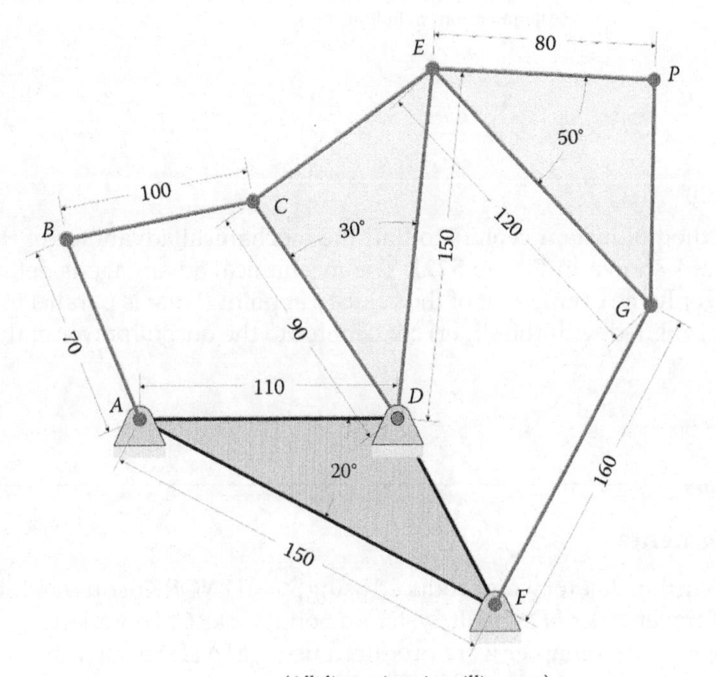

FIGURE 5.119
Problem 5.18.

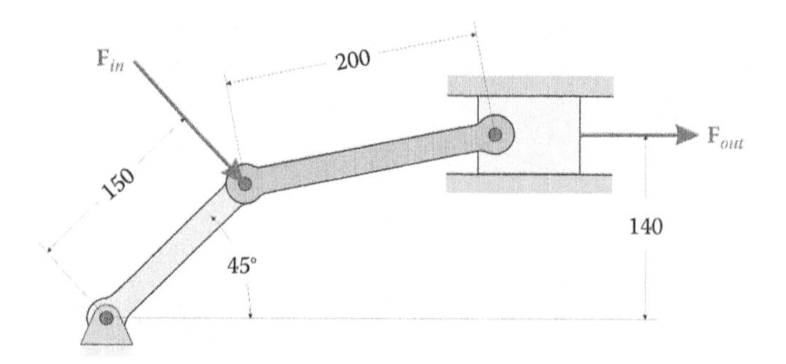

FIGURE 5.120
Problem 5.19.

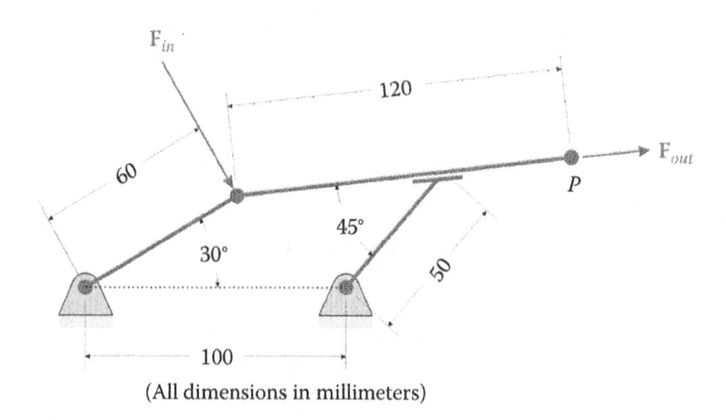

FIGURE 5.121
Problem 5.20.

Problem 5.20

Use the method of instant centers to find the mechanical advantage of the inverted slider-crank shown in Figure 5.121. The mechanical advantage is defined as F_{out}/F_{in}. Hint: only the component of the velocity at point *P* that is parallel to the output force (i.e., aligned with the slider) contributes to the output power of the linkage.

Acknowledgments

Several images in this chapter were produced using SOLIDWORKS software. SOLIDWORKS is a registered trademark of Dassault Systèmes SolidWorks Corporation.

Several images in this chapter were produced using MATLAB software. MATLAB is a registered trademark of The MathWorks, Inc.

6

Acceleration Analysis of Linkages

6.1 Introduction to Acceleration Analysis

The next logical step in our development is to study the acceleration of points on a moving linkage. According to Newton's Second Law, forces are proportional to accelerations. Thus, if we know the accelerations of every point on the linkage, we can predict the forces that the linkage must withstand. This can be very important in cases where the linkage must move at high speed, for example, in an automobile engine.

To begin our analysis, consider the single link shown in Figure 6.1. It has length b, which is rotating with angular velocity ω, and has angular acceleration α. If we write the position vector of point B in the usual way, we have

$$\mathbf{r}_B = b\mathbf{e} \tag{6.1}$$

To find the velocity at point B, we differentiate $\mathbf{r}_B$ with respect to time, as before.

$$\mathbf{v}_B = \frac{d\mathbf{r}_B}{dt} = b\omega\mathbf{n} \tag{6.2}$$

where we have used

$$\frac{d\mathbf{e}}{dt} = \omega\mathbf{n}$$

as found in Section 4.2.5. To find the acceleration of point B, simply differentiate once more with respect to time.

$$\mathbf{a} = \frac{d\mathbf{v}}{dt} = \frac{d}{dt}(b\omega\mathbf{n}) \tag{6.3}$$

Before proceeding, let us consider which terms are constant, and which terms vary with time.

Constant	b
Time-varying	$\omega, \theta, \mathbf{n}$

Since ω may not be constant, we must employ the product rule for differentiation

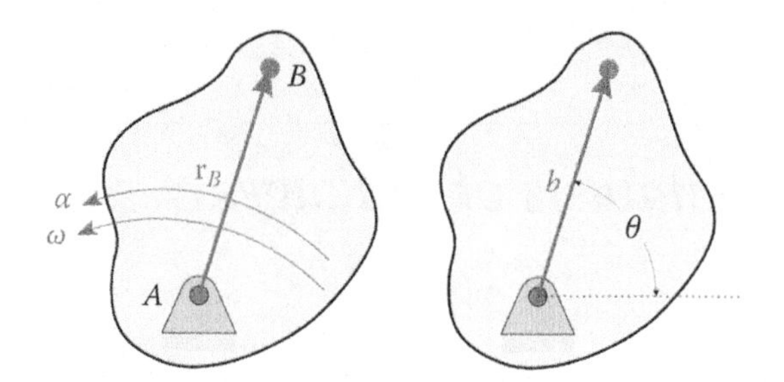

FIGURE 6.1
A rigid body in pure rotation about a point pinned to ground.

$$\mathbf{a} = \frac{d}{dt}(b\omega\mathbf{n}) = b\frac{d\omega}{dt}\mathbf{n} + b\omega\frac{d\mathbf{n}}{dt} \tag{6.4}$$

Define the time derivative of angular velocity ω as angular acceleration, α. Thus

$$\mathbf{a} = b\alpha\mathbf{n} - b\omega^2\mathbf{e} \tag{6.5}$$

where we have used

$$\frac{d\mathbf{n}}{dt} = -\omega\mathbf{e} \tag{6.6}$$

for the time derivative of the unit normal.

From Equation (6.5), it is apparent that the acceleration has two components. These components are shown in Figure 6.2 where, the first component ($b\alpha\mathbf{n}$) is tangential to the motion of point B, and is called the *tangential acceleration*. Note that if the link is rotating at constant angular velocity, then the angular acceleration and tangential acceleration are both zero.

The second component ($-b\omega^2\mathbf{e}$) points toward the center of rotation and is called *centripetal acceleration*. Centripetal acceleration is a byproduct of rotation – in order to rotate, a point must accelerate towards the center of rotation. If the link does not rotate, then the angular velocity and centripetal acceleration are both zero. The tangential acceleration points in a

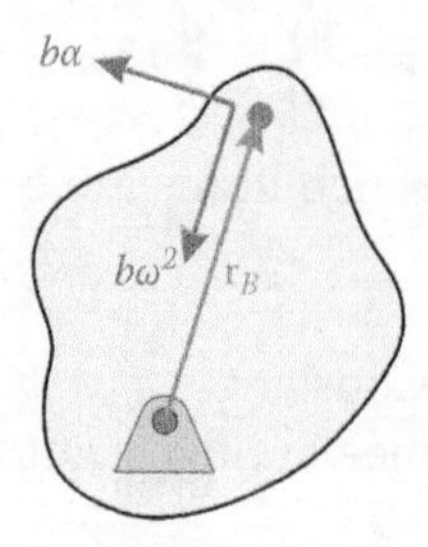

FIGURE 6.2
Acceleration of a point fixed on a link in pure rotation.

direction normal to the unit vector **e** and the centripetal acceleration points in the direction opposite the unit vector **e**.

6.1.1 Acceleration of a Moving Point on a Moving Link

Let us now examine a more general case. As seen in Figure 6.3, the slider is free to move inside a slot in the rotating link. The point B is attached to the slider, and we wish to find an expression for the acceleration of this point. The position vector to the slider is

$$\mathbf{r}_B = b\mathbf{e} \tag{6.7}$$

Note that length b is not constant for this case, so we must use the product rule to find velocity. To find the velocity at point B, we differentiate $\mathbf{r}_B$ with respect to time, as before.

$$\mathbf{v}_B = \frac{d}{dt}(b\mathbf{e}) \tag{6.8}$$

using the product rule

$$\mathbf{v}_B = \dot{b}\mathbf{e} + b\omega\mathbf{n} \tag{6.9}$$

To find acceleration, simply differentiate once more with respect to time.

$$\mathbf{a}_B = \frac{d\mathbf{v}_B}{dt} = \frac{d}{dt}\left(\dot{b}\mathbf{e} + b\omega\mathbf{n}\right) \tag{6.10}$$

Now consider which terms are constant, and which terms vary with time.

Constant	*nothing*!
Time-varying	$b, \dot{b}, \theta, \omega, \mathbf{n}, \mathbf{e}$

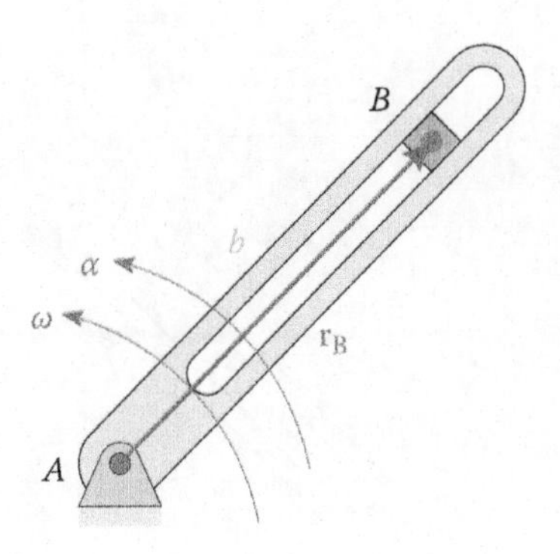

FIGURE 6.3
The link rotates with angular velocity ω and angular acceleration α, while the distance, b, to the slider is a function of time.

Thus, we must apply the product rule twice for each term

$$\mathbf{a}_B = \frac{d\dot{b}}{dt}\mathbf{e} + \dot{b}\frac{d\mathbf{e}}{dt} + \frac{d\omega}{dt}b\mathbf{n} + \omega\frac{db}{dt}\mathbf{n} + \omega b\frac{d\mathbf{n}}{dt} \tag{6.11}$$

$$\mathbf{a}_B = \ddot{b}\mathbf{e} + \dot{b}\omega\mathbf{n} + b\alpha\mathbf{n} + \dot{b}\omega\mathbf{n} - b\omega^2\mathbf{e} \tag{6.12}$$

$$\mathbf{a}_B = \ddot{b}\mathbf{e} + 2\dot{b}\omega\mathbf{n} + b\alpha\mathbf{n} - b\omega^2\mathbf{e} \tag{6.13}$$

Each of these terms has a common name:

$\alpha b\mathbf{n}$	Tangential acceleration
$-\omega^2 b\mathbf{e}$	Centripetal acceleration
$\ddot{b}\mathbf{e}$	Acceleration of slip
$2\dot{b}\omega\mathbf{n}$	Coriolis acceleration

The direction of each component is shown in Figure 6.4. The tangential and centripetal acceleration are the same as for the rigid rotating link. The acceleration of slip gives the acceleration of the slider relative to the slot. If you were sitting on the link as it rotated, you would see the slider moving toward or away from you with acceleration $\ddot{a}$. The Coriolis term arises from the fact that an object must accelerate as it changes distance to the center of rotation if it is to maintain a constant angular velocity. If the velocity inward or outward from the center of rotation is zero, then the Coriolis term vanishes. To visualize the effect that Coriolis acceleration might have, imagine a child standing 2 m from the center of a spinning merry-go-round. For convenience, assume that the angular velocity of the merry-go-round is a constant 1 rad/s. The tangential velocity of the child is

$$v_{t1} = (2\,\text{m})\left(1\frac{\text{rad}}{\text{s}}\right) = 2\frac{\text{m}}{\text{s}} \tag{6.14}$$

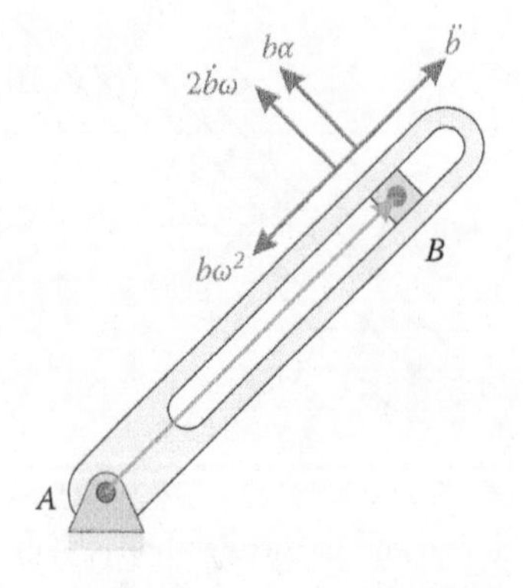

FIGURE 6.4
The four components of acceleration for the rotating link with slider.

TABLE 6.1

The Four Different Types of Translational Acceleration

Quantity	Formula	Units
Tangential acceleration	Angular acceleration × length	rad · m/s² = m/s²
Centripetal acceleration	Angular velocity² × length	(rad/s)² · m = m/s²
Acceleration of slip	Translational acceleration	m/s²
Coriolis acceleration	2 × translational velocity × angular velocity	(m/s) · (rad/s) = m/s²

If the child wishes to join her friend standing 4 m from the center, she must match her tangential velocity, which is

$$v_{t2} = (4\,\text{m})\left(1\frac{\text{rad}}{\text{s}}\right) = 4\frac{\text{m}}{\text{s}} \tag{6.15}$$

Thus, to join her friend, the child must *accelerate* in the tangential direction. But remember that the Coriolis term depends upon the inward or outward *velocity*: if the child moves very slowly toward her friend on the outside of the merry-go-round she might not notice the sideways acceleration. But if she tries to move rapidly from the inner part of the merry-go-round to the outer, the sideways acceleration (and the force that accompanies it) will likely knock her off balance *sideways*. Thus, the effects of Coriolis acceleration can be somewhat counterintuitive, but are always present when a body subject to rotation moves inward or outward from the center of rotation.

We will encounter tangential and centripetal acceleration as we analyze the four-bar linkage. The other two types of acceleration will appear during the analysis of the threebar and inverted slider-crank. To recap, the total acceleration is made up of the four components shown in Table 6.1. Note that the units of translational acceleration are all the same: m/s². Thus, if you encounter a term with a length multiplied by an angular velocity squared, it is very likely to be a centripetal acceleration. Likewise, the appearance of a factor of 2 in an acceleration formula is an almost certain sign of a Coriolis acceleration.

To summarize, we can see that acceleration analysis is considerably more complicated than velocity analysis owing to the presence of centripetal and Coriolis acceleration. Luckily, the vector loop procedure we have developed for velocity analysis can still be applied to acceleration analysis.

6.2 Acceleration Analysis of the Threebar Slider-Crank

To conduct the acceleration analysis on the threebar linkage shown in Figure 6.5 we proceed in the same manner as we did for velocity analysis. We assume that we have performed a full position and velocity analysis as described in the previous sections, and wish to find the acceleration of one or more points on the linkage. Once we have found the slider's angular acceleration, α_3, and acceleration of slip, $\ddot{b}$, it is simple to use the expressions derived in the preceding section to find the desired accelerations. The input to the linkage is, as always, the crank and we assume that the angular acceleration of the crank is known.

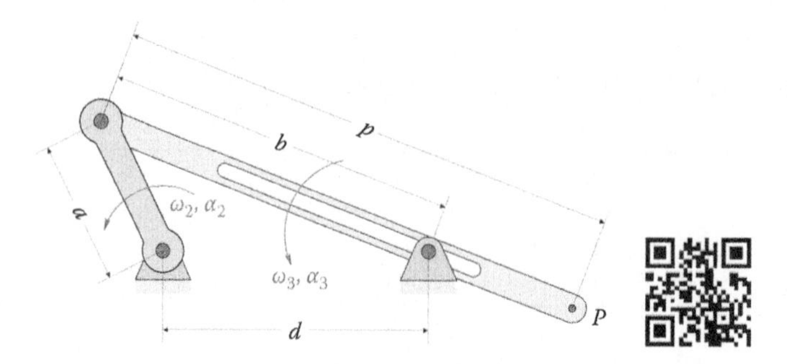

FIGURE 6.5
Diagram of the threebar linkage showing angular velocities and accelerations.

We start by listing the known and unknown quantities in the analysis

$$\text{Known} = a, b, d, \theta_2, \theta_3, \omega_2, \omega_3, \dot{b}, \alpha_2$$

$$\text{Unknown} = \alpha_3, \ddot{b}$$

Begin by writing the vector loop equation, as shown in Figure 6.6.

$$\mathbf{r}_2 + \mathbf{r}_3 - \mathbf{r}_1 = 0 \tag{6.16}$$

Or, in unit vector form

$$a\mathbf{e}_2 + b\mathbf{e}_3 - d\mathbf{e}_1 = 0 \tag{6.17}$$

Differentiate this with respect to time

$$a\omega_2\mathbf{n}_2 + \dot{b}\mathbf{e}_3 + b\omega_3\mathbf{n}_3 = 0 \tag{6.18}$$

Then, differentiate again. For clarity, we will proceed term by term. The first term differentiates as

$$\frac{d}{dt}(a\omega_2\mathbf{n}_2) = a\frac{d\omega_2}{dt}\mathbf{n}_2 + a\omega_2\frac{d}{dt}(\mathbf{n}_2) = a\alpha_2\mathbf{n}_2 - a\omega_2^2\mathbf{e}_2 \tag{6.19}$$

The second term is a product of two functions of time, so that

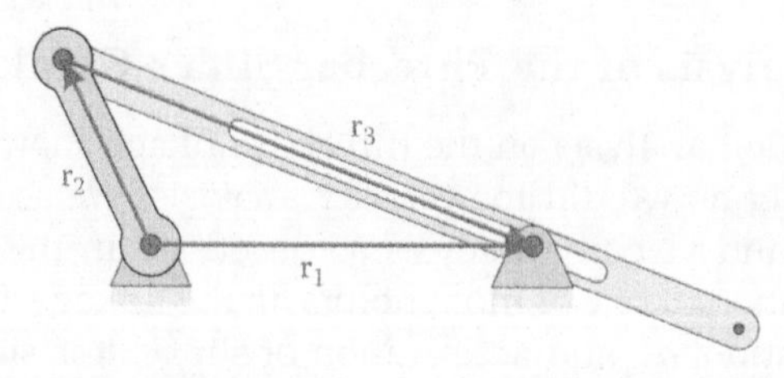

FIGURE 6.6
Vector loop diagram for the threebar linkage.

$$\frac{d}{dt}(\dot{b}\mathbf{e}_3) = \frac{d\dot{b}}{dt}\mathbf{e}_3 + \dot{b}\frac{d\mathbf{e}_3}{dt} = \ddot{b}\mathbf{e}_3 + \dot{b}\omega_3\mathbf{n}_3 \tag{6.20}$$

The third term is a product of *three* functions of time

$$\frac{d}{dt}(b\omega_3\mathbf{n}_3) = \frac{db}{dt}\omega_3\mathbf{n}_3 + b\frac{d\omega_3}{dt}\mathbf{n}_3 + b\omega_3\frac{d\mathbf{n}_3}{dt} = \dot{b}\omega_3\mathbf{n}_3 + b\alpha_3\mathbf{n}_3 - b\omega_3^2\mathbf{e}_3 \tag{6.21}$$

Adding the results together gives the acceleration equation for the threebar slider-crank.

$$a\alpha_2\mathbf{n}_2 - a\omega_2^2\mathbf{e}_2 + \ddot{b}\mathbf{e}_3 + 2\dot{b}\omega_3\mathbf{n}_3 + b\alpha_3\mathbf{n}_3 - b\omega_3^2\mathbf{e}_3 = 0 \tag{6.22}$$

The reader will recognize several familiar terms, including centripetal accelerations ($a\omega_2^2\mathbf{e}_2$, $b\omega_3^2\mathbf{e}_3$), tangential accelerations ($a\alpha_2\mathbf{n}_2$,$b\alpha_3\mathbf{n}_3$), and even a Coriolis acceleration ($2\dot{b}\omega_3\mathbf{n}_3$). The term $\ddot{b}$ is known as the *acceleration of slip*, since it gives the acceleration of the slider relative to the fixed pin *D*. As before, we move the known quantities to the right side of the equation, and the unknowns to the left

$$\ddot{b}\mathbf{e}_3 + b\alpha_3\mathbf{n}_3 = -a\alpha_2\mathbf{n}_2 + a\omega_2^2\mathbf{e}_2 - 2\dot{b}\omega_3\mathbf{n}_3 + b\omega_3^2\mathbf{e}_3 \tag{6.23}$$

And finally, put this into matrix form for solution in MATLAB®

$$\mathbf{C}\boldsymbol{\alpha} = \mathbf{d} \tag{6.24}$$

where

$$\mathbf{C} = \begin{bmatrix} b\mathbf{n}_3 & \vdots & \mathbf{e}_3 \end{bmatrix} \tag{6.25}$$

$$\boldsymbol{\alpha} = \begin{Bmatrix} \alpha_3 \\ \ddot{b} \end{Bmatrix} \qquad \mathbf{d} = -a\alpha_2\mathbf{n}_2 + a\omega_2^2\mathbf{e}_2 - 2\dot{b}\omega_3\mathbf{n}_3 + b\omega_3^2\mathbf{e}_3 \tag{6.26}$$

You may have noticed that the **C** matrix is identical to the **A** matrix found earlier for the velocity analysis. This is no coincidence – this matrix is called the *Jacobian matrix*. It is formed by taking the first derivative of the constraint equations in (6.18). We will encounter a similar phenomenon when we conduct the acceleration analysis of the remaining linkages. Jacobian matrices and constraint equations are the subjects of more advanced courses in kinematics and are beyond the scope of this text, although they do appear, and are briefly discussed, in the Newton–Raphson algorithm in Section 4.15.

6.2.1 Computing the Accelerations Using MATLAB®

We will now implement the acceleration matrix equation in MATLAB in order to calculate the acceleration of any point on the threebar linkage. Start by opening your threebar velocity analysis code. Use `Save As` to save the file as `Threebar_Acceleration_Analysis.m`. As an example, we will consider the same linkage we have used for the position and velocity analysis codes in earlier sections (see Figure 6.7). We will assume a

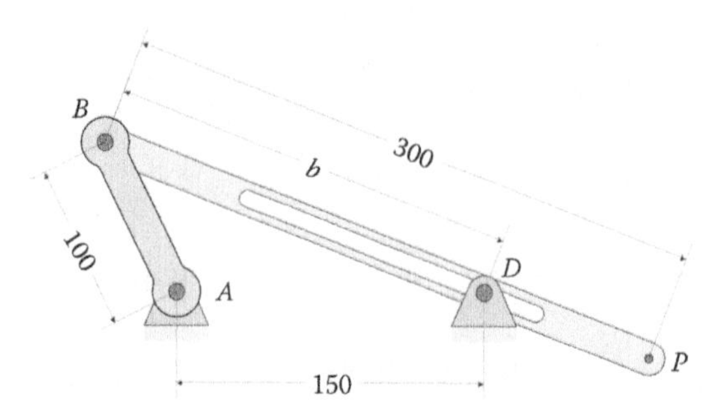

(All dimensions in millimeters)

Crank length: 100
Slider length: 300
Distance between ground pins: 150
Crank angular velocity: 10 rad/s
Crank angular acceleration: 0 rad/s^2

FIGURE 6.7
Example threebar linkage used to develop the MATLAB code.

constant angular velocity for the crank, so that its angular acceleration is zero. Add the `a0` and `alpha2` variables to their respective positions as shown below.

```
% Threebar_Acceleration_Analysis.m
% Conducts an acceleration analysis on the threebar crank-slider linkage
% and plots the acceleration of point P
% by Eric Constans, June 1, 2017

% Prepare Workspace
clear variables; close all; clc;

% Linkage dimensions
a = 0.100;            % crank length (m)
d = 0.150;            % length between ground pins (m)
p = 0.300;            % slider length (m)

% ground pins
x0 = [0;0];  % ground pin at A (origin)
xD = [d;0];  % ground pin at D
v0 = [0;0];  % velocity of origin
a0 = [0;0];  % accel of origin

% Angular velocity and acceleration of crank
omega2 = 10;          % angular velocity of crank (rad/sec)
alpha2 = 0;           % angular acceleration of crank (rad/sec2)
```

Now preallocate memory for the new variables associated with acceleration.

```
N = 361;    % number of times to perform position calculations
[xB,xP] = deal(zeros(2,N)); % allocate space for position of B,P
[vB,vP] = deal(zeros(2,N)); % allocate space for velocity of B,P
```

```
[aB,aP] = deal(zeros(2,N)); % allocate space for accel of B,P
[theta2,theta3] = deal(zeros(1,N)); % allocate space for link angles
[omega3,alpha3] = deal(zeros(1,N)); % allocate space for vel and accel

[b,bdot,bddot] = deal(zeros(1,N));  % allocate space for b, bdot, bddot
```

Inside the main loop, immediately after the velocity calculations, add the following lines of code:

```
% conduct acceleration analysis to solve for alpha3 and bddot
  ac = a*omega2^2;              % centripetal accel
  at = a*alpha2;                       % tangential accel
  bC = 2*bdot(i)*omega3(i);            % Coriolis accel
  bc = b(i)*omega3(i)^2;   % centripetal accel

  C_Mat = A_Mat;
  d_Vec = -at*n2 + ac*e2 - bC*n3 + bc*e3;
  alpha_Vec = C_Mat\d_Vec;                    % solve for angular accelerations

  alpha3(i) = alpha_Vec(1);
  bddot(i)  = alpha_Vec(2);
```

Here we have defined the variables ac, at, bC, and bc to compute the centripetal, tangential, and Coriolis components of the accelerations that are on the right side of the matrix equation (the known side). This makes the form of d _ Vec much simpler to type and debug. Immediately after the main loop, we will use the Derivative _ Plot function to validate our solution for alpha3 and alpha4.

```
dt = 2*pi/((N-1)*omega2);  % time increment between calculations
Derivative_Plot(theta2, omega3, alpha3, dt)
```

Figure 6.8 shows the computed and estimated angular acceleration of the slider for the example problem we have been considering. Figure 6.9 shows the same plot for the acceleration of slip for the slider ($\ddot{b}$). Make sure that your plots are the same as those shown in Figures 6.8 and 6.9 before continuing with the analysis.

6.2.2 Acceleration at the Pins

We will employ the relative acceleration formula to compute the translational acceleration at the pins. As with velocity and position, we will create a separate function to compute acceleration. Open up a new MATLAB script and enter the following function:

```
% Function FindAcc.m
% Calculates the translational acceleration of a point on the linkage
% using the relative acceleration formula
%
% a0    = acceleration of first point
% L     = length of vector to second point on the link
% omega = angular velocity of link
% alpha = angular acceleration of link
```

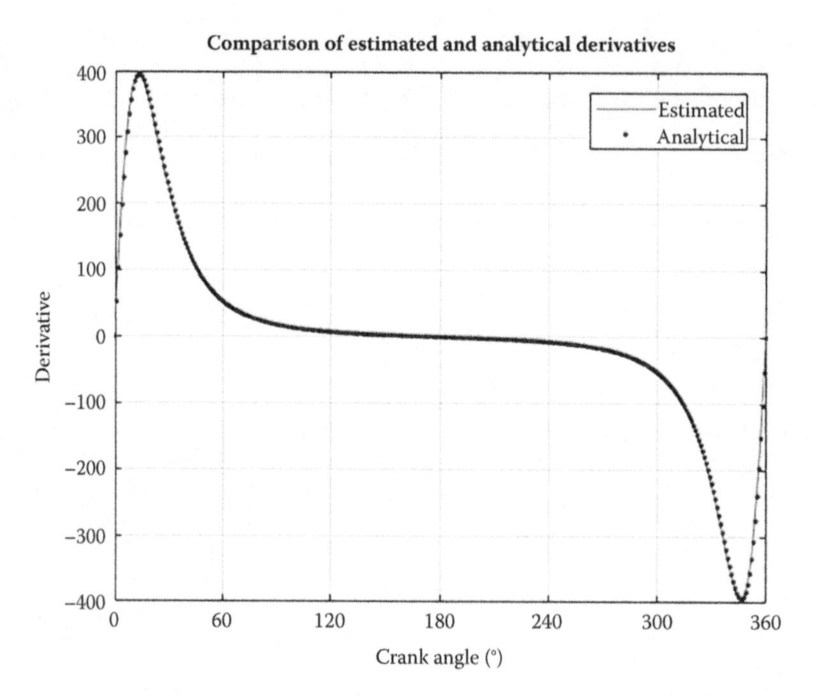

FIGURE 6.8
The estimated and analytical values of α_3.

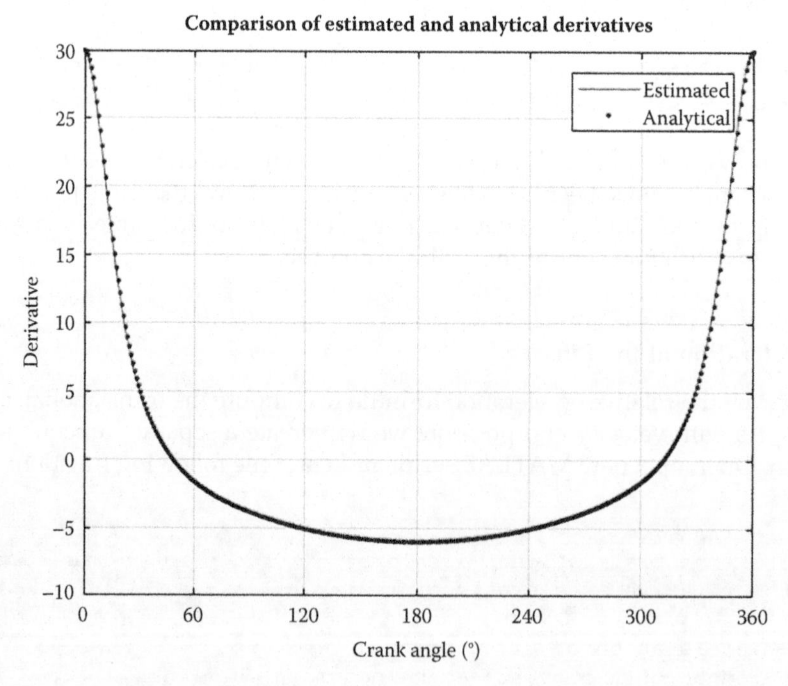

FIGURE 6.9
The estimated and analytical values of $\ddot{b}$.

```
% e      = unit vector btw first and second points
% n      = unit normal to vector btw first and second points
% a      = acceleration of second point

function a = FindAcc(a0, L, omega, alpha, e, n)

a = a0 + L*alpha*n - L*omega^2*e;
```

As before, this is a very simple function, but it will make the main program much neater and easier to follow. In the main program, use the following lines of code to calculate translational accelerations

```
% calculate acceleration at important points on linkage
  aB(:,i) = FindAcc(       a0,     a,      omega2,      alpha2,     e2,     n2);
  aP(:,i) = FindAcc(aB(:,i),     p, omega3(i), alpha3(i),     e3,     n3);
```

Of course, you should check your work for each of the accelerations using the `Derivative_Plot` routine.

Figure 6.10 shows the x and y accelerations of the point P on the example threebar linkage. As we have seen, conducting the full acceleration analysis on the threebar linkage requires only a few additional lines of code once the position and velocity analyses are complete. This will also be the case for the other linkages, which we will analyze in the next few sections.

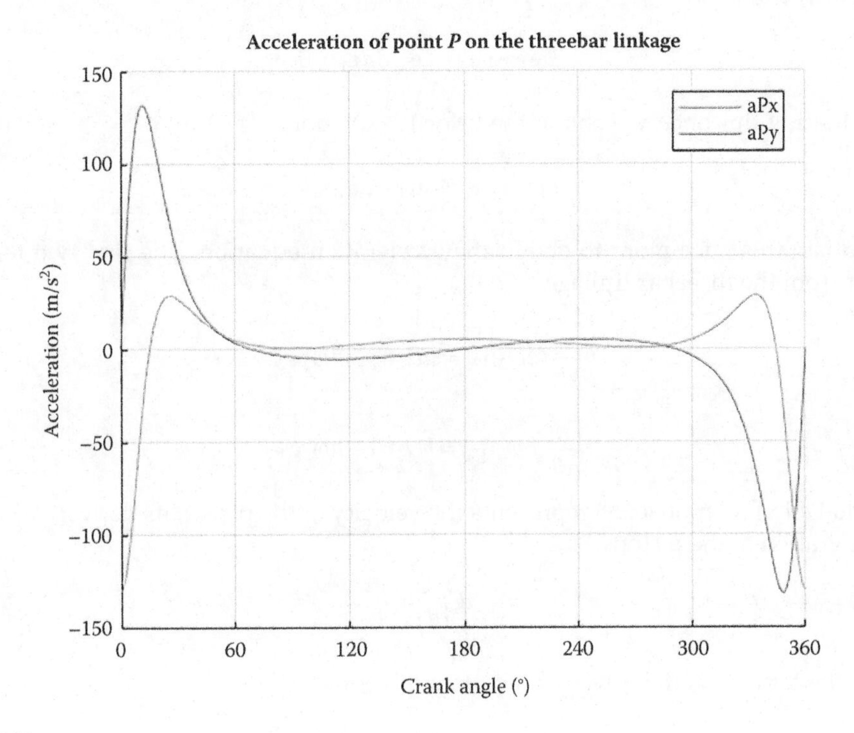

FIGURE 6.10
Acceleration of point P on the example threebar linkage.

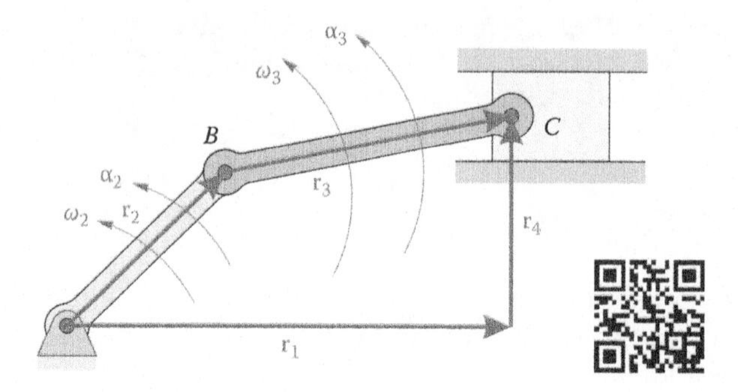

FIGURE 6.11
Vector loop diagram for the slider-crank linkage showing angular velocities and accelerations.

6.3 Acceleration Analysis of the Slider-Crank

The acceleration analysis for the slider-crank shown in Figure 6.11 proceeds in a similar manner as the threebar. We begin, as usual, with the vector loop equation for the slider-crank.

$$\mathbf{r}_2 + \mathbf{r}_3 - \mathbf{r}_4 - \mathbf{r}_1 = 0 \tag{6.27}$$

or, in unit vector form

$$a\mathbf{e}_2 + b\mathbf{e}_3 - c\mathbf{e}_4 - d\mathbf{e}_1 = 0 \tag{6.28}$$

Differentiating this once, we obtain the velocity equation.

$$a\omega_2\mathbf{n}_2 + b\omega_3\mathbf{n}_3 - \dot{d}\mathbf{e}_1 = 0 \tag{6.29}$$

Differentiate this once more to obtain the acceleration equation. The first two terms are familiar from the threebar linkage

$$\frac{d}{dt}(a\omega_2\mathbf{n}_2) = a\alpha_2\mathbf{n}_2 - a\omega_2^2\mathbf{e}_2 \tag{6.30}$$

$$\frac{d}{dt}(b\omega_3\mathbf{n}_3) = b\alpha_3\mathbf{n}_3 - b\omega_3^2\mathbf{e}_3 \tag{6.31}$$

Recall that $\dot{d}$ in Equation (6.29) represents the velocity of the piston; its derivative is simply the acceleration of the piston.

$$\frac{d}{dt}\left(\dot{d}\right) = \ddot{d} \tag{6.32}$$

Adding the terms together gives the acceleration equation

$$a\alpha_2\mathbf{n}_2 - a\omega_2^2\mathbf{e}_2 + b\alpha_3\mathbf{n}_3 - b\omega_3^2\mathbf{e}_3 - \ddot{d}\mathbf{e}_1 = 0 \tag{6.33}$$

It is useful again at this point to take stock of what variables we know, and which are unknown. As before, we assume that a full position and velocity analysis has been performed, so that we know

$$\text{Known}: a, b, c, d, \theta_2, \theta_3, \omega_2, \omega_3, \alpha_2$$

Let the crank be driven by a motor with known speed and angular acceleration, ω_2, α_2. The only unknown quantities in Equation (6.33) are

$$\text{Unknown}: \alpha_3, \ddot{d}$$

Before assembling the matrices, let us move the known quantities to the right side of the equation, and the unknown quantities to the left.

$$b\alpha_3\mathbf{n}_3 - \ddot{d}\mathbf{e}_1 = -a\alpha_2\mathbf{n}_2 + a\omega_2^2\mathbf{e}_2 + b\omega_3^2\mathbf{e}_3 \tag{6.34}$$

Now rearrange this equation into matrix form, so that we can use MATLAB to solve it.

$$\mathbf{C}\boldsymbol{\alpha} = \mathbf{d} \tag{6.35}$$

where

$$\mathbf{C} = \begin{bmatrix} b\mathbf{n}_3 & \vdots & -\mathbf{e}_1 \end{bmatrix} \qquad \boldsymbol{\alpha} = \begin{Bmatrix} \alpha_3 \\ \ddot{d} \end{Bmatrix} \tag{6.36}$$

and

$$\mathbf{d} = \left\{ -a\alpha_2\mathbf{n}_2 + a\omega_2^2\mathbf{e}_2 + b\omega_3^2\mathbf{e}_3 \right\} \tag{6.37}$$

6.3.1 Slider-Crank with Constant Crank Angular Velocity

Figure 6.12 shows the dimensions of the slider-crank mechanism we have used in earlier chapters. Copy your slider-crank velocity analysis code to a file named

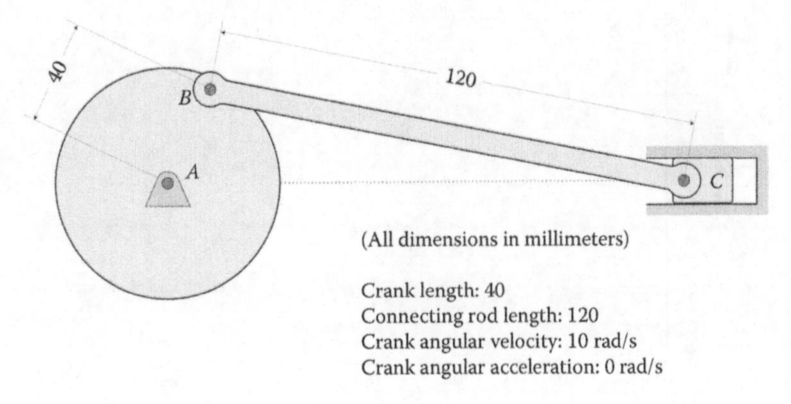

FIGURE 6.12
Dimensions of the slider-crank used in the example calculations.

`SliderCrank _ Acceleration _ Analysis.m` and modify it with the matrix equations given above. If you use the Derivative_Plot function to verify your values for α_3 you should obtain the plot shown in Figure 6.13. The acceleration of the piston for this example is shown in Figure 6.14.

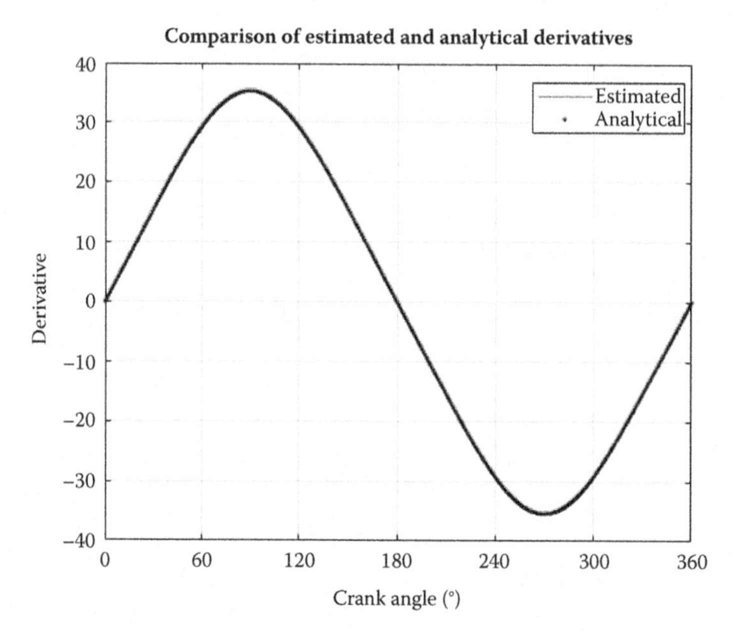

FIGURE 6.13
Verification of the angular acceleration of the connecting rod, α_3, for Section 6.3.1.

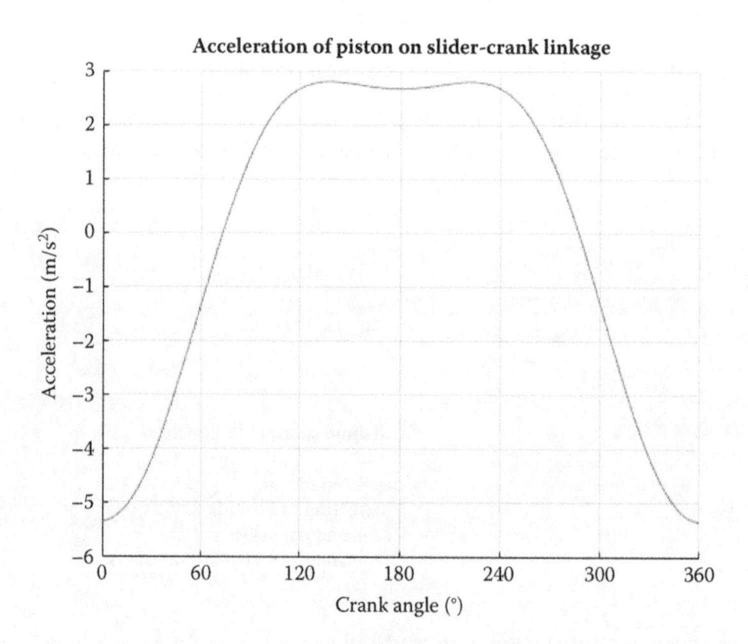

FIGURE 6.14
Acceleration of the piston in Section 6.3.1.

6.3.2 A Note on the Angular Acceleration of the Crank

In the preceding section, and in the sections that follow, we have assumed that the crank rotates at a constant angular velocity, that is, $\alpha_2 = 0$. This would be the case if the linkage is operating in the "steady state"; that is, the motor driving the crank has reached its operating speed and is no longer accelerating. In some situations, however, we might be interested in calculating the accelerations and forces on the linkage during the time the crank is "ramping up" to its desired speed. In this situation, the crank (and the rest of the linkage) would start from rest, and begin accelerating as soon as the crank motor is turned on. In this case, the angular acceleration of the crank is no longer zero, and we must make a few modifications to the code to reflect this. Let us consider now the case where the linkage starts from rest, and turning on the crank motor results in a short sinusoidal pulse of crank acceleration. The pulse of acceleration lasts for a period T, after which the crank moves at a constant angular velocity. Let the *amplitude* of the acceleration pulse be A, so that

$$\alpha_2 = A\left(1-\cos\left(\frac{2\pi t}{T}\right)\right) \tag{6.38}$$

for $t < T$ and

$$\alpha_2 = 0 \tag{6.39}$$

afterward.

An example of such a pulse is shown in Figure 6.15, where the amplitude is 1 rad/s^2 and the period is 4 s. Note that the total height of the pulse is twice the amplitude, since the cosine function ranges between –1 and 1.

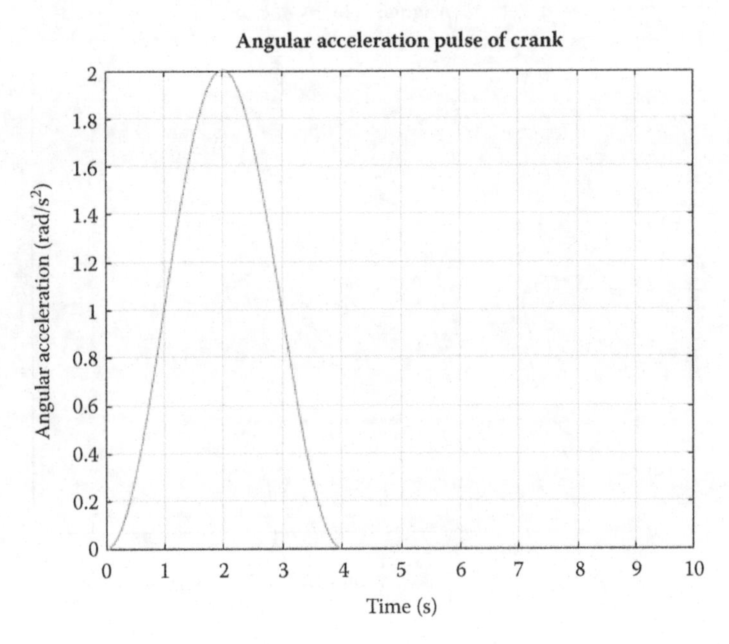

FIGURE 6.15
The crank is given a sinusoidal pulse of angular acceleration, starting from rest.

To find angular velocity, we must integrate Equations (6.38) and (6.39). Equation (6.38) is integrated over a period $t = 0$ to $t = T$, and the constant of integration is zero because the angular velocity of the crank is zero when $t = 0$.

$$\omega_2 = At - \frac{T}{2\pi}\sin\left(\frac{2\pi t}{T}\right) \tag{6.40}$$

for $t < T$ and

$$\omega_2 = AT \tag{6.41}$$

for $t > T$.

The resulting angular velocity is shown in Figure 6.16. Note that ω_2 starts at zero, and steadily increases until it reaches its steady-state value of 4 rad/s (assuming that A = 1 rad/s). We now integrate once more to find the crank angle

$$\theta_2 = \frac{1}{2}At^2 + \frac{T^2}{4\pi^2}\left(\cos\left(\frac{2\pi t}{T}\right) - 1\right) \tag{6.42}$$

for $t < T$ and

$$\theta_2 = ATt - \frac{T^2 A}{2} \tag{6.43}$$

Constants of integration are required in Equations (6.42) and (6.43) to ensure that the crank angle starts at zero at $t = 0$ and that it takes on the value $\theta_2 = AT^2/2$ when $t = T$.

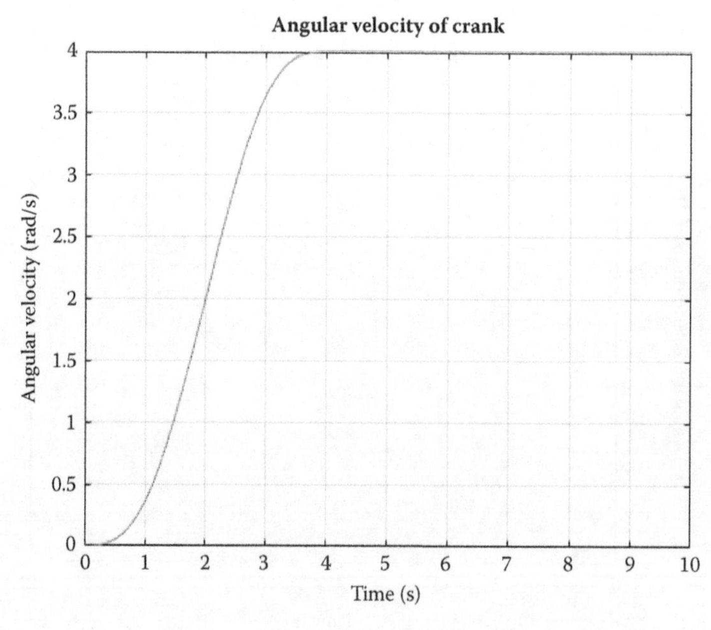

FIGURE 6.16
The angular velocity of the crank resulting from the pulse of angular acceleration described above.

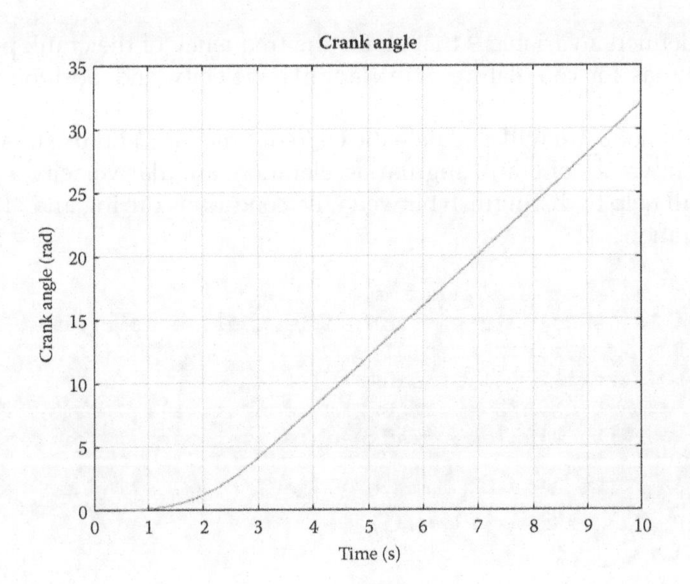

FIGURE 6.17
Crank angle resulting from the acceleration pulse described above.

The crank angle function given in Equations (6.42) and (6.43) is shown in Figure 6.17. As expected, it starts at zero, and increases linearly after $t = T$.

Up to now, we have calculated and plotted positions, velocities, and accelerations for a single revolution of the crank. Since we are now considering non-steady-state condition, it will make more sense to perform our calculations for a given length of time. Let us repeat the analysis of the linkage in the example using the sinusoidal crank acceleration given in Figure 6.15. We will perform the calculations for 1000 time steps. Use `Save As` to rename your code `SliderCrank_Nonsteady_Acceleration_Analysis.m`, then modify the upper part of your code with the following statements.

```
N = 1000;    % number of time steps to calculate
```

Since the angular velocity and acceleration of the crank are not constant, we must allocate space in memory for their calculated values. Additionally, we will plot positions, velocities, and accelerations versus *time*, instead of crank angle as we did before. Add the following lines of code to allocate the necessary memory.

```
omega2 = zeros(1,N); % angular velocity of crank (rad/sec)
alpha2 = zeros(1,N); % angular acceleration of crank (rad/sec2)
t = zeros(1,N);      % time (sec)
```

Of course, we must also change `omega2` to `omega2(i)` and `alpha2` to `alpha2(i)` inside the main loop. Now let us define the crank acceleration pulse and time increment:

```
A = 1;               % amp of crank angular accel pulse (rad/sec2)
T = 4;               % period of crank angular acceleration (sec)
B = 2*pi/T;          % freq of crank angular accel pulse (1/sec)
dt = 0.01;           % time increment
```

We have also defined a variable B that stores the frequency of the crank pulse – this will make the formulas for calculating crank angular velocity and position simpler in the main loop.

Inside the main loop, we will calculate the current time first, and use this value to determine the current values of crank angular acceleration, angular velocity and angle. Some simple logic will help to distinguish between the conditions during and after the pulse of angular acceleration.

```
t(i) = i*dt;           % calculate time
if (t(i) <= T)   % calculate crank angle, vel, accel for t < T
  alpha2(i) = A*(1-cos(2*pi*t(i)/T));
  omega2(i) = A*t(i) - (1/B)*sin(B*t(i));
  theta2(i) = A*t(i)^2/2 + (1/(B^2))*(cos(B*t(i))-1);
else                   % calculate crank angle, vel, accel for t > T
  alpha2(i) = 0;
  omega2(i) = A*T;
  theta2(i) = A*T*t(i) - T^2*A/2;
end
```

That's it! You should now change the legends and title of your plot statements to reflect the fact that the position, velocity, and acceleration of the piston are being plotted with respect to time, instead of crank angle.

Figure 6.18 shows the position of the piston versus crank angle for the case of constant angular velocity (this is the solution from the example in Chapter 4). Figures 6.19 and 6.20 show the piston position versus time resulting from the angular acceleration pulse discussed above. Figure 6.20 is a "zoomed in" version of Figure 6.19 showing one cycle of the

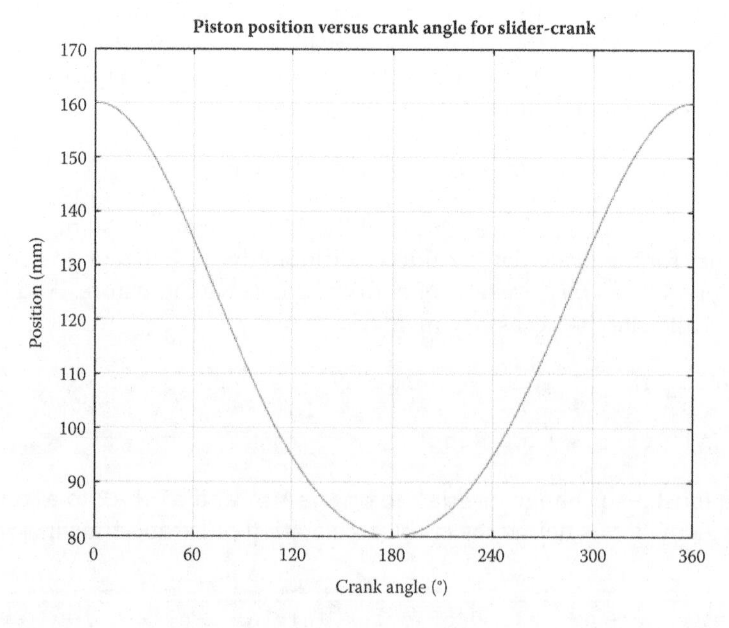

FIGURE 6.18
Position of piston versus crank angle for constant angular velocity (repeated from Figure 4.39).

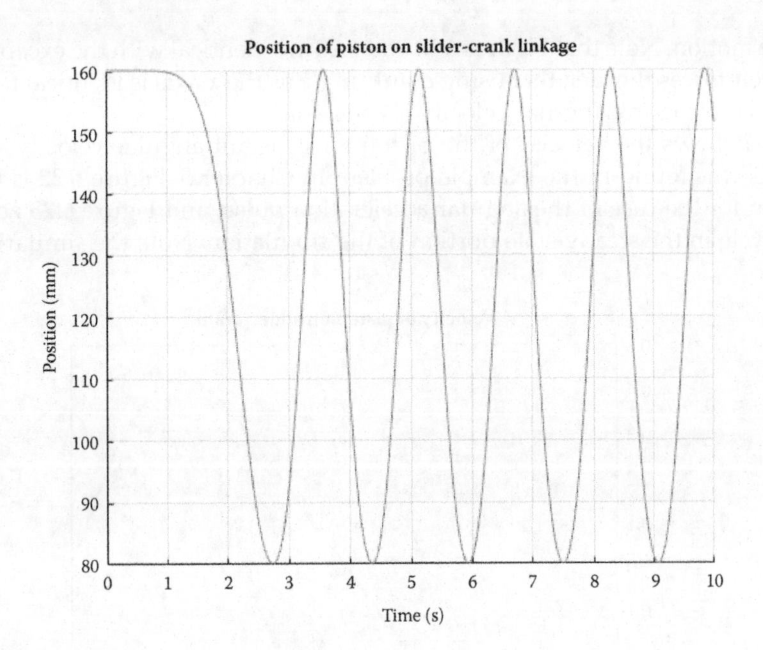

FIGURE 6.19
Position of piston versus time for angular acceleration pulse.

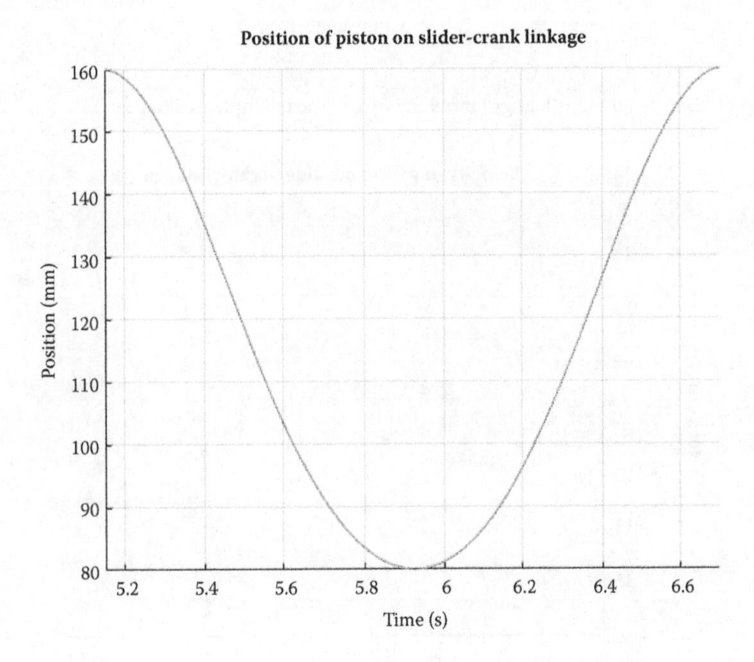

FIGURE 6.20
Zoomed in view of piston position versus time for angular acceleration pulse during steady-state portion. Note similarity to Figure 6.18.

steady-state motion. Note that Figures 6.18 and 6.20 are identical with the exception of the x axis – this reinforces the idea that using crank angle as the x axis is identical to using time on the x axis if the crank angular velocity is constant.

Figure 6.21 shows the velocity of the piston for constant angular velocity – this is the same plot as was found in the example problem in Chapter 5. Figure 6.22 is the velocity of the piston for the case of the angular acceleration pulse, and Figure 6.23 zoomed-in to show one cycle in the steady-state portion of the simulation. Note the similarity between

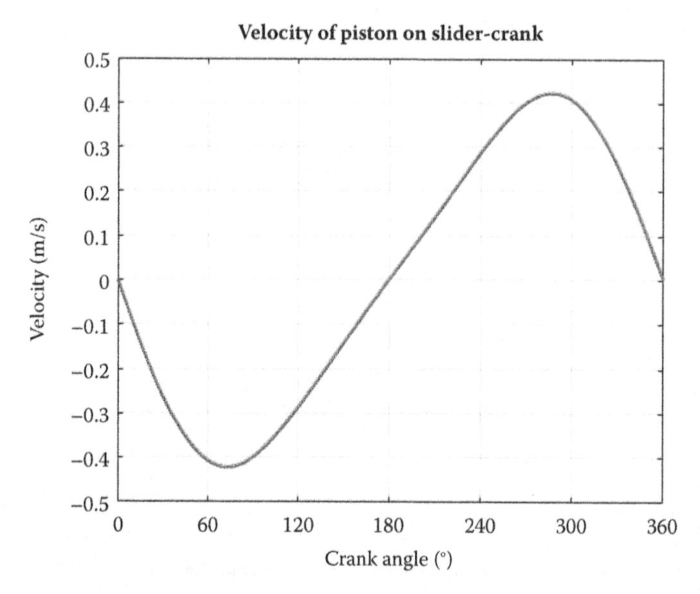

FIGURE 6.21
Velocity of piston with constant crank angular velocity from the example in Chapter 5.

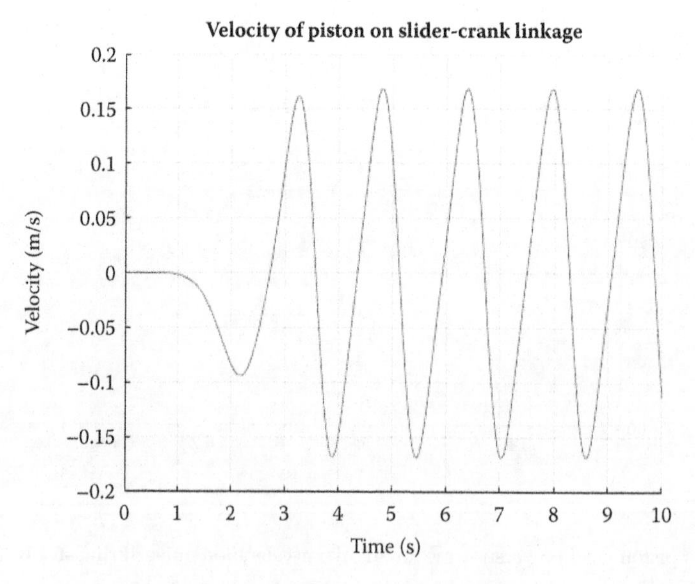

FIGURE 6.22
Velocity of piston versus time for angular acceleration pulse.

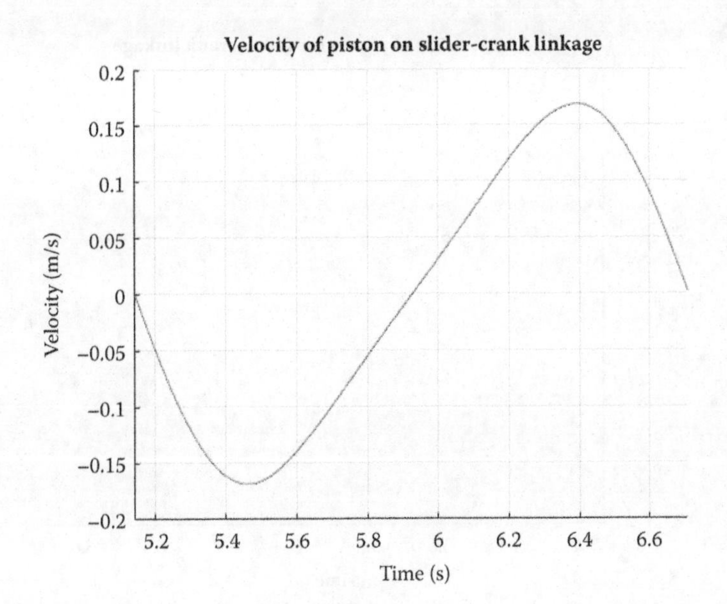

FIGURE 6.23
Zoomed in view of piston velocity versus time for angular acceleration pulse during steady-state portion. Note similarity to Figure 6.21.

Figures 6.21 and 6.23. The absolute magnitude of the velocity is different in each case because the crank angular velocity was set to 10 rad/s in the constant angular velocity case, but it only reaches 4 rad/s in the case of the acceleration pulse.

Finally, Figures 6.24 and 6.25 show the acceleration of the piston for the angular acceleration pulse. As before, Figures 6.14 and 6.25 are off by a scaling factor because of the difference in steady-state angular velocity of the crank.

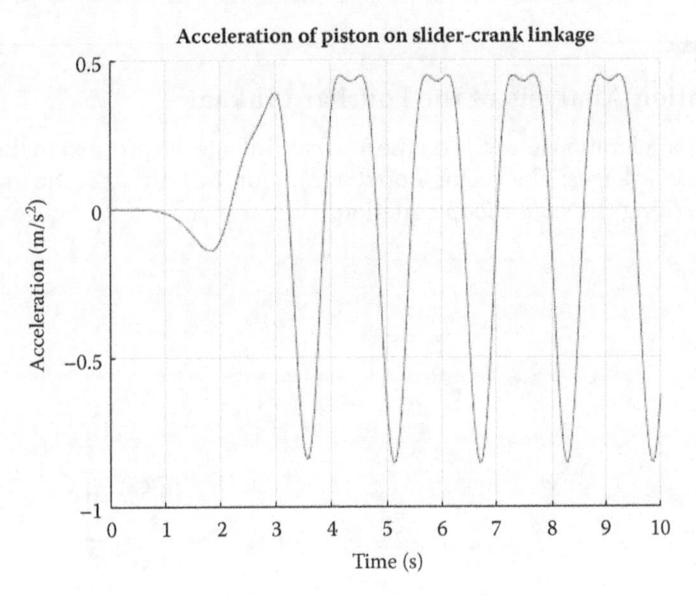

FIGURE 6.24
Acceleration of piston for the crank angular acceleration pulse case.

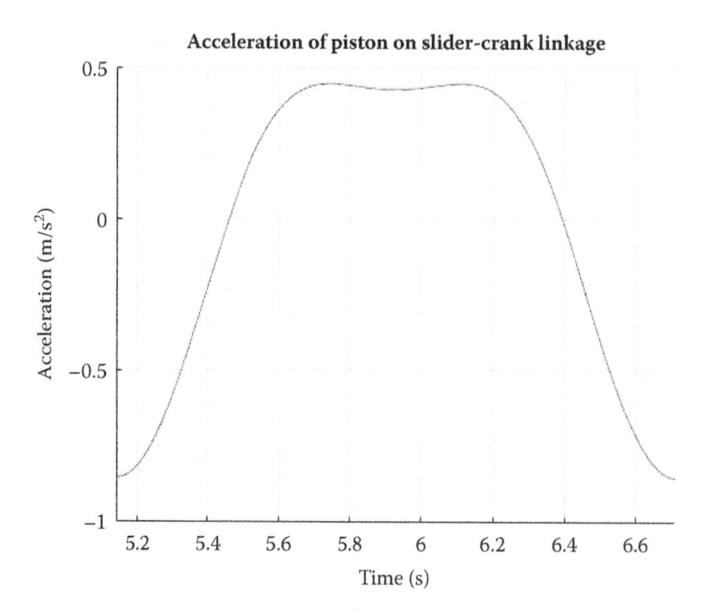

FIGURE 6.25
Zoomed in view of piston velocity versus time for angular acceleration pulse during steady-state portion. Note similarity to Figure 6.14.

To summarize, we have developed a method for calculating positions, velocities, and accelerations for the case when the crank has a non-constant angular velocity. To keep things simple, we will assume a constant crank angular velocity in the sections that follow. As we have seen, it is a relatively simple matter to convert to non-zero angular acceleration if that is required.

6.4 Acceleration Analysis of the Fourbar Linkage

To conduct the acceleration analysis on the fourbar linkage we proceed in the same manner as the preceding linkages. The vector loop diagram for the fourbar is shown in Figure 6.26. We begin by writing the vector loop equation

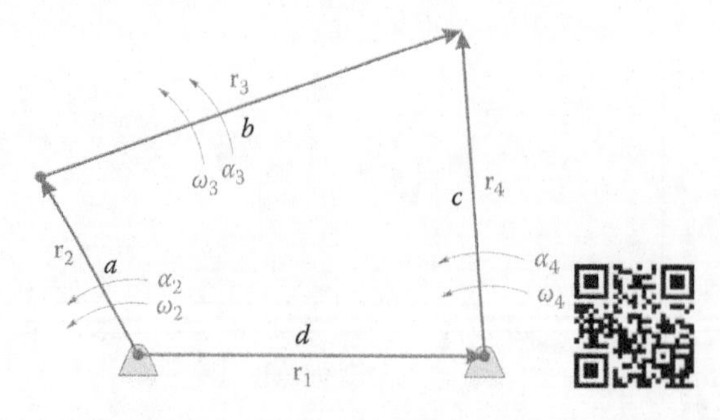

FIGURE 6.26
Vector loop diagram of the fourbar linkage showing angular velocities and accelerations.

$$\mathbf{r}_2 + \mathbf{r}_3 - \mathbf{r}_4 - \mathbf{r}_1 = 0 \tag{6.44}$$

Or, in unit vector form

$$a\mathbf{e}_2 + b\mathbf{e}_3 - c\mathbf{e}_4 - d\mathbf{e}_1 = 0 \tag{6.45}$$

Differentiate this with respect to time

$$\omega_2 a\mathbf{n}_2 + \omega_3 b\mathbf{n}_3 - \omega_4 c\mathbf{n}_4 = 0 \tag{6.46}$$

Then, differentiate again. The first term differentiates as

$$\frac{d}{dt}(\omega_2 a\mathbf{n}_2) = \frac{d\omega_2}{dt} a\mathbf{n}_2 + \omega_2 a \frac{d}{dt}(\mathbf{n}_2) = \alpha_2 a\mathbf{n}_2 - \omega_2^2 a\mathbf{e}_2 \tag{6.47}$$

The other terms proceed in a similar manner, so that the acceleration equation is

$$\alpha_2 a\mathbf{n}_2 - \omega_2^2 a\mathbf{e}_2 + \alpha_3 b\mathbf{n}_3 - \omega_3^2 b\mathbf{e}_3 - \alpha_4 c\mathbf{n}_4 + \omega_4^2 c\mathbf{e}_4 = 0 \tag{6.48}$$

As before, move the known quantities to the right side of the equation, and the unknowns to the left

$$\alpha_3 b\mathbf{n}_3 - \alpha_4 c\mathbf{n}_4 = -\alpha_2 a\mathbf{n}_2 + \omega_2^2 a\mathbf{e}_2 + \omega_3^2 b\mathbf{e}_3 - \omega_4^2 c\mathbf{e}_4 \tag{6.49}$$

And finally, put this into matrix form for solution in MATLAB

$$\mathbf{C}\boldsymbol{\alpha} = \mathbf{d} \tag{6.50}$$

where

$$\mathbf{C} = \begin{bmatrix} b\mathbf{n}_3 & \vdots & -c\mathbf{n}_4 \end{bmatrix} \tag{6.51}$$

$$\boldsymbol{\alpha} = \begin{Bmatrix} \alpha_3 \\ \alpha_4 \end{Bmatrix} \qquad \mathbf{d} = -\alpha_2 a\mathbf{n}_2 + \omega_2^2 a\mathbf{e}_2 + \omega_3^2 b\mathbf{e}_3 - \omega_4^2 c\mathbf{e}_4 \tag{6.52}$$

As before, the **C** matrix is identical to the **A** matrix found earlier for the velocity analysis.

6.4.1 Computing the Accelerations Using MATLAB®

Start by opening your fourbar velocity analysis code. Use `Save As` to save the file as `Fourbar_Acceleration_Analysis.m`. As an example, we will consider the same linkage we have used for the position and velocity analysis codes in earlier sections, as shown in Figure 6.27. We will assume a constant angular velocity for the crank, so that its angular acceleration is zero. Modify your code to perform the matrix calculations given above, then check the angular accelerations of the coupler and rocker using the `Derivative_Plot` function.

Figure 6.28 shows the computed and estimated angular acceleration of the coupler for the example problem we have been considering. Figure 6.29 shows the same plot for the

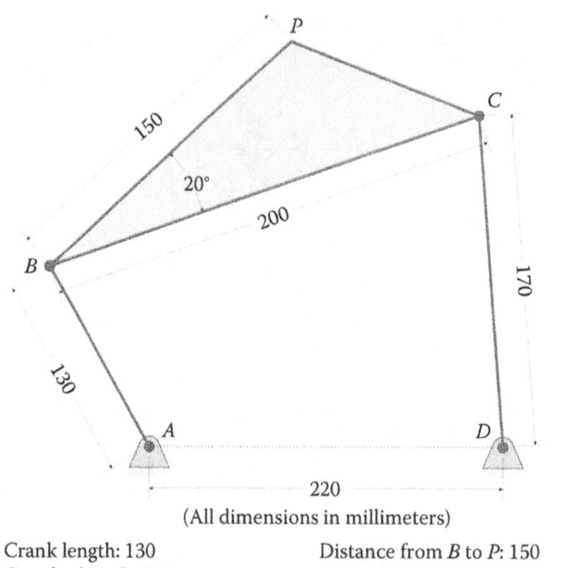

FIGURE 6.27
Fourbar linkage used in the example calculations.

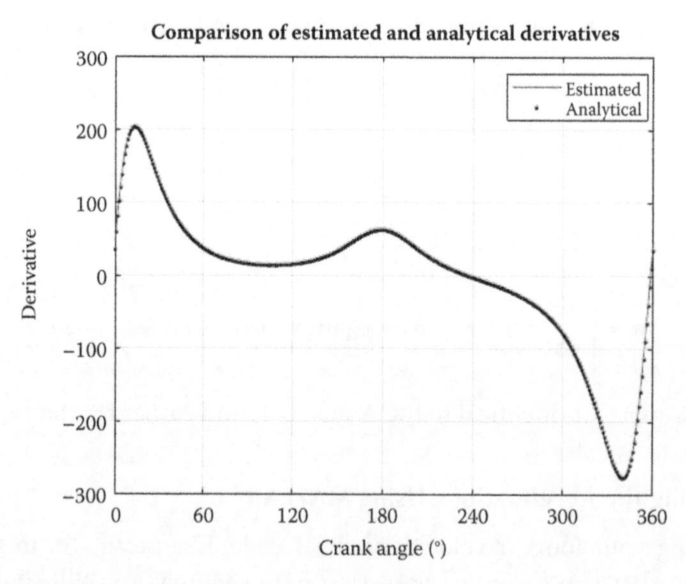

FIGURE 6.28
The estimated and analytical values of α_3.

angular acceleration of the rocker. Make sure that your plots are the same as those shown in Figures 6.28 and 6.29 before continuing with the analysis.

To find the acceleration of the point P on the coupler use the `FindAcc` function defined earlier. Figure 6.30 shows the x and y components of the acceleration of point P for the

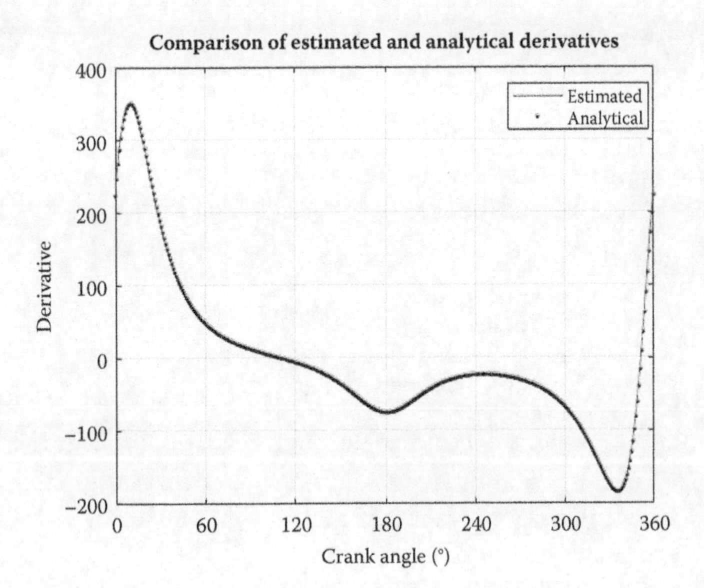

FIGURE 6.29
Estimated and analytical values of α_4.

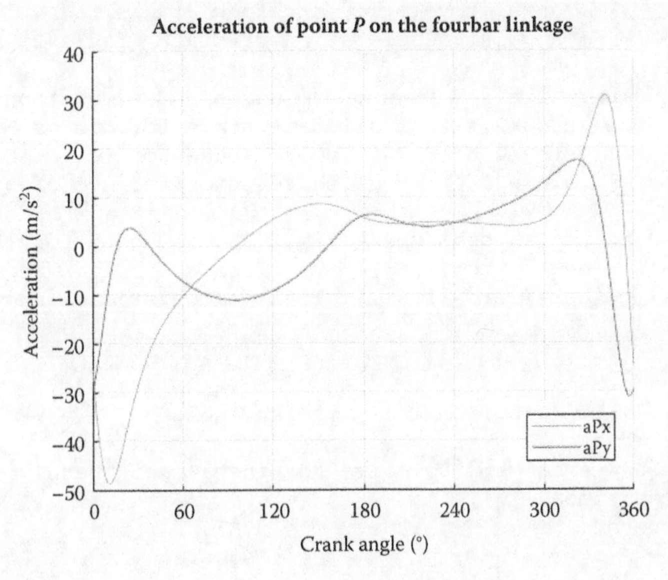

FIGURE 6.30
Acceleration of the point P on the coupler.

example problem. It is interesting to note that the acceleration is relatively large, with peaks above −4 g, even for a relatively low crank angular velocity. If the example fourbar were intended to move something delicate, the large acceleration might be hazardous. This is one of the main reasons for performing an acceleration analysis – we must always ensure that the objects being moved do not experience excessive acceleration, especially if they are human beings! The completed code for fourbar acceleration analysis is presented below.

```
% Fourbar_Acceleration_Analysis.m
% Conducts an acceleration analysis on the fourbar linkage
% and plots the acceleration of point P
% by Eric Constans, June 14, 2017

% Prepare Workspace
clear variables; close all; clc;

% Linkage dimensions
a = 0.130;          % crank length (m)
b = 0.200;          % coupler length (m)
c = 0.170;          % rocker length (m)
d = 0.220;          % length between ground pins (m)
p = 0.150;          % length from B to P (m)
gamma3 = 20*pi/180; % angle between BP and coupler (converted to rad)

% ground pins
x0 = [0;0];  % ground pin at A (origin)
xD = [d;0];  % ground pin at D
v0 = [0;0];  % velocity of origin
a0 = [0;0];  % accel of origin

% Angular velocity and acceleration of crank
omega2 = 10;        % angular velocity of crank (rad/s)
alpha2 = 0;         % angular acceleration of crank (rad/s^2)

N = 361;   % number of times to perform position calculations
[xB,xC,xP] = deal(zeros(2,N)); % allocate space for pos of B, C, P
[vB,vC,vP] = deal(zeros(2,N)); % allocate space for vel of B, C, P
[aB,aC,aP] = deal(zeros(2,N)); % allocate space for acc of B, C, P

[theta2,theta3,theta4] = deal(zeros(1,N));  % allocate for link angles
[omega3,omega4] = deal(zeros(1,N));  % allocate for link angular vel
[alpha3,alpha4] = deal(zeros(1,N));  % allocate for link angular acc

% Main Loop
for i = 1:N
  theta2(i) = (i-1)*(2*pi)/(N-1);  % crank angle

% conduct position analysis to solve for theta3 and theta4
  r = d - a*cos(theta2(i));
  s = a*sin(theta2(i));
  f2 = r^2 + s^2;                      % f squared
  delta = acos((b^2+c^2-f2)/(2*b*c)); % angle between coupler and rocker

  g = b - c*cos(delta);
  h = c*sin(delta);

  theta3(i) = atan2((h*r - g*s),(g*r + h*s));
  theta4(i) = theta3(i) + delta;

% calculate unit vectors
  [e2,n2] = UnitVector(theta2(i));
```

```
  [e3,n3] = UnitVector(theta3(i));
  [e4,n4] = UnitVector(theta4(i));
  [eBP,nBP] = UnitVector(theta3(i) + gamma3);

% solve for positions of points B, C and P on the linkage
  xB(:,i) = FindPos(      x0,    a,    e2);
  xC(:,i) = FindPos(      xD,    c,    e4);
  xP(:,i) = FindPos(xB(:,i),    p,   eBP);

% conduct velocity analysis to solve for omega3 and omega4
  A_Mat = [b*n3  -c*n4];
  b_Vec = -a*omega2*n2;
  omega_Vec = A_Mat\b_Vec;  % solve for angular velocities

  omega3(i) = omega_Vec(1);     % decompose omega_Vec into
  omega4(i) = omega_Vec(2);     % individual components

% calculate velocity at important points on linkage
  vB(:,i) = FindVel(      v0,    a,     omega2,    n2);
  vC(:,i) = FindVel(      v0,    c, omega4(i),    n4);
  vP(:,i) = FindVel(vB(:,i),    p, omega3(i),   nBP);

% conduct acceleration analysis to solve for alpha3 and alpha4
  ac = a*omega2^2;
  bc = b*omega3(i)^2;
  cc = c*omega4(i)^2;
  pc = p*omega3(i)^2;

  C_Mat = A_Mat;
  d_Vec = ac*e2 + bc*e3 - cc*e4;
  alpha_Vec = C_Mat\d_Vec;  % solve for angular accelerations

  alpha3(i) = alpha_Vec(1);
  alpha4(i) = alpha_Vec(2);

% find acceleration of pins
  aB(:,i) = FindAcc(      a0, a,     omega2,     alpha2,  e2,  n2);
  aC(:,i) = FindAcc(      a0, c, omega4(i), alpha4(i),  e4,  n4);
  aP(:,i) = FindAcc(aB(:,i), p, omega3(i), alpha3(i), eBP, nBP);
end

% plot the acceleration of point P
figure; hold on
plot(theta2*180/pi,aP(1,:),'Color',[153/255 153/255 153/255])
plot(theta2*180/pi,aP(2,:),'Color',[0 110/255 199/255])
legend('aPx','aPy','Location','Southeast')
title('Acceleration of point P on the fourbar linkage')
xlabel('Crank angle (degrees)')
ylabel('Acceleration (m/s^2)')
grid on
set(gca,'xtick',0:60:360)
xlim([0 360])
```

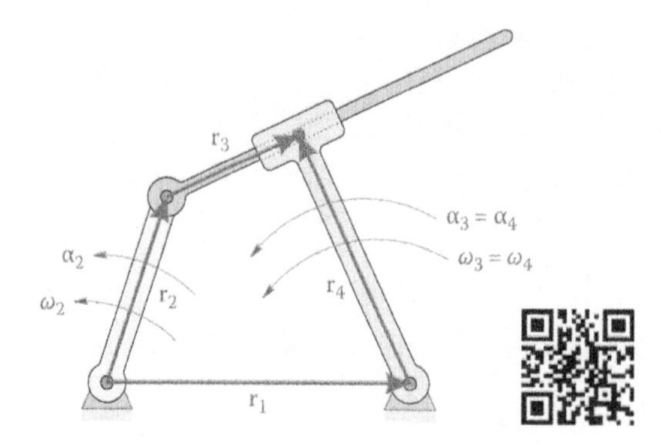

FIGURE 6.31
Vector loop diagram of the inverted slider-crank showing angular velocities and accelerations.

6.5 Acceleration Analysis of the Inverted Slider-Crank

The important dimensions of the inverted slider-crank are shown in Figure 6.31. We begin, as always, with the vector loop equation

$$\mathbf{r}_2 + \mathbf{r}_3 - \mathbf{r}_4 - \mathbf{r}_1 = 0 \tag{6.53}$$

In unit vector form

$$a\mathbf{e}_2 + b\mathbf{e}_3 - c\mathbf{e}_4 - d\mathbf{e}_1 = 0 \tag{6.54}$$

Recall that $\theta_4 = \theta_3 + \delta$ and thus $\omega_3 = \omega_4$. Substituting this, and differentiating we have

$$a\omega_2\mathbf{n}_2 + \dot{b}\mathbf{e}_3 + b\omega_3\mathbf{n}_3 - c\omega_3\mathbf{n}_4 = 0 \tag{6.55}$$

To obtain the accelerations, we differentiate the above equation again, using the chain and product rules where appropriate. The first term is simple

$$\frac{d}{dt}(a\omega_2\mathbf{n}_2) = a\alpha_2\mathbf{n}_2 - a\omega_2^2\mathbf{e}_2 \tag{6.56}$$

The second term is slightly more complicated

$$\frac{d}{dt}\left(\dot{b}\mathbf{e}_3\right) = \ddot{b}\mathbf{e}_3 + \dot{b}\omega_3\mathbf{n}_3 \tag{6.57}$$

The third term is similar to the second

$$\frac{d}{dt}(b\omega_3\mathbf{n}_3) = \dot{b}\omega_3\mathbf{n}_3 + b\alpha_3\mathbf{n}_3 - b\omega_3^2\mathbf{e}_3 \tag{6.58}$$

And the fourth term is similar to the first term

$$\frac{d}{dt}(c\omega_3\mathbf{n}_4) = c\alpha_3\mathbf{n}_4 - c\omega_3^2\mathbf{e}_4 \tag{6.59}$$

Adding these together, we obtain the acceleration equation for the inverted slider-crank

$$a\alpha_2\mathbf{n}_2 - a\omega_2^2\mathbf{e}_2 + \ddot{b}\mathbf{e}_3 + 2\dot{b}\omega_3\mathbf{n}_3 + b\alpha_3\mathbf{n}_3 - b\omega_3^2\mathbf{e}_3 - c\alpha_3\mathbf{n}_4 + c\omega_3^2\mathbf{e}_4 = 0 \tag{6.60}$$

The reader will recognize several familiar terms including three centripetal accelerations, three tangential acceleration terms, an acceleration of slip and a Coriolis acceleration term. Now place the knowns on the right side of the equation and the unknowns on the left.

$$\ddot{b}\mathbf{e}_3 + b\alpha_3\mathbf{n}_3 - c\alpha_3\mathbf{n}_4 = -a\alpha_2\mathbf{n}_2 + a\omega_2^2\mathbf{e}_2 - 2\dot{b}\omega_3\mathbf{n}_3 + b\omega_3^2\mathbf{e}_3 - c\omega_3^2\mathbf{e}_4 \tag{6.61}$$

Collecting terms, and placing this into matrix form, we have

$$\mathbf{C}\boldsymbol{\alpha} = \mathbf{d} \tag{6.62}$$

where

$$\mathbf{C} = \begin{bmatrix} b\mathbf{n}_3 - c\mathbf{n}_4 & \vdots & \mathbf{e}_3 \end{bmatrix}$$

$$\boldsymbol{\alpha} = \begin{Bmatrix} \alpha_3 \\ \ddot{b} \end{Bmatrix} \qquad \mathbf{d} = \left\{-a\alpha_2\mathbf{n}_2 + a\omega_2^2\mathbf{e}_2 - 2\dot{b}\omega_3\mathbf{n}_3 + b\omega_3^2\mathbf{e}_3 - c\omega_3^2\mathbf{e}_4\right\}$$

6.5.1 Computing the Accelerations Using MATLAB®

Figure 6.32 shows the example inverted slider-crank that we have used for position and velocity analysis. Modify your inverted slider-crank velocity analysis code to perform the acceleration calculations described above.

Make sure to check your `alpha3` and `bddot` calculations using the `Derivative_Plot` function. If all goes well, your plot for the *x* and *y* components of the acceleration at point *P* should look like Figure 6.33.

6.6 Acceleration Analysis of the Geared Fivebar Linkage

Our next acceleration analysis will be conducted on the geared fivebar linkage, whose vector loop diagram is shown in Figure 6.34. Recall that the angle of link 5 (the second gear) is known as a function of the crank angle, through the gear ratio

$$\theta_5 = -\rho\theta_2 + \varphi \tag{6.63}$$

where

$$\rho = \frac{N_1}{N_2} \tag{6.64}$$

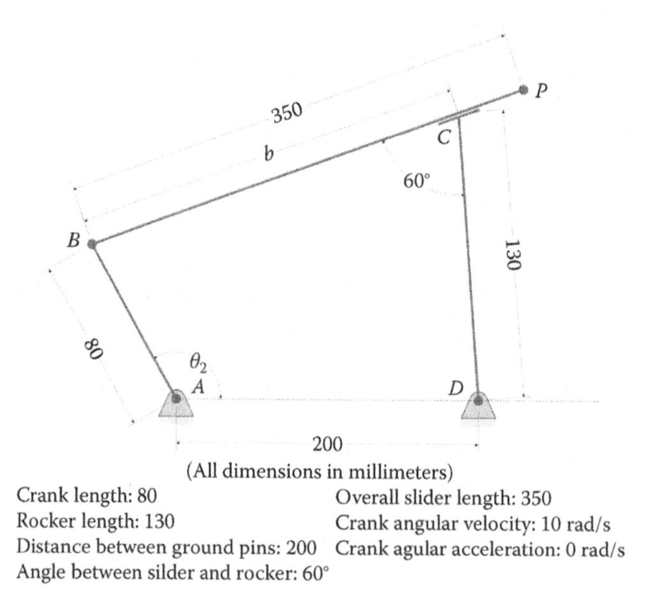

FIGURE 6.32
Dimensions of the inverted slider-crank used in the example calculations.

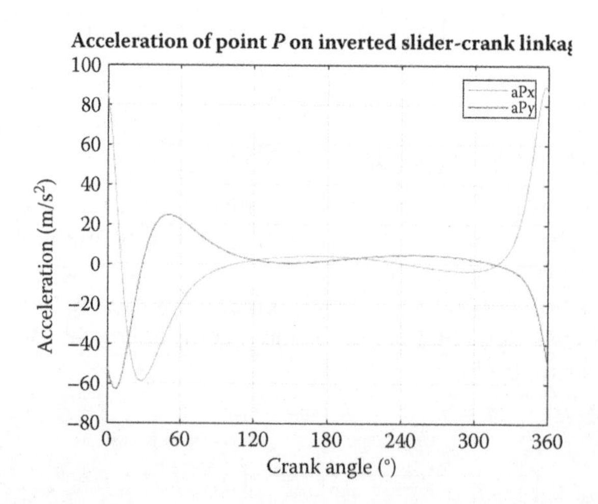

FIGURE 6.33
Acceleration of point P on the example inverted slider-crank linkage.

is the gear ratio. If we differentiate this expression twice with respect to time, we arrive at the acceleration ratio between the two gears

$$\alpha_5 = -\rho\alpha_2 \tag{6.65}$$

The vector loop equation in unit vector form is

$$a\mathbf{e}_2 + b\mathbf{e}_3 - c\mathbf{e}_4 - u\mathbf{e}_5 - d\mathbf{e}_1 = 0 \tag{6.66}$$

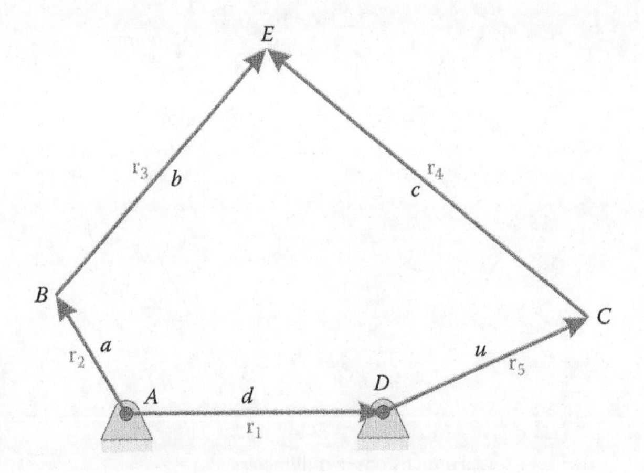

FIGURE 6.34
Vector loop diagram of the geared fivebar linkage.

Differentiating this twice with respect to time gives

$$a\alpha_2\mathbf{n}_2 - a\omega_2^2\mathbf{e}_2 + b\alpha_3\mathbf{n}_3 - b\omega_3^2\mathbf{e}_3 - c\alpha_4\mathbf{n}_4 + c\omega_4^2\mathbf{e}_4 - u\alpha_5\mathbf{n}_5 + u\omega_5^2\mathbf{e}_5 = 0 \tag{6.67}$$

Collecting all known terms on the right-hand side of the equation and rearranging into matrix form results in a familiar formula

$$\mathbf{C}\boldsymbol{\alpha} = \mathbf{d} \tag{6.68}$$

with

$$\mathbf{A} = \begin{bmatrix} b\mathbf{n}_3 & \vdots & -c\mathbf{n}_4 \end{bmatrix} \tag{6.69}$$

$$\mathbf{d} = \begin{bmatrix} -a\alpha_2\mathbf{n}_2 + a\omega_2^2\mathbf{e}_2 + b\omega_3^2\mathbf{e}_3 - c\omega_4^2\mathbf{e}_4 + u\alpha_5\mathbf{n}_5 - u\omega_5^2\mathbf{e}_5 \end{bmatrix} \tag{6.70}$$

and

$$\boldsymbol{\alpha} = \begin{Bmatrix} \alpha_3 \\ \alpha_4 \end{Bmatrix} \tag{6.71}$$

As with the other linkages, the **C** matrix is identical to the **A** matrix for the fivebar.

6.6.1 Computing the accelerations using MATLAB®

Figure 6.35 shows the dimensions of the geared fivebar linkage that we have used for position and velocity analysis. Modify your fivebar velocity code to perform the matrix calculations above and use it to plot the acceleration of point *P*.

Figure 6.36 shows the x and y components of the acceleration at point *P* for the example linkage. Next, modify the dimensions of the linkage such that a = 150 mm, u = 150 mm, and d = 200 mm. You might recall that we obtained discontinuities in the velocity plot with

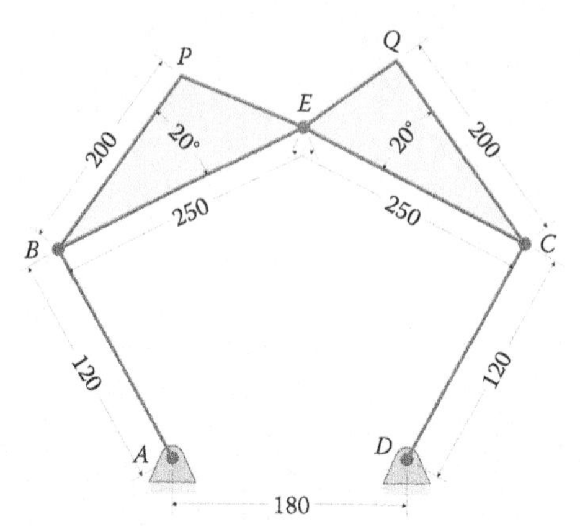

Length of link on gear 1: 120
Coupler 1 length: 250
Coupler 2 length: 250
Length of link on gear 2: 120
Distance between ground pins: 180
Length from *B* to *P*: 200
Length from *C* to *Q*: 200
Teeth on gear 1: 48
Teeth on gear 2: 24
Angular velocity of gear 1: 10 rad/s
Angular acceleration of gear 1: 0 rad/s

FIGURE 6.35
Dimensions of the geared fivebar used in the example calculations.

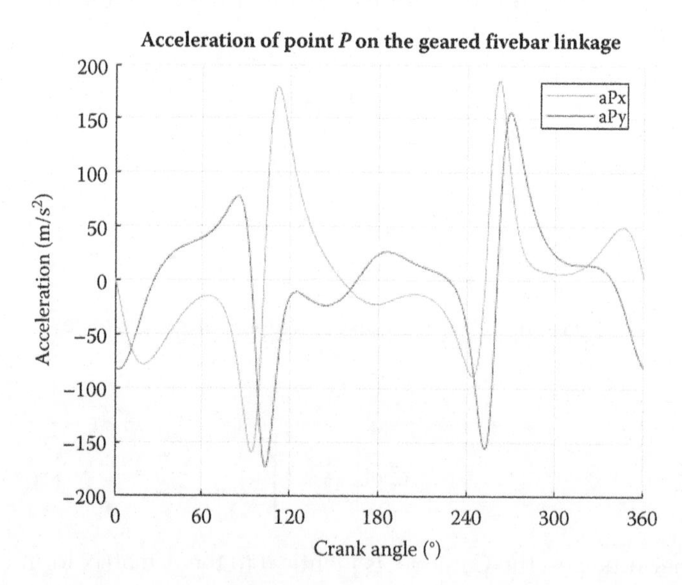

FIGURE 6.36
Acceleration of point *P* for the example geared fivebar linkage.

these dimensions reflecting the fact that the point *P* made a sudden change in direction when $\theta_2 = 180°$.

The resulting acceleration plot is shown in Figure 6.37. Observe the near infinite acceleration of the point *P* when $\theta_2 = 180°$. Because acceleration is proportional to force, this behavior will result in a sudden jolt in all of the pins when the linkage passes through this point. Clearly, this is a design to be avoided!

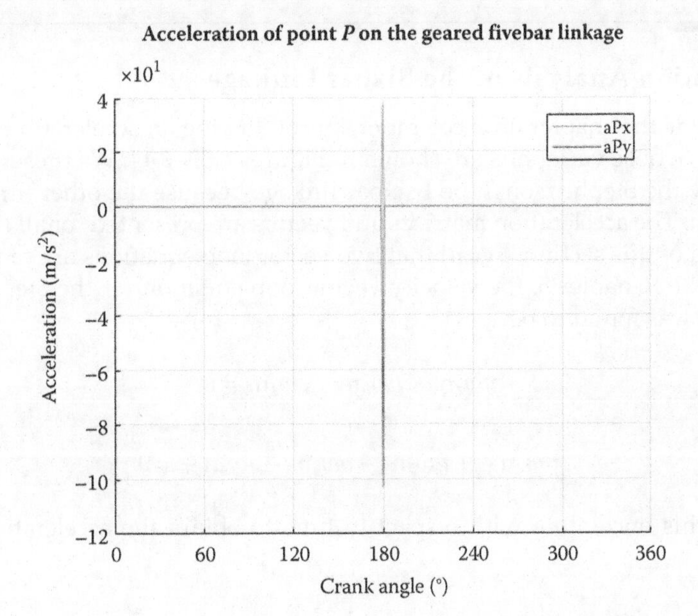

FIGURE 6.37
Acceleration of point P with the modified dimensions. The discontinuities in the velocity of point P have resulted in near infinite accelerations when $\theta_2 = 180°$.

If we "zoom in" on the acceleration plot by using `ylim([-400 400])` we see that the acceleration of point P seems very ordinary except where $\theta_2 = 180°$ (see Figure 6.38). This demonstrates the benefit of performing a kinematic analysis for the entire range of motion of the linkage. If we had simply done the acceleration calculations at a few crank angles, we might have missed the disaster that occurs when $\theta_2 = 180°$.

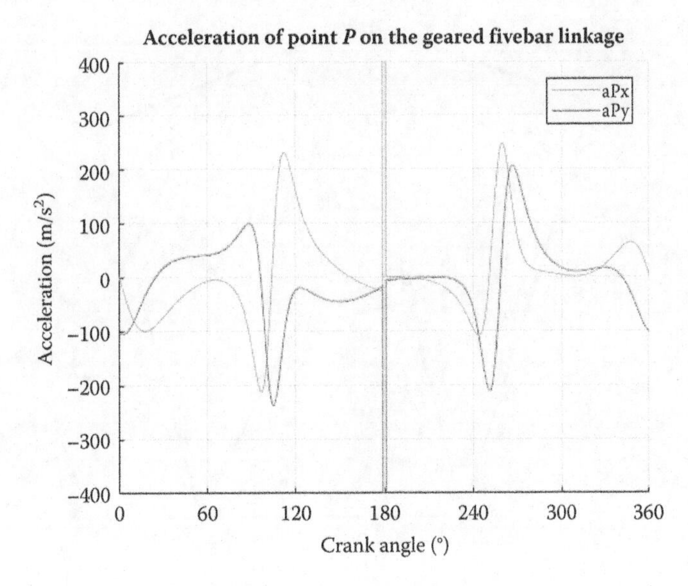

FIGURE 6.38
The acceleration of point P is "normal" except at $\theta_2 = 180°$.

6.7 Acceleration Analysis of the Sixbar Linkage

We will conclude the chapter on acceleration by conducting an acceleration analysis of the sixbar linkage, whose variations are shown in Figures 6.39–6.44. We present the full derivation for only the Stephenson Type I sixbar linkage because the other linkages follow a similar pattern. The acceleration matrices and vectors are presented for all five sixbar linkages at the end of this section. Recall that two vector loop equations are required for each sixbar linkage. In Chapter 5, the velocity vector loop equation for the Stephenson Type I sixbar linkage was found to be

$$a\omega_2\mathbf{n}_2 + b\omega_3\mathbf{n}_3 - c\omega_4\mathbf{n}_4 = 0$$
$$p\omega_2\mathbf{n}_{AE} + u\omega_5\mathbf{n}_5 - v\omega_6\mathbf{n}_6 - q\omega_4\mathbf{n}_{DF} = 0 \tag{6.72}$$

Differentiate this once more with respect to time to obtain the acceleration vector loop equations.

$$a\alpha_2\mathbf{n}_2 - a\omega_2^2\mathbf{e}_2 + b\alpha_3\mathbf{n}_3 - b\omega_3^2\mathbf{e}_3 - c\alpha_4\mathbf{n}_4 + c\omega_4^2\mathbf{e}_4 = 0 \tag{6.73}$$

$$p\alpha_2\mathbf{n}_{AE} - p\omega_2^2\mathbf{e}_{AE} + u\alpha_5\mathbf{n}_5 - u\omega_5^2\mathbf{e}_5 - v\alpha_6\mathbf{n}_6 + v\omega_6^2\mathbf{e}_6 - q\alpha_4\mathbf{n}_{DF} + q\omega_4^2\mathbf{e}_{DF} = 0$$

Since each link is rigid, we do not have any Coriolis or slip acceleration terms. Instead, each moving link provides a centripetal and tangential acceleration term. Thus, there are two acceleration terms for each moving link in a given loop. Now collect all unknown terms on the right side of the equations, and the known terms on the left.

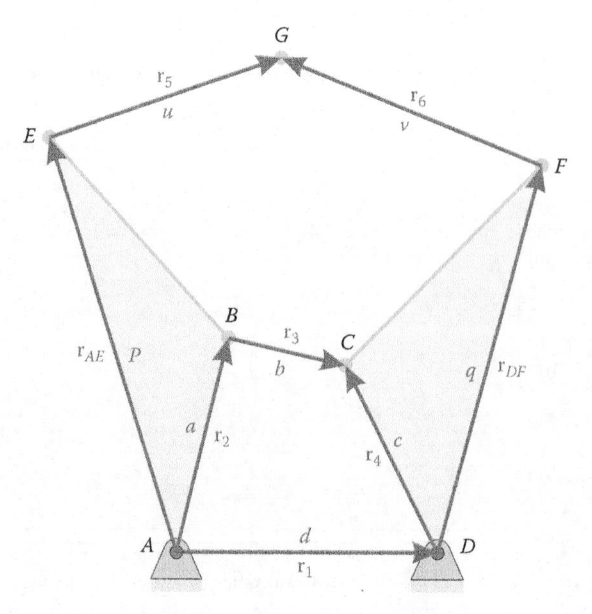

FIGURE 6.39
Vector loop diagram for the Stephenson Type I sixbar linkage.

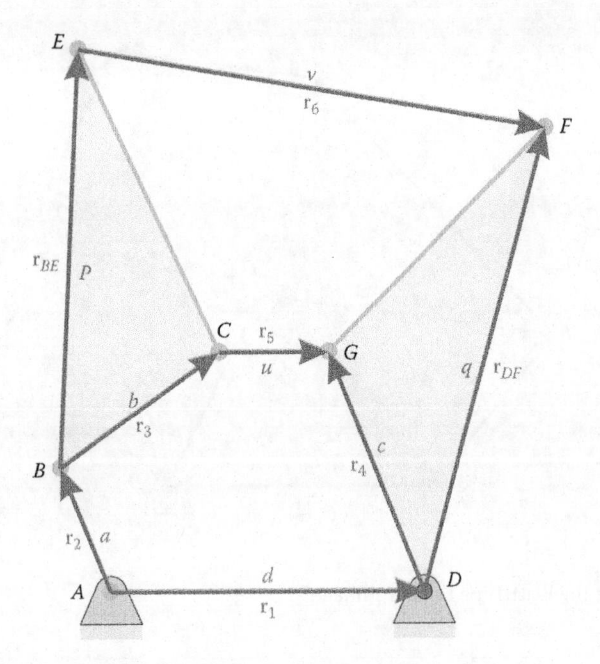

FIGURE 6.40
Vector loop diagram for the Stephenson Type II sixbar linkage.

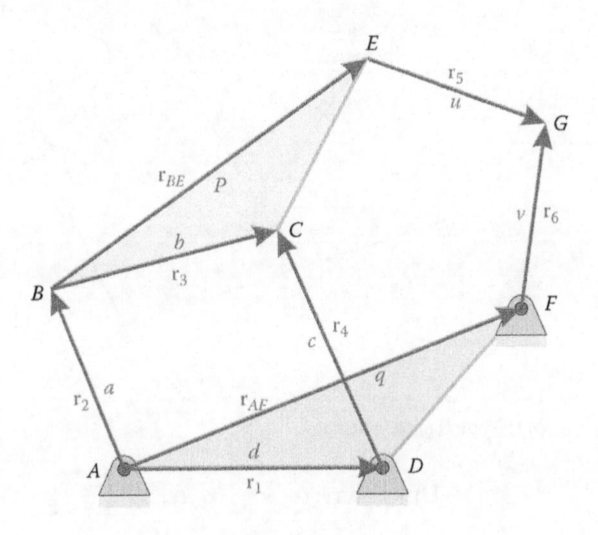

FIGURE 6.41
Vector loop diagram for the Stephenson Type III sixbar linkage.

$$b\alpha_3\mathbf{n}_3 - c\alpha_4\mathbf{n}_4 = -a\alpha_2\mathbf{n}_2 + a\omega_2^2\mathbf{e}_2 + b\omega_3^2\mathbf{e}_3 - c\omega_4^2\mathbf{e}_4 \tag{6.74}$$

$$-q\alpha_4\mathbf{n}_{DF} + u\alpha_5\mathbf{n}_5 - v\alpha_6\mathbf{n}_6 = -p\alpha_2\mathbf{n}_{AE} + p\omega_2^2\mathbf{e}_{AE} + u\omega_5^2\mathbf{e}_5 - v\omega_6^2\mathbf{e}_6 - q\omega_4^2\mathbf{e}_{DF} \tag{6.75}$$

We now have four equations, with four unknowns

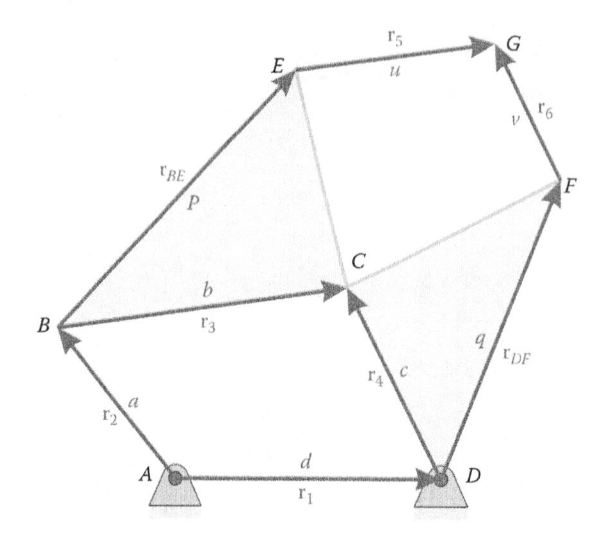

FIGURE 6.42
Vector loop diagram for the Watt Type I sixbar linkage.

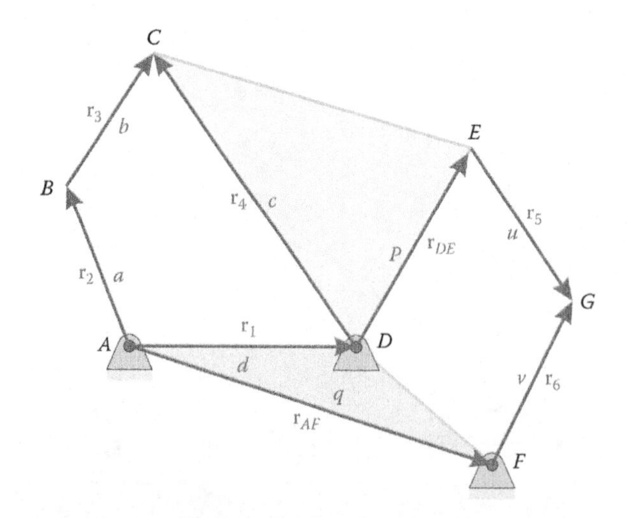

FIGURE 6.43
Vector loop diagram for the Watt Type II sixbar linkage.

$$\text{Unknown} : \alpha_3, \alpha_4, \alpha_5, \alpha_6 \tag{6.76}$$

and we can arrange the equations into matrix form as before

$$\mathbf{C}\alpha = \mathbf{d} \tag{6.77}$$

where

$$\mathbf{A} = \begin{bmatrix} b\mathbf{n}_3 & \vdots & -c\mathbf{n}_4 & \vdots & 0_{21} & \vdots & 0_{21} \\ 0_{21} & \vdots & -q\mathbf{n}_{DF} & \vdots & u\mathbf{n}_5 & \vdots & -v\mathbf{n}_6 \end{bmatrix} \tag{6.78}$$

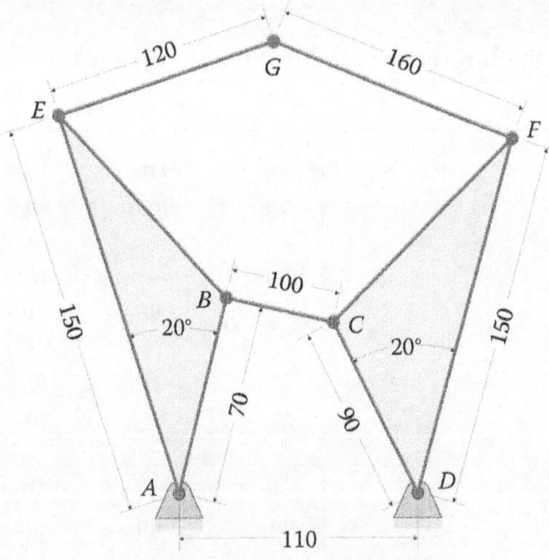

(All dimensions in millimeters)

Crank length: 70	Rocker length: 90
Length AE on crank: 150	Length DF on rocker: 150
Internal angle of crank: 20°	Internal angle of rocker: –20°
Coupler length: 100	Length of link 5: 120
Distance between ground pins: 110	Length of link 6: 160
Crank angular velocity: 10 rad/s	Crank angular acceleration: 0 rad/s

FIGURE 6.44
Dimensions of the example Stephenson Type I sixbar linkage.

$$\mathbf{d} = \left\{ \begin{array}{c} -a\alpha_2\mathbf{n}_2 + a\omega_2^2\mathbf{e}_2 + b\omega_3^2\mathbf{e}_3 - c\omega_4^2\mathbf{e}_4 \\ -p\alpha_2\mathbf{n}_{AE} + p\omega_2^2\mathbf{e}_{AE} + u\omega_5^2\mathbf{e}_5 - v\omega_6^2\mathbf{e}_6 - q\omega_4^2\mathbf{e}_{DF} \end{array} \right\} \tag{6.79}$$

$$\boldsymbol{\alpha} = \left\{ \begin{array}{cccc} \alpha_3 & \alpha_4 & \alpha_5 & \alpha_6 \end{array} \right\}^T \tag{6.80}$$

Of course, this set of equations is only valid for the Stephenson Type I sixbar linkage. A similar analysis can be carried out for the other sixbar linkages, the results are seen in Tables 6.2 and 6.3.

6.7.1 Some Example Solutions for the Sixbar Linkage

Figures 6.45–6.53 show the dimensions and acceleration plots for the example sixbar linkages. In each case, the angular velocity of the crank is a constant 10 rad/s. Be sure to check your answers against these plots before completing the homework problems.

6.8 Summary

This concludes our study of acceleration analysis of single degree of freedom linkages. As we have seen, acceleration analysis is a simple extension of velocity analysis, although the

TABLE 6.2

C Matrix for One Degree of Freedom Sixbar Linkages

Linkage	**C**
Stephenson I	$\begin{bmatrix} b\mathbf{n}_3 & \vdots & -c\mathbf{n}_4 & \vdots & \mathbf{0}_{21} & \vdots & \mathbf{0}_{21} \\ \mathbf{0}_{21} & \vdots & -q\mathbf{n}_{DF} & \vdots & u\mathbf{n}_5 & \vdots & -v\mathbf{n}_6 \end{bmatrix}$
Stephenson II	$\begin{bmatrix} b\mathbf{n}_3 & \vdots & -c\mathbf{n}_4 & \vdots & u\mathbf{n}_5 & \vdots & \mathbf{0}_{21} \\ p\mathbf{n}_{BE} & \vdots & -q\mathbf{n}_{DF} & \vdots & \mathbf{0}_{21} & \vdots & v\mathbf{n}_6 \end{bmatrix}$
Stephenson III	$\begin{bmatrix} b\mathbf{n}_3 & \vdots & -c\mathbf{n}_4 & \vdots & \mathbf{0}_{21} & \vdots & \mathbf{0}_{21} \\ p\mathbf{n}_{BE} & \vdots & \mathbf{0}_{21} & \vdots & u\mathbf{n}_5 & \vdots & -v\mathbf{n}_6 \end{bmatrix}$
Watt I	$\begin{bmatrix} b\mathbf{n}_3 & \vdots & -c\mathbf{n}_4 & \vdots & \mathbf{0}_{21} & \vdots & \mathbf{0}_{21} \\ p\mathbf{n}_{BE} & \vdots & -q\mathbf{n}_{DF} & \vdots & u\mathbf{n}_5 & \vdots & -v\mathbf{n}_6 \end{bmatrix}$
Watt II	$\begin{bmatrix} b\mathbf{n}_3 & \vdots & -c\mathbf{n}_4 & \vdots & \mathbf{0}_{21} & \vdots & \mathbf{0}_{21} \\ \mathbf{0}_{21} & \vdots & p\mathbf{n}_{DE} & \vdots & u\mathbf{n}_5 & \vdots & -v\mathbf{n}_6 \end{bmatrix}$

TABLE 6.3

d Vector for the One Degree of Freedom Sixbar Linkages

Linkage	**d**
Stephenson I	$\left\{ \begin{matrix} -a\alpha_2\mathbf{n}_2 + a\omega_2^2\mathbf{e}_2 + b\omega_3^2\mathbf{e}_3 - c\omega_4^2\mathbf{e}_4 \\ -p\alpha_2\mathbf{n}_{AE} + p\omega_2^2\mathbf{e}_{AE} + u\omega_5^2\mathbf{e}_5 - v\omega_6^2\mathbf{e}_6 - q\omega_4^2\mathbf{e}_{DF} \end{matrix} \right\}$
Stephenson II	$\left\{ \begin{matrix} -a\alpha_2\mathbf{n}_2 + a\omega_2^2\mathbf{e}_2 + b\omega_3^2\mathbf{e}_3 - c\omega_4^2\mathbf{e}_4 + u\omega_5^2\mathbf{e}_5 \\ -a\alpha_2\mathbf{n}_2 + a\omega_2^2\mathbf{e}_2 + p\omega_3^2\mathbf{e}_{BE} - q\omega_4^2\mathbf{e}_{DF} + v\omega_6^2\mathbf{e}_6 \end{matrix} \right\}$
Stephenson III	$\left\{ \begin{matrix} -a\alpha_2\mathbf{n}_2 + a\omega_2^2\mathbf{e}_2 + b\omega_3^2\mathbf{e}_3 - c\omega_4^2\mathbf{e}_4 \\ -a\alpha_2\mathbf{n}_2 + a\omega_2^2\mathbf{e}_2 + p\omega_3^2\mathbf{e}_{BE} + u\omega_5^2\mathbf{e}_5 - v\omega_6^2\mathbf{e}_6 \end{matrix} \right\}$
Watt I	$\left\{ \begin{matrix} -a\alpha_2\mathbf{n}_2 + a\omega_2^2\mathbf{e}_2 + b\omega_3^2\mathbf{e}_3 - c\omega_4^2\mathbf{e}_4 \\ -a\alpha_2\mathbf{n}_2 + a\omega_2^2\mathbf{e}_2 + p\omega_3^2\mathbf{e}_{BE} - q\omega_4^2\mathbf{e}_{DF} + u\omega_5^2\mathbf{e}_5 - v\omega_6^2\mathbf{e}_6 \end{matrix} \right\}$
Watt II	$\left\{ \begin{matrix} -a\alpha_2\mathbf{n}_2 + a\omega_2^2\mathbf{e}_2 + b\omega_3^2\mathbf{e}_3 - c\omega_4^2\mathbf{e}_4 \\ p\omega_4^2\mathbf{e}_{DE} + u\omega_5^2\mathbf{e}_5 - v\omega_6^2\mathbf{e}_6 \end{matrix} \right\}$

matrix equations are considerably more complicated. Now that we can find the acceleration of any point on a linkage we are ready to begin calculating forces and torques. The study of position, velocity, and acceleration provides a *kinematic* analysis of each linkage. In the next chapter, we will add forces and torques, giving us a *dynamic* analysis. We will see that dynamic analysis is considerably more complicated than kinematic analysis, but by following a systematic procedure, we will be able to tackle each linkage in turn.

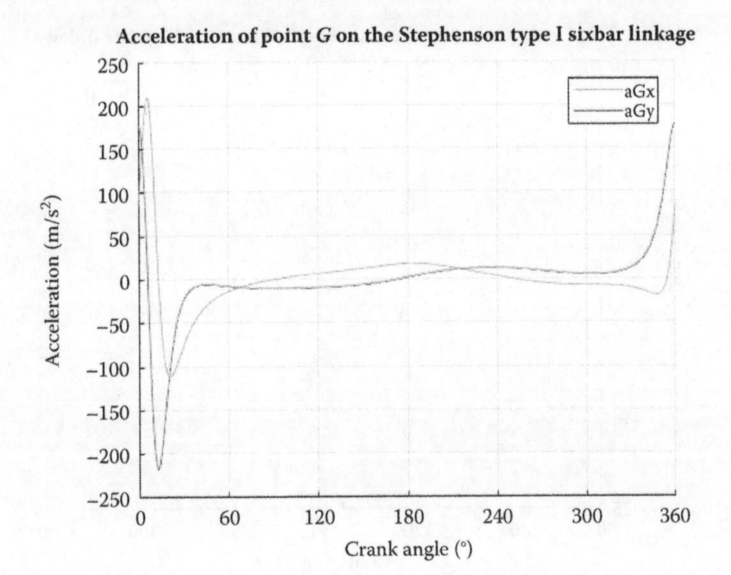

FIGURE 6.45
Acceleration of point G on the example Stephenson Type I linkage.

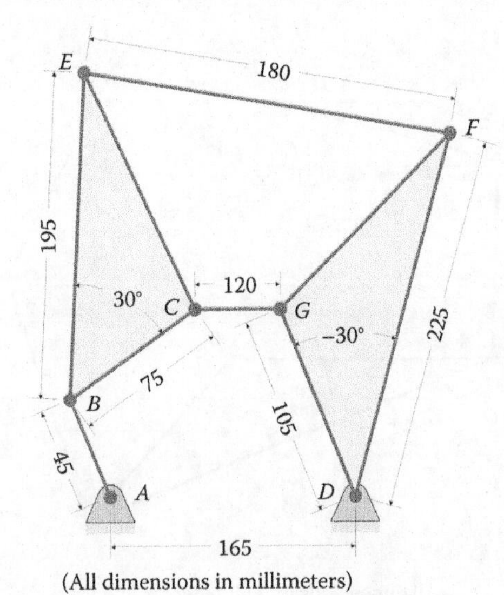

Crank length: 45	Rocker length: 105
Coupler length: 75	Length *DF* on rocker: 225
Length *BE* on coupler: 195	Internal angle of rocker: −30°
Internal angle of coupler: 30°	Length of link 5: 120
Distance between ground pins: 165	Length of link 6: 180
Crank angular velocity: 10 rad/s	Crank angular acceleration: 0 rad/s

FIGURE 6.46
Dimensions of the example Stephenson Type II sixbar linkage.

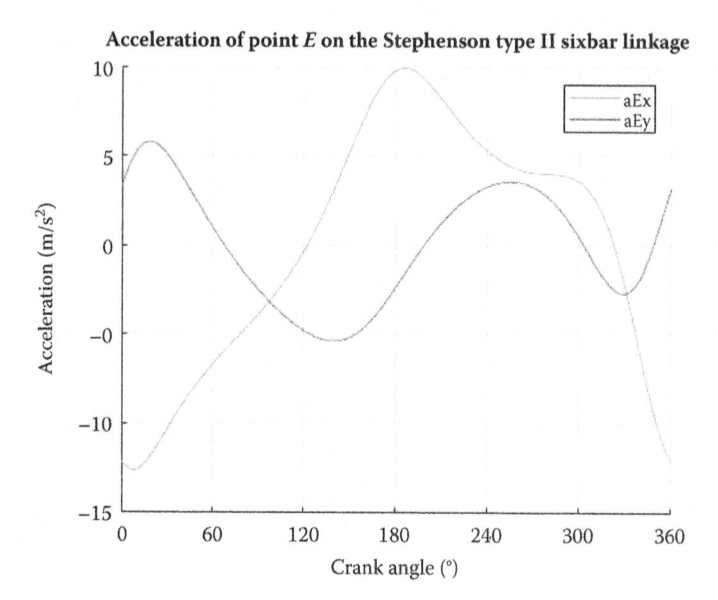

FIGURE 6.47
Acceleration of point E on the example Stephenson Type II linkage.

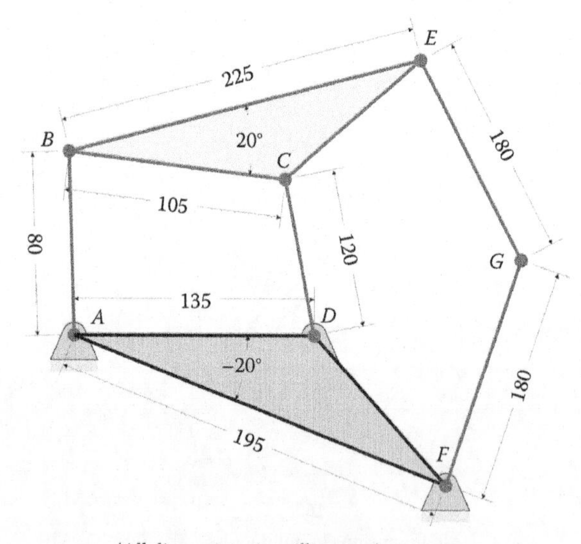

Crank length: 80
Coupler length: 105
Internal angle of coupler: 20°
Length BE on coupler: 225
Distance between ground pins: 135
Crank angular velocity: 10 rad/s
Internal angle of ground: −20°
Length AF on ground: 195
Rocker length: 120
Length of link 5: 180
Length of link 6: 180
Crank angular acceleration: 0 rad/s

FIGURE 6.48
Dimensions of the example Stephenson Type III sixbar linkage.

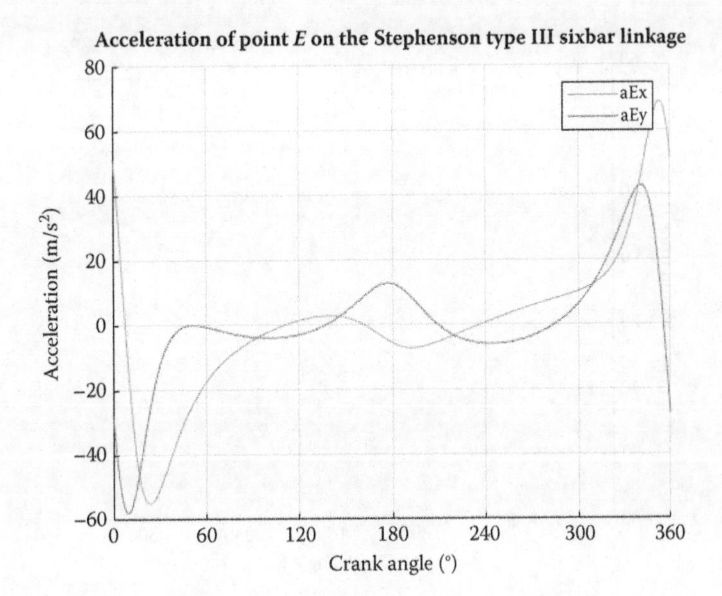

FIGURE 6.49
Acceleration of point E on the example Stephenson Type III linkage.

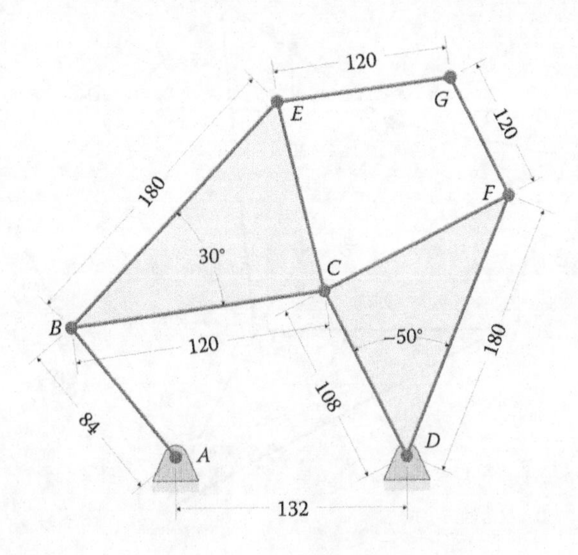

(All dimensions in millimeters)

Crank length: 84	Rocker length: 108
Coupler length: 120	Internal angle of rocker: −50°
Internal angle of coupler: 30°	Length DF on rocker: 180
Length BE on coupler: 180	Length of link 5: 120
Distance between ground pins: 132	Length of link 6: 120
Crank angular velocity: 10 rad/s	Crank angular acceleration: 0 rad/s

FIGURE 6.50
Dimensions of the example Watt Type I linkage.

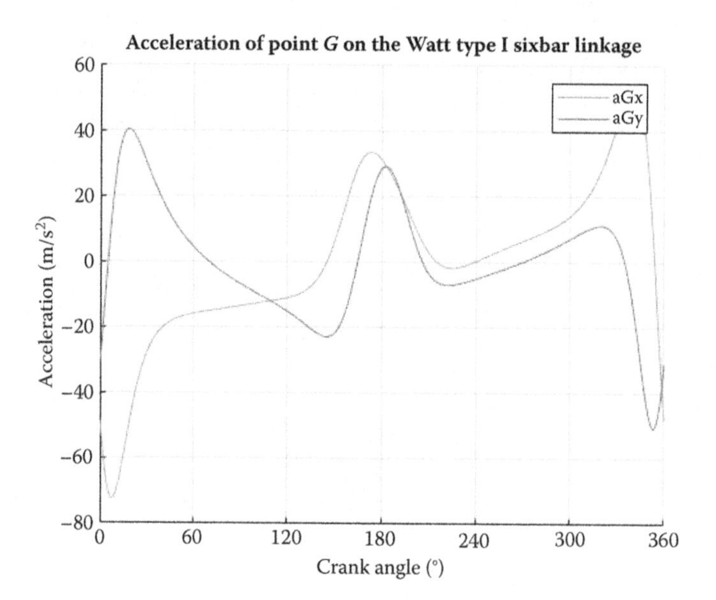

FIGURE 6.51
Acceleration of point G on the example Watt Type I linkage.

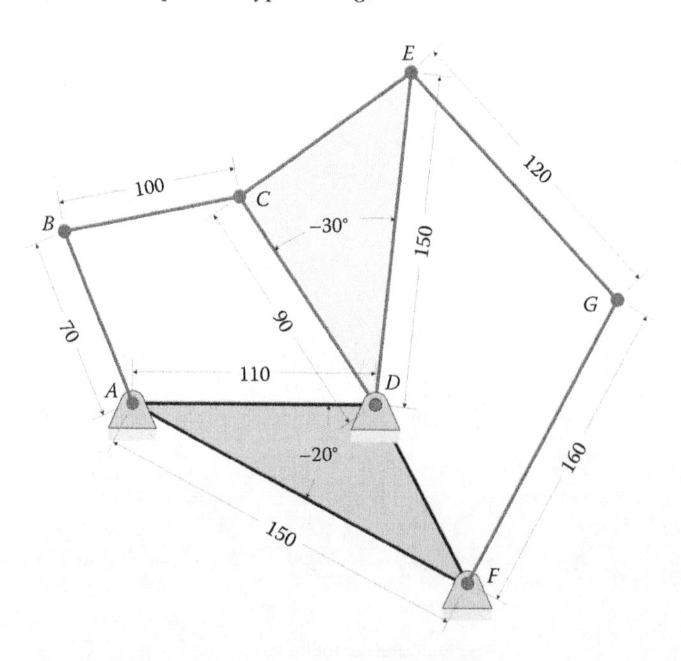

(All dimensions in millimeters)

Crank length: 70	Rocker length: 90
Coupler length: 100	Internal angle of rocker: −30°
Distance between ground pins: 110	Length DE on rocker: 150
Internal angle of ground: −20°	Length of link 5: 120
Length AF on ground: 150	Length of link 6: 160
Crank angular velocity: 10 rad/s	Crank angular acceleration: 0 rad/:

FIGURE 6.52
Dimensions of the example Watt Type II sixbar linkage.

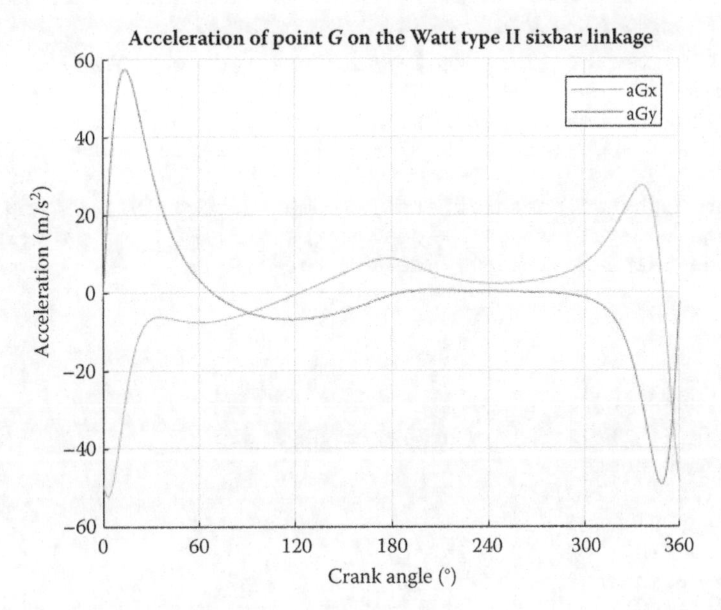

FIGURE 6.53
Acceleration of point G for example Watt Type II linkage.

```
% Sixbar_S1_Acceleration_Analysis.m
% Conducts an acceleration analysis on the Stephenson Type I sixbar
% linkage and plots the acceleration of point G
% by Eric Constans, June 14, 2017

% Prepare Workspace
clear variables; close all; clc;

% Linkage dimensions
a = 0.070;                % crank length (m)
b = 0.100;                % coupler length (m)
c = 0.090;                % rocker length (m)
d = 0.110;                % length between ground pins (m)
p = 0.150;                % length to third pin on crank triangle (m)
q = 0.150;                % length to third pin on rocker triangle (m)
u = 0.120;                % length of link 5 (m)
v = 0.160;                % length of link 6 (m)
gamma2 = 20*pi/180;   % internal angle of crank triangle
gamma4 = -20*pi/180;  % internal angle of rocker triangle

% Ground pins
x0 = [ 0; 0];     % ground pin at A (origin)
xD = [ d; 0];     % ground pin at D
v0 = [ 0; 0];     % velocity of origin
a0 = [ 0; 0];     % acceleration of origin
Z21 = zeros(2,1);

% Angular velocity and acceleration of crank
omega2 = 10;        % angular velocity of crank (rad/sec)
alpha2 = 0;         % angular acceleration of crank (rad/sec^2)
```

```
% allocate space for variables
N = 361;   % number of times to perform position calculations
[xB,xC,xE,xF,xG] = deal(zeros(2,N));  % position of B, C, E, F, G
[vB,vC,vE,vF,vG] = deal(zeros(2,N));  % velocity of B, C, E, F, G
[aB,aC,aE,aF,aG] = deal(zeros(2,N));  % acceleration of B, C, E, F, G

[theta2,theta3,theta4,theta5,theta6] = deal(zeros(1,N)); % angles
[omega3,omega4,omega5,omega6] = deal(zeros(1,N));         % velocities
[alpha3,alpha4,alpha5,alpha6] = deal(zeros(1,N));         % accelerations

% Main Loop
for i = 1:N

% solve lower fourbar linkage
  theta2(i) = (i-1)*(2*pi)/(N-1); % crank angle
  r = d - a*cos(theta2(i));
  s = a*sin(theta2(i));
  f2 = r^2 + s^2;
  delta = acos((b^2+c^2-f2)/(2*b*c));
  g = b - c*cos(delta);
  h = c*sin(delta);

  theta3(i) = atan2((h*r - g*s),(g*r + h*s)); % coupler angle
  theta4(i) = theta3(i) + delta;               % rocker angle

% calculate unit vectors
  [e2,n2] = UnitVector(theta2(i));
  [e3,n3] = UnitVector(theta3(i));
  [e4,n4] = UnitVector(theta4(i));
  [eAE,nAE] = UnitVector(theta2(i) + gamma2);
  [eDF,nDF] = UnitVector(theta4(i) + gamma4);

% solve for positions of points B, C, E, F
  xB(:,i) = FindPos(x0,a, e2);
  xC(:,i) = FindPos(xD,c, e4);
  xE(:,i) = FindPos(x0,p,eAE);
  xF(:,i) = FindPos(xD,q,eDF);

% solve upper fourbar linkage
  xFB = xF(1,i) - xB(1,i);   yFB = xF(2,i) - xB(2,i);
  xEB = xE(1,i) - xB(1,i);   yEB = xE(2,i) - xB(2,i);
  beta =  atan2(yFB, xFB);
  alpha = atan2(yEB, xEB);
  aPrime = sqrt(xEB^2 + yEB^2);
  dPrime = sqrt(xFB^2 + yFB^2);
  theta2Prime = alpha - beta; % virtual crank angle on upper fourbar

  r = dPrime - aPrime*cos(theta2Prime);
  s = aPrime*sin(theta2Prime);
  f2 = r^2 + s^2;
  delta = acos((u^2+v^2-f2)/(2*u*v));
  g = u - v*cos(delta);
  h = v*sin(delta);
```

```
  theta5Prime = atan2((h*r - g*s),(g*r + h*s)); % coupler and rocker angles
  theta6Prime = theta5Prime + delta;            % on upper fourbar
  theta5(i) = theta5Prime + beta;               % return angles to fixed
  theta6(i) = theta6Prime + beta;               % fixed CS

% calculate remaining unit vectors
  [e5,n5]    = UnitVector(theta5(i));
  [e6,n6]    = UnitVector(theta6(i));

% calculate position of point G
  xG(:,i) = FindPos(xE(:,i),   u,   e5);

% Conduct velocity analysis to solve for omega3 - omega6
  A_Mat = [b*n3    -c*n4    Z21    Z21;
             Z21   -q*nDF   u*n5  -v*n6];
  b_Vec = [-a*omega2*n2; -p*omega2*nAE];
  omega_Vec = A_Mat\b_Vec;

  omega3(i) = omega_Vec(1);
  omega4(i) = omega_Vec(2);
  omega5(i) = omega_Vec(3);
  omega6(i) = omega_Vec(4);

% Calculate velocity at important points on linkage
  vB(:,i) = FindVel(      v0,    a,     omega2,    n2);
  vC(:,i) = FindVel(      v0,    c, omega4(i),    n4);
  vE(:,i) = FindVel(      v0,    p,     omega2,   nAE);
  vF(:,i) = FindVel(      v0,    q, omega4(i),   nDF);
  vG(:,i) = FindVel(vE(:,i),    u, omega5(i),    n5);

% Conduct acceleration analysis to solve for alpha3 - alpha6
  ac = a*omega2^2;     at = a*alpha2;
  bc = b*omega3(i)^2; cc = c*omega4(i)^2;
  pt = p*alpha2;       pc = p*omega2^2;
  uc = u*omega5(i)^2; vc = v*omega6(i)^2;
  qc = q*omega4(i)^2;

  C_Mat = A_Mat;
  d_Vec = [ -at*n2 +  ac*e2 + bc*e3 - cc*e4;
           -pt*nAE + pc*eAE + uc*e5 - vc*e6 - qc*eDF];
  alpha_Vec = C_Mat\d_Vec;

  alpha3(i) = alpha_Vec(1);
  alpha4(i) = alpha_Vec(2);
  alpha5(i) = alpha_Vec(3);
  alpha6(i) = alpha_Vec(4);

% Calculate acceleration at important points on linkage
  aB(:,i) = FindAcc(      a0,    a,     omega2,      alpha2,    e2,    n2);
  aC(:,i) = FindAcc(      a0,    c, omega4(i), alpha4(i),    e4,    n4);
  aE(:,i) = FindAcc(      a0,    p,     omega2,      alpha2,   eAE,   nAE);
  aF(:,i) = FindAcc(      a0,    q, omega4(i), alpha4(i),   eDF,   nDF);
```

```
  aG(:,i) = FindAcc(aE(:,i),    u, omega5(i), alpha5(i),    e5,    n5);
end

% plot the acceleration of point G
figure; hold on
plot(theta2*180/pi,aG(1,:),'Color',[153/255 153/255 153/255])
plot(theta2*180/pi,aG(2,:),'Color',[0 110/255 199/255])
legend('aGx','aGy')
title('Acceleration of point G on the Stephenson Type I sixbar linkage')
xlabel('Crank angle (degrees)')
ylabel('Acceleration (m/s2)')
grid on
set(gca,'xtick',0:60:360)
xlim([0 360])
```

6.9 Practice Problems

Problem 6.1

a. A child runs from the center of a rotating merry-go-round to the outer edge and stumbles sideways along the way. What type of acceleration causes the stumble?

b. A child is pressed against the outer rail of a merry-go-round as it rotates at a constant speed. What type of acceleration is the child experiencing?

c. An armored vehicle fires a shell as its turret is rotating at a constant speed. What types of acceleration does the shell experience?

Problem 6.2

Figure 6.54 shows an ant crawling at a speed of 10 mm/s outward on a link that is rotating at 2 rad/s in the counterclockwise direction. At the time of interest, the ant is located at a distance of 50 mm from the center of rotation. What is the total acceleration of the ant?

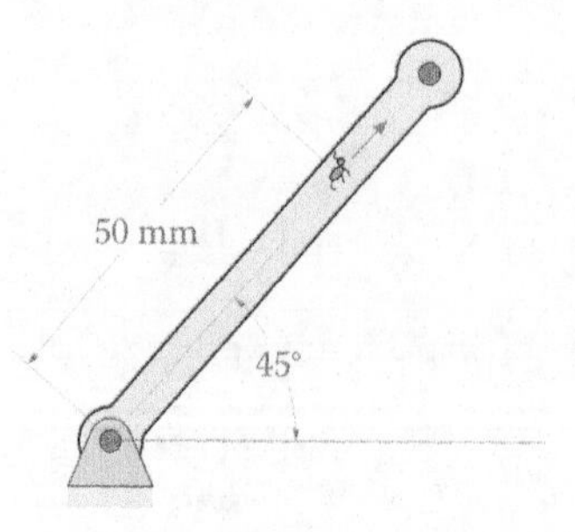

FIGURE 6.54
Problem 6.2.

Problem 6.3

In the linkage shown in Figure 6.55, the length AB and BC are both 200 mm. Link AB rotates at 100 rpm clockwise and link BC rotates at 200 rpm counterclockwise. Both links are slowing down at a rate of 10 rad/s^2. Find the total acceleration at point C.

Problem 6.4

Figure 6.56 shows the ant from Problem 6.2 is now crawling at a speed of 10 mm/s from point C to point B and speeding up at a rate of 5 mm/s^2. The length AB and BC are both 200 mm. Link AB rotates at 10 rpm clockwise and link BC rotates at 20 rpm counterclockwise and both links are speeding up at a rate of 1 rad/s^2. At the time of interest the ant is 150 mm from point B. Find the total acceleration of the ant.

Problem 6.5

Figure 6.57 shows the rider of an amusement park ride sitting at point D at the end of arm CD. To enhance his enjoyment of the ride, he is attempting to bring a sandwich toward his mouth at 20 mm/s. The length AB is 10 m, the length BC is 5 m, and the length CD is 2 m. What is the overall acceleration of the sandwich?

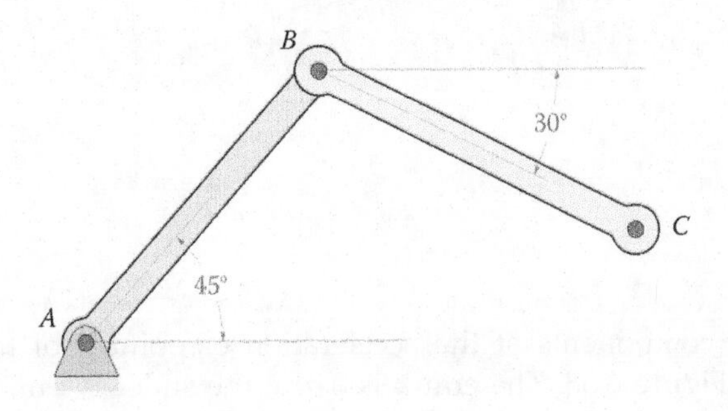

FIGURE 6.55
Problem 6.3.

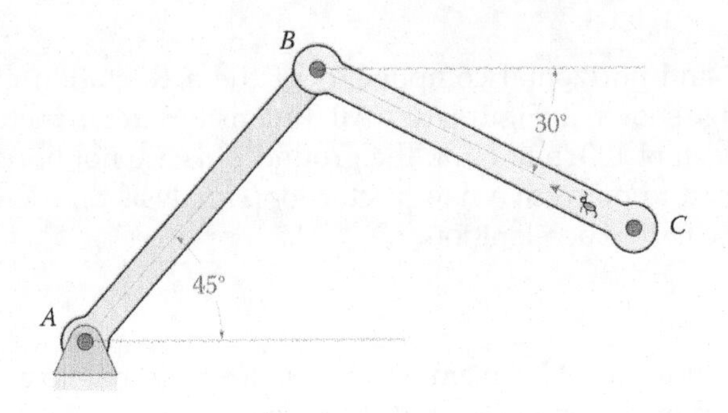

FIGURE 6.56
Problem 6.4.

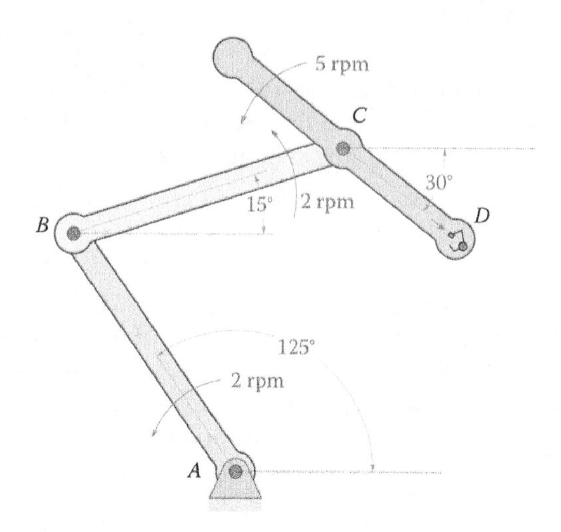

FIGURE 6.57
Problem 6.5.

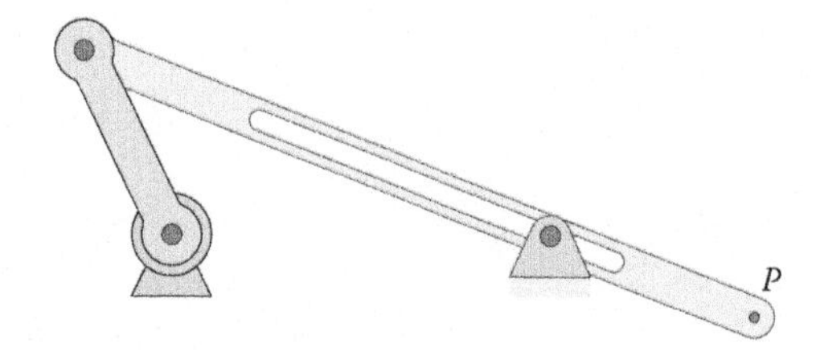

FIGURE 6.58
Problem 6.6.

Problem 6.6

Plot the x and y components of the acceleration of point P of the threebar linkage shown in Figure 6.58. The crank is 0.5 m, the slider is 2 m, and the distance between the ground pins is 1 m. The crank has a constant angular velocity of 8 rad/s. What is the slider's maximum angular acceleration? Verify your results using the `Derivative_Plot` function.

Problem 6.7

Plot the vertical and horizontal components of the acceleration of point P on the threebar linkage shown in Figure 6.59. All dimensions are in meters and the crank rotates at a constant 100 rpm. Hint: the ground pins are not horizontally aligned, so you will need to perform a new vector loop analysis on the linkage, and differentiate twice to find acceleration.

Problem 6.8

The crank in the threebar of Problem 6.6 accelerates from rest to a speed of 100 rpm in 1 s. Plot the x and y components of the acceleration of point P at the end of the

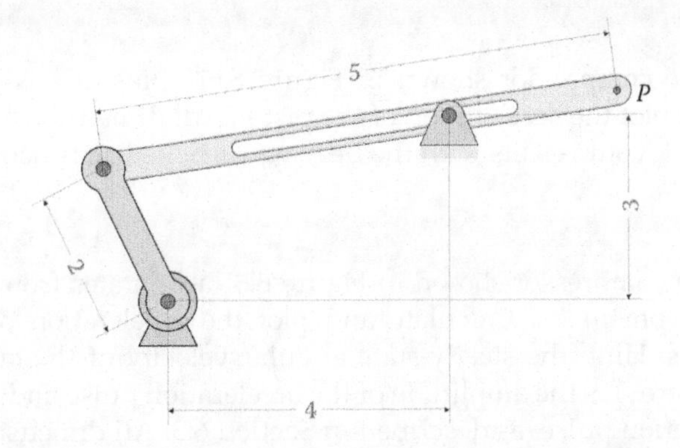

FIGURE 6.59
Problem 6.7.

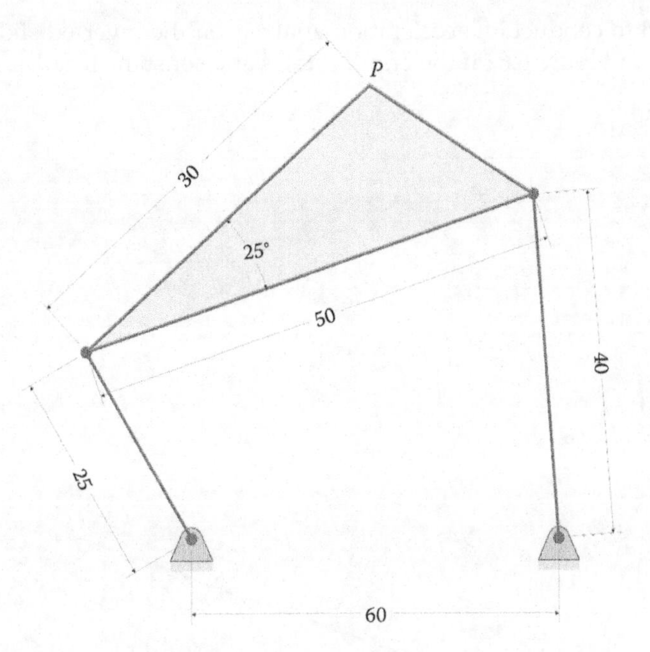

FIGURE 6.60
Problem 6.9.

slider for a period of 2 s. Hint: the steady-state angular velocity of the crank is given by $\omega_{2ss} = AT$, where A is the amplitude of the acceleration pulse and T is the duration of the acceleration pulse, as described in Section 6.3.

Problem 6.9

Plot the x and y components of acceleration of point P for the fourbar linkage shown in Figure 6.60. All dimensions in the figure are in centimeters. The crank rotates at a constant 10 rad/s. Check your results with the `Derivative _ Plot` function.

Problem 6.10

The crank of the compressor shown in Figure 6.61 spins at a constant 1700 rpm. Calculate and plot the acceleration of the piston. All dimensions are given in millimeters. Check your results with the `Derivative_Plot` function.

Problem 6.11

The crank of the compressor shown in Figure 6.61 accelerates from rest to a steady speed of 500 rpm in 2 s. Calculate and plot the acceleration of the piston for a period of 3 s. Hint: the steady-state angular velocity of the crank is given by $\omega_{2ss} = AT$, where A is the amplitude of the acceleration pulse and T is the duration of the acceleration pulse, as described in Section 6.3. All dimensions are given in millimeters.

Problem 6.12

Use MATLAB to conduct an acceleration analysis on the inverted slider-crank linkage shown in Figure 6.62 if the crank rotates at a constant 10 rad/s. Plot the x and

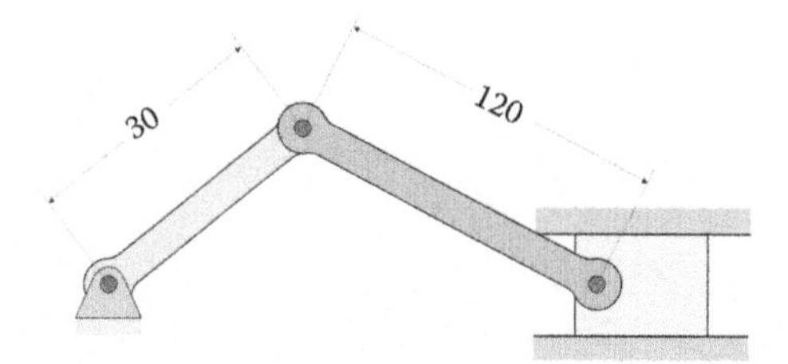

FIGURE 6.61
Problems 6.10 and 6.11.

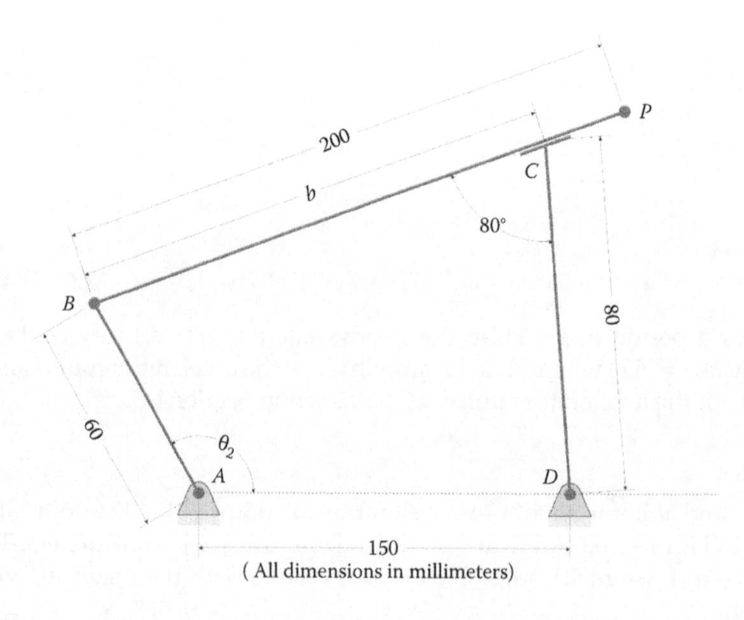

FIGURE 6.62
Problem 6.12.

y components of the acceleration at point P. Also, plot the acceleration of slip ($\ddot{b}$) between the slider and rocker.

Problem 6.13

Use MATLAB to conduct an acceleration analysis on the geared fivebar linkage shown in Figure 6.63 if the crank rotates at a constant 10 rad/s. The gear at point A has 50 teeth and the gear at point D has 25 teeth. Gear D is rotated 180° when gear A is at 0°. Plot the x and y components of the acceleration at point P. Check your answer using the `Derivative_Plot.m` function.

Problem 6.14

Use MATLAB to conduct an acceleration analysis on the sixbar linkage shown in Figure 6.64 if the crank rotates at a constant 10 rad/s. Plot the x and y components of the acceleration at point P for one revolution of the crank. Check your answer using the `Derivative_Plot.m` function.

Problem 6.15

Use MATLAB to conduct an acceleration analysis on the sixbar linkage shown in Figure 6.65 if the crank rotates at a constant 10 rad/s. Plot the x and y components of the acceleration at point P for one revolution of the crank. Check your answer using the `Derivative_Plot.m` function.

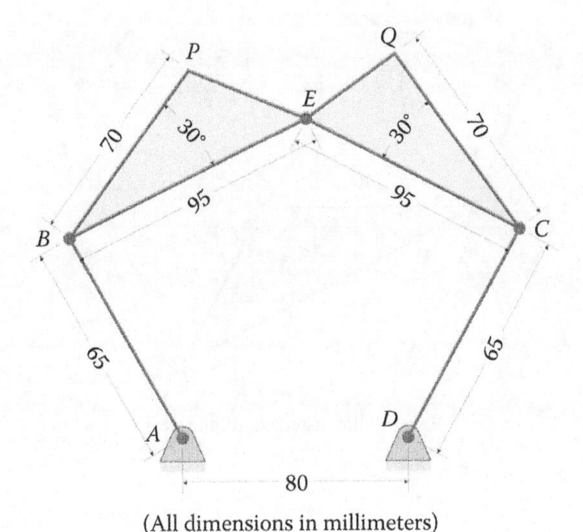

FIGURE 6.63
Problem 6.13.

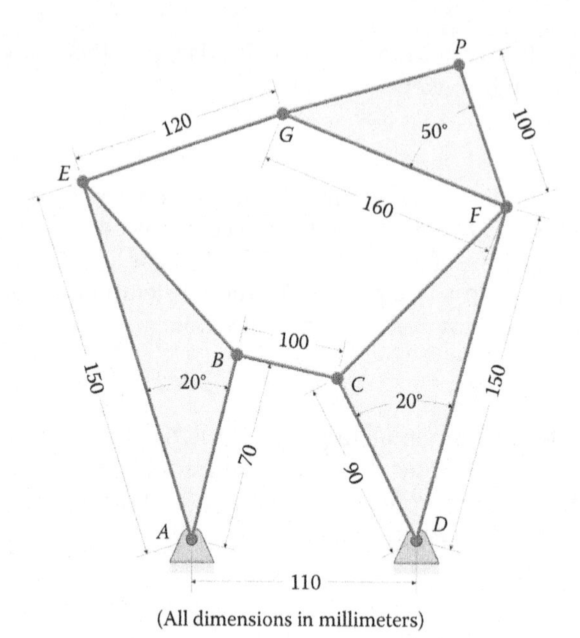

FIGURE 6.64
Problem 6.14.

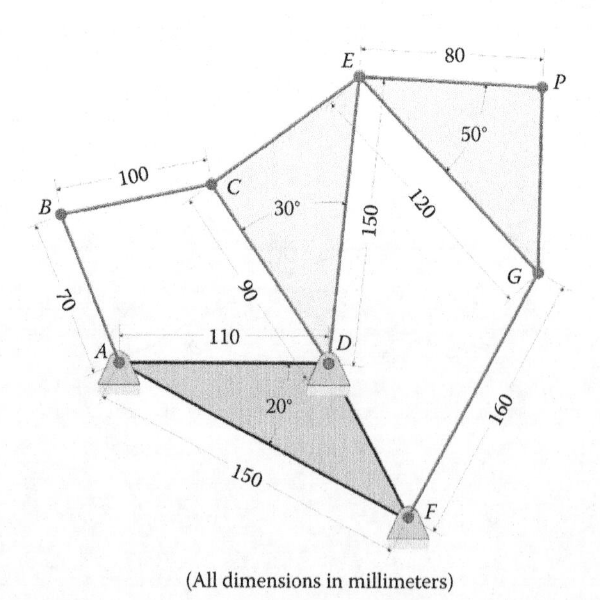

FIGURE 6.65
Problem 6.15.

Acknowledgments

Several images in this chapter were produced using SOLIDWORKS® software. SOLIDWORKS is a registered trademark of Dassault Systèmes SolidWorks Corporation.

Several images in this chapter were produced using MATLAB software.

MATLAB is a registered trademark of The MathWorks, Inc.

7

Force Analysis on Linkages

7.1 Fundamentals of Dynamics

Until now, we have discussed kinematics, which is the study of position, velocity, and acceleration without regard to forces. We assumed that a "magic motor" turned the crank on a linkage at a constant speed, and plotted the resulting motion of the rest of the linkage. We will now turn our attention to the forces and torques required to produce such a motion. Recall the definition of *dynamics*, given earlier:

Dynamics: the study of forces in a moving system

In general, we may encounter two types of dynamics problems:

1. Given a set of forces and moments on a system, find the accelerations, velocities, and positions of the bodies in the system as a function of time.
2. Given the position, velocity, and acceleration of each body within a system, find the forces and torques (or moments) required to produce this motion.

The first problem requires the derivation and solution of a set of nonlinear differential equations, which can be quite challenging for nontrivial systems. For this reason, the general *forward dynamics* problem is usually reserved for senior or graduate level classes, except for very simple systems (e.g. the block on an inclined plane). The second problem, which is often called the *inverse dynamic problem,* requires only the solution of a set of algebraic equations, which is quite straightforward in MATLAB®. We will be concerned with the inverse dynamic problem in this text.

7.1.1 Dynamic Models

To develop the inverse dynamic model of a mechanism we must first create a dynamic model of each component in the system. For each link, we must know the critical dimensions (e.g. the distances between pins, etc.) and also the primary inertial properties, which are:

1. Mass – m
2. Center of mass – $\overline{x},\overline{y}$
3. Mass moment of inertia – I

Each of these quantities is easy to calculate using the **Mass Properties** feature in SOLIDWORKS®, but we will also show a means of finding the properties "by hand" in the section that follows.

7.1.1.1 Mass

First, recall that mass is not equal to weight! Mass is a measure of the quantity of matter in an object, while weight reflects the gravitational attraction between two bodies. In this text, we will use the metric unit for mass, the kilogram (kg). Although we will not use them in this text, you may encounter problems presented in English units. It is the authors' recommendation to always convert the problem to SI units before starting the problem. The English units of mass are uniformly terrible. The most common English mass unit is the *slug*, which is equivalent to

$$1 \text{ slug} = 1\frac{\text{lbf} \cdot \text{s}^2}{\text{ft}}$$

Unfortunately, feet are an inconvenient unit in mechanical design, and most machine tools are calibrated in inches. There is, unfortunately, no standardized inch-pound-second unit of mass. Even worse is the unit *pound-mass*, which is the quantity of mass required to exert one pound-force in Earth's gravitational field. The opportunities for confusion are simply too great with the pound-mass, and its use should always be avoided! To be safe, we will perform all of our calculations in SI units. If necessary, you can always convert your final answers to inches and pounds after all of the dynamics calculations have been completed.

7.1.1.2 Center of Mass

The center of mass is the "average" location of all the mass in a body. It is normally indicated in a drawing by the "target" symbol shown in Figure 7.1. As a first approximation, let us pretend that the body in Figure 7.2 is made up of point masses, each of which has mass m_i and coordinates (x_i, y_i). All the masses are rigidly connected and cannot move relative to each other. To compute the coordinates for the center of mass of the body we use a "weighted average" formula

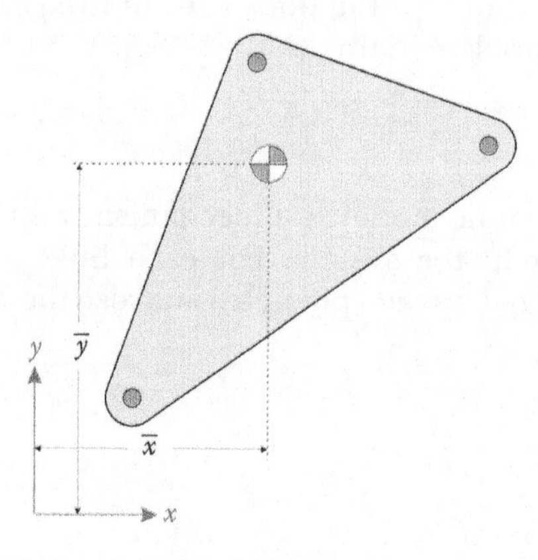

FIGURE 7.1
A typical link showing the location of the center of mass.

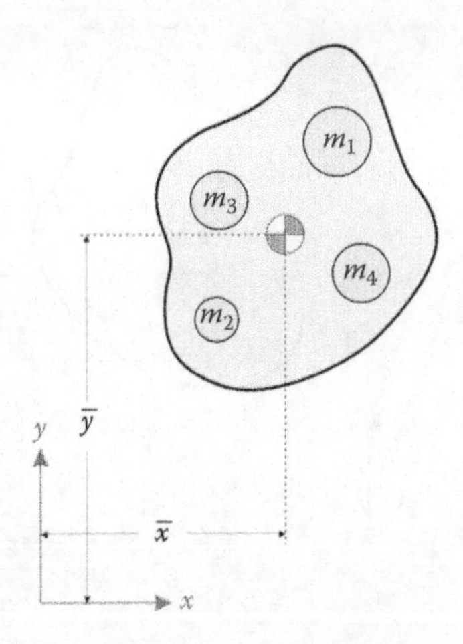

FIGURE 7.2
The "space potato" above is composed of point masses m_1, m_2, m_3, and m_4.

$$\bar{x} = \frac{m_1x_1 + m_2x_2 + m_3x_3 + m_4x_4}{m} \tag{7.1}$$

$$\bar{y} = \frac{m_1y_1 + m_2y_2 + m_3y_3 + m_4y_4}{m} \tag{7.2}$$

where

$$m = m_1 + m_2 + m_3 + m_4 \tag{7.3}$$

is the total mass of the body. It is far simpler to write the center of mass coordinates in summation notation:

$$\bar{x} = \frac{1}{m}\sum_i m_i x_i \quad \bar{y} = \frac{1}{m}\sum_i m_i y_i \tag{7.4}$$

If the mass is continuously distributed throughout the body (as is the case with most links, which are usually made from homogeneous materials like aluminum or steel) we convert the summations to integrals

$$\bar{x} = \frac{1}{m}\int x\, dm \quad \bar{y} = \frac{1}{m}\int y\, dm \tag{7.5}$$

where (x, y) gives the coordinates of each differential mass, dm. As we will see below, the summation formulas will work for links that are composed of simple shapes, and the centers of mass for more complicated shapes can be computed using SOLIDWORKS.

The center of mass has two useful properties that we should consider when designing linkages:

1. If we pin a body at its center of mass (as shown in Figure 7.3) it will remain balanced in a gravitational field; that is, it will not rotate about the pivot.
2. If we pin a body to ground at its center of mass and spin it using a motor, the unbalance force on the pivot will be zero. If the unbalance force is zero, then

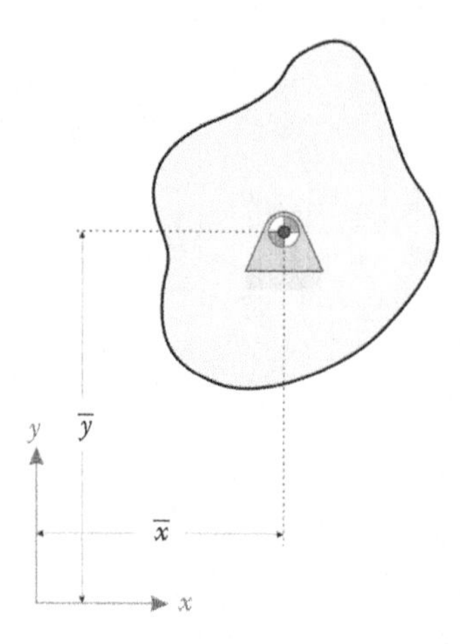

FIGURE 7.3
A link is statically balanced if we pin it to ground at its center of mass.

spinning the link will not create vibratory forces at the pivot. This is very useful when balancing a linkage – if we can get the center of mass to remain stationary when the linkage moves, then there will be no unbalanced forces at the pivots.

Example 7.1

The object in Figure 7.4 is made up of a 5 cm square attached to a 10 cm square. Find the center of mass of the object.

Solution

A coordinate system is attached to the center of the left side of the object. The center of mass of the 5 cm square is 2.5 cm in the positive x direction and the center of mass of the 10 cm square is 10 cm in the positive x direction. The y coordinate of both squares is zero. Thus,

$$x_1 = 2.5\text{ cm} \quad x_2 = 10\text{ cm}$$

The mass of the first square is

$$m_1 = (5\text{ cm})^2 t\rho = 25\, t\rho$$

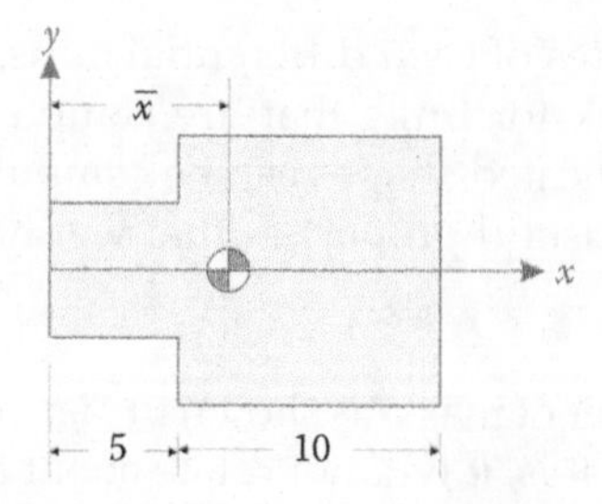

FIGURE 7.4
This object is composed of a 5 cm square and 10 cm square attached together.

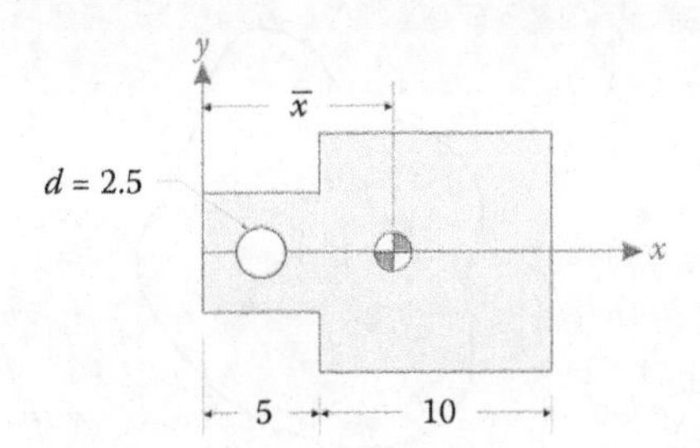

FIGURE 7.5
A 2.5 cm hole has been drilled in the object of Example 7.1.

where t is the thickness of the object and ρ is its mass density (given in kg/cm³). The mass of the second square is 100 $t\rho$. Thus, the x coordinate of the center of mass is

$$\bar{x} = \frac{2.5 \cdot 25\, t\rho + 10 \cdot 100\, t\rho}{25\, t\rho + 100\, t\rho} = 8.5 \text{ cm}$$

Of course, the y coordinate of the center of mass is zero.

Example 7.2

A 2.5 cm hole has been drilled at the center of the small square on the left side of the object in Example 7.1, as shown in Figure 7.5. Find the new center of mass.

Solution
The total mass of the object (before drilling the hole) was 125 $t\rho$ and its center of mass was located at 8.5 cm in the positive x direction. We will find the new center of mass by *subtracting* the mass that has been removed from the hole.

$$m_3 = -\frac{\pi (2.5 \text{ cm})^2\, t\rho}{4} = -4.909 t\rho \text{ kg}$$

and the center of the circle is at

$$x_3 = 2.5 \text{ cm}$$

Thus, the location of the new center of mass is

$$\bar{x} = \frac{125 t\rho \cdot 8.5 \text{ cm} - 4.909 t\rho \cdot 2.5 \text{ cm}}{125 t\rho - 4.909 t\rho} = 8.745 \text{ cm}$$

7.1.1.3 Mass Moment of Inertia

The mathematical definition of mass moment of inertia is

$$I = \int \left(x^2 + y^2\right) dm = \int r^2 dm$$

where r is the distance from the origin to the differential mass, dm, as shown in Figure 7.6. While this definition can be used to calculate the mass moment of inertia, it is not very

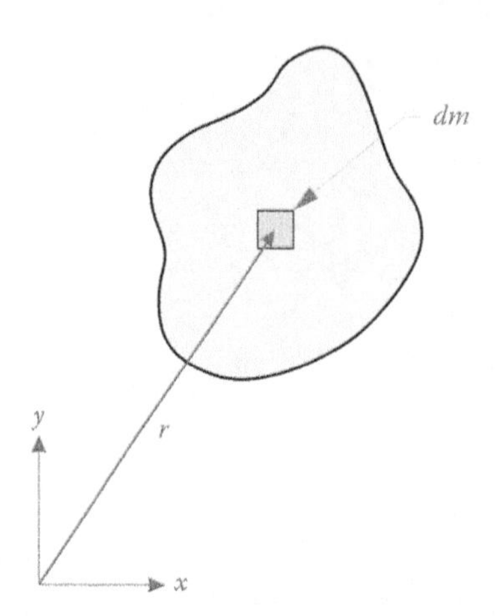

FIGURE 7.6
The mass moment of inertia gives a measure of how mass is distributed about a given axis.

helpful in gaining an intuitive understanding of its meaning. Instead, consider the following definition:

> Mass moment of inertia gives the resistance of a body to an increase or decrease in rotational speed.

This is analogous to the role that mass plays in translational motion. An object with a large mass is more difficult to accelerate than an object with a small mass.

Consider the situation shown in Figure 7.7. Both links consist of two point masses attached to rigid rods that rotate about a pivot. A point mass is a body whose mass is assumed to be concentrated at a single point. They are shown as circles in Figure 7.7 as an aid to visualization, but the mass of each circle is concentrated at a point at the end of the rod. The masses are at equal distances from the pivot, but the masses in Link 1 are greater than those in Link 2. Imagine attaching a motor to the pivot in order to cause the speed of rotation of the masses to increase. As you might guess, it takes more torque to cause a speed increase in Link 1 than in Link 2. This leads us to believe that moment of inertia has something to do with the mass of the system being rotated.

Now consider the second pair of systems in Figure 7.8. The masses in both systems are equal, but the masses in Link 1 are attached farther from the pivot than those of Link 2.

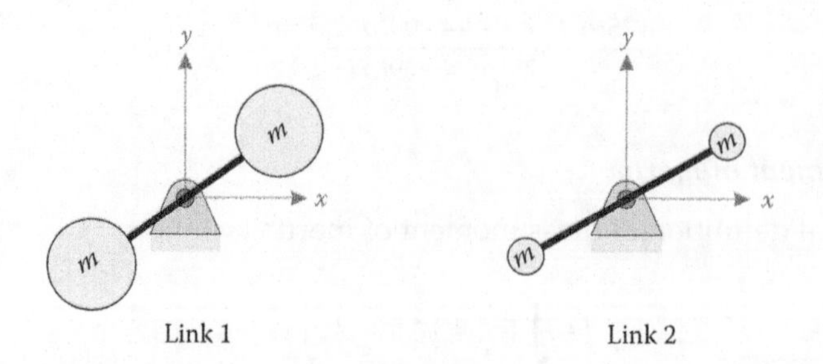

FIGURE 7.7
The link with the larger masses (at left) has a higher mass moment of inertia than the link with the smaller masses.

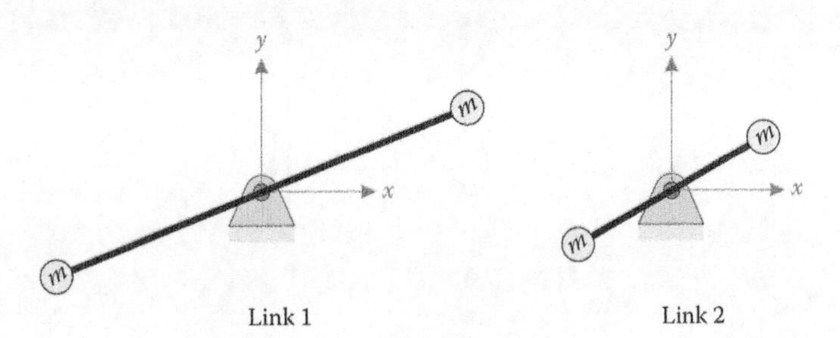

FIGURE 7.8
Both links have the same mass, but the mass is farther from the center of rotation in Link 1.

Again, it takes more torque to cause a speed increase in Link 1 than in Link 2. Thus, moment of inertia involves not only *mass*, but *distance* as well. For each point mass at the end of a rod, the moment of inertia is

$$I = mr^2 \tag{7.6}$$

so that the total moment of inertia for the systems shown in Figures 7.7 and 7.8 are

$$I_{total} = 2mr^2 \tag{7.7}$$

For many common shapes (e.g. rectangles, disks, etc.) we can use the tables in Appendix A to find moments of inertia. For more complicated shapes, it is sometimes necessary to resort to the mathematical definition and integrate. In most instances, however, we can simply break the object into several simpler pieces, and use the *parallel axis theorem* to find the moment of inertia, or we use the **Mass Properties** feature in SOLIDWORKS to calculate the moment of inertia directly.

7.1.2 The Parallel Axis Theorem

The tables in Appendix A give the moments of inertia of common shapes *about their centers of mass.* Most parts of a linkage, however, do not rotate about their center of mass. We can transfer inertial properties from one axis to another using the parallel axis theorem

$$I_A = I_g + md^2 \tag{7.8}$$

where

- I_A is the moment of inertia about the desired axis located at point A.
- I_g is the moment of inertia about the center of mass (found using the tables in Appendix A).
- m is the mass of the body.
- d is the distance between the point A and the center of mass.

Figure 7.9 shows a 3D view of a 2D object. The parallel axis theorem enables us to calculate the moment of inertia about the axis at point A, given the moment of inertia about the center of mass. To see how this all fits together, let us perform a simple example.

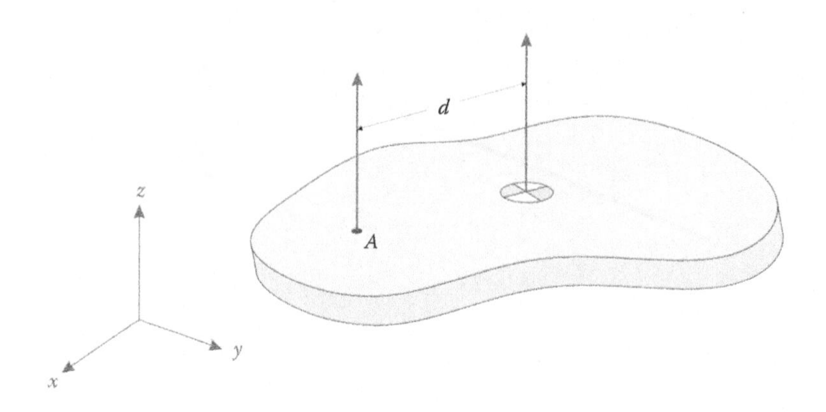

FIGURE 7.9
We wish to find the mass moment of inertia about point A which is located a distance d from the center of mass.

Example 7.3: A Rectangular Link

The link shown in Figure 7.10 has two holes of differing size, and thickness t. The dimensions are as follows:

$$W = 0.1\,\text{m} \quad L = 1\,\text{m} \quad t = 0.02\,\text{m}$$

$$r_1 = 0.025\,\text{m} \quad r_2 = 0.01\,\text{m} \quad l = 0.8\,\text{m}$$

The link is made of aluminum, with density $\rho = 2700\,\text{kg/m}^3$. Find the mass, center of mass, and moment of inertia about the center of mass of the link. A coordinate system has been attached to the center of the large hole on the left.

Solution
To find the requested quantities, we will adopt the following procedure

1. Break the link into simpler components like the ones in Appendix A.
2. Find the mass of each component.
3. Find the center of mass of each component.
4. Find the center of mass of the entire link.
5. Find the moment of inertia for each component about its center of mass.

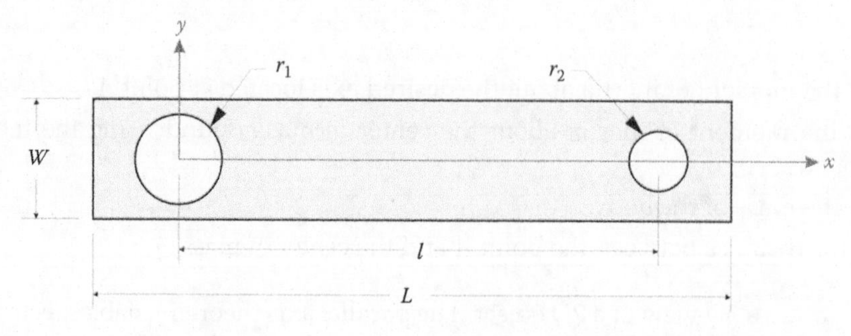

FIGURE 7.10
The link used in Example 7.3.

6. Transfer each moment of inertia to the center of mass of the link using the parallel axis theorem.
7. Add these together to find the mass moment of inertia of the entire link.

Step 1: The link consists of a rectangle prism and two disks, as shown in Figure 7.11. Let us denote the large circle as component 1, the rectangular prism as component 2, and the small circle as component 3. The **Mass Properties** of each of these shapes are given in Appendix A.

Step 2: We will now find the mass of each component.

$$m_1 = \rho t \pi r_1^2 = \left(2700 \frac{\text{kg}}{\text{m}^3}\right)(0.02\ \text{m})\pi(0.025\ \text{m})^2 = 0.106\ \text{kg}$$

$$m_2 = \rho tLW = \left(2700 \frac{\text{kg}}{\text{m}^3}\right)(0.02\ \text{m})\pi(1\ \text{m})(0.1\ \text{m}) = 5.4\ \text{kg}$$

$$m_3 = \rho t \pi r_2^2 = \left(2700 \frac{\text{kg}}{\text{m}^3}\right)(0.02\ \text{m})\pi(0.01\ \text{m})^2 = 0.017\ \text{kg}$$

The total mass is then

$$m = -m_1 + m_2 - m_3 = 5.277\ \text{kg}$$

Step 3: Find the center of mass of each part. The center of mass of the large hole is (0, 0) since the xy coordinate system is located at the center of this hole.

$$\bar{x}_1, \bar{y}_1 = (0,0)$$

The center of mass of the rectangle is located at the midpoint between the centers of the two holes.

$$\bar{x}_2, \bar{y}_2 = (0,0) = \left(\frac{l}{2}, 0\right) = (0.4, 0)\ \text{m}$$

And the center of the third hole has coordinates:

$$\bar{x}_3, \bar{y}_3 = (l, 0) = (0.8, 0)\ \text{m}$$

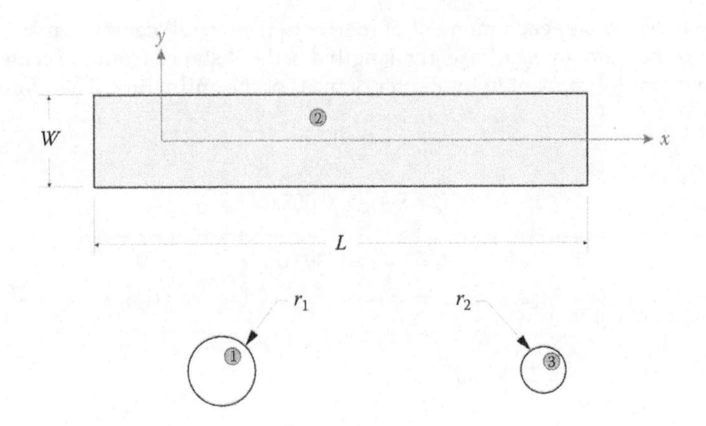

FIGURE 7.11
The link in the example has been broken into its three constituent parts: a rectangle and two circles.

Step 4: Use the formula for center of mass to find the center of mass for the entire link. In the x direction we have:

$$\bar{x} = \frac{1}{m}\left(-m_1\bar{x}_1 + m_2\bar{x}_2 - m_3\bar{x}_3\right)$$

$$= \frac{1}{5.277\,\text{kg}}\left[(-0.106\,\text{kg})(0\,\text{m}) + (5.4\,\text{kg})(0.4\,\text{m}) - (0.017\,\text{kg})(0.8\,\text{m})\right]$$

$$= 0.407\,\text{m}$$

And in the y direction we have:

$$\bar{y} = \frac{1}{m}\left(-m_1\bar{y}_1 + m_2\bar{y}_2 - m_3\bar{y}_3\right)$$

$$= \frac{1}{5.277\,\text{kg}}\left[(-0.106\,\text{kg})(0\,\text{m}) + (5.5\,\text{kg})(0\,\text{m}) - (0.017\,\text{kg})(0\,\text{m})\right]$$

$$= 0\,\text{m}$$

As expected. The center of mass of the entire link is then:

$$(\bar{x}, \bar{y}) = (0.407,\ 0)\,\text{m}$$

Step 5: The formulas for the moments of inertia of each component about its own center of mass are given in Appendix A:

$$I_{1g} = \frac{m_1 r_1^2}{2} = \frac{(0.106\,\text{kg})(0.025\,\text{m})^2}{2} = 3.313 \times 10^{-5}\,\text{kg}\cdot\text{m}^2$$

$$I_{2g} = \frac{m_2\left(L^2 + W^2\right)}{12} = \frac{(5.4\,\text{kg})\left(1^2 + 0.1^2\right)\text{m}^2}{12} = 0.455\,\text{kg}\cdot\text{m}^2$$

$$I_{3g} = \frac{m_3 r_2^2}{2} = \frac{(0.017\,\text{kg})(0.01\,\text{m})^2}{2} = 8.5 \times 10^{-7}\,\text{kg}\cdot\text{m}^2$$

Step 6: We now transfer each moment of inertia to the overall center of mass using the parallel axis theorem. In each case, the length d is the distance from the center of mass of the individual component to the center of mass of the entire link. Therefore,

$$d_1 = \bar{x} = 0.407\,\text{m}$$

$$d_2 = \bar{x} - \frac{l}{2} = 0.007\,\text{m}$$

$$d_3 = l - \bar{x} = 0.393\,\text{m}$$

And the moments of inertia are

$$I_1 = I_{1g} + m_1 d_1^2$$

$$= 3.313 \times 10^{-5}\,\text{kg}\cdot\text{m}^2 + (0.106\,\text{kg})(0.407\,\text{m})^2$$

$$= 0.0176\,\text{kg}\cdot\text{m}^2$$

$$I_2 = I_{2g} + m_2 d_2^2$$

$$= 0.455\,\text{kg}\cdot\text{m}^2 + (5.4\,\text{kg})(0.007\,\text{m})^2$$

$$= 0.4552\,\text{kg}\cdot\text{m}^2$$

$$I_3 = I_{3g} + m_3 d_3^2$$

$$= 8.5\times10^{-7}\,\text{kg}\cdot\text{m}^2 + (0.017\,\text{kg})(0.393\,\text{m})^2$$

$$= 0.0026\,\text{kg}\cdot\text{m}^2$$

Step 7: Finally, we add these together to find the overall moment of inertia of the link.

$$I = -I_1 + I_2 - I_3$$

$$= 0.4754\,\text{kg}\cdot\text{m}^2$$

It is important to note that moments of inertia can only be added together if they are taken about the same axis. If we had added the moments of inertia of each body taken about its own center of mass, we would have obtained an incorrect result.

7.1.3 Using SOLIDWORKS® to Calculate Moment of Inertia

The procedure given above was rather lengthy and prone to error. A much simpler method is to use SOLIDWORKS to calculate the **Mass Properties** of each link in a mechanism. There are a few simple steps to follow in this method.

1. Set the appropriate material in the *Material* property in the Feature Tree. The SOLIDWORKS default material is plastic, and by changing the material, you will set the mass density to the appropriate value.
2. Draw the link in the ordinary manner.
3. In the **Evaluate** tab, select **Mass Properties.**
4. Click the **Options** button and select **Use Custom Settings** to change the length unit to meters, the mass unit to kilograms and the volume unit to cubic meters. You can also change the accuracy to its highest setting, since it will not slow the calculations appreciably for simple links.

You will now be presented with a dialog box containing a multitude of mass quantities for the link, as shown in Figure 7.12. At the top of the box the mass and volume are given. Toward the middle, the coordinates of the center of mass are presented. Note that these coordinates are given relative to the origin that you used to draw the link (also called the *Output Coordinate System*). Finally, there are three tables of moments of inertia. For our purposes we are interested in the middle table, which gives the moments of inertia relative to the center of mass of the object and aligned with the output coordinate system. The pink triad on the link shows the location of the center of mass, and also the directions of the coordinate axes. Since the profile of the link was sketched on the *Front Plane*, the z direction is the axis of rotation. Therefore, we use L_{zz} to find the correct moment of inertia about the center of mass. If you draw your base sketch on a different plane, be sure you select the correct value from the table.

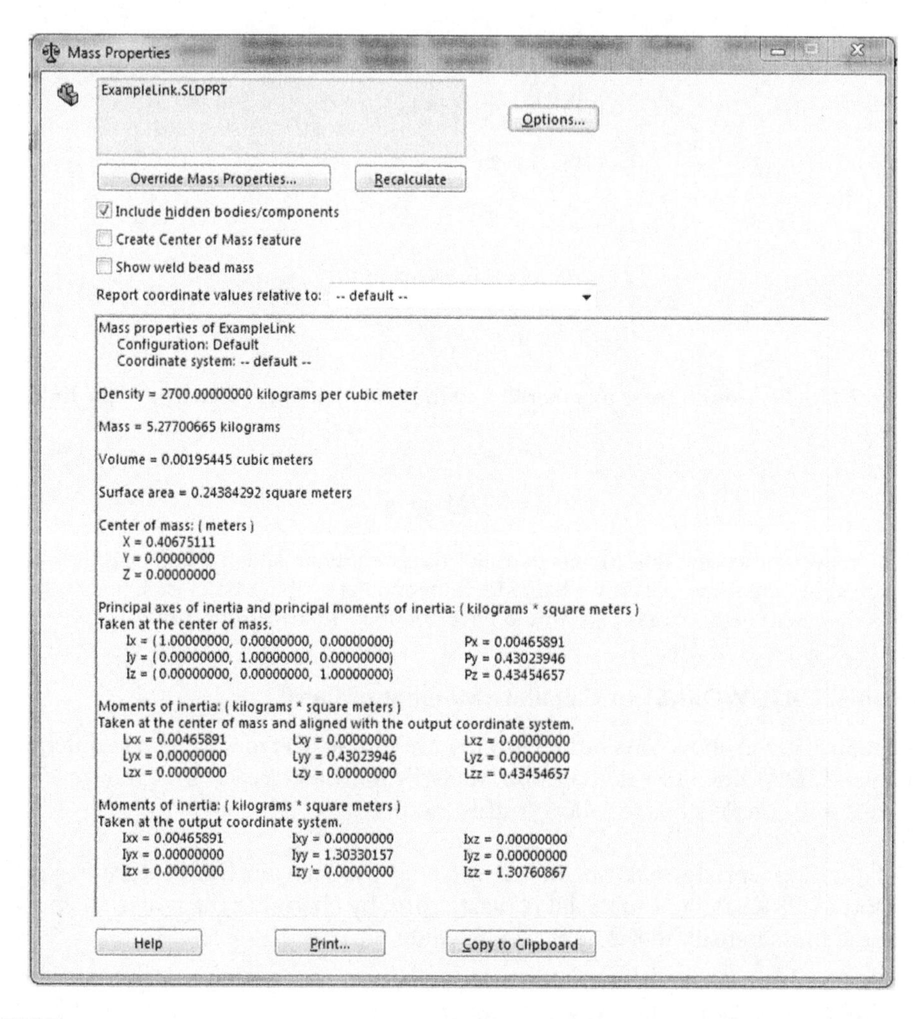

FIGURE 7.12
Screen shot of the **Mass Properties** dialog box in SOLIDWORKS for the link in Example 7.3. Note that the moment of inertia L_{zz} is identical with the results obtained above.

The lowest table in the dialog box gives the moments of inertia about the output coordinate system; that is, the moments of inertia about the origin of the part. Since the origin of the part is located at the center of the large circle, we can use the parallel axis theorem to find the moment of inertia about this axis:

$$I' = I + md^2$$

$$I' = 0.4345\,\text{kg}\cdot\text{m}^2 + (5.277\,\text{kg})\cdot(0.407\,\text{m})^2$$

$$I' = 1.3076\,\text{kg}\cdot\text{m}^2$$

which is identical to the entry for I_{zz} in the dialog box.

For the remainder of the text, we will use SOLIDWORKS to calculate the **Mass Properties** of the links that we design. This method is much simpler and less error-prone than performing the calculations by hand. We will also adopt the convention of using the moment of inertia about the center of mass of an object, and not the *Output Coordinate System*. This will have the effect of making the equations of motion simpler and more uniform than would be the case if we adopted a different convention for each link.

7.2 Newtonian Kinetics of a Rigid Body

Now that we have the definitions of the important inertial properties of a rigid body we turn our attention to developing the equations of motion for the body. We will use Newton's three laws of motion to formulate our equations. Since Newton formulated his laws for point masses, we will need a few tricks to extend them to rigid bodies. In employing these tricks, the appropriate equations of motion will "fall out" automatically.

Consider the two point masses in 3D space shown in Figure 7.13. The two masses might be independent bodies (e.g. planets) or they might be two particles within a rigid body. Each point mass has an externally applied force acting on it (shown as $\mathbf{F}_1$ and $\mathbf{F}_2$), and there are also internal forces acting *between* the masses ($\mathbf{f}_{21}$ and $\mathbf{f}_{12}$). The numbering convention we adopt here is as follows:

$$\mathbf{f}_{12} = \text{force from body 1 on body 2}$$

$$\mathbf{f}_{21} = \text{force from body 2 on body 1}$$

The internal forces may be caused by gravitational attraction, electrostatic attraction (or repulsion), or may simply be the forces that hold the body together, if m_1 and m_2 form parts of a rigid body. The forces cause each particle to accelerate (shown as $\mathbf{a}_1$ and $\mathbf{a}_2$ in the figure). We can paraphrase Newton's laws for the pair of particles as:

1. Both masses will remain at a constant velocity (or at rest) in the absence of the internal and external forces.
2. The internal forces between the particles have equal magnitude but opposite direction. In other words, $\mathbf{f}_{21} = -\mathbf{f}_{12}$.

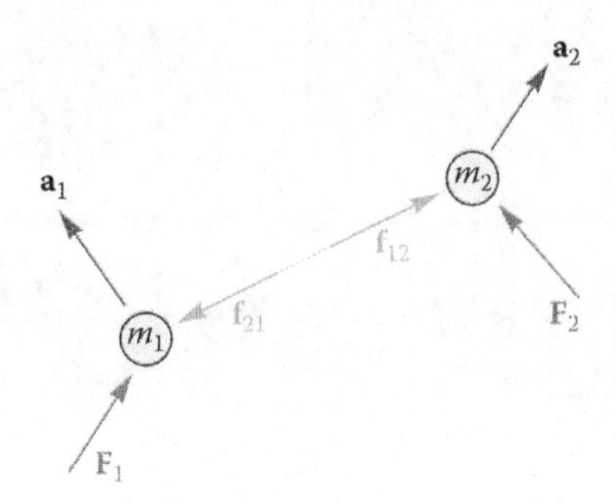

FIGURE 7.13
Two point masses in 3D space. An external force ($\mathbf{F}_1$, $\mathbf{F}_2$) acts on each point mass, and internal forces ($\mathbf{f}_{21}$ and $\mathbf{f}_{12}$) act between the masses.

3. The *net force* on a particle causes it to accelerate in proportion with its mass. The net force is found by vectorially adding together all the forces on the particle. In other words:

$$\mathbf{F}_1 + \mathbf{f}_{21} = m_1\mathbf{a}_1 \quad \mathbf{F}_2 + \mathbf{f}_{12} = m_2\mathbf{a}_2$$

It is our goal now to apply these laws to an arbitrary set of rigid bodies in two dimensions (i.e. linkages). To do this, we will assume that a rigid body is composed of an infinite number of infinitesimally small point masses. It will be easiest to stick with the notion of point masses as we proceed through the derivation. At the end of the derivation, we will convert each mass, m, to an infinitesimally small differential mass dm and integrate over the rigid body.

7.2.1 Equations of Motion for the Rigid Body

Figure 7.14 shows three of the infinitesimally small point masses that make up a rigid body. Of course, the rigid body consists of an infinite number of these point masses, but we will use these three as representative of the entire body. A rigid body, by definition, consists of particles whose location relative to each other remains fixed. The internal forces $\mathbf{f}_{12}$, $\mathbf{f}_{21}$, ... can be thought of as the interatomic forces holding the body together. According to Newton's Second Law, any two internal forces have the relationship

$$\mathbf{f}_{ij} = -\mathbf{f}_{ji} \tag{7.9}$$

Since the internal forces come in equal and opposite pairs, they add to zero:

$$\mathbf{f}_{ij} + \mathbf{f}_{ji} = 0 \tag{7.10}$$

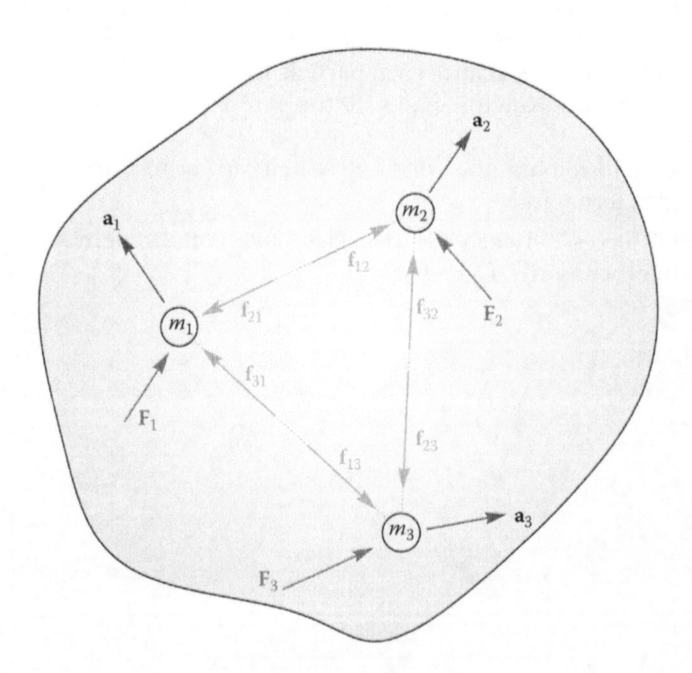

FIGURE 7.14
A rigid body consists of an infinite number of infinitesimally small point masses. Three of the point masses are shown here.

For mass 1, we may use Newton's third law to calculate its acceleration

$$\mathbf{F}_1 + \mathbf{f}_{21} + \mathbf{f}_{31} = m_1\mathbf{a}_1 \tag{7.11}$$

We can write similar equations for each of the other two particles. If we add the resulting equations together we obtain

$$\mathbf{F}_1 + \mathbf{F}_2 + \mathbf{F}_3 + \mathbf{f}_{21} + \mathbf{f}_{31} + \mathbf{f}_{12} + \mathbf{f}_{32} + \mathbf{f}_{13} + \mathbf{f}_{23} = m_1\mathbf{a}_1 + m_2\mathbf{a}_2 + m_3\mathbf{a}_3 \tag{7.12}$$

But from Equation (7.10) the internal forces cancel each other and we are left with

$$\sum_i \mathbf{F}_i = \sum_i m_i\mathbf{a}_i \tag{7.13}$$

The acceleration is defined as the second derivative of the position vector

$$\mathbf{a}_i = \frac{d^2\mathbf{r}_i}{dt^2} \tag{7.14}$$

so that we may write

$$\sum_i \mathbf{F}_i = \sum_i m_i \frac{d^2\mathbf{r}_i}{dt^2} \tag{7.15}$$

Since the mass of each particle is assumed constant, we may shift the derivative operator outside the summation symbol

$$\sum_i \mathbf{F}_i = \frac{d^2}{dt^2}\sum_i m_i\mathbf{r}_i \tag{7.16}$$

If we add together the masses of each particle, we obtain the total mass of the rigid body.

$$m = \sum_i m_i \tag{7.17}$$

In the previous section, we defined the *center of mass* to be

$$\mathbf{r}_G = \frac{1}{m}\sum_i m_i\mathbf{r}_i \tag{7.18}$$

Then Equation (7.16) simplifies to

$$\sum_i \mathbf{F}_i = \frac{d^2}{dt^2}(m\mathbf{r}_G) \tag{7.19}$$

Since the mass of the rigid body is constant, it can be moved outside the derivative operator to obtain

$$\sum_i \mathbf{F}_i = m\mathbf{a}_G \tag{7.20}$$

In other words, we may use Newton's Third Law to treat a rigid body exactly like a point mass, as long as we use the acceleration of the center of mass in our computations. The summation of forces on the left side of Equation (7.20) is called the *net force* on the body. If all of the external forces on the body cancel out, then the net force (and therefore acceleration) is zero and we are left with the *Statics* problem. In the case of a moving linkage, however, the net force on a link will almost never be zero.

7.2.2 Rotational Equations of Motion

The major difference between a rigid body and a point mass is that the rigid body can rotate. In most cases, the forces (and moments) applied to the rigid body will cause both translational and rotational motion. Consider the rigid body shown in Figure 7.15. Here we have attached the origin to an arbitrary point P on the body. The vector from the origin to m_1 is $\mathbf{r}_{P1}$, and we define similar vectors for the other point masses. The rotational effect of the forces on the body can be determined by computing the *moment* of the forces about a point P. The moment is defined as the cross-product of the vector from P to the point of application of the force with the force itself. Clearly, the location of the point P will affect the moment that is calculated, and we must choose P carefully. For now, let us assume that P is an arbitrary point on the body. In a later section, we will discuss the most suitable locations for P from a computational point of view.

Before proceeding with our example, let us take a moment to review a simple moment calculation when only one force is acting on a point. Examine the two instances of a single link shown in Figure 7.16. In each instance, the force is applied at a different angle with respect to the link. On the left, the force is applied almost perpendicular to the vector from the center of rotation, $\mathbf{r}_B$, which would cause a large angular acceleration. On the right the

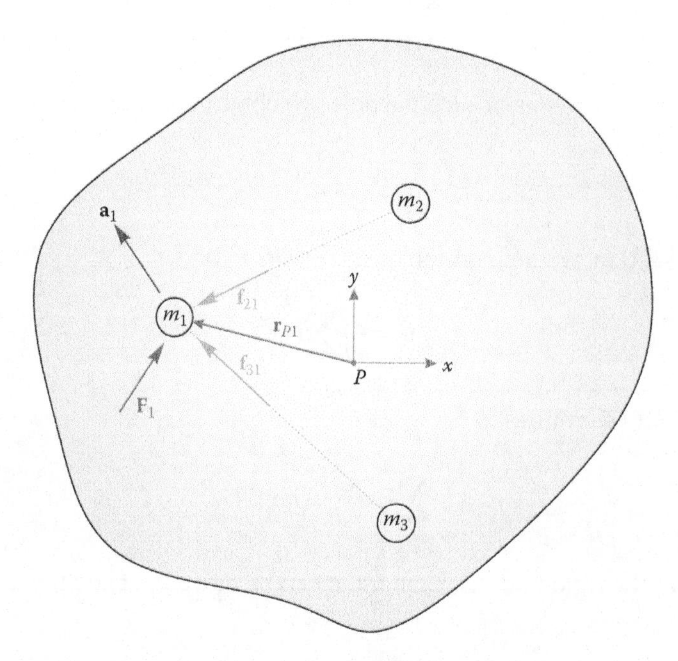

FIGURE 7.15
A rigid body with point masses m_1, m_2, and m_3 shown. The origin is chosen at an arbitrary point P on the body.

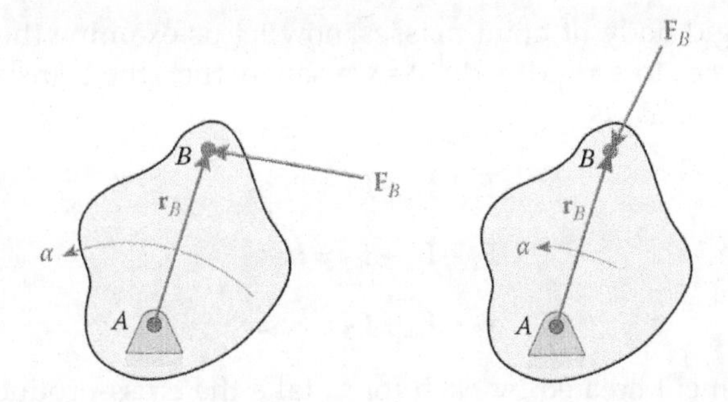

FIGURE 7.16
The angular acceleration caused by a force depends upon the direction of the force relative to the vector from the center of rotation.

force is almost parallel to $\mathbf{r}_B$, which would cause a small angular acceleration. The moment about the ground pin at A created by the force in each case is

$$\mathbf{M}_A = \mathbf{r}_B \times \mathbf{F}_B \tag{7.21}$$

Since we are confining our analysis to the two-dimensional case, we can write

$$\mathbf{r}_B = a \begin{Bmatrix} \cos\theta \\ \sin\theta \end{Bmatrix} \tag{7.22}$$

and the cross-product yields

$$\mathbf{M}_A = \begin{vmatrix} a\cos\theta & a\sin\theta & 0 \\ F_{Bx} & F_{By} & 0 \end{vmatrix} = \begin{Bmatrix} 0 \\ 0 \\ -aF_{Bx}\sin\theta + aF_{By}\cos\theta \end{Bmatrix} \tag{7.23}$$

For this two-dimensional case, this may be rewritten in terms of the unit normal as

$$\mathbf{M}_A = a(\mathbf{n}\cdot\mathbf{F}_B)\hat{k} \tag{7.24}$$

or

$$M_A = a(\mathbf{n}\cdot\mathbf{F}_B) \tag{7.25}$$

For the two-dimensional case, the moment is always in the $\hat{k}$ direction, out of the page. From the definition of the dot product, we see that the moment is at a maximum when the force is aligned with the unit normal, and is zero if the force is perpendicular to the unit normal (i.e. when it is parallel to the link itself). This is intuitively obvious: if we apply the force along the length of the link, the force will be directed into the pin, and will not cause it to rotate. Conversely, applying the force perpendicular to the link will cause the maximum possible rotational motion for a given force.

Returning to our rigid body of point masses, now let us examine the rotational effect of applying a set of forces to a rigid body. As we saw earlier, the translational equations of motion for each point mass is

$$\begin{aligned} \mathbf{F}_1 + \mathbf{f}_{21} + \mathbf{f}_{31} &= m_1\mathbf{a}_1 \\ \mathbf{F}_2 + \mathbf{f}_{32} + \mathbf{f}_{12} &= m_2\mathbf{a}_2 \\ \mathbf{F}_3 + \mathbf{f}_{13} + \mathbf{f}_{23} &= m_3\mathbf{a}_3 \end{aligned} \tag{7.26}$$

To calculate the moment created by each force, take the cross-product of Equation (7.26) with the position vectors.

$$\begin{aligned} \mathbf{r}_{P1} \times (\mathbf{F}_1 + \mathbf{f}_{21} + \mathbf{f}_{31}) &= \mathbf{r}_{P1} \times m_1\mathbf{a}_1 \\ \mathbf{r}_{P2} \times (\mathbf{F}_2 + \mathbf{f}_{32} + \mathbf{f}_{12}) &= \mathbf{r}_{P2} \times m_2\mathbf{a}_2 \\ \mathbf{r}_{P3} \times (\mathbf{F}_3 + \mathbf{f}_{13} + \mathbf{f}_{23}) &= \mathbf{r}_{P3} \times m_3\mathbf{a}_3 \end{aligned} \tag{7.27}$$

If we examine the moments created by the internal forces we see that the forces share a common line of action; thus, their moment arm, h, is the same, as shown in Figure 7.17. Since the forces are equal and opposite we see that:

$$\mathbf{r}_{P1} \times \mathbf{f}_{21} + \mathbf{r}_{P2} \times \mathbf{f}_{12} = 0 \tag{7.28}$$

Thus, the moments created by the internal forces cancel when all three equations are added together. Adding the three equations in (7.19) gives

$$\sum_i \mathbf{r}_{Pi} \times \mathbf{F}_i = \sum_i \mathbf{r}_{Pi} \times m_i\mathbf{a}_i \tag{7.29}$$

For convenience, let us define the sum of moments created by the external forces as

$$\mathbf{M}_P \equiv \sum_i \mathbf{r}_{Pi} \times \mathbf{F}_i \tag{7.30}$$

Then the rotational equation of motion becomes

$$\mathbf{M}_P = \sum_i \mathbf{r}_{Pi} \times m_i\mathbf{a}_i \tag{7.31}$$

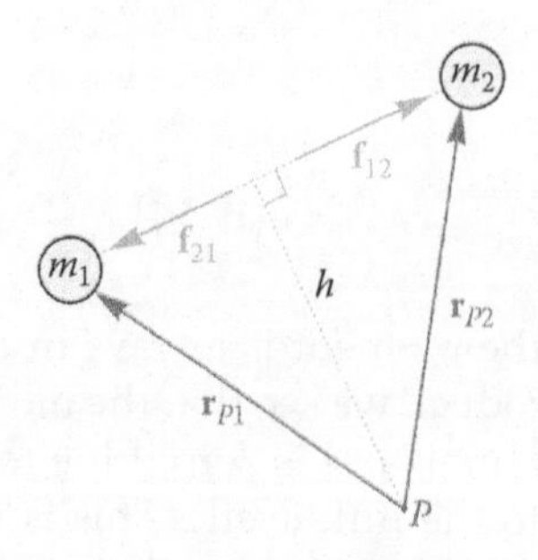

FIGURE 7.17
The moments created by internal forces cancel, since the internal forces are equal and opposite, and act along a common line of action.

Now rewrite the vector from the point P to each mass in unit vector form, as shown in Figure 7.18.

$$\mathbf{r}_{Pi} = r_{Pi}\mathbf{e}_{Pi} \tag{7.32}$$

where r_{Pi} is the distance from point P to the point mass and $\mathbf{e}_{Pi}$ is the unit vector pointing in the direction of $\mathbf{r}_{Pi}$. Then

$$\mathbf{M}_P = \sum_i r_{Pi}\mathbf{e}_{Pi} \times m_i\mathbf{a}_i \tag{7.33}$$

We may use the relative acceleration formula given in Chapter 6 to rewrite the acceleration of each point mass i as

$$\mathbf{a}_i = \mathbf{a}_P + r_{Pi}\alpha\mathbf{n}_{Pi} - r_{Pi}\omega^2\mathbf{e}_{Pi} \tag{7.34}$$

where $\mathbf{a}_P$ is the acceleration of point P. Substituting this into Equation (7.31) gives

$$\mathbf{M}_P = \sum_i m_i r_{Pi}\mathbf{e}_{Pi} \times \left(\mathbf{a}_P + r_{Pi}\alpha\mathbf{n}_{Pi} - r_{Pi}\omega^2\mathbf{e}_{Pi}\right) \tag{7.35}$$

Now separate the cross-product into its three terms

$$\mathbf{M}_P = \sum_i m_i r_{Pi}\mathbf{e}_{Pi} \times \mathbf{a}_P + \sum_i m_i r_{Pi}\mathbf{e}_{Pi} \times \rho_{Pi}\alpha\mathbf{n}_{Pi} - \sum_i m_i r_{Pi}\mathbf{e}_{Pi} \times r_{Pi}\omega^2\mathbf{e}_{Pi} \tag{7.36}$$

Since r_{Pi}, ω and α are scalars, we may factor them outside the cross-products

$$\mathbf{M}_P = \sum_i m_i\rho_{Pi}\mathbf{e}_{Pi} \times \mathbf{a}_P + \sum_i m_i\rho_{Pi}^2\alpha(\mathbf{e}_{Pi} \times \mathbf{n}_{Pi}) - \sum_i m_i\rho_{Pi}^2\omega^2(\mathbf{e}_{Pi} \times \mathbf{e}_{Pi}) \tag{7.37}$$

The third term vanishes because

$$\mathbf{e}_{Pi} \times \mathbf{e}_{Pi} = 0 \tag{7.38}$$

Also, the cross-product in the second term may be reduced by noting that

$$\mathbf{e}_{Pi} \times \mathbf{n}_{Pi} = \hat{k} \tag{7.39}$$

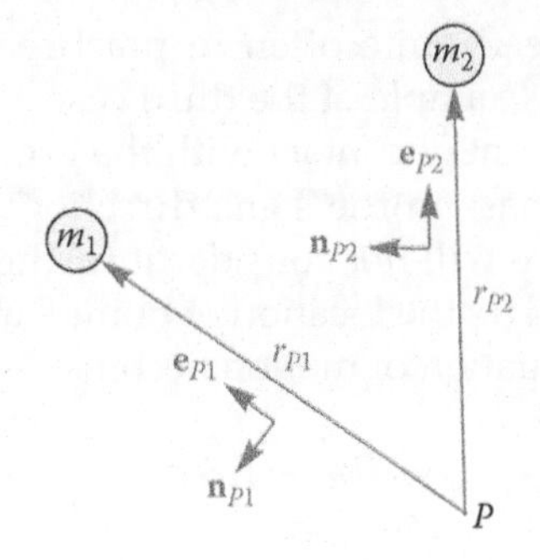

FIGURE 7.18
The vector from point P to each mass may be written in terms of the unit vector $\mathbf{e}_{Pi}$ and length r_{Pi}.

The resulting equation can be simplified by noting that $\mathbf{a}_P$, the acceleration of point P, is uniform across the body, as is the angular acceleration, α.

$$\mathbf{M}_P = \left(\sum_i m_i r_{Pi}\mathbf{e}_{Pi}\right) \times \mathbf{a}_P + \alpha \sum_i m_i r_{Pi}^2 \hat{k} \tag{7.40}$$

We may use the definition of center of mass given in Equation (7.18) to simplify the first term

$$\mathbf{M}_P = m\mathbf{r}_g \times \mathbf{a}_P + \alpha \sum_i m_i r_{Pi}^2 \hat{k} \tag{7.41}$$

where $\mathbf{r}_g$ is the vector from the point P to the center of mass of the body. Note that

$$r_{Pi}^2 = x_{Pi}^2 + y_{Pi}^2 \tag{7.42}$$

is the squared distance from the point P to a point mass i. If we convert the summation in Equation (7.41) to an integral by changing the point masses to differential masses we have

$$\mathbf{M}_P = m\mathbf{r}_g \times \mathbf{a}_P + \alpha \hat{k} \int \left(x_P^2 + y_P^2\right) dm \tag{7.43}$$

which the alert reader will recognize as the definition of mass moment of inertia. Thus, we have

$$\mathbf{M}_P = m\mathbf{r}_g \times \mathbf{a}_P + I_P \alpha \hat{k} \tag{7.44}$$

We will now discuss the proper choice of the location of point P. Judicious selection of this point will enable us to eliminate the first term in Equation (7.44), leading to a considerable simplification of the rotational equations of motion. There are three choices that enable us to eliminate this term:

1. $\mathbf{r}_g = 0 \rightarrow$ the point P is at the center of mass of the body
2. $\mathbf{a}_P = 0 \rightarrow$ the point P is pinned to ground, as in a ground pivot
3. $\mathbf{r}_g \parallel \mathbf{a}_P \rightarrow$ since the cross-product of parallel vectors is zero

The first two conditions are used quite often in practice, while the third is difficult to achieve in most situations. One example of the third case is a disk rolling on a flat surface. If the point P is chosen at the point of contact with the ground, its acceleration is directed upward toward the center of mass in the same direction as $\mathbf{r}_g$. Since this situation is so rarely encountered, however, we will not consider it further. For analyzing the motion of linkages, it is straightforward to fix the location of point P at the center of mass of the link. If this is done, the rotational equation of motion becomes

$$M_g = I_g \alpha \tag{7.45}$$

where I_g is the moment of inertia of the link about its center of mass. It cannot be emphasized too strongly that this equation is only valid when moments are summed about the center of mass, and not an arbitrary point on the link. We obtain a similarly simple equation of motion if we sum moments about a ground pivot

$$M_P = I_P \alpha \tag{7.46}$$

but we must calculate the moment of inertia about the ground pivot, and not the center of mass. Since this equation will work only with links that are pinned to ground, we choose to employ only Equation (7.45) for the analysis that follows. This will enable us to systematically develop a uniform set of equations of motion for a linkage, without having to worry about whether a link is pinned to ground or not. In the geared fivebar linkage, however, it will prove simplest to sum moments about the ground pivots.

This was a very long-winded way to arrive at a very simple equation of motion for rotation. The development was necessary in order to emphasize the importance of choosing the correct point P to sum moments about. We will attach the point P to the center of mass for most of the remainder of this text because it results in relatively simple equations of motion, and *it always works*!

7.2.3 A Digression on Moments, Torques, and Couples

We will ordinarily drive a link in rotation by attaching a motor to it, as shown in Figure 7.19. The electromagnetic properties of the motor create a force $\mathbf{F}_e/2$ at the top and bottom of the motor. Here we have assumed a two-pole motor, but the principle is the same for motors with more than two poles. The electromagnetic forces are equal in magnitude but opposite in direction and form a *couple*. Since the forces are equal and opposite the net force created by a couple is zero. The vector from the ground pivot to the top force is $\mathbf{r}_{AB}$ and to the bottom force is $\mathbf{r}_{AC}$, and these vectors are also equal and opposite. The net *torque* created by the forces is

$$T = \mathbf{r}_{AB} \times \frac{\mathbf{F}_e}{2} + \mathbf{r}_{AC} \times \frac{-\mathbf{F}_e}{2} = \mathbf{r}_{AB} \times \mathbf{F}_e$$

There is no such thing as a pure torque in nature – all torques are created by force couples as in this example.

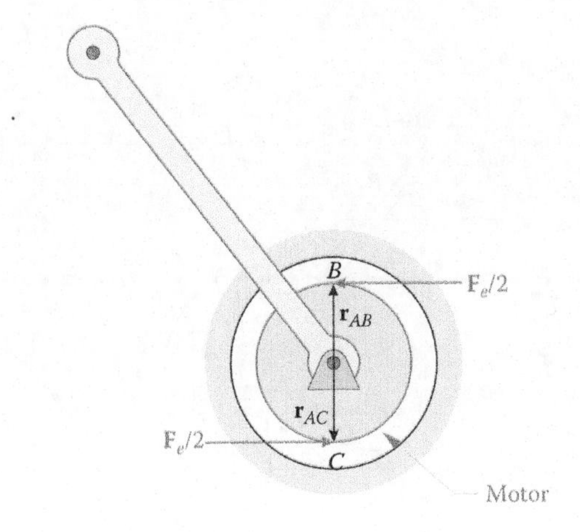

FIGURE 7.19
A motor is used to drive the link.

Now consider the problem of finding the net moment created by the couple on an arbitrary point P on the link as shown in Figure 7.20.

$$M = \mathbf{r}_{PB} \times \frac{\mathbf{F}_e}{2} + \mathbf{r}_{PC} \times \frac{-\mathbf{F}_e}{2} \tag{7.47}$$

But $\mathbf{r}_{PB}$ can be written as the sum of

$$\mathbf{r}_{PB} = \mathbf{r}_{PA} + \mathbf{r}_{AB}$$

and $\mathbf{r}_{PC}$ is

$$\mathbf{r}_{PC} = \mathbf{r}_{PA} + \mathbf{r}_{AC}$$

But $\mathbf{r}_{AC}$ is equal and opposite to $\mathbf{r}_{AB}$, so that

$$\mathbf{r}_{PC} = \mathbf{r}_{PA} - \mathbf{r}_{AB}$$

Substituting this into Equation (7.47) gives

$$M = (\mathbf{r}_{PA} + \mathbf{r}_{AB}) \times \frac{\mathbf{F}_e}{2} + (\mathbf{r}_{PA} - \mathbf{r}_{AB}) \times \frac{-\mathbf{F}_e}{2}$$

Using the distributive property of the cross-product results in

$$M = \mathbf{r}_{PA} \times \frac{\mathbf{F}_e}{2} + \mathbf{r}_{AB} \times \frac{\mathbf{F}_e}{2} - \mathbf{r}_{PA} \times \frac{\mathbf{F}_e}{2} + \mathbf{r}_{AB} \times \frac{\mathbf{F}_e}{2}$$

$$M = \mathbf{r}_{AB} \times \mathbf{F}_e = T$$

Notice that all reference to point P has vanished! The conclusion is that it does not matter where we apply a torque to the link; the moment created by the torque is unaffected by its location. Henceforth, we will assume that the motor provides a pure torque to the link,

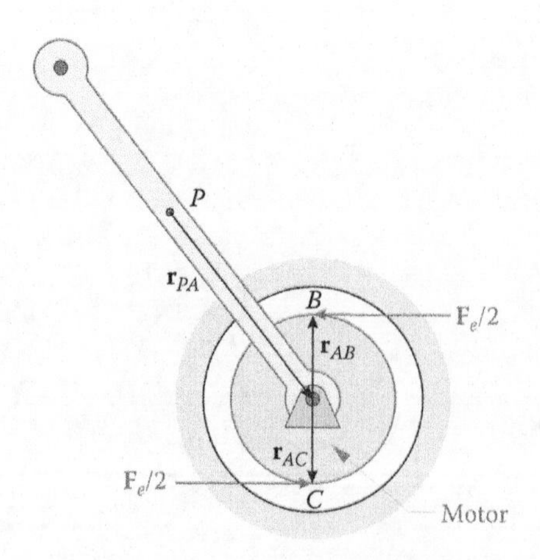

FIGURE 7.20
Finding the moment created by the force couple on an arbitrary point P on the link.

and we can apply this torque to any point on the link that is convenient. When we sum moments about the center of mass of a link (as in Case 1, above) we will apply the motor torque to the center of mass, and when we sum moments about the ground pivot (as in Case 2, above) we will apply the motor torque to the ground pivot.

7.3 Force Analysis on a Single Link

We now have enough background material to begin analyzing forces on a linkage. Recall that we will be conducting *inverse dynamics*; that is, given a set of positions, velocities, and accelerations, we wish to find the forces that would be necessary to produce this motion. To conduct our analysis, we will employ Newton's second law:

$$\sum_i \mathbf{F}_i = m\mathbf{a}_g \quad \sum_i M_i = I_g \alpha \tag{7.48}$$

The first equation states that the sum of all forces on the link is equal to its mass multiplied by the acceleration of the *center of mass* of the link. In the second equation, we sum moments (or torques) also *about the center of mass*. The result of this summation is the mass moment of inertia – also taken about the center of mass – multiplied by the angular acceleration of the link.

Since we have already performed a position/velocity/acceleration analysis, the right-hand sides of Equations (7.48) are known, and we will be solving for the forces on the left-hand side of the equations. Let us begin with a simple example: the force analysis of a single link. Figure 7.21 shows a single link that has been pinned to ground at point A and is free to rotate. A known force, $\mathbf{F}_B$, has been applied at B, and we wish to solve for the torque, T, that is necessary to drive the link. We would also like to solve for the force, $\mathbf{F}_A$, that keeps the link pinned to the ground. Below are definitions of some of the important quantities for this problem:

$\mathbf{F}_B$ = force on pin B (an externally applied force, assumed known)
$\mathbf{F}_A$ = force on pin A (the reaction force from the ground pivot)
T = applied torque (usually from a motor attached to the link)
$\mathbf{a}_g$ = acceleration at center of mass of link

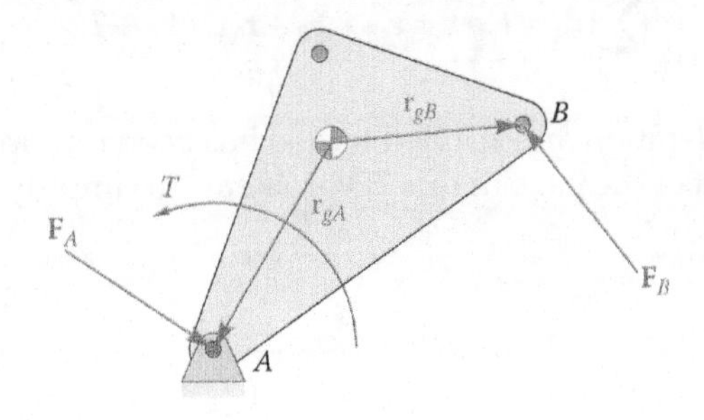

FIGURE 7.21
Diagram of a single link showing pin forces and applied torque.

α = angular acceleration of the link
$\mathbf{r}_{gA}$ = vector from the center of mass to point A
$\mathbf{r}_{gB}$ = vector from the center of mass to point B

Since the vectors $\mathbf{r}_{gA}$ and $\mathbf{r}_{gB}$ are attached to the link, we assume that these are known. In a later section, we will discuss a method for solving for these vectors explicitly. As usual, we start by making a list of knowns and unknowns

$$\text{known:} \quad \theta, \omega, \alpha, m, I_g, \mathbf{a}_g, \mathbf{r}_{gA}, \mathbf{r}_{gB}, \mathbf{F}_B$$

$$\text{find:} \quad \mathbf{F}_A, T$$

First, apply Newton's second law for forces

$$\sum_i \mathbf{F}_i = m\mathbf{a}_g = \mathbf{F}_B + \mathbf{F}_A \tag{7.49}$$

Rearrange this equation so that the knowns are on the right side, and the unknowns are on the left.

$$\mathbf{F}_A = m\mathbf{a}_g - \mathbf{F}_B \tag{7.50}$$

Remember that this is actually two equations, one for the x component and one for the y component. Next, we will examine the rotational equation of motion,

$$\sum_i M_i = I_G\alpha \tag{7.51}$$

The sum of moments on the left side of the equation derive from two sources: the moments created by the forces acting on the link and the applied torque, T. As described earlier, we must calculate each moment about the center of mass of the link. The moment created by the force acting on point A can be calculated using the cross-product of the vector from the center of mass to point A with the force at point A

$$\mathbf{M}_A = \mathbf{r}_{gA} \times \mathbf{F}_A \tag{7.52}$$

and a similar expression can be found for the force at point B. Adding all of the torques and moments on the link gives

$$\sum_i M_i = I_g\alpha\hat{k} = \mathbf{r}_{gB} \times \mathbf{F}_B + \mathbf{r}_{gA} \times \mathbf{F}_A + T\hat{k} \tag{7.53}$$

where the angular acceleration and applied torque has been multiplied by $\hat{k}$ to account for the fact that it points out of the page in this 2D problem. To simplify this equation, evaluate the cross-products

$$\mathbf{r}_{gB} \times \mathbf{F}_B = \begin{vmatrix} \hat{i} & \hat{j} & \hat{k} \\ r_{gBx} & r_{gBy} & 0 \\ F_{Bx} & F_{By} & 0 \end{vmatrix} = \left(r_{gBx}F_{By} - r_{gBy}F_{Bx}\right)\hat{k}$$

$$\mathbf{r}_{gA} \times \mathbf{F}_A = \begin{vmatrix} \hat{i} & \hat{j} & \hat{k} \\ r_{gAx} & r_{gAy} & 0 \\ F_{Ax} & F_{Ay} & 0 \end{vmatrix} = \left(r_{gAx}F_{Ay} - r_{gAy}F_{Ax}\right)\hat{k} \tag{7.54}$$

Let us define a new set of vectors that are normal to the position vectors $\mathbf{r}_{gA}$ and $\mathbf{r}_{gB}$. We can use the formula for finding a perpendicular vector to write

$$\mathbf{s}_{gA} = \begin{Bmatrix} -r_{gAy} \\ r_{gAx} \end{Bmatrix} \quad \mathbf{s}_{gB} = \begin{Bmatrix} -r_{gBy} \\ r_{gBx} \end{Bmatrix} \tag{7.55}$$

Then the cross-product terms become

$$\mathbf{r}_{gA} \times \mathbf{F}_A = \left(\mathbf{s}_{gA} \cdot \mathbf{F}_A\right)\hat{k} \quad \mathbf{r}_{gB} \times \mathbf{F}_B = \left(\mathbf{s}_{gB} \cdot \mathbf{F}_B\right)\hat{k} \tag{7.56}$$

Thus, we have converted the cross-products in Equation (7.53) to dot products.

If the preceding seems a little "mathemagical" to you, here is an alternative approach. The scalar triple product is often used in mathematics to calculate the volume of a parallelpiped (a solid body of which each face is parallelogram).

$$V = \mathbf{a} \cdot (\mathbf{b} \times \mathbf{c}) \tag{7.57}$$

where **a**, **b**, and **c** are vectors along the sides of the parallelpiped. One property of the scalar triple product is that

$$\mathbf{a} \cdot (\mathbf{b} \times \mathbf{c}) = \mathbf{b} \cdot (\mathbf{c} \times \mathbf{a}) \tag{7.58}$$

Since $\mathbf{r}_{gA}$ and $\mathbf{F}_A$ lie in the same 2D plane, their cross-product points out of the plane in the $\hat{k}$ direction. To find the *magnitude* of the moment created by $\mathbf{F}_A$, we can take the dot product of $\hat{k}$ with $\mathbf{r}_{gA} \times \mathbf{F}_A$

$$M_A = \hat{k} \cdot \left(\mathbf{r}_{gA} \times \mathbf{F}_A\right) \tag{7.59}$$

But, according to Equation (7.58) this can be rearranged to form

$$M_A = \mathbf{F}_A \cdot \left(\hat{k} \times \mathbf{r}_{gA}\right) \tag{7.60}$$

The cross-product $\hat{k} \times r_{gA}$ results in a vector perpendicular to both $\hat{k}$ and $\mathbf{r}_{gA}$, in other words, $\mathbf{s}_{gA}$ (see Figure 7.22). Thus

$$M_A = \mathbf{F}_A \cdot \mathbf{s}_{gA} = \mathbf{s}_{gA} \cdot \mathbf{F}_A \tag{7.61}$$

as we found earlier. Having dot products instead of cross-products will make our equations of motion much simpler to write out. Please note that this conversion only works for 2D systems; for 3D systems, you will need to perform the full cross-product operations.

Substituting Equation (7.56) into Equation (7.53) and dividing out the $\hat{k}$ gives

$$I_g\alpha = \mathbf{s}_{gB} \cdot \mathbf{F}_B + \mathbf{s}_{gA} \cdot \mathbf{F}_A + T \tag{7.62}$$

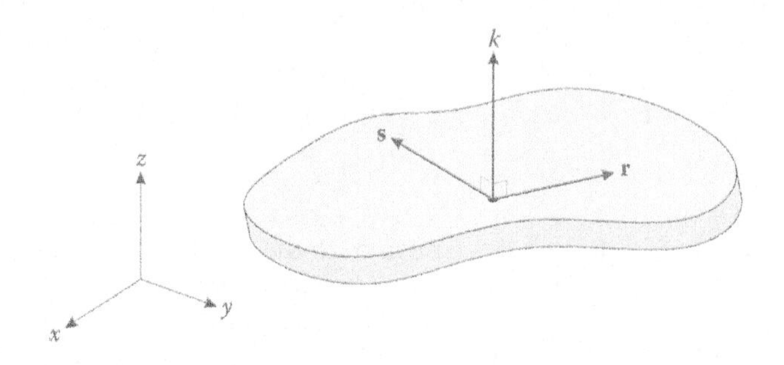

FIGURE 7.22
The vectors **r**, **s**, and $\hat{k}$ are mutually perpendicular.

Rearrange this so that the knowns are on the right side and the unknowns are on the left.

$$\mathbf{s}_{gA} \cdot \mathbf{F}_A + T = I_g\alpha - \mathbf{s}_{gB} \cdot \mathbf{F}_B \tag{7.63}$$

Thus, the three equations of motion for the link are

$$\begin{aligned} \mathbf{F}_A &= m\mathbf{a}_g - \mathbf{F}_B \\ \mathbf{s}_{gA} \cdot \mathbf{F}_A + T &= I_g\alpha - \mathbf{s}_{gB} \cdot \mathbf{F}_B \end{aligned} \tag{7.64}$$

Remember that the dot product can also be written in matrix form as

$$\mathbf{s}_{gA} \cdot \mathbf{F}_A = \mathbf{s}_{gA}^T \mathbf{F}_A \tag{7.65}$$

Thus, we can rewrite the equations of motion in matrix form as

$$\begin{bmatrix} \mathbf{U}_2 & \mathbf{Z}_{21} \\ \mathbf{s}_{gA}^T & 1 \end{bmatrix} \begin{Bmatrix} \mathbf{F}_A \\ T \end{Bmatrix} = \begin{Bmatrix} m\mathbf{a}_g - \mathbf{F}_B \\ I_g\alpha - \mathbf{s}_{gB} \cdot \mathbf{F}_B \end{Bmatrix} \tag{7.66}$$

where

$$\mathbf{U}_2 = \begin{bmatrix} 1 & 0 \\ 0 & 1 \end{bmatrix} \tag{7.67}$$

is the 2×2 identity matrix and

$$\mathbf{Z}_{21} = \begin{bmatrix} 0 \\ 0 \end{bmatrix} \tag{7.68}$$

is a 2×1 matrix of zeros. The true form of the matrix equation may be a little obscure, since there are so many vectors within the matrix and vector of knowns. Writing out the equation in full gives

$$\begin{bmatrix} 1 & 0 & 0 \\ 0 & 1 & 0 \\ s_{gAx} & s_{gAy} & 1 \end{bmatrix} \begin{Bmatrix} F_{Ax} \\ F_{Ay} \\ T \end{Bmatrix} = \begin{Bmatrix} ma_{gx} - F_{Bx} \\ ma_{gy} - F_{By} \\ I_g\alpha - \mathbf{s}_{gB} \cdot \mathbf{F}_B \end{Bmatrix} \tag{7.69}$$

Remember that the dot product $\mathbf{s}_{gB} \cdot \mathbf{F}_B$ produces a scalar. As before, we can write this equation more compactly in matrix form

$$\mathbf{Sf} = \mathbf{t} \tag{7.70}$$

where

$$\mathbf{S} = \begin{bmatrix} \mathbf{U}_2 & \mathbf{Z}_{21} \\ \mathbf{s}_{gA}^T & 1 \end{bmatrix} \qquad \mathbf{f} = \begin{Bmatrix} \mathbf{F}_A \\ T \end{Bmatrix} \tag{7.71}$$

$$\mathbf{t} = \begin{Bmatrix} m\mathbf{a}_g - \mathbf{F}_B \\ I_g\alpha - \mathbf{s}_{gB} \cdot \mathbf{F}_B \end{Bmatrix} \tag{7.72}$$

As with the kinematic equations, this can easily be solved using MATLAB. The compact form of Equations (7.67) and (7.68) may take some time to get used to, but the resulting code will be simple to write and to scan for errors. Spend a little time now making sure that you understand how Equation (7.66) translates into Equation (7.69) – it will be worth it in the end!

7.3.1 Another Useful MATLAB® Function

Performing the dynamic analysis requires that we know the vectors that point from the center of mass of the link to each pin, shown as $\mathbf{r}_{gA}$, $\mathbf{r}_{gB}$, $\mathbf{r}_{gC}$ in Figure 7.23. In addition, we must find the normal to each of these vectors, $\mathbf{s}_{gA}$, $\mathbf{s}_{gB}$, and $\mathbf{s}_{gC}$. It is interesting to note that we do not require the vectors $\mathbf{r}_{gA}$, $\mathbf{r}_{gB}$, $\mathbf{r}_{gC}$ for our force computations, but we will end up calculating them anyway as part of the process of calculating $\mathbf{s}_{gA}$, $\mathbf{s}_{gB}$, and $\mathbf{s}_{gC}$. Since we must calculate these vectors for each link, it is clear that we have a lengthy and repetitive task ahead. The simplest approach is to define a new MATLAB function that will return the normal vectors based on a few link parameters that we send it.

Additionally, we must compute the acceleration at the center of mass for each link in order to assemble the **t** vector in Equation (7.72). If we know the acceleration at pin A (from the kinematic analysis performed earlier) then we may use the relative acceleration formula

$$\mathbf{a}_g = \mathbf{a}_A + l_{Ag}\alpha\mathbf{n}_{Ag} - l_{Ag}\omega^2\mathbf{e}_{Ag} \tag{7.73}$$

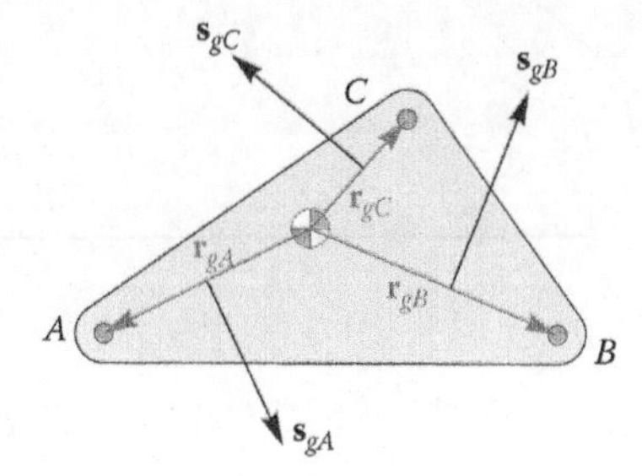

FIGURE 7.23
We must now find the set of vectors that point from the center of mass of the link to each pin.

to compute the acceleration at the center of mass. The function `FindAcc` will come in handy here. Note that $\mathbf{e}_{Ag}$ points from pin A to the center of mass, instead of the other way around – see Figure 7.24.

A generic three-pin link is shown in Figure 7.25. The distance from A to B is a, and the distance from A to C is c; both of these are known from the geometry of the linkage, as is the angle γ. It is a simple matter to use the **Mass Properties** feature in SOLIDWORKS to arrive at the coordinates of the center of mass of the link, $(\bar{x},\bar{y})$. Along the way, we can also determine the link's mass and moment of inertia (about its center of mass). Here we assume that the origin of the part has been defined at point A. In the analysis that follows, we will use a *local* coordinate system that has the x axis aligned with AB and the y axis pointed upward. Vectors in the local coordinate system will be indicated with a prime. We will return to the global coordinate system at the end by adding the angle of the link, θ, as shown in Figure 7.26.

The input and output structure of our new function should be

$$\text{Inputs:} \quad a, c, \gamma, \bar{x}, \bar{y}, \theta$$

$$\text{Outputs:} \quad \mathbf{s}_{gA}, \mathbf{s}_{gB}, \mathbf{s}_{gC}, l_{Ag}, \mathbf{e}_{Ag}, \mathbf{n}_{Ag}$$

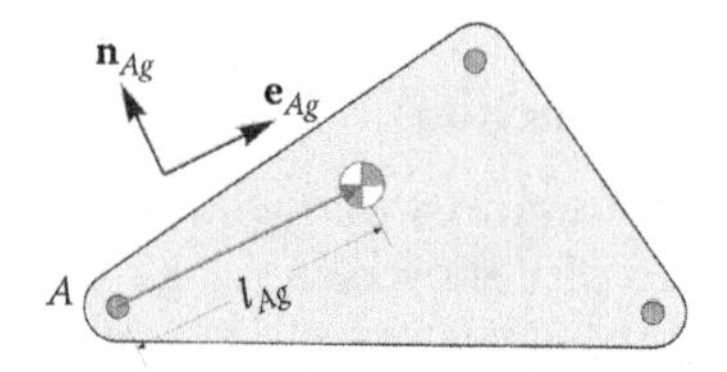

FIGURE 7.24
The vector $\mathbf{r}_{Ag}$ extends from point A to the center of mass. We use this, along with the unit vectors $\mathbf{e}_{Ag}$ and $\mathbf{n}_{Ag}$ to find the acceleration of the center of mass.

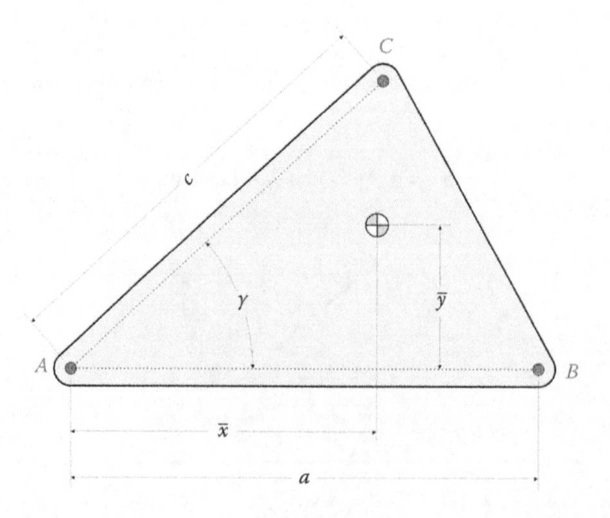

FIGURE 7.25
Input parameters for the `LinkCG` function. Most of these are known from the geometry of the linkage, $\bar{x}$ and $\bar{y}$ are calculated using the **Mass Properties** feature in SOLIDWORKS.

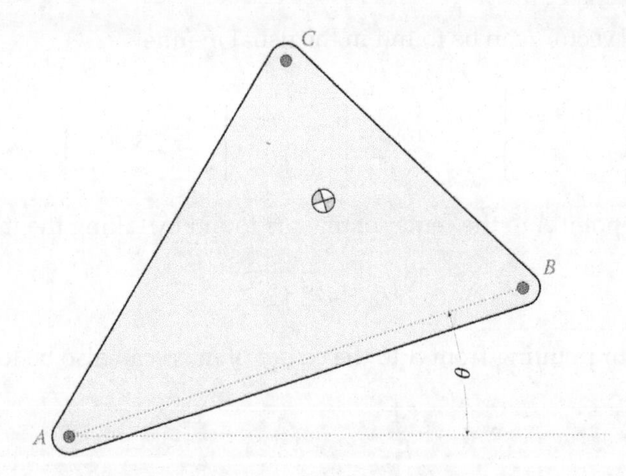

FIGURE 7.26
The angle θ is measured from the horizontal to the line AB.

As shown in Figure 7.27, the local vector from A to B is found to be

$$\mathbf{r}'_{AB} = a\begin{Bmatrix} 1 \\ 0 \end{Bmatrix} \tag{7.74}$$

Also, the local vector from the center of mass to A is

$$\mathbf{r}'_{gA} = -\begin{Bmatrix} \bar{x} \\ \bar{y} \end{Bmatrix} \tag{7.75}$$

We can use simple vector addition to find $\mathbf{r}'_{gB}$ and $\mathbf{r}'_{gC}$.

$$\begin{aligned} \mathbf{r}'_{gB} &= \mathbf{r}'_{AB} + \mathbf{r}'_{gA} \\ \mathbf{r}'_{gC} &= \mathbf{r}'_{AC} + \mathbf{r}'_{gA} \end{aligned} \tag{7.76}$$

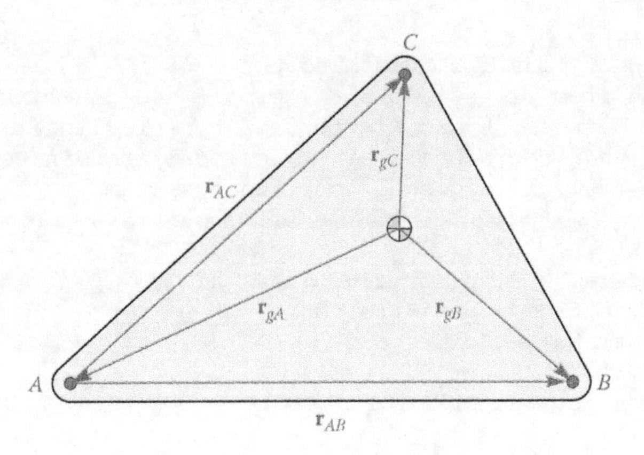

FIGURE 7.27
The local x axis is aligned with AB, $\mathbf{r}'_{AB}$ points from A to B and $\mathbf{r}'_{AC}$ points from A to C.

The perpendicular vectors can be found in the usual manner

$$\mathbf{s}'_{gA} = \begin{Bmatrix} -r'_{gAy} \\ r'_{gAx} \end{Bmatrix} \qquad \mathbf{s}'_{gB} = \begin{Bmatrix} -r'_{gBy} \\ r'_{gBx} \end{Bmatrix} \qquad \mathbf{s}'_{gC} = \begin{Bmatrix} -r'_{gCy} \\ r'_{gCx} \end{Bmatrix} \tag{7.77}$$

The distance from point A to the center of mass is found by taking the magnitude of $\mathbf{r}_{gA}$.

$$l_{Ag} = \sqrt{\bar{x}^2 + \bar{y}^2} \tag{7.78}$$

The local unit vector pointing from A to the center of mass can also be found using $\mathbf{r}_{gA}$

$$\mathbf{e}'_{Ag} = -\frac{1}{l_{Ag}}\mathbf{r}'_{gA} \quad \mathbf{n}'_{Ag} = -\frac{1}{l_{Ag}}\mathbf{s}'_{gA} \tag{7.79}$$

Finally, we must transform these vectors from the local coordinate system to the global coordinate system by rotating each one through the angle θ. Recall the definition of the rotation matrix from Chapter 4:

$$\mathbf{R} = \begin{bmatrix} \cos\theta & -\sin\theta \\ \sin\theta & \cos\theta \end{bmatrix} \tag{7.80}$$

The vectors in the global coordinate system can be written

$$\mathbf{s}_{gA} = \mathbf{R}\mathbf{s}'_{gA} \quad \mathbf{s}_{gB} = \mathbf{R}\mathbf{s}'_{gB} \quad \mathbf{s}_{gC} = \mathbf{R}\mathbf{s}'_{gC} \tag{7.81}$$

And a similar transformation can be performed on the unit vectors. We are now ready to begin defining the function `LinkCG` that will execute these calculations for any three-pin link. If a link has only two pins we can enter zero for the length c and for the angle γ. Open a new MATLAB script and type the following header:

```
% LinkCG.m
% calculates the vectors associated with the center of mass of a
% two or three pin link.
%
% *** Inputs ***
% a      = length of link from pin A to pin B
% c      = length from pin A to pin C (zero if two-pin link)
% gamma = angle between AB and AC (zero if two-pin link)
% xbar   = [xbar,ybar]coordinates of CM in local coordinate system
% theta = angle of link in global coordinate system
%
% *** Outputs ***
% eAg,nAg = unit vector and normal point A to CM
% LAg      = length from point A to CM
% sgA      = normal to vector from CM to A
% sgB      = normal to vector from CM to B
% sgC      = normal to vector from CM to C

function [eAg,nAg,LAg,sgA,sgB,sgC] = LinkCG(a,c,gamma,xbar,theta)
```

First, calculate the vectors from the center of mass to the pins in the local coordinate system

```
rAB = a*[           1; 0];           % local vector from A to B
rAC = c*[cos(gamma); sin(gamma)]; % local vector from A to C
rgA =    [  -xbar(1);    -xbar(2)]; % local vector from CM to A
rgB = rAB + rgA;                     % local vector from CM to B
rgC = rAC + rgA;                     % local vector from CM to C
```

Next, we compute the length from the point A to the center of mass, and the unit vector and normal associated with the vector $\mathbf{r}_{Ag}$.

```
LAg = norm(rgA);                     % length from A to CM
eAg = -(1/LAg)*rgA;                  % unit vector from A to CM
nAg = [-eAg(2); eAg(1)];             % unit normal of vector from A to CM
```

The `norm` function in MATLAB calculates the magnitude of a vector. Dividing a vector by its length gives a unit vector. Now calculate the normal vectors $\mathbf{s}'_{gA}$, $\mathbf{s}'_{gB}$, $\mathbf{s}'_{gC}$.

```
sgA = [-rgA(2); rgA(1)];    % local vector normal to rGA
sgB = [-rgB(2); rgB(1)];    % local vector normal to rGB
sgC = [-rgC(2); rgC(1)];    % local vector normal to rGC
```

Finally, define the rotation matrix and transform the vectors into the global coordinate system

```
R = [cos(theta) -sin(theta);      % rotation matrix
     sin(theta)  cos(theta)];

% transform to global coordinate system
sgA = R*sgA;    sgB = R*sgB;    sgC = R*sgC;
eAg = R*eAg;    nAg = R*nAg;
```

That's it! This was the most complicated function we have developed so far, but it will save us quite a bit of time and effort in performing force analysis. This function will work for two or three-pin linkages. If you encounter a four-pin linkage, you must extend the function to handle the fourth pin. Luckily, none of the linkages that we will study have more than three pins.

We now have all the tools we need to begin conducting force analysis on our family of linkages. We will begin with the humble threebar linkage, and conclude with the sixbar. We will perform a general force analysis for each linkage, and then conduct a case study of a specific, practical linkage for each linkage type. Because of the large number of quantities that must be computed, our force analysis programs will become rather long and complicated. If you keep in mind that each program consists of a set of small, discrete modules (position, velocity, acceleration, and force analysis) then the process will seem less intimidating.

7.3.2 Force Analysis of a Threebar Linkage using MATLAB®

We will now employ the techniques that we have developed to conduct a force analysis of the threebar linkage shown in Figure 7.28. For this simple example we will use arbitrary values for the masses and moments of inertia of the links. The next section will demonstrate using SOLIDWORKS to calculate these values. The crank and slider are assumed to be symmetric, so that their centers of mass lie at the geometric center of each link. Thus,

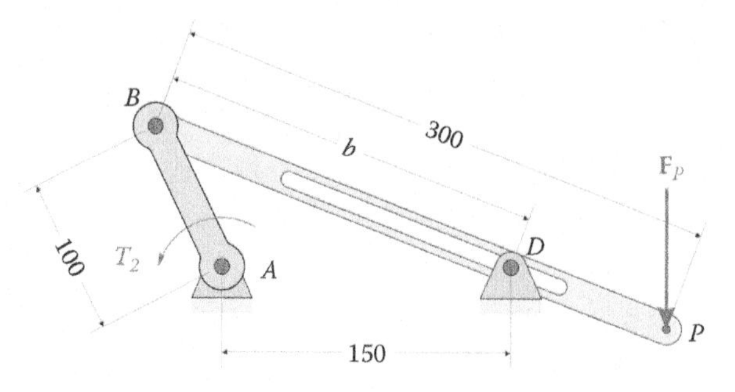

(All dimensions in millimeters)

Crank length: 100
Slider length: 300
Distance between ground pins: 150

Crank angular velocity: 10 rad/s
Crank angular acceleration: 0 rad/s²
Force at pin *P*: 100 N downward

FIGURE 7.28
The threebar linkage used in the example problem.

$$\bar{x}_2 = 50\text{ mm} \quad \bar{x}_3 = 150\text{ mm}$$

$$\bar{y}_2 = 0\text{ mm} \quad \bar{y}_3 = 0\text{ mm}$$

The (arbitrary) values for the mass and moment of inertia of each link are

$$m_2 = 0.1\text{ kg} \quad m_3 = 0.3\text{ kg}$$

$$I_2 = 0.0002\text{ kg}\cdot\text{m}^2 \quad I_3 = 0.0020\text{kg}\cdot\text{m}^2$$

The applied force is 100N vertically downward at the point *P*. The goal of the exercise is to find the driving torque, T_2, necessary to move the crank through one rotation and to find the resulting pin forces.

We are now ready to begin modifying our MATLAB code to perform the force analysis on the threebar linkage. Begin by using **Save As** to save a copy of your threebar acceleration analysis code. Add the code below to the program at the beginning.

```
% Threebar_Force_Analysis.m
% Conducts a force analysis on the threebar crank-slider linkage
% and calculates the driving torque and pin forces
% by Eric Constans, June 16, 2017

% Prepare Workspace
clear variables; close all; clc;

% Linkage dimensions
a = 0.100;              % crank length (m)
d = 0.150;              % length between ground pins (m)
p = 0.300;              % slider length (m)

% ground pins
x0 = [0;0];   % ground pin at A (origin)
xD = [d;0];   % ground pin at D
v0 = [0;0];   % velocity of origin
a0 = [0;0];   % accel of origin
Z2 = zeros(2); Z21 = zeros(2,1); Z12 = zeros(1,2); U2 = eye(2);
```

```
% CM locations
xbar2 = [0.050; 0];   % CM of crank from point A
xbar3 = [0.150; 0];   % CM of slider from point B
```

Here we have initialized the matrices `Z`$_2$, `Z`$_{21}$, `Z`$_{12}$, and `U`$_2$ for use in the force matrix later in the program. Next, define the inertial properties of each link:

```
% Inertial properties
m2 = 0.1;             % mass of crank (kg)
m3 = 0.3;             % mass of slider (kg)
I2 = 0.0002;          % moment of inertia of crank about CM (kg-m2)
I3 = 0.0020;          % moment of inertia of slider about CM (kg-m2)
```

For this example, the load applied at point P is 100 N downward, so we define the applied loads as

```
% Applied loads
FP = [0; -100];   % force at point P (N)
```

Next, allocate memory for the driving torque and pin forces, as well as the accelerations of the center of mass of each link.

```
% Angular velocity and acceleration of crank
omega2 = 10;          % angular velocity of crank (rad/sec)
alpha2 = 0;           % angular acceleration of crank (rad/sec2)

N = 361;   % number of times to perform position calculations
[xB,xP]         = deal(zeros(2,N)); % allocate space for positions
[vB,vP]         = deal(zeros(2,N)); % allocate space for velocities
[aB,aP,a2,a3]   = deal(zeros(2,N)); % allocate space for accelerations
[FA,FB,FC]      = deal(zeros(2,N)); % pin forces

[theta2,theta3]     = deal(zeros(1,N)); % link angles
[omega3,alpha3]     = deal(zeros(1,N)); % angular vel and accel
[b,bdot,bddot,T2]   = deal(zeros(1,N)); % b, bdot, bddot, driving torque
```

Of course, each force has an x and y component, but the torque has only a single component in the $\hat{k}$ direction. Inside the main loop, we use the `LinkCG` function to calculate the vectors used in the force analysis.

```
for i = 1:N
  theta2(i) = (i-1)*(2*pi)/(N-1);
  theta3(i) = atan2(-a*sin(theta2(i)),d - a*cos(theta2(i)));
  b(i) = (d - a*cos(theta2(i)))/cos(theta3(i));

% calculate unit vectors
  [e2,n2]   = UnitVector(theta2(i));
  [e3,n3]   = UnitVector(theta3(i));
  [eA2,nA2,LA2,s2A,s2B,~]   = LinkCG(   a,0,0,xbar2,theta2(i));
  [eB3,nB3,LB3,s3B,s3D,s3P] = LinkCG(b(i),p,0,xbar3,theta3(i));
```

The crank is a two-pin link, so we need only supply the length a and the location of the center of mass. The tilde (~) is used to indicate that we do not need the final return value from the `LinkCG` function, since there is no third pin. The slider is a three-pin link, with pins *B*, *D* and *P*. Since all three pins lie in a straight line, the internal angle γ is zero. The distance from pin *B* to pin *D* is given by the length *b* that is calculated in the position analysis. We next perform a velocity and acceleration analysis on the linkage, as usual, and we then calculate the acceleration of the center of mass of each link.

```
% conduct velocity analysis to solve for omega3 and omega4
  A_Mat = [b(i)*n3   e3];
  b_Vec = -a*omega2*n2;
  omega_Vec = A_Mat\b_Vec;  % solve for angular velocities

  omega3(i) = omega_Vec(1);     % decompose omega_Vec into
  bdot(i)   = omega_Vec(2);     % individual components

% calculate velocity at important points on linkage
  vB(:,i) = FindVel(      v0,    a,     omega2,  n2);
  vP(:,i) = FindVel(vB(:,i),   p, omega3(i),  n3);

% conduct acceleration analysis to solve for alpha3 and bddot
  ac = a*omega2^2;
  at = a*alpha2;
  bC = 2*bdot(i)*omega3(i);
  bc = b(i)*omega3(i)^2;

  C_Mat = A_Mat;
  d_Vec = -at*n2 + ac*e2 - bC*n3 + bc*e3;
  alpha_Vec = C_Mat\d_Vec;  % solve for angular accelerations

  alpha3(i) = alpha_Vec(1);
  bddot(i)  = alpha_Vec(2);

% calculate acceleration at important points on linkage
  aB(:,i) = FindAcc(      a0,   a,     omega2,    alpha2,   e2,  n2);
  aP(:,i) = FindAcc(aB(:,i),   p, omega3(i), alpha3(i),   e3,  n3);
  a2(:,i) = FindAcc(      a0, LA2,     omega2,    alpha2,  eA2, nA2);
  a3(:,i) = FindAcc(aB(:,i), LB3, omega3(i), alpha3(i),  eB3, nB3);
```

We now define the **S** matrix and **t** vector as

```
% Conduct force analysis
  S_Mat = [       U2          -U2          Z2    Z21;
                  Z2           U2         -U2    Z21;
                 s2A'        -s2B'        Z12      1;
                  Z12          s3B'      -s3D'     0;
                  Z12          Z12         e3'     0];

  t_Vec = [m2*a2(:,i);
           m3*a3(:,i) - FP;
           I2*alpha2;
           I3*alpha3(i) - dot(s3P,FP);
           0];
```

Remember that the prime operator (') gives the transpose of a matrix or vector. And finally, we solve for the pin forces and driving torque:

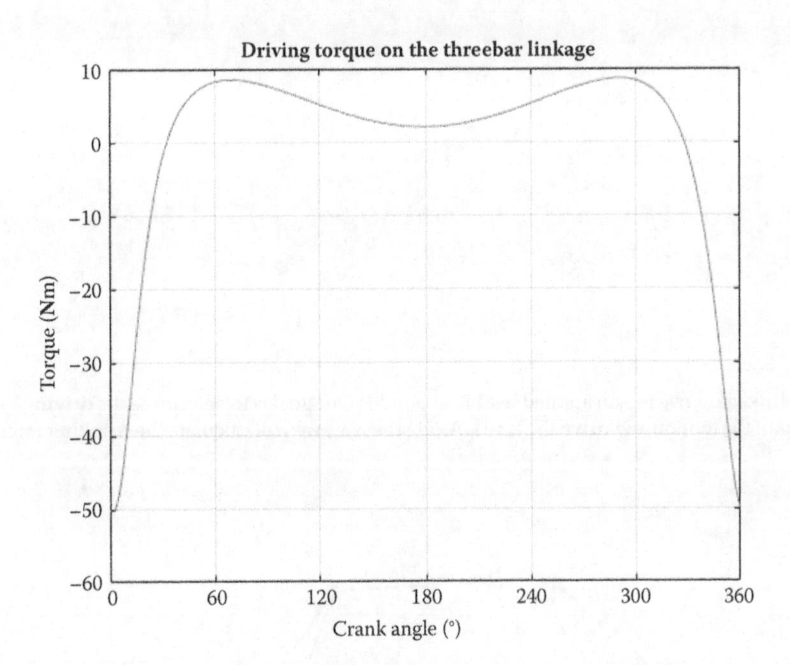

FIGURE 7.29
Driving torque for the example threebar linkage.

```
f_Vec = S_Mat\_Vec;
FA(:,i) = [f_Vec(1); f_Vec(2)];
FB(:,i) = [f_Vec(3); f_Vec(4)];
FD(:,i) = [f_Vec(5); f_Vec(6)];
T2(i)   = f_Vec(7);
```

The driving torque for the example linkage is shown in Figure 7.29.

7.4 Force Analysis of the Threebar Slider-Crank

We are now prepared to turn our attention to a more interesting problem: the force analysis of a threebar linkage. A general threebar linkage is shown in Figure 7.30. For this case, we assume that a known load ($\mathbf{F}_P$) is applied to point P on the linkage. The goal of our analysis is to find the motor torque, T_2, required to drive the crank through a rotation. Along the way we will solve for the forces at all of the pins. As before, we assume that a complete position, velocity, and acceleration analysis has been conducted earlier.

To perform the force analysis, we first draw a free-body diagram of each link, as shown in Figure 7.31. It is important to remember that the pin force at B is equal and opposite when applied to the crank and slider. For example, the force at pin B on the slider has been denoted $\mathbf{F}_B$, while the same force on the crank is denoted $-\mathbf{F}_B$. We have arranged the minus signs such that each link receives a "positive" and "negative" force – this will lend symmetry to the force and moment equations that will make them easy to scan for errors. This sign convention will, of course, produce the correct results as long as we remain consistent throughout the derivation. Summing forces and moments for the crank gives

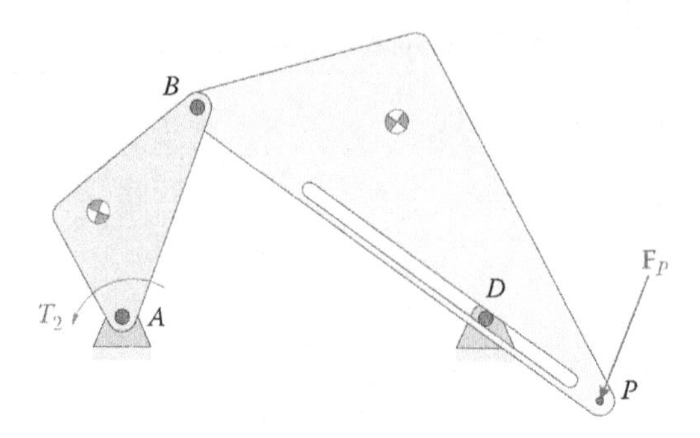

FIGURE 7.30
The threebar linkage above has an applied load $\mathbf{F}_P$ at point P. Our goal is to determine the driving torque, T_2, necessary to sustain the motion and drive the loads. Along the way, we will calculate the resulting forces at the pins.

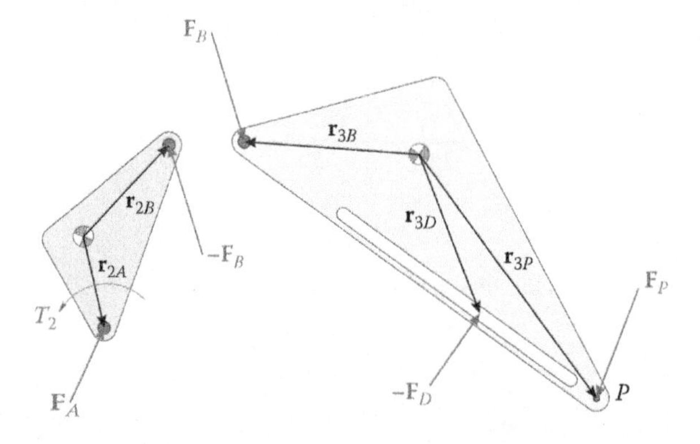

FIGURE 7.31
Free-body diagram of each link in the threebar linkage. Note the equal and opposite force placed at the pin joint B.

$$\mathbf{F}_A - \mathbf{F}_B = m_2\mathbf{a}_2 \tag{7.82}$$

$$\mathbf{r}_{2A} \times \mathbf{F}_A + \mathbf{r}_{2B} \times -\mathbf{F}_B + \mathbf{T}_2 = I_2\alpha_2 \tag{7.83}$$

Doing the same for the slider results in

$$\mathbf{F}_B - \mathbf{F}_D + \mathbf{F}_P = m_3\mathbf{a}_3 \tag{7.84}$$

$$\mathbf{r}_{3B} \times \mathbf{F}_B + \mathbf{r}_{3D} \times -\mathbf{F}_D + \mathbf{r}_{3P} \times \mathbf{F}_P = I_3\alpha_3 \tag{7.85}$$

Remember that the accelerations on the right-hand side of the equations are taken at the center of mass of each link, and the moments of inertia are also calculated about the center of mass. The g subscript has been dropped from each quantity to make the equations neater and more compact. We may use the vector triple product identity to simplify the moment equations in (7.83) and (7.85).

$$\mathbf{s}_{2A} \cdot \mathbf{F}_A - \mathbf{s}_{2B} \cdot \mathbf{F}_B + T_2 = I_2\alpha_2 \tag{7.86}$$

$$\mathbf{s}_{3B} \cdot \mathbf{F}_B - \mathbf{s}_{3D} \cdot \mathbf{F}_D + \mathbf{s}_{3P} \cdot \mathbf{F}_P = I_3\alpha_3 \tag{7.87}$$

If we rearrange the equations to place the unknowns on the left-hand side and the knowns on the right, we obtain

$$\begin{aligned}
\mathbf{F}_A - \mathbf{F}_B &= m_2\mathbf{a}_2 \\
\mathbf{F}_B - \mathbf{F}_D &= m_3\mathbf{a}_3 - \mathbf{F}_P \\
\mathbf{s}_{2A}\cdot\mathbf{F}_A - \mathbf{s}_{2B}\cdot\mathbf{F}_B + T_2 &= I_2\alpha_2 \\
\mathbf{s}_{3B}\cdot\mathbf{F}_B - \mathbf{s}_{3D}\cdot\mathbf{F}_D &= I_3\alpha_3 - \mathbf{s}_{3P}\cdot\mathbf{F}_P
\end{aligned} \tag{7.88}$$

These are six equations (remember that each force equation has an x and y component) but we have seven unknowns: the x and y components of $\mathbf{F}_A$, $\mathbf{F}_B$, $\mathbf{F}_D$ plus T_2, which is a scalar. We require one more equation to be able to solve for the system of forces and the driving torque.

Recall that a half-slider joint removes only one degree of freedom: the slider cannot move in the direction perpendicular to the slot. It follows that the force at pin D must be perpendicular to the slot at all times, and there can be no component of $\mathbf{F}_D$ that is parallel to the slot. The dot product of two vectors is zero if the vectors are orthogonal to each other, so that we can write

$$\mathbf{F}_D\cdot\mathbf{e}_3 = 0 \tag{7.89}$$

where $\mathbf{e}_3$ is the unit vector directed along the slot, which we have assumed runs between point B and point P. Equation (7.89) gives the seventh equation, and we are now ready to build our force matrix equation.

To place these equations into matrix form, we must first modify the dot product notation slightly. The reader may verify through direct computation that

$$\mathbf{s}\cdot\mathbf{F} = s_xF_x + s_yF_y = \begin{Bmatrix} s_x & s_y \end{Bmatrix}\begin{Bmatrix} F_x \\ F_y \end{Bmatrix} = \mathbf{s}^T\mathbf{F} \tag{7.90}$$

since the normal has been defined as a column vector. The matrix form of the force equations in its compact form is

$$\mathbf{Sf} = \mathbf{t}$$

where

$$\mathbf{S} = \begin{bmatrix} \mathbf{U}_2 & -\mathbf{U}_2 & \mathbf{0}_2 & \mathbf{0}_{21} \\ \mathbf{0}_2 & \mathbf{U}_2 & -\mathbf{U}_2 & \mathbf{0}_{21} \\ \mathbf{s}_{2A}^T & -\mathbf{s}_{2B}^T & \mathbf{0}_{12} & 1 \\ \mathbf{0}_{12} & \mathbf{s}_{3B}^T & -\mathbf{s}_{3D}^T & 0 \\ \mathbf{0}_{12} & \mathbf{0}_{12} & \mathbf{e}_3^T & 0 \end{bmatrix} \tag{7.91}$$

$$\mathbf{f} = \begin{Bmatrix} \mathbf{F}_A \\ \mathbf{F}_B \\ \mathbf{F}_D \\ T_2 \end{Bmatrix} \quad \mathbf{t} = \begin{Bmatrix} m_2\mathbf{a}_2 \\ m_3\mathbf{a}_3 - \mathbf{F}_P \\ I_2\alpha_2 \\ I_3\alpha_3 - \mathbf{s}_{3P}^T\mathbf{F}_P \\ 0 \end{Bmatrix} \tag{7.92}$$

where

$$\mathbf{U}_2 = \begin{bmatrix} 1 & 0 \\ 0 & 1 \end{bmatrix} \quad \mathbf{0}_2 = \begin{bmatrix} 0 & 0 \\ 0 & 0 \end{bmatrix} \quad \mathbf{0}_{12} = \begin{Bmatrix} 0 & 0 \end{Bmatrix} \quad \mathbf{0}_{21} = \begin{Bmatrix} 0 \\ 0 \end{Bmatrix} \tag{7.93}$$

Since each of the forces has both an x and y component, the **S** matrix has dimension 7×7, and the **f** and **t** vectors have dimension 7×1. We must take special care to ensure that the number of zeros in a given location keeps the dimension of the matrix correct. The reader may wish to write out the **S** matrix in its 7×7 glory to ensure understanding.

7.4.1 Code Verification

Verifying the code for the force analysis is quite a bit trickier than it was for acceleration and velocity analysis. For force analysis, there are no obvious derivatives to compute with the `Derivative _ Plot` routine. We will take two very different approaches to verifying the code: one static and the other dynamic.

7.4.1.1 Static Verification

If we consider the threebar mechanism as a single unit, we may draw a free-body diagram of the linkage as shown in Figure 7.32. The only external forces that support the linkage and keep the ground pins from moving are $\mathbf{F}_A$ and $\mathbf{F}_D$. Adding external forces on the linkage gives:

$$\mathbf{F}_P + \mathbf{F}_A - \mathbf{F}_D = m\mathbf{a} \tag{7.94}$$

where m is the mass of the entire linkage and **a** is the acceleration of the composite center of mass of the linkage. Let us assume that the angular velocity of the crank is very small so that we have, in effect, a static system. In this case, the acceleration of the composite center of mass is also very small, so that

$$\mathbf{F}_A - \mathbf{F}_D \approx -\mathbf{F}_P \tag{7.95}$$

Figure 7.33 shows a comparison of the sum of the vertical ground pin forces with the load applied at point P. The crank angular velocity was set to 0.001 rad/s, which makes the accelerations negligible. We could also arrive at this result by setting all of the masses and moments of inertia to zero temporarily.

7.4.1.2 Verifying the Code using the Energy Method

A second method for verifying the code uses the principle of Conservation of Energy. This method is more complicated than the static method, but can be used for non-negligible

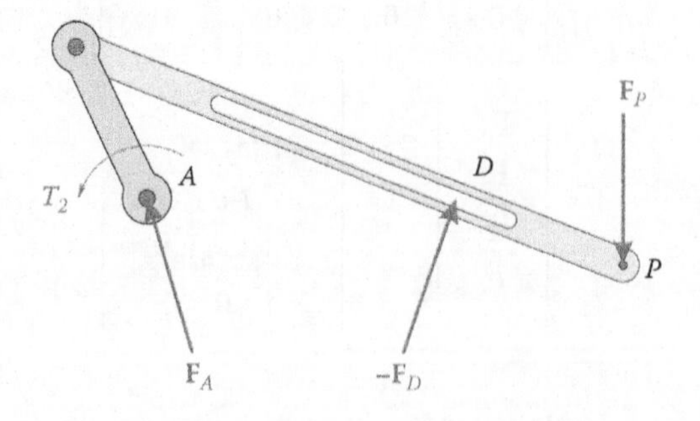

FIGURE 7.32
Free-body diagram of threebar mechanism as a unit. Here we are performing a static force analysis on the linkage by assuming that the crank angular velocity is very small.

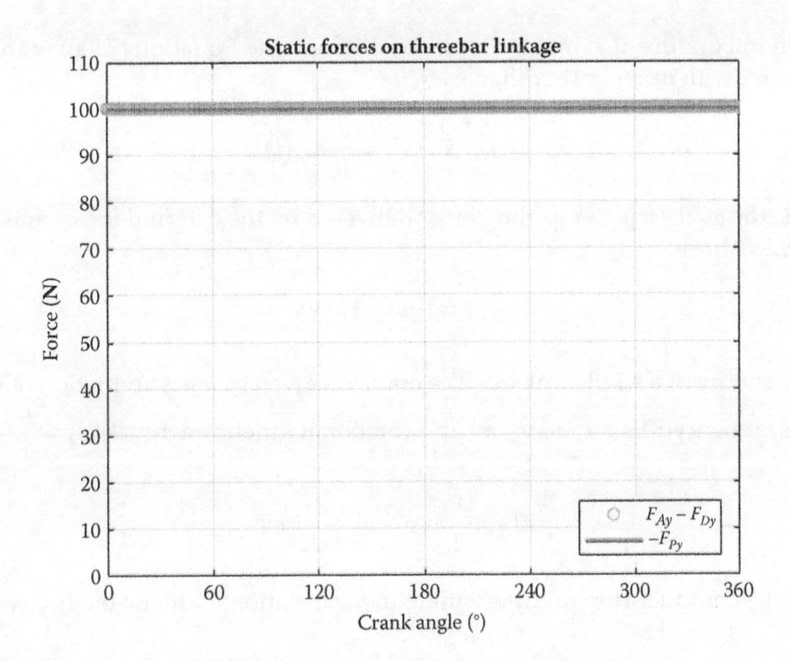

FIGURE 7.33
The sum of the vertical ground pin forces compared with the vertical load applied at point P for the example problem. The two curves are identical.

crank speeds. From the start of the simulation to an arbitrary time t, the work done by the external torques and external loads must be balanced by the total change in kinetic energy of the linkage.

$$W_{ext} = \Delta KE \tag{7.96}$$

The work done by the crank torque is

$$W_{T2} = \int_0^t T_2 \omega_2 \, dt \tag{7.97}$$

The work done by the external load is

$$W_{FP} = \int_0^t \mathbf{F}_P \cdot \mathbf{v}_P \, dt \tag{7.98}$$

while the change in kinetic energy is

$$\Delta KE = KE(t) - KE_0 \tag{7.99}$$

where $KE(t)$ is the total kinetic energy of the linkage at the current time and KE_0 is the kinetic energy at $t = 0$. Equating the external work with the change in kinetic energy gives:

$$\int_0^t T_2 \omega_2 \, dt + \int_0^t \mathbf{F}_P \cdot \mathbf{v}_P \, dt = KE(t) - KE_0 \tag{7.100}$$

Rather than integrating, it is much simpler to differentiate Equation (7.100) with respect to time and work with *power*, rather than *energy*.

$$T_2\omega_2 + \mathbf{F}_P \cdot \mathbf{v}_P = \frac{d}{dt}KE(t) \tag{7.101}$$

If we define the *external power* as the power delivered by the external forces and torques to the linkage, we have

$$P_{ext} = T_2\omega_2 + \mathbf{F}_P \cdot \mathbf{v}_P \tag{7.102}$$

The kinetic energy of a single link (e.g. the crank) is given by the summation of its translational kinetic energy $KE_{T2} = \frac{1}{2}m_2\mathbf{v}_2 \cdot \mathbf{v}_2$ and rotational kinetic energy $KE_{R2} = \frac{1}{2}I_2\omega_2^2$

$$KE_2 = \frac{1}{2}m_2\mathbf{v}_2 \cdot \mathbf{v}_2 + \frac{1}{2}I_2\omega_2^2 \tag{7.103}$$

We employ the product rule to differentiate the translational kinetic energy with respect to time

$$\frac{d}{dt}\left(\frac{1}{2}m_2\mathbf{v}_2 \cdot \mathbf{v}_2\right) = \frac{1}{2}m_2\mathbf{a}_2 \cdot \mathbf{v}_2 + \frac{1}{2}m_2\mathbf{v}_2 \cdot \mathbf{a}_2 = m_2\mathbf{a}_2 \cdot \mathbf{v}_2 \tag{7.104}$$

Since the rotational kinetic energy is expressed without vectors, we use the chain rule to differentiate

$$\frac{d}{dt}\left(\frac{1}{2}I_2\omega_2^2\right) = I_2\omega_2\frac{d}{dt}(\omega_2) = I_2\alpha_2\omega_2$$

The *inertial power* of the crank is then

$$P_2 = m_2\mathbf{a}_2 \cdot \mathbf{v}_2 + I_2\alpha_2\omega_2 \tag{7.105}$$

and the total inertial power of the linkage at a given time is

$$P_{kin} = \sum_i (m_i\mathbf{a}_i \cdot \mathbf{v}_i + I_i\alpha_i\omega_i) \tag{7.106}$$

Thus, the instantaneous power relation is

$$P_{ext} = P_{kin} \tag{7.107}$$

Since we can compute the inertial power of each link using the same equation, we will define a new function, `InertialPower` to perform the calculations.

```
% InertialPower.m
% Computes the total inertial power of a link
% m      = mass of link
% I      = moment of inertia of link
% v      = velocity of CM of link
```

```
% a     = acceleration of CM of link
% omega = angular velocity of link
% alpha = angular acceleration of link

function P = InertialPower(m, I, v, a, omega, alpha)

P = m*dot(v,a) + I*omega*alpha; % inertial power of link
```

After computing the forces and torques on the linkage, add the following lines of code just before the final end statement.

```
  P2 = InertialPower(m2,I2,v2(:,i),a2(:,i),    omega2,    alpha2);
  P3 = InertialPower(m3,I3,v3(:,i),a3(:,i),omega3(i),alpha3(i));
  PKin(i) = P2 + P3;      % Inertial power
  PF = dot(FP,vP(:,i));   % Power from external force FP
  PT = T2(i) * omega2;    % Power from driving torque T2
  PExt(i) = PF + PT;      % Total external power
```

Finally, add the following plotting commands to compare the computed kinetic power with the external power. Both plots should overlay each other exactly, regardless of crank speed. The power plot for the example is shown in Figure 7.34.

```
% plot the kinetic and external powers
plot(theta2*180/pi,PKin,'o','Color',[153/255 153/255 153/255])
hold on
plot(theta2*180/pi,PExt,'Color',[0 110/255 199/255])
legend('Kinetic','External','Location','Southeast')
```

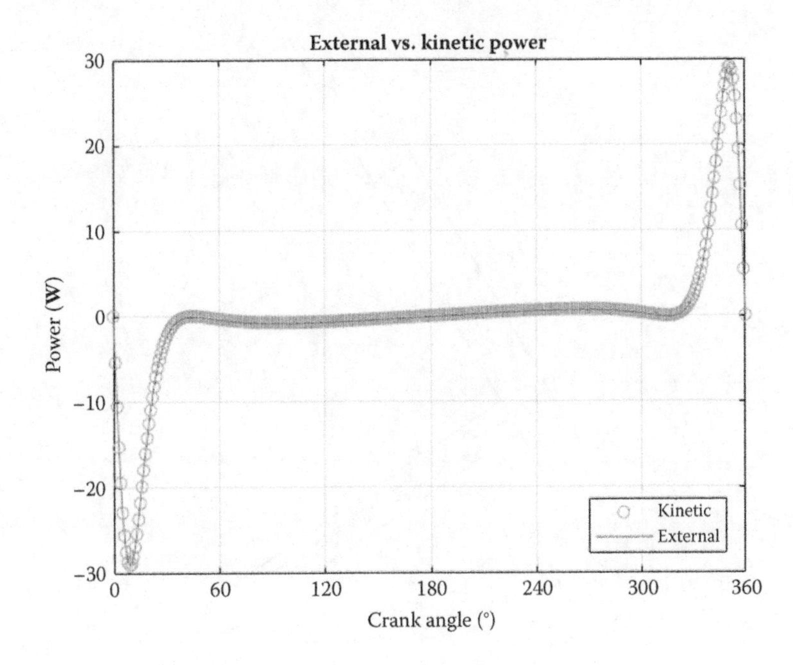

FIGURE 7.34
Comparison of external and kinetic power for the example problem. Both traces overlay each other exactly, and we conclude that the driving torque is being calculated correctly.

```
title('External vs. Kinetic Power')
xlabel('Crank angle (degrees)')
ylabel('Power (W)')
grid on
set(gca,'xtick',0:60:360)
xlim([0 360])
```

7.4.2 Summary

We have developed a method for conducting the force analysis on a simple linkage – the threebar. In addition, we have also demonstrated two methods for checking our calculations: one static and the other dynamic. Our next step will be to tackle a more realistic threebar problem, the door closing mechanism. Afterwards, we will conduct force analysis on more complicated mechanisms. Although the matrix equations will become more involved as we proceed, the underlying method of constructing free-body diagrams and summing forces and torques will remain the same.

7.5 Force Analysis Example 1 – The Threebar Door Closing Mechanism

As a first force analysis case study we choose a simple mechanism: the threebar door closer. Figure 7.35 shows a schematic of the mechanism. A heavy steel door pivots on its

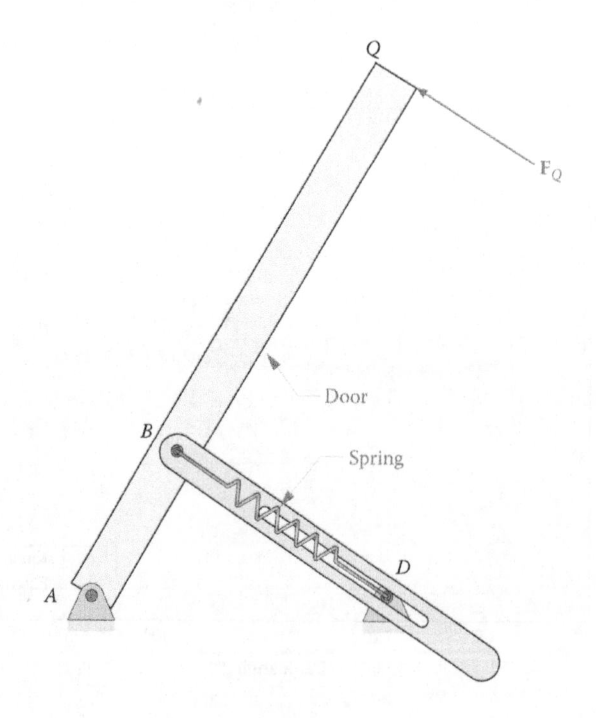

FIGURE 7.35
A threebar linkage has been selected for use as a door closing mechanism. The user pushes the door at its end with a force $\mathbf{F}_Q$ and a spring pulls the door closed when not in use.

hinges at point A. An extension spring is attached between points B and D. As the door is opened, the length of the spring increases, which exerts a tensile force between B and D. We assume that the spring is unstretched when the door is in its closed position. A person wishing to open the door pushes with a force $\mathbf{F}_Q$ at the end of the door, and $\mathbf{F}_Q$ is always perpendicular to the plane of the door.

Figure 7.36 shows the dimensions of the steel door, which is a heavy commercial unit used for fire protection – hence, the desire to keep it closed when not in use. Much of the force exerted by the user will go into accelerating the door from its resting position.

The dimensions of the door closing mechanism are shown in Figure 7.37. The spring has a constant of 1000 N/meter of stretch. That is, if we stretch the spring so that it is 1 m longer than its natural length then it will exert a tensile force of 1000 N.

Finally, Figure 7.38 shows the door closing mechanism in the closed position. We assume that the spring is unstretched in this position and exerts no tensile force. As seen in the figure, the unstretched length of the spring is

$$b_0 = d - a \tag{7.108}$$

Thus, we may compute the amount that the spring has stretched, x, as

$$x = b - b_0 \tag{7.109}$$

The force exerted by the spring is calculated using the spring law

$$|\mathbf{F}_k| = kx \tag{7.110}$$

where k is the spring constant (1000 N/m in this case).

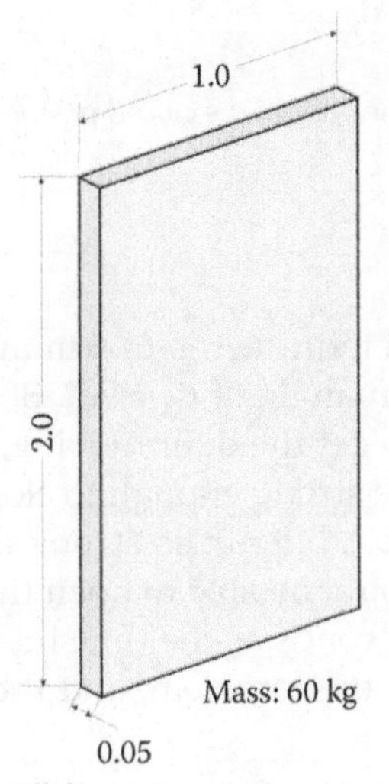

FIGURE 7.36
Dimensions of the door. The door is hollow with steel cladding and has a mass of 60 kg.

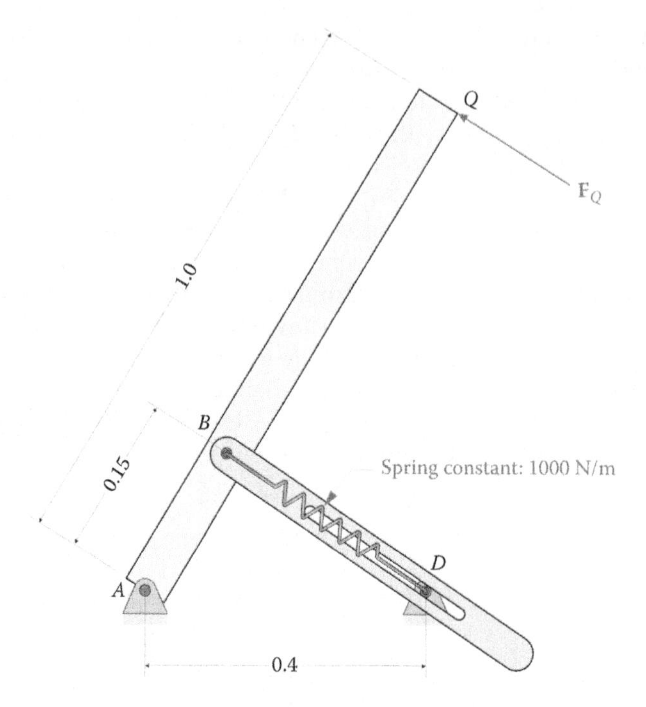

FIGURE 7.37
Dimensions of the door closing mechanism. The spring has a constant of 1000 N/m of stretch.

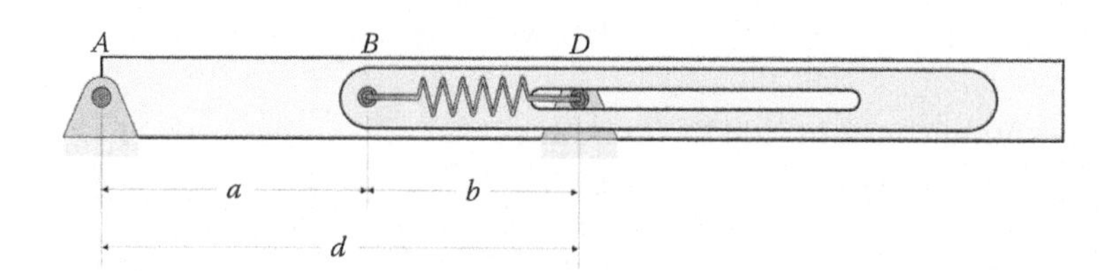

FIGURE 7.38
View of the door closer mechanism with the door in the closed position. The spring is at its natural, unstretched length in this position and exerts no force.

7.5.1 The Problem Statement

The goal of this case study is to determine the maximum amount of force, $\mathbf{F}_Q$, exerted by the user in opening the door to an angle of $\theta_2 = 90°$. Because the door is so massive, the user must exert a pushing force to get the door moving, and then a pulling force to get it to stop at 90°. The spring should be strong enough to close the door (and possibly to assist the user in stopping the door at 90°), but not so strong that it prevents the user from easily opening the door. The user should be able to open the door in a reasonable amount of time: for this study, we choose 2 seconds as the time interval between $\theta_2 = 0°$ and $\theta_2 = 90°$. We will follow a "recipe" to solve this problem, and the same recipe will be used in the remaining case studies.

1. Determine the critical dimensions of the linkage.
2. Calculate the inertial properties of each body in the mechanism.

3. Determine the nature of the external forces acting on the mechanism.
4. Draw free-body diagrams of each link in the mechanism showing internal and external forces and moments.
5. Determine the nature of the movement of crank. Does it move with constant angular velocity, or does it receive an acceleration pulse?
6. Use the matrix equation given in the previous section to solve for the pin forces and driving torque (or force) and plot the desired results.

Use **Save As** to save the threebar force analysis code to the file `ThreebarDoorCloser.m`. We will modify this code to conduct the force analysis on the door closer mechanism.

7.5.1.1 Critical Dimensions of the Linkage

There are only two important dimensions for this linkage, the crank length and distance between ground pins. We will also need the dimensions of the door for calculating its moment of inertia, and the unstretched length of the spring will come in handy as well. We have chosen an arbitrary total length of the slider to be 0.3 m; this is only used to find the location of its center of mass. Modify the top of your code as follows:

```
% Threebar_Force_Analysis_DoorCloser.m
% Conducts a force analysis on the threebar crank-slider linkage
% used in the door-closer example
% and calculates the driving force and pin forces
% by Eric Constans, June 20, 2017

% Prepare Workspace
clear variables; close all; clc;

% Linkage dimensions
a = 0.150;              % crank length (m)
d = 0.400;              % length between ground pins (m)
p = 0.300;              % total length of slider (m)
H = 2.0;                % height of door (m)
W = 1.0;                % width of door (m)
L = 0.050;              % thickness of door (m)
b0 = d - a;             % unstretched spring length
k = 1000;               % spring constant (N/m)

% ground pins
x0 = [0;0];   % ground pin at A (origin)
xD = [d;0];   % ground pin at D
v0 = [0;0];   % velocity of origin
a0 = [0;0];   % accel of origin
Z2 = zeros(2); Z21 = zeros(2,1); Z12 = zeros(1,2); U2 = eye(2);
```

7.5.1.2 Inertial Properties of the Mechanism

There are two moving bodies in the mechanism: the door and the slider. The slider is a relatively small piece of stamped steel, and its mass and moment of inertia may be neglected relative to that of the door. The door is a rectangular parallelpiped with mass 60 kg. From the tables in the Appendix we see that the centroidal moment of inertia of the door is

$$I_2 = \frac{m}{12}\left(W^2 + L^2\right) \tag{7.111}$$

It is easiest to have MATLAB perform this calculation, especially if we wish to use the code to model doors with different dimensions.

```
% Inertial properties
m2 = 60;                    % mass of crank (kg)
m3 = 0.0;                   % mass of slider (kg)
I2 = (m2/12)*(W^2+L^2);     % moment of inertia of crank about CM (kg-m2)
I3 = 0.0;                   % moment of inertia of slider about CM (kg-m2)

% CM locations
xbar2 = [W/2; 0];  % CM of crank from point A
xbar3 = [p/2; 0];  % CM of slider from point B
```

Since we have a force, $\mathbf{F}_Q$, acting at the end of the crank we must "change" the crank to a three-pin link using our `LinkCG` function:

```
  [eA2,nA2,LA2,s2A,s2B,s2Q] = LinkCG(  a,W,0,xbar2,theta2(i));
```

7.5.1.3 External Forces Acting on the Mechanism

Figure 7.39 shows the external forces acting on the bodies of the mechanism. The force exerted by the user, $\mathbf{F}_Q$, creates a torque on the door given by

$$T_2 = \mathbf{r}_Q \times \mathbf{F}_Q \cdot \hat{k} = \mathbf{s}_{2Q} \cdot \mathbf{F}_Q \tag{7.112}$$

where $r_Q = W\mathbf{e}_2$ and W is the width of the door – the length from A to Q. The spring force acts in the direction of the slider such that

$$\mathbf{F}_k = k\left(b - b_0\right)\mathbf{e}_3 \tag{7.113}$$

The spring exerts a force of equal magnitude and opposite direction on pin D, but this is not important for our analysis since pin D is grounded.

7.5.1.4 Free-Body Diagrams of each Link in the Mechanism

The free-body diagrams of the door and slider can be seen in Figure 7.40. They are identical to the diagrams of the threebar linkage from the previous section with the exception that the driving torque, T_2, has been replaced by an unknown driving force, $\mathbf{F}_Q$, and a spring force has been added. The force equation for the door is

$$\mathbf{F}_A - \mathbf{F}_B + \mathbf{F}_Q = m_2\mathbf{a}_2 \tag{7.114}$$

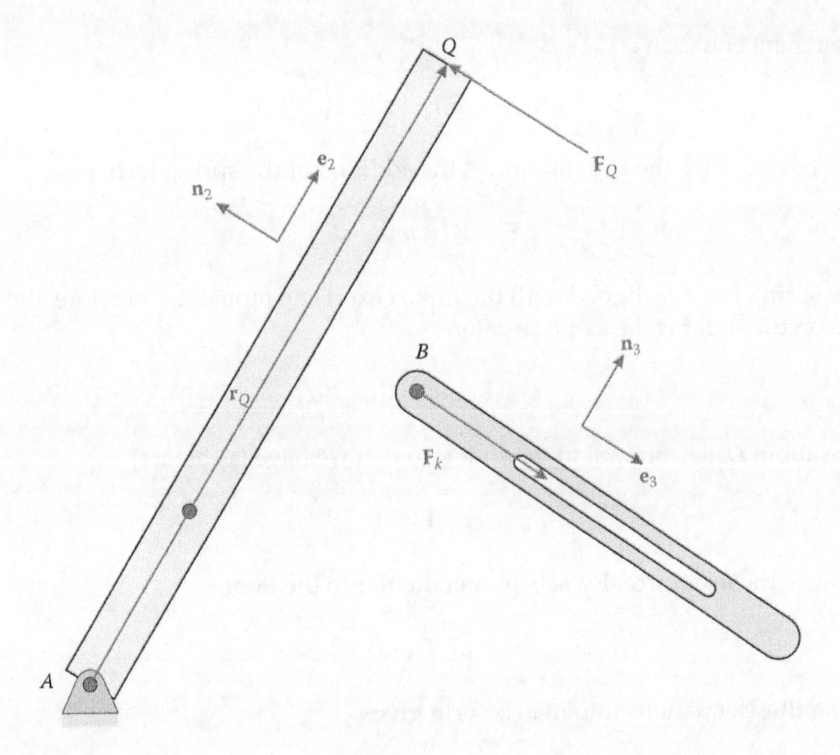

FIGURE 7.39
External forces acting on the door closer mechanism.

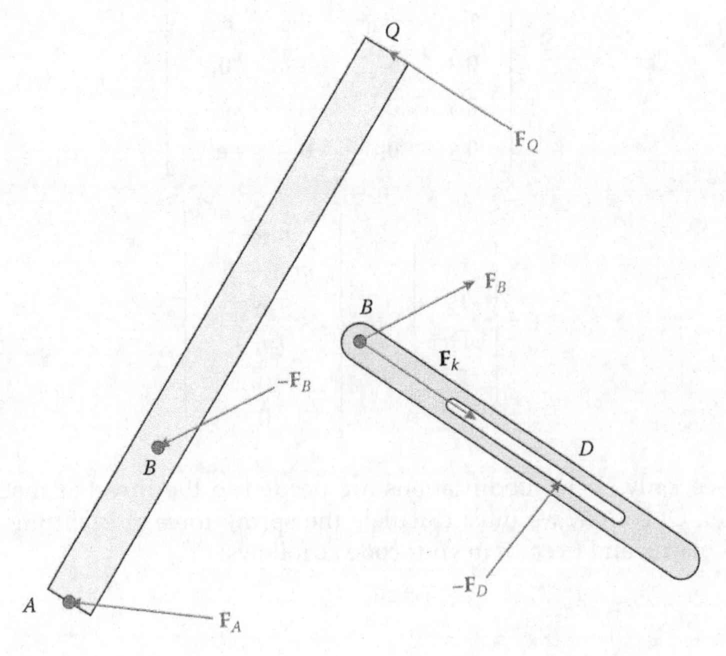

FIGURE 7.40
Free-body diagrams of the door and slider.

and its moment equation is

$$\mathbf{s}_{2A} \cdot \mathbf{F}_A - \mathbf{s}_{2B} \cdot \mathbf{F}_B + \mathbf{s}_{2Q} \cdot \mathbf{F}_Q = I_2\alpha_2 \tag{7.115}$$

The force equation for the slider requires the addition of the spring force

$$\mathbf{F}_B - \mathbf{F}_D = m_3\mathbf{a}_3 - \mathbf{F}_k \tag{7.116}$$

Since the spring force is aligned with the link, it exerts no moment. Therefore, the moment equation on the slider is the same as before.

$$\mathbf{s}_{3B} \cdot \mathbf{F}_B - \mathbf{s}_{3D} \cdot \mathbf{F}_D = I_3\alpha_3 \tag{7.117}$$

The force at pin D acts normal to the slider, so that we have (as before)

$$\mathbf{e}_3 \cdot \mathbf{F}_D = 0 \tag{7.118}$$

And because the user force, $\mathbf{F}_Q$, acts perpendicular to the door

$$\mathbf{e}_2 \cdot \mathbf{F}_Q = 0 \tag{7.119}$$

Arranging these equations into matrix form gives

$$\mathbf{S} = \begin{bmatrix} \mathbf{U}_2 & -\mathbf{U}_2 & \mathbf{0}_2 & \mathbf{U}_2 \\ \mathbf{0}_2 & \mathbf{U}_2 & -\mathbf{U}_2 & \mathbf{0}_2 \\ \mathbf{s}_{2A}^T & -\mathbf{s}_{2B}^T & \mathbf{0}_{12} & \mathbf{s}_{2Q}^T \\ \mathbf{0}_{12} & \mathbf{s}_{3B}^T & -\mathbf{s}_{3D}^T & \mathbf{0}_{12} \\ \mathbf{0}_{12} & \mathbf{0}_{12} & \mathbf{e}_3^T & \mathbf{0}_{12} \\ \mathbf{0}_{12} & \mathbf{0}_{12} & \mathbf{0}_{12} & \mathbf{e}_2^T \end{bmatrix} \tag{7.120}$$

$$\mathbf{f} = \begin{Bmatrix} \mathbf{F}_A \\ \mathbf{F}_B \\ \mathbf{F}_D \\ \mathbf{F}_Q \end{Bmatrix} \quad \mathbf{t} = \begin{Bmatrix} m_2\mathbf{a}_2 \\ m_3\mathbf{a}_3 - \mathbf{F}_k \\ I_2\alpha_2 \\ I_3\alpha_3 \\ 0 \\ 0 \end{Bmatrix} \tag{7.121}$$

As you can see, only slight modifications are needed to the threebar matrix equations derived earlier. Of course, we must calculate the spring force at each time step as well. Modify the **S** matrix and **t** vector in your code as follows:

```
% Conduct force analysis
  Fk = k*(b(i) - b0)*e3;
  S_Mat = [        U2          -U2          Z2       U2;
                   Z2           U2         -U2       Z2;
                 s2A'         -s2B'        Z12     s2Q';
```

```
        Z12         s3B'        -s3C'       Z12;
        Z12         Z12          e3'        Z12;
        Z12         Z12         Z12          e2'];

t_Vec = [m2*a2(:,i);
         m3*a3(:,i) - Fk;
         I2*alpha2(i);
         I3*alpha3(i);
         0;
         0];
```

7.5.1.5 Motion of the Crank

We wish the crank to start from a resting state at $\theta_2 = 0°$ and end in a resting state at $\theta_2 = 90°$, so the angular velocity of the crank will be non-constant. Let us use the technique of the angular acceleration pulse developed earlier in Section 6.3. The opening of the door lasts 2 seconds, so that

$$T = 2\text{ s} \tag{7.122}$$

If the angular acceleration pulse starts and ends at zero, a suitable function is

$$\alpha_2 = A\sin\left(\frac{2\pi}{T}t\right) \tag{7.123}$$

where A is the (unknown) amplitude of the acceleration pulse. The reader can easily confirm that this takes on the value of zero for $t = 0$ sec and $t = T$. Let us define the constant $\lambda = 2\pi/T$ such that

$$\alpha_2 = A\sin\lambda t \tag{7.124}$$

Integrate this once to find the angular velocity

$$\omega_2 = -\frac{A}{\lambda}\cos\lambda t + C_1 \tag{7.125}$$

We require that the door be at rest when $t = T$, so that

$$C_1 = \frac{A}{\lambda} \tag{7.126}$$

Integrate again to find the crank angle, θ_2

$$\theta_2 = -\frac{A}{\lambda^2}\sin\lambda t + \frac{A}{\lambda}t + C_2 \tag{7.127}$$

Since the crank angle begins at zero, $C_2 = 0$. The crank angle is $\pi/2$ at $t = T$ so that

$$A = \frac{\lambda\pi}{2T} = \frac{\lambda^2}{4} \tag{7.128}$$

Thus, the crank angle function is

$$\theta_2 = \frac{1}{4}(\lambda t - \sin \lambda t)$$

$$\omega_2 = \frac{\lambda}{4}(1 - \cos \lambda t) \tag{7.129}$$

$$\alpha_2 = \frac{\lambda^2}{4}\sin \lambda t$$

Immediately after defining the center of mass locations, enter the following (constant) parameters for the acceleration pulse.

```
% Acceleration pulse parameters
T = 2;                % time it takes to open door
lambda = 2*pi/T;      % door opening "frequency"
N = 1001;             % number of time steps
dt = T/(N-1);         % time increment
t = 0:dt:T;           % vector of simulation time
```

The crank angle, angular velocity, and angular acceleration must also be changed within the main loop.

```
  theta2(i) = 0.25*(lambda*t(i) - sin(lambda*t(i)));
  omega2(i) = 0.25*lambda*(1-cos(lambda*t(i)));
  alpha2(i) = 0.25*lambda^2*sin(lambda*t(i));
```

7.5.1.6 Solving for the Pin Forces and Plotting Results

Solving the force matrix equations proceeds in the same manner as in the previous section; however, you will need to update your code to "retrieve" values from your solution and assign them to the appropriate variables.

```
  f_Vec = S_Mat\_Vec;
  FA(:,i) = [f_Vec(1); f_Vec(2)];
  FB(:,i) = [f_Vec(3); f_Vec(4)];
  FD(:,i) = [f_Vec(5); f_Vec(6)];
  FQ(:,i) = [f_Vec(7); f_Vec(8)];
```

A plot of the force exerted by the user is shown in Figure 7.41. The force is relatively large (50 N) at the beginning of the motion when the user must accelerate the door. Less force is required to decelerate the door because of the assistance of the spring. The force at the end of the motion (–25 N) is needed to overcome the tensile force of the spring that is trying to close the door. A 50 N force may be considered excessive for opening a door, and it is likely that the requirement of opening such a heavy door in 2 seconds is too severe. Increasing the length of time for opening the door to 4 secs reduces the user force to a more manageable 12 N. Similarly, the force required to hold open the door at 90° would probably defeat any doorstop device (e.g. a rubber wedge) used to hold it open. Cutting the spring constant in half reduces this force by half, and may be a better design. As with all mechanical design problems, there is no single "right" answer – the judgment of the engineer plays the most critical role. Now that we have developed a

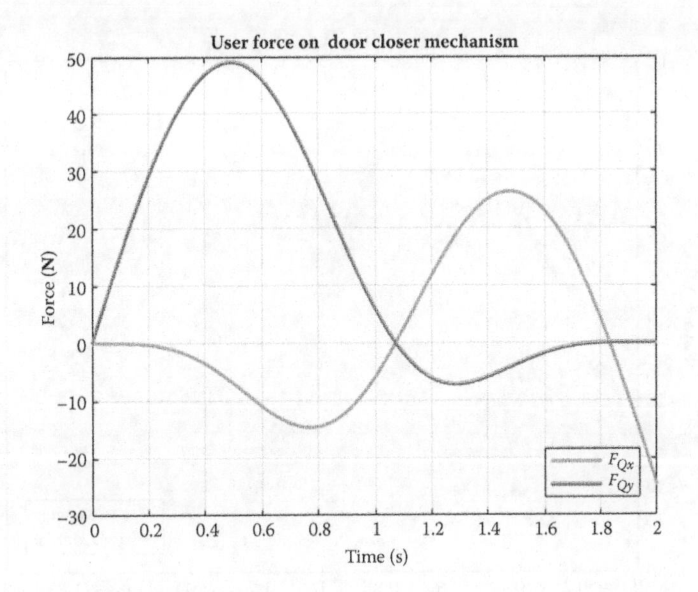

FIGURE 7.41
Force exerted by the user in opening the door. The largest force occurs near the beginning of the motion when the user must accelerate the door and the force at the end of the motion is required to overcome the spring.

force analysis code for the mechanism, it is easy to tweak the design until a feasible solution has been found.

7.5.2 Verification of the Code

We can use the conservation of power to verify the code as we did in the previous section. There are two external forces acting on the door: $\mathbf{F}_Q$ and $\mathbf{F}_k$.

$$P_Q = \mathbf{F}_Q \cdot \mathbf{v}_Q \quad P_k = \mathbf{F}_k \cdot \mathbf{v}_B \tag{7.130}$$

The velocity at point B has already been found in the motion analysis of the linkage, but we must add a line of code to find the velocity at point Q.

```
vQ(:,i) = FindVel(      v0,    W, omega2(i),  n2);
```

The external power is then

```
Pk = dot(Fk,vB(:,i));
PQ = dot(FQ(:,i),vQ(:,i));
PExt(i) = PQ + Pk;
```

A plot of the kinetic and external power is shown in Figure 7.42. Since the two curves match exactly, we can have some confidence that the code is producing accurate results.

A second method for verification makes use of the fact that the spring stores potential energy as it stretches. The potential energy stored in a spring is given by

$$PE_k = \frac{1}{2}kx^2 \tag{7.131}$$

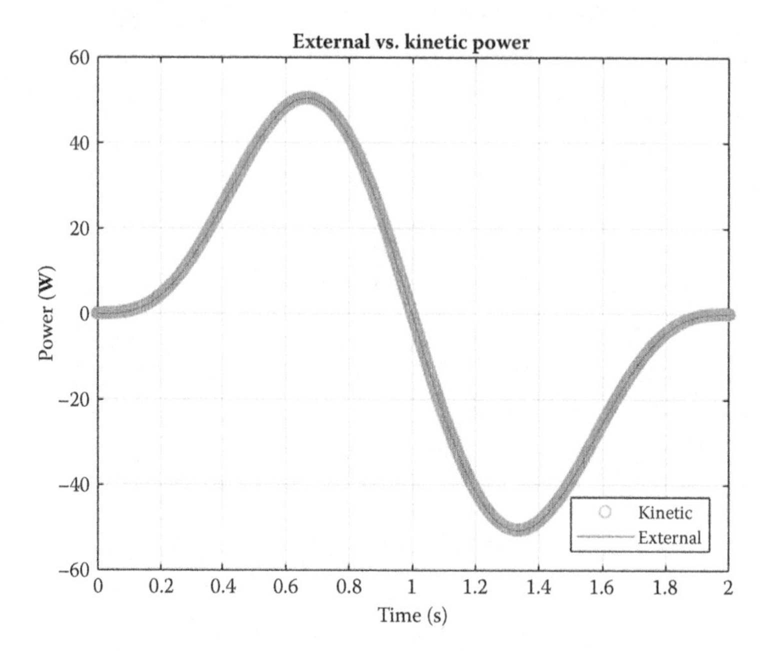

FIGURE 7.42
Kinetic and external power for the door closer mechanism. The two curves are identical.

where x is the amount that the spring has stretched from its neutral position, given by Equation (7.109)

$$x = b - b_0 \tag{7.132}$$

To convert this to a power, take the derivative with respect to time

$$\begin{aligned} P_k &= \frac{d}{dt}\left(\frac{1}{2}k(b-b_0)^2\right) \\ &= k(b-b_0)\cdot\frac{d}{dt}(b-b_0) \\ &= k(b-b_0)\dot{b} \end{aligned} \tag{7.133}$$

where we have again used the chain rule to perform the differentiation. Since the potential energy of the spring is internal to the system, it should be added to the kinetic power and the power caused by the spring force should be removed.

```
P2 = InertialPower(m2,I2,v2(:,i),a2(:,i),omega2(i),alpha2(i));
P3 = InertialPower(m3,I3,v3(:,i),a3(:,i),omega3(i),alpha3(i));
Pk = k*bdot(i)*(b(i)-b0);
PKin(i) = P2 + P3 + Pk;

 PQ = dot(FQ(:,i),vQ(:,i));
PExt(i) = PQ;
```

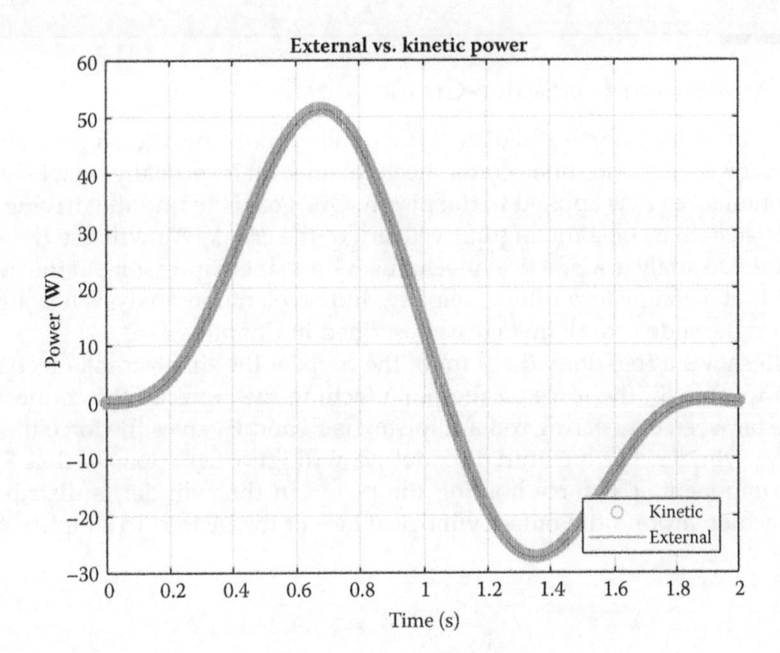

FIGURE 7.43
Kinetic and external power for the door closer mechanism. Here we have added the power from the spring potential energy to the kinetic power and removed it from the external power.

The resulting power plot is shown in Figure 7.43. It is slightly different from the plot in Figure 7.42 since we have added the spring power to the kinetic power (and removed it from the external power).

7.5.3 Summary

We have used a simple door closer mechanism to demonstrate a general technique for performing the force analysis on a practical mechanism. Some of the mechanisms in the sections that follow will be rather complicated, and it is easy to become overwhelmed with the code – especially when it can run to 200 lines or more! It is therefore a good idea to have a systematic procedure for conducting the analysis, one that will work for any mechanism. The steps we took in analyzing the door closer are repeated in Table 7.1. Review the steps and make sure you understand each one before proceeding to the more complicated mechanisms.

TABLE 7.1

General Procedure for Force Analysis of a Mechanism

1. Determine the critical dimensions of the linkage.
2. Calculate the inertial properties of each body in the linkage.
3. Determine the magnitude and direction of the external forces acting on the linkage.
4. Draw free-body diagrams of each link showing internal and external forces and moments.
5. Determine the nature of the movement of crank. Does it move with constant angular velocity, or does it receive an acceleration pulse?
6. Use the matrix equation to solve for the pin forces and driving torque (or force) and plot the desired results.

7.6 Force Analysis of the Slider-Crank

We are now in a position to calculate forces and torques on the slider-crank linkage. Figure 7.44 shows a generic slider-crank mechanism with a vertically offset slider. A constant horizontal force $\mathbf{F}_P$ is applied to the piston. Our goal is to find the driving torque, T_2, necessary to sustain a constant angular velocity of the crank. We will use the techniques developed here to analyze a practical mechanism – an air compressor – in the next section. We assume that a complete position, velocity, and acceleration analysis has already been performed on the slider-crank linkage as described in Chapters 4–6.

Figure 7.45 shows a free-body diagram of the parts of the slider-crank mechanism. The ground pin force is $\mathbf{F}_A$, the force on the pin attaching the crank to the connecting rod is $\mathbf{F}_B$, the force between connecting rod and piston is $\mathbf{F}_C$ and $\mathbf{F}_D$ gives the force that holds the piston in the cylinder. For our initial model we will ignore friction, so that $\mathbf{F}_D$ has only a vertical component. The force holding the piston in the cylinder is distributed in an unknown fashion around the outer cylindrical face of the piston. To simplify matters we

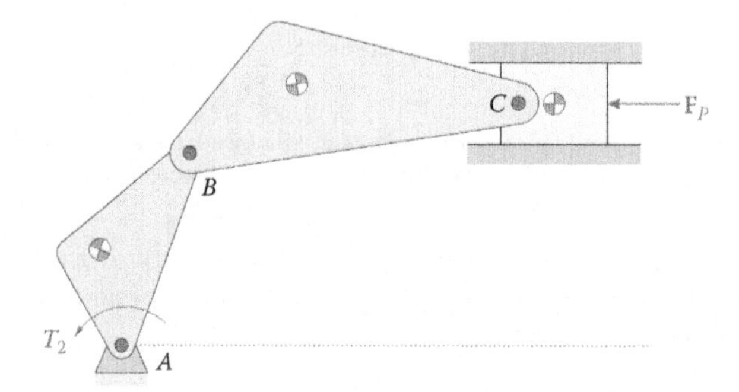

FIGURE 7.44
The slider-crank mechanism has a horizontal force, $\mathbf{F}_P$, applied to the piston. Our goal is to find the driving torque, T_2, necessary to sustain a constant crank angular velocity, ω_2.

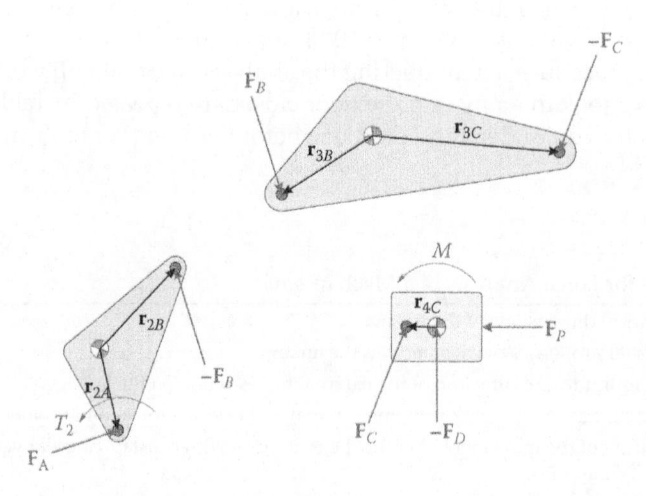

FIGURE 7.45
Free-body diagram of the parts of the slider-crank linkage.

will assume that the resultant force is applied directly to the center of mass of the piston. The cylinder also exerts an unknown moment, M, on the piston which keeps it from rotating.

Summing forces and torques on the crank is the same as for the threebar linkage

$$\mathbf{F}_A - \mathbf{F}_B = m_2\mathbf{a}_2 \tag{7.134}$$

$$\mathbf{s}_{2A} \cdot \mathbf{F}_A - \mathbf{s}_{2B} \cdot \mathbf{F}_B + T_2 = I_2\alpha_2 \tag{7.135}$$

The connecting rod forces and torques are similar to those for the threebar slider, minus the applied force at point P.

$$\mathbf{F}_B - \mathbf{F}_C = m_3\mathbf{a}_3 \tag{7.136}$$

$$\mathbf{s}_{3B} \cdot \mathbf{F}_B - \mathbf{s}_{3C} \cdot \mathbf{F}_C = I_3\alpha_3 \tag{7.137}$$

Finally, we sum forces on the piston.

$$\mathbf{F}_C - \mathbf{F}_D + \mathbf{F}_P = m_4\mathbf{a}_4 \tag{7.138}$$

To sum moments, we note that the cylinder force, $\mathbf{F}_D$, acts through the center of mass and creates no moment. The external force, $\mathbf{F}_P$, also acts through the center of mass. Since the piston cannot rotate, its angular acceleration is zero.

$$\mathbf{s}_{4C} \cdot \mathbf{F}_C + M = 0 \tag{7.139}$$

Since we have assumed that there is no friction, the horizontal force on the piston, $\mathbf{F}_{Dx}$, is zero. We could eliminate $\mathbf{F}_{Dx}$ from the list of unknowns, but we will retain it for now. When we model friction in the cylinder in a later section $\mathbf{F}_{Dx}$ will not be zero.

So far, we have ten unknowns ($\mathbf{F}_A$, $\mathbf{F}_B$, $\mathbf{F}_C$, $\mathbf{F}_D$, T_2, and M) but only nine equations. If we stipulate that the cylinder force, $\mathbf{F}_D$, act only in the vertical direction then we have

$$\mathbf{e}_1 \cdot \mathbf{F}_D = 0 \tag{7.140}$$

This provides the tenth equation. Summarizing the equations that we have developed so far, and moving the known quantities to the right-hand sides, gives

$$
\begin{aligned}
&\mathbf{F}_A - \mathbf{F}_B = m_2\mathbf{a}_2 \\
&\mathbf{F}_B - \mathbf{F}_C = m_3\mathbf{a}_3 \\
&\mathbf{F}_C - \mathbf{F}_D = m_4\mathbf{a}_4 - \mathbf{F}_P \\
&\mathbf{s}_{2A} \cdot \mathbf{F}_A - \mathbf{s}_{2B} \cdot \mathbf{F}_B + \mathrm{T}_2 = \mathrm{I}_2\alpha_2 \\
&\mathbf{s}_{3B} \cdot \mathbf{F}_B - \mathbf{s}_{3C} \cdot \mathbf{F}_C = \mathrm{I}_3\alpha_3 \\
&\mathbf{s}_{4C} \cdot \mathbf{F}_C + M = 0 \\
&\mathbf{e}_1 \cdot \mathbf{F}_D = 0
\end{aligned}
\tag{7.141}
$$

Writing Equation (7.141) in matrix form gives

$$\mathbf{S} = \begin{bmatrix} \mathbf{U}_2 & -\mathbf{U}_2 & \mathbf{0}_2 & \mathbf{0}_2 & \mathbf{0}_{21} & \mathbf{0}_{21} \\ \mathbf{0}_2 & \mathbf{U}_2 & -\mathbf{0}_2 & \mathbf{0}_2 & \mathbf{0}_{21} & \mathbf{0}_{21} \\ \mathbf{0}_2 & \mathbf{0}_2 & \mathbf{U}_2 & -\mathbf{U}_2 & \mathbf{0}_{21} & \mathbf{0}_{21} \\ \mathbf{s}_{2A}^T & -\mathbf{s}_{2B}^T & \mathbf{0}_{12} & \mathbf{0}_{12} & 0 & 1 \\ \mathbf{0}_{12} & \mathbf{s}_{3B}^T & -\mathbf{s}_{3C}^T & \mathbf{0}_{12} & 0 & 0 \\ \mathbf{0}_{12} & \mathbf{0}_{12} & \mathbf{s}_{4C} & \mathbf{0}_{12} & 1 & 0 \\ \mathbf{0}_{12} & \mathbf{0}_{12} & \mathbf{0}_{12} & \mathbf{e}_1^T & 0 & 0 \end{bmatrix} \tag{7.142}$$

$$\mathbf{f} = \begin{Bmatrix} \mathbf{F}_A \\ \mathbf{F}_B \\ \mathbf{F}_C \\ \mathbf{F}_D \\ M \\ T_2 \end{Bmatrix} \quad \mathbf{t} = \begin{Bmatrix} m_2\mathbf{a}_2 \\ m_3\mathbf{a}_3 \\ m_4\mathbf{a}_4 - \mathbf{F}_P \\ I_2\alpha_2 \\ I_3\alpha_3 \\ 0 \\ 0 \end{Bmatrix} \tag{7.143}$$

This matrix equation is remarkably similar to the one for the threebar linkage. There are 10 unknowns and the **S** matrix has dimension 10×10.

7.6.1 Force Analysis of the Example Linkage

Let us now add some specific numbers to our example slider-crank. The dimensions of the example linkage are shown in Figure 7.46. Here we will assume that the crank is a steel disk with diameter 100 mm and thickness 25 mm. The length from ground pin to point *B* is 40 mm. The connecting rod length is 120 mm. Starting with the slider-crank acceleration code, enter the following dimensions at the top of the file.

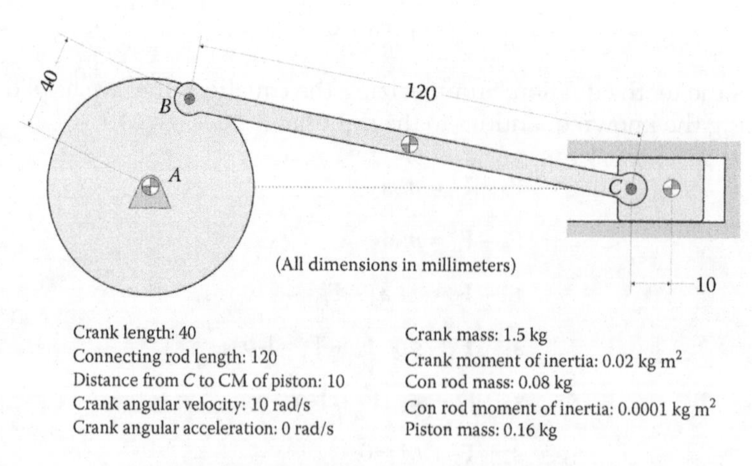

FIGURE 7.46
Dimensions of the example slider-crank linkage.

```
% Linkage dimensions
a = 0.040;              % crank length (m)
b = 0.120;              % connecting rod length (m)
c = 0.0;                % vertical slider offset (m)
p = 0.01;               % distance from point C to CM of piston (m)
```

Next, enter the inertial properties of the crank, connecting rod, and piston.

```
% Inertial properties
m2 = 1.5;               % mass of crank (kg)
m3 = 0.08;              % mass of con rod (kg)
m4 = 0.16;              % mass of piston (kg)
I2 = 0.002;             % moment of inertia of crank about CM (kg-m2)
I3 = 0.0001;            % moment of inertia of con rod about CM (kg-m2)

% CM locations
xbar2 = [   0; 0]; % CM of crank
xbar3 = [ b/2; 0]; % CM of coupler
xbar4 = [   p; 0]; % CM of piston
```

Much of the remaining code can be copied directly from the threebar-force analysis program, although the force analysis matrices must be modified appropriately. A full listing of the code is given at the end of this section, if you get stuck.

Figure 7.47 shows a plot of the driving torque for the example mechanism. It looks surprisingly like a sine wave, but isn't really. To prove this, try increasing the crank speed to 100 rad/s. You should obtain the decidedly non-sinusoidal plot shown in Figure 7.48.

This is a good time to conduct the conservation of power check discussed in the previous section. The driving torque is the only source of external power, and we must

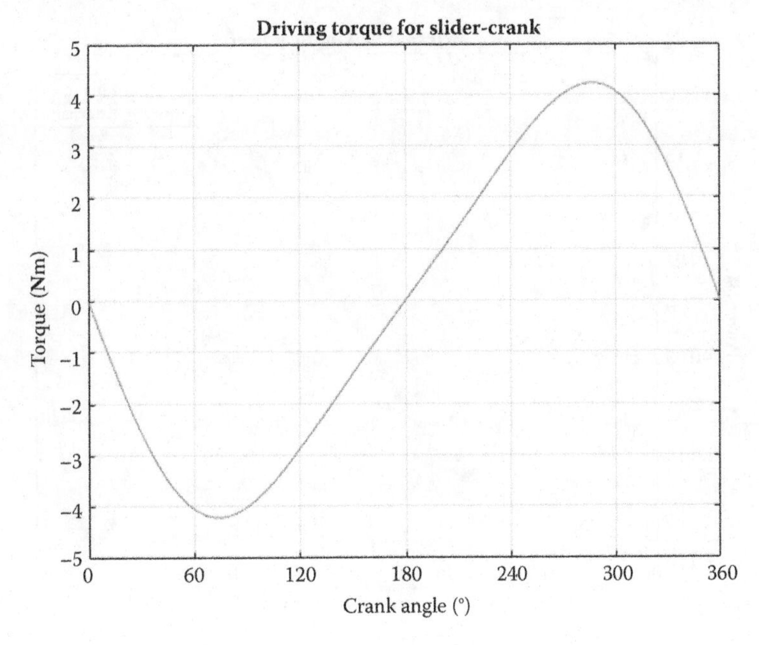

FIGURE 7.47
Driving torque for the example problem with crank speed 10 rad/s.

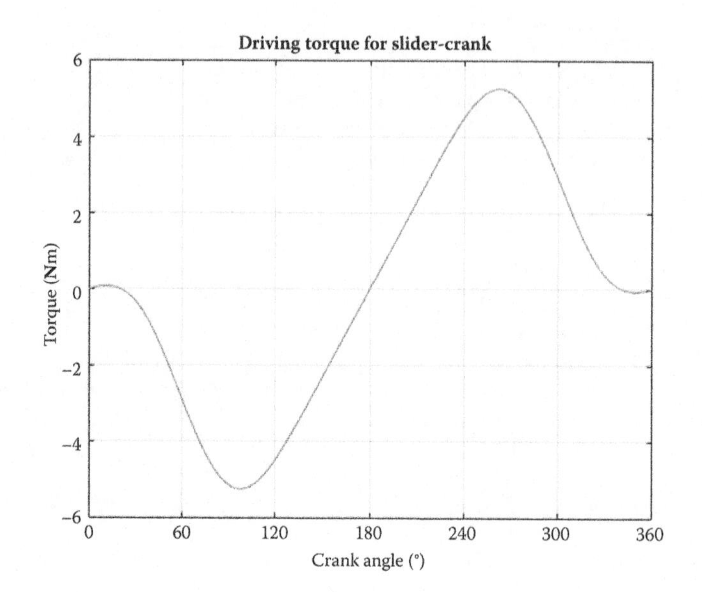

FIGURE 7.48
Driving torque for the example problem with crank speed 100 rad/s.

calculate kinetic power for the crank, connecting rod, and piston. The crank has no translational kinetic energy since its center of mass does not move, and the piston has no rotational kinetic energy since it is not permitted to rotate.

Figure 7.49 shows the external and kinetic power for the example compressor without pressure force or friction. As you can see, the two curves overlay each other exactly. Make sure that your power curves match these before proceeding to the next section.

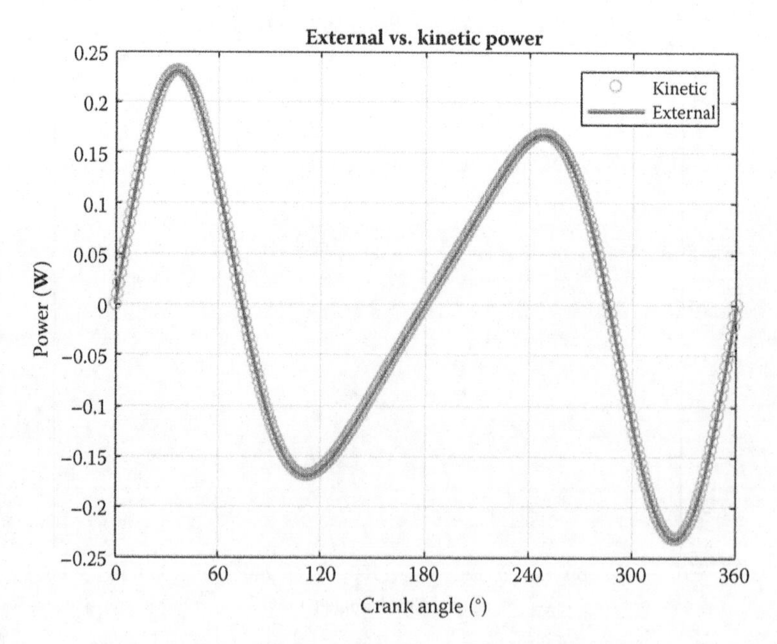

FIGURE 7.49
External and kinetic power for the example slider-crank with crank speed 10 rad/s.

Finally, it is interesting to plot the cylinder moment, M, for crank speeds 10 rad/s and 100 rad/s. As seen in Figures 7.50 and 7.51, the shape of the curve matches that of the driving torque, although the magnitudes are lower. This is the moment that the cylinder must exert in order to keep the piston oriented horizontally. Since it changes direction with every cycle, it will contribute to the vibration forces caused by the movement of the linkage.

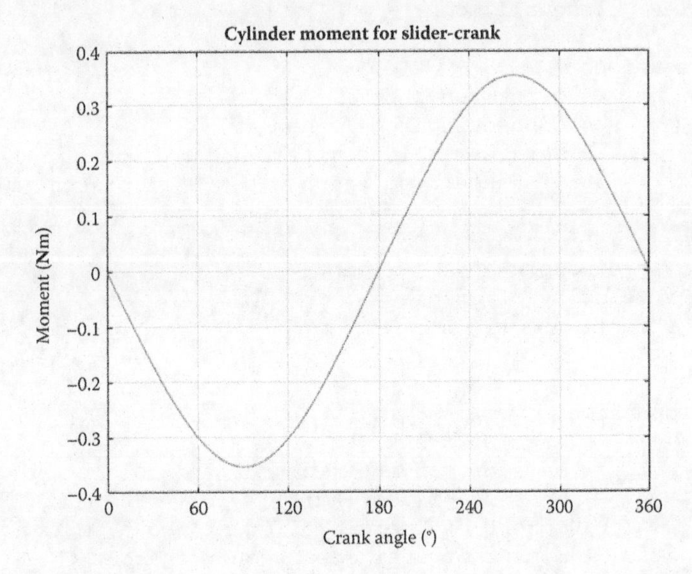

FIGURE 7.50
Cylinder moment for example slider-crank with crank speed 10 rad/s. The shape of the curve matches the driving torque, but the magnitude is lower.

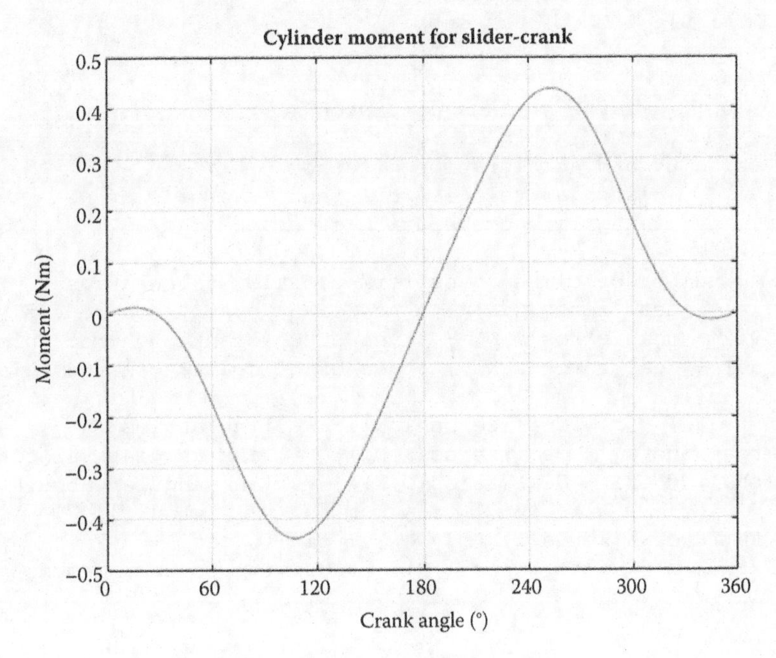

FIGURE 7.51
Cylinder moment for example slider-crank with crank speed 100 rad/s.

```
% SliderCrank_Force_Analysis.m
% performs a force analysis on the slider-crank mechanism and
% plots the driving torque
% by Eric Constans, June 20, 2017

% Prepare Workspace
clear variables; close all; clc;

% Linkage dimensions
a = 0.040;            % crank length (m)
b = 0.120;            % connecting rod length (m)
c = 0.0;              % vertical slider offset (m)
p = 0.01;             % distance from point C to CM of piston (m)

% ground pins
x0 = [0;0];  % ground pin at A (origin)
v0 = [0;0];  % velocity of origin
a0 = [0;0];  % accel of origin
Z2 = zeros(2); Z21 = zeros(2,1); Z12 = zeros(1,2); U2 = eye(2);

% Inertial properties
m2 = 1.5;             % mass of crank (kg)
m3 = 0.08;            % mass of con rod (kg)
m4 = 0.16;            % mass of piston (kg)
I2 = 0.002;           % moment of inertia of crank about CM (kg-m2)
I3 = 0.0001;          % moment of inertia of con rod about CM (kg-m2)

% CM locations
xbar2 = [   0; 0]; % CM of crank
xbar3 = [ b/2; 0]; % CM of coupler
xbar4 = [   p; 0]; % CM of piston

% external loads
FP = [-100; 0];  % force on face of piston

% Angular velocity and acceleration of crank
omega2 = 10;     % angular velocity of crank (rad/s)
alpha2 = 0;      % angular acceleration of crank (rad/s^2)

N = 361;   % number of times to perform position calculations
[xB,x2,x3,x4] = deal(zeros(2,N)); % pos of B, crank, con rod, piston
[vB,v2,v3,v4] = deal(zeros(2,N)); % vel of B, crank, con rod, piston
[aB,a2,a3,a4] = deal(zeros(2,N)); % acc of B, crank, con rod, piston

[theta2,theta3,d]      = deal(zeros(1,N));  % link angles
[omega3,omega4,ddot]  = deal(zeros(1,N));  % link angular velocities
[alpha3,alpha4,dddot] = deal(zeros(1,N));  % link angular accelerations

[FA,FB,FC,FD] = deal(zeros(2,N));       % pin forces
[P,M,T2,PExt,PKin] = deal(zeros(1,N)); % driving torque, moment and
powers

% Main loop
for i = 1:N
```

```
  theta2(i) = (i-1)*(2*pi)/(N-1);
  theta3(i) = asin((c - a*sin(theta2(i)))/b);
  d(i) = a*cos(theta2(i)) + b*cos(theta3(i));

% calculate unit vectors
  [e1,n1] = UnitVector(0);
  [e2,n2] = UnitVector(theta2(i));
  [e3,n3] = UnitVector(theta3(i));
  [eA2,nA2,LA2,s2A,s2B,~] = LinkCG(a,0,0,xbar2,theta2(i));
  [eB3,nB3,LB3,s3B,s3C,~] = LinkCG(b,0,0,xbar3,theta3(i));
  [eC4,nC4,LC4,s4C,s4D,~] = LinkCG(p,0,0,xbar4,0);

% solve for position of point B on the linkage
  xB(:,i) = FindPos(     x0,   a,  e2);

% conduct velocity analysis to solve for omega3 and omega4
  A_Mat = [b*n3 -e1];
  b_Vec = -a*omega2*n2;
  omega_Vec = A_Mat\b_Vec;  % solve for angular velocities

  omega3(i) = omega_Vec(1);    % decompose omega_Vec into
  ddot(i)   = omega_Vec(2);    % individual components

% calculate velocity at important points on linkage
  vB(:,i) = FindVel(     v0,   a,    omega2,  n2);
  v2(:,i) = FindVel(     v0, LA2,    omega2, nA2);
  v3(:,i) = FindVel(vB(:,i), LB3, omega3(i), nB3);
  v4(:,i) = FindVel(vB(:,i),   b, omega3(i),  n3);

% conduct acceleration analysis to solve for alpha3 and dddot
  ac = a*omega2^2;
  at = a*alpha2;
  bc = b*omega3(i)^2;

  C_Mat = A_Mat;
  d_Vec = -at*n2 + ac*e2 + bc*e3;
  alpha_Vec = C_Mat\d_Vec;  % solve for angular accelerations

  alpha3(i) = alpha_Vec(1);
  dddot(i)  = alpha_Vec(2);

% find acceleration of pins
  aB(:,i) = FindAcc(     a0,   a,    omega2,    alpha2,  e2,   n2);
  a2(:,i) = FindAcc(     a0, LA2,    omega2,    alpha2, eA2,  nA2);
  a3(:,i) = FindAcc(aB(:,i), LB3, omega3(i), alpha3(i), eB3,  nB3);
  a4(:,i) = FindAcc(aB(:,i),   b, omega3(i), alpha3(i),  e3,   n3);

% conduct force analysis
  S_Mat = [      U2         -U2          Z2          Z2  Z21   Z21;
                 Z2          U2         -U2          Z2  Z21   Z21;
                 Z2          Z2          U2         -U2  Z21   Z21;
               s2A'       -s2B'         Z12         Z12    0     1;
                Z12        s3B'       -s3C'         Z12    0     0;
                Z12         Z12        s4C'         Z12    1     0;
                Z12         Z12         Z12         e1'    0     0];
```

```
  t_Vec = [m2*a2(:,i);
           m3*a3(:,i);
           m4*a4(:,i) - FP;
           I2*alpha2;
           I3*alpha3(i);
           0;
           0];

  f_Vec = S_Mat      _Vec;
  FA(:,i) = [f_Vec(1); f_Vec(2)];
  FB(:,i) = [f_Vec(3); f_Vec(4)];
  FC(:,i) = [f_Vec(5); f_Vec(6)];
  FD(:,i) = [f_Vec(7); f_Vec(8)];
  M(i)   = f_Vec(9);
  T2(i)  = f_Vec(10);

  P2 = InertialPower(m2,I2,v2(:,i),a2(:,i),omega2,alpha2);
  P3 = InertialPower(m3,I3,v3(:,i),a3(:,i),omega3(i),alpha3(i));
  P4 = InertialPower(m4, 0,v4(:,i),a4(:,i),        0,        0);
  PKin(i) = P2 + P3 + P4;           % kinetic power
  PF = dot(FP,v4(:,i));             % power from external force
  PT = T2(i) * omega2;              % power from crank torque
  PExt(i) =  PF + PT;               % total external power
end

% plot the driving torque
plot(theta2*180/pi,T2,'Color',[0 110/255 199/255])
title('Driving Torque for Slider-Crank')
xlabel('Crank angle (degrees)')
ylabel('Torque (N-m)')
grid on
set(gca,'xtick',0:60:360)
xlim([0 360])
```

7.7 Force Analysis Example 2 – The Air Compressor Mechanism

We will now conduct a dynamic analysis of a specific device: an air compressor. Along the way we will learn several techniques that are useful in modeling dynamic systems. We will start with a very simple model and add complexity (and fidelity) as we proceed.

Figure 7.52 shows a sketch of the air compressor mechanism. The purpose of an air compressor, obviously, is to compress air. As the piston moves to the right in the cylinder, the volume diminishes and the air inside the cylinder is compressed. The pressurized air creates a force, $\mathbf{F}_P$, on the piston, which is given by

$$\mathbf{F}_P = \left\{ \begin{array}{c} -P \cdot A \\ 0 \end{array} \right\} \tag{7.144}$$

where A is the cross-sectional area of the piston and P is the gauge pressure (above atmospheric) inside the cylinder. The negative sign is included because the pressure force acts

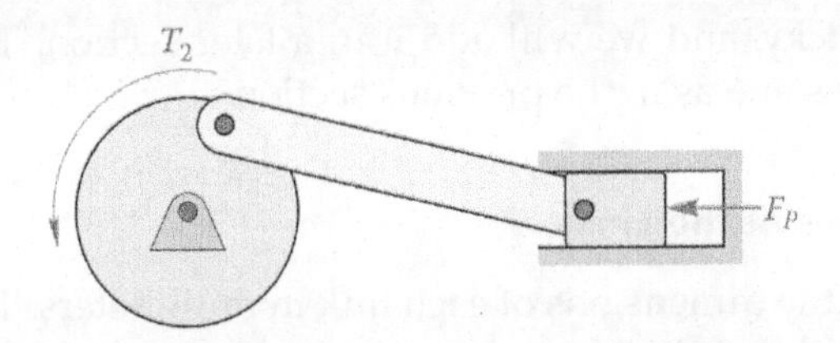

FIGURE 7.52
The air compressor mechanism. The crank is link 2, the connecting rod is link 3, and the piston is link 4.

in the negative x direction. A motor drives the crank with torque T_2. The goals of making the compressor model are:

- To determine the motor torque, T_2, necessary to drive the compressor and achieve the desired air pressure in the cylinder.
- To determine the forces acting on the pins and the wall of the cylinder.

Once the driving torque and speed are known, we can choose an appropriate motor from an industrial supplier or motor manufacturer. Knowledge of the pin forces will enable us to choose the proper materials for the pins, and also to choose bearings rated for the given loads.

7.7.1 First, a Simple Model

To begin, we will create a dynamic model of the mechanism alone, ignoring friction and the force from the high-pressure air. This model will simulate the situation where the top of the cylinder is left open, and no air is compressed. The dimensions of the mechanism are given in Figure 7.53. Note that the vertical offset of the cylinder (given as c in previous chapters) is set to zero. The length of the crank is a = 40 mm and the length of the connecting rod is b = 120 mm. In this mechanism, the crank takes the form of a flywheel – a heavy disk with large moment of inertia. The center of mass of the flywheel is located at the ground pin so that the translational acceleration of the crank is zero.

The free-body diagram of the compressor is the same as it was for the slider-crank in the previous section, although the force on the face of the piston will be more complicated this time. For our initial model we will ignore friction, so that $\mathbf{F}_D$ has only a vertical component.

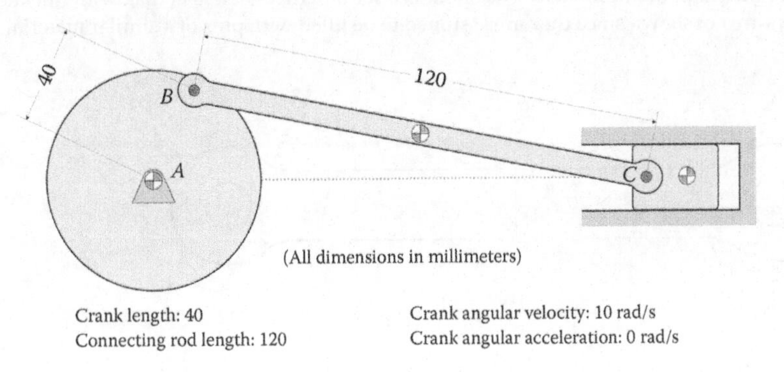

FIGURE 7.53
Dimensions of the compressor mechanism. The vertical offset of the cylinder is zero.

Modeling friction is tricky, and we will add it in a later section. The matrix equation of motion will also be the same as in the previous section.

7.7.2 Inertial Properties of the Links

Figures 7.54–7.56 show the dimensions of each link in millimeters. The crank and connecting rod are made of steel and the piston is made of aluminum. Holes have been drawn at the pin locations, but we will find the **Mass Properties** of each link without the holes, since we assume each hole is filled by a pin. If the pins are of a material different from the link then you should fill each hole with a cylinder of the proper material. Here we assume that the pins are made of steel.

Since the crank is a uniform cylinder, its center of mass lies at its center, which is also the origin. Thus

$$\bar{x}_2 = 0\text{ mm} \quad \bar{y}_2 = 0\text{ mm}$$

The connecting rod is symmetric in the x and y directions so its center of mass lies at its geometric center

$$\bar{x}_3 = 60\text{ mm} \quad \bar{y}_3 = 0\text{ mm}$$

The piston has a nonuniform shape, so we use the **Mass Properties** feature in SOLIDWORKS to find the coordinates of its center of mass (with the holes suppressed).

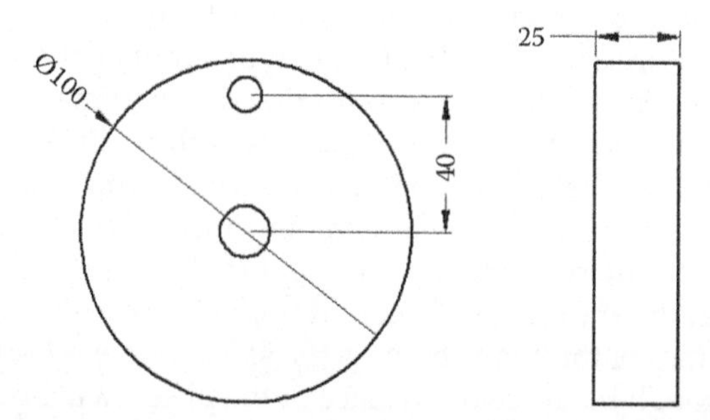

FIGURE 7.54
Dimensions of the crank in millimeters. The crank is a 100 mm diameter steel disk with thickness 25 mm. The hole diameters are not shown since they are assumed to be filled with pins of a similar material to the links.

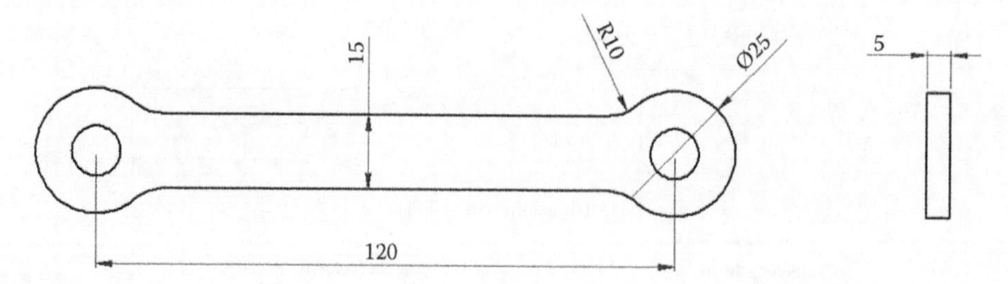

FIGURE 7.55
Dimensions of the connecting rod in millimeters. The connecting rod is made of steel and is 5 mm thick. All corners have radius 10 mm.

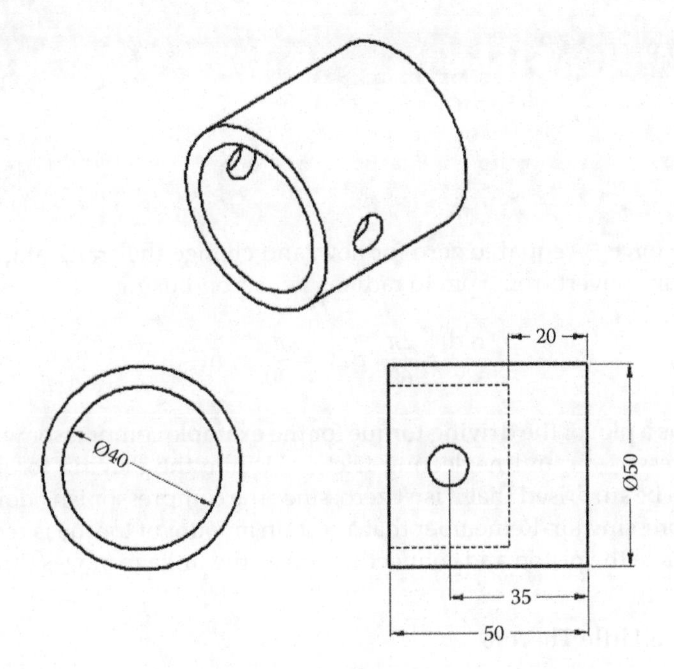

FIGURE 7.56
Dimensions of the piston in millimeters. The piston is a hollowed-out aluminum cylinder of diameter 50 mm.

The x coordinate of the center of mass is 19 mm away from the top face of the piston, which means that it is (35 mm − 19 mm = 16 mm) from the center of the hole. Thus,

$$\bar{x}_4 = 16\text{ mm} \quad \bar{y}_4 = 0\text{ mm}$$

Using the **Mass Properties** feature in SOLIDWORKS, we obtain the following inertial properties of each link (with the holes suppressed).

$$m_2 = 1.551\text{ kg} \qquad m_3 = 0.096\text{ kg} \qquad m_4 = 0.163\text{ kg}$$

$$I_2 = 0.001939\text{ kg m}^2 \qquad I_3 = 0.000189\text{ kg m}^2$$

7.7.3 Driving Torque without Pressure Force

Use **Save As** to save your slider-crank force analysis code to the file `SliderCrankCompressor.m`. Enter the following dimensions at the top of the file.

```
% Linkage dimensions
a = 0.040;              % crank length (m)
b = 0.120;              % connecting rod length (m)
c = 0.0;                % vertical slider offset (m)
p = 0.016;              % distance from pin C to CM of piston (m)
```

Next, enter the inertial properties into the code as shown.

```
% Inertial properties
m2 = 1.551;             % mass of crank (kg)
m3 = 0.096;             % mass of con rod (kg)
m4 = 0.163;             % mass of piston (kg)
I2 = 0.001939;          % moment of inertia of crank about CM (kg-m2)
I3 = 0.000189;          % moment of inertia of coupler about CM (kg-m2)
```

Set the pressure force `FP` equal to zero for now, and change the crank angular velocity to 1200 rpm. You can convert from rpm to radians per second using

$$\frac{\text{rad}}{\text{sec}} = \frac{2\pi}{60} \cdot \text{rpm} = \frac{\pi}{30} \cdot \text{rpm} \tag{7.145}$$

Figure 7.57 shows a plot of the driving torque for the example compressor without the force from the compressed air and neglecting friction. While the driving torque is relatively small, you might be surprised that it isn't zero, since the compressor isn't doing any "useful work" like compressing air. Remember that a certain amount of torque is required to accelerate and decelerate the piston and connecting rod as the linkage moves through the cycle.

7.7.4 And Now, a Little Thermo

Our next step is to calculate the pressure inside the cylinder as a function of the piston position. We will make a few assumptions in order to keep the model simple:

- The compression process happens so quickly that no heat from the compressed air is transmitted to the cylinder. That is, the compression process is assumed to be *adiabatic*.

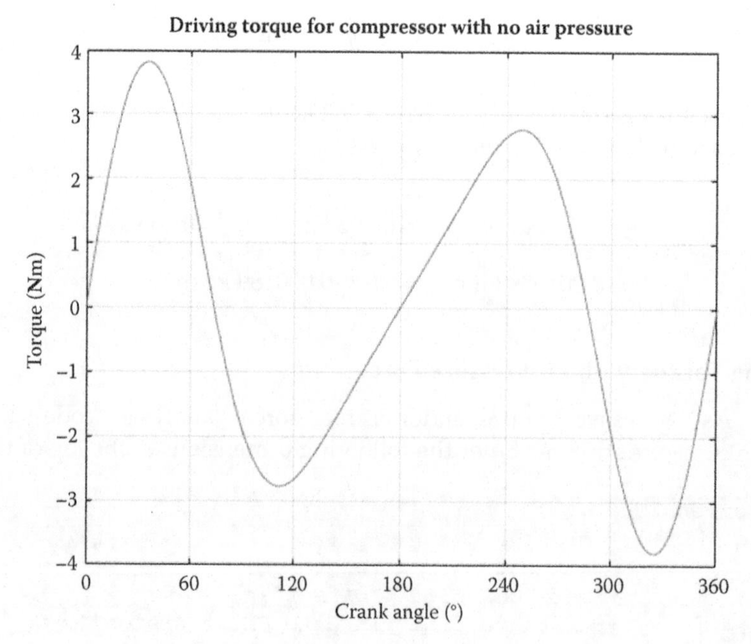

FIGURE 7.57
Driving torque for the example problem without pressure force and friction

- The linkage must exert a certain amount of *mechanical* work to create the *thermodynamic* work of compressing the air. We assume that the thermodynamic work can be converted back to mechanical work without loss of energy. In other words, the compressed air could be used to "back drive" the linkage.
- Air behaves as an ideal gas and obeys the ideal gas law.

The combination of these three assumptions leads to the conclusion that the compression process is *isentropic*; that is, entropy remains constant through the compression process. For isentropic compression, we obtain the following relationship between pressure and volume (see [1, pp. 282–283]).

$$P_2 = P_1\left(\frac{V_1}{V_2}\right)^{1.4} \tag{7.146}$$

where P_1 and V_1 are the pressure and volume at state 1 and P_2 and V_2 are the pressure and volume at state 2, respectively.

Figure 7.58 shows the dimensions of the compressor that are needed to calculate the volume inside the cylinder. We will define the term "Bottom Dead Center" (BDC) to denote the position of the piston when it is at its left-most extreme and "Top Dead Center" (TDC) when it is furthest to the right. We will assume that the air inside the cylinder is at atmospheric pressure when the piston is at BDC; this is also the position where the volume inside the cylinder is at its maximum.

$$P_1 = P_{atm} = 101\text{ kPa} \tag{7.147}$$

The piston does not entirely reach the top of the cylinder at TDC and there is a small head space, h, between the top of the piston and the inside of the cylinder. If the head space were not present (i.e. if the volume inside the cylinder were zero), the pressure inside the cylinder would reach infinity at TDC. Examining Figure 7.58, we see that

$$a+b+H+h=d+H+x+h \tag{7.148}$$

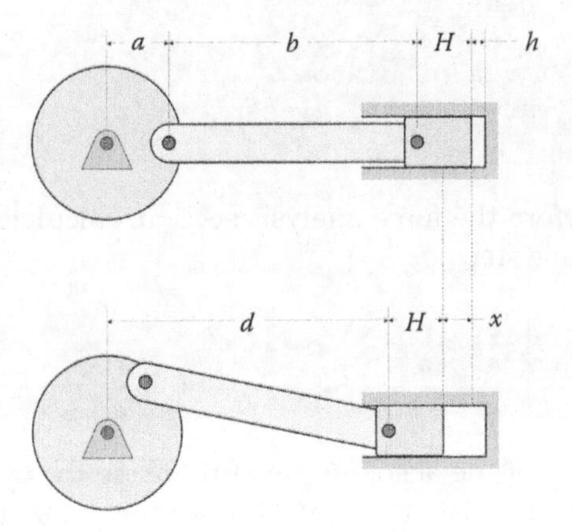

FIGURE 7.58
Dimensions of the compressor used to calculate the volume inside the cylinder.

where x is the displacement of the piston from TDC. Solving for x gives

$$x = a + b - d \tag{7.149}$$

Since d is known from conducting the position analysis, we conclude that the volume inside the cylinder at an arbitrary time is

$$V_2 = A(a + b - d + h) \tag{7.150}$$

The maximum displacement of the piston from TDC is

$$x_{BDC} = 2a \tag{7.151}$$

So that the volume at BDC (the maximum volume) is

$$V_1 = A(2a + h) \tag{7.152}$$

Thus, using Equation (7.146) we find that the pressure inside the cylinder at an arbitrary point during the stroke is

$$P_2 = P_{atm}\left(\frac{2a + h}{a + b - d + h}\right)^{1.4} \tag{7.153}$$

This equation gives the *absolute* pressure in the cylinder. Since the backside of the piston is always subject to atmospheric pressure, we must use the *net* pressure to calculate the pressure force.

$$F_P = \begin{Bmatrix} -A(P_2 - P_{atm}) \\ 0 \end{Bmatrix} \tag{7.154}$$

Add the following lines of code near the top of the file to define some of the constants associated with the pressure calculations

```
% Pressure constants
h = 0.015;            % head space (m)
D = 0.05;             % diameter of piston (m)
A = pi*D^2/4;         % area of piston
Patm = 101e3;         % atmospheric pressure
```

Inside the main loop, before the force analysis section, calculate the pressure inside the cylinder at the current time step:

```
% calculate pressure force
  P(i) = Patm*((2*a+h)/(a+b-d(i)+h))^1.4;
  FP = [-A*(P(i)-Patm); 0];
```

If you run the code and plot the absolute pressure inside the cylinder, the result should be similar to Figure 7.59. The piston is at TCD when the crank angle is 0° or 360°, so the pressure reaches a maximum at these crank angles. When the piston is at BDC (crank angle = 180°) the air pressure is at one atmosphere. With our current design, the compressor

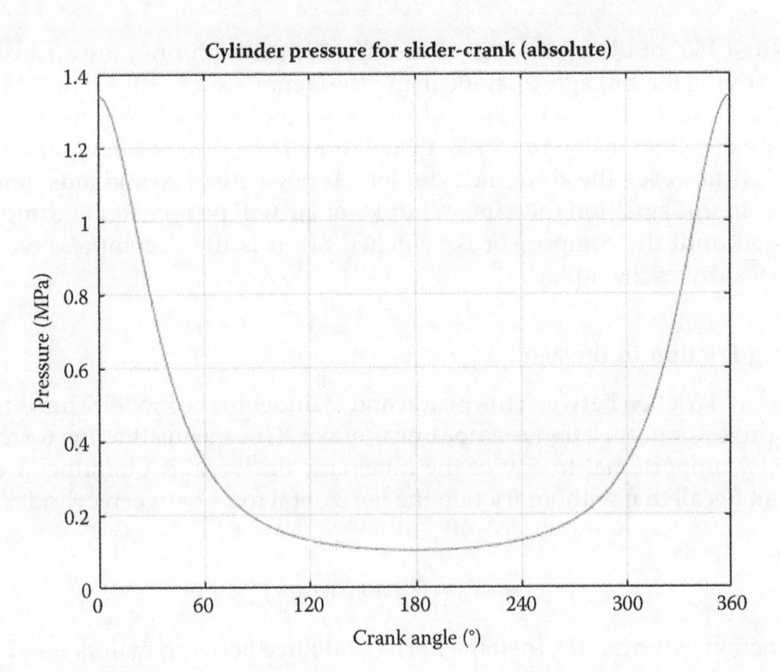

FIGURE 7.59
Pressure inside the cylinder for the example problem.

will achieve a net pressure of 1.24 MPa (or roughly 180 psi, for those who insist on using Imperial units!)

If you now use the code to plot the driving torque necessary to compress the air, the result should resemble Figure 7.60. You might question the presence of negative torque

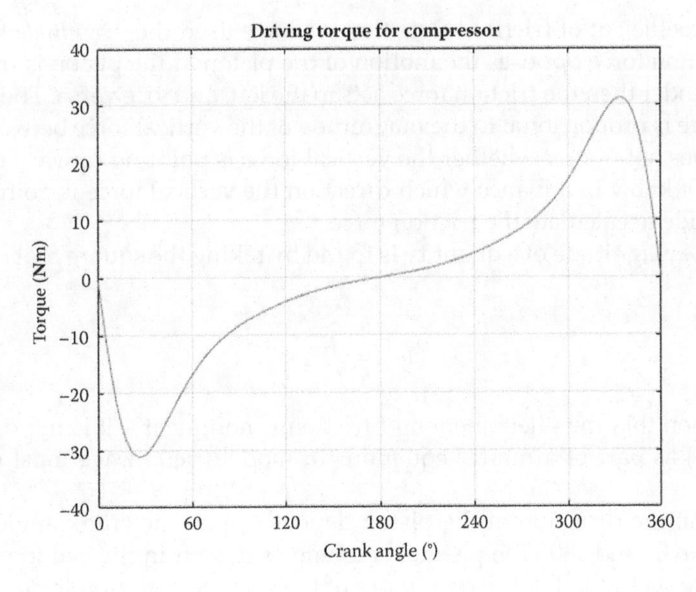

FIGURE 7.60
Driving torque needed to compress air in the example problem. Note the presence of negative torque in the first half of the plot.

during the first 180° of the cycle – this is the effect of the high-pressure air inside the cylinder back-driving the linkage. Imagine using the compressor to fill a tank of compressed air, as would be the case for a portable air-powered tool setup. If the pressure inside the cylinder is greater than inside the tank, air will flow from the cylinder to the tank. When the tank is full, however, the air in the cylinder has nowhere to go and must remain inside the cylinder. In this situation the same "chunk" of air will be repeatedly compressed and decompressed until the compressor is switched off. It is the decompression part of the cycle that back-drives the linkage.

7.7.5 Adding Friction to the Model

We will now add friction between the piston and cylinder to the model. This is the trickiest part of the project, since friction is a nonlinear effect. This means that the force of friction is not directly proportional to any of the kinematic or dynamic quantities we have discussed so far. Recall that without friction the horizontal force between cylinder and piston was zero

$$\mathbf{e}_1 \cdot \mathbf{F}_D = 0 \quad \text{(no friction)} \tag{7.155}$$

If we add friction to the model, then the horizontal force between cylinder and piston will not be zero any more:

$$\mathbf{e}_1 \cdot \mathbf{F}_D = F_f \tag{7.156}$$

where F_f is the friction force. The friction force is proportional to the normal force between piston and cylinder:

$$|F_f| = \mu |\mathrm{F}_{Dy}| \tag{7.157}$$

where μ is the coefficient of friction. Note that we have used the *magnitude* for each of the forces. The friction force opposes the motion of the piston: if the piston is traveling to the right in the cylinder then the friction force acts to the left, and vice versa. The magnitude of the friction force is proportional to the magnitude of the vertical force between piston and cylinder – it does not matter whether the vertical force is pointing upward or downward. Since we cannot know in advance which direction the vertical force is pointing, we must use its magnitude to calculate the friction force.

Recall that the magnitude of a quantity is found by taking the square root of the quantity squared

$$|\mathrm{F}_{Dy}| = \sqrt{\mathrm{F}_{Dy}^2} \tag{7.158}$$

This is the reason that the phenomenon of friction is nonlinear – it is impossible to write Equation (7.158) as part of a matrix equation. To model friction, we must employ a few tricks.

First, note that the direction of the piston depends upon the crank angle. If the crank angle is between 0° and 180°, the piston is moving to the left in the cylinder, and the friction force acts to the right. For the remainder of the cycle, the friction force acts to the left. Therefore, we have

$$\begin{aligned} &\text{if } 0° < \theta_2 < 180° \quad F_f = \mu|F_{Dy}| \\ &\text{if } 180° < \theta_2 < 360° \quad F_f = -\mu|F_{Dy}| \end{aligned} \tag{7.159}$$

But F_{Dy} is one of the variables being solved for in the matrix equation, so we cannot place the quantity $\mu|F_{Dy}|$ in the vector of known quantities, **t**. Instead, we will make the assumption that the normal force, F_{Dy}, does not change a great deal from one time step to the next. We can then use the normal force from the previous time step in calculating the friction force.

$$\begin{aligned} &\text{if } 0° < \theta_2 < 180° \quad F_f(i) = \mu|F_{Dy}(i-1)| \\ &\text{if } 180° < \theta_2 < 360° \quad F_f(i) = -\mu|F_{Dy}(i-1)| \end{aligned} \tag{7.160}$$

Near the top of the code, add the definition of the coefficient of friction (assumed to be 0.1 for our example). The friction force must be initialized to zero in order to have a value for the first time step (when $i - 1 = 0$).

```
mu = 0.1;                    % coefficient of friction btw piston and cylinder
Ff = 0;                      % friction force starts at zero
```

Inside the main loop, add the following code to calculate the friction force.

```
% Calculate Friction Force
if (i > 1)
  if (theta2 < pi)
    Ff = mu*abs(FD(2,i-1));
  else
    Ff = -mu*abs(FD(2,i-1));
  end
end
```

The first `if` statement is necessary because we are using the normal force from the previous time step, and $F_{Dy}(0)$ does not exist – MATLAB doesn't allow zero or negative indices. This is why we initialized F_f to be zero before the main loop. Next, modify the **t** vector to include the friction force.

```
t_Vec = [m2*a2(:,i);
         m3*a3(:,i);
         m4*a4(:,i) - FP;
         I2*alpha2;
         I3*alpha3(i);
         0;
         Ff];
```

Figure 7.61 shows the normal and friction forces acting on the piston for the example problem. The friction force acts in the positive x direction for the first 180° of crank rotation – as the piston is moving in the negative x direction.

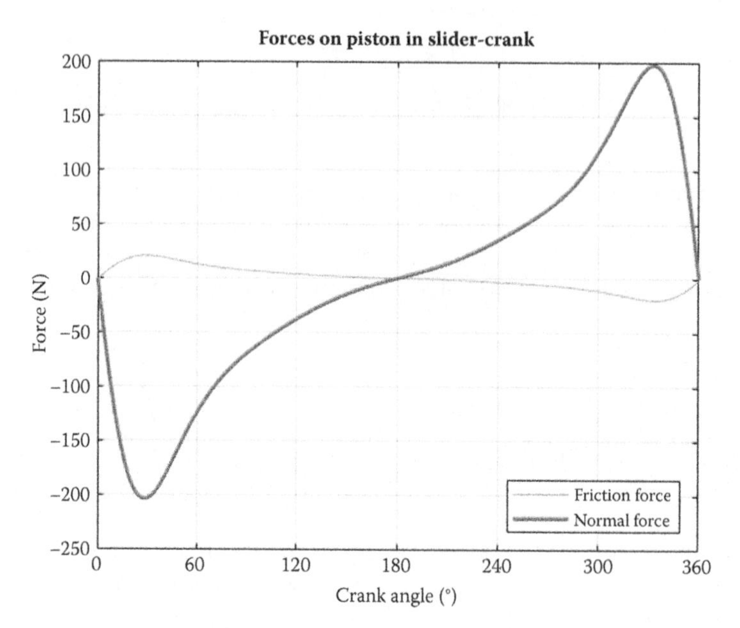

FIGURE 7.61
Horizontal and vertical forces on piston with friction added to the model.

If we next create a conservation of power plot for the compressor with friction the result should resemble Figure 7.62. The curves do not match, and it appears as though we have violated the law of Conservation of Energy!

To make the curves match, we must remember to account for the power consumed by friction as the compressor moves through its cycle. The frictional power can be calculated as

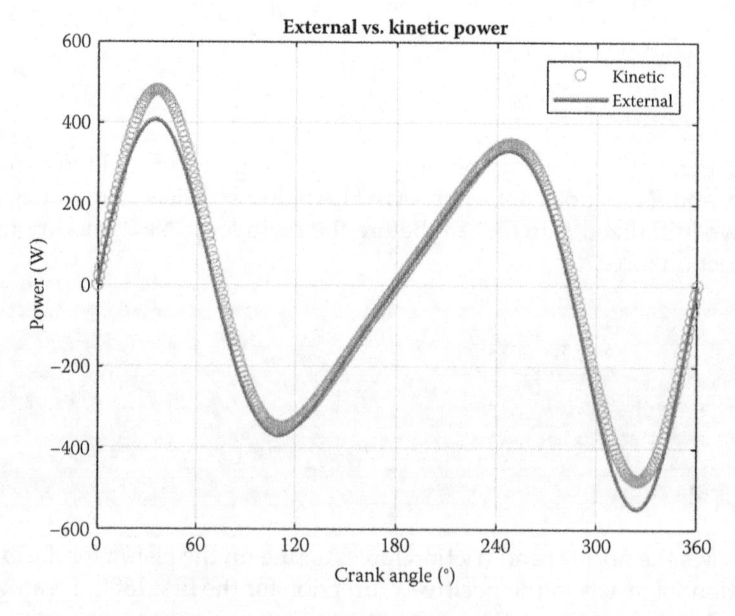

FIGURE 7.62
Conservation of power plot for the compressor with friction added. Power does not appear to be conserved!

$$P_f = F_D \cdot v_4 \tag{7.161}$$

Since the velocity of the piston has a zero y component, the y component of the dot product will be zero, and only frictional power will be computed. We must remember to *subtract* the power consumed by friction since it does *negative* work on the system. Subtracting the frictional power from the overall external power produces the plot shown in Figure 7.63. Happily, power appears to be conserved once again.

7.7.6 Potential Energy of Air Inside the Cylinder

We may also create a pair of conservation of power curves by noting that compressing the air inside the cylinder increases its potential energy, much like a spring. If the absolute pressure inside the cylinder is P, then the *net* force on the piston is

$$F_{net} = A(P - P_{atm}) \tag{7.162}$$

where we must subtract the atmospheric pressure because it acts on both faces of the piston. If we move the piston forward by a distance dx, then the work done is [1, pp. 87–92]

$$dW_{air} = -A(P - P_{atm})dx \tag{7.163}$$

The sign is negative because the pressure force acts to oppose the motion. The work done to compress the air increases its potential energy, such that

$$dPE_{air} = -(P - P_{atm})A\ dx \tag{7.164}$$

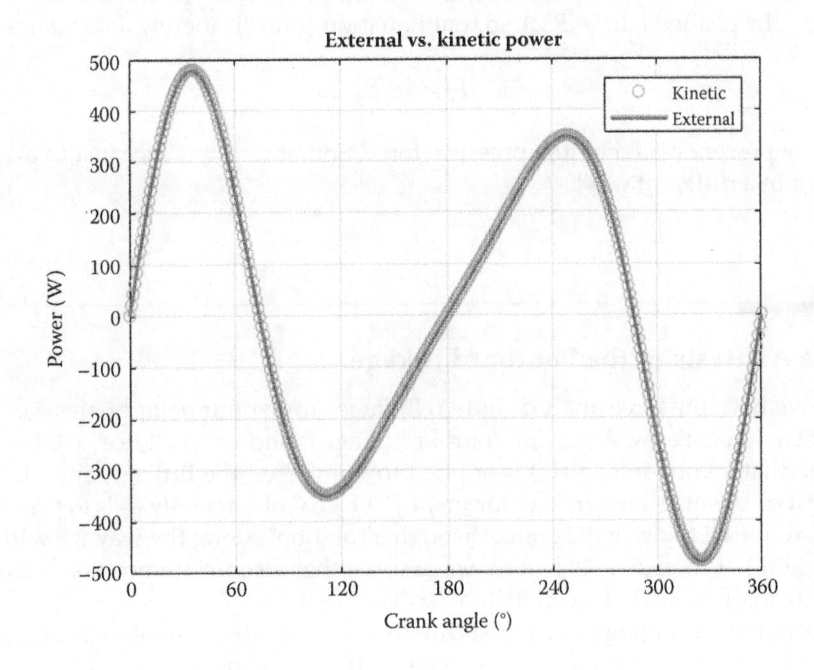

FIGURE 7.63
Power curves with the power consumed by friction taken into account. The curves are now identical, and Conservation of Energy is preserved.

The quantity $A\, dx$ is equal to a differential change in volume, so that

$$dPE_{air} = -(P - P_{atm})dV \tag{7.165}$$

The total increase in potential energy is found by adding (integrating) the individual, differential changes in potential energy.

$$PE_{air} = \int_0^t -(P - P_{atm})\, dV = \int_0^t -(P - P_{atm})\frac{dV}{dt}dt \tag{7.166}$$

The volume of air inside the cylinder is, from Equation (7.150)

$$V = A(a + b + h - d) \tag{7.167}$$

Thus, the change in volume with respect to time is

$$\frac{dV}{dt} = -A\dot{d} \tag{7.168}$$

The power consumed in compressing the air is then

$$P_{air} = \frac{d}{dt}PE_{air} = A\dot{d}(P - P_{atm}) \tag{7.169}$$

In retrospect, this result should have been obvious! The velocity of the piston is $\dot{d}$, and the net force on the piston is $A(P - P_{atm})$, so that Equation (7.169) is merely a restatement of

$$P_F = F_P \cdot v_4 \tag{7.170}$$

which is the power created by the pressure force. Sometimes it is nice just to arrive at the same result in a different way!

7.8 Force Analysis of the Fourbar Linkage

Now that we have the force analysis of two linkages under our belts, analysis of the four-bar will seem almost easy. A general fourbar linkage is shown in Figure 7.64. For this case we assume that a known load ($\mathbf{F}_P$) is applied to point P on the linkage, and also that the rocker must overcome a known load torque, T_4. The goal of our analysis is to find the motor torque, T_2, required to drive the crank through a rotation. Along the way we will solve for the forces at all of the pins. As before, we assume that a complete position, velocity, and acceleration analysis has been conducted earlier.

To perform the force analysis, we first draw a free-body diagram of each link, as shown in Figure 7.65. Summing forces and moments for the crank gives

$$\mathbf{F}_A - \mathbf{F}_B = m_2\mathbf{a}_2 \tag{7.171}$$

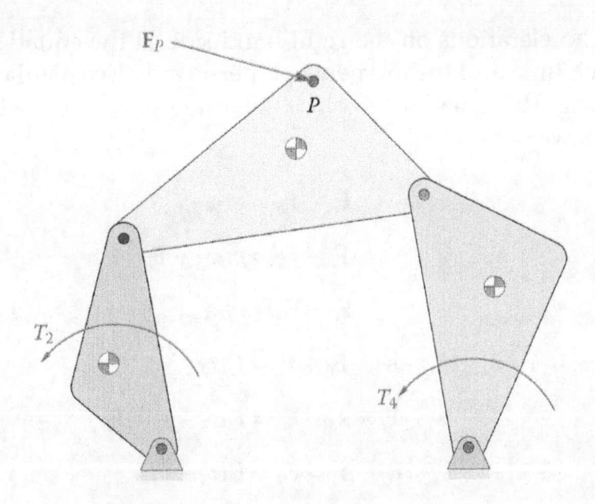

FIGURE 7.64
The fourbar linkage above has an applied load $\mathbf{F}_P$ and an applied torque T_4. Our goal is to determine the driving torque, T_2, necessary to sustain the motion and drive the loads.

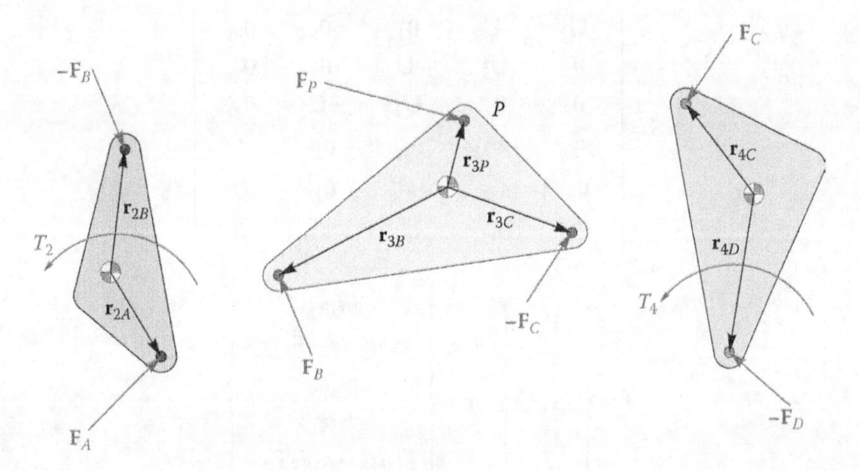

FIGURE 7.65
Free-body diagram of each link in the fourbar linkage.

$$\mathbf{s}_{2A} \cdot \mathbf{F}_A - \mathbf{s}_{2B} \cdot \mathbf{F}_B + T_2 = I_2\alpha_2 \tag{7.172}$$

Doing the same for the coupler results in

$$\mathbf{F}_B - \mathbf{F}_C + \mathbf{F}_P = m_3\mathbf{a}_3 \tag{7.173}$$

$$\mathbf{s}_{3B} \cdot \mathbf{F}_B - \mathbf{s}_{3C} \cdot \mathbf{F}_C + \mathbf{s}_{3P} \cdot \mathbf{F}_P = I_3\alpha_3 \tag{7.174}$$

And finally, the equations for the rocker are

$$\mathbf{F}_C - \mathbf{F}_D = m_4\mathbf{a}_4 \tag{7.175}$$

$$\mathbf{s}_{4C} \cdot \mathbf{F}_C - \mathbf{s}_{4D} \cdot \mathbf{F}_D + T_4 = I_4\alpha_4 \tag{7.176}$$

Remember that the accelerations on the right-hand side of the equations are taken at the center of mass of each link, and the moments of inertia are also calculated about the center of mass. If we rearrange the equations to place the unknowns on the left-hand side and the knowns on the right, we obtain

$$\begin{aligned} \mathbf{F}_A - \mathbf{F}_B &= m_2\mathbf{a}_2 \\ \mathbf{F}_B - \mathbf{F}_C &= m_3\mathbf{a}_3 - \mathbf{F}_P \\ \mathbf{F}_C - \mathbf{F}_D &= m_4\mathbf{a}_4 \\ \mathbf{s}_{2A} \cdot \mathbf{F}_A - \mathbf{s}_{2B} \cdot \mathbf{F}_B + T_2 &= I_2\alpha_2 \\ \mathbf{s}_{3B} \cdot \mathbf{F}_B - \mathbf{s}_{3C} \cdot \mathbf{F}_C &= I_3\alpha_3 - \mathbf{s}_{3P} \cdot \mathbf{F}_P \\ \mathbf{s}_{4C} \cdot \mathbf{F}_C - \mathbf{s}_{4D} \cdot \mathbf{F}_D &= I_4\alpha_4 - T_4 \end{aligned} \tag{7.177}$$

There are nine unknowns ($\mathbf{F}_A$, $\mathbf{F}_B$, $\mathbf{F}_C$, $\mathbf{F}_D$, and T_2) and nine equations, so we have all of the information we need.

$$\mathbf{S} = \begin{bmatrix} \mathbf{U}_2 & -\mathbf{U}_2 & \mathbf{0}_2 & \mathbf{0}_2 & \mathbf{0}_{21} \\ \mathbf{0}_2 & \mathbf{U}_2 & -\mathbf{U}_2 & \mathbf{0}_2 & \mathbf{0}_{21} \\ \mathbf{0}_2 & \mathbf{0}_2 & \mathbf{U}_2 & -\mathbf{U}_2 & \mathbf{0}_{21} \\ \mathbf{s}_{2A}^T & -\mathbf{s}_{2B}^T & \mathbf{0}_{12} & \mathbf{0}_{12} & 1 \\ \mathbf{0}_{12} & \mathbf{s}_{3B}^T & -\mathbf{s}_{3C}^T & \mathbf{0}_{12} & 0 \\ \mathbf{0}_{12} & \mathbf{0}_{12} & \mathbf{s}_{4C}^T & -\mathbf{s}_{4D}^T & 0 \end{bmatrix} \tag{7.178}$$

$$\mathbf{f} = \begin{Bmatrix} \mathbf{F}_A \\ \mathbf{F}_B \\ \mathbf{F}_C \\ \mathbf{F}_D \\ T_2 \end{Bmatrix} \quad \mathbf{t} = \begin{Bmatrix} m_2\mathbf{a}_2 \\ m_3\mathbf{a}_3 - \mathbf{F}_P \\ m_4\mathbf{a}_4 \\ I_2\alpha_2 \\ I_3\alpha_3 - \mathbf{s}_{3P} \cdot \mathbf{F}_P \\ I_4\alpha_4 - T_4 \end{Bmatrix} \tag{7.179}$$

Since each of the forces has both an x and y component, the **S** matrix has dimension 9×9, and the **f** and **t** vectors have dimension 9×1.

7.8.1 Force Analysis of the Sample Linkage

The dimensions and **Mass Properties** of our example fourbar linkage are shown in Figure 7.66. We are now ready to begin modifying our fourbar acceleration analysis code to perform the force analysis on the fourbar linkage. Add the code below near to the beginning of the file.

```
% Inertial properties
m2 = 0.124;             % mass of crank (kg)
m3 = 0.331;             % mass of coupler (kg)
m4 = 0.157;             % mass of rocker (kg)
```

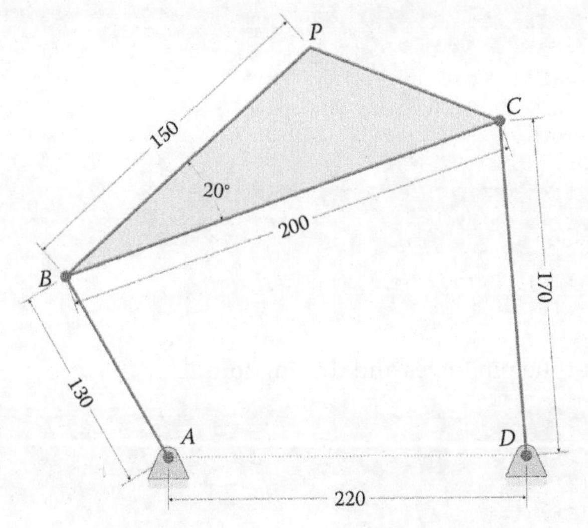

Crank length: 130
Coupler length: 200
Rocker length: 170
Distance betweet ground pins: 220
Distance from *B* to *P*: 150
Angle *PBC*: 20°
Crank angular velocity: 10 rad/s
Crank angular acceleration: 0 rad/s

Crank mass: 0.124 kg
Coupler mass: 0.331 kg
Rocker mass: 0.157 kg
Crank moment of inertia: 0.000255 kg m^2
Coupler moment of inertia: 0.001188 kg m^2
Rocker moment of inertia: 0.000503 kg m^2
Coupler *CM*: (107.1, 14.8)

FIGURE 7.66
The fourbar linkage used in the example. There is a 100 N downward force applied at point *P*.

```
I2 = 0.000255;          % moment of inertia of crank about CM (kg-m2)
I3 = 0.001188;          % moment of inertia of coupler about CM (kg-m2)
I4 = 0.000503;          % moment of inertia of rocker about CM (kg-m2)
```

Since the crank and rocker are symmetric, we use $a/2$ and $c/2$ to define the center of mass locations for each. The coupler center of mass is found geometrically or using SOLIDWORKS.

```
% CM locations
xbar2 = [   a/2;          0]; % CM of crank
xbar3 = [0.1071; 0.0148]; % CM of coupler
xbar4 = [   c/2;          0]; % CM of rocker
```

For this example, we will assume that the load applied at point *P* is 100 N downward, so we define the applied loads as

```
% Applied Loads
FP = [0; -100];         % force at point P (N)
T4 = 0;                 % torque applied to rocker (N-m)
```

The applied torque on the rocker is assumed to be zero for this example. Next, define the **S** matrix and **t** vector as

```
   S_Mat = [    U2      -U2       Z2       Z2    Z21;
                Z2       U2      -U2       Z2    Z21;
```

```
          Z2       Z2       U2      -U2    Z21;
         s2A'    -s2B'      Z12      Z12      1;
          Z12      s3B'    -s3C'      Z12      0;
          Z12      Z12      s4C'    -s4D'      0];

t_Vec = [m2*a2(:,i);
         m3*a3(:,i) - FP;
         m4*a4(:,i);
         I2*alpha2;
         I3*alpha3(i) - dot(FP,s3P);
         I4*alpha4(i) - T4];
```

And finally, we solve for the pin forces and driving torque:

```
f_Vec = S_Mat       _Vec;
FA(:,i) = [f_Vec(1); f_Vec(2)];
FB(:,i) = [f_Vec(3); f_Vec(4)];
FC(:,i) = [f_Vec(5); f_Vec(6)];
FD(:,i) = [f_Vec(7); f_Vec(8)];
T2(i) = f_Vec(9);
```

The driving torque for the example linkage is shown in Figure 7.67. In this example, the crank has a constant angular velocity of $\omega_2 = 10\,\text{rad/s}$. Finally, add the following plotting commands to compare the computed kinetic power with the external power. Both plots should overlay each other exactly, regardless of crank speed. The power plot for the example is shown in Figure 7.68. The complete code for the fourbar force analysis is given below. In the next section, we will pursue a full example with the fourbar linkage from design through force analysis.

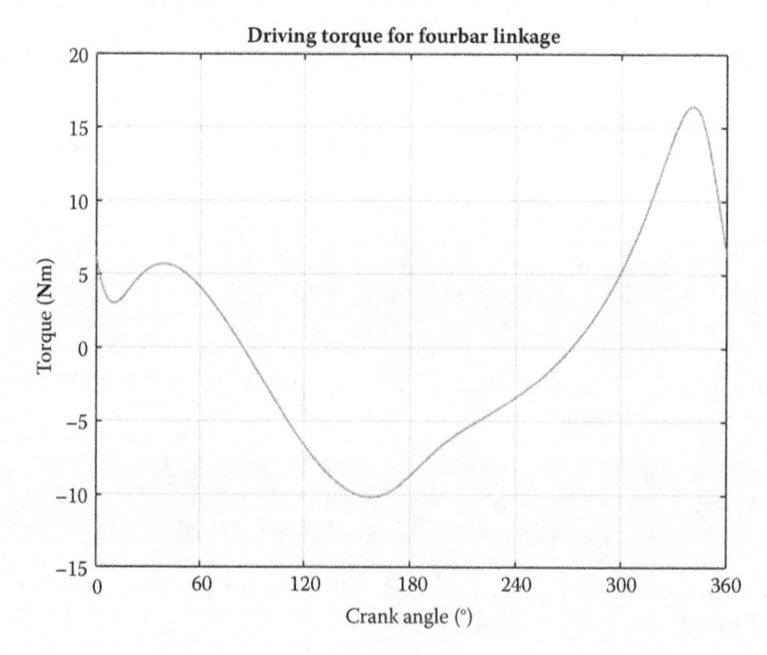

FIGURE 7.67
Driving torque for the example fourbar linkage. The angular velocity of the crank is 10 rad/s and the angular acceleration of the crank is zero.

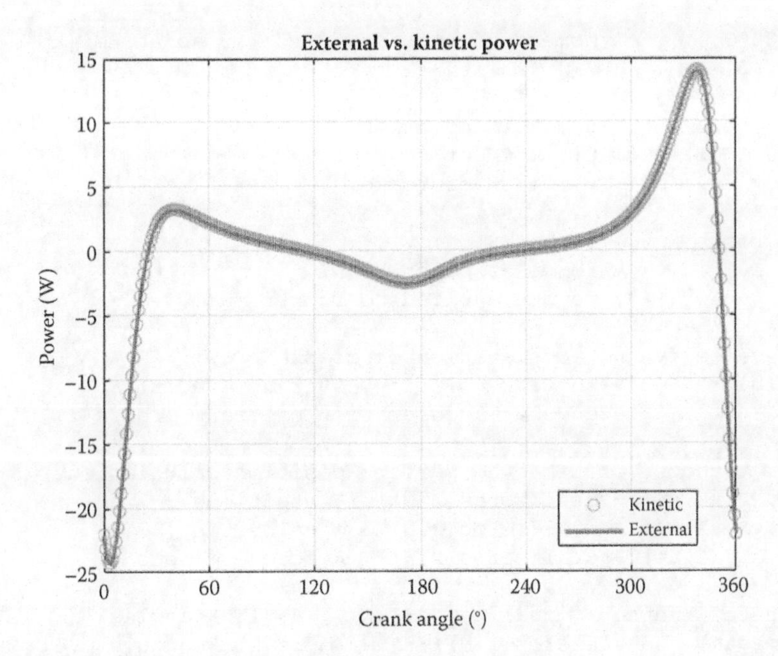

FIGURE 7.68
Comparison of external and kinetic power for the example problem. Both traces overlay each other exactly, and we conclude that the driving torque is being calculated correctly.

```
% Fourbar_Force_Analysis.m
% solves for the pin forces and driving torques of the fourbar linkage
% by Eric Constans, June 21, 2017

% Prepare Workspace
clear variables; close all; clc;

% Linkage dimensions
a = 0.130;            % crank length (m)
b = 0.200;            % coupler length (m)
c = 0.170;            % rocker length (m)
d = 0.220;            % length between ground pins (m)
p = 0.150;            % length from B to P (m)
gamma3 = 20*pi/180; % angle between BP and coupler (converted to rad)

% ground pins
x0 = [0;0];           % ground pin at A (origin)
xD = [d;0];           % ground pin at D
v0 = [0;0];           % velocity of origin
a0 = [0;0];           % accel of origin
Z2 = zeros(2); Z21 = zeros(2,1); Z12 = zeros(1,2); U2 = eye(2);

% Inertial properties
m2 = 0.124;           % mass of crank (kg)
m3 = 0.331;           % mass of coupler (kg)
m4 = 0.157;           % mass of rocker (kg)
I2 = 0.000255;        % moment of inertia of crank about CM (kg-m2)
I3 = 0.001188;        % moment of inertia of coupler about CM (kg-m2)
```

```
I4 = 0.000503;       % moment of inertia of rocker about CM (kg-m2)

% CM locations
xbar2 = [   a/2;       0]; % CM of crank
xbar3 = [0.1071; 0.0148]; % CM of coupler
xbar4 = [   c/2;       0]; % CM of rocker

% Applied Loads
FP = [0; -100];       % force at point P (N)
T4 = 0;               % torque applied to rocker (N-m)

% Angular velocity and acceleration of crank
omega2 = 10;          % angular velocity of crank (rad/s)
alpha2 = 0;           % angular acceleration of crank (rad/s^2)

N = 361;   % number of times to perform position calculations
[xB,xC,xP,x2,x3,x4] = deal(zeros(2,N)); % positions
[vB,vC,vP,v2,v3,v4] = deal(zeros(2,N)); % velocities
[aB,aC,aP,a2,a3,a4] = deal(zeros(2,N)); % accelerations

[theta2,theta3,theta4] = deal(zeros(1,N));  % link angles
[omega3,omega4] = deal(zeros(1,N));         % link angular vel
[alpha3,alpha4] = deal(zeros(1,N));         % link angular acc

[FA,FB,FC,FD] = deal(zeros(2,N));        % pin forces
[T2,PExt,PKin] = deal(zeros(1,N));       % driving torque and powers

% Main Loop
for i = 1:N
  theta2(i) = (i-1)*(2*pi)/(N-1);  % crank angle

% conduct position analysis to solve for theta3 and theta4
  r = d - a*cos(theta2(i));
  s = a*sin(theta2(i));
  f2 = r^2 + s^2;                        % f squared
  delta = acos((b^2+c^2-f2)/(2*b*c)); % angle between coupler and rocker

  g = b - c*cos(delta);
  h = c*sin(delta);

  theta3(i) = atan2((h*r - g*s),(g*r + h*s));
  theta4(i) = theta3(i) + delta;

% calculate unit vectors
  [e2,n2] = UnitVector(theta2(i));
  [e3,n3] = UnitVector(theta3(i));
  [e4,n4] = UnitVector(theta4(i));
  [eBP,nBP] = UnitVector(theta3(i) + gamma3);
  [eA2,nA2,LA2,s2A,s2B,  ~] = LinkCG(        a,0,      0,xbar2,theta2(i));
  [eB3,nB3,LB3,s3B,s3C,s3P] = LinkCG(        b,p,gamma3,xbar3,theta3(i));
  [eD4,nD4,LD4,s4D,s4C,  ~] = LinkCG(        c,0,      0,xbar4,theta4(i));

% solve for positions of points B, C and P on the linkage
  xB(:,i) = FindPos(      x0,    a,    e2);
```

```
  xC(:,i) = FindPos(      xD,   c,   e4);
  xP(:,i) = FindPos(xB(:,i),   p,  eBP);
  x2(:,i) = FindPos(      x0, LA2,  eA2);
  x3(:,i) = FindPos(xB(:,i), LB3,  eB3);
  x4(:,i) = FindPos(      xD, LD4,  eD4);

% conduct velocity analysis to solve for omega3 and omega4
  A_Mat = [b*n3 -c*n4];
  b_Vec = -a*omega2*n2;
  omega_Vec = A_Mat\b_Vec;  % solve for angular velocities

  omega3(i) = omega_Vec(1); % decompose omega_Vec into
  omega4(i) = omega_Vec(2); % individual components

% calculate velocity at important points on linkage
  vB(:,i) = FindVel(      v0,   a,    omega2,   n2);
  vC(:,i) = FindVel(      v0,   c, omega4(i),   n4);
  vP(:,i) = FindVel(vB(:,i),   p, omega3(i),  nBP);
  v2(:,i) = FindVel(      v0, LA2,    omega2,  nA2);
  v3(:,i) = FindVel(vB(:,i), LB3, omega3(i),  nB3);
  v4(:,i) = FindVel(      v0, LD4, omega4(i),  nD4);

% conduct acceleration analysis to solve for alpha3 and alpha4
  ac = a*omega2^2;
  bc = b*omega3(i)^2;
  cc = c*omega4(i)^2;
  pc = p*omega3(i)^2;

  C_Mat = A_Mat;
  d_Vec = ac*e2 + bc*e3 - cc*e4;
  alpha_Vec = C_Mat\d_Vec;  % solve for angular accelerations

  alpha3(i) = alpha_Vec(1);
  alpha4(i) = alpha_Vec(2);

% find acceleration of pins
  aB(:,i) = FindAcc(      a0, a,    omega2,    alpha2,  e2,  n2);
  aC(:,i) = FindAcc(      a0, c, omega4(i), alpha4(i),  e4,  n4);
  aP(:,i) = FindAcc(aB(:,i), p, omega3(i), alpha3(i), eBP, nBP);

% solve for accelerations at centers of mass of each link
  a2(:,i) = FindAcc(      a0, LA2,    omega2,    alpha2, eA2, nA2);
  a3(:,i) = FindAcc(aB(:,i), LB3, omega3(i), alpha3(i), eB3, nB3);
  a4(:,i) = FindAcc(      a0, LD4, omega4(i), alpha4(i), eD4, nD4);

  S_Mat = [   U2    -U2     Z2     Z2   Z21;
              Z2     U2    -U2     Z2   Z21;
              Z2     Z2     U2    -U2   Z21;
            s2A'  -s2B'    Z12    Z12     1;
             Z12   s3B'  -s3C'    Z12     0;
             Z12    Z12   s4C'  -s4D'     0];

  t_Vec = [m2*a2(:,i);
           m3*a3(:,i) - FP;
```

```
            m4*a4(:,i);
            I2*alpha2;
            I3*alpha3(i) - dot(FP,s3P);
            I4*alpha4(i) - T4];

  f_Vec = S_Mat      _Vec;
  FA(:,i) = [f_Vec(1); f_Vec(2)];
  FB(:,i) = [f_Vec(3); f_Vec(4)];
  FC(:,i) = [f_Vec(5); f_Vec(6)];
  FD(:,i) = [f_Vec(7); f_Vec(8)];
  T2(i) = f_Vec(9);

  P2 = InertialPower(m2,I2,v2(:,i),a2(:,i),omega2,alpha2);
  P3 = InertialPower(m3,I3,v3(:,i),a3(:,i),omega3(i),alpha3(i));
  P4 = InertialPower(m4,I4,v4(:,i),a4(:,i),omega4(i),alpha4(i));
  PKin(i) = P2 + P3 + P4;     % kinetic power
  PF = dot(FP,vP(:,i));       % power from external force
  PT = T2(i) * omega2;        % power from crank torque
  PExt(i) =  PF + PT;         % total external power
end

% plot the driving torque
plot(theta2*180/pi,T2,'Color',[0 110/255 199/255])
title('Driving Torque for Fourbar Linkage')
xlabel('Crank angle (degrees)')
ylabel('Torque (N-m)')
grid on
set(gca,'xtick',0:60:360)
xlim([0 360])

% plot the external vs. kinetic power
figure
plot(theta2*180/pi,PKin,'o','Color',[153/255 153/255 153/255])
hold on
plot(theta2*180/pi,PExt,'Color',[0 110/255 199/255],'LineWidth',2)

title('External vs. Kinetic Power')
xlabel('Crank angle (degrees)')
ylabel('Power (W)')
legend('Kinetic','External')
grid on
set(gca,'xtick',0:60:360)
xlim([0 360])
```

7.9 Force Analysis Example 3 – The Grill Lid Lifting Mechanism

Some high-end outdoor grills, called "Kamado Grills" are made of solid ceramic, instead of the traditional sheet metal. The thick ceramic walls enable the grills to maintain a steady temperature for long periods of time, which is essential for proper slow-cooking of meats. One major drawback for this type of grill is its weight – some Kamado grills can weigh over 100 kg! Since the lid is made of the same solid ceramic material, it is also very heavy. It is common for these grills to have some type of lifting assistance mechanism instead of an ordinary hinge, in order to help the user raise the heavy lid.

The goal of this example problem is to design and analyze a fourbar mechanism for helping to raise the lid of a typical Kamado grill. The lid weighs 60 kg, and we wish to raise it to an angle of 70° from horizontal. The lid should also move a distance of 65 mm away from the base of the grill in order to provide a bigger open area over the cooking surface. Figure 7.69 shows a diagram of the grill and the desired motion of the lid.

Figure 7.70 shows a very rough conceptual design of the lifting mechanism. At this stage of the design process, it is helpful to make several sketches on some scrap paper (*not* on a computer!) This will help you to eliminate many of the inevitable "silly" ideas that arise during brainstorming. In our conceptual design, we attach the lid to the coupler of the fourbar linkage, since the lid must translate and rotate. The base of the grill will serve as the ground. For ease of manufacture, both of the ground pins are situated on a line that is 40 mm away from the base. We will use a spring connected between the coupler and ground to keep the lid in the raised position after lifting. Part of our design exercise is to find out how strong to make the spring, and where to attach it to the ground and coupler.

The ground pins are vertically aligned in Figure 7.70, but all of our analyses have assumed that the ground pins are horizontally aligned. This means that we will need to rotate the mechanism by 90° after the design is complete in order to perform the force calculations – gravity will be acting horizontally in our analysis. We will keep the ground pins vertically aligned during the design process to aid with visualization.

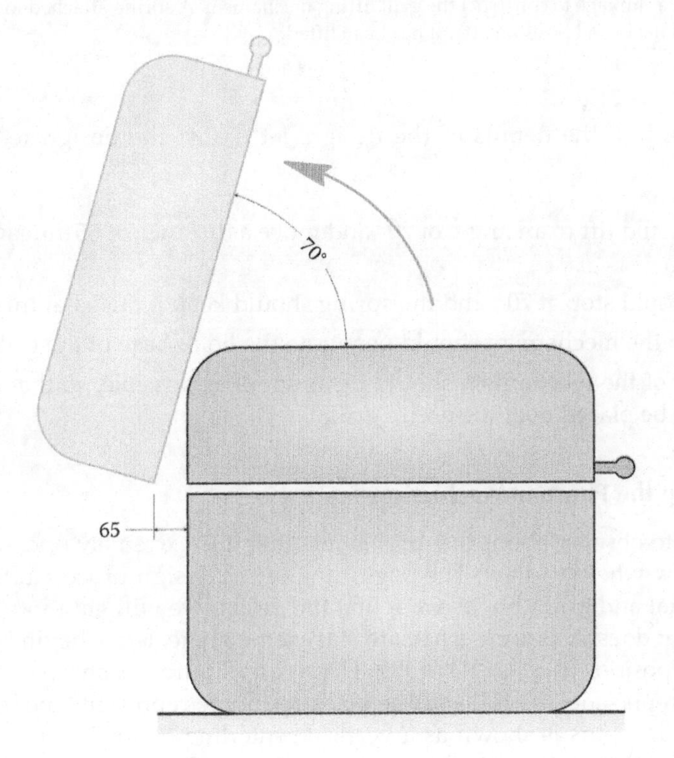

FIGURE 7.69
Desired motion for the lid of the grill. The lid should move 65 mm horizontally and raise to an angle of 70°.

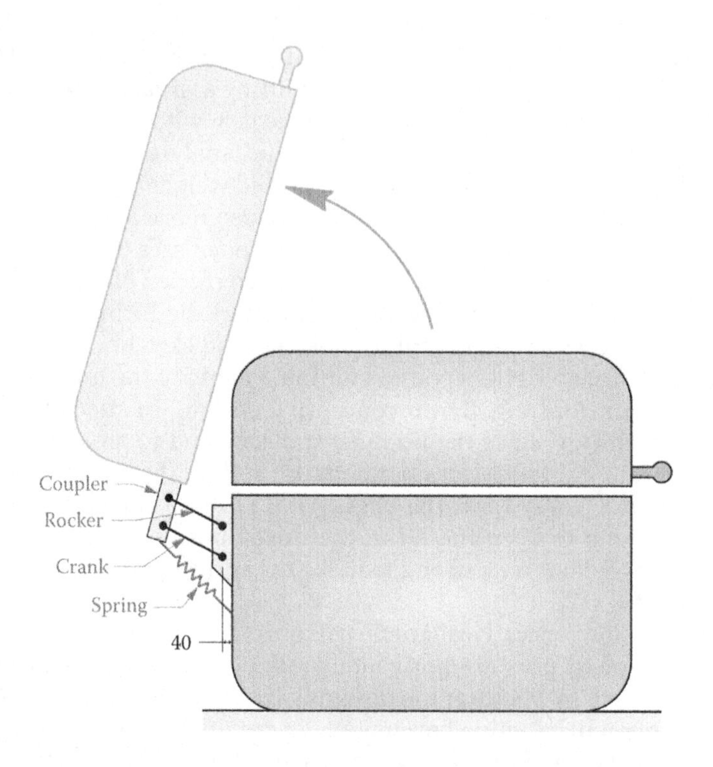

FIGURE 7.70
We will use a fourbar linkage to construct the grill lifting mechanism. A spring attached to the base of the grill will keep the lid in the raised position after it has been lifted.

Before diving into the details of the design, let us list the design requirements and constraints:

1. The lid should lift to an angle of 70° and move a distance of 65 mm away from the base.
2. The lid should stop at 70°, and the spring should keep it raised at this height.
3. No part of the mechanism should penetrate the lid or base of the grill.
4. The parts of the mechanism should be as small as possible, so that a cloth cover can easily be placed over the entire grill.

7.9.1 Designing the Fourbar Mechanism

The first thing to observe about this problem is that it is extremely open-ended, and it is difficult to know where to start. The beginning of the design process often consists of a great deal of trial and error, but as we refine the design we will gain intuition as to what works, and what doesn't. Since we have to start somewhere, let us begin by sketching the lid in its closed position in a SOLIDWORKS Drawing. Figure 7.71 shows such a sketch. The blue rectangle represents the lid, and the black rectangle represents the body of the grill. The line of ground pins is shown as a vertical centerline 40 mm away from the body of the grill. The lid is attached to the lifting mechanism with a 15 mm wide metal band that wraps around the circumference of the lid; thus, the centerline of this band is 7.5 mm above

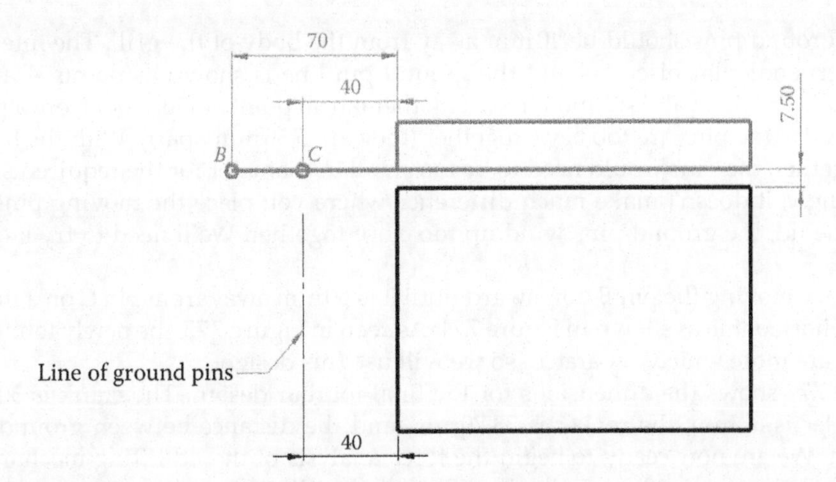

FIGURE 7.71
A schematic drawing of the lid in its closed position. We will first try aligning the moving pins (*B* and *C*) with the bottom of the lid.

the top of the body of the grill. What is shown as the "bottom" of the lid in the figure is actually the centerline of the band.

As a first guess, let us place the moving pins in line with the centerline of the band, as shown in Figure 7.71 as points *B* and *C*. Our next step would be to draw the lid in its raised position, as shown in Figure 7.72. It is helpful to use the **Layers** feature in SOLIDWORKS to keep each position (and the dimensions) on their own layers.

With both lid positions visible, draw perpendicular bisectors between $B_1 - B_2$ and $C_1 - C_2$. Remember that anywhere along these lines represents a valid ground point that will achieve the desired motion for the lid. In the design requirements, however, we stipulated

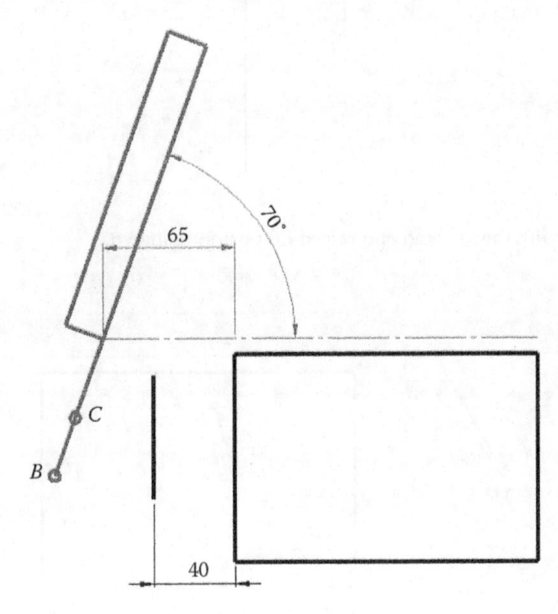

FIGURE 7.72
The lid shown in the raised position, with the moving pins (*B* and *C*) added.

that the ground pins should lie 40 mm away from the body of the grill. The intersection of the perpendicular bisectors and the ground pin line is shown as points A and D in Figure 7.73. This is a valid solution from a mathematical point of view, but from a practical point of view the pins are too close together (they are 5.56 mm apart). With the holes this close together, the pins would need to be too small in diameter for the required strength. Interestingly, it doesn't make much difference where you place the moving pins on the line of the lid, the ground pins wind up too close together. We'll need to try a different solution.

Let us try moving the pin B downward until it is 60 mm away from pin C on a line −120° from the horizontal, as shown in Figure 7.74. As seen in Figure 7.75, the newly found points A and D are more widely separated, so we will use this design.

Figure 7.76 shows the dimensions for the final fourbar design. The crank is 30.04 mm, the coupler is 60 mm, the rocker is 38.70 mm, and the distance between ground pins is 14.81 mm. We are now ready to begin the force analysis of the lid-lifting mechanism. To keep ourselves organized, we will follow Steps 1–6 outlined in Section 7.5.

7.9.2 Determine the Critical Dimensions of the Linkage

We have found the dimensions using graphical linkage synthesis (with some trial and error.) For the crank we have a = 30.04 mm, for the coupler we have b = 60 mm, for the rocker we have c = 38.70 mm and the distance between ground pins is d = 14.81 mm. As

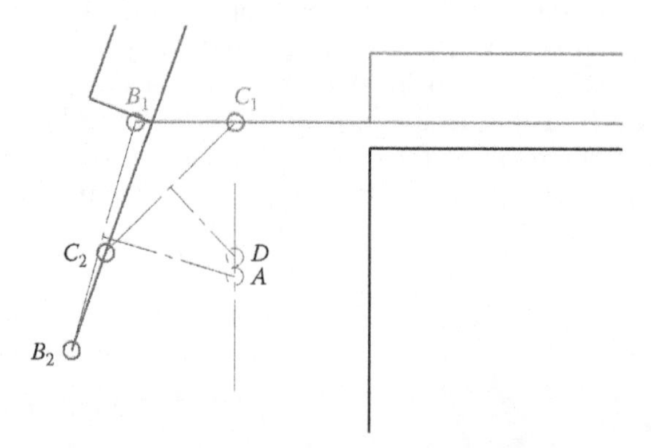

FIGURE 7.73
The first design iteration, with the lowered and raised lid positions shown.

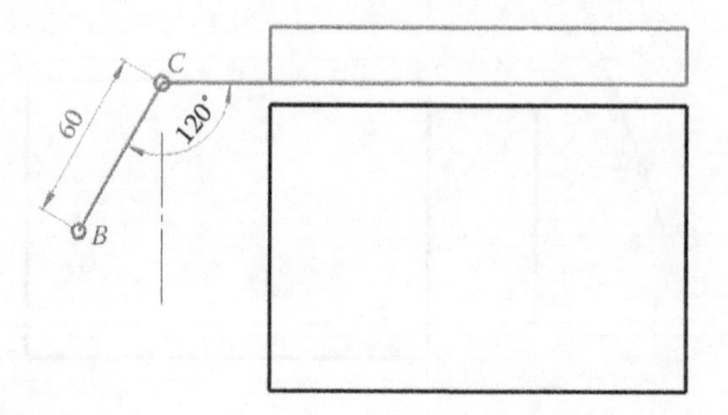

FIGURE 7.74
For our second design, we will move the pin B downward until it is 60 mm away from pin C along a 120° line.

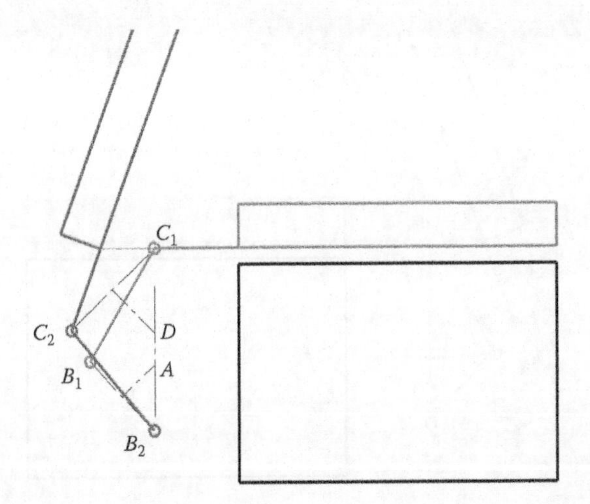

FIGURE 7.75
Points A and D found using the new design for the coupler.

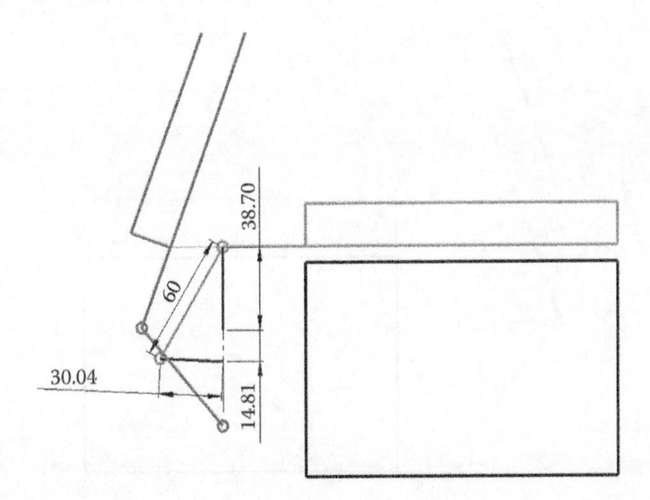

FIGURE 7.76
Final design for the grill lifter mechanism.

seen in Figure 7.77 the crank is at an angle of 87.04° when the lid is closed and moves to an angle of 179.78° when the lid is in the open position (see Figure 7.78). Use **Save As** to save your fourbar force analysis code to the file `Fourbar_GrillLifter.m`. Change the linkage dimensions as follows, and add the crank limiting angles.

```
% Linkage dimensions
a = 0.03004;              % crank length (m)
b = 0.06000;              % coupler length (m)
c = 0.03870;              % rocker length (m)
d = 0.01481;              % length between ground pins (m)
p = 0.59730;              % length from B to P (m)
gamma3 = -55*pi/180;      % angle between BP and coupler (converted to rad)
theta2min = 87.04*pi/180;   % crank position with lid closed
theta2max = 179.78*pi/180;  % crank position with lid open
```

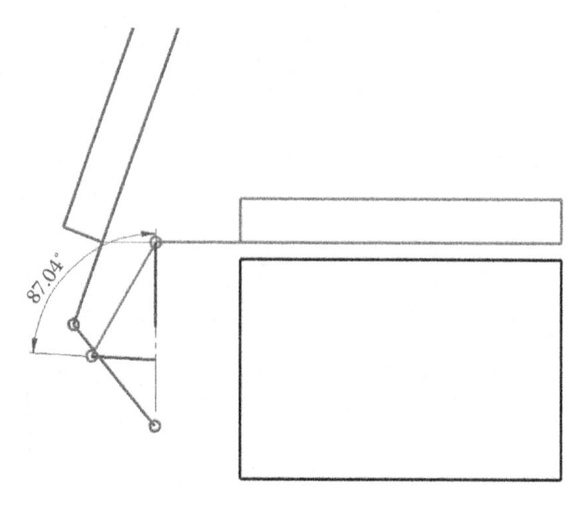

FIGURE 7.77
The crank begins its rotation at 87.04° from the horizontal.

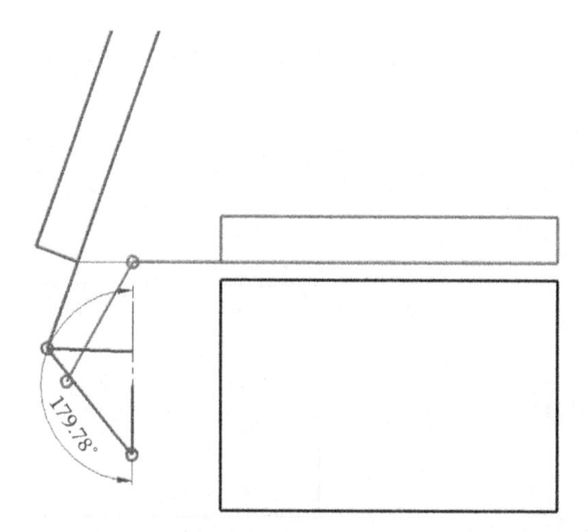

FIGURE 7.78
The ending angle of the crank is 179.78° from the horizontal.

7.9.3 Determine the Inertial Properties of Each Body in the Mechanism

Next, we will analyze the inertial properties of the links. The lid, which has a mass of 60 kg, is attached to the coupler. The rocker and crank are made of relatively thin stamped steel, and their masses and moments of inertia may safely be neglected in comparison with the lid. The lid is 450 mm wide and 220 mm high and its moment of inertia and location of center of mass were found using the **Mass Properties** feature in SOLIDWORKS.

$$m_3 = 60 \text{ kg} \quad I_3 = 1.35 \text{ kg} \cdot \text{m}^2$$

Figure 7.79 shows the dimensions of the coupler and attached lid. In designing the coupler we located pin C a distance of 40 mm away from the lid. The center of mass of the lid is 110 mm above the line *CP* and the handle is located at point *P*, which is 75 mm from the front of the lid. In our previous analyses of the fourbar we placed the origin of the coupler

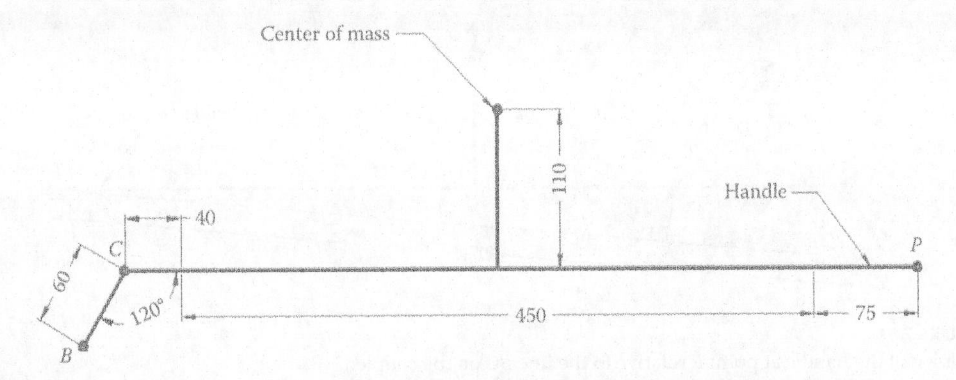

FIGURE 7.79
Dimensions of the coupler, which is attached to the lid of the grill. The location of the center of mass has been found using **Mass Properties.** The user lifts the lid with the handle at point *P*.

at point *B* and measured the angle of the coupler from the line *BC*. We must, therefore, find the location of the other points on the coupler (the center of mass and the handle at point *P*) relative to the line *BC*. Figure 7.80 shows the location of the center of mass relative to the line *BC*.

$$\bar{x}_3 = 287.8 \text{ mm} \quad \bar{y}_3 = -174.5 \text{ mm}$$

The user lifts the lid with the handle located at point *P*, as shown in Figure 7.81. From the figure, we see that

$$p = 597.3 \text{ mm} \quad \gamma_3 = -55°$$

We will assume that the centers of mass of the crank and rocker lie at their geometric centers

$$\bar{x}_2 = \frac{a}{2} \quad \bar{y}_2 = 0$$

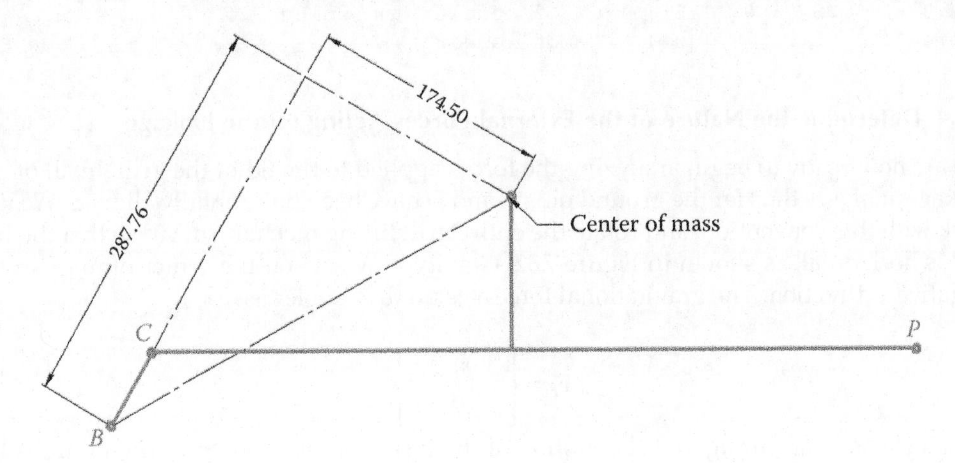

FIGURE 7.80
Location of the center of mass relative to the line *BC* on the coupler.

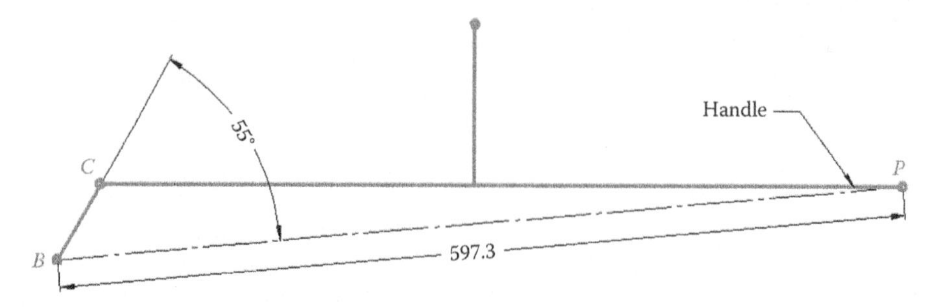

FIGURE 7.81
Location of the handle at point P relative to the line BC on the coupler.

$$\bar{x}_4 = \frac{c}{2} \quad \bar{y}_4 = 0$$

The code giving the inertial properties is shown below. Also given is the gravitational force (which is constant) and the spring stiffness and unstretched length. Both of these are design variables that we can "tweak" once we have a working force analysis code. For now, we have entered a spring constant of zero.

```
% Inertial properties
m2 = 0.0;            % mass of crank (kg)
m3 = 60.0;           % mass of coupler including lid (kg)
m4 = 0.0;            % mass of rocker (kg)
I2 = 0.0;            % moment of inertia of crank about CM (kg-m2)
I3 = 1.35;           % moment of inertia of coupler incl lid about CM (kg-m2)
I4 = 0.0;            % moment of inertia of rocker about CM (kg-m2)
g = 9.81;            % acceleration of gravity (m/s2)
Fg = [-m3*g; 0];     % force of gravity on the lid
xk0 = 0.01;          % unstretched length of spring
k = 0;               % spring constant (N/m)

% CM locations
xbar2 = [   a/2;         0]; % CM of crank
xbar3 = [0.2878;-0.1745]; % CM of coupler including lid
xbar4 = [   c/2;         0]; % CM of rocker
```

7.9.4 Determine the Nature of the External Forces Acting on the Linkage

We are now ready to begin analyzing the forces applied to the lid of the grill. In all of our linkage analyses thus far the ground pins A and D have been horizontally aligned. We will stick with this convention, and rotate the entire grill lifting mechanism 90° so that the line AD is horizontal, as shown in Figure 7.82. Gravity now acts on the center of mass in the negative x direction. The gravitational force acts on the coupler as

$$\mathbf{F}_g = \begin{Bmatrix} -m_3 g \\ 0 \end{Bmatrix}$$

which is constant throughout the motion of the linkage. Also shown in the figure is the spring, which is connected between point Q on the ground and the pin B. Since the pin B attaches the crank to the coupler we can choose to have the spring force act on either of

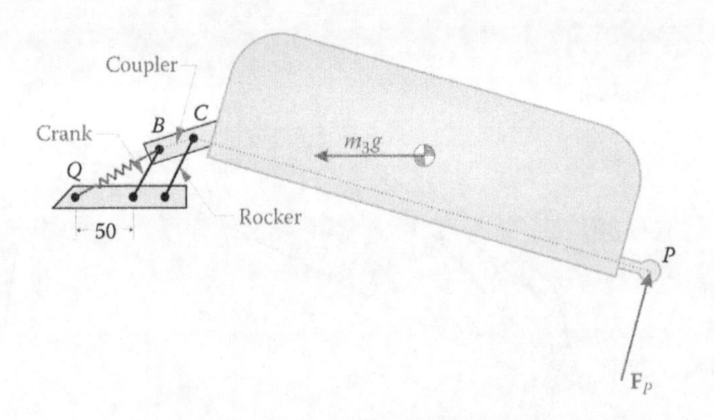

FIGURE 7.82
The grill has been rotated 90° to make the ground pins aligned with the horizontal.

these links. To keep the crank equations simple, we will assume that the spring force acts on the coupler. The point Q is 50 mm (an arbitrary number!) away from point A on the line AD. Let us define a vector $\mathbf{r}_{BQ}$ that extends from point B to point Q.

$$\mathbf{r}_{BQ} = \mathbf{x}_B - \mathbf{x}_Q$$

The spring force is then

$$\mathbf{F}_k = k\left(\left|\mathbf{r}_{BQ}\right| - x_0\right)\mathbf{e}_{BQ}$$

where k is the spring constant, x_0 is the unstretched length of the spring and

$$\mathbf{e}_{BQ} = \frac{\mathbf{r}_{BQ}}{\left|\mathbf{r}_{BQ}\right|}$$

is the unit vector acting in the direction of the spring.

The user lifts the lid with unknown force $\mathbf{F}_P$, which we assume acts perpendicular to the line CP. We can write a vector from point C to point P as

$$\mathbf{r}_{CP} = \mathbf{x}_P - \mathbf{x}_C$$

Then the user force $\mathbf{F}_P$ must satisfy

$$\mathbf{F}_P \cdot \mathbf{e}_{CP} = 0$$

where

$$\mathbf{e}_{CP} = \frac{\mathbf{r}_{CP}}{\left|\mathbf{r}_{CP}\right|}$$

is the unit vector pointing from C to P.

7.9.5 Draw Free-body Diagrams of Each Link in the Mechanism

The free-body diagrams of each link in the lid-lifting mechanism are shown in Figure 7.83. The crank has only the two-pin forces acting on it, so its equations of motion are

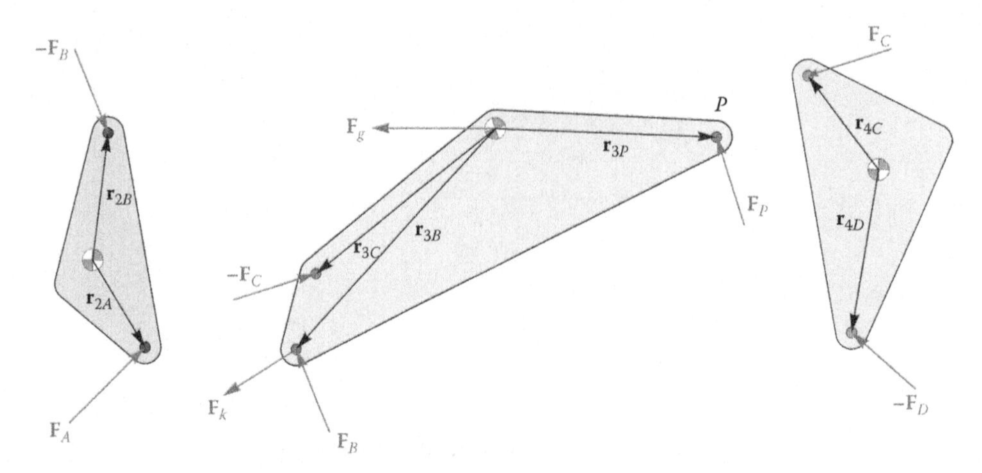

FIGURE 7.83
Free-body diagrams of each link in the lid-lifting mechanism

$$\mathbf{F}_A - \mathbf{F}_B = m_2\mathbf{a}_2$$

$$\mathbf{s}_{2A} \cdot \mathbf{F}_A - \mathbf{s}_{2B} \cdot \mathbf{F}_B = I_2\alpha_2$$

The coupler is more complicated, since it has the spring force, the gravitational force, and the force from the user acting on it. Since it acts at point B, moment arm of the spring force is the same as for the force $\mathbf{F}_B$. Gravity exerts no moment on the coupler since it acts at the center of mass. The equations of motion for the coupler are then

$$\mathbf{F}_B - \mathbf{F}_C + \mathbf{F}_P + \mathbf{F}_k + \mathbf{F}_g = m_3\mathbf{a}_3$$

$$\mathbf{s}_{3B} \cdot \mathbf{F}_B - \mathbf{s}_{3C} \cdot \mathbf{F}_C + \mathbf{s}_{3P} \cdot \mathbf{F}_P + \mathbf{s}_{3B} \cdot \mathbf{F}_k = I_3\alpha_3$$

Finally, the equations of motion for the rocker are

$$\mathbf{F}_C - \mathbf{F}_D = m_4\mathbf{a}_4$$

$$\mathbf{s}_{4C} \cdot \mathbf{F}_C - \mathbf{s}_{4D} \cdot \mathbf{F}_D = I_4\alpha_4$$

since it experiences only the two forces at its pins. The unknown forces in these equations are $\mathbf{F}_A$, $\mathbf{F}_B$, $\mathbf{F}_C$, $\mathbf{F}_D$, and $\mathbf{F}_P$ – a total of 10 unknowns. We assume that the spring force is known once the position analysis has been carried out. The matrix force equation is mostly unchanged from that of the generic fourbar

$$\mathbf{S} = \begin{bmatrix} \mathbf{U}_2 & -\mathbf{U}_2 & \mathbf{0}_2 & \mathbf{0}_2 & \mathbf{0}_2 \\ \mathbf{0}_2 & \mathbf{U}_2 & -\mathbf{U}_2 & \mathbf{0}_2 & \mathbf{U}_2 \\ \mathbf{0}_2 & \mathbf{0}_2 & \mathbf{U}_2 & -\mathbf{U}_2 & \mathbf{0}_2 \\ \mathbf{s}_{2A}^T & -\mathbf{s}_{2B}^T & \mathbf{0}_{12} & \mathbf{0}_{12} & \mathbf{0}_{12} \\ \mathbf{0}_{12} & \mathbf{s}_{3B}^T & -\mathbf{s}_{3C}^T & \mathbf{0}_{12} & \mathbf{s}_{3P}^T \\ \mathbf{0}_{12} & \mathbf{0}_{12} & \mathbf{s}_{4C}^T & -\mathbf{s}_{4D}^T & \mathbf{0}_{12} \\ \mathbf{0}_{12} & \mathbf{0}_{12} & \mathbf{0}_{12} & \mathbf{0}_{12} & \mathbf{e}_{CP}^T \end{bmatrix}$$

$$\mathrm{f} = \begin{Bmatrix} \mathbf{F}_A \\ \mathbf{F}_B \\ \mathbf{F}_C \\ \mathbf{F}_D \\ \mathbf{F}_P \end{Bmatrix} \quad \mathrm{t} = \begin{Bmatrix} m_2\mathbf{a}_2 \\ m_3\mathbf{a}_3 - \mathbf{F}_k - \mathbf{F}_g \\ m_4\mathbf{a}_4 \\ I_2\alpha_2 \\ I_3\alpha_3 - \mathbf{s}_{3B}\cdot\mathbf{F}_k \\ I_4\alpha_4 \\ 0 \end{Bmatrix}$$

The force analysis portion of the code is

```
% conduct force analysis
  Fk = -k*(norm(xB(:,i)-xQ) - xk0)*eBQ;

  S_Mat = [   U2     -U2     Z2     Z2      Z2;
              Z2      U2    -U2     Z2      U2;
              Z2      Z2     U2    -U2      Z2;
             s2A'   -s2B'   Z12    Z12     Z12;
             Z12     s3B'  -s3C'   Z12     s3P';
             Z12     Z12    s4C'  -s4D'    Z12;
             Z12     Z12    Z12    Z12     eCP'];

  t_Vec = [m2*a2(:,i);
           m3*a3(:,i) - Fk - Fg;
           m4*a4(:,i);
           I2*alpha2;
           I3*alpha3(i) - dot(Fk,s3B);
           I4*alpha4(i);
           0];

  f_Vec = S_Mat\t_Vec
  FA(:,i) = [f_Vec(1); f_Vec(2)];
  FB(:,i) = [f_Vec(3); f_Vec(4)];
  FC(:,i) = [f_Vec(5); f_Vec(6)];
  FD(:,i) = [f_Vec(7); f_Vec(8)];
  FP(:,i) = [f_Vec(9); f_Vec(10)];
```

Remember that the `norm` function gives the magnitude of a vector – in this case the total length of the spring.

7.9.6 Determine the Nature of the Motion of the Crank

We will use the same acceleration pulse technique that we used for the threebar door closing mechanism in order to open the grill lid. We will assume that the user takes 2 seconds to open the grill. The only difference between the grill lifter and the door closer is that the crank of the grill lid begins its rotation at 87.04° instead of zero. The crank equations of motion for the grill lifter are

$$\alpha_2 = A\sin\lambda t$$

$$\omega_2 = \frac{A}{\lambda}(1-\cos\lambda t)$$

$$\theta_2 = \frac{A}{\lambda}t - \frac{A}{\lambda^2}\sin\lambda t + \theta_{2\min}$$

where

$$A = \frac{\lambda}{T}(\theta_{2\max} - \theta_{2\min})$$

The code that initializes the acceleration pulse parameters is

```
% Acceleration pulse parameters
T = 2;                                  % time it takes to open door
lambda = 2*pi/T;                        % door opening "frequency"
N = 1001;                               % number of time steps
dt = T/(N-1);                           % time increment
t = 0:dt:T;                             % vector of simulation time
A = lambda*(theta2max - theta2min)/T;   % amplitude of acceleration pulse
```

Inside the main loop, define the crank angle, angular velocity, and angular acceleration as

```
  theta2(i) = (A/lambda)*t(i) - (A/lambda^2)*sin(lambda*t(i)) + theta2min;
  omega2 = (A/lambda)*(1-cos(lambda*t(i)));
  alpha2 = A*sin(lambda*t(i));
```

This completes the definition of the crank motion.

7.9.7 Solve the Equations of Motion and Plot the Desired Results

Before we begin using our code to find the best spring constant we should do a few quick verification calculations to make sure we are getting meaningful results. First, make sure that the spring constant is set to zero. Lifting the lid in the presence of gravity will increase its potential energy, where the gravitational potential energy of an object is

$$PE = mgh$$

and h is the height that the object has been lifted. To convert this to a power, differentiate with respect to time

$$P_g = \frac{d}{dt}(PE) = \frac{d}{dt}mgh = mg\frac{dh}{dt}$$

The quantity (dh/dt) is the velocity of the object in the direction of the gravitational field, and mg is the gravitational force, so that we have:

$$P_g = \mathbf{F}_g \cdot \mathbf{v}$$

where $\mathbf{v}$ is the velocity of the center of mass of the object. This is the same power expression as we have used for calculating the power from any other external force. We can use the same logic to find the power consumed in stretching the spring

$$P_k = \mathbf{F}_k \cdot \mathbf{v}_B$$

since the point Q at the other end of the spring is fixed. The power computations are then

```
P2 = InertialPower(m2,I2,v2(:,i),a2(:,i),omega2,alpha2);
P3 = InertialPower(m3,I3,v3(:,i),a3(:,i),omega3(i),alpha3(i));
P4 = InertialPower(m4,I4,v4(:,i),a4(:,i),omega4(i),alpha4(i));
PKin(i) = P2 + P3 + P4;      % kinetic power

Pg = dot(Fg,v3(:,i));         % gravitational power
Pk = dot(Fk,vB(:,i));         % spring power
PF = dot(FP(:,i),vP(:,i));  % power from handle force
PExt(i) = PF+Pg;              % total external power
```

The conservation of power plot with no spring force is shown in Figure 7.84. As you can see, the two curves match exactly. Now change the spring constant to $k = 10{,}000\,\text{N/m}$ and plot the power curves again. Interestingly, nothing seems to have changed! The additional energy stored in the stretched spring goes directly into raising the potential energy of the lid, so the net effect on the power curve is zero. Since our power curves line up so nicely, we conclude that the code is providing meaningful results.

7.9.8 Using the Code to Improve the Design

Our primary design goal, as originally stated, is to develop a linkage and spring that would assist the user in lifting the heavy lid. Remember that gravity is acting horizontally in our simulation. To make the plots easier to interpret we define

$$\mathbf{F}_P' = \begin{Bmatrix} -F_{Py} \\ F_{Px} \end{Bmatrix}$$

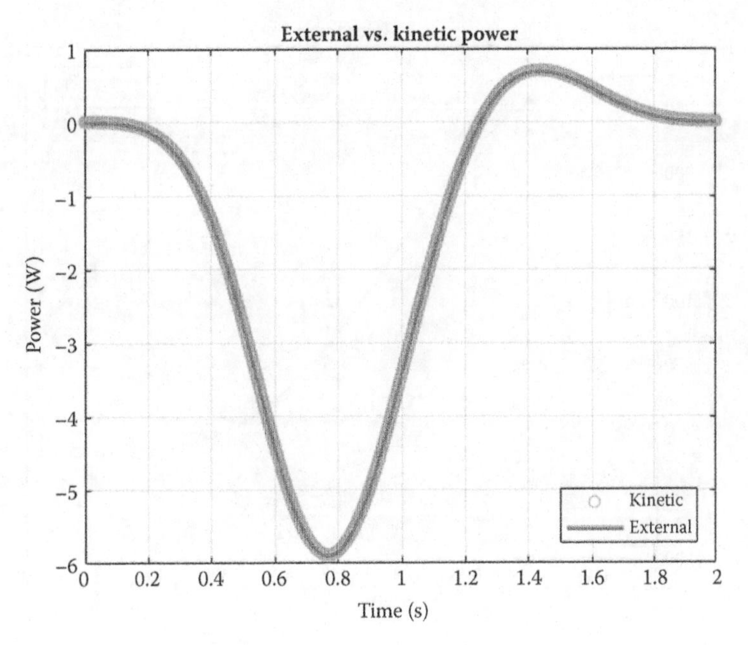

FIGURE 7.84
Conservation of power plot for the linkage without the force from the spring.

That is, we have rotated the axes 90° for plotting purposes. Change the spring constant back to zero, and plot the lid lifting force, $\mathbf{F}_P$. The result is shown in Figure 7.85. At the beginning of the motion the user must exert a 275 N vertical force to lift the lid – too much for most users. Clearly, we will need the spring force to assist the user.

A secondary goal was to ensure that the lid remains in the upright position after it has been raised so that the user does not need to keep the lid raised manually. Since the force has two components, it is difficult to know whether we have achieved this goal. Let us instead plot the *torque* created by the handle force

$$T_3 = \mathbf{F}_P \cdot \mathbf{s}_{3P}$$

At the beginning of the motion, the torque will be positive since the user is moving the lid counterclockwise and raising it in the gravitational field. If the torque is negative at the end of the motion it means that gravity is assisting the user in moving the lid, and that the lid will stay open by itself after being raised.

The torque exerted by the user in order to lift the lid without the spring force is shown in Figure 7.86. As expected, the torque starts out positive and ends negative. Now try increasing the spring constant to 50,000 N/m.

As shown in Figure 7.87, the maximum required torque has been reduced to 50 N-m. For the best user experience, we should try to reduce the peak torque level (positive or negative) to the minimum possible value.

The required torque and handle forces for a spring rate of 90,000 N/m are shown in Figure 7.88 and Figure 7.89. The peak force has been reduced from 275 N to less than 80 N, which is well within the capabilities of most adults. Thus, in a relatively short time we have achieved a "pretty good" design through modifying only one of the design variables. We have written the code in such a way that it would be easy to choose other variables to optimize (e.g. the location of the fixed end of the spring, the unstretched length of the spring,

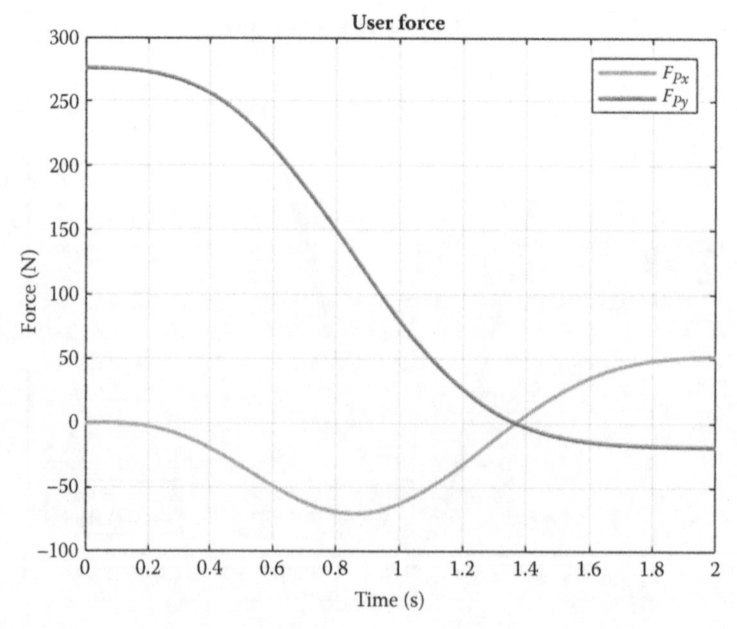

FIGURE 7.85
Handle force exerted by the user in lifting the lid.

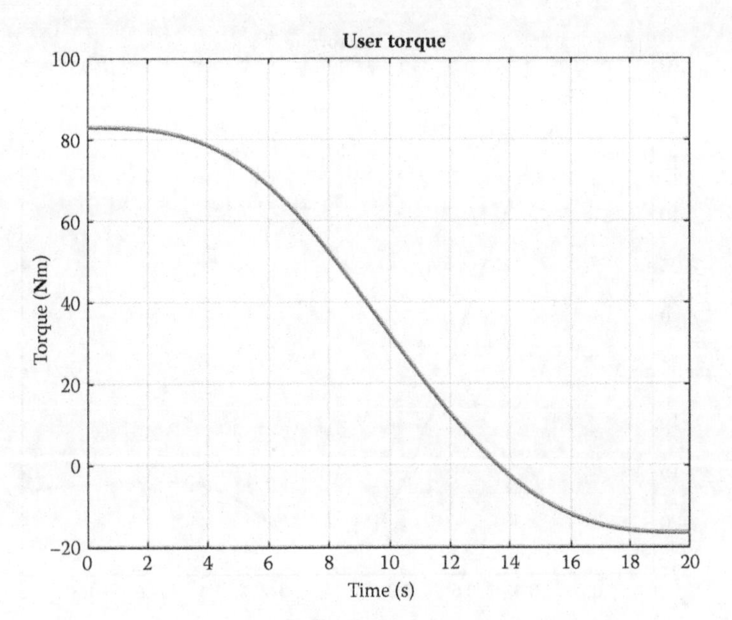

FIGURE 7.86
Torque exerted by the user to lift the lid without the spring force.

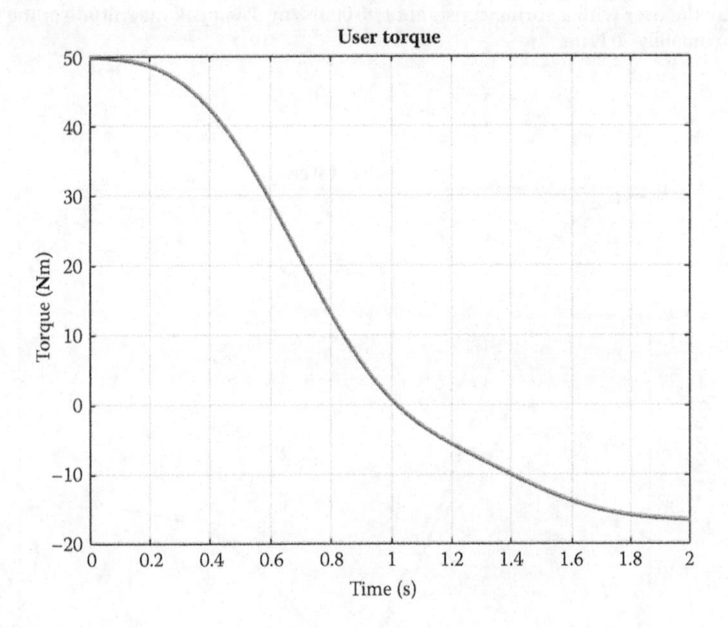

FIGURE 7.87
User torque with a spring constant of 50,000 N/m. The maximum required torque has been reduced to 50 N m.

etc.) but we leave this as an exercise for the reader. It should be noted that some of the design variables – the link lengths – are quite a bit more difficult to optimize. If we change any of the link lengths we must return to our SOLIDWORKS drawing to ensure that we retain the desired motion and we must recalculate the crank starting and ending angles. In addition, the center of mass location relative to the line *BC* will change. This is the reason that path optimization of linkages is still an area of active research in kinematics today.

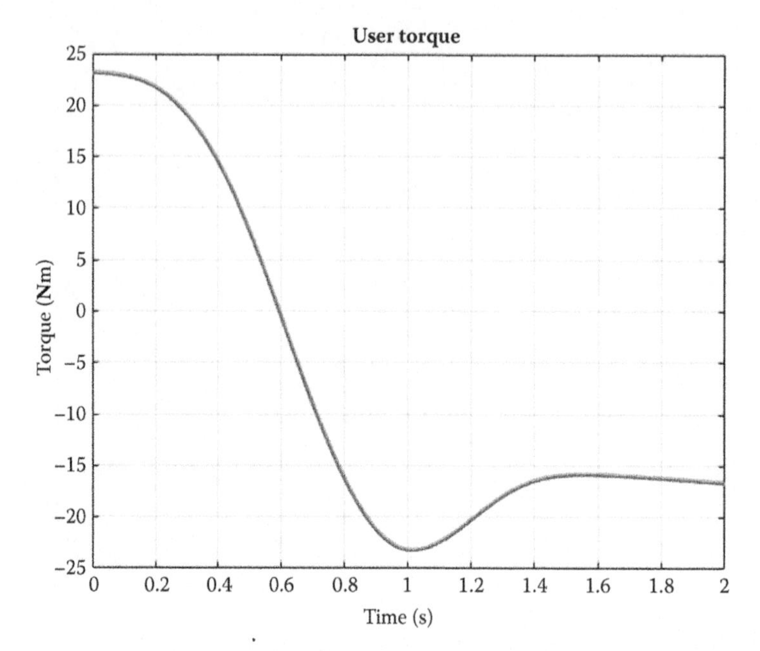

FIGURE 7.88
Torque exerted by the user with a spring constant of 90,000 N/m. The peak magnitude of the torque has been reduced to approximately 20 N m.

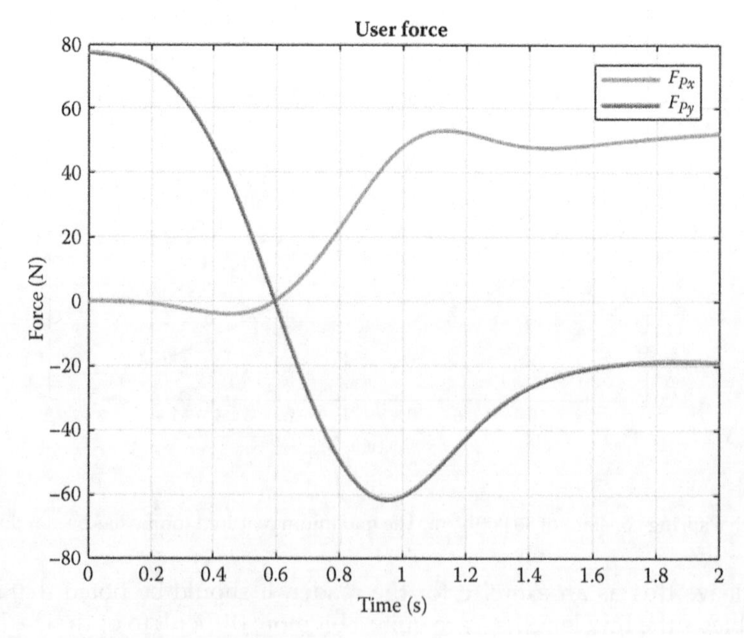

FIGURE 7.89
Force exerted by the user with a spring of 90,000 N/m. The peak force is slightly less than 80 N, which is achievable by most users.

7.9.9 Summary

This section has presented a practical design problem, a lifting mechanism for a heavy grill lid. We began by using graphical linkage synthesis techniques to find a suitable fourbar linkage for raising the lid to its desired position. We used the SOLIDWORKS **Mass Properties** feature to calculate the center of mass and inertial properties of the lid. Next we modified the fourbar force analysis code to account for the user force input at the handle, instead of a crank motor torque. Finally, we demonstrated the use of the code for finding a suitable value of one of the design variables – the spring constant. By using a systematic approach, we were able to tackle this complicated design problem in a straightforward way. In the end, we reached the conclusion that some design variable are quite simple to modify, once the simulation code is complete, but others (e.g. link lengths) are more challenging.

7.10 Force Analysis of the Inverted Slider-Crank

Force analysis on the inverted slider-crank is a little more complicated than it was for the slider-crank. We proceed in the same manner as before, drawing a free-body diagram of each link and summing forces and torques. Figure 7.90 shows a diagram of the slider-crank with a known force, $\mathbf{F}_P$, applied at the end of the slider. The problem statement is the same as for the fourbar and slider-crank: find the driving torque necessary to actuate the linkage at the desired speed. Along the way, we will solve for the pin forces and the *moment* between the slider and rocker.

The free-body diagram of the crank is the same as it was for the fourbar, see Figure 7.91. The main distinguishing feature of the inverted slider-crank is the moving full-slider joint. If we assume that friction is negligible, then the force between the slider and the rocker can have no component parallel to the slider. Furthermore, rotating the slider will

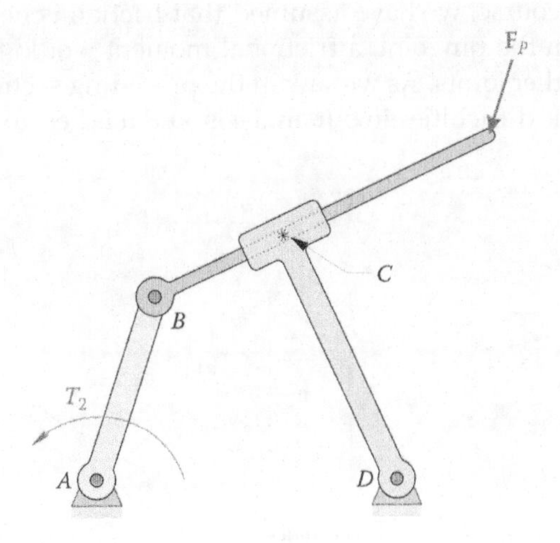

FIGURE 7.90
The inverted slider-crank linkage.

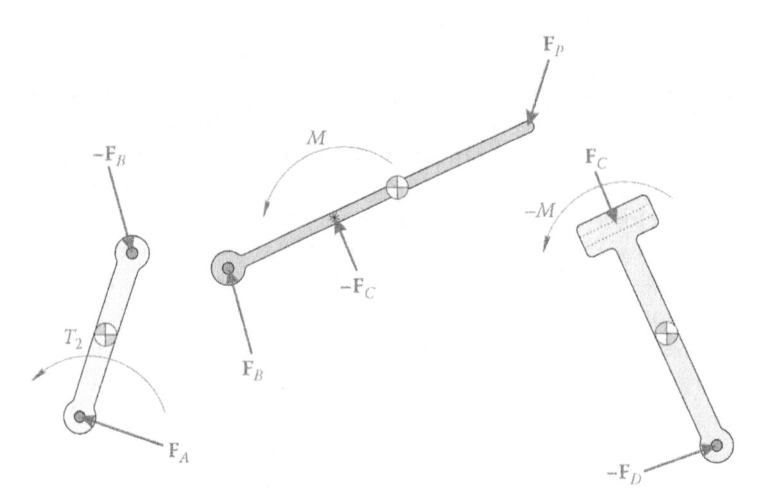

FIGURE 7.91
Free-body diagram of each link in the inverted slider-crank.

create a *moment* that causes the rocker to rotate by the same amount. Remember that one of the important features of the inverted slider-crank is that the slider and rocker have the same rotation, angular velocity, and angular acceleration. The moment between the two bodies is what enforces the rotational part of the full-slider constraint. Because the full-slider makes no constraint along the axis of the slider, the force is zero in this direction.

This discussion provides a new way to look at the various joints (or constraints) that we discussed in Chapter 1. Each joint requires a set of forces or moments to enforce the constraint condition. As seen in Figure 7.92, the pin joint requires two forces, one in the x direction and one in the y direction. The full-slider requires a normal force and a moment. The normal force keeps the slider in the slot, and the moment prevents the slider from rotating relative to the slot. Finally, the half-slider has only a normal force that keeps the pin inside the slot. Of course, we have assumed that friction is negligible in the joints. If friction were present in the pin joint, a frictional moment would be created, and similar results obtain for the other joints. As we saw in the preceding section, the presence of friction creates considerable difficulties in our analysis, and it is best to ignore it for a first-pass analysis.

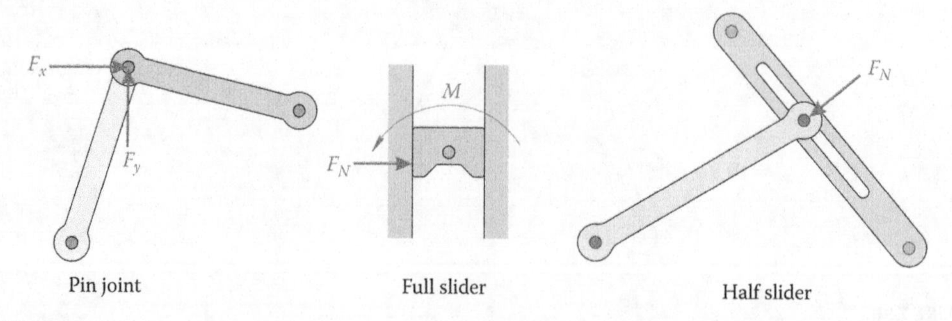

FIGURE 7.92
The forces and moments required to enforce constraint conditions. The pin joint requires two forces (x and y), the full-slider requires a normal force and a moment, and the half-slider requires only a normal force.

The equations of motion for the crank are the same as those for the fourbar:

$$\mathbf{F}_A - \mathbf{F}_B = m_2\mathbf{a}_2$$

$$\mathbf{s}_{2A}\cdot\mathbf{F}_A - \mathbf{s}_{2B}\cdot\mathbf{F}_B + T_2 = I_2\alpha_2$$

Adding forces and moments on the slider gives

$$\mathbf{F}_B - \mathbf{F}_C + \mathbf{F}_P = m_3\mathbf{a}_3$$

$$\mathbf{s}_{3B}\cdot\mathbf{F}_B - \mathbf{s}_{3C}\cdot\mathbf{F}_C + \mathbf{s}_{3P}\cdot\mathbf{F}_P + M = I_3\alpha_3$$

where M is the unknown moment created by the full-slider joint. The equations of motion for the rocker are

$$\mathbf{F}_C - \mathbf{F}_D = m_4\mathbf{a}_4$$

$$\mathbf{s}_{4C}\cdot\mathbf{F}_C - \mathbf{s}_{4D}\cdot\mathbf{F}_D - M = I_4\alpha_4$$

Note that we have used $-M$ in the equation above to account for the fact that the unknown moment acts in the opposite direction as on the slider. We now have nine equations (two force and one moment equation for each link) and ten unknowns ($\mathbf{F}_A$, $\mathbf{F}_B$, $\mathbf{F}_C$, $\mathbf{F}_D$, M, T_2). We will obtain the final equation by considering the action of the full-slider joint on the force $\mathbf{F}_C$. Since $\mathbf{F}_C$ is perpendicular to the slider, we may write

$$\mathbf{F}_C\cdot\mathbf{e}_3 = 0$$

because $\mathbf{e}_3$ is parallel to the slider. This provides our final equation. Collecting the equations into matrix form gives

$$\mathbf{S} = \begin{bmatrix} \mathbf{U}_2 & -\mathbf{U}_2 & \mathbf{0}_2 & \mathbf{0}_2 & \mathbf{0}_{21} & \mathbf{0}_{21} \\ \mathbf{0}_2 & \mathbf{U}_2 & -\mathbf{U}_2 & \mathbf{0}_2 & \mathbf{0}_{21} & \mathbf{0}_{21} \\ \mathbf{0}_2 & \mathbf{0}_2 & \mathbf{U}_2 & -\mathbf{U}_2 & \mathbf{0}_{21} & \mathbf{0}_{21} \\ \mathbf{s}_{2A}^T & -\mathbf{s}_{2B}^T & \mathbf{0}_{12} & \mathbf{0}_{12} & 0 & 1 \\ \mathbf{0}_{12} & \mathbf{s}_{3B}^T & -\mathbf{s}_{3C}^T & \mathbf{0}_{12} & 1 & 0 \\ \mathbf{0}_{12} & \mathbf{0}_{12} & \mathbf{s}_{4C}^T & -\mathbf{s}_{4D}^T & -1 & 0 \\ \mathbf{0}_{12} & \mathbf{0}_{12} & \mathbf{e}_3^T & \mathbf{0}_{12} & 0 & 0 \end{bmatrix}$$

$$\mathbf{f} = \begin{Bmatrix} \mathbf{F}_A \\ \mathbf{F}_B \\ \mathbf{F}_C \\ \mathbf{F}_D \\ M \\ T_2 \end{Bmatrix} \quad \mathbf{t} = \begin{Bmatrix} m_2\mathbf{a}_2 \\ m_3\mathbf{a}_3 - \mathbf{F}_P \\ m_4\mathbf{a}_4 \\ I_2\alpha_2 \\ I_3\alpha_3 - \mathbf{s}_{3P}\cdot\mathbf{F}_P \\ I_4\alpha_4 \\ 0 \end{Bmatrix}$$

Modify your code for the inverted slider crank acceleration analysis to perform the force analysis.

Example 7.5

Calculate the driving torque for an inverted slider-crank with the dimensions and inertial properties shown in Figure 7.93. A downward vertical force of 100 N is applied at the end of the slider.

Solution

After modifying the inverted slider crank acceleration code to perform force analysis, we should first conduct a power balance to ensure that our code is functioning correctly. Figure 7.94 shows the kinetic and external power for the example linkage. Because the plots match exactly, we can proceed with the analysis.

Figure 7.95 shows the driving torque for the example linkage. Make sure your torque plot matches this before proceeding with the homework problems.

```
% InvSlider_Force_Analysis.m
% Conducts a force analysis on the inverted slider-crank linkage
% and plots the driving torque
% by Eric Constans, June 26, 2017

% Prepare Workspace
```

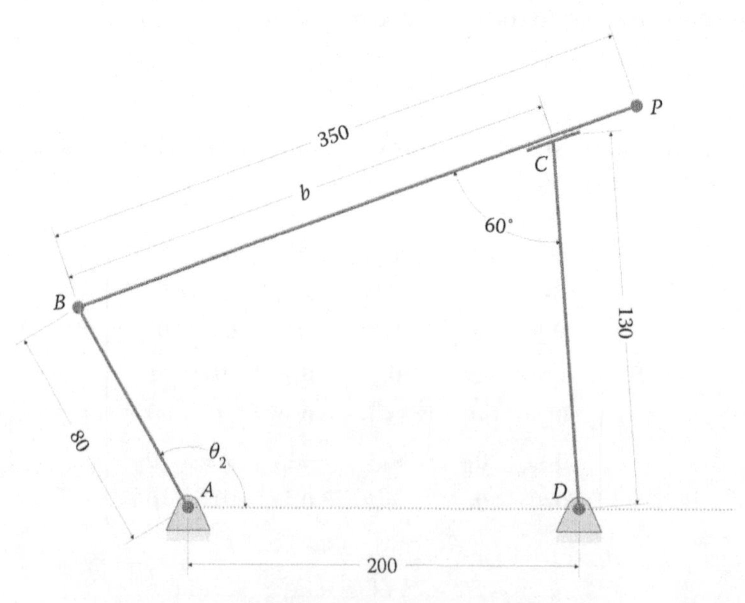

(All dimensions in millimeters)

Crank length: 80
Rocker length: 130
Distance betweet ground pins: 200
Angle between slider and rocker: 60°
Overall slider length: 350
Crank angular velocity: 10 rad/s
Crank angular acceleration: 0 rad/s

Crank mass: 0.1 kg
Slider mass: 0.3 kg
Rocker mass: 0.15 kg
Crank moment of inertia: 0.0004 kg m^2
Slidermoment of inertia: 0.0010 kg m^2
Rocker moment of inertia: 0.0007 kg m^2

FIGURE 7.93
Example inverted slider-crank linkage.

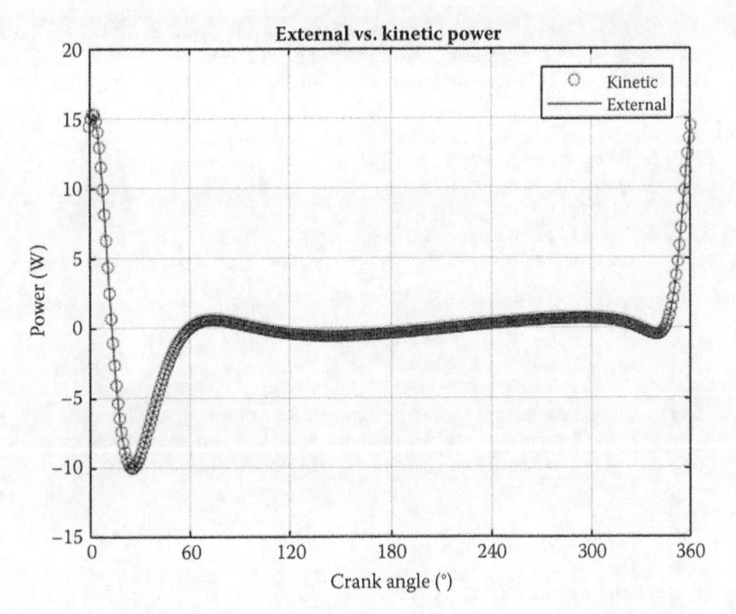

FIGURE 7.94
Power balance plot for the example inverted slider-crank.

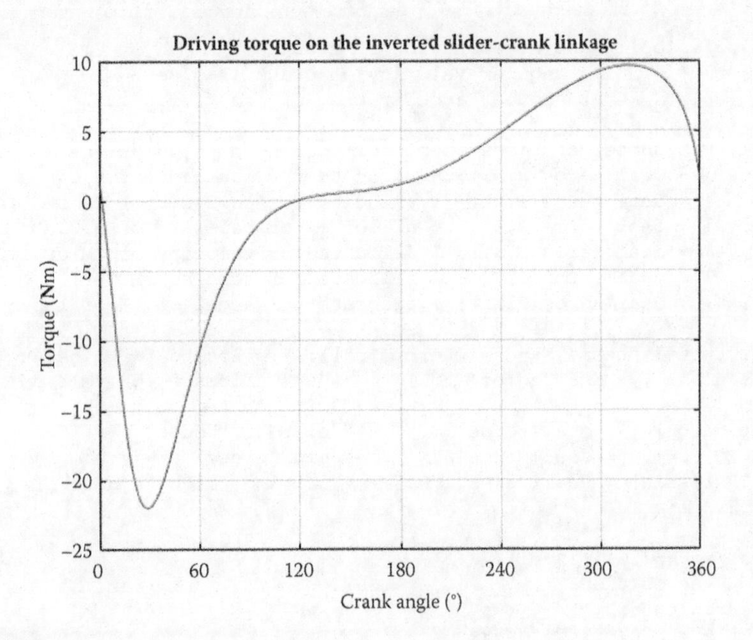

FIGURE 7.95
Driving torque for the example inverted slider-crank linkage.

```
clear variables; close all; clc;

% Linkage dimensions
a = 0.080;              % crank length (m)
c = 0.130;              % rocker length (m)
d = 0.200;              % length between ground pins (m)
p = 0.350;              % slider length (m)
```

```
delta = 60*pi/180;   % angle between slider and rocker (converted to rad)
gamma3 = 0;          % angle to point P on slider
h = c*sin(delta);    % h is a constant, only calculate it once

% ground pins
x0 = [0;0];  % ground pin at A (origin)
xD = [d;0];  % ground pin at D
v0 = [0;0];  % velocity of origin
a0 = [0;0];  % accel of origin
Z2 = zeros(2); Z21 = zeros(2,1); Z12 = zeros(1,2); U2 = eye(2);

% Inertial properties
m2 = 0.1;            % mass of crank (kg)
m3 = 0.3;            % mass of slider (kg)
m4 = 0.15;           % mass of rocker (kg)
I2 = 0.0004;         % moment of inertia of crank about CM (kg-m2)
I3 = 0.0010;         % moment of inertia of slider about CM (kg-m2)
I4 = 0.0007;         % moment of inertia of rocker about CM (kg-m2)

% CM locations
xbar2 = [a/2; 0];  % CM of crank
xbar3 = [p/2; 0];  % CM of slider
xbar4 = [c/2; 0];  % CM of rocker

% Applied loads
FP = [0; -100];  % force at point P (N)

% Angular velocity and acceleration of crank
omega2 = 10;         % angular velocity of crank (rad/sec)
alpha2 = 0;          % angular acceleration of crank (rad/sec2)

N = 361;   % number of times to perform position calculations
[xB,xC,xP] = deal(zeros(2,N)); % allocate for position of B,C,P
[x2,x3,x4] = deal(zeros(2,N)); % allocate for position of CM of links
[vB,vC,vP] = deal(zeros(2,N)); % allocate for velocity of B,C,P
[v2,v3,v4] = deal(zeros(2,N)); % allocate for velocity of CM of links
[aB,aC,aP] = deal(zeros(2,N)); % allocate for accel of B,C,P
[a2,a3,a4] = deal(zeros(2,N)); % allocate for accel of CM of links

[theta2,theta3,theta4] = deal(zeros(1,N)); % allocate for link angles
[omega3,alpha3] = deal(zeros(1,N));        % allocate for vel and accel

[b,bdot,bddot]    = deal(zeros(1,N));  % b, bdot, bddot
[FA,FB,FC,FD]     = deal(zeros(2,N));  % pin forces
[M,T2,PExt,PKin] = deal(zeros(1,N));  % driving torque and powers

for i = 1:N
  theta2(i) = (i-1)*(2*pi)/(N-1);
  r = d - a*cos(theta2(i));
  s = a*sin(theta2(i));
  f2 = r^2 + s^2;  % f squared

  b(i) = c * cos(delta) + sqrt(f2 - h^2);
  g = b(i) - c*cos(delta);

  theta3(i) = atan2((h*r - g*s),(g*r + h*s));
  theta4(i) = theta3(i) + delta;

% calculate unit vectors
  [e1,n1]   = UnitVector(0);
  [e2,n2]   = UnitVector(theta2(i));
```

```
    [e3,n3]    = UnitVector(theta3(i));
    [e4,n4]    = UnitVector(theta4(i));
    [eBP,nBP] = UnitVector(theta3(i) + gamma3);
    [eA2,nA2,LA2,s2A,s2B,  ~] = LinkCG(   a,0,      0,xbar2,theta2(i));
    [eB3,nB3,LB3,s3B,s3C,s3P] = LinkCG(b(i),p,gamma3,xbar3,theta3(i));
    [eD4,nD4,LD4,s4D,s4C,  ~] = LinkCG(   c,0,      0,xbar4,theta4(i));

  % solve for positions of points B, C and P on the linkage
    xB(:,i) = FindPos(      x0,a,e2);
    xC(:,i) = FindPos(      xD,c,e4);
    xP(:,i) = FindPos(xB(:,i),p,eBP);
    x2(:,i) = FindPos(      x0,LA2,eA2);
    x3(:,i) = FindPos(xB(:,i),LB3,eB3);
    x4(:,i) = FindPos(      xD,LD4,eD4);

  % conduct velocity analysis to solve for omega3 and omega4
    A_Mat = [b(i)*n3-c*n4   e3];
    b_Vec = -a*omega2*n2;
    omega_Vec = A_Mat\b_Vec;  % solve for angular velocities

    omega3(i) = omega_Vec(1);     % decompose omega_Vec into
    bdot(i)   = omega_Vec(2);     % individual components

  % calculate velocity at important points on linkage
    vB(:,i) = FindVel(      v0,   a,    omega2, n2);
    vC(:,i) = FindVel(      v0,   c, omega3(i), n4);
    vP(:,i) = FindVel(vB(:,i),   p, omega3(i), nBP);
    v2(:,i) = FindVel(      v0, LA2,    omega2, nA2);
    v3(:,i) = FindVel(vB(:,i), LB3, omega3(i), nB3);
    v4(:,i) = FindVel(      v0, LD4, omega3(i), nD4);

  % conduct acceleration analysis to solve for alpha3 and bddot
    ac = a*omega2^2;
    at = a*alpha2;
    bC = 2*bdot(i)*omega3(i);
    bc = b(i)*omega3(i)^2;
    cc = c*omega3(i)^2;

    C_Mat = A_Mat;
    d_Vec = -at*n2 + ac*e2 - bC*n3 + bc*e3 - cc*e4;
    alpha_Vec = C_Mat\d_Vec;  % solve for angular accelerations

    alpha3(i) = alpha_Vec(1);
    bddot(i)  = alpha_Vec(2);

  % calculate acceleration at important points on linkage
    aB(:,i) = FindAcc(      a0,   a,    omega2,    alpha2,  e2,  n2);
    aC(:,i) = FindAcc(      a0,   c, omega3(i), alpha3(i),  e4,  n4);
    aP(:,i) = FindAcc(aB(:,i),   p, omega3(i), alpha3(i), eBP, nBP);
    a2(:,i) = FindAcc(      a0, LA2,    omega2,    alpha2, eA2, nA2);
    a3(:,i) = FindAcc(aB(:,i), LB3, omega3(i), alpha3(i), eB3, nB3);
    a4(:,i) = FindAcc(      a0, LD4, omega3(i), alpha3(i), eD4, nD4);

  % Conduct force analysis
    S_Mat = [   U2   -U2    Z2    Z2   Z21  Z21;
                Z2    U2   -U2    Z2   Z21  Z21;
                Z2    Z2    U2   -U2   Z21  Z21;
              s2A' -s2B'   Z12   Z12     0    1;
               Z12  s3B' -s3C'   Z12     1    0;
               Z12   Z12  s4C' -s4D'    -1    0;
               Z12   Z12   e3'   Z12     0    0];
```

```
    t_Vec = [m2*a2(:,i);
             m3*a3(:,i) - FP;
             m4*a4(:,i);
             I2*alpha2;
             I3*alpha3(i) - dot(s3P,FP);
             I4*alpha3(i);
             0];

    f_Vec = S_Mat\t_Vec
    FA(:,i) = [f_Vec(1); f_Vec(2)];
    FB(:,i) = [f_Vec(3); f_Vec(4)];
    FC(:,i) = [f_Vec(5); f_Vec(6)];
    FD(:,i) = [f_Vec(7); f_Vec(8)];
    M(i)    = f_Vec(9);
    T2(i)   = f_Vec(10);

    P2 = InertialPower(m2,I2,v2(:,i),a2(:,i),omega2,alpha2);
    P3 = InertialPower(m3,I3,v3(:,i),a3(:,i),omega3(i),alpha3(i));
    P4 = InertialPower(m4,I4,v4(:,i),a4(:,i),omega3(i),alpha3(i));
    PKin(i) = P2 + P3 + P4;
    PF = dot(FP,vP(:,i));
    PT = T2(i) * omega2;
    PExt(i) = PF + PT;
end

% plot the driving torque
plot(theta2*180/pi,T2,'Color',[0 110/255 199/255])
title('Driving Torque on the Inverted Slider-Crank Linkage')
xlabel('Crank angle (degrees)')
ylabel('Torque (N-m)')
grid on
set(gca,'xtick',0:60:360)
xlim([0 360])
```

7.11 Force Analysis Example 4 – The Bicycle Air Pump

A very commonly encountered inverted slider-crank mechanism is the foot-operated bicycle pump shown in Figure 7.96. The user operates the mechanism by pressing downward on the pedal.

Figure 7.97 shows the bicycle pump in the retracted position, with the crank at 90° from the horizontal. The air inside the cylinder is at atmospheric pressure in this position, and the volume inside the cylinder is at its maximum value. Figure 7.98 shows what happens to the linkage when the user steps on the pedal: the cylinder volume decreases and the air is compressed. For this mechanism the crank can rotate from a starting point of 90° in the *clockwise* direction until it reaches a final value of 20°. This means that the angular velocity of the crank will be *negative*, since the crank angle decreases with time.

At first glance, the bicycle pump mechanism might appear to be quite different from the inverted slider-crank; there is no obvious link corresponding to the rocker. If we take the length of the rocker, c, to be zero and the angle between slider and rocker, δ, to be 180° then we obtain the situation shown in Figure 7.99. The unit vectors for slider and cylinder are collinear, but point in opposite directions. Since the two links are collinear there is no obvious location for point C, which was the point of intersection between slider and

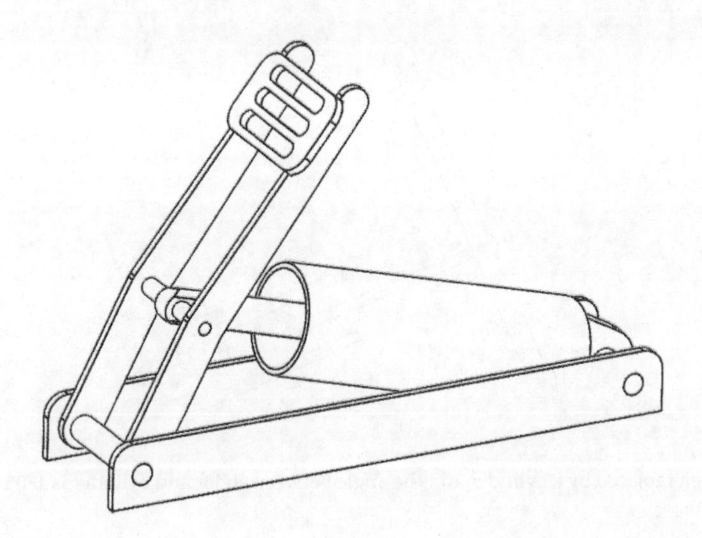

FIGURE 7.96
A common example of the inverted slider-crank is a bicycle pump, shown here. The crank is attached to the foot pedal, the slider is the piston and the rocker is the cylinder. The rocker length is zero in this mechanism.

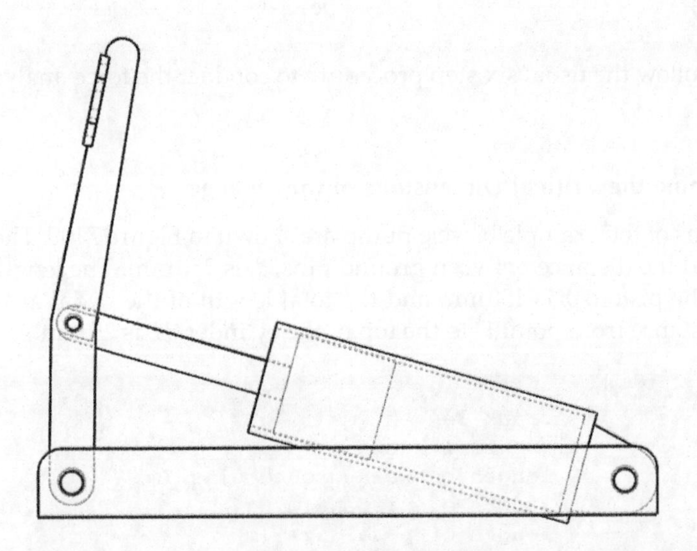

FIGURE 7.97
The bicycle pump in the retracted position, with the crank at 90°. The air inside the cylinder is at atmospheric pressure.

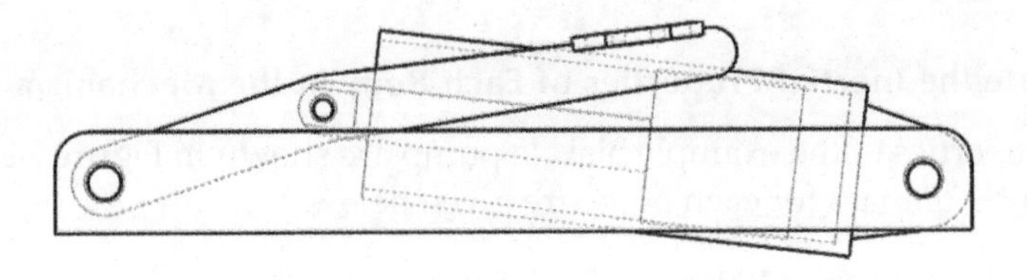

FIGURE 7.98
The bicycle pump in the extended position with the crank at 20°. The air inside the cylinder has been compressed.

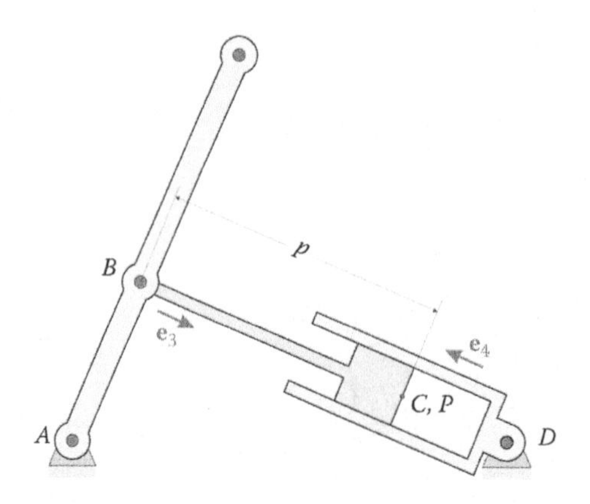

FIGURE 7.99
The rocker has been replaced by a cylinder and the angle between slider and cylinder is 180°.

rocker. For this analysis we will merge the points C and P on the front face of the piston such that

$$\mathbf{r}_{BP} = p\mathbf{e}_3$$

We will now follow the usual six-step procedure to conduct the force analysis on the bicycle pump.

7.11.1 Determine the Critical Dimensions of the Linkage

The dimensions of the example bicycle pump are shown in Figure 7.100. The crank length, a, is 70 mm and the distance between ground pins, d, is 250 mm. The length from point B to the top of the piston, p, is 150 mm and the total length of the pedal arm, q, is 200 mm. Finally, the distance from point D to the top of the cylinder, H, is 25 mm.

```
% Linkage dimensions
a = 0.070;              % crank length (m)
c = 0.0;                % rocker length (m)
d = 0.250;              % length between ground pins (m)
p = 0.150;              % length of piston (m)
q = 0.200;              % length of pedal arm (m)
delta = 180*pi/180; % angle between piston and cylinder (converted to rad)
gamma3 = 0;             % angle to point P on piston
H = 0.025;              % distance btw point D and top of cylinder
h = c*sin(delta);       % h is a constant, only calculate it once
```

7.11.2 Calculate the Inertial Properties of Each Body in the Mechanism

The inertial properties of the example bicycle pump are shown in Figure 7.100. The coordinates of the center of mass for each body are given below.

$$\bar{x}_2 = 100\,\text{mm} \quad \bar{x}_3 = 100\,\text{mm} \quad \bar{x}_4 = 55\,\text{mm}$$

$$\bar{y}_2 = 0\,\text{mm} \quad \bar{y}_3 = 0\,\text{mm} \quad \bar{y}_4 = 0\,\text{mm}$$

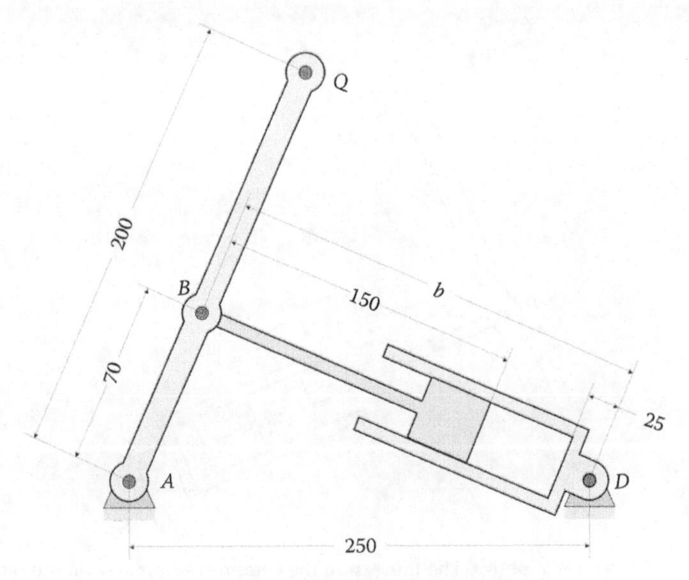

(All dimensions in millimeters)

Crank length: 70	Crank mass: 0.090 kg
Distance between ground pins: 250	Slider/piston mass: 0.240 kg
Length of piston: 150	Rocker/cylinder mass: 0.200 kg
Cylinder head height: 25	Crank moment of inertia: 0.0004 kg m^2
Pedal arm length: 200	Slider/piston moment of inertia: 0.0003 kg m^2
Piston diameter: 45	Rocker/cylinder moment of inertia: 0.0006 kg m^2

FIGURE 7.100
Dimensions of the bicycle pump.

```
% Inertial properties
m2 = 0.090;          % mass of crank (kg)
m3 = 0.240;          % mass of slider (kg)
m4 = 0.200;          % mass of rocker (kg)
I2 = 0.0004;         % moment of inertia of crank about CM (kg-m2)
I3 = 0.0003;         % moment of inertia of slider about CM (kg-m2)
I4 = 0.0006;         % moment of inertia of rocker about CM (kg-m2)

% CM locations
xbar2 = [0.100; 0];  % CM of crank
xbar3 = [0.100; 0];  % CM of slider
xbar4 = [0.055; 0];  % CM of rocker
```

7.11.3 Determine the External Forces

As shown in Figure 7.101, there are two external forces acting on the bicycle pump: the user applies an unknown force $\mathbf{F}_Q$ to the pedal and the compressed air applies a known force $\mathbf{F}_P$ to the face of the piston. We will assume that the pedal force acts perpendicular to the pedal link at all times. The pressure force is determined by the instantaneous volume inside the cylinder.

We will make the same assumption of isentropic compression as we did for the slider-crank compressor mechanism.

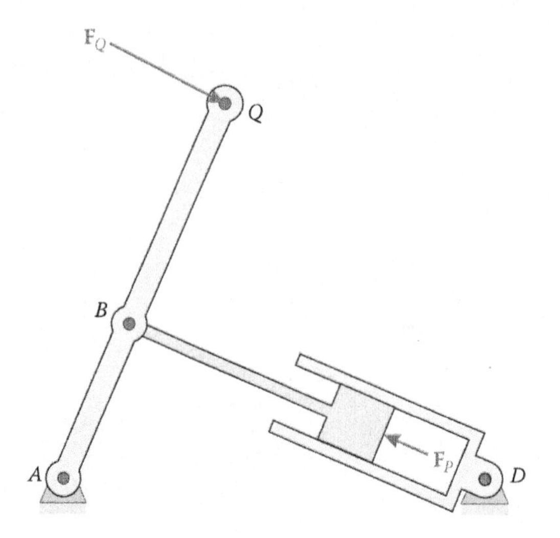

FIGURE 7.101
External forces acting on the bicycle pump. The force from the compressed air acts on the piston face and the user generates a force $\mathbf{F}_Q$ at the pedal.

$$P_2 = P_1\left(\frac{V_1}{V_2}\right)^{1.4}$$

The maximum cylinder volume, V_1, occurs when the crank is at 90°, as seen in Figure 7.102. In this position, the distance from point B to point D can be found using the Pythagorean theorem, since the linkage forms a right triangle.

$$(p + x_1 + H)^2 = a^2 + d^2$$

$$x_1 = \sqrt{a^2 + d^2} - p - H$$

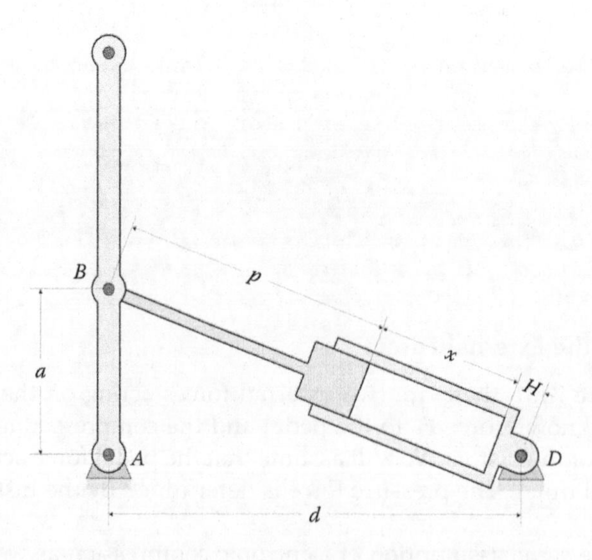

FIGURE 7.102
The dimensions used in calculating the maximum cylinder volume.

Thus, the maximum volume inside the cylinder is

$$V_1 = A\left(\sqrt{a^2 + d^2} - p - H\right)$$

where A is the cross-sectional area of the cylinder. As shown in Figure 7.100, the instantaneous volume inside the cylinder is given by

$$V_2 = A\left(b - p - H\right)$$

Thus, the instantaneous pressure is

$$P_2 = P_{atm}\left(\frac{V_1}{A\left(b - p - H\right)}\right)^{1.4}$$

Before the main loop enter the constants for the pressure calculations. Make sure to allocate memory for P, the instantaneous (absolute) air pressure inside the cylinder.

```
% Pressure constants
D = 0.045;                              % diameter of piston (m)
A = pi*D^2/4;                           % area of piston
Patm = 101e3;                           % atmospheric pressure
V1 = A*(sqrt(a^2 + d^2) - p - H);       % maximum cylinder volume
```

Enter the instantaneous pressure calculations inside the main loop, before the force analysis section.

```
% Pressure calculations
  V2 = A*(b(i) - p - H);   % current volume in cylinder
  P(i) = Patm*(V1/V2)^1.4;
  FP = -A*(P(i)-Patm)*e3;
```

7.11.4 Draw Free-Body Diagrams of Each Link in the Mechanism

A free-body diagram of each link in the bicycle pump mechanism is shown in Figure 7.103. The forces and torques on the crank are the same as before, with the addition of the pedal force $\mathbf{F}_Q$.

$$\mathbf{F}_A - \mathbf{F}_B + \mathbf{F}_Q = m_2\mathbf{a}_2$$

$$\mathbf{s}_{2A} \cdot \mathbf{F}_A - \mathbf{s}_{2B} \cdot \mathbf{F}_B + \mathbf{s}_{2Q} \cdot \mathbf{F}_Q = I_2\alpha_2$$

The equations of motion for the piston are a little trickier. In the previous section, the normal force between piston and cylinder, $\mathbf{F}_C$ was placed at point C – the intersection of the slider and rocker. In this case, the piston is much shorter in length than the slider. For convenience we choose to place the normal force $\mathbf{F}_C$ at point P at the end of the piston, since we have already defined a dimension, p, that specifies this location. In reality, the force $\mathbf{F}_C$ is distributed in an unknown fashion around the cylindrical outer face of the piston, so there is no simple method for finding its exact location, anyway. The equations of motion for the piston are then

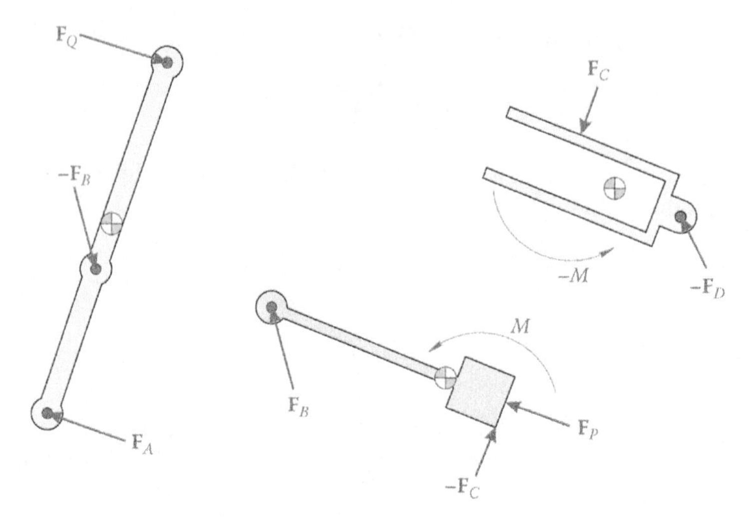

FIGURE 7.103
Free-body diagrams of each link in the bicycle pump mechanism.

$$\mathbf{F}_B - \mathbf{F}_C = m_3\mathbf{a}_3 - \mathbf{F}_P$$

$$\mathbf{s}_{3B} \cdot \mathbf{F}_B - \mathbf{s}_{3P} \cdot \mathbf{F}_C + M = I_3\alpha_3 - \mathbf{s}_{3P} \cdot \mathbf{F}_P$$

where M is the unknown moment created by the full-slider joint. The equations of motion for the rocker are

$$\mathbf{F}_C - \mathbf{F}_D = m_4\mathbf{a}_4$$

$$\mathbf{s}_{4P} \cdot \mathbf{F}_C - \mathbf{s}_{4D} \cdot \mathbf{F}_D - M = I_4\alpha_4$$

The challenge with $\mathbf{F}_C$ is that its position moves with piston. In all cases thus far, the position vectors from the center of mass of a link to a pin joint have remained constant. To address this problem we treat the cylinder as a two-pin link with one pin at D and the other pin at P, on the face of the piston. Defining the vector from point D to point P as

$$\mathbf{r}_{DP} = (b - p)\mathbf{e}_4$$

allows us to use the `LinkCG` function can find $\mathbf{s}_{4P}$ in the usual manner.

```
[eA2,nA2,LA2,s2A,s2B,s2Q] = LinkCG(     a, q, 0,xbar2,theta2(i));
[eB3,nB3,LB3,s3B,s3P,  ~] = LinkCG(     p, 0, 0,xbar3,theta3(i));
[eD4,nD4,LD4,s4D,s4P,  ~] = LinkCG(b(i)-p, 0, 0,xbar4,theta4(i));
```

There are eleven unknowns ($\mathbf{F}_A$, $\mathbf{F}_B$, $\mathbf{F}_C$, $\mathbf{F}_D$, $\mathbf{F}_Q$, M) but thus far we have found only nine equations. The remaining two can be found by noting that $\mathbf{F}_Q$ is always perpendicular to the crank and $\mathbf{F}_C$ is always perpendicular to the slider.

$$\mathbf{F}_Q \cdot \mathbf{e}_2 = 0$$

$$\mathbf{F}_C \cdot \mathbf{e}_3 = 0$$

The matrix equation of motion is then

$$\mathbf{S} = \begin{bmatrix} \mathbf{U}_2 & -\mathbf{U}_2 & \mathbf{0}_2 & \mathbf{0}_2 & \mathbf{U}_2 & \mathbf{0}_{21} \\ \mathbf{0}_2 & \mathbf{U}_2 & -\mathbf{U}_2 & \mathbf{0}_2 & \mathbf{0}_2 & \mathbf{0}_{21} \\ \mathbf{0}_2 & \mathbf{0}_2 & \mathbf{U}_2 & -\mathbf{U}_2 & \mathbf{0}_2 & \mathbf{0}_{21} \\ \mathbf{s}_{2A}^T & -\mathbf{s}_{2B}^T & \mathbf{0}_{12} & \mathbf{0}_{12} & \mathbf{s}_{2Q}^T & 0 \\ \mathbf{0}_{12} & \mathbf{s}_{3B}^T & -\mathbf{s}_{3P}^T & \mathbf{0}_{12} & \mathbf{0}_{12} & 1 \\ \mathbf{0}_{12} & \mathbf{0}_{12} & \mathbf{s}_{4P}^T & -\mathbf{s}_{4D}^T & \mathbf{0}_{12} & -1 \\ \mathbf{0}_{12} & \mathbf{0}_{12} & \mathbf{e}_3^T & \mathbf{0}_{12} & \mathbf{0}_{12} & 0 \\ \mathbf{0}_{12} & \mathbf{0}_{12} & \mathbf{0}_{12} & \mathbf{0}_{12} & \mathbf{e}_2^T & 0 \end{bmatrix}$$

with the vector of unknowns

$$\mathbf{f} = \begin{Bmatrix} \mathbf{F}_A \\ \mathbf{F}_B \\ \mathbf{F}_C \\ \mathbf{F}_D \\ \mathbf{F}_Q \\ M \end{Bmatrix}$$

and the vector of knowns

$$\mathbf{t} = \begin{Bmatrix} m_2\mathbf{a}_2 \\ m_3\mathbf{a}_3 - \mathbf{F}_P \\ m_4\mathbf{a}_4 \\ I_2\alpha_2 \\ I_3\alpha_3 - \mathbf{s}_{3P} \cdot \mathbf{F}_P \\ I_4\alpha_4 \\ 0 \\ 0 \end{Bmatrix}$$

7.11.5 Determine the Nature of the Movement of Crank

The crank receives an angular acceleration pulse, starting at 90° and ending at 20°. We may use the same procedure as we did for the grill lifter mechanism to calculate the crank angle, angular velocity and angular acceleration. Before the main loop define the acceleration parameters:

```
% Acceleration pulse parameters
theta2min = 90*pi/180;           % crank position at top of stroke
theta2max = 20*pi/180;           % crank position at bottom of stroke
T = 2;                           % time it takes to open door
lambda = 2*pi/T;                 % door opening "frequency"
```

```
N = 1001;                             % number of time steps
dt = T/(N-1);                         % time increment
t = 0:dt:T;                           % vector of simulation time
B = lambda*(theta2max - theta2min)/T; % amplitude of acceleration pulse
```

and then at the beginning of the main loop calculate the crank angle, etc.

```
theta2(i) = (B/lambda)*t(i) - (B/lambda^2)*sin(lambda*t(i)) + theta2min;
omega2 = (B/lambda)*(1-cos(lambda*t(i)));
alpha2 = B*sin(lambda*t(i));
```

7.11.6 Solve for the Pin Forces and Driving Force

The solution procedure is the same as for the inverted slider-crank, although we must solve for the pedal force, $\mathbf{F}_Q$.

```
f_Vec = S_Mat         _Vec;
FA(:,i) = [f_Vec(1); f_Vec(2)];
FB(:,i) = [f_Vec(3); f_Vec(4)];
FC(:,i) = [f_Vec(5); f_Vec(6)];
FD(:,i) = [f_Vec(7); f_Vec(8)];
FQ(:,i) = [f_Vec(9); f_Vec(10)];
M(i)    = f_Vec(11);
```

To verify our calculations, we should first create a conservation of power plot. The inertial powers are calculated in the normal fashion, and we have two external powers caused by the pedal force and the air pressure force.

```
P2 = InertialPower(m2,I2,v2(:,i),a2(:,i),omega2,alpha2);
P3 = InertialPower(m3,I3,v3(:,i),a3(:,i),omega3(i),alpha3(i));
P4 = InertialPower(m4,I4,v4(:,i),a4(:,i),omega3(i),alpha3(i));
PKin(i) = P2 + P3 + P4;
PP = dot(FP,vP(:,i));
PQ = dot(FQ(:,i),vQ(:,i));
PExt(i) = PP + PQ;
```

Figure 7.104 shows the conservation of power plot for the bicycle pump in its current configuration. Since the curves match, there is a good chance that we have calculated the correct results. The kinetic power is very low because the links have very little mass and are moving slowly.

Next, plot the air pressure inside the cylinder, as shown in Figure 7.105. The final pressure is approximately 1.7 MPa. The maximum recommended tire pressure for a road bicycle is around 900 kPa, so our current design is something of an overachiever. Luckily, we have written the code in such a way that tweaking the design is simple. To lower the maximum pressure produced by the pump, we can simply lower the value of p, which will increase the volume of air inside the cylinder.

Try reducing the value of p to 145 mm; the results are shown in Figure 7.106. The maximum gauge pressure developed by the pump is now 1.05 MPa, which is much more reasonable. For completeness, we should also modify the CM location of the piston accordingly, although this will likely have a negligible effect on our calculations. When changing variables be aware of the impact it may have on the rest of the design. If we shorten p too

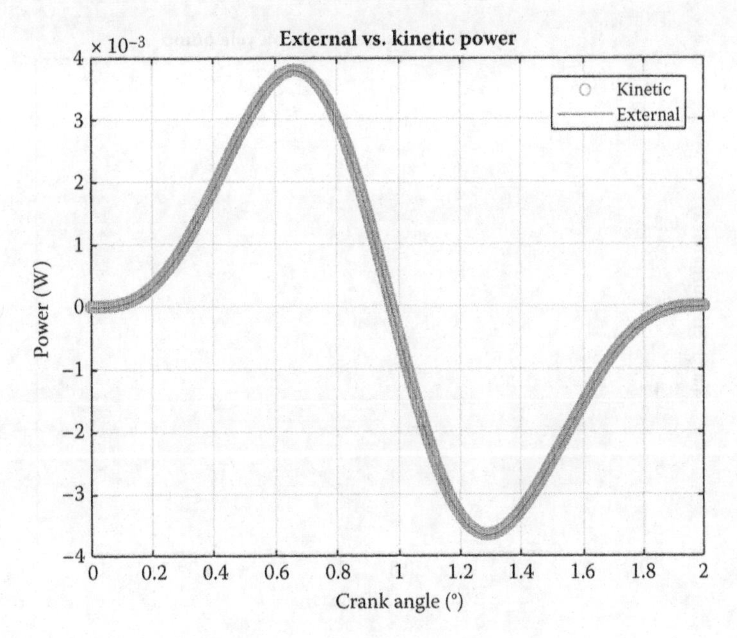

FIGURE 7.104
Conservation of power plot for the bicycle pump mechanism.

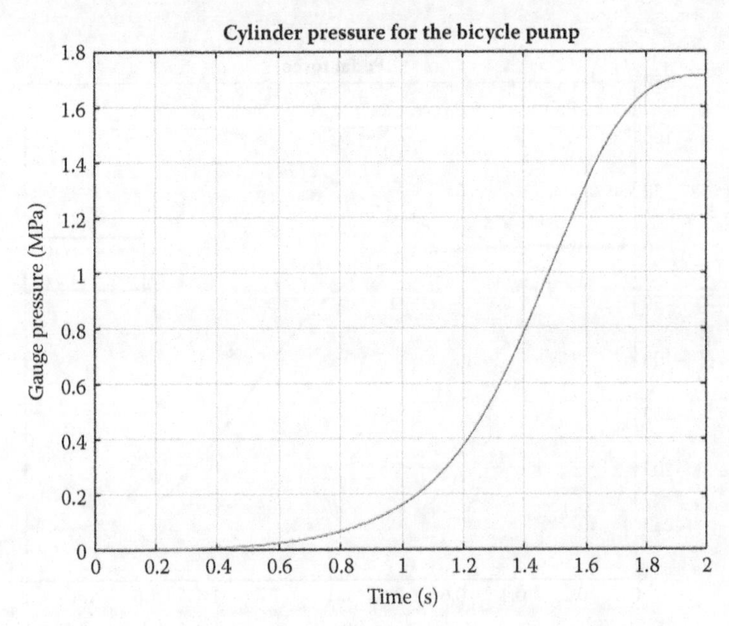

FIGURE 7.105
Gauge pressure inside the cylinder.

much, we may have to reconsider the length of the cylinder to ensure the piston does not "pop" out the back.

Finally, we should plot the force $\mathbf{F}_Q$ exerted by the user to operate the pump. Figure 7.107 shows the force plot with the shortened value of p taken into account. The maximum

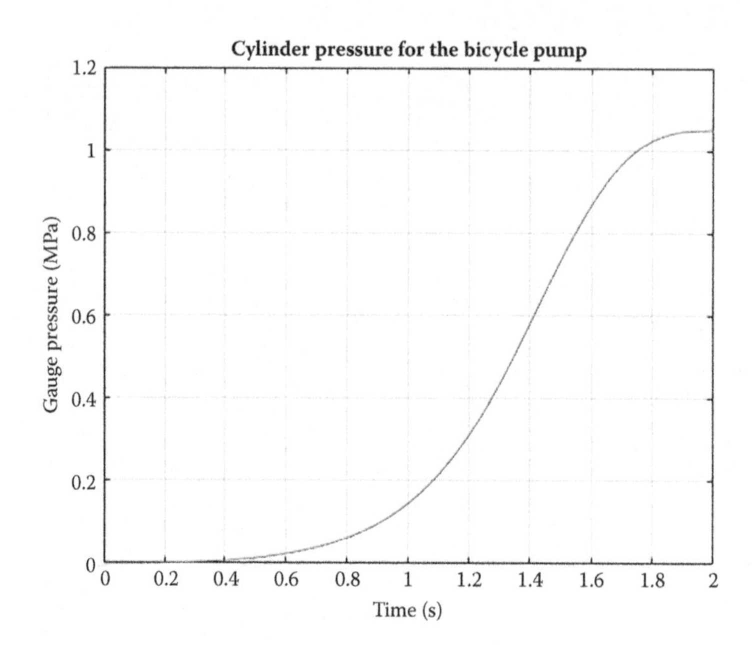

FIGURE 7.106
By reducing the length p to 145 mm we have lowered the maximum air pressure to 1.05 MPa, a much safer value.

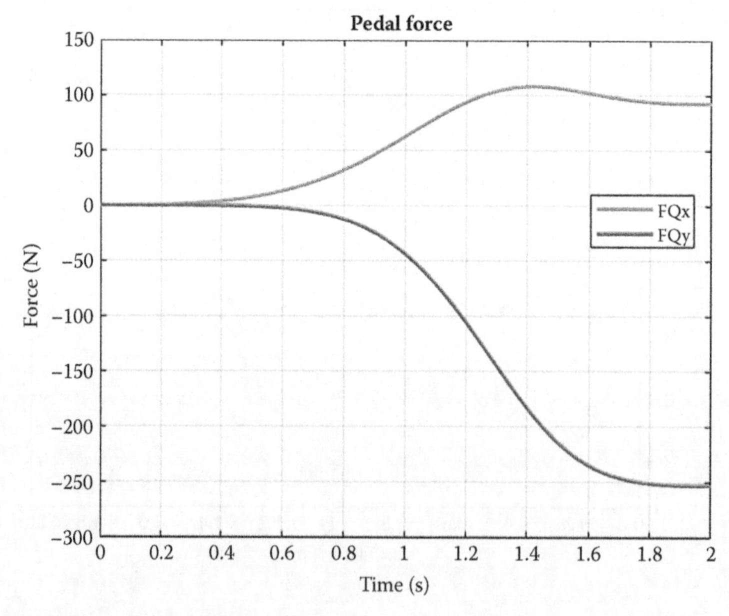

FIGURE 7.107
Pedal force exerted by the user.

magnitude of the force required by the user is 268.7 N, which is a reasonable amount of force to apply with the legs. Consider that the gravitational force exerted by a person with a weight of 40 kg is 392.4 N; by standing on the pump even a lightweight person can achieve maximum pressure.

As a final exercise, it is interesting to plot the mechanical advantage of the bicycle pump as a function of the crank angle. Here, we define mechanical advantage to be the output force divided by the input force:

$$MA = \frac{|\mathbf{F}_P|}{|\mathbf{F}_Q|}$$

A mechanical advantage plot for our current design is shown in Figure 7.108. It is fairly low when the crank is vertical, but reaches a value of 6.2 when the crank is at 20°. This is desirable because the maximum air pressure also occurs near the bottom of the stroke. Since the value seems to increase as crank angle diminishes, we suspect that even greater mechanical advantage would be achieved if we could rotate the crank below 20°. Unfortunately, the pedal makes contact with the cylinder shortly below 20°, so we would need to modify the crank arm to achieve this.

7.12 Force Analysis of the Geared Fivebar Linkage

Force analysis on the geared fivebar is relatively straightforward, although there are a few new aspects to it that are worth discussing. Figure 7.109 shows the geared fivebar linkage with the links labeled. The problem statement is the usual one: find the motor torque, T_2, required to drive the load $\mathbf{F}_Q$ with a given fivebar linkage and solve for the forces at the pins.

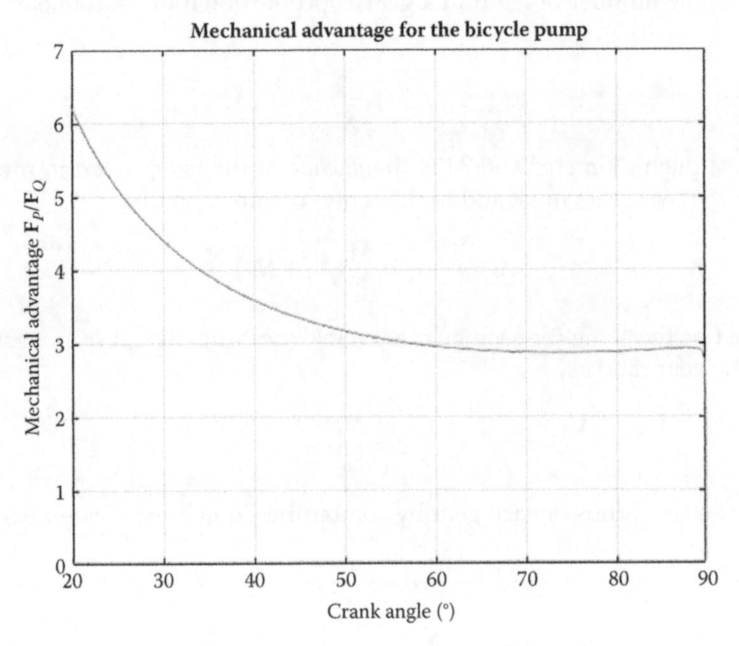

FIGURE 7.108
Mechanical advantage of the bicycle pump as a function of crank angle.

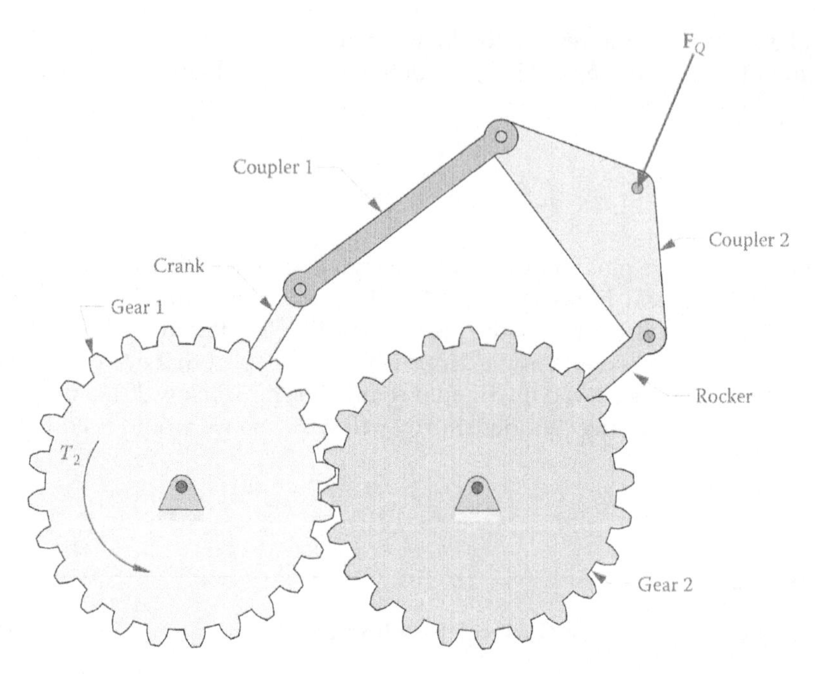

FIGURE 7.109
The geared fivebar linkage.

7.12.1 Some Gear Geometry

Let the number of teeth in the first gear be N_1 and the number of teeth in the second gear be N_2. The distance between the ground pins, d, is also known as the *center distance* between the two gears. The number of teeth in a gear is proportional to its diameter, through the relation

$$N = \frac{D}{M} \tag{7.180}$$

where D is the pitch diameter and M is the *module* of the gear, given in mm/tooth. The pitch radii of the two gears must add to the center distance, so that

$$d = r_1 + r_2 = \frac{M}{2}(N_1 + N_2) \tag{7.181}$$

where M must be the same for each gear to enable meshing. Recall from Section 4.14 that we defined the gear ratio as

$$\rho = \frac{N_1}{N_2} \tag{7.182}$$

We can solve for the radius of each gear by combining Equations (7.180–7.182)

$$r_1 = \frac{d\rho}{\rho + 1}$$

$$r_2 = \frac{d}{\rho + 1}$$

Thus, by knowing the center distance and the gear ratio, we can calculate the pitch radius of each gear. This will be important for determining the moment caused by the contact force between the gears.

Figure 7.110 shows a magnified view of the two gears in contact. The line between the centers of the gears is called, sensibly enough, the *line of centers*. The line perpendicular to the line of centers is the *common tangent*, or *pitch line*. If you picture the two gears as two circles making contact at a single point, then the common tangent is tangent to both circles at this point. Because of the shape of the gear teeth (which we shall have much to say about in Chapter 8 on gears) the force between the two gears is at an angle α with the common tangent, as shown in the figure. The angle α is known as the *pressure angle* and takes on certain standard values (e.g. 14.5° or 20°) in real gears. Thus, the gear force, $\mathbf{F}_g$, has a vertical and horizontal component. The vertical component transmits the torque between the two gears, and is the component that does the actual work of the gearset. The horizontal component transmits no torque, and instead increases the load on the pin at point A.

Now imagine a situation where the second gear "back-drives" the first gear. This might happen when the applied load on the linkage tries to rotate gear 2 faster than it would be driven by the motor on gear 1. In this case, as shown in Figure 7.111, the vertical (torque-inducing) component of the gear force has changed direction, but the horizontal (useless)

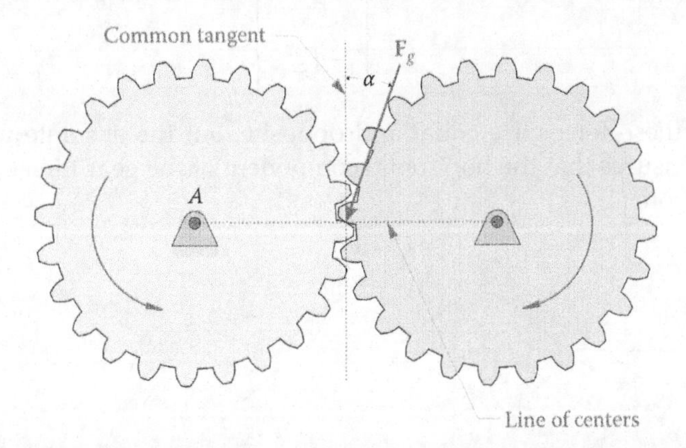

FIGURE 7.110
Torque is transmitted between gears with the force $\mathbf{F}_g$. The gear force is at an angle α with the common tangent between the two gears.

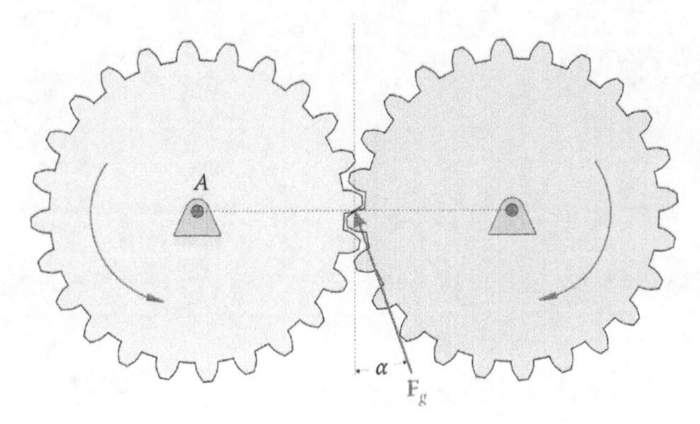

FIGURE 7.111
The second gear is "back-driving" the first gear. The horizontal component of the gear force $\mathbf{F}_g$ always points toward the pin at A.

component is still pointing toward pin A. This phenomenon will add some difficulty to our force analysis of the geared fivebar.

A free-body diagram of each link in the geared fivebar is shown in Figure 7.112. Here we have separated the x and y components of the gear force. The x component of the gear force always points toward the ground pin on a given gear, while the y force may change direction depending upon which gear is driving, and which is driven. The magnitude of the x component of each gear force is

$$F_{gx} = |F_{gy}| \tan \alpha \tag{7.183}$$

The force from gear 2 acting on gear 1 is

$$\mathbf{F}_{g1} = \begin{Bmatrix} -F_{gx} \\ F_{gy} \end{Bmatrix} \tag{7.184}$$

and the force from gear 1 acting on gear 2 is

$$\mathbf{F}_{g2} = \begin{Bmatrix} F_{gx} \\ -F_{gy} \end{Bmatrix} \tag{7.185}$$

As you can see, these forces are equal and opposite, but the absolute value operator in Equation (7.183) ensures that the horizontal component of the gear forces always points in the correct direction.

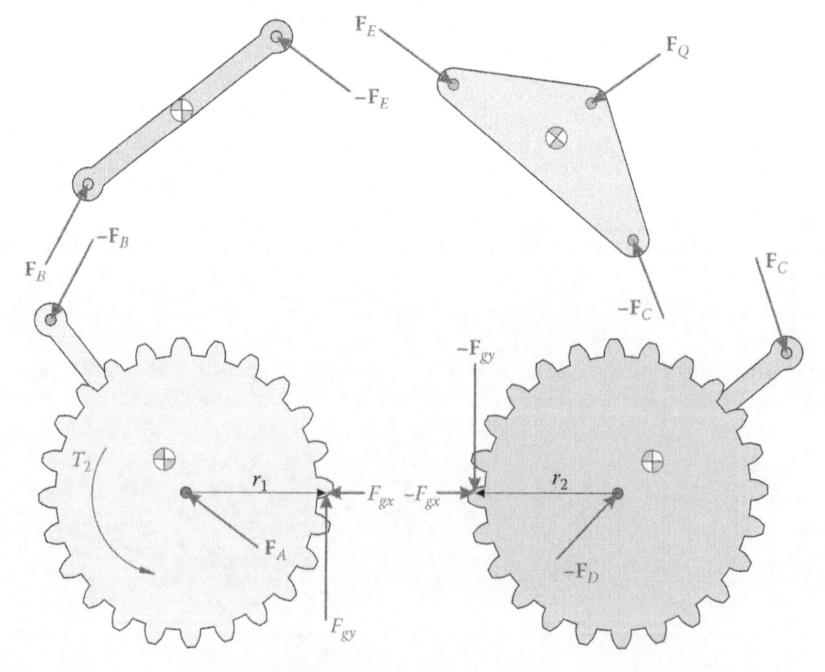

FIGURE 7.112
Free-body diagram of each link in the geared fivebar.

The procedure for summing forces on each link is the same as it was for the previous linkages.

$$\mathbf{F}_A - \mathbf{F}_B + \mathbf{F}_{g1} = m_2\mathbf{a}_2 \tag{7.186}$$

$$\mathbf{F}_B - \mathbf{F}_E = m_3\mathbf{a}_3 \tag{7.187}$$

$$\mathbf{F}_E - \mathbf{F}_C = m_4\mathbf{a}_4 - \mathbf{F}_Q \tag{7.188}$$

$$\mathbf{F}_C - \mathbf{F}_D + \mathbf{F}_{g2} = m_5\mathbf{a}_5 \tag{7.189}$$

The moment equations for the two-coupler links are also derived in the usual manner

$$\mathbf{s}_{3B} \cdot \mathbf{F}_B - \mathbf{s}_{3E} \cdot \mathbf{F}_E = I_3\alpha_3$$

$$\mathbf{s}_{4E} \cdot \mathbf{F}_E - \mathbf{s}_{4C} \cdot \mathbf{F}_C = I_4\alpha_4 - \mathbf{s}_{4Q} \cdot \mathbf{F}_Q$$

Recall from Section 7.2 that there are two possible points to sum moments about on a link: the center of mass or a grounded pivot. For the two gears we choose to sum moments about the grounded pivots at A and D, since this will eliminate $\mathbf{F}_A$ and $\mathbf{F}_D$ from these equations. In addition, the horizontal component of the gear force will not be present in the equations, since it creates no moment about the ground pivot. The moment equation for gear 1 is

$$-\mathbf{s}_{AB} \cdot \mathbf{F}_B + r_1F_{gy} + T_2 = I_{2A}\alpha_2$$

where $\mathbf{s}_{AB}$ is the normal to the vector from point A to point B. The moment equation for gear 2 is

$$\mathbf{s}_{DC} \cdot \mathbf{F}_C + r_2F_{gy} = I_{5D}\alpha_5$$

Note that in both equations we have used the moment of inertia *about the ground pin*. Using the moment of inertia about the center of mass would be invalid since we have summed moments about the ground pins. We will use the parallel axis theorem to calculate the moments of inertia about the ground pins.

$$I_{2A} = I_2 + m_2\bar{\mathbf{x}}_2 \cdot \bar{\mathbf{x}}_2$$

$$I_{5D} = I_5 + m_5\bar{\mathbf{x}}_5 \cdot \bar{\mathbf{x}}_5$$

where we have made use of the fact that

$$\bar{\mathbf{x}} \cdot \bar{\mathbf{x}} = \begin{Bmatrix} \bar{x} \\ \bar{y} \end{Bmatrix} \cdot \begin{Bmatrix} \bar{x} \\ \bar{y} \end{Bmatrix} = \bar{x}^2 + \bar{y}^2$$

Since we have computed the unit vectors and normals for each link during the position analysis, calculating $\mathbf{s}_{AB}$ and $\mathbf{s}_{DC}$ is quite simple.

$$\mathbf{s}_{AB} = a\mathbf{n}_2$$

$$\mathbf{s}_{DC} = u\mathbf{n}_5$$

We now have a total of 12 unknowns ($\mathbf{F}_A$, $\mathbf{F}_B$, $\mathbf{F}_C$, $\mathbf{F}_D$, $\mathbf{F}_E$, F_{gy}, T_2) and 12 equations (four force equations and four moment equations). Unfortunately, Equations (184) and (185) are nonlinear since $\mathbf{F}_{g1}$ and $\mathbf{F}_{g2}$ contain the absolute values of F_{gy}. Because of this nonlinearity, we cannot place all 12 equations into a matrix and solve them simultaneously. But because we summed moments about the ground pins for the two gears, the quantities $\mathbf{F}_A$ and $\mathbf{F}_D$ appear only in Equations (186) and (189), so we can solve these equations separately. Let us collect the other equations

$$\mathbf{F}_B - \mathbf{F}_E = m_3\mathbf{a}_3$$

$$\mathbf{F}_E - \mathbf{F}_C = m_4\mathbf{a}_4 - \mathbf{F}_Q$$

$$-\mathbf{s}_{AB} \cdot \mathbf{F}_B + r_1 F_{gy} + T_2 = I_{2A}\alpha_2$$

$$\mathbf{s}_{3B} \cdot \mathbf{F}_B - \mathbf{s}_{3E} \cdot \mathbf{F}_E = I_3\alpha_3$$

$$\mathbf{s}_{4E} \cdot \mathbf{F}_E - \mathbf{s}_{4C} \cdot \mathbf{F}_C = I_4\alpha_4 - \mathbf{s}_{4Q} \cdot \mathbf{F}_Q$$

$$\mathbf{s}_{DC} \cdot \mathbf{F}_C + r_2 F_{gy} = I_{5D}\alpha_5$$

These are now eight equations and eight unknowns ($\mathbf{F}_B$, $\mathbf{F}_C$, $\mathbf{F}_E$, F_{gy}, T_2) and all of the equations are linear. We may, therefore, collect them into matrix form as

$$\mathbf{S} = \begin{bmatrix} \mathbf{U}_2 & \mathbf{0}_2 & -\mathbf{U}_2 & \mathbf{0}_{21} & \mathbf{0}_{21} \\ \mathbf{0}_2 & -\mathbf{U}_2 & \mathbf{U}_2 & \mathbf{0}_{21} & \mathbf{0}_{21} \\ -\mathbf{s}'_{AB} & \mathbf{0}_{12} & \mathbf{0}_{12} & r_1 & 1 \\ -\mathbf{s}'_{3B} & \mathbf{0}_{12} & -\mathbf{s}'_{3E} & 0 & 0 \\ \mathbf{0}_{12} & -\mathbf{s}'_{4C} & -\mathbf{s}'_{4E} & 0 & 0 \\ \mathbf{0}_{12} & -\mathbf{s}'_{DC} & \mathbf{0}_{12} & r_2 & 0 \end{bmatrix}$$

$$\mathbf{f} = \begin{Bmatrix} \mathbf{F}_B \\ \mathbf{F}_C \\ \mathbf{F}_E \\ F_{gy} \\ T_2 \end{Bmatrix}$$

$$\mathbf{t} = \begin{Bmatrix} m_3\mathbf{a}_3 \\ m_4\mathbf{a}_4 - \mathbf{F}_Q \\ I_{2A}\alpha_2 \\ I_3\alpha_3 \\ I_4\alpha_4 - \mathbf{s}_{4Q} \cdot \mathbf{F}_Q \\ I_{5D}\alpha_5 \end{Bmatrix}$$

Once the matrix equation has been solved, we can use Equations (186) and (189) to solve for the ground pin forces

$$\mathbf{F}_A = \mathbf{F}_B + m_2\mathbf{a}_2 - \mathbf{F}_{g1}$$

$$\mathbf{F}_D = \mathbf{F}_C - m_5\mathbf{a}_5 + \mathbf{F}_{g2}$$

where $\mathbf{F}_{g1}$ and $\mathbf{F}_{g2}$ have been defined in Equations (184) and (185).

Now is a good time to modify your code for the fivebar acceleration analysis to perform the force analysis using the matrix equation above.

Example 7.6

Calculate the driving torque for a geared fivebar linkage with the dimensions and inertial properties shown in Figure 7.113. A downward vertical force of 100 N is applied at point Q.

Solution

After modifying the geared fivebar acceleration code to perform force analysis, we should first conduct a power balance to ensure that our code is functioning correctly. Figure 7.114 shows the kinetic and external power for the example linkage. Because the plots match exactly, we can proceed with the analysis.

Another way to check the code is to perform a static balance on the entire linkage, as we did for the threebar. This is especially useful in making sure that we have defined

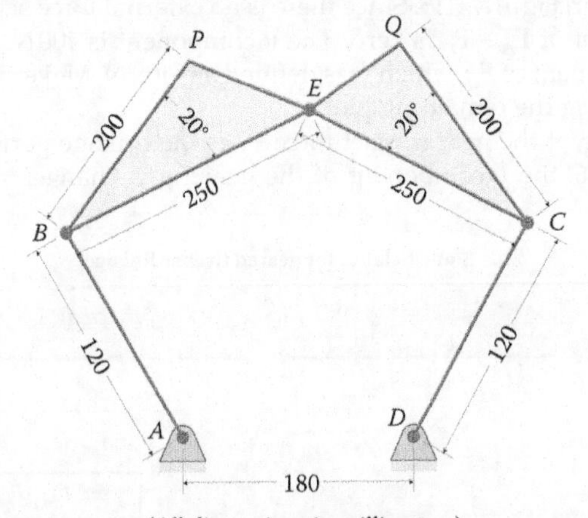

(All dimensions in millimeters)

Length of link on gear 1: 120	Mass of gear 1: 0.100 kg
Coupler 1 length: 250	Mass of coupler 1: 0.200 kg
Coupler 2 length: 250	Mass of coupler 2: 0.200 kg
Length of link on gear 2: 120	Mass of gear 2: 0.100 kg
Distance between ground pins: 180	Moment of inertia of gear 1: 0.0001 kg m^2
Length from B to P: 200	Moment of inertia of coupler 1: 0.0002 kg m^2
Length from C to Q: 200	Moment of inertia of coupler 2: 0.0002 kg m^2
Teeth on gear 1: 48	Moment of inertia of gear 2: 0.0001 kg m^2
Teeth on gear 2: 24	$\overline{x}_2 = (60, 0)$ $\overline{x}_3 = (125, 0)$
Angular velocity of gear 1: 10 rad/s	$\overline{x}_4 = (125, 0)$ $\overline{x}_5 = (60, 0)$
Angular acceleration of gear 1: 0 rad/s	

FIGURE 7.113
Example geared fivebar linkage.

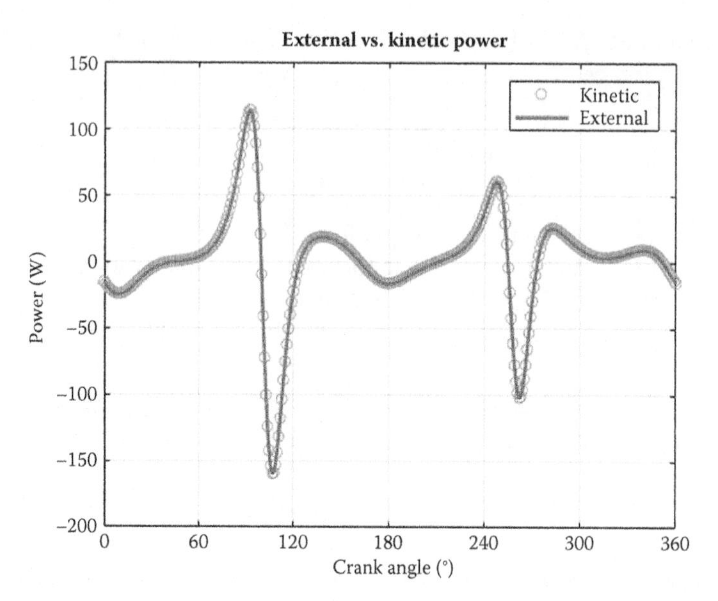

FIGURE 7.114
Power balance plot for the example geared fivebar linkage.

the internal gear forces in the correct direction. Set the crank angular velocity to a very small number (e.g. 0.0001 rad/s) and plot the resultant of $\mathbf{F}_A - \mathbf{F}_D$ in both the x and y directions, as shown in Figure 7.115. Since there is no external force acting in the x direction, the x component of $\mathbf{F}_A - \mathbf{F}_D$ is zero. The y component is 100 N, since this exactly balances the y component of $\mathbf{F}_Q$, which was defined as –100 N. Make sure to return ω_2 to 10 rad/s before creating the remaining plots.

It is interesting to plot the gear contact force $\mathbf{F}_g$ as the linkage performs its cycle. As shown in Figure 7.116, the y component of the gear force changes sign several times

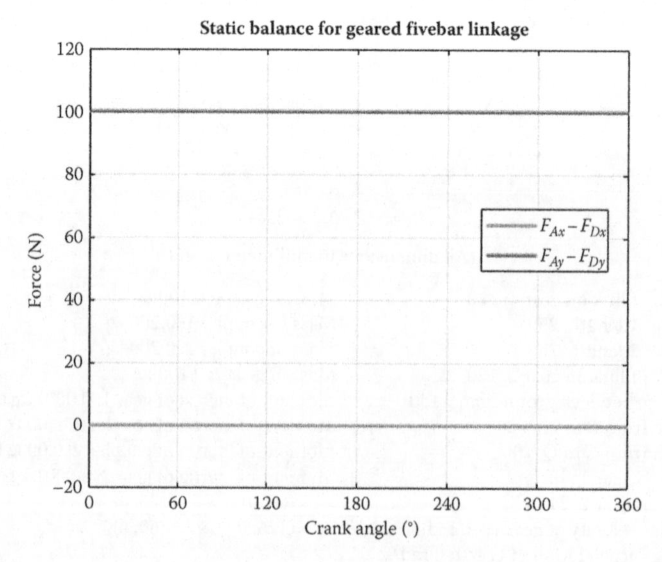

FIGURE 7.115
Static balance for the geared fivebar linkage. The pin forces in the x direction cancel out and the pin forces in the y direction add up to the force $\mathbf{F}_Q$.

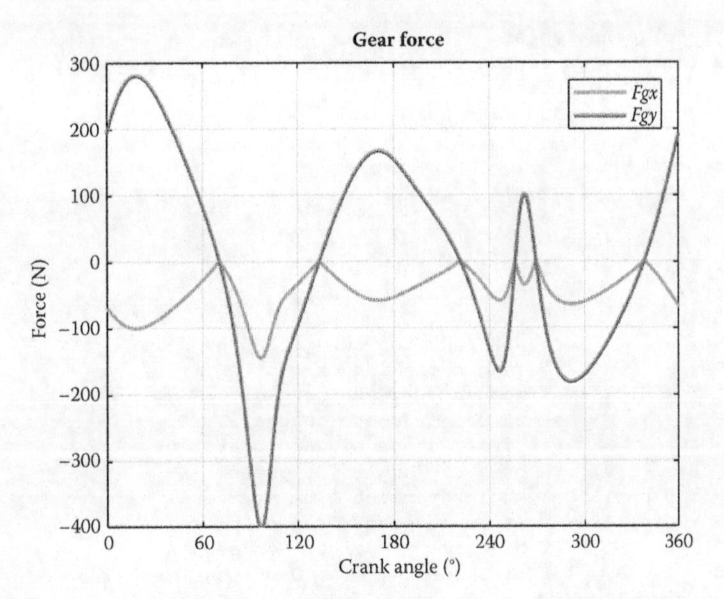

FIGURE 7.116
Gear contact force for the geared fivebar linkage. Notice that the x component is always negative – it is always directed inward toward the ground pin.

during the cycle, which means that gear 1 is alternately driving and being driven by gear 2. The horizontal component of the gear force, on the other hand, is always negative – always directed inward toward the ground pin A. This must be the case, since the two gears cannot pull on one another, they can only push.

Finally, Figure 7.117 shows the driving torque for the example linkage. Make sure your torque plot matches this before proceeding with the homework problems.

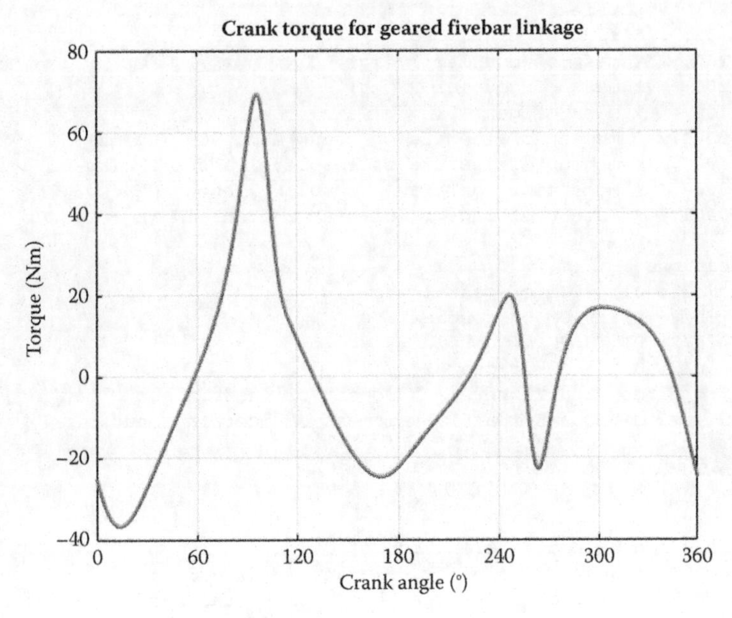

FIGURE 7.117
Driving torque for the example geared fivebar linkage.

```
% Fivebar_Force_Analysis.m
% solves for the pin forces on the geared fivebar linkage and
% plots the driving torque
% by Eric Constans, June 28, 2017

% Prepare Workspace
clear variables; close all; clc;

% Linkage dimensions
a = 0.120;              % crank length (m)
b = 0.250;              % coupler 1 length (m)
c = 0.250;              % coupler 2 length (m)
d = 0.180;              % distance between ground pins (m)
u = 0.120;              % length of link on gear 2 (m)
N1 = 48;                % number of teeth on gear 1
N2 = 24;                % number of teeth on gear 2
phi = 0;                % offset angle between gears
alpha = 20*pi/180;      % pressure angle for gears
gamma3 =  20*pi/180;    % angle to point P on coupler 1
gamma4 = -20*pi/180;    % angle to point Q on coupler 2
p = 0.200;              % distance to point P on coupler 1
q = 0.200;              % distance to point Q on coupler 2

% some gear calculations
rho = N1/N2;                        % gear ratio
r1 = rho*d/(rho+1);                 % radius of gear 1
r2 =     d/(rho+1);                 % radius of gear 2

% ground pins
x0 = [0;0];  % ground pin at A (origin)
xD = [d;0];  % ground pin at D
v0 = [0;0];  % velocity of origin
a0 = [0;0];  % accel of origin
Z2 = zeros(2); Z21 = zeros(2,1); Z12 = zeros(1,2); U2 = eye(2);

% Inertial properties
m2 = 0.100;   % mass of crank (kg)
m3 = 0.200;   % mass of coupler 1 (kg)
m4 = 0.200;   % mass of coupler 2 (kg)
m5 = 0.100;   % mass of gear 2(kg)
I2 = 0.0001;  % moment of inertia of crank about CM (kg-m2)
I3 = 0.0002;  % moment of inertia of coupler 1 about CM (kg-m2)
I4 = 0.0002;  % moment of inertia of coupler 2 about CM (kg-m2)
I5 = 0.0001;  % moment of inertia of gear 2 about CM (kg-m2)

% CM locations
xbar2 = [a/2; 0]; % CM of gear 1
xbar3 = [b/2; 0]; % CM of coupler 1
xbar4 = [c/2; 0]; % CM of coupler 2
xbar5 = [u/2; 0]; % CM of gear 2
I2A = I2 + m2*dot(xbar2,xbar2); % moment of inertia about pin A
I5D = I5 + m5*dot(xbar5,xbar5); % moment of inertia about pin D

% External forces
FQ = [0;-100];

% Angular velocity and acceleration of crank
omega2 = 10;  % angular velocity of crank (rad/sec)
alpha2 = 0;   % angular acceleration of crank (rad/sec^2)
omega5 = -omega2 * rho; % angular velocity of second gear
alpha5 = -alpha2 * rho; % angular accel of second gear
```

```
% Main Loop
N = 361;   % number of times to perform position calculations
[xB,xC,xE,xP,xQ] = deal(zeros(2,N)); % pin positions
[x2,x3,x4,x5]    = deal(zeros(2,N)); % CM positions
[vB,vC,vE,vP,vQ] = deal(zeros(2,N)); % pin velocities
[v2,v3,v4,v5]    = deal(zeros(2,N)); % CM velocities
[aB,aC,aE,aP,aQ] = deal(zeros(2,N)); % pin accelerations
[a2,a3,a4,a5]    = deal(zeros(2,N)); % CM accelerations

[theta2,theta3,theta4] = deal(zeros(1,N));  % link angles
[omega3,omega4] = deal(zeros(1,N));         % link angular vels
[alpha3,alpha4] = deal(zeros(1,N));         % link angular accels

[FA,FB,FC,FD,FE,Fg] = deal(zeros(2,N));  % pin forces
[T2,PKin,PExt] = deal(zeros(1,N));       % torque and powers

for i = 1:N
  theta2(i) = (i-1)*(2*pi)/(N-1);  % crank angle
  theta5 = -rho*theta2(i) + phi;   % angle of second gear
  [e2,n2] = UnitVector(theta2(i)); % unit vector for crank
  [e5,n5] = UnitVector(theta5);    % unit vector for second gear

  xC(:,i) = FindPos(xD, u, e5);    % coords of pin C

  dprime = sqrt(xC(1,i)^2 + xC(2,i)^2);  % distance to pin C
  beta = atan2(xC(2,i),xC(1,i));         % angle to pin C

  r = dprime - a*cos(theta2(i) - beta);
  s = a*sin(theta2(i) - beta);
  f2 = r^2 + s^2;                        % f squared

  delta = acos((b^2+c^2-f2)/(2*b*c));
  g = b - c*cos(delta);
  h = c*sin(delta);

  theta3(i) = atan2((h*r - g*s),(g*r + h*s)) + beta;
  theta4(i) = theta3(i) + delta;

  [e3,n3] = UnitVector(theta3(i));  % unit vector for first coupler
  [e4,n4] = UnitVector(theta4(i));  % unit vector for second coupler

  [eBP,nBP] = UnitVector(theta3(i) + gamma3); % unit vec from B to P
  [eCQ,nCQ] = UnitVector(theta4(i) + gamma4); % unit vec from C to Q

  [eA2,nA2,LA2,s2A,s2B,  ~] = LinkCG( a, 0,     0,xbar2,theta2(i));
  [eB3,nB3,LB3,s3B,s3E,s3P] = LinkCG( b, p,gamma3,xbar3,theta3(i));
  [eC4,nC4,LC4,s4C,s4E,s4Q] = LinkCG( c, q,gamma4,xbar4,theta4(i));
  [eD5,nD5,LD5,s5D,s5C,  ~] = LinkCG( u, 0,     0,xbar5,theta5);

  xB(:,i) = FindPos(     x0,   a,  e2);
  xE(:,i) = FindPos(xB(:,i),   b,  e3);
  xP(:,i) = FindPos(xB(:,i),   p, eBP);
  xQ(:,i) = FindPos(xC(:,i),   q, eCQ);
  x2(:,i) = FindPos(     x0, LA2, eA2);
  x3(:,i) = FindPos(xB(:,i), LB3, eB3);
  x4(:,i) = FindPos(xC(:,i), LC4, eC4);
  x5(:,i) = FindPos(     xD, LD5, eD5);

% conduct velocity analysis to solve for omega3 and omega4
  A_Mat = [b*n3 -c*n4];
```

```
    b_Vec = -a*omega2*n2 + u*omega5*n5;
    omega_Vec = A_Mat\b_Vec;    % solve for angular velocities

    omega3(i) = omega_Vec(1);  % decompose omega_Vec into
    omega4(i) = omega_Vec(2);  % individual components

  % calculate velocity at important points on the linkage
    vB(:,i) = FindVel(     v0,   a,    omega2, n2);
    vC(:,i) = FindVel(     v0,   u,    omega5, n5);
    vE(:,i) = FindVel(vB(:,i),   b, omega3(i), n3);
    vP(:,i) = FindVel(vB(:,i),   p, omega3(i), nBP);
    vQ(:,i) = FindVel(vC(:,i),   q, omega4(i), nCQ);
    v2(:,i) = FindVel(     v0, LA2,    omega2, nA2);
    v3(:,i) = FindVel(vB(:,i), LB3, omega3(i), nB3);
    v4(:,i) = FindVel(vC(:,i), LC4, omega4(i), nC4);
    v5(:,i) = FindVel(     v0, LD5,    omega5, nD5);

  % conduct acceleration analysis to solve for alpha3 and alpha4
    ac = a*omega2^2;
    at = a*alpha2;
    bc = b*omega3(i)^2;
    cc = c*omega4(i)^2;
    uc = u*omega5^2;
    ut = u*alpha5;

    C_Mat = A_Mat;
    d_Vec = -at*n2 + ac*e2 + bc*e3 - cc*e4 + ut*n5 - uc*e5;
    alpha_Vec = C_Mat\d_Vec;  % solve for angular accelerations

    alpha3(i) = alpha_Vec(1);
    alpha4(i) = alpha_Vec(2);

  % find accelerations of important points on linkage
    aB(:,i) = FindAcc(     a0,   a,    omega2,    alpha2,   e2,  n2);
    aC(:,i) = FindAcc(     a0,   u,    omega5,    alpha5,   e5,  n5);
    aE(:,i) = FindAcc(aB(:,i),   b, omega3(i), alpha3(i),   e3,  n3);
    aP(:,i) = FindAcc(aB(:,i),   p, omega3(i), alpha3(i),  eBP, nBP);
    aQ(:,i) = FindAcc(aC(:,i),   q, omega4(i), alpha4(i),  eCQ, nCQ);

  % find accelerations at centers of mass of each link
    a2(:,i) = FindAcc(     a0, LA2,    omega2,    alpha2, eA2, nA2);
    a3(:,i) = FindAcc(aB(:,i), LB3, omega3(i), alpha3(i), eB3, nB3);
    a4(:,i) = FindAcc(aC(:,i), LC4, omega4(i), alpha4(i), eC4, nC4);
    a5(:,i) = FindAcc(     a0, LD5,    omega5,    alpha5, eD5, nD5);

    sAB = a*n2;  % normal to vector from A to B
    sDC = u*n5;  % normal to vector from D to C

  % conduct force analysis
    S_Mat = [  U2      Z2     -U2     Z21  Z21;
               Z2     -U2      U2     Z21  Z21;
             -sAB'    Z12     Z12      r1    1;
              s3B'    Z12   -s3E'       0    0;
               Z12   -s4C'    s4E'      0    0;
               Z12    sDC'    Z12      r2    0];

    t_Vec = [m3*a3(:,i);
             m4*a4(:,i) - FQ;
             I2A*alpha2;
             I3*alpha3(i);
             I4*alpha4(i) - dot(s4Q,FQ);
             I5D*alpha5];
```

```
    f_Vec = S_Mat\t_Vec
    FB(:,i) = [f_Vec(1); f_Vec(2)];
    FC(:,i) = [f_Vec(3); f_Vec(4)];
    FE(:,i) = [f_Vec(5); f_Vec(6)];
    Fg(2,i) = f_Vec(7);                    % y component of gear force
    Fg(1,i) = -abs(Fg(2,i))*tan(alpha); % x component of gear force
    T2(i)   = f_Vec(8);

    FA(:,i) = FB(:,i) + m2*a2(:,i) - Fg(:,i);
    FD(:,i) = FC(:,i) - m5*a5(:,i) - Fg(:,i);

    P2 = InertialPower(m2,I2,v2(:,i),a2(:,i),    omega2,alpha2);
    P3 = InertialPower(m3,I3,v3(:,i),a3(:,i),omega3(i),alpha3(i));
    P4 = InertialPower(m4,I4,v4(:,i),a4(:,i),omega4(i),alpha4(i));
    P5 = InertialPower(m5,I5,v5(:,i),a5(:,i),    omega5,alpha5);
    PKin(i) = P2 + P3 + P4 + P5;
    PF = dot(FQ,vQ(:,i));
    PT = T2(i)*omega2;
    PExt(i) = PF + PT;
end

% plot the crank torque
plot(180*theta2/pi,T2,'Color',[0 110/255 199/255],'LineWidth',2)
title('Crank Torque for Geared Fivebar Linkage')
xlabel('Crank Angle (deg)')
ylabel('Torque (N-m)')
grid on
set(gca,'xtick',0:60:360)
xlim([0 360])
```

7.13 Force Analysis of the Sixbar Linkage

Force analysis on the sixbar linkage is straightforward, although somewhat tedious. Figure 7.118 shows a Stephenson Type I sixbar linkage with the links labeled. The problem statement is the usual one: find the motor torque, T_2, required to drive the load $\mathbf{F}_W$ with a given sixbar linkage and solve for the forces at the pins.

A free-body diagram of each link in the Stephenson Type I sixbar is shown in Figure 7.119. We can immediately begin writing force and moment equations for each link. For the crank we have

$$\mathbf{F}_A - \mathbf{F}_B - \mathbf{F}_E = m_2\mathbf{a}_2$$

$$\mathbf{s}_{2A} \cdot \mathbf{F}_A - \mathbf{s}_{2B} \cdot \mathbf{F}_B - \mathbf{s}_{2E} \cdot \mathbf{F}_E + T_2 = I_2\alpha_2$$

For the coupler

$$\mathbf{F}_B - \mathbf{F}_C = m_3\mathbf{a}_3$$

$$\mathbf{s}_{3B} \cdot \mathbf{F}_B - \mathbf{s}_{3C} \cdot \mathbf{F}_C = I_3\alpha_3$$

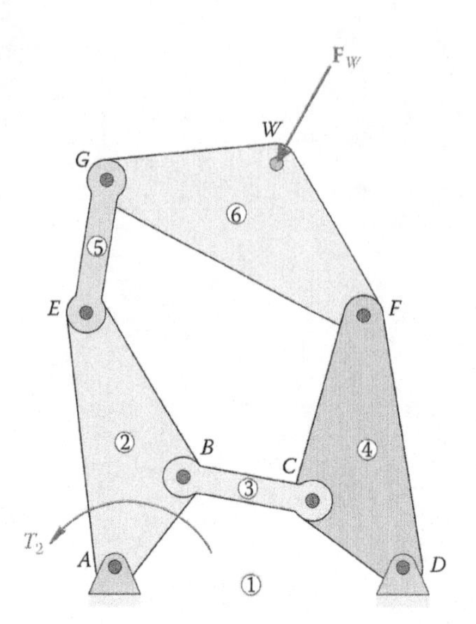

FIGURE 7.118
The Stephenson Type I sixbar linkage.

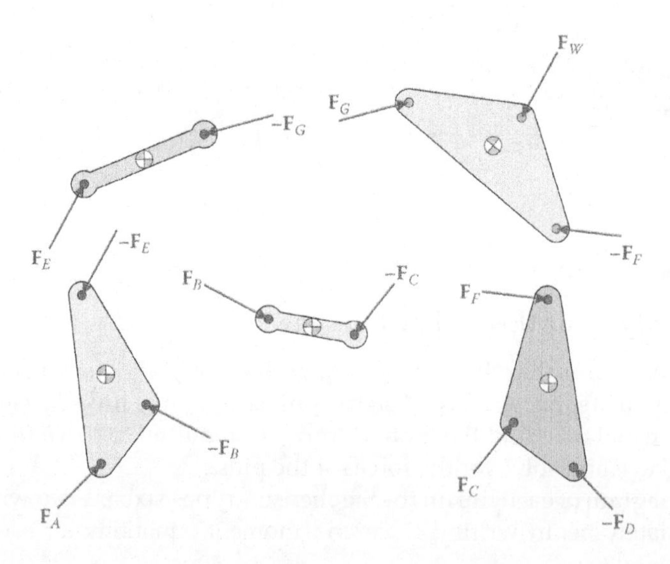

FIGURE 7.119
Free-body diagram of each link in the geared fivebar.

For the rocker

$$\mathbf{F}_C - \mathbf{F}_D + \mathbf{F}_F = m_4\mathbf{a}_4$$

$$\mathbf{s}_{4C} \cdot \mathbf{F}_C - \mathbf{s}_{4D} \cdot \mathbf{F}_D + \mathbf{s}_{4F} \cdot \mathbf{F}_F = I_4\alpha_4$$

For link 5

$$\mathbf{F}_E - \mathbf{F}_G = m_5\mathbf{a}_5$$

$$\mathbf{s}_{5E} \cdot \mathbf{F}_E - \mathbf{s}_{5G} \cdot \mathbf{F}_G = I_5\alpha_5$$

and finally, link 6

$$\mathbf{F}_G - \mathbf{F}_F = m_6\mathbf{a}_6 - \mathbf{F}_W$$
$$\mathbf{s}_{6G} \cdot \mathbf{F}_G - \mathbf{s}_{6F} \cdot \mathbf{F}_F = I_6\alpha_6 - \mathbf{s}_{6W} \cdot \mathbf{F}_W$$

We now have a total of 15 unknowns ($\mathbf{F}_A$, $\mathbf{F}_B$, $\mathbf{F}_C$, $\mathbf{F}_D$, $\mathbf{F}_E$, $\mathbf{F}_F$, $\mathbf{F}_G$ T_2) and 15 equations (five force equations and five moment equations). We may, therefore, collect them into matrix form as

$$\mathbf{S} = \begin{bmatrix} \mathbf{U}_2 & -\mathbf{U}_2 & \mathbf{0}_2 & \mathbf{0}_2 & -\mathbf{U}_2 & \mathbf{0}_2 & \mathbf{0}_2 & \mathbf{0}_{21} \\ \mathbf{0}_2 & \mathbf{U}_2 & -\mathbf{U}_2 & \mathbf{0}_2 & \mathbf{0}_2 & \mathbf{0}_2 & \mathbf{0}_2 & \mathbf{0}_{21} \\ \mathbf{0}_2 & \mathbf{0}_2 & \mathbf{U}_2 & -\mathbf{U}_2 & \mathbf{0}_2 & \mathbf{U}_2 & \mathbf{0}_2 & \mathbf{0}_{21} \\ \mathbf{0}_2 & \mathbf{0}_2 & \mathbf{0}_2 & \mathbf{0}_2 & \mathbf{U}_2 & \mathbf{0}_2 & -\mathbf{U}_2 & \mathbf{0}_{21} \\ \mathbf{0}_2 & \mathbf{0}_2 & \mathbf{0}_2 & \mathbf{0}_2 & \mathbf{0}_2 & -\mathbf{U}_2 & \mathbf{U}_2 & \mathbf{0}_{21} \\ \mathbf{s}_{2A}^T & -\mathbf{s}_{2B}^T & \mathbf{0}_{12} & \mathbf{0}_{12} & -\mathbf{s}_{2E}^T & \mathbf{0}_{12} & \mathbf{0}_{12} & 1 \\ \mathbf{0}_{12} & \mathbf{s}_{3B}^T & -\mathbf{s}_{3C}^T & \mathbf{0}_{12} & \mathbf{0}_{12} & \mathbf{0}_{12} & \mathbf{0}_{12} & 0 \\ \mathbf{0}_{12} & \mathbf{0}_{12} & \mathbf{s}_{4C}^T & -\mathbf{s}_{4D}^T & \mathbf{0}_{12} & \mathbf{s}_{4F}^T & \mathbf{0}_{12} & 0 \\ \mathbf{0}_{12} & \mathbf{0}_{12} & \mathbf{0}_{12} & \mathbf{0}_{12} & \mathbf{s}_{5E}^T & \mathbf{0}_{12} & -\mathbf{s}_{5G}^T & 0 \\ \mathbf{0}_{12} & \mathbf{0}_{12} & \mathbf{0}_{12} & \mathbf{0}_{12} & \mathbf{0}_{12} & -\mathbf{s}_{6F}^T & \mathbf{s}_{6G}^T & 0 \end{bmatrix}$$

$$\mathbf{f} = \begin{Bmatrix} \mathbf{F}_A & \mathbf{F}_B & \mathbf{F}_C & \mathbf{F}_D & \mathbf{F}_E & \mathbf{F}_F & \mathbf{F}_G & T_2 \end{Bmatrix}^T$$

$$\mathbf{t} = \begin{Bmatrix} m_2\mathbf{a}_2 \\ m_3\mathbf{a}_3 \\ m_4\mathbf{a}_4 \\ m_5\mathbf{a}_5 \\ m_6\mathbf{a}_6 - \mathbf{F}_W \\ I_2\alpha_2 \\ I_3\alpha_3 \\ I_4\alpha_4 \\ I_5\alpha_5 \\ I_6\alpha_6 - \mathbf{s}_{6W} \cdot \mathbf{F}_W \end{Bmatrix}$$

The **S** matrices for the remaining sixbar linkages can be derived in the same manner, and are shown at the end of this section. It is interesting to note that the **t** vector is identical for all five types of sixbar linkage as long as the external load is applied at point *W*. If it is applied at another point then the **t** vector must be modified accordingly. Modify your code for the Stephenson Type I sixbar acceleration analysis to perform the force analysis using the matrix equation above.

Example 7.7

Calculate the driving torque for the Stephenson Type I sixbar linkage with the dimensions shown in Figure 7.120. The inertial properties of the linkage are given in Table 7.2. A downward vertical force of 100 N is applied at point *W*.

Solution

After modifying the geared sixbar acceleration code to perform force analysis, we should first conduct a power balance to ensure that our code is functioning correctly. Figure 7.121 shows the kinetic and external power for the example linkage. Because the plots match exactly, we can proceed with the analysis.

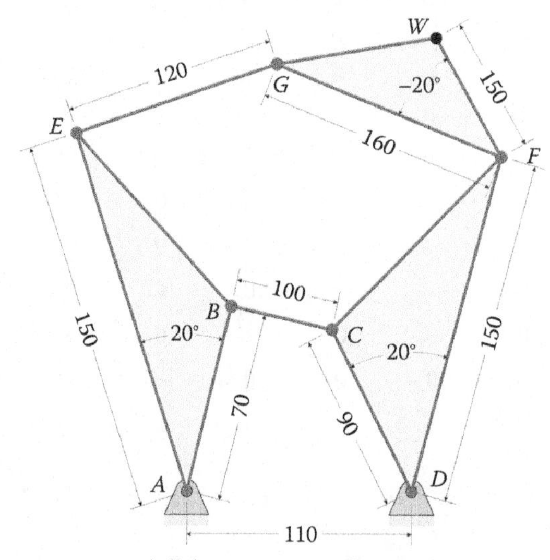

Crank length: 70
Length *AE* on crank: 150
Internal angle of crank: 20°
Coupler length: 100
Distance between ground pins: 110
Crank angular velocity: 10 rad/s

Rocker length: 90
Length *DF* on rocker: 150
Internal angle of rocker: –20°
Length of link 5: 120
Length of link 6: 160
Crank angular acceleration: 0 rad/s

FIGURE 7.120
Dimensions of the example Stephenson Type I linkage.

TABLE 7.2
Inertial Properties of Example Stephenson Type I Sixbar Linkage

	Mass (kg)	Moment of Inertia (kg m^2)	$\bar{x}$ (m)	$\bar{y}$ (m)
Crank	0.200	0.0002	0.035	0.0175
Coupler	0.100	0.0001	0.050	0.0
Rocker	0.200	0.0002	0.045	–0.0225
Link 5	0.100	0.0001	0.060	0.0
Link 6	0.150	0.00015	0.080	0.04

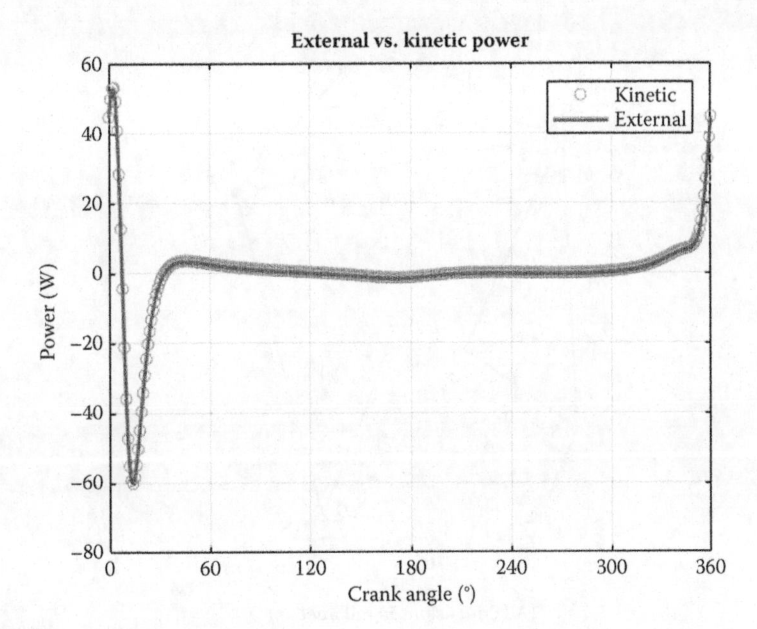

FIGURE 7.121
Power balance plot for the example Stephenson Type I sixbar linkage.

Finally, Figure 7.122 shows the driving torque for the example linkage. Make sure your torque plot matches this before proceeding with the homework problems. The remainder of this section gives example driving torque plots for the other sixbar linkages. In each case, a downward vertical force of 100 N is applied to the point W on link 6 (Figures 7.123–7.130 and Tables 7.3–7.6).

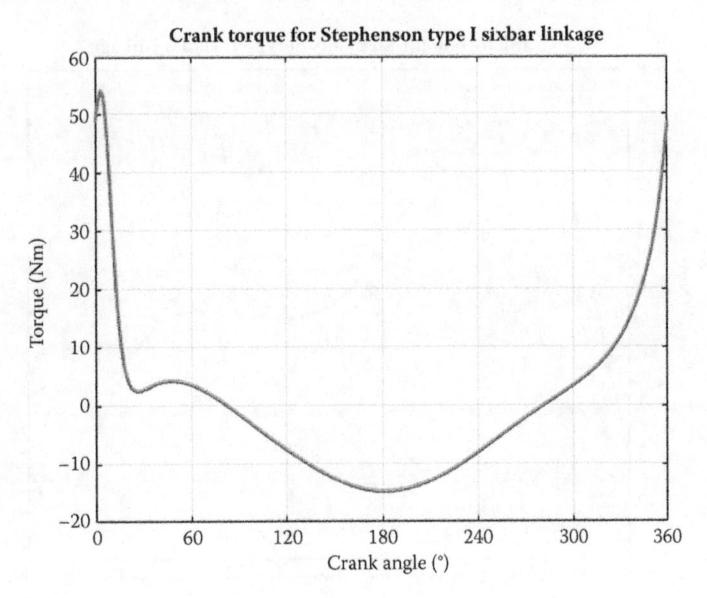

FIGURE 7.122
Driving torque for the example Stephenson Type I sixbar linkage.

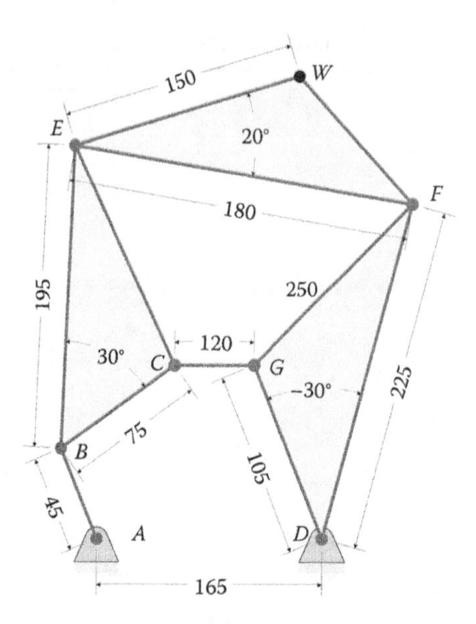

(All dimensions in millimeters)

Crank length: 45
Coupler length: 75
Length BE on coupler: 195
Internal angle of coupler: 30°
Distance between ground pins: 165
Crank angular velocity: 10 rad/s

Rocker length: 105
Length DF on rocker: 225
Internal angle of rocker: −30°
Length of link 5: 120
Length of link 6: 180
Crank angular acceleration: 0 rad/s

FIGURE 7.123
Dimensions of the example Stephenson Type II sixbar linkage.

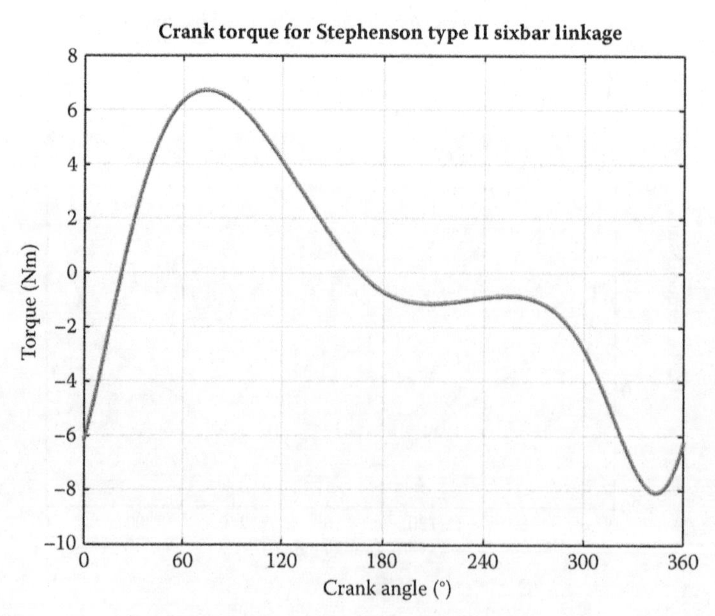

FIGURE 7.124
Driving torque for the example Stephenson Type II sixbar linkage.

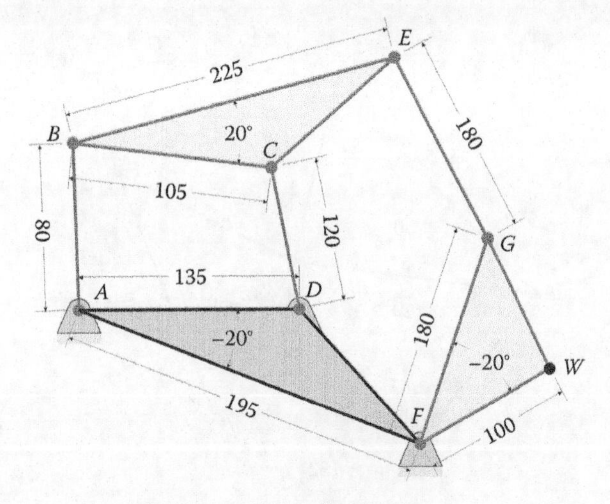

(All dimensions in millimeters)

Crank length: 80	Internal angle of ground: –20°
Coupler length: 105	Length *AF* on ground: 195
Internal angle of coupler: 20°	Rocker length: 120
Length *BE* on coupler: 225	Length of link 5: 180
Distance between ground pins: 135	Length of link 6: 180
Crank angular velocity: 10 rad/s	Crank angular acceleration: 0 rad/s

FIGURE 7.125
Dimensions of the example Stephenson Type III sixbar linkage.

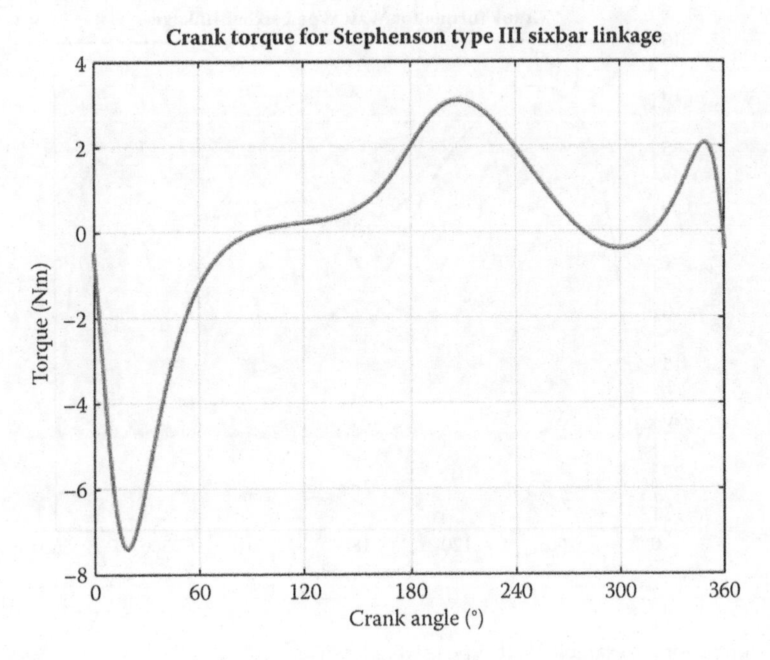

FIGURE 7.126
Driving torque for the example Stephenson Type III sixbar linkage.

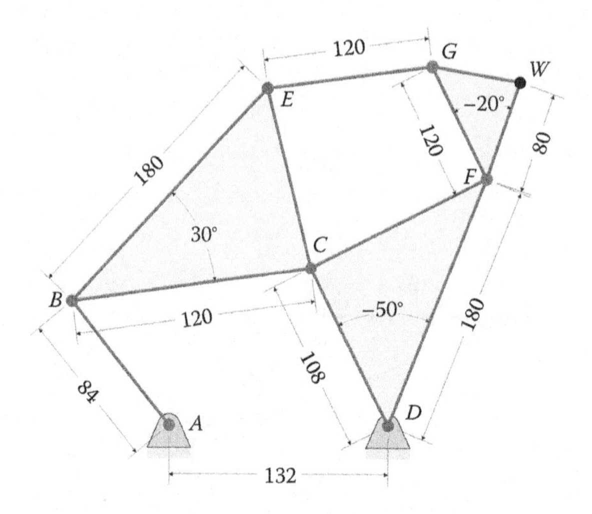

(All dimensions in millimeters)

Crank length: 84
Coupler length: 120
Internal angle of coupler: 30°
Length BE on coupler: 180
Distance between ground pins: 132
Crank angular velocity: 10 rad/s

Rocker length: 108
Internal angle of rocker: −50°
Length DF on rocker: 180
Length of link 5: 120
Length of link 6: 120
Crank angular acceleration: 0 rad/s

FIGURE 7.127
Dimensions of the example Watt Type I sixbar linkage.

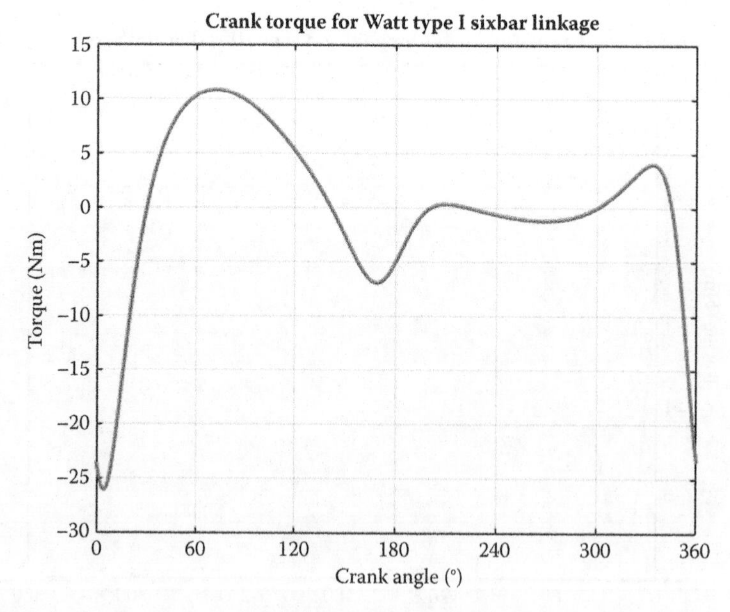

FIGURE 7.128
Driving torque for the example Watt Type I sixbar linkage.

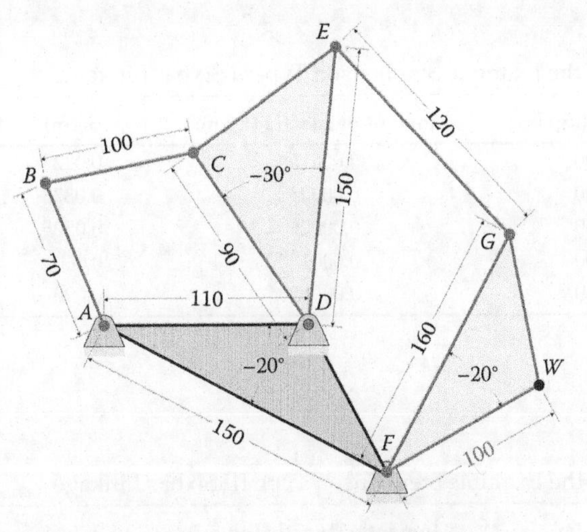

(All dimensions in millimeters)

Crank length: 70
Coupler length: 100
Distance between ground pins: 110
Internal angle of ground: −20°
Length *AF* on ground: 150
Crank angular velocity: 10 rad/sec

Rocker length: 90
Internal angle of rocker: −30°
Length *DE* on rocker: 150
Length of link 5: 120
Length of link 6: 160
Crank angular acceleration: 0 rad/sec

FIGURE 7.129
Dimensions of the example Watt Type II sixbar linkage.

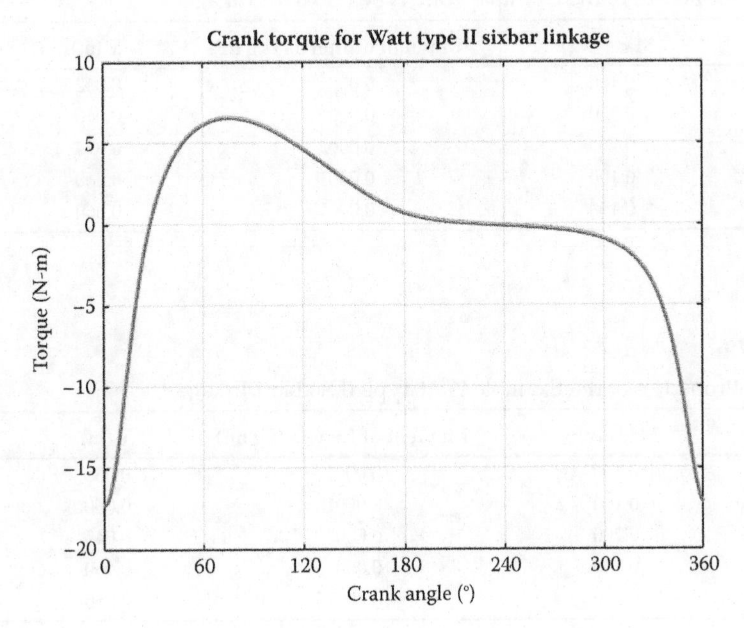

FIGURE 7.130
Driving torque for the example Watt Type I sixbar linkage.

TABLE 7.3

Inertial Properties of the Example Stephenson Type II Sixbar Linkage

	Mass (kg)	Moment of Inertia (kg m^2)	$\bar{x}$ (m)	$\bar{y}$ (m)
Crank	0.100	0.0001	0.0225	0.0
Coupler	0.200	0.0002	0.0375	0.01875
Rocker	0.250	0.00025	0.0525	−0.02625
Link 5	0.100	0.0001	0.060	0.0
Link 6	0.150	0.00015	0.090	0.045

TABLE 7.4

Inertial Properties of the Example Stephenson Type III Sixbar Linkage

	Mass (kg)	Moment of Inertia (kg m^2)	$\bar{x}$ (m)	$\bar{y}$ (m)
Crank	0.100	0.0001	0.040	0.0
Coupler	0.200	0.0002	0.0525	0.02625
Rocker	0.100	0.0001	0.060	0.0
Link 5	0.100	0.0001	0.090	0.0
Link 6	0.150	0.00015	0.090	−0.045

TABLE 7.5

Inertial Properties of the Example Watt Type I Sixbar Linkage

	Mass (kg)	Moment of Inertia (kg m^2)	$\bar{x}$ (m)	$\bar{y}$ (m)
Crank	0.100	0.0001	0.042	0.0
Coupler	0.200	0.0002	0.060	0.030
Rocker	0.200	0.0002	0.054	−0.027
Link 5	0.100	0.0001	0.060	0.0
Link 6	0.150	0.00015	0.060	−0.030

TABLE 7.6

Inertial Properties of the Example Watt Type II Sixbar Linkage

	Mass (kg)	Moment of Inertia (kg m^2)	$\bar{x}$ (m)	$\bar{y}$ (m)
Crank	0.100	0.0001	0.035	0.0
Coupler	0.100	0.0001	0.050	0.0
Rocker	0.200	0.0002	0.045	0.0225
Link 5	0.100	0.0001	0.060	0.0
Link 6	0.150	0.00015	0.080	−0.040

7.13.1 Force Matrices for Sixbar Linkages

$$\text{Stephenson Type I} = \begin{bmatrix} \mathbf{U}_2 & -\mathbf{U}_2 & \mathbf{0}_2 & \mathbf{0}_2 & -\mathbf{U}_2 & \mathbf{0}_2 & \mathbf{0}_2 & \mathbf{0}_{21} \\ \mathbf{0}_2 & \mathbf{U}_2 & -\mathbf{U}_2 & \mathbf{0}_2 & \mathbf{0}_2 & \mathbf{0}_2 & \mathbf{0}_2 & \mathbf{0}_{21} \\ \mathbf{0}_2 & \mathbf{0}_2 & \mathbf{U}_2 & -\mathbf{U}_2 & \mathbf{0}_2 & \mathbf{U}_2 & \mathbf{0}_2 & \mathbf{0}_{21} \\ \mathbf{0}_2 & \mathbf{0}_2 & \mathbf{0}_2 & \mathbf{0}_2 & \mathbf{U}_2 & \mathbf{0}_2 & -\mathbf{U}_2 & \mathbf{0}_{21} \\ \mathbf{0}_2 & \mathbf{0}_2 & \mathbf{0}_2 & \mathbf{0}_2 & \mathbf{0}_2 & -\mathbf{U}_2 & \mathbf{U}_2 & \mathbf{0}_{21} \\ \mathbf{s}_{2A}^T & -\mathbf{s}_{2B}^T & \mathbf{0}_{12} & \mathbf{0}_{12} & -\mathbf{s}_{2E}^T & \mathbf{0}_{12} & \mathbf{0}_{12} & 1 \\ \mathbf{0}_{12} & \mathbf{s}_{3B}^T & -\mathbf{s}_{3C}^T & \mathbf{0}_{12} & \mathbf{0}_{12} & \mathbf{0}_{12} & \mathbf{0}_{12} & 0 \\ \mathbf{0}_{12} & \mathbf{0}_{12} & \mathbf{s}_{4C}^T & -\mathbf{s}_{4D}^T & \mathbf{0}_{12} & \mathbf{s}_{4F}^T & \mathbf{0}_{12} & 0 \\ \mathbf{0}_{12} & \mathbf{0}_{12} & \mathbf{0}_{12} & \mathbf{0}_{12} & \mathbf{s}_{5E}^T & \mathbf{0}_{12} & -\mathbf{s}_{5G}^T & 0 \\ \mathbf{0}_{12} & \mathbf{0}_{12} & \mathbf{0}_{12} & \mathbf{0}_{12} & \mathbf{0}_{12} & -\mathbf{s}_{6F}^T & \mathbf{s}_{6G}^T & 0 \end{bmatrix}$$

$$\text{Stephenson Type II} = \begin{bmatrix} \mathbf{U}_2 & -\mathbf{U}_2 & \mathbf{0}_2 & \mathbf{0}_2 & \mathbf{0}_2 & \mathbf{0}_2 & \mathbf{0}_2 & \mathbf{0}_{21} \\ \mathbf{0}_2 & \mathbf{U}_2 & -\mathbf{U}_2 & \mathbf{0}_2 & -\mathbf{U}_2 & \mathbf{0}_2 & \mathbf{0}_2 & \mathbf{0}_{21} \\ \mathbf{0}_2 & \mathbf{0}_2 & \mathbf{0}_2 & -\mathbf{U}_2 & \mathbf{0}_2 & \mathbf{U}_2 & \mathbf{U}_2 & \mathbf{0}_{21} \\ \mathbf{0}_2 & \mathbf{0}_2 & \mathbf{U}_2 & \mathbf{0}_2 & \mathbf{0}_2 & \mathbf{0}_2 & -\mathbf{U}_2 & \mathbf{0}_{21} \\ \mathbf{0}_2 & \mathbf{0}_2 & \mathbf{0}_2 & \mathbf{0}_2 & \mathbf{U}_2 & -\mathbf{U}_2 & \mathbf{0}_2 & \mathbf{0}_{21} \\ \mathbf{s}_{2A}^T & -\mathbf{s}_{2B}^T & \mathbf{0}_{12} & \mathbf{0}_{12} & \mathbf{0}_{12} & \mathbf{0}_{12} & \mathbf{0}_{12} & 1 \\ \mathbf{0}_{12} & \mathbf{s}_{3B}^T & -\mathbf{s}_{3C}^T & \mathbf{0}_{12} & -\mathbf{s}_{3E}^T & \mathbf{0}_{12} & \mathbf{0}_{12} & 0 \\ \mathbf{0}_{12} & \mathbf{0}_{12} & \mathbf{0}_{12} & -\mathbf{s}_{4D}^T & \mathbf{0}_{12} & \mathbf{s}_{4F}^T & \mathbf{s}_{4G}^T & 0 \\ \mathbf{0}_{12} & \mathbf{0}_{12} & \mathbf{s}_{5C}^T & \mathbf{0}_{12} & \mathbf{0}_{12} & \mathbf{0}_{12} & -\mathbf{s}_{5G}^T & 0 \\ \mathbf{0}_{12} & \mathbf{0}_{12} & \mathbf{0}_{12} & \mathbf{0}_{12} & \mathbf{s}_{6E}^T & -\mathbf{s}_{6F}^T & \mathbf{0}_{12} & 0 \end{bmatrix}$$

$$\text{Stephenson Type III} = \begin{bmatrix} \mathbf{U}_2 & -\mathbf{U}_2 & \mathbf{0}_2 & \mathbf{0}_2 & \mathbf{0}_2 & \mathbf{0}_2 & \mathbf{0}_2 & \mathbf{0}_{21} \\ \mathbf{0}_2 & \mathbf{U}_2 & -\mathbf{U}_2 & \mathbf{0}_2 & -\mathbf{U}_2 & \mathbf{0}_2 & \mathbf{0}_2 & \mathbf{0}_{21} \\ \mathbf{0}_2 & \mathbf{0}_2 & \mathbf{U}_2 & -\mathbf{U}_2 & \mathbf{0}_2 & \mathbf{0}_2 & \mathbf{0}_2 & \mathbf{0}_{21} \\ \mathbf{0}_2 & \mathbf{0}_2 & \mathbf{0}_2 & \mathbf{0}_2 & \mathbf{U}_2 & \mathbf{0}_2 & -\mathbf{U}_2 & \mathbf{0}_{21} \\ \mathbf{0}_2 & \mathbf{0}_2 & \mathbf{0}_2 & \mathbf{0}_2 & \mathbf{0}_2 & -\mathbf{U}_2 & \mathbf{U}_2 & \mathbf{0}_{21} \\ \mathbf{s}_{2A}^T & -\mathbf{s}_{2B}^T & \mathbf{0}_{12} & \mathbf{0}_{12} & \mathbf{0}_{12} & \mathbf{0}_{12} & \mathbf{0}_{12} & 1 \\ \mathbf{0}_{12} & \mathbf{s}_{3B}^T & -\mathbf{s}_{3C}^T & \mathbf{0}_{12} & -\mathbf{s}_{3E}^T & \mathbf{0}_{12} & \mathbf{0}_{12} & 0 \\ \mathbf{0}_{12} & \mathbf{0}_{12} & \mathbf{s}_{4C}^T & -\mathbf{s}_{4D}^T & \mathbf{0}_{12} & \mathbf{0}_{12} & \mathbf{0}_{12} & 0 \\ \mathbf{0}_{12} & \mathbf{0}_{12} & \mathbf{0}_{12} & \mathbf{0}_{12} & \mathbf{s}_{5E}^T & \mathbf{0}_{12} & -\mathbf{s}_{5G}^T & 0 \\ \mathbf{0}_{12} & \mathbf{0}_{12} & \mathbf{0}_{12} & \mathbf{0}_{12} & \mathbf{0}_{12} & -\mathbf{s}_{6F}^T & \mathbf{s}_{6G}^T & 0 \end{bmatrix}$$

$$\text{Watt Type I} = \begin{bmatrix} \mathbf{U}_2 & -\mathbf{U}_2 & \mathbf{0}_2 & \mathbf{0}_2 & \mathbf{0}_2 & \mathbf{0}_2 & \mathbf{0}_2 & \mathbf{0}_{21} \\ \mathbf{0}_2 & \mathbf{U}_2 & -\mathbf{U}_2 & \mathbf{0}_2 & -\mathbf{U}_2 & \mathbf{0}_2 & \mathbf{0}_2 & \mathbf{0}_{21} \\ \mathbf{0}_2 & \mathbf{0}_2 & \mathbf{U}_2 & -\mathbf{U}_2 & \mathbf{0}_2 & \mathbf{U}_2 & \mathbf{0}_2 & \mathbf{0}_{21} \\ \mathbf{0}_2 & \mathbf{0}_2 & \mathbf{0}_2 & \mathbf{0}_2 & \mathbf{U}_2 & \mathbf{0}_2 & -\mathbf{U}_2 & \mathbf{0}_{21} \\ \mathbf{0}_2 & \mathbf{0}_2 & \mathbf{0}_2 & \mathbf{0}_2 & \mathbf{0}_2 & -\mathbf{U}_2 & \mathbf{U}_2 & \mathbf{0}_{21} \\ \mathbf{s}_{2A}^T & -\mathbf{s}_{2B}^T & \mathbf{0}_{12} & \mathbf{0}_{12} & \mathbf{0}_{12} & \mathbf{0}_{12} & \mathbf{0}_{12} & 1 \\ \mathbf{0}_{12} & \mathbf{s}_{3B}^T & -\mathbf{s}_{3C}^T & \mathbf{0}_{12} & -\mathbf{s}_{3E}^T & \mathbf{0}_{12} & \mathbf{0}_{12} & 0 \\ \mathbf{0}_{12} & \mathbf{0}_{12} & \mathbf{s}_{4C}^T & -\mathbf{s}_{4D}^T & \mathbf{0}_{12} & \mathbf{s}_{4F}^T & \mathbf{0}_{12} & 0 \\ \mathbf{0}_{12} & \mathbf{0}_{12} & \mathbf{0}_{12} & \mathbf{0}_{12} & \mathbf{s}_{5E}^T & \mathbf{0}_{12} & -\mathbf{s}_{5G}^T & 0 \\ \mathbf{0}_{12} & \mathbf{0}_{12} & \mathbf{0}_{12} & \mathbf{0}_{12} & \mathbf{0}_{12} & -\mathbf{s}_{6F}^T & \mathbf{s}_{6G}^T & 0 \end{bmatrix}$$

$$\text{Watt Type II} = \begin{bmatrix} \mathbf{U}_2 & -\mathbf{U}_2 & \mathbf{0}_2 & \mathbf{0}_2 & \mathbf{0}_2 & \mathbf{0}_2 & \mathbf{0}_2 & \mathbf{0}_{21} \\ \mathbf{0}_2 & \mathbf{U}_2 & -\mathbf{U}_2 & \mathbf{0}_2 & \mathbf{0}_2 & \mathbf{0}_2 & \mathbf{0}_2 & \mathbf{0}_{21} \\ \mathbf{0}_2 & \mathbf{0}_2 & \mathbf{U}_2 & -\mathbf{U}_2 & -\mathbf{U}_2 & \mathbf{0}_2 & \mathbf{0}_2 & \mathbf{0}_{21} \\ \mathbf{0}_2 & \mathbf{0}_2 & \mathbf{0}_2 & \mathbf{0}_2 & \mathbf{U}_2 & \mathbf{0}_2 & -\mathbf{U}_2 & \mathbf{0}_{21} \\ \mathbf{0}_2 & \mathbf{0}_2 & \mathbf{0}_2 & \mathbf{0}_2 & \mathbf{0}_2 & -\mathbf{U}_2 & \mathbf{U}_2 & \mathbf{0}_{21} \\ \mathbf{s}_{2A}^T & -\mathbf{s}_{2B}^T & \mathbf{0}_{12} & \mathbf{0}_{12} & \mathbf{0}_{12} & \mathbf{0}_{12} & \mathbf{0}_{12} & 1 \\ \mathbf{0}_{12} & \mathbf{s}_{3B}^T & -\mathbf{s}_{3C}^T & \mathbf{0}_{12} & \mathbf{0}_{12} & \mathbf{0}_{12} & \mathbf{0}_{12} & 0 \\ \mathbf{0}_{12} & \mathbf{0}_{12} & \mathbf{s}_{4C}^T & -\mathbf{s}_{4D}^T & -\mathbf{s}_{4E}^T & \mathbf{0}_{12} & \mathbf{0}_{12} & 0 \\ \mathbf{0}_{12} & \mathbf{0}_{12} & \mathbf{0}_{12} & \mathbf{0}_{12} & \mathbf{s}_{5E}^T & \mathbf{0}_{12} & -\mathbf{s}_{5G}^T & 0 \\ \mathbf{0}_{12} & \mathbf{0}_{12} & \mathbf{0}_{12} & \mathbf{0}_{12} & \mathbf{0}_{12} & -\mathbf{s}_{6F}^T & \mathbf{s}_{6G}^T & 0 \end{bmatrix}$$

```
% Sixbar_S1_Force_Analysis.m
% conducts a force analysis on the Stephenson Type I linkage and
% plots the driving torque.
% by Eric Constans, June 29, 2017

% Prepare Workspace
clear variables; close all; clc;

% Linkage dimensions
a = 0.070;              % crank length (m)
b = 0.100;              % coupler length (m)
c = 0.090;              % rocker length (m)
d = 0.110;              % length between ground pins (m)
p = 0.150;              % length to third pin on crank triangle (m)
q = 0.150;              % length to third pin on rocker triangle (m)
u = 0.120;              % length of link 5 (m)
v = 0.160;              % length of link 6 (m)
w = 0.150;              % length from pin F to point W (m)
gamma2 = 20*pi/180;     % internal angle of crank triangle
gamma4 = -20*pi/180;    % internal angle of rocker triangle
gamma6 = -20*pi/180;    % internal angle of link 6

% Ground pins
x0 = [ 0; 0];      % ground pin at A (origin)
xD = [ d; 0];      % ground pin at D
v0 = [ 0; 0];      % velocity of origin
a0 = [ 0; 0];      % acceleration of origin
Z2 = zeros(2); Z21 = zeros(2,1); Z12 = zeros(1,2); U2 = eye(2);

% Inertial properties
m2 = 0.200;        % mass of crank (kg)
m3 = 0.100;        % mass of coupler (kg)
m4 = 0.200;        % mass of rocker (kg)
m5 = 0.100;        % mass of link 5 (kg)
m6 = 0.150;        % mass of link 6 (kg)
I2 = 0.0002;       % moment of inertia of crank about CM (kg-m^2)
I3 = 0.0001;       % moment of inertia of coupler about CM (kg-m^2)
I4 = 0.0002;       % moment of inertia of rocker about CM (kg-m^2)
```

```
I5 = 0.0001;      % moment of inertia of link 5 about CM (kg-m^2)
I6 = 0.00015;     % moment of inertia of link 6 about CM (kg-m^2)

% CM Locations
xbar2 = [a/2;  a/4]; % CM of crank
xbar3 = [b/2;    0]; % CM of coupler
xbar4 = [c/2; -c/4]; % CM of rocker
xbar5 = [u/2;    0]; % CM of link 5
xbar6 = [v/2;  v/4]; % CM of link 6

% Applied loads
FW = [0; -100];      % force at point W (N)

% Angular velocity and acceleration of crank
omega2 = 10;         % angular velocity of crank (rad/sec)
alpha2 = 0;          % angular acceleration of crank (rad/sec^2)

% allocate space for variables
N = 361;   % number of times to perform position calculations
[xB,xC,xE,xF,xG,xW] = deal(zeros(2,N));  % position of B, C, E, F, G, W
[x2,x3,x4,x5,x6]    = deal(zeros(2,N));  % CM position of links
[vB,vC,vE,vF,vG,vW] = deal(zeros(2,N));  % velocity of B, C, E, F, G, W
[v2,v3,v4,v5,v6]    = deal(zeros(2,N));  % CM velocity of links
[aB,aC,aE,aF,aG,aW] = deal(zeros(2,N));  % acceleration of B, C, E, F, G
[a2,a3,a4,a5,a6]    = deal(zeros(2,N));  % CM acceleration of links

[theta2,theta3,theta4,theta5,theta6] = deal(zeros(1,N)); % angles
[omega3,omega4,omega5,omega6] = deal(zeros(1,N));        % velocities
[alpha3,alpha4,alpha5,alpha6] = deal(zeros(1,N));        % accelerations

[FA,FB,FC,FD,FE,FF,FG] = deal(zeros(2,N)); % pin forces
[T2,PExt,PKin] = deal(zeros(1,N));         % driving torque and powers

% Main Loop
for i = 1:N

% solve lower fourbar linkage
  theta2(i) = (i-1)*(2*pi)/(N-1); % crank angle
  r = d - a*cos(theta2(i));
  s = a*sin(theta2(i));
  f2 = r^2 + s^2;
  delta = acos((b^2+c^2-f2)/(2*b*c));
  g = b - c*cos(delta);
  h = c*sin(delta);

  theta3(i) = atan2((h*r - g*s),(g*r + h*s)); % coupler angle
  theta4(i) = theta3(i) + delta;              % rocker angle

% calculate unit vectors
  [e2,n2] = UnitVector(theta2(i));
  [e3,n3] = UnitVector(theta3(i));
  [e4,n4] = UnitVector(theta4(i));
  [eAE,nAE] = UnitVector(theta2(i) + gamma2);
  [eDF,nDF] = UnitVector(theta4(i) + gamma4);
```

```
% solve for positions of points B, C, E, F
  xB(:,i) = FindPos(x0,a, e2);
  xC(:,i) = FindPos(xD,c, e4);
  xE(:,i) = FindPos(x0,p,eAE);
  xF(:,i) = FindPos(xD,q,eDF);

% solve upper fourbar linkage
  xFB = xF(1,i) - xB(1,i);   yFB = xF(2,i) - xB(2,i);
  xEB = xE(1,i) - xB(1,i);   yEB = xE(2,i) - xB(2,i);
  beta =  atan2(yFB, xFB);
  alpha = atan2(yEB, xEB);
  aStar = sqrt(xEB^2 + yEB^2);
  dStar = sqrt(xFB^2 + yFB^2);
  theta2star = alpha - beta; % virtual crank angle on upper fourbar

  r = dStar - aStar*cos(theta2star);
  s = aStar*sin(theta2star);
  f2 = r^2 + s^2;
  delta = acos((u^2+v^2-f2)/(2*u*v));
  g = u - v*cos(delta);
  h = v*sin(delta);

  theta5star = atan2((h*r - g*s),(g*r + h*s)); % coupler and rocker angles
  theta6star = theta5star + delta;             % on upper fourbar
  theta5(i) = theta5star + beta;               % return angles to fixed
  theta6(i) = theta6star + beta;               % fixed CS

% calculate remaining unit vectors
  [e5,n5]   = UnitVector(theta5(i));
  [e6,n6]   = UnitVector(theta6(i));
  [eFW,nFW] = UnitVector(theta6(i) + gamma6);

  [eA2,nA2,LA2,s2A,s2B,s2E] = LinkCG( a, p,gamma2,xbar2,theta2(i));
  [eB3,nB3,LB3,s3B,s3C,  ~] = LinkCG( b, 0,     0,xbar3,theta3(i));
  [eD4,nD4,LD4,s4D,s4C,s4F] = LinkCG( c, q,gamma4,xbar4,theta4(i));
  [eE5,nE5,LE5,s5E,s5G,  ~] = LinkCG( u, 0,     0,xbar5,theta5(i));
  [eF6,nF6,LF6,s6F,s6G,s6W] = LinkCG( v, w,gamma6,xbar6,theta6(i));

% calculate position of pins
  xG(:,i) = FindPos(xE(:,i),   u,  e5);
  xW(:,i) = FindPos(xF(:,i),   w, eFW);
  x2(:,i) = FindPos(      x0, LA2, eA2);
  x3(:,i) = FindPos(xB(:,i), LB3, eB3);
  x4(:,i) = FindPos(      x0, LD4, eD4);
  x5(:,i) = FindPos(xE(:,i), LE5, eE5);
  x6(:,i) = FindPos(xF(:,i), LF6, eF6);

% Conduct velocity analysis to solve for omega3 - omega6
  A_Mat = [b*n3  -c*n4  Z21 Z21;
           Z21 -q*nDF u*n5 -v*n6];
  b_Vec = [-a*omega2*n2; -p*omega2*nAE];
  omega_Vec = A_Mat\b_Vec;
```

```
  omega3(i) = omega_Vec(1);
  omega4(i) = omega_Vec(2);
  omega5(i) = omega_Vec(3);
  omega6(i) = omega_Vec(4);

% Calculate velocity at important points on linkage
  vB(:,i) = FindVel(      v0,   a,    omega2,  n2);
  vC(:,i) = FindVel(      v0,   c, omega4(i),  n4);
  vE(:,i) = FindVel(      v0,   p,    omega2, nAE);
  vF(:,i) = FindVel(      v0,   q, omega4(i), nDF);
  vG(:,i) = FindVel(vE(:,i),   u, omega5(i),  n5);
  vW(:,i) = FindVel(vF(:,i),   w, omega6(i), nFW);
  v2(:,i) = FindVel(      v0, LA2,    omega2, nA2);
  v3(:,i) = FindVel(vB(:,i), LB3, omega3(i), nB3);
  v4(:,i) = FindVel(      v0, LD4, omega4(i), nD4);
  v5(:,i) = FindVel(vE(:,i), LE5, omega5(i), nE5);
  v6(:,i) = FindVel(vF(:,i), LF6, omega6(i), nF6);

% Conduct acceleration analysis to solve for alpha3 - alpha6
  ac = a*omega2^2;     at = a*alpha2;
  bc = b*omega3(i)^2; cc = c*omega4(i)^2;
  pt = p*alpha2;       pc = p*omega2^2;
  uc = u*omega5(i)^2; vc = v*omega6(i)^2;
  qc = q*omega4(i)^2;

  C_Mat = A_Mat;
  d_Vec = [-at*n2 + ac*e2 + bc*e3 - cc*e4;
          -pt*nAE + pc*eAE + uc*e5 - vc*e6 - qc*eDF];
  alpha_Vec = C_Mat\d_Vec;

  alpha3(i) = alpha_Vec(1);
  alpha4(i) = alpha_Vec(2);
  alpha5(i) = alpha_Vec(3);
  alpha6(i) = alpha_Vec(4);

% Calculate acceleration at important points on linkage
  aB(:,i) = FindAcc(      a0,   a,    omega2,    alpha2,  e2,  n2);
  aC(:,i) = FindAcc(      a0,   c, omega4(i), alpha4(i),  e4,  n4);
  aE(:,i) = FindAcc(      a0,   p,    omega2,    alpha2, eAE, nAE);
  aF(:,i) = FindAcc(      a0,   q, omega4(i), alpha4(i), eDF, nDF);
  aG(:,i) = FindAcc(aE(:,i),   u, omega5(i), alpha5(i),  e5,  n5);
  aW(:,i) = FindAcc(aF(:,i),   w, omega6(i), alpha6(i), eFW, nFW);
  a2(:,i) = FindAcc(      a0, LA2,    omega2,    alpha2, eA2, nA2);
  a3(:,i) = FindAcc(aB(:,i), LB3, omega3(i), alpha3(i), eB3, nB3);
  a4(:,i) = FindAcc(      a0, LD4, omega4(i), alpha4(i), eD4, nD4);
  a5(:,i) = FindAcc(aE(:,i), LE5, omega5(i), alpha5(i), eE5, nE5);
  a6(:,i) = FindAcc(aF(:,i), LF6, omega6(i), alpha6(i), eF6, nF6);

% Conduct force analysis
  S_Mat = [ U2    -U2    Z2    Z2   -U2    Z2    Z2  Z21;
            Z2     U2   -U2    Z2    Z2    Z2    Z2  Z21;
            Z2     Z2    U2   -U2    Z2    U2    Z2  Z21;
            Z2     Z2    Z2    Z2    U2    Z2   -U2  Z21;
```

```
             Z2     Z2     Z2     Z2     Z2    -U2     U2  Z21;
           s2A'  -s2B'    Z12    Z12  -s2E'    Z12    Z12    1;
            Z12   s3B'  -s3C'    Z12    Z12    Z12    Z12    0;
            Z12    Z12   s4C'  -s4D'    Z12   s4F'    Z12    0;
            Z12    Z12    Z12    Z12   s5E'    Z12  -s5G'    0;
            Z12    Z12    Z12    Z12    Z12  -s6F'   s6G'    0];

  t_Vec = [m2*a2(:,i);
           m3*a3(:,i);
           m4*a4(:,i);
           m5*a5(:,i);
           m6*a6(:,i) - FW;
           I2*alpha2;
           I3*alpha3(i);
           I4*alpha4(i);
           I5*alpha5(i);
           I6*alpha6(i) - dot(s6W,FW)];

  f_Vec = S_Mat\t_Vec

  FA(:,i) = [f_Vec(1);  f_Vec(2)];
  FB(:,i) = [f_Vec(3);  f_Vec(4)];
  FC(:,i) = [f_Vec(5);  f_Vec(6)];
  FD(:,i) = [f_Vec(7);  f_Vec(8)];
  FE(:,i) = [f_Vec(9);  f_Vec(10)];
  FF(:,i) = [f_Vec(11); f_Vec(12)];
  FG(:,i) = [f_Vec(13); f_Vec(14)];
  T2(i) = f_Vec(15);

% Calculate inertial powers
  P2 = InertialPower(m2,I2,v2(:,i),a2(:,i),   omega2,   alpha2);
  P3 = InertialPower(m3,I3,v3(:,i),a3(:,i),omega3(i),alpha3(i));
  P4 = InertialPower(m4,I4,v4(:,i),a4(:,i),omega4(i),alpha4(i));
  P5 = InertialPower(m5,I5,v5(:,i),a5(:,i),omega5(i),alpha5(i));
  P6 = InertialPower(m6,I6,v6(:,i),a6(:,i),omega6(i),alpha6(i));
  PKin(i) = P2 + P3 + P4 + P5 + P6;

% Calculate external power
  PF = dot(FW,vW(:,i));
  PT = T2(i)*omega2;
  PExt(i) = PF + PT;
end

% plot the crank torque
plot(180*theta2/pi,T2,'Color',[0 110/255 199/255],'LineWidth',2)
title('Crank Torque for Stephenson Type I Sixbar Linkage')
xlabel('Crank Angle (deg)')
ylabel('Torque (N-m)')
grid on
set(gca,'xtick',0:60:360)
xlim([0 360])
```

7.14 Practice Problems

Problem 7.1

Determine the centers of mass of objects (a), (b), and (c) in Figure 7.131. All dimensions are in centimeters.

Problem 7.2

Determine the mass moments of inertia of the objects in Problem 7.1 about the z axis if each object is made of AISI 1020 steel with a thickness of 1 cm. Use hand calculations and confirm your results using the **Mass Properties** feature in SOLIDWORKS.

Problem 7.3

Find the mass moments of inertia about the z axis of objects (a), (b), and (c) in Figure 7.132. Use hand calculations and the inertial properties shown in Appendix A. The objects in Parts (b) and (c) are made of aluminum.

Problem 7.4

Determine the mass, center of mass, and moment of inertia about the z axis of the acrylic link shown in Figure 7.133 if its thickness is 1.0 cm. All dimensions are in centimeters. Perform the calculations by hand, and confirm your answer using the **Mass Properties** feature in SOLIDWORKS.

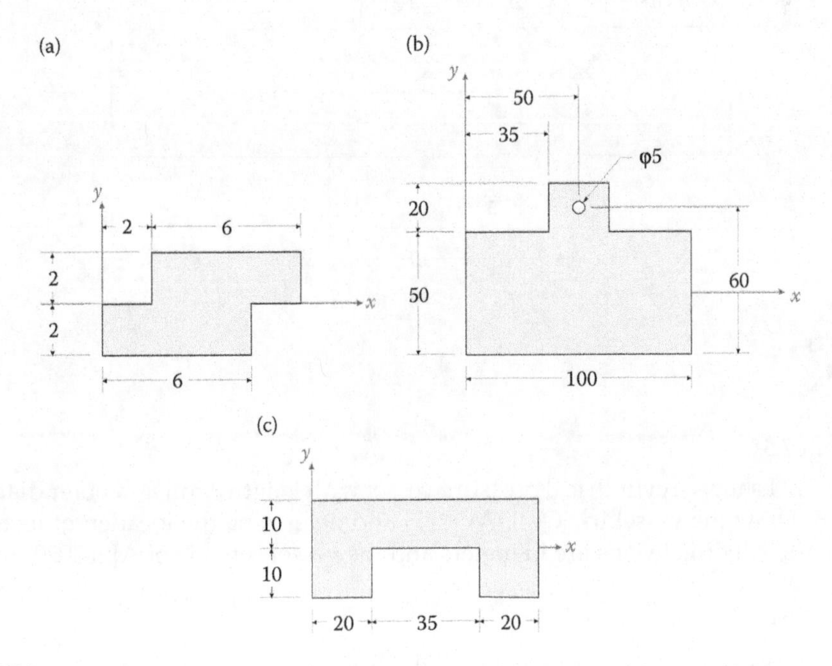

FIGURE 7.131
Problem 7.1.

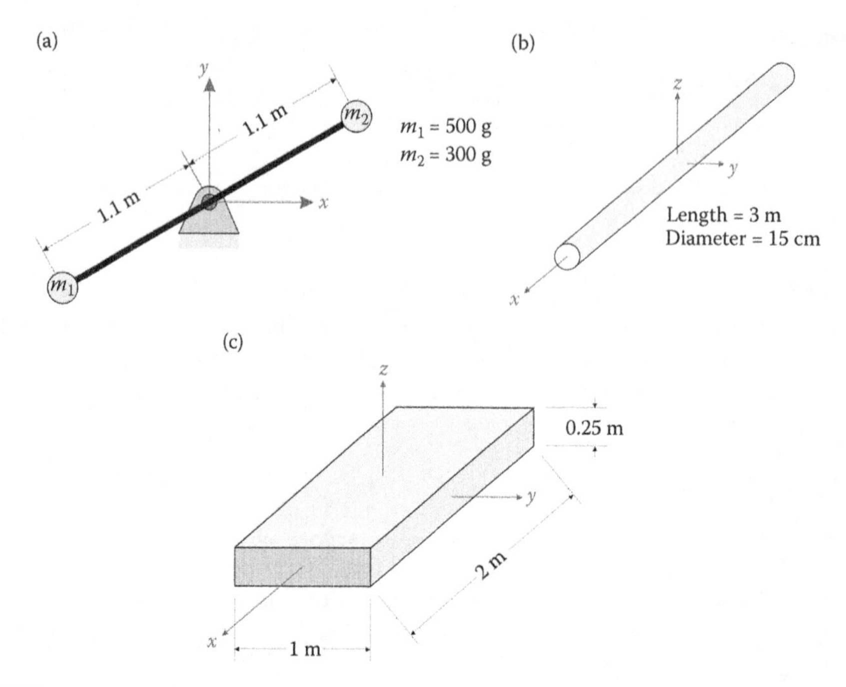

FIGURE 7.132
Problem 7.3.

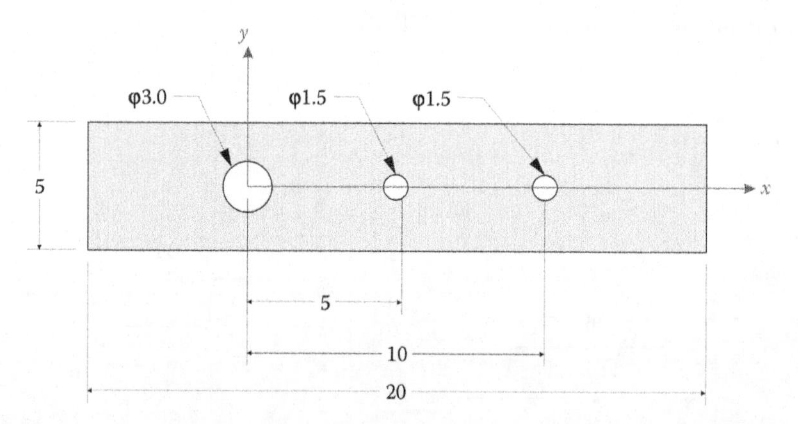

FIGURE 7.133
Problem 7.4.

Problem 7.5

Figure 7.134 shows a cylindrical pressure vessel with length 3.5 m and outer diameter 2.0 m. Draw the vessel in SOLIDWORKS and determine the location of its center of mass. All dimensions are in meters and the vessel is made of AISI 1020 steel.

Problem 7.6

Use the **Mass Properties** feature in SOLIDWORKS to determine the mass moment of inertia of the pressure vessel in Problem 7.5 about the *z* axis.

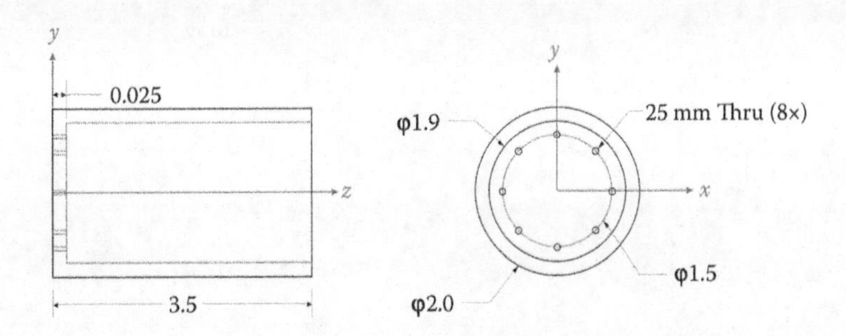

FIGURE 7.134
Problem 7.5.

Problem 7.7

a. Determine the moment about the pivot that is created by the 5N force in Figure 7.135a.
b. Determine the moment about the pivot that is created by the 5N force when it is applied at an angle of 30° from the horizontal as shown in Figure 7.135b.
c. For the configuration presented in Figure 7.135b determine the force that would be required to produce the same moment calculated in Part (a).

Problem 7.8

Determine the angular acceleration of the link in Problem 7.7 (a) if

a. The link is a cylindrical rod of diameter 20mm, length 0.25m, and mass 0.5kg.
b. The link is a massless rod with a point mass of 0.5kg at the end of the rod where the force is applied.

Problem 7.9

A 5mm thick link with the dimensions presented in Figure 7.136 made of 6061-T6 aluminum has a 10kg mass fixed to its end. Use MATLAB to plot the driving torque of the link, which rotates at 100rpm. Hint: use SOLIDWORKS to calculate the inertial properties of the link without the point mass, then use MATLAB to add the inertial properties of the mass to the link. Do not neglect gravity.

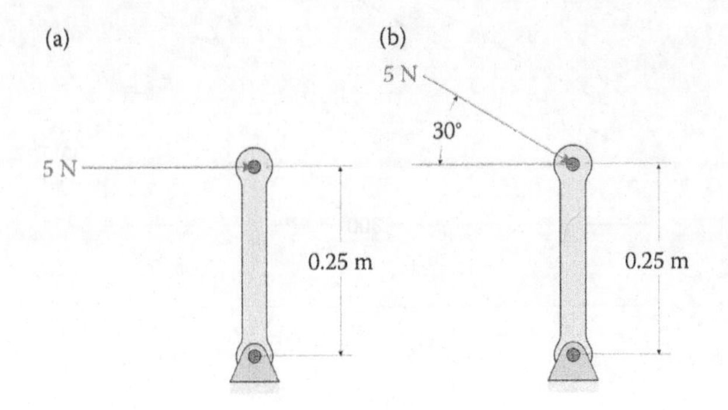

FIGURE 7.135
Problem 7.7.

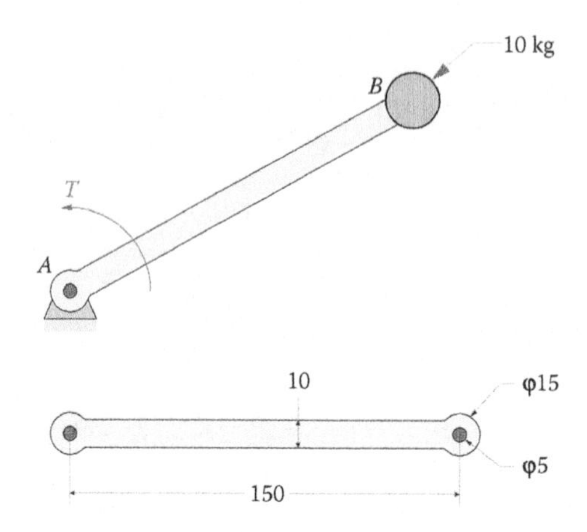

FIGURE 7.136
Problem 7.9.

Problem 7.10

A 5mm thick link shown in Figure 7.137 is made of 6061-T6 aluminum and has a spring attached at its end. Use MATLAB to plot the driving torque of the link if the spring constant is $k = 1{,}000$ N/m. The spring is in its neutral position when the crank angle is 0° and the link rotates at 100 rpm. Use SOLIDWORKS to calculate the inertial properties of the link.

Problem 7.11

Use MATLAB to plot the driving torque of the threebar linkage shown in Figure 7.138 if a 100N vertical force is applied at point P and the crank rotates at 30 rad/s. The masses and moments of inertia of the links are given below, and all dimensions are in millimeters.

$$m_2 = 0.10 \text{ kg} \quad m_3 = 0.15 \text{ kg}$$

$$I_2 = 0.000075 \text{ kg} \cdot \text{m}^2 \quad I_3 = 0.000175 \text{ kg} \cdot \text{m}^2$$

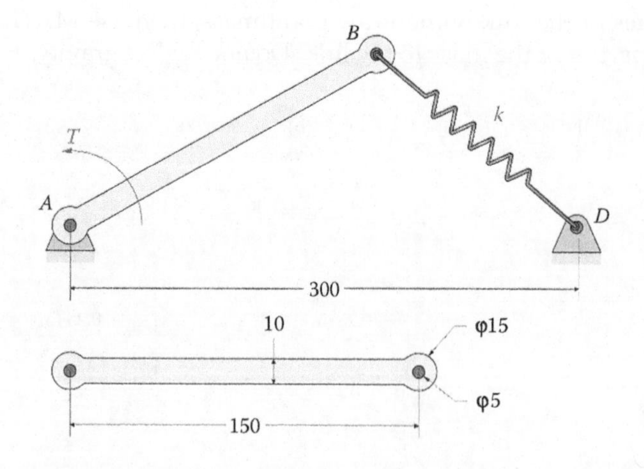

FIGURE 7.137
Problem 7.10.

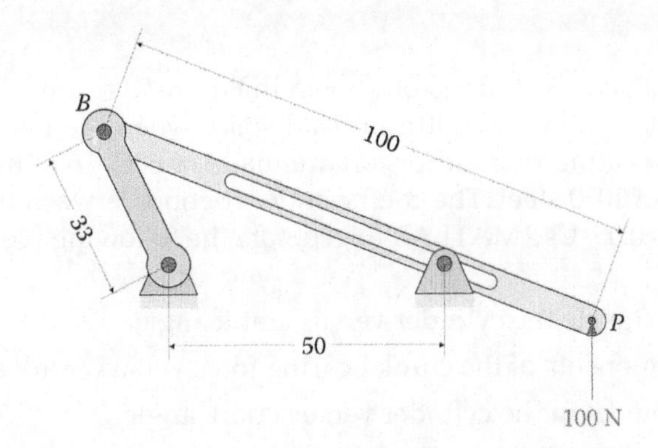

FIGURE 7.138
Problem 7.11.

Problem 7.12

Verify the code created in Problem 7.11 by plotting the external and kinetic power consumed by the linkage versus crank angle.

Problem 7.13

A slider-crank mechanism has a crank length of 50 mm, connecting rod length 150 mm, and vertical offset –25 mm. The mechanism is driven by a motor with constant angular velocity 10,000 rpm. All links are flat and made of aluminum, with thickness 5 mm. The dimensions of the links are shown in Figure 7.139. Calculate and plot:

a. The x and y forces on the ground pivot versus crank angle.
b. The driving torque versus crank angle needed to accomplish this motion.
c. The reaction moment on the cylinder versus crank angle.
d. The normal force on the cylinder versus crank angle.

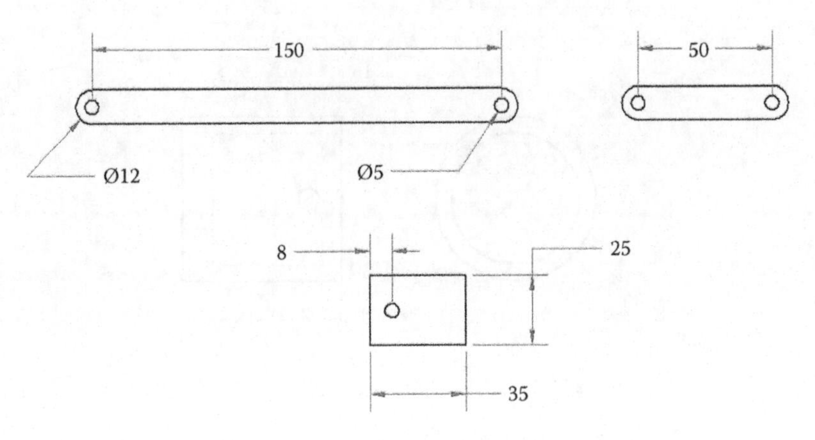

FIGURE 7.139
Problem 7.13.

Problem 7.14

The connecting rod, crank, and piston shown in Figure 7.140 are used to build an air compressor. The cylinder has 10 mm head space when the piston is at top dead center. The connecting rod and piston are made of 6061-T6 aluminum, while the crank is made of 1020 steel. The coefficient of friction between the piston and the cylinder wall is 0.05. Use MATLAB to generate the following plots:

a. air pressure inside the cylinder versus crank angle.
b. x and y components of the crank bearing force versus crank angle.
c. reaction moment in the cylinder versus crank angle.
d. driving torque versus crank angle.

Confirm your results by creating a kinetic/external power plot for one rotation of the crank.

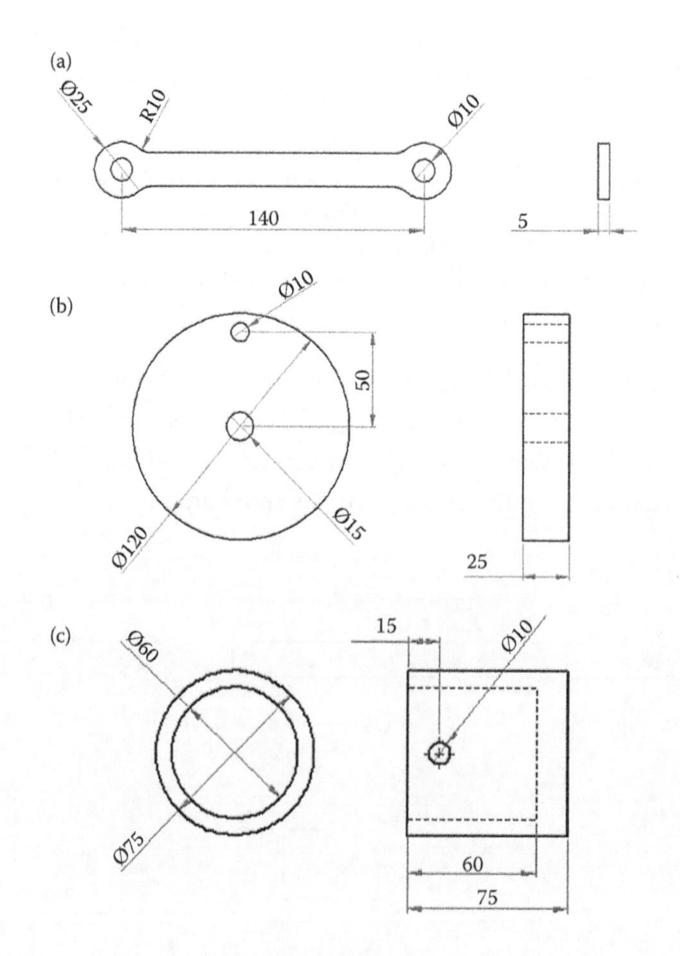

FIGURE 7.140
Problem 7.14.

Problem 7.15

A sadistic carnival operator has designed a ride based on a fourbar linkage, shown in Figure 7.141. The passenger sits at point P while the crank rotates at a constant 40 rpm. The thickness of all links is 25 mm, and all links are made of aluminum. Assume that the rider is a point mass of 100 kg located at point P. The distance between ground pins is 1200mm. Calculate and plot the forces in the x and y direction at the crank pivot, and the driving torque required to operate the ride. *Do not ignore gravity.*

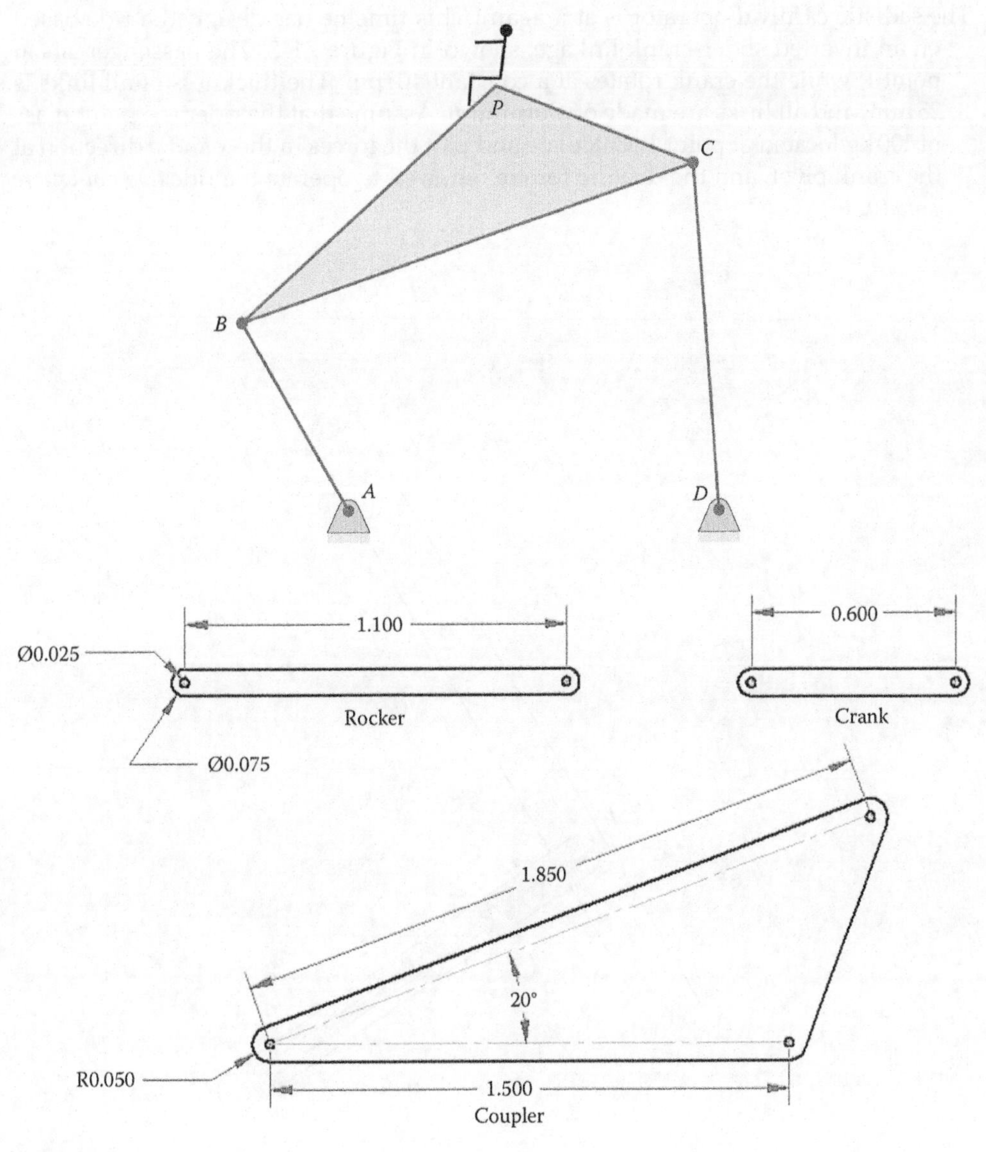

FIGURE 7.141
Problem 7.15.

Problem 7.16

A threebar linkage and spring have been designed to ensure that a solid aluminum door panel shuts properly after it has been opened. The door has a width and height of 2 m and is 25 mm thick as shown in Figure 7.142. Use MATLAB to calculate and plot the x and y components of the force $\mathbf{F}_Q$ used to open the door. Assume that the door takes 5 s to open and that the spring is in its neutral position when the door is closed.

Problem 7.17

The sadistic carnival operator is at it again! This time he has designed a ride based on an inverted slider-crank linkage, shown in Figure 7.143. The passenger sits at point P while the crank rotates at a constant 40 rpm. The thickness of all links is 25 mm, and all links are made of aluminum. Assume that the rider is a point mass of 100 kg located at point P. Calculate and plot the forces in the x and y direction at the crank pivot, and the driving torque required to operate the ride. *Do not ignore gravity.*

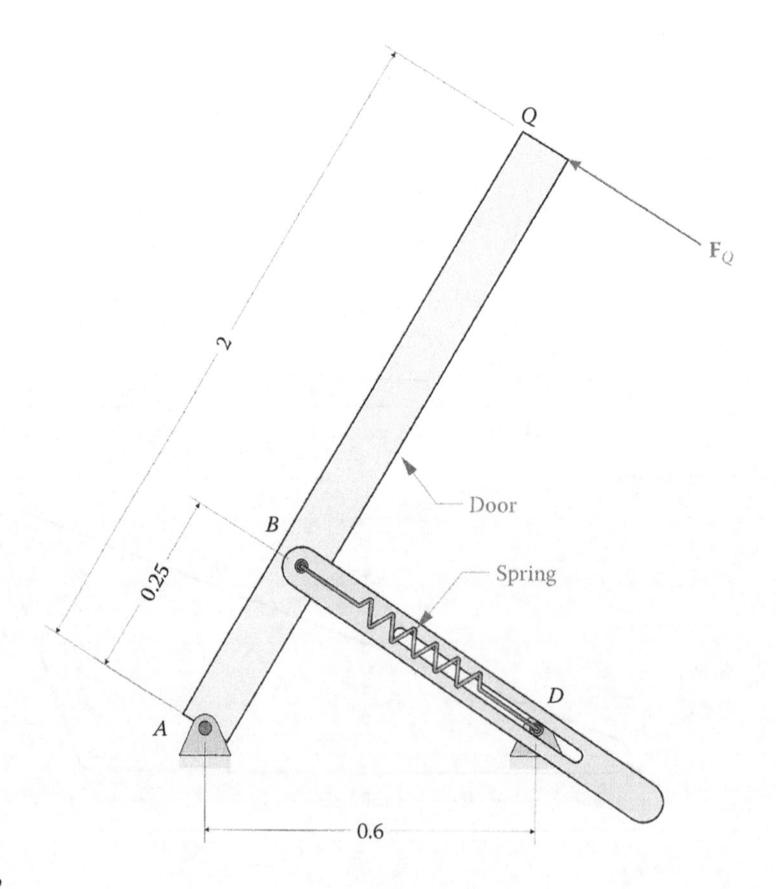

FIGURE 7.142
Problem 7.16.

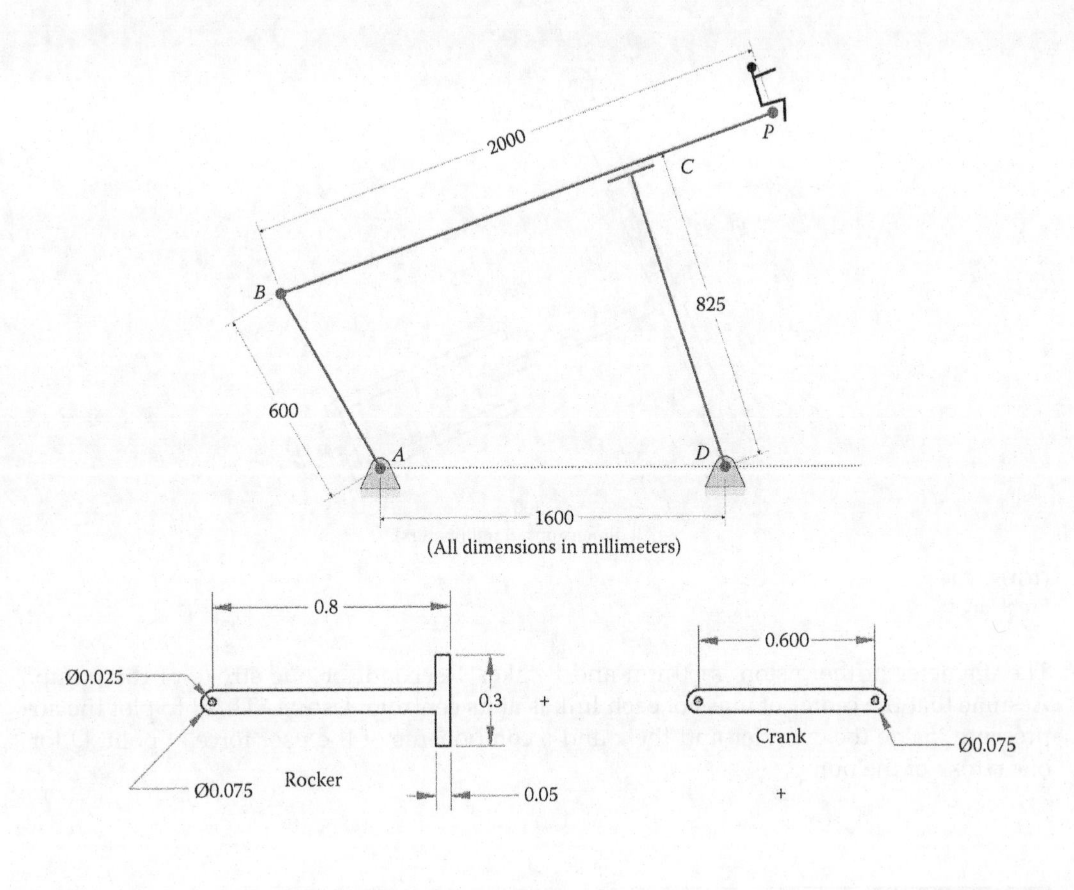

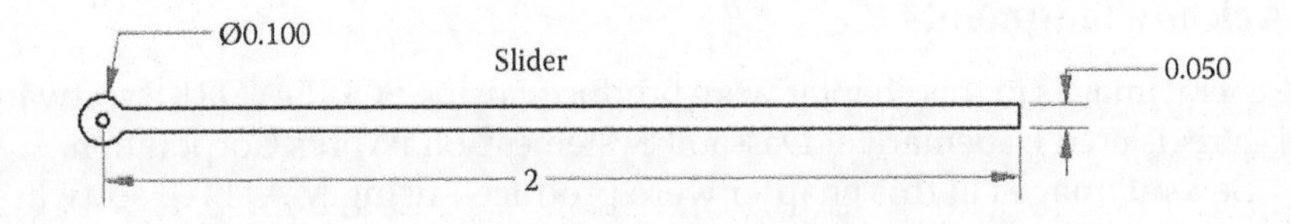

FIGURE 7.143
Problem 7.17.

Problem 7.18

An inverted slider-crank linkage is being employed as a bicycle pump, as shown in Figure 7.144. The air inside the cylinder is at atmospheric pressure when the crank angle is at 90°, and reaches its maximum when the crank angle is at 10° from horizontal. The links have the following inertial properties:

$$m_2 = 0.1\text{ kg} \qquad I_2 = 0.0005\text{ kg}\cdot\text{m}^2$$

$$m_3 = 0.250\text{ kg} \qquad I_3 = 0.0004\text{ kg}\cdot\text{m}^2$$

$$m_4 = 0.15\text{ kg} \qquad I_4 = 0.0005\text{ kg}\cdot\text{m}^2$$

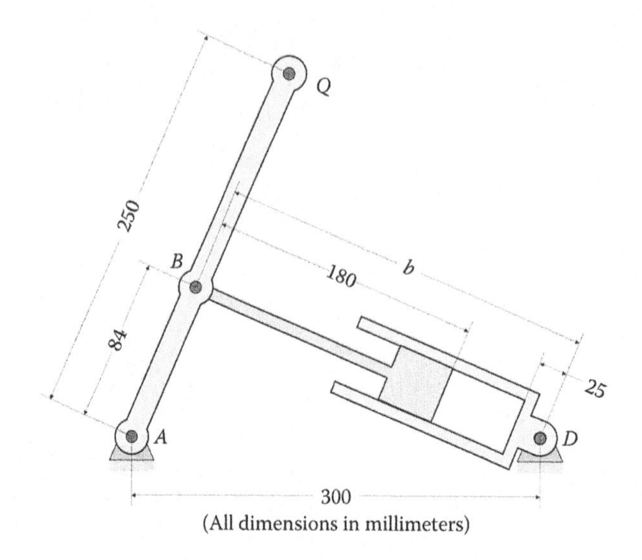

FIGURE 7.144
Problem 7.18.

The diameter of the piston is 50 mm and it takes 1 second for one stroke of the pump. Assume that the center of mass of each link is at its centroid. Use MATLAB to plot the air pressure inside the cylinder and the x and y components of the user force at point Q for one stroke of the pump.

Acknowledgments

Several images in this chapter were produced using SOLIDWORKS software. SOLIDWORKS is a registered trademark of Dassault Systèmes SolidWorks Corporation.

Several images in this chapter were produced using MATLAB software.

MATLAB is a registered trademark of The MathWorks, Inc.

Work Cited

1. Y. Cengel and M. Boles, *Thermodynamics - An Engineering Approach*, New York: McGraw-Hill, 1989.

8

Gears and Gear Trains

8.1 Introduction to Gears

Gears are used to transmit motion and torque from one shaft to another. In this section, we will discuss the kinematics of gears; that is, the motion relationships between gears. In a later section, we will learn how to conduct force and torque analysis on gears in order to calculate bearing loads and the like. There are several different types of gears to choose from, depending upon the application and the available budget.

8.1.1 Spur Gears

Spur gears transmit motion between parallel shafts, as shown in Figure 8.1. They are the simplest gears to manufacture, and are the most commonly encountered in practice. They are also relatively noisy and weak, compared to helical gears. The "reverse" gear in a manual transmission is composed of spur gears – hence the distinctive noise when you back your car up. Spur gears are made of almost every material, from the softest plastic to the hardest steel.

8.1.2 Helical Gears

Like spur gears, helical gears can transmit motion between parallel axes. The teeth of a helical gear are set at an angle, called the helix angle. A common value for the helix angle is 45° as shown in Figure 8.2. Because of the helix angle, the teeth engage gradually (and therefore quietly) instead of suddenly, as in the case of spur gears. All forward gears in a manual transmission are helical, which is the main reason that manual transmissions are almost inaudible at highway speeds. Helical gears have more teeth in contact than spur gears; this distributes the load more evenly and makes helical gears, on average, stronger than spur gears.

When purchasing helical gears you must remember to order one left-handed gear and one right-handed gear, where the "handedness" refers to the direction of the helix, see Figure 8.3. If you order two left-handed gears (or two right-handed gears) they will mesh as shown in Figure 8.4. As you can see, the two right-handed helical gears are meshing on non-parallel, non-intersecting shafts. This is kinematically quite interesting, but is impractical in most situations since the high friction levels would make for an inefficient drivetrain.

8.1.3 Bevel Gears

Bevel gears transmit motion between non-parallel, intersecting shafts, in most cases the shafts are perpendicular to one another. The two most common types of bevel gears are straight bevel gears, shown Figure 8.5 left, and spiral bevel gears, shown in Figure 8.5 on

FIGURE 8.1
Spur gears transmit motion between parallel shafts.

FIGURE 8.2
Helical gears are available in left-handed and right-handed configurations.

FIGURE 8.3
Mating a left-handed helical gear with a right-handed helical gear transmits motion between parallel shafts.

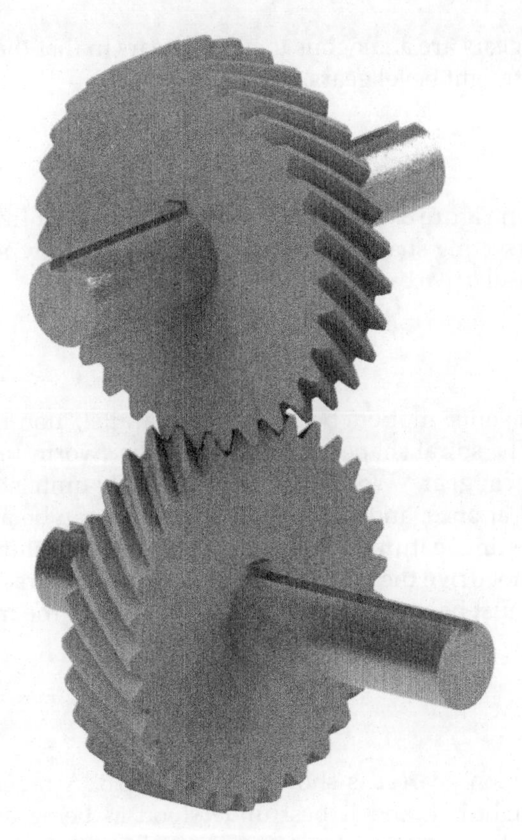

FIGURE 8.4
Mating two left-handed (or two right-handed) helical gears will transmit motion between perpendicular, non-intersecting shafts.

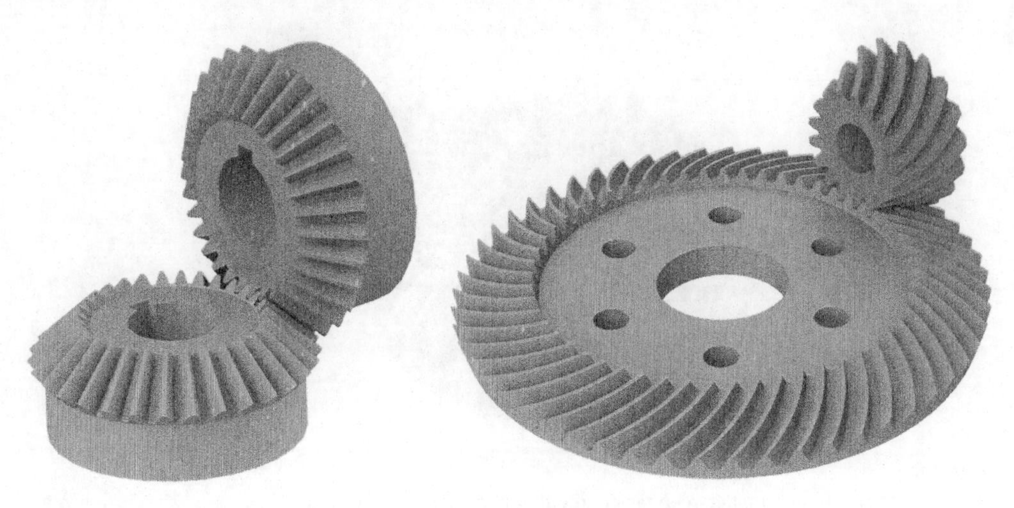

FIGURE 8.5
Bevel gears can have straight teeth or spiral teeth.[1]

the right. Spiral bevel gears are analogous to helical gears in that they have stronger teeth, and are quieter than straight bevel gears.

8.1.4 Hypoid Gears

Hypoid gears, shown in Figure 8.6, are similar to bevel gears, but transmit torque between non-parallel, non-intersecting shafts. A common application of hypoid gears is in the rear differential in rear-wheel drive cars.

8.1.5 Worm Gears

A worm gearset transmits motion between non-parallel, non-intersecting shafts as shown in Figure 8.7. The spiral shaped gear is called the "worm" and the gear it meshes with is called the "worm gear." Worm gears are typically quite strong, since they have several teeth engaged at once, and very large reductions can be achieved in a compact space. One very interesting feature of most worm gearsets is that the worm can drive the gear, but the gear cannot drive the worm! This makes worm gearsets valuable in applications where the load must be prevented from "back-driving" the motor, as in the case of a winch or crane.

8.1.6 Rack

A typical rack and pinion gearset is shown in Figure 8.8. A rack is a set of gear teeth machined onto a straight bar, and is best understood as being an ordinary spur gear with infinite diameter. Racks are most commonly used to change rotary motion to linear motion. In a "rack and pinion" steering system, rotary motion from the steering wheel is converted to linear motion of the steering linkage.

FIGURE 8.6
Hypoid gears transmit motion between perpendicular, non-intersecting shafts.[1]

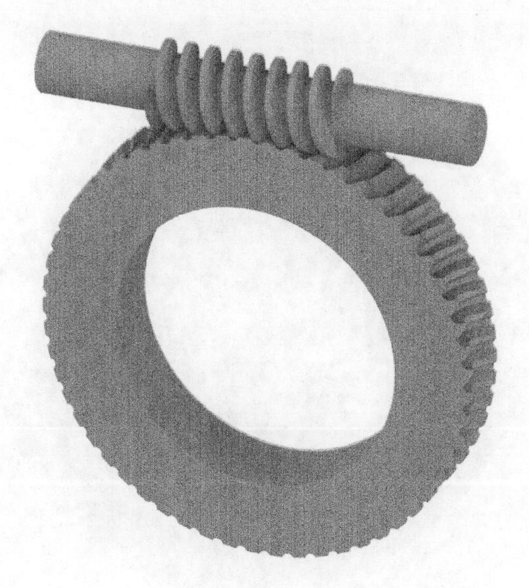

FIGURE 8.7
Worm gearsets transmit motion between perpendicular, non-intersecting shafts. Most worm gearsets cannot be "back-driven"; that is, trying to turn the gear will not rotate the worm.[1]

FIGURE 8.8
A rack is a straight gear that meshes with an ordinary, circular gear. It is used to convert rotary motion to linear motion, or vice versa.

8.1.7 Internal Gears

Internal gears have teeth machined into the interior of a circle, as shown in Figure 8.9. They are often used where the output motion must be in the same direction as the input motion, or where a large speed reduction in a small space is needed. We will encounter internal gears quite a bit during our discussion of planetary gearsets.

FIGURE 8.9
Internal gears can be used to achieve large speed reductions in a small space.

8.2 Properties of the Involute Curve

One of the most noticeable features of gear teeth is that they are curved, rather than straight. This is done to ensure *constant velocity meshing*; that is, to ensure that the speed of the driven gear does not vary as teeth come in and out of contact. Constant-velocity meshing implies that the speed of the driven gear depends only upon the speed of the driving gear, and does not depend upon the angle of rotation of either gear. In mathematical terms, we would be written

$$\omega_2 = -\rho\omega_1$$

where ω_1 is the angular velocity of the driving gear, ω_2 is the angular velocity of the driven gear and ρ is the *gear ratio*, which must be constant for constant velocity meshing. The negative sign occurs because the driven gear rotates in the opposite direction from the driving gear for external gears.

The utility of constant-velocity meshing is probably obvious – we do not want the driven gear to speed up and slow down when the driving gear is rotating at a constant speed. The fact that the gear teeth must have a particular shape is not so obvious. To demonstrate this, let us examine the worst-case scenario: a pair of gears with straight teeth. Figure 8.10 shows two of these "primitive" gears in mesh with each other. Let us assume that the gear on the left is driving the gear on the right. As shown in the figure, the left gear has positive angular velocity and the right gear has negative angular velocity.

A zoomed-in view of the two gears is shown in Figure 8.11. The center of the driving gear is at point A and the center of the driven gear is at point B. The driving gear makes contact with the driven gear at the tip of its tooth, which is shown as point C.

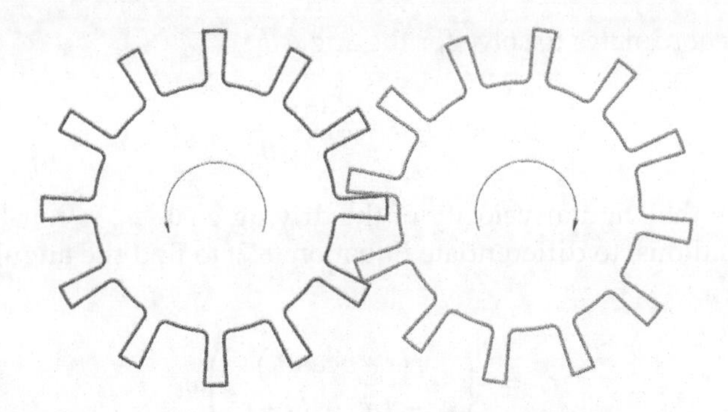

FIGURE 8.10
Two primitive gears in mesh with straight-sided teeth.

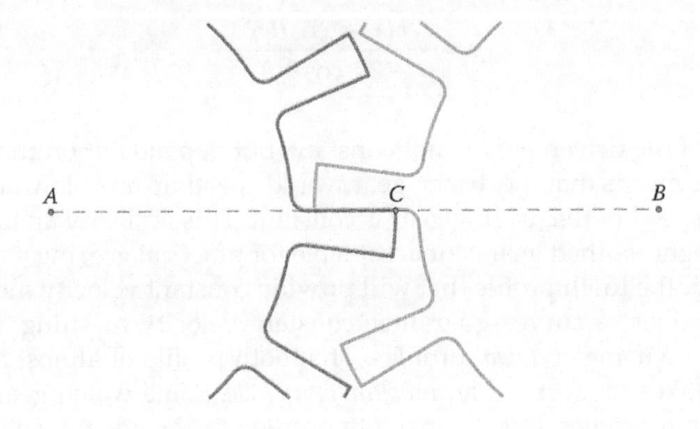

FIGURE 8.11
Zoomed-in view of the straight-toothed gears. The driving gear makes contact with the driven gear at point C, which is the tip of its tooth.

Now rotate the driving gear through an angle θ_1. As seen in Figure 8.12, the driven gear rotates through an unknown angle θ_2. Let the *center distance* (the distance between points A and B) be a constant c. The distance from the center of the driving gear to the tip of its teeth (the distance between points A and C) is r. The coordinates of point C are then

$$\mathbf{x}_C = r\begin{Bmatrix} \cos\theta_1 \\ \sin\theta_1 \end{Bmatrix} \tag{8.1}$$

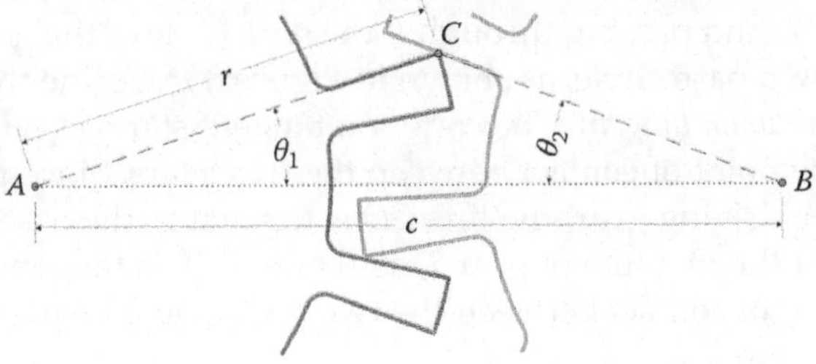

FIGURE 8.12
The driving gear has been rotated by an angle θ_1, which rotates the driven gear through an angle θ_2.

We can use these coordinates to solve for the angle θ_2.

$$\tan\theta_2 = \frac{r\sin\theta_1}{c - r\cos\theta_1} \tag{8.2}$$

Let us assume that the angular velocity of the driving gear, ω_1, is constant. It is straightforward (though tedious) to differentiate Equation (8.2) to find the angular velocity of the driven gear

$$\omega_2 = -\left(\frac{r(r - c\cos\theta_1)}{c^2 - 2rc\cos\theta_1 + r^2}\right)\omega_1 \tag{8.3}$$

The gear ratio is then:

$$\rho = \frac{r(r - c\cos\theta_1)}{c^2 - 2rc\cos\theta_1 + r^2} \tag{8.4}$$

Thus, the speed of the driven gear is not constant, but depends upon the angle, θ_1, of the driving gear. This means that the driven gear would speed up and slow down as it rotates, even though the speed of the driving gear is constant. This is clearly an undesirable situation, and the straight-toothed gear is unacceptable for practical gearing. Our goal is to find a type of curve for the tooth profile that will provide constant velocity meshing.

Only a few families of curves guarantee constant velocity meshing, and many older gearsets were made using *cycloidal* profiles. The tooth profile of almost all modern commercial gearing takes the form of an *involute of a circle.* Some watch gears are still made with cycloidal tooth profiles, but the involute profile has the advantage that slight variations in center distance will not impair the constant velocity meshing. It is entirely possible (and common) to be able to design gearsets without understanding the constant velocity meshing of the involute tooth profile, but the proof is interesting in its own regard.

The involute of a circle is shown in Figure 8.13. You can generate your own involute by wrapping a string around an aluminum can and tying a pencil to the end of the string. As you unwind the string from the can – making sure to keep the string taut – the resulting curve is an involute. The outside surface of the can is called the *base circle* of the involute. Because the string is always tangent to the outside surface of the can, every segment of the involute is perpendicular to a line tangent to the base circle. This will be important as we consider contact forces later in this section.

Now consider the two gears in mesh with involute teeth as shown in Figure 8.14. The gear teeth make contact at point C. Draw a line from C that is tangent to the base circle and label the intersection *D*. The line *CD* intersects the *line of centers* at point *P*. We will denote the circle centered at *A* and passing through *P* the *pitch circle* of the gear.

Each gear has its own base circle, as shown in Figure 8.15. A line that is tangent to both circles is called the *common tangent*. The common tangent starts at point *D* and extends to point *E*. Point C is the point of contact between the two gears. Since point C is part of the involute of gear 1 (the driving gear) the line *CD* is tangent to the base circle of gear 1. The point C is also part of the involute of gear 2, so the line *CE* is tangent to the base circle of gear 2. Thus, the point of contact between the two gears must lie on the common tangent between the two base circles.

A magnified view of the zone of contact between the two gears is shown in Figure 8.16. As can be seen, all of the contact points lie on the common tangents between the two base

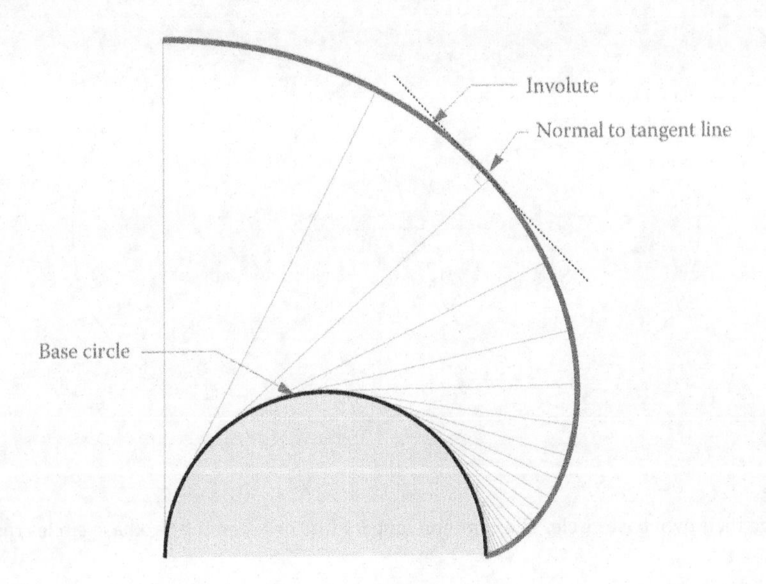

FIGURE 8.13
The involute of a circle can be generated by unwinding a string that is wrapped around the base circle.

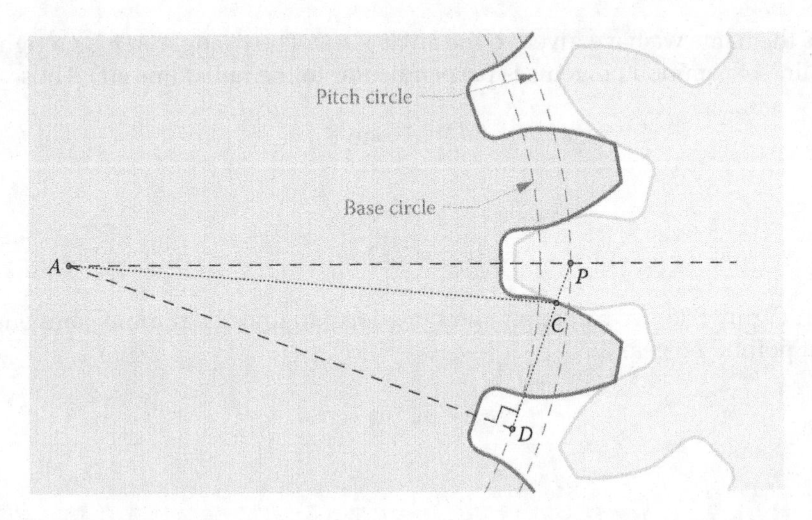

FIGURE 8.14
Two gears in mesh. The gears make contact at point C.

circles. Since there are two common tangent lines between any two circles, the location of the contact points depends upon which direction the driving gear is rotating. The important thing to note is that all contact points lie in a straight line – the common tangent line.

We can calculate the velocity on gear 1 of point C by using the angular velocity formula

$$|\mathbf{v}_{C1}| = AC \cdot \omega_1$$

where ω_1 is the angular velocity of gear 1 – see Figure 8.17. Note that this velocity is perpendicular to the line AC. The component of this velocity that is normal to the face of the tooth is

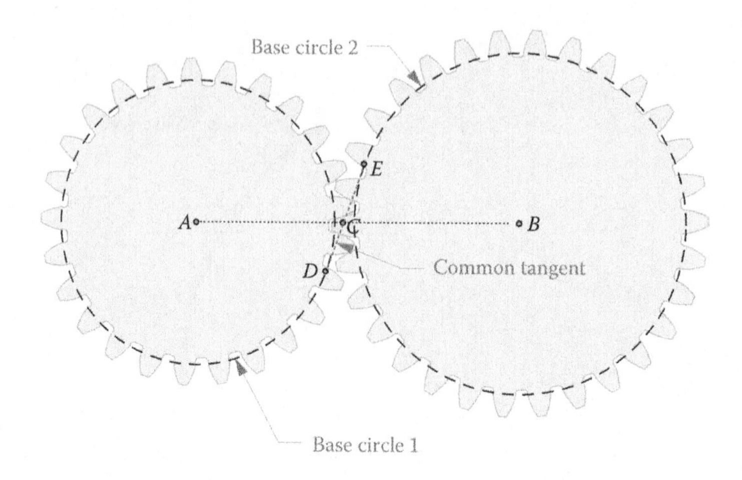

FIGURE 8.15
Both gears have their own base circle. The common tangent line between the two base circles runs from point D to point E.

$$|\mathbf{v}_{Cn}| = AC \cdot \omega_1 \cdot \cos\gamma$$

where γ is the name we have given to the angle CAD. The triangle ADC is a right triangle since the line of common tangents is perpendicular to the radial line AD. Thus

$$AD = AC\cos\gamma \tag{8.5}$$

and

$$|\mathbf{v}_{Cn}| = AD \cdot \omega_1$$

As seen in Figure 8.18, we can apply the same logic to find the normal component of the velocity at point C on gear 2

$$|\mathbf{v}_{Cn}| = BC \cdot \omega_2 \cdot \cos\delta$$

or

$$|\mathbf{v}_{Cn}| = BE \cdot \omega_2$$

The tangential component for the two gears will, in general, be different because the gear teeth slide past each other as the gears mesh. The normal component, however, must be the same. Thus, we can write

$$\frac{\omega_2}{\omega_1} = \frac{AD}{BE}$$

This ratio must be constant for constant velocity meshing. It is interesting to note that the normal to **both** gear teeth at the point C lies in the same direction, along the line DE. Because of how the involute was constructed, this line is tangent to **both** base circles and is commonly called the *line of action*.

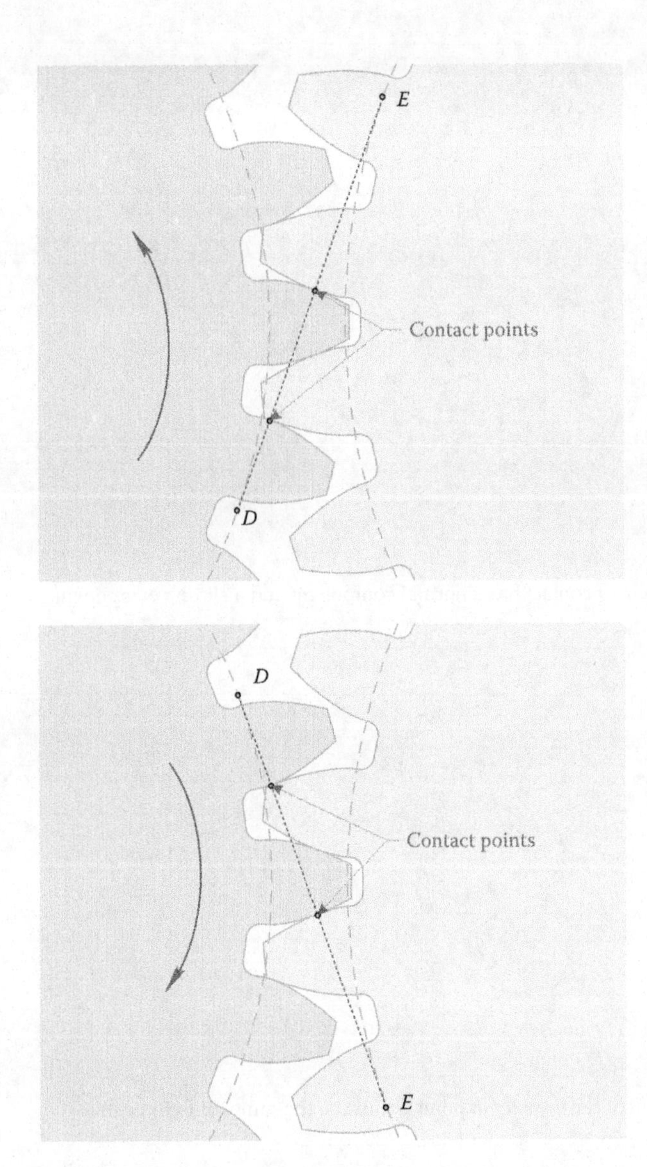

FIGURE 8.16
A zoomed-in view of the area of contact between the two gears. All points of contact between the gears lie on the common tangents between the base circles.

Now define the point P where the line of action crosses the line of centers, as seen in Figure 8.19. This is an important point in gear train design and is called the *pitch point*. Since DE and AB are both straight lines, the angles DPA and EPB are equal. Angles ADP and BEP are both right angles, so the triangles ADP and BEP are similar. Because of this we can write

$$\frac{AD}{BE} = \frac{AP}{BP}$$

Thus, the velocity ratio between the two gears is

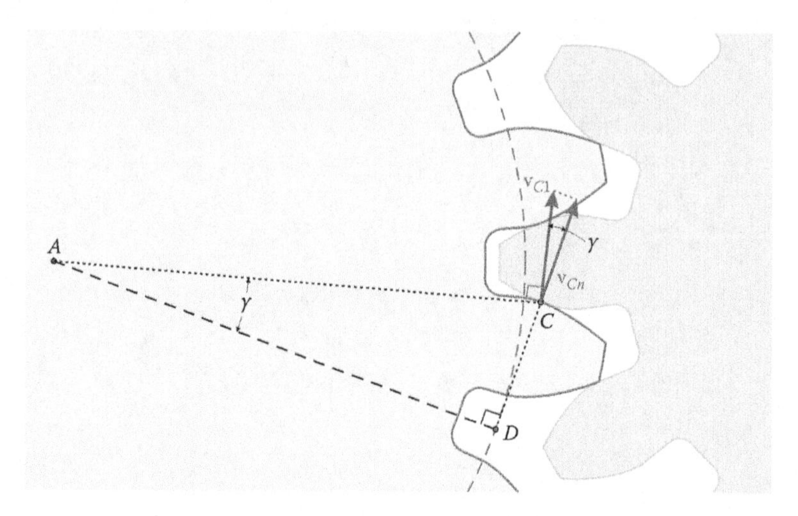

FIGURE 8.17
The velocity at the point of contact has a normal component and a sliding component.

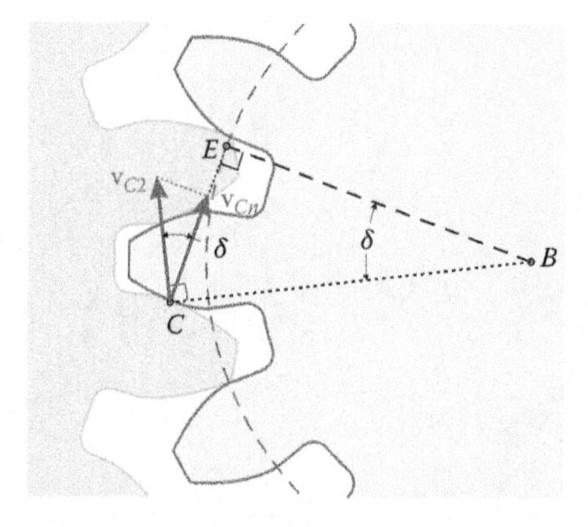

FIGURE 8.18
The normal component of the velocity at point C must be the same for both gears.

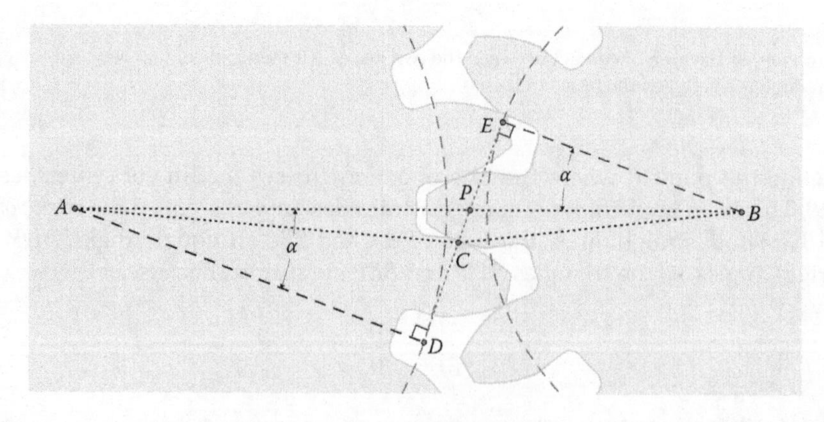

FIGURE 8.19
The line of action crosses the line of centers at the pitch point.

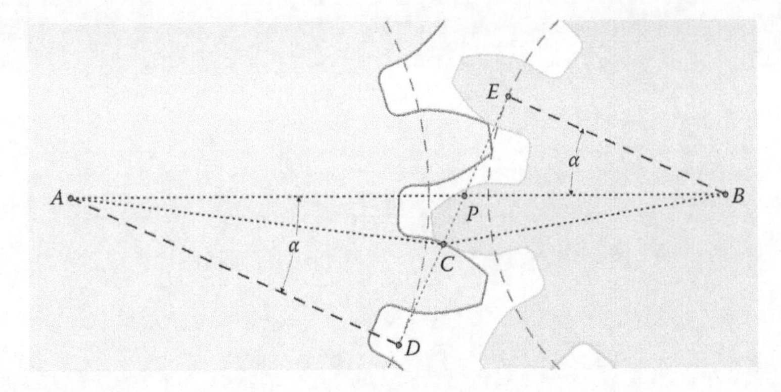

FIGURE 8.20
Changing the center distance between the gears does not change the velocity ratio.

$$\frac{\omega_2}{\omega_1} = \frac{AP}{BP} \tag{8.6}$$

For the velocity ratio to be constant, the point P must be stationary. Since it lies at the intersection of the line of common tangents and the line of centers, this is indeed the case. Thus, we have proved that involute gear teeth produce constant velocity meshing.

It is interesting to observe what happens if the center distance between the two gears is changed slightly, as might be the case if the shafts holding the two gears were located incorrectly. As shown in Figure 8.20 the geometry of the situation is basically unchanged. Although the lengths AP and BP have increased, the ratio

$$\frac{AD}{BE} = \frac{AP}{BP}$$

still holds and the velocity ratio is the same as it was before (since the base circle radii are fixed). Thus, slight errors in center distance have no effect on the velocity ratio or the constant velocity meshing. This (along with manufacturing considerations) is the reason that involute gearing is so widely used in practice.

Note that using improperly spaced gears will introduce backlash into the gear train. If the gears continue to spin in the same direction, then constant velocity meshing will be maintained. However, if the driving gear changes direction, there will be a brief period of time when the gear teeth lose contact with one another, which is known as backlash. Once the teeth regain contact in their new direction constant meshing velocity is achieved once again.

8.2.1 Base Circles and Pitch Circles

Despite its importance in generating the involute tooth shape the base circle is almost never used in designing a gear train. The most fundamental quantity is instead the *pitch circle*. As shown in Figure 8.21 each pitch circle passes through the pitch point, P, and both are tangent to each other. Referring to Figure 8.19, we see that the base circle radius and pitch radius are related to each other by

$$r_b = r_p \cos\alpha \tag{8.7}$$

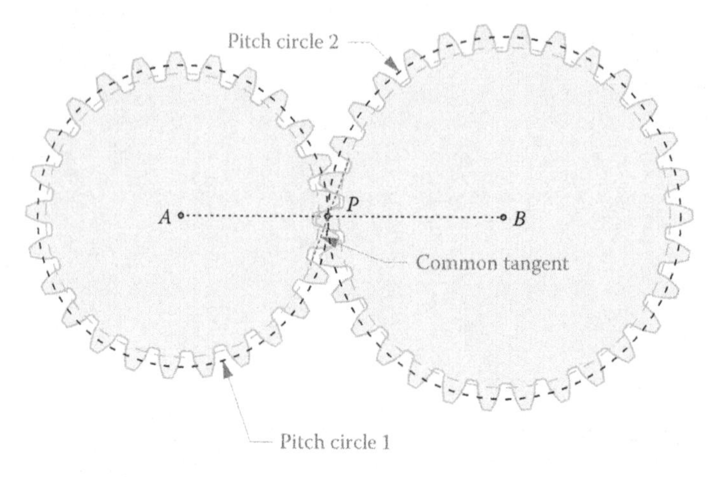

FIGURE 8.21
Pitch circles are used in designing gearsets.

where the angle α is called the *pressure angle*, for reasons that will become clear in the next section. It is more common to specify the *pitch diameter* (the diameter of the pitch circle) for a particular gear. If the pitch diameter of gear 1 is d_1 and the pitch diameter of gear 2 is d_2 then the center distance is

$$c = \frac{d_1 + d_2}{2}$$

Figure 8.22 shows two gears that have the same base circle diameter but different numbers of teeth. If a gear has a small number of teeth, each tooth will span a relatively large angle, allowing more of the involute curve to be traced out. Gears with large tooth numbers have relatively straight teeth. As we will see in the next section, the teeth of a rack (which is just a gear with infinite diameter) are perfectly straight.

8.2.2 Force Analysis on Involute Gears

Figure 8.23 shows the force that gear 2 exerts on gear 1 as torque is transmitted from gear 1 to gear 2. If we assume that the sliding friction force on the gear teeth is negligible, then the resulting force must be normal to the gear teeth; that is, directed along the line of action.

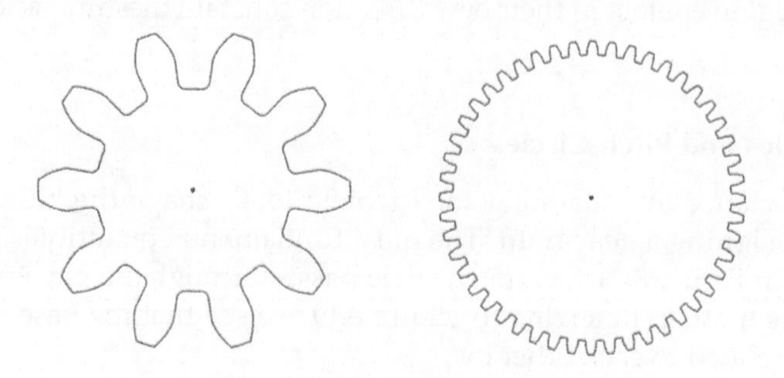

FIGURE 8.22
For a given base circle, the lower the number of teeth, the more curvature on the tooth profile.

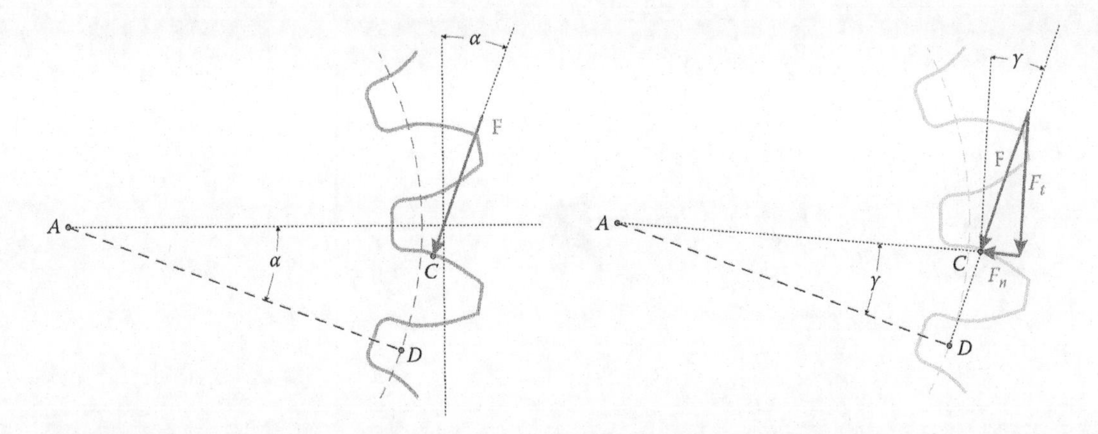

FIGURE 8.23
Torque is transmitted between the two gears through the contact force, **F**. The contact force acts at an angle α with the vertical.

This means that the contact force is always oriented at an angle α with the vertical. This is why the angle α is called the pressure angle.

The contact force has two components. The component F_t that is normal to the line AC does the job of transmitting the torque between gears. The other component, F_n, is collinear with AC. Since this component is directed inward toward the center of the gear, it does no useful work, and merely increases the load on the bearing that supports the gear shaft. The torque-transmitting component can be calculated as

$$F_t = |\mathbf{F}|\cos\gamma$$

This force creates a torque on gear 1 given by

$$T_1 = AC \times F_t$$

Substituting Equation 8.5 into this relation gives

$$T_1 = AD \times |\mathbf{F}|$$

Recall that AD is the base circle radius. Since we normally work with the pitch radius (or diameter) we can use Equation 8.7 to obtain

$$T_1 = \frac{d_1}{2}|\mathbf{F}|\cos\alpha$$

It is somewhat inconvenient to work with the force **F** acting on the point of contact C, since C moves as the gear rotates. Instead, let us calculate a statically equivalent set of forces acting at point P, which is fixed, as shown in Figure 8.24. This is allowed since the force **F** is always directed along the line of action and never changes direction. The torque-producing component of **F** is vertical at point P, and the radius to point P is the pitch radius. The torque produced by **F** at point P is

$$T_1 = \frac{d_1}{2}|\mathbf{F}|\cos\alpha$$

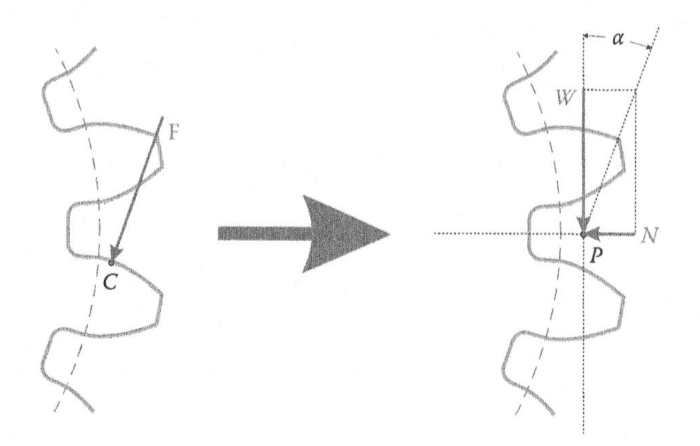

FIGURE 8.24
The force components W and N acting at P are statically equivalent to the force F acting at C.

Thus, the component of **F** that creates torque is

$$W = |\mathbf{F}|\cos\alpha$$

and

$$T_1 = \frac{d_1 W}{2}$$

The remaining component of **F** that is orthogonal to W is

$$N = |\mathbf{F}|\sin\alpha$$

Since **F** is never needed directly for calculating torque or bearing reaction load, it is common to eliminate **F** entirely and use

$$N = W\tan\alpha$$

We can conduct the same analysis on gear 2 after drawing the free-body diagram shown in Figure 8.25. Since the torque-transmitting force is equal and opposite on both gears, we have

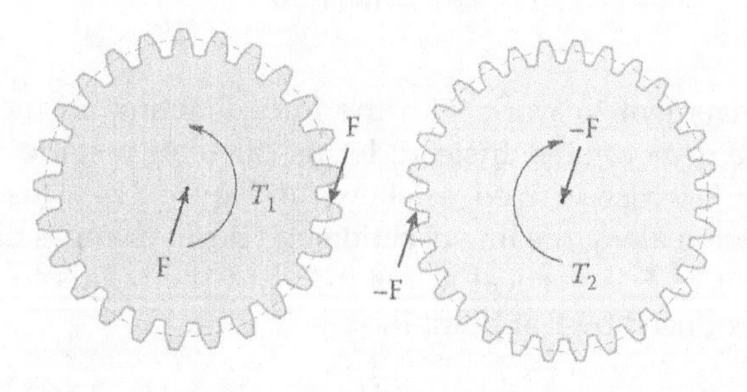

FIGURE 8.25
Free-body diagram of both gears.

$$T_1 = \frac{d_1 W}{2} \quad T_2 = \frac{d_2 W}{2}$$

or, written as a torque ratio

$$\frac{T_2}{T_1} = \frac{d_2}{d_1}$$

Thus, using a small gear to drive a large gear will result in an increase in torque, and vice versa. Comparing this with the speed ratio equation in Equation (8.6) we see that

$$\frac{T_2}{T_1} = \frac{\omega_1}{\omega_2}$$

An increase in torque is accompanied by a decrease in speed. In fact, this is the purpose of most gear trains – to decrease the speed and increase the torque of a particular motor. In fact, almost all forward gears in an automotive transmission are used to reduce the speed of the engine and increase its torque.

8.2.3 Summary

To conclude, we have demonstrated that involute gear profiles provide constant-velocity meshing, even when the center distance is slightly out of specification. The force transmitted between two gears acts along a line of action, which is directed at an angle α (the pressure angle) from the vertical. Finally, a statically equivalent force system located at the fixed point P was derived. It is customary to use the W, N force system to calculate transmitted torques and bearing loads.

8.3 Gear Terminology

In this section, we will describe the set of gear parameters that are important for design. Most of these parameters can be found in gear catalogs, and must be specified when ordering a set of gears. Figure 8.26 shows a typical spur gear. The *circular pitch*, p_c, is the arc distance between two adjacent teeth. The pitch diameter, d, is the nominal diameter of the gear that is used in calculating center distances and speed/torque ratios. Circular pitch is not often used in design, but pitch diameter is very important.

Figure 8.27 shows two spur gears in mesh. The smaller gear in a gear set is often called the *pinion*, while the larger gear is simply denoted the *gear*. As we found in the previous section, the center distance between the two gears can be found using the pitch diameter of each gear.

$$c = \frac{d_1 + d_2}{2} \tag{8.8}$$

The number of teeth on a gear is proportional to its diameter. There are two systems for specifying the ratio of gear teeth to diameter. In the United States, it is common to use the

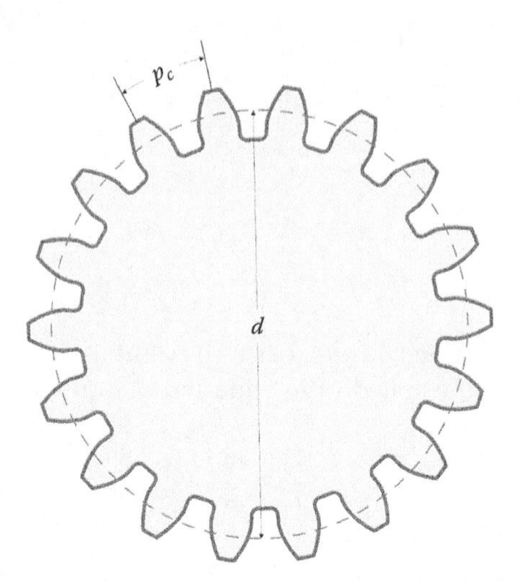

FIGURE 8.26
A typical spur gear showing the circular pitch, p_c, and pitch diameter, d.

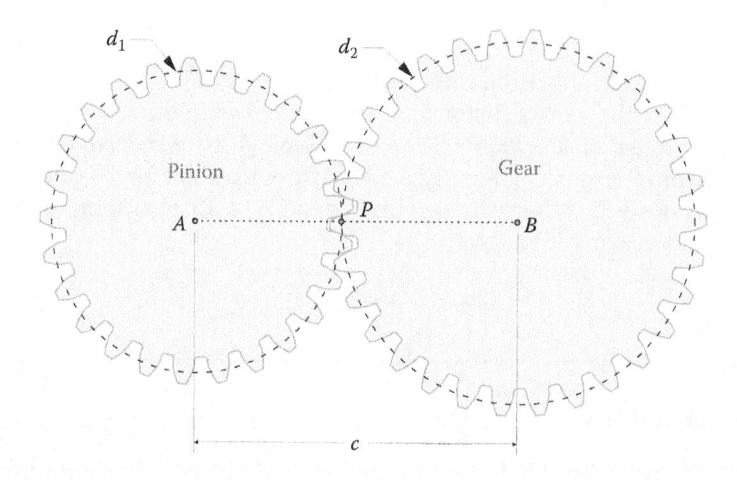

FIGURE 8.27
The center distance between two gears can be found using the pitch diameters.

diametral pitch, P, which is given in *teeth per diametral inch.* In this system the relationship between number of teeth and pitch diameter is

$$N = Pd \tag{8.9}$$

Thus, a gear with a diametral pitch of 10 teeth/in and a pitch diameter of 2 in would have 20 teeth. In the metric system, we specify a *module* in *diametral millimeters per tooth.* The number of teeth on a metric gear is found using

$$N = \frac{d}{m} \tag{8.10}$$

TABLE 8.1

A Selection of Standard Diametral Pitches and Modules

Pitch (teeth/in)	64	48	32	24	20	16	12	10	8	6	5	4
Module (mm/tooth)	0.4	0.5	0.75	1.0	1.25	1.5	2.0	2.5	3.0	4.0	5.0	6.0

Thus, a gear with a diameter of 50 mm and a module of 2.5 mm/tooth would have 20 teeth. A list of some standard diametral pitches and modules is given in Table 8.1. This list is not exhaustive, but shows some of the more common tooth pitches seen in practice and found in gear catalogs. You should choose a standard pitch (or module) whenever possible in order to keep manufacturing costs at a minimum. Both gears in a mating pair must have the same module (or diametral pitch) in order to mesh. This is analogous to the situation with threaded fasteners – a fine-pitch nut will not thread onto a coarse-pitch screw. We can use the module (or pitch) to find the center distance between gears, given the number of teeth by combining Equations (8.8) and (8.9).

$$c = \frac{N_1 + N_2}{2P} \tag{8.11}$$

or Equations (8.8) and (8.10)

$$c = \frac{N_1 + N_2}{2} m \tag{8.12}$$

Example 8.1

Figure 8.28 shows three different gear pairs. If the space between holes on a construction brick beam is 8 mm, find the module of the gears.

Solution

There are three different gears shown in the figure: one with 8 teeth, one with 24 teeth, and one with 40 teeth. Since all three mesh together, they must all have the same module. Rearrange Equation (8.10) to solve for the module

$$m = \frac{2c}{N_1 + N_2}$$

For the first gear pair we have a center distance of 3×8 mm = 24 mm (remember to count the spaces between holes, not the holes themselves!) so that

$$m = \frac{2(24\text{ mm})}{8 + 40} = 1\frac{\text{mm}}{\text{tooth}}$$

For the second case we have a center distance of 2×8 mm = 16 mm so that

$$m = \frac{2(16\text{ mm})}{8 + 24} = 1\frac{\text{mm}}{\text{tooth}}$$

And finally, for the third case we have

$$m = \frac{2(32\text{ mm})}{24 + 40} = 1\frac{\text{mm}}{\text{tooth}}$$

As expected, the module is identical for all three cases. Since all gears are compatible with each other, they are all made with a module of 1 mm/tooth. The involute profile is most clearly seen on the eight-tooth gear since it has the smallest base circle, while the teeth on the 40 tooth gear are nearly straight.

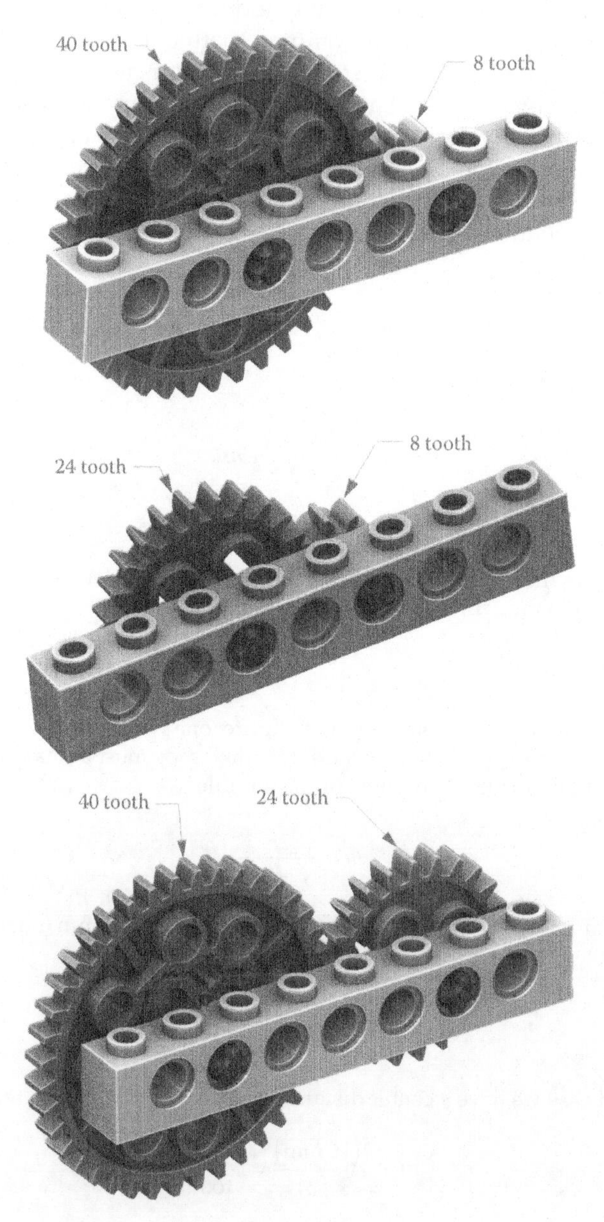

FIGURE 8.28
The construction brick gearsets used in the example problem.[2]

8.3.1 Parts of the Gear Tooth

A typical involute gear tooth is shown in Figure 8.29. The portion above the tooth that is outside the pitch circle is called the *addendum*, while the portion of the tooth inside the pitch circle is called the *dedendum*. The American Gear Manufacturers Association (AGMA) has defined the standard radii for the addendum and dedendum circles as [1]

$$r_a = r_p + \frac{1}{P}$$

$$r_d = r_p - \frac{1.25}{P} \text{ or } r_d = r_p - \frac{1.35}{P}$$

for US gears or

$$r_a = r_p + m$$

$$r_d = r_p - 1.25 \text{ m or } r_d = r_p - 1.35 \text{ m}$$

for metric gears. In these equations, r_p is the pitch radius. The tooth profile that exists outside the base circle follows an involute curve, but the involute is undefined inside the base circle. If the dedendum extends below the base circle, a radial line is commonly used for this part of the tooth. At the root of the tooth is the root fillet, which decreases the bending stress concentration created by the sharp corner there. According to the AGMA standard [1], the radius of the root fillet is

$$f = \frac{0.3}{P} \text{ or } 0.3 \text{ m} \tag{8.13}$$

In practice, the radius of this fillet will be determined by the geometry of the cutter used to form the gear teeth.

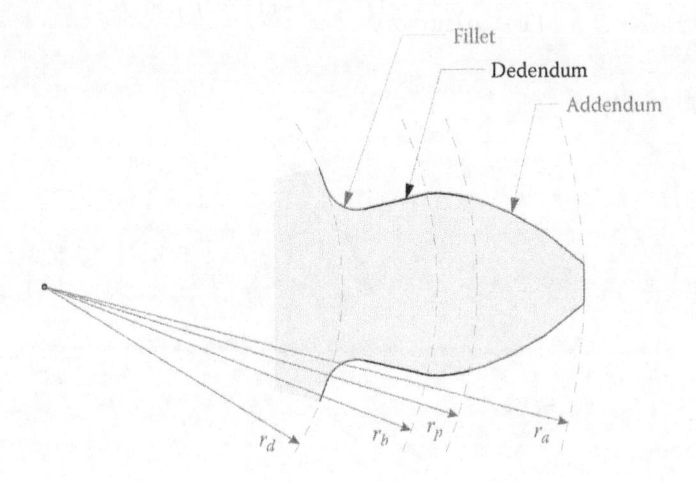

FIGURE 8.29
The main parts of the tooth profile are the addendum, dedendum, and base fillet.

8.3.2 Pressure Angle

Another important parameter in choosing a set of gears is the pressure angle. Pressure angles are also standardized, and gears with pressure angles of 14.5° and 20° are easy to find. The other standard pressure angles, 22.5° and 25°, are more rare. Three 10 tooth gears with differing pressure angles are shown in Figure 8.30. The difference is subtle, but the 25° pressure angle gears have thicker teeth at the base and are capable of transmitting heavier loads. The tradeoff is a higher normal force – the component of the contact force that does not transmit torque. Since

$$N = W \tan\alpha$$

as found earlier, a higher pressure angle will result in higher forces on the bearings that support the gear shaft. Thus, we have a tradeoff: higher pressure angles result in stronger teeth but higher bearing forces, while low pressure angles give weaker teeth but lower bearing forces. As we will see in the next section, low pressure angle gears have another disadvantage that has led to 20° becoming the most common pressure angle in practice. As with diametral pitch, two gears must have the same pressure angle in order to mesh properly.

8.3.3 Interference

If a pinion with a very small number of teeth is used to drive a much larger gear then the tips of the teeth on the gear may "undercut" the radial flanks of the pinion, as shown in Figure 8.31. In the figure, an eight-tooth pinion is in mesh with a 24 tooth gear, and both gears have 14.5° pressure angle. To avoid interference, the pinion must have a certain *minimum* number of teeth. This minimum number (for full depth teeth) is given by

$$N_p = \frac{2}{(1+2\rho)\sin^2\alpha}\left(\rho + \sqrt{\rho^2 + (1+2\rho)\sin^2\alpha}\right) \tag{8.14}$$

where ρ is the gear ratio.

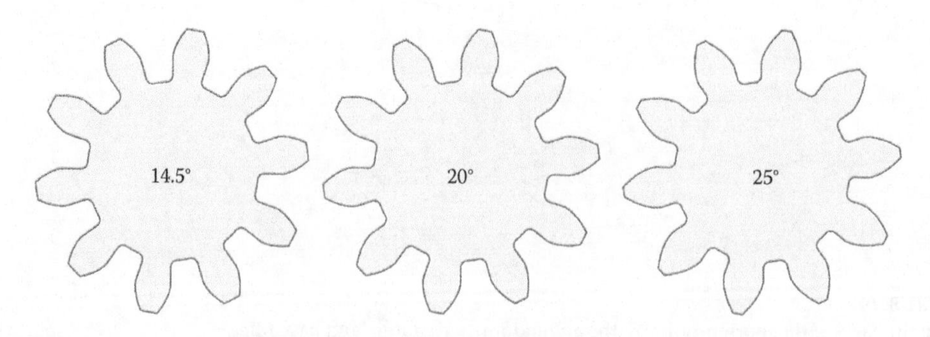

FIGURE 8.30
A 10 tooth gear in three of the standard pressure angles. The 25° teeth are thicker at the base and therefore stronger.

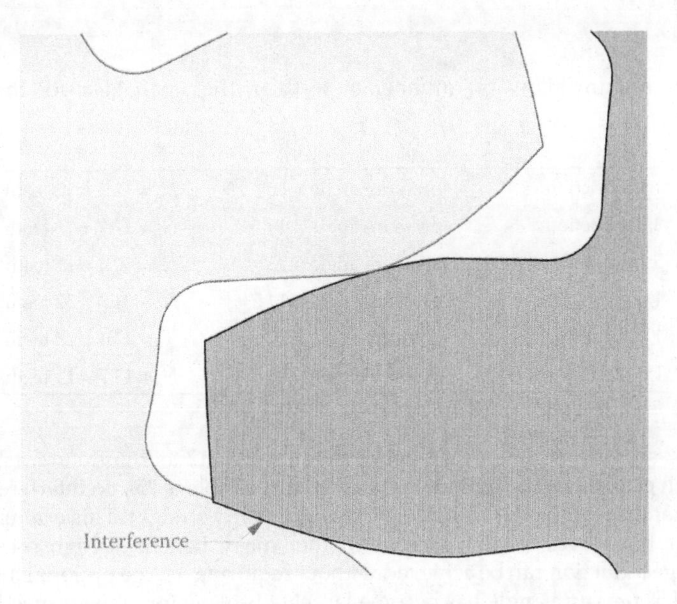

FIGURE 8.31
Gears with low numbers of teeth may exhibit interference between the tip of the large gear's tooth and the radial line of the small gear.

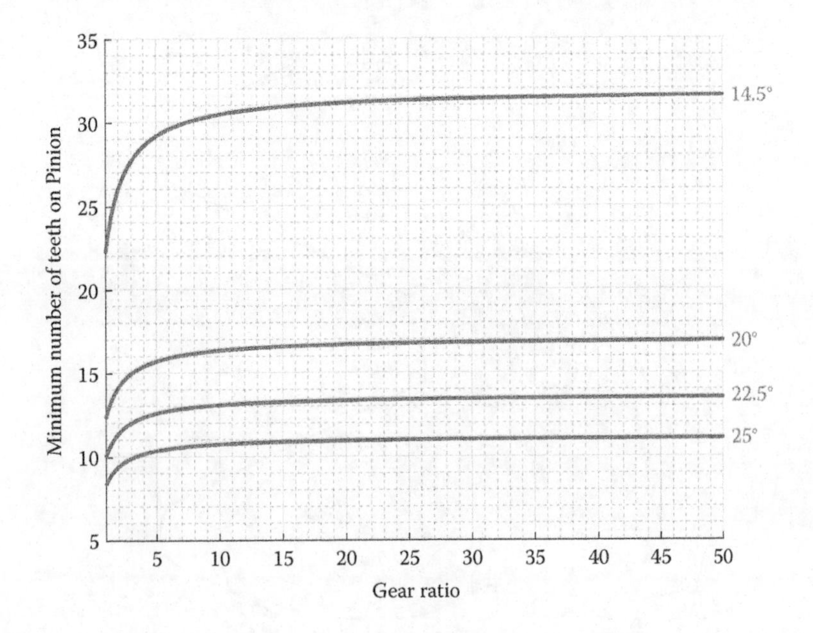

FIGURE 8.32
Minimum number of pinion teeth needed to avoid interference or undercutting.

The formula in Equation (8.14) has been plotted for a variety of gear ratios and pressure angles in Figure 8.32. Larger pressure angles allow a smaller number of teeth to be used in the pinion, which may result in a smaller gear train overall.

Example 8.2

Find the minimum allowable number of teeth in the small gear for the following examples

1:1 reduction	pressure angle = 20°	$N_p = 12.3 = 13$ teeth
3:1 reduction	pressure angle = 14.5°	$N_p = 27.7 = 28$ teeth
5:1 reduction	pressure angle = 25°	$N_p = 10.4 = 11$ teeth
1,000:1 reduction	pressure angle = 14.5°	$N_p = 31.9 = 32$ teeth
1,000:1 reduction	pressure angle = 20°	$N_p = 17.1 = 18$ teeth
1,000:1 reduction	pressure angle = 22.5°	$N_p = 13.7 = 14$ teeth

Figure 8.33 shows the somewhat fanciful situation of a 12 tooth pinion meshing with a 12,000 tooth gear. Since the pressure angle in this example is 25°, no interference occurs. Observe that the teeth in the 12,000 tooth gear are nearly straight. This example is clearly impractical, but serves to demonstrate another major benefit of high pressure angle gears: a large reduction can be achieved without requiring a large pinion. This illustrates why the 14.5° pressure angle has become largely obsolete: for a given speed reduction, the number of teeth on the pinion (and therefore on the gear) can be excessively large.

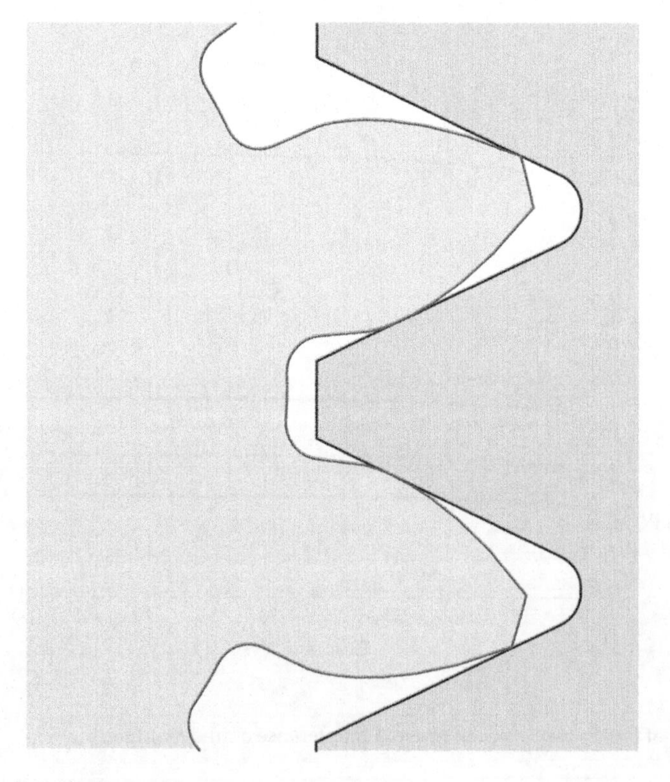

FIGURE 8.33
A 12 tooth pinion meshing with a 12,000 tooth gear. The pressure angle is 25 and no interference occurs.

8.4 Speed Reduction using Gear Trains

In Section 8.3, we learned to deduce the speed reduction in a pair of gears using the pitch diameter of each gear. We will now discuss the more common method of using the number of teeth to calculate speed reduction and also the effect of using multiple stages to form a compound speed reducer. Since the number of teeth on a gear is proportional to its pitch diameter (through the diametral pitch or module) we can use the number of teeth as a "stand-in" in our equations. Consider the single-stage gear reducer shown in Figure 8.34, which has an eight-tooth pinion driving a 24 tooth gear. Because 24/8 = 3, the pinion must make three revolutions for every one revolution of the gear. Thus, we can write

$$n_2 = -\frac{N_1}{N_2} n_1 \tag{8.15}$$

where n_1 and n_2 are the speeds of gears 1 and 2 in revolutions per minute. In this equation, we have used n instead of ω for the angular velocity of each gear. It is more common to specify gear speed in revolutions per minute (rpm) than radians per second so we will use n to represent angular velocity in rpm, reserving ω for angular velocity in radians per second. We can easily switch between the two representations using

$$\omega = \frac{2\pi n}{60} = \frac{\pi n}{30} \frac{\text{rad}}{\text{s}}$$

Remember that the minus sign was needed to account for the fact that the gear rotates in the opposite direction from the pinion. Earlier we defined the gear ratio as the ratio of the diameters of the two gears

$$\rho = \frac{d_2}{d_1}$$

But since the number of teeth is proportional to diameter, we can redefine the gear ratio in terms of tooth numbers

$$\rho = \frac{N_2}{N_1}$$

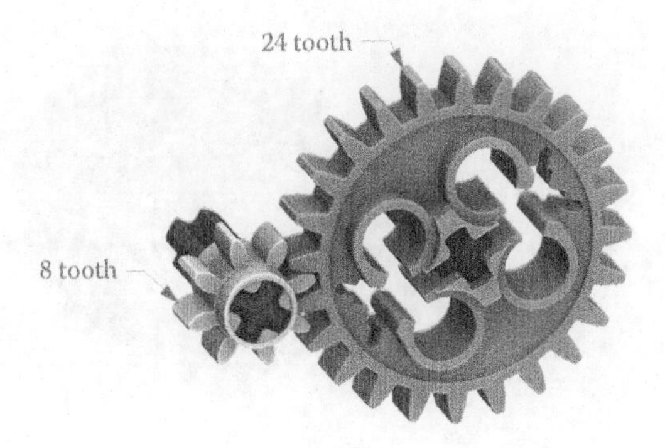

FIGURE 8.34
The eight-tooth pinion is driving the 24 tooth gear.[2]

The speed of gear 2 can be written in terms of the gear ratio as

$$n_2 = -\frac{n_1}{\rho}$$

Since most gear trains are designed to reduce speed, the gear ratio will ordinarily be greater than one. For the example in Figure 8.34 the gear ratio is 24/8 = 3. Let us now try a more complicated example:

Figure 8.35 shows a two-stage speed reducer. The first stage contains gears 1 and 2, and the second stage includes gears 3 and 4. It is important to note that gears 2 and 3 are mounted to the same shaft and spin at the same speed. In other words:

$$n_2 = n_3 \tag{8.16}$$

Let us find the speed of gear 4 (the output of the speed reducer) if the speed of gear 1 is 1,000 rpm. First, the speed of gear 2 can be found using Equation (8.15)

$$n_2 = -\frac{N_1}{N_2}n_1 = -\frac{8}{24}(1{,}000\text{ rpm}) = -333.3\text{ rpm}$$

Since gear 2 is attached to gear 3, we have

$$n_3 = -333.3\text{ rpm}$$

Finally, the speed of gear 4 is

$$n_4 = -\frac{N_3}{N_4}n_3 = -\frac{8}{40}(-333.3\text{ rpm}) = 66.7\text{ rpm}$$

Since there are two stages, the output rotates in the same direction as the input. By combining Equations (8.15) and (8.16) we can find the overall reduction in one step

$$n_4 = \left(-\frac{N_3}{N_4}\right)\left(-\frac{N_1}{N_2}\right)n_1$$

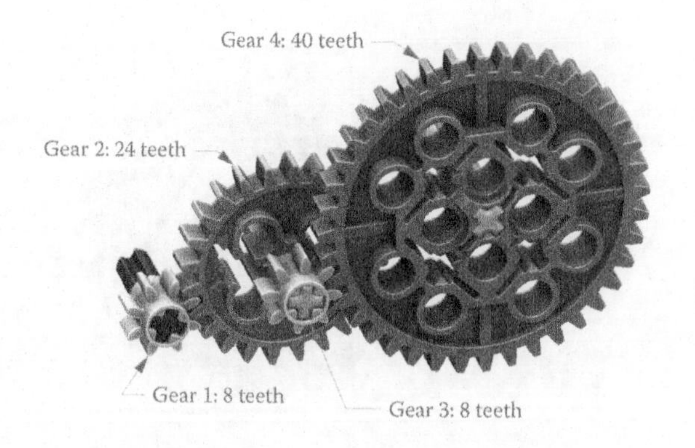

FIGURE 8.35
This speed reducer has two stages, with the first stage consisting of gears 1 and 2 and the second stage consisting of gears 3 and 4. Gears 2 and 3 are mounted on the same shaft, so they spin at the same speed.[2]

In general, if ρ_1 is the reduction in stage 1 and ρ_2 is the reduction in stage 2, then the overall reduction is

$$\rho_{total} = (-\rho_1) \times (-\rho_2)$$

For this example we have

$$\rho_{total} = \left(-\frac{N_2}{N_1}\right) \times \left(-\frac{N_4}{N_3}\right) = \left(-\frac{24}{8}\right) \times \left(-\frac{40}{8}\right) = 15$$

Thus, this speed reducer has an overall gear ratio of 15, which means that the output speed is a factor of 15 lower than the input speed:

$$\frac{1{,}000 \text{ rpm}}{15} = 66.7 \text{ rpm}$$

In Section 8.2, we also learned that the output torque is increased by the same factor as the speed is reduced, so that

$$T_2 = -\rho T_1$$

Thus, if we reduce speed by $1/\rho$, we increase torque by ρ. For the two-stage speed reducer the torque would be increased in the same ratio as speed is reduced:

$$T_4 = (-\rho_1)(-\rho_2) T_1 = \rho_{total} T_1$$

Example 8.3: Design of a Gear Reducer

Design a gear reducer with an input speed of 1,764 rpm and an output speed of 7 rpm. Use 20° pressure angle gears with a module of 1 mm/tooth. Find the reduction in each stage, and specify the number of teeth and pitch diameter of each gear.

Solution

The overall speed reduction must be

$$\rho_{total} = \frac{1764}{7} = 252$$

The problem statement does not specify how many stages are to be used, so we will need to use a trial and error approach. Let us try to achieve the reduction in one stage. Looking at Figure 8.32 in Section 8.3 the minimum number of pinion teeth to avoid interference is slightly over 17, so we choose 18 teeth. The gear would need to have

$$N_g = \rho N_p = 252 \cdot 18 = 4536 \text{ teeth}$$

Aside from the fact that we are unlikely to find a gear with over 4,000 teeth, the diameter of the gear would be

$$d_g = (4536 \text{ teeth}) \left(\frac{1 \text{ mm}}{\text{tooth}} \right) = 4.536 \text{ m}$$

This design is clearly impractical, so we must achieve the reduction in multiple stages. What about two stages? Since we know that the overall reduction is given by

$$\rho_{total} = \rho_1 \rho_2$$

we can get an initial estimate for the reduction in each stage using

$$\rho_1 \approx \sqrt{\rho_{total}} = 15.9$$

For the 20° pressure angle, the smallest number of teeth on the pinion is

$$N_p = 17 \text{ teeth}$$

so that the number of gear teeth is (17·15.9) = 270 teeth. This gear still seems a little too large to be practical, so let us try three stages.

$$\rho_1 \approx \sqrt[3]{\rho_{total}} = 6.3$$

This is a more reasonable reduction to achieve in a single stage. Let us try a reduction of 7 in the first stage. Then

$$\frac{\rho_{total}}{\rho_1} = \frac{252}{7} = 36$$

This is lucky! If we let the remaining two stages have reductions of 6 each, we will achieve our goal exactly:

$$m_{total} = 7 \cdot 6 \cdot 6 = 252$$

The minimum number of pinion teeth to avoid interference is still 17 for the first stage

$$N_1 = 17 \text{ teeth} \quad d_1 = 17 \text{ mm}$$

while the minimum numbers in stages 2 and 3 are

$$N_3 = 16 \text{ teeth} \quad d_3 = 16 \text{ mm}$$

$$N_5 = 16 \text{ teeth} \quad d_5 = 16 \text{ mm}$$

Thus, we can calculate the numbers of teeth for the gear in each stage as

$$N_2 = 119 \text{ teeth} \quad d_2 = 119 \text{ mm}$$

$$N_4 = 96 \text{ teeth} \quad d_4 = 96 \text{ mm}$$

$$N_6 = 96 \text{ teeth} \quad d_6 = 96 \text{ mm}$$

Example 8.4: Speed Reduction for a Clock

Design the appropriate speed reduction between the minute and second hands of a clock using a combination of gears that have 8, 12, 16, 20, 24, 36, 40, and 56 teeth. Determine the number of stages, the reduction in each stage, and the numbers of teeth in each gear.

TABLE 8.2

Four-Stage Gear Reducer for Use in a Clock Mechanism

	Pinion	Gear
Stage 1	8	40
Stage 2	8	24
Stage 3	8	16
Stage 4	8	16

Solution

The speed ratio between the minute and second hand of a clock is 60, so this is our overall reduction. The largest single-stage reduction that is achievable using the prescribed gears is

$$\rho = \frac{56}{8} = 7$$

so we know that we need at least three stages, since $7 \cdot 7 = 49$, which is less than 60. The cube root of 60 is 3.9, so our first guess might be to try a reduction of 4 in the first stage. Oddly enough, there is no combination of prescribed gears that produces a reduction of 4! Instead, let us try a reduction of 5 in the first stage by combining an eight-tooth gear with a 40 tooth gear. The remaining required reduction is

$$\rho = \frac{60}{5} = 12$$

Since a reduction of 4 is impossible, we will use reductions of 3, 2, and 2 for the remaining stages, which gives us a four stage gear reducer. A workable clock gear reducer is shown in Table 8.2.

Of course, there are many valid ways to achieve a gear reduction of 2 from the set of gears, so other solutions are possible.

8.5 Efficiency of Gear Trains

The topic of gear train efficiency has been much studied in the past 100 years, and no simple, all-encompassing theory has emerged. One of the main reasons for this is the large number of factors that influence the efficiency of a gear train, including lubrication and friction, the precisions of the gears, the types (and quality) of bearings used, temperature, loading and other factors. To obtain an *absolute* estimate of the efficiency of a given gear train would expand this text beyond reasonable limits, so we will confine ourselves to estimating the *relative* efficiency of a gear train. That is, we will attempt to solve the problem of which of a pair of competing gear train designs is likely to be more efficient. Once a gear train has been selected and manufactured, laboratory testing must be carried out if an estimate of absolute efficiency is required. Thus, the methods presented here are *approximate*, and should only be used for the purpose of comparing one design with another. The most common method for presenting the efficiency of a pair of gears is to calculate the power loss:

$$P_{out} = E_0 \cdot P_{in} \tag{8.17}$$

where P_{in} is the power input into the gear pair, P_{out} is the power available to be transmitted "downstream" and E_0 is called the "basic efficiency," which is always less than one. Most of the power loss in a gear pair is through the rubbing (friction) of the gear teeth as the motion is transmitted, although some power is also lost from friction in the bearings. The efficiency of a pair of gears can be calculated as

$$E_0 = 1 - \mu L \tag{8.18}$$

where μ is the coefficient of friction and L is called the *tooth loss factor*. The full derivation of the tooth loss factor is beyond the scope of this text, but an excellent explanation can be found in [2]. For a pair of spur gears with pressure angle 20°, 22.5°, or 25° the tooth loss factor can be calculated as

$$L = \left(\frac{\rho+1}{\rho}\right)\left(\frac{1}{d_b}\right)\left[\frac{Z_a^2 + Z_r^2}{p_b} + p_b - Z_r - Z_a\right]$$

where

$$d_b = d\cos\alpha$$

is the base diameter of the pinion,

$$p_b = \frac{\pi\cos\alpha}{P_d} = m\pi\cos\alpha$$

is the base pitch, ρ is the gear ratio and

$$Z_a = \frac{1}{2P_d}\left[\sqrt{(N_g+2)^2 - (N_g\cos\alpha)^2} - N_g\sin\alpha\right]$$

$$Z_r = \frac{1}{2P_d}\left[\sqrt{(N_p+2)^2 - (N_p\cos\alpha)^2} - N_p\sin\alpha\right]$$

are the approach and recess portions of the line of action, respectively. If internal gears are used, then the ratio $(\rho + 1)/\rho$ must be replaced with $(\rho - 1)/\rho$.

Given the approximate nature of our calculations, it is much simpler to estimate the tooth loss factor using the charts presented in Figures 8.36 and 8.37. Some additional rules given by Molian [3] are as follows:

- For a pair of external spur gears, use the tooth loss factors shown in Figure 8.36 for a 20° pressure angle and Figure 8.37 for a 25° pressure angle.
- For an external spur gear mating with an internal spur gear, multiply the tooth loss factor found in Figure 8.36 or Figure 8.37 by $(\rho - 1)/(\rho + 1)$.
- For helical gears the tooth loss factor must be multiplied by 0.8 cos ψ, where ψ is the helix angle.

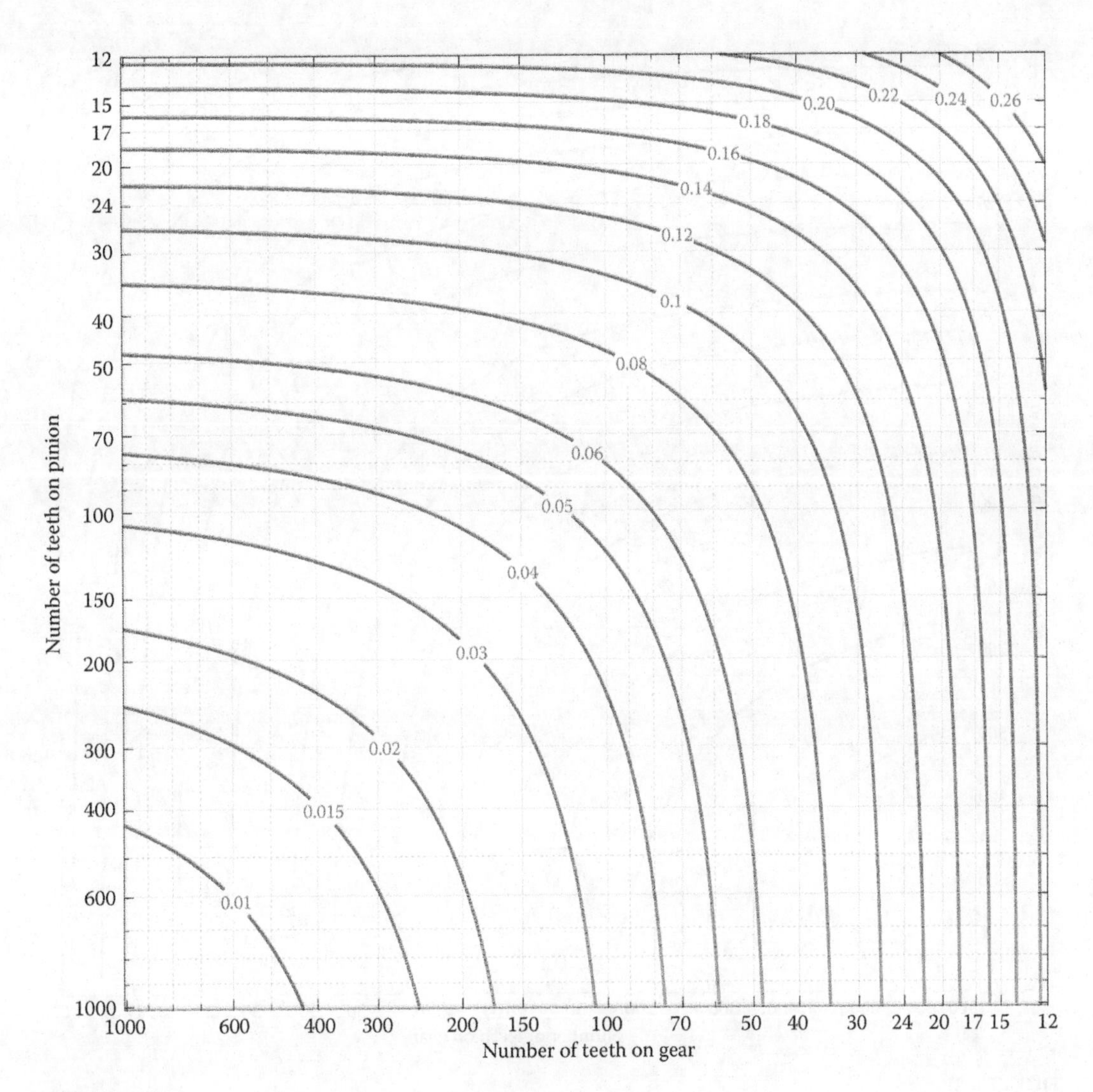

FIGURE 8.36
Tooth loss factor for external gear pair with pressure angle 20°.

Looking at the second rule, we see that an internal gear pair will have lower losses than an external pair if the gear ratio is small. This is one reason that planetary gearsets commonly use internal gears in their design (the other being compactness).

The coefficient of friction between pairs of gear teeth depends upon several factors including the gear materials, the lubrication used, temperature, accuracy of the tooth profile, surface finish, and many others. Merritt [4] suggests a value of 0.08 for pairs of steel gears while Tso [2] uses more recent data to propose.

$$\mu = 0.04 \text{ for precision steel gears}$$

$$\mu = 0.05 \text{ for accurately cut steel gears.}$$

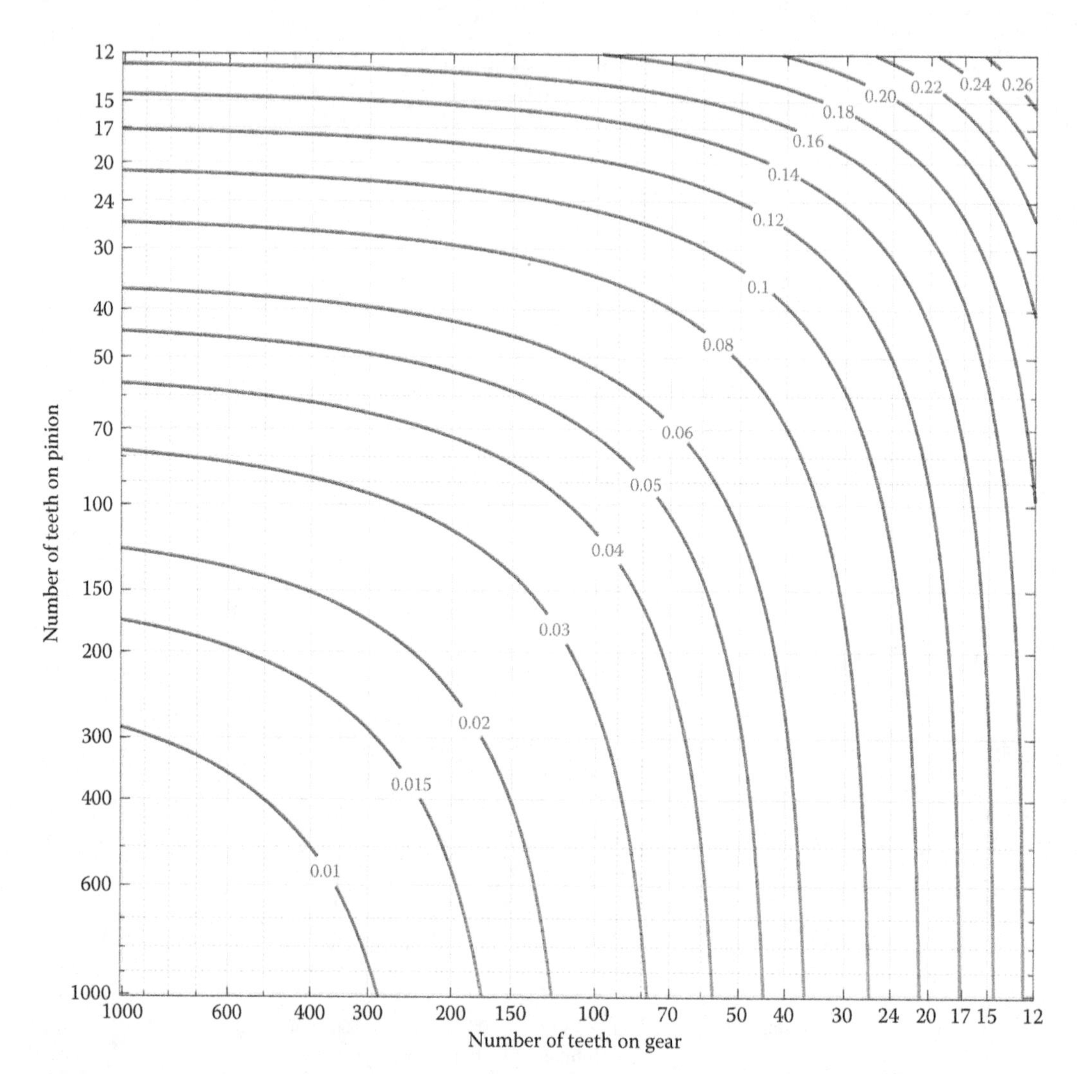

FIGURE 8.37
Tooth loss factor for external gear pair with pressure angle 25°.

If a gear train is composed of more than one pair, the efficiencies of each pair are multiplied to obtain the overall efficiency. Thus, for a compound gear train with two pairs, the overall efficiency is

$$E_0 = E_1 E_2 \tag{8.19}$$

$$E_0 = (1 - \mu_1 L_1)(1 - \mu_2 L_2) \tag{8.20}$$

Example 8.5

Calculate the efficiency of an external spur gear pair where the pinion has 16 teeth and the gear has 32 teeth. Repeat the calculation if the 32 tooth gear is internal. Assume a pressure angle of 20° and a coefficient of friction of 0.05.

Solution

Examining Figure 8.36 it appears that the tooth loss factor is approximately 0.2 so that the efficiency is

$$E_0 = (1 - 0.05 \cdot 0.2) = 0.99$$

or 99%. For the internal pair we have $\rho = 2$, so that

$$L_{\text{int}} = \frac{\rho - 1}{\rho + 1} L_{ext}$$

$$L_{\text{int}} = \frac{2 - 1}{2 + 1} \cdot 0.2$$

$$L_{\text{int}} = 0.0667$$

and the efficiency is

$$E_0 = (1 - 0.05 \cdot 0.0667) = 0.9967$$

or 99.67%. Clearly, the internal gearset is much more efficient than the external gearset, although both are fairly efficient.

Example 8.6

Compare the efficiencies of the following two gearsets:

1. One-stage reduction with $N_p = 12$ and $N_g = 48$.
2. Two-stage reduction with $N_{p1} = N_{p2} = 12$ and $N_{g1} = N_{g2} = 24$.

Assume both gearsets have pressure angle 25° and coefficient of friction 0.10.

Solution

Examining Figure 8.37 we have $L = 0.19$ so that

$$E_0 = 100 - 0.1 \cdot 0.19 = 0.981$$

for the single-stage reduction. For the two-stage reduction, each stage has $L = 0.225$ so that

$$E_0 = (1 - 0.1 \cdot 0.225)(1 - 0.1 \cdot 0.225) = 0.956$$

The two-stage reducer is much less efficient, although space constraints may prevent us from using the single-stage reducer.

Example 8.7

Compare the efficiency of the following single-stage reducers. Assume a pressure angle of 20° and a coefficient of friction of 0.05.

1. $N_p = 12, N_g = 48$
2. $N_p = 100, N_g = 400$

For the first gearset we have $L = 0.225$ so that

$$E_0 = 1 - 0.05 \cdot 0.225 = 0.989$$

For the second, we have $L = 0.035$ so that

$$E_0 = 1 - 0.05 \cdot 0.035 = 0.998$$

Clearly, the second stage is more efficient. Examining Figures 8.36 and 8.37 we see that gears with larger numbers of teeth tend to have lower tooth loss factors and are generally more efficient. The tradeoff, of course, is the size of the gears. If a large amount of power is to be transmitted, we may require a set of gears with large teeth; that is, gears with a large module or small diametral pitch. Since the gear diameter is proportional to the number of teeth, size constraints may prevent us from using the gearset that is "optimal" from an efficiency standpoint. In general, however, the designer should select gears with the highest number of teeth within strength and packaging requirements.

Example 8.8

Figure 8.38 shows an example of the most common type of planetary gearset. We will give planetary gearsets a thorough treatment in Chapter 9, but for now, we wish to find the *basic efficiency* of the gearset. The basic efficiency gives the power loss assuming that

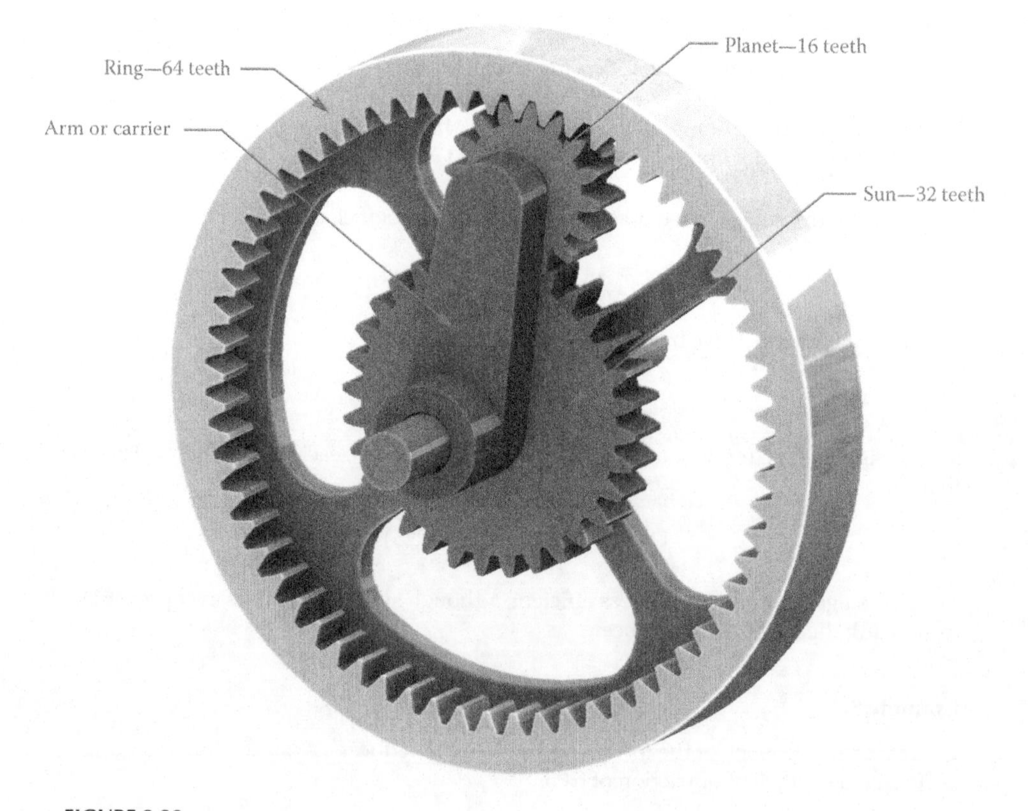

FIGURE 8.38
A common type of planetary gearset. For this example, the arm is fixed and the mechanism behaves like an ordinary gearset.

the arm is fixed so that the planetary gearset behaves like ordinary gearset. Find the basic efficiency of the gearset shown in Figure 8.38 assuming a pressure angle of 20° and a coefficient of friction of 0.05.

Solution

Let us take the sun and planet as the first gear pair. From Figure 8.37 we have $L_1 = 0.205$. The second gear pair has a ratio of 4 so that we calculate the tooth loss factor as

$$L_2 = \frac{4-1}{4+1} \cdot 0.18 = 0.108$$

Thus, the overall basic efficiency of this gearset is

$$E_0 = (1 - 0.05 \cdot 0.205)(1 - 0.05 \cdot 0.108) = 0.984$$

or 98.4%.

Example 8.9

Figure 8.39 shows another common type of planetary gearset with two suns and two planets. The problem statement is the same as before: calculate the basic efficiency of this gearset (the efficiency with the carrier fixed) assuming a pressure angle of 20° and coefficient of friction 0.05.

Solution

Let us denote the first stage as the pair of 24 tooth gears. The tooth loss factor for the first stage is 0.18 and for the second stage is 0.20. Thus, the basic efficiency of this gearset is

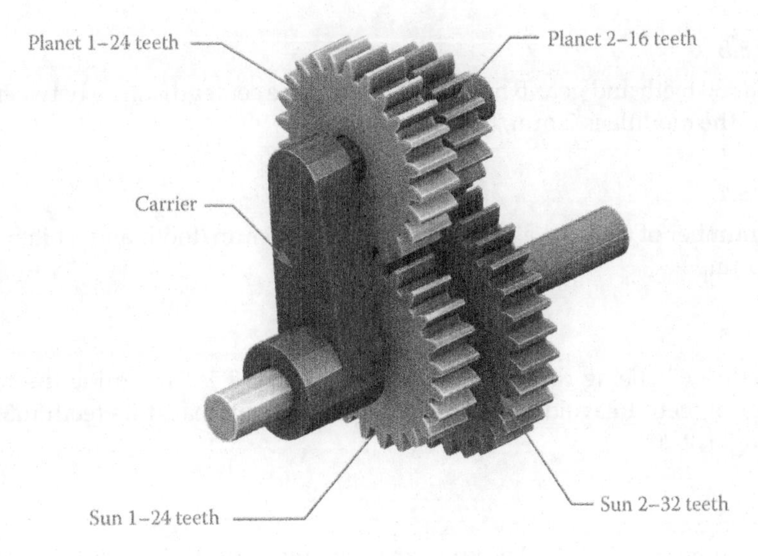

FIGURE 8.39
Another common type of planetary gearset. Again, we assume that the carrier is fixed when calculating the basic efficiency.

$$E_0 = (1-0.05\cdot 0.18)(1-0.05\cdot 0.2) = 0.981$$

or 98.1%. Thus, the planetary gearset in Example 8.8 is slightly more efficient if the carrier is fixed. As we will see in Chapter 9, however, calculating the efficiency of planetary gearsets is quite a bit more complicated when the arm is allowed to rotate.

8.5.1 Summary

This section has presented a method for estimating the efficiency of a simple (non-epicyclic) gearset. As we have seen, the efficiency of spur gears can be quite high, and we can maximize efficiency by increasing tooth count to the maximum permitted by strength and packaging considerations. It must be emphasized that the methods given in this section provide *very approximate estimates* and should not be used to calculate the absolute efficiency of a given gearset. Instead, they should be employed for *comparing* two or more potential designs with each other. Once a design has been selected, laboratory testing is usually necessary to obtain the absolute efficiency of the gearset.

8.6 Practice Problems

Problem 8.1

A 30 tooth gear spins clockwise at 10 rpm and drives a 90 tooth gear. Find the speed and rotational direction of the driven gear.

Problem 8.2

Gear A has 80 teeth and gear B has 40 teeth. If gear B is being driven at 20 rpm clockwise, find the speed of gear A.

Problem 8.3

Gear A has 30 teeth and gear B has 60 teeth. Find the center distance between the two gears if the module is 2 mm/tooth.

Problem 8.4

Find the number of teeth on a gear with module of 5 mm/tooth and a pitch diameter of 100 mm.

Problem 8.5

A gear reducer is being designed for a reduction of 2.5. Determine the minimum number of teeth to avoid interference for the pinion gear if its teeth have a pressure angle of 20°.

Problem 8.6

As shown in Figure 8.40, a 20 tooth gear transmits 1,000 W to a 40 tooth gear. Both gears have module 4 mm/tooth. What is the torque produced by gear 2 if gear 1 rotates at 1,000 rpm?

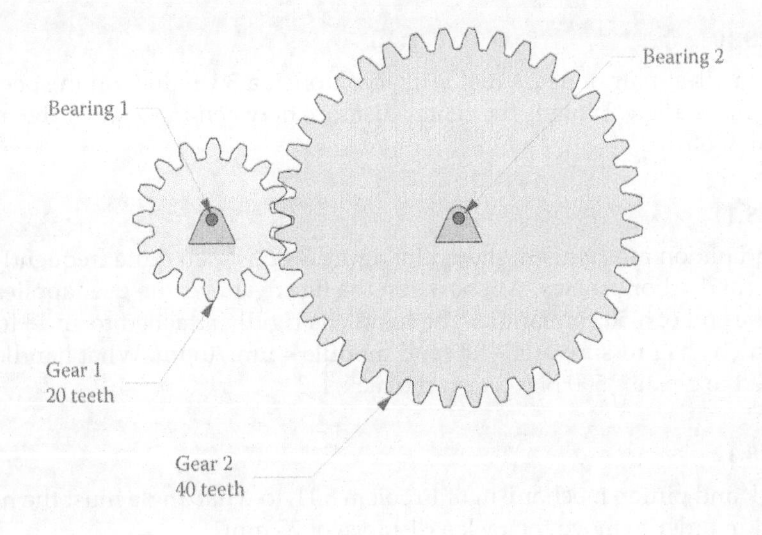

FIGURE 8.40
Problem 8.6.

Problem 8.7

Using the results of Problem 8.6, what is the vertical and horizontal load on bearing 2 if the pressure angle of the gears is 14.5°?

Problem 8.8

Using the results of Problems 8.6 and 8.7, find the vertical and horizontal loads on bearing 2 if the pressure angle of the gears is 25°.

Problem 8.9

In Figure 8.41, if gear 1 spins at 3,600 rpm clockwise, what is the speed of gear 4?

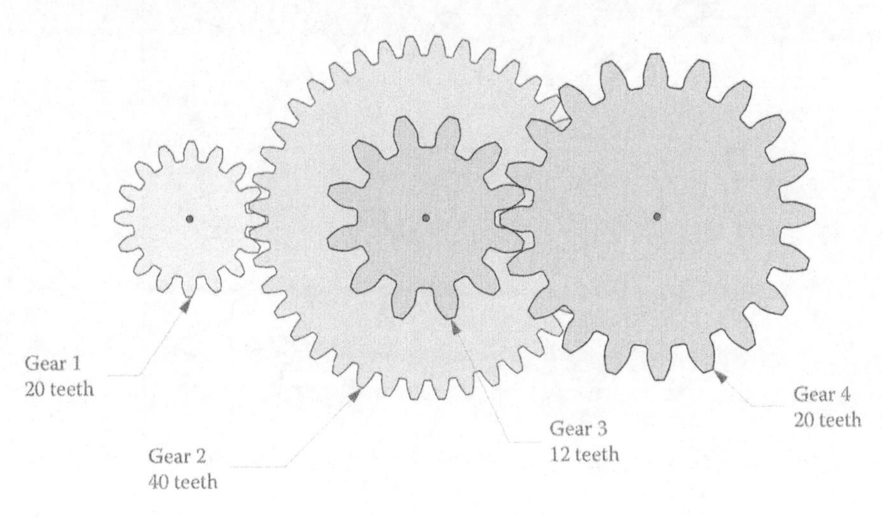

FIGURE 8.41
Problem 8.9.

Problem 8.10

Find the smallest pair of gears that will accomplish a 3:1 reduction in speed with a 20° pressure angle. What is the center distance between the gears if the module is 0.5 mm/tooth?

Problem 8.11

A rack and pinion mechanism, shown in Figure 8.42, is seen quite frequently in drill presses and arbor presses. As shown in the figure above, the user applies a force, F_h, at the end of a 50 cm handle. The handle is rigidly attached to an 18 tooth pinion, which has pressure angle 20° and module 4 mm/tooth. What handle force is required to create a 500 N force on the rack, F_d?

Problem 8.12

In the rack and pinion mechanism of Problem 8.11, to what angle must the handle be rotated in order to move the rack a distance of 25 mm?

Problem 8.13

For the rack and pinion mechanism in Problem 8.11, what angular velocity must the pinion have in order for the rack to move 10 mm/s? Give your answer in revolutions per minute.

Problem 8.14

In the gearset of Problem 8.9, gear 1 spins at 3450 rpm and transmits 750 W. What torque is present at gear 4, assuming 100% efficiency?

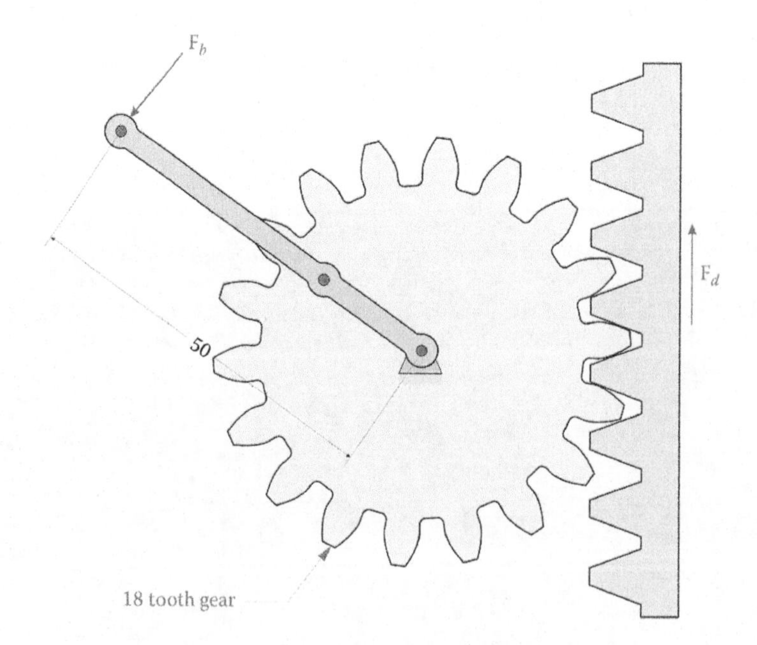

FIGURE 8.42
Problem 8.11.

Problem 8.15

In problem 8.9, the module of the first pair of gears is 1.5 mm/tooth and the second pair of gears is 4 mm/tooth. Find the center distance between gears 1 and 4 if the gears are aligned horizontally, as shown in the figure.

Problem 8.16

In the gearset of Problem 8.9, the speed of gear 1 is 3,600 rpm. Keeping all other tooth numbers constant, how many teeth should gear 4 have in order for its speed to be 600 rpm?

Problem 8.17

Design a two-stage gear reducer of minimum size for an input speed of 3,600 rpm and an output speed of 1,000 rpm. Both stages should have pressure angle 20° and module 2 mm/tooth.

Problem 8.18

See Figure 8.43. You are given the task of designing a gearbox for a small lathe. Three different output speeds are required: 100 rpm, 250 rpm, and 500 rpm. The motor spins at a constant 3,600 rpm. Each reduction unit should have three stages, but the first two stages are shared between all three speeds (i.e. only the third stage in each unit will vary among the three speeds.) A clutch mechanism allows the operator to choose which of the reductions in the third stage is active (gears 5–6, gears 7–8 or gears 9–10). The center distance for all three-gear pairs in the third stage must be identical, as seen in the figure. Find the numbers of teeth to accomplish the required reductions, assuming a pressure angle of 20°.

Problem 8.19

A battery-powered motor spins at 12,000 rpm. Design a five-stage gear reduction unit that achieves an output speed of 60 rpm. Use gears with a pressure angle of 20°, and try to minimize the tooth count of each gear.

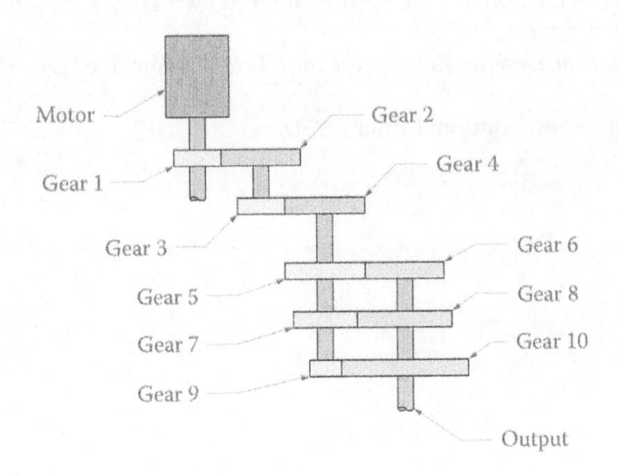

FIGURE 8.43
Problem 8.18.

Problem 8.20

Estimate the overall efficiency of the gear train in Problem 8.9 if the pressure angle of all gears is 20° and the coefficient of friction is 0.05. What is the torque at gear 4 if gear 1 is driven by a 500 W motor that spins at 1,750 rpm?

Acknowledgments

Several images in this chapter were produced using SOLIDWORKS® software. SOLIDWORKS is a registered trademark of Dassault Systèmes SolidWorks Corporation.

Notes

1. Dr. Lunin has uploaded several models of gears with complex tooth profiles under the user name dr.lunin-1 on www.grabcad.com, these may be downloaded with a free user account. Dr. Lunin generated these models using his software available at www.spiralbevel.com.
2. The construction brick images used in this text were made with the SOLIDWORKS® models created by the user Yauhen on www.grabcad.com. Yauhen has uploaded hundreds of first-rate models of construction bricks and gears that can be downloaded with a free user account.

Works Cited

1. AGMA, *Gear Nomenclature, Definitions of Terms with Symbols.* ANSI/AGMA Standard 1012-F90, 1500 King St., Suite 201, Alexandria, VA: American Gear Manufacturers Association, 1990.
2. L.-N. Tso, A study of friction loss for spur gear teeth (MS Thesis), Monterey, CA: United States Naval Postgraduate School, 1961.
3. S. Molian, *Mechanism Design: An Introductory Text,* Cambridge, UK: Cambridge University Press, 1982.
4. H.E. Merritt, *Gear Trains,* London: Pitman, 1947.

9

Planetary Gear Trains

9.1 Introduction to Planetary Gearsets

The gearsets we have considered thus far have been relatively simple in having only one input and one output, and thus only one overall gear ratio. In this section, we introduce the *planetary gearset,* which is much more versatile. As we will see, this versatility comes at a price: planetary gearsets are much more complicated to design and analyze. They often behave in very counterintuitive ways, which makes their use all the more difficult (but interesting!). As with ordinary gearsets, however, a few simple rules can be used to determine the performance and efficiency of any planetary gearset.

The most common type of planetary gearset is shown in Figure 9.1. It consists of four major components: the *sun,* one or more *planets,* a *ring* gear, and a *carrier* (often called the *arm* or the *spider*). The major difference between planetary gearsets and ordinary gearsets is that the planets "orbit" the sun; that is, their shafts are not fixed in space. It is this *epicyclic motion* that permits such interesting behavior.

Like all planetary gearsets, the gearset in Figure 9.1 has two inputs and one output. We may freely choose which shafts are the inputs and which is the output, as shown in the Table 9.1. It is this free choice that makes determining an overall ratio so interesting.

9.1.1 Types of Planetary Gearsets

Zoltan Levai [1] described a total of 34 different types of planetary gearsets that can be constructed; of these, only a few are used in practice. A few of the more common ones are described in the section that follows; the 12 shown by Norton [2] are given in Table 9.2. For clarity, only one planet gear is shown for each type; more planets can be added for balancing and to increase the torque capacity of the gearset as shown in Figure 9.2.

9.1.2 Sun, Ring, and Planet

The most basic type of planetary gearset is shown in Figure 9.1. In this geartrain, inputs and output can be taken from the carrier, ring, and sun gears, and only the planet experiences epicyclic motion. This is the most common type of planetary gearset (with the exception of the differential) and it finds application in automatic transmissions, hybrid-electric transmissions (e.g. the Toyota Prius) and even battery-powered drills.

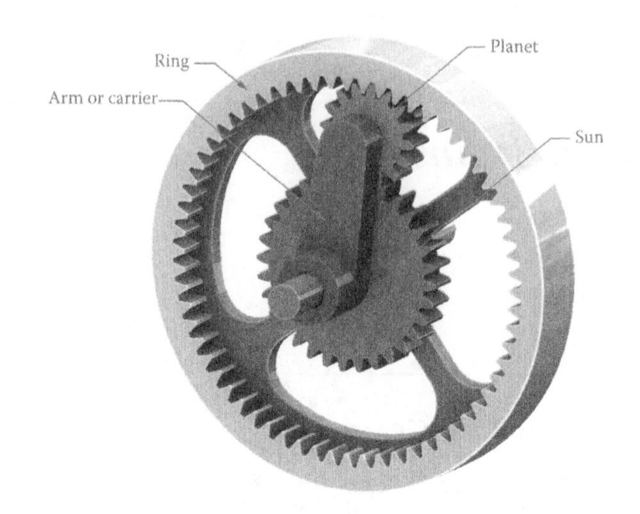

FIGURE 9.1
The most basic planetary gearset consists of a sun, arm (or carrier), ring, and planet.

TABLE 9.1
Possible Inputs and Outputs from the Planetary Gearset in Figure 9.1

Option	Inputs	Output
1	Sun and carrier	Ring
2	Ring and sun	Carrier
3	Ring and carrier	Sun

TABLE 9.2
Twelve of the 34 Possible Types of Planetary Gearset as Described by Levai [1] and Norton [2]

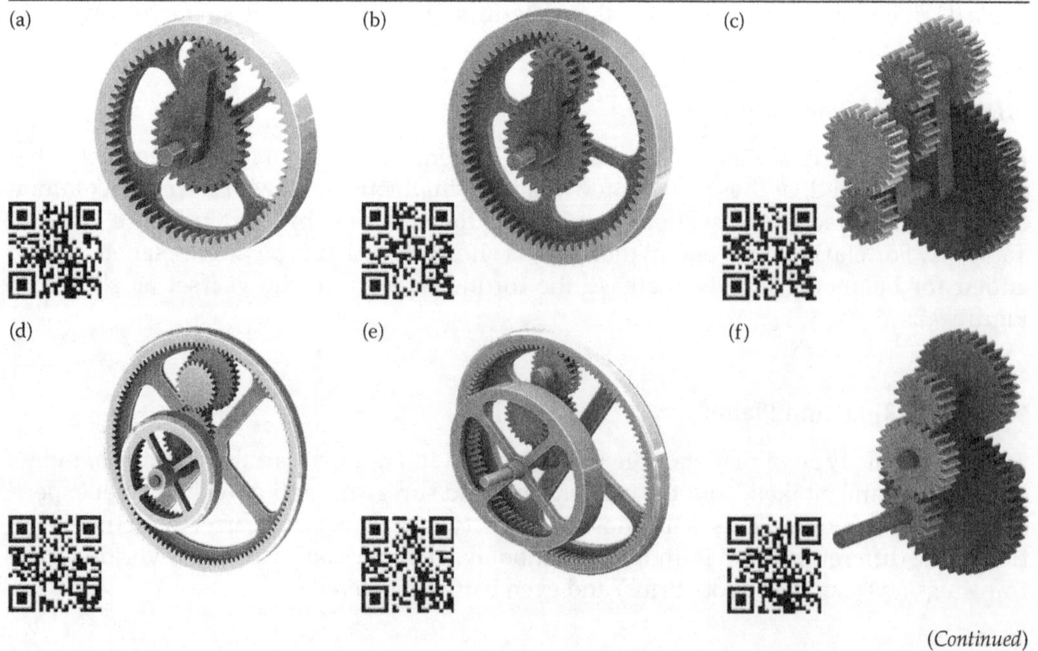

(*Continued*)

TABLE 9.2 (*Continued*)

Twelve of the 34 Possible Types of Planetary Gearset as Described by Levai [1] and Norton [2]

FIGURE 9.2
This planetary gearset has three planets instead of just one. The additional planets are used to balance the gearset and to provide additional torque-transmitting capacity.

As seen in Figure 9.3, the diameters of each gear are not entirely independent of one another. In order for this gearset to fit together, the diameters must follow

$$d_r = 2d_p + d_s \tag{9.1}$$

Since all three must have the same diametral pitch (or module) in order to mesh together, we can write a similar relationship for the numbers of teeth.

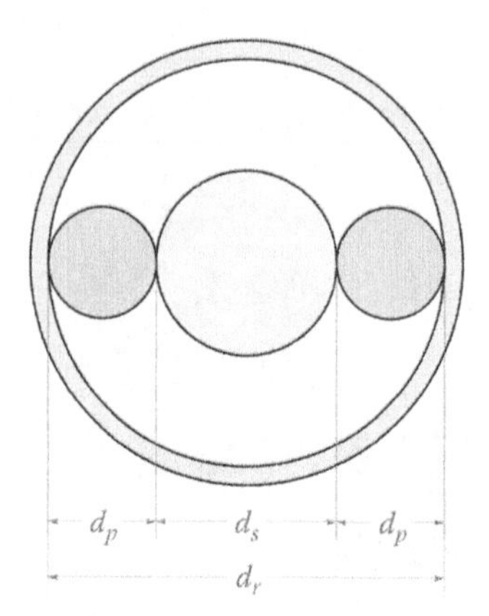

FIGURE 9.3
The dimensions (and number of teeth) of the gears are not entirely independent of each other.

$$N_r = 2N_p + N_s \tag{9.2}$$

where N_r is the number of teeth on the ring, N_p is the number of teeth on the planet and N_s is the number of teeth on the sun.

9.1.3 Two Suns and Two Planets

The gearset shown in Figure 9.4 has two sun gears, and the two planet gears rotate as a single unit. The sun gears can rotate independently of one another. The inputs and output can be selected from both sun gears and the carrier. Very high-speed reductions can be achieved with this unit, but it can suffer from low efficiency if not designed correctly.

As seen in Figure 9.5, the pitch diameters are not independent of one another, and must follow the relationship

$$d_{s1} + d_{p1} = d_{s2} + d_{p2} \tag{9.3}$$

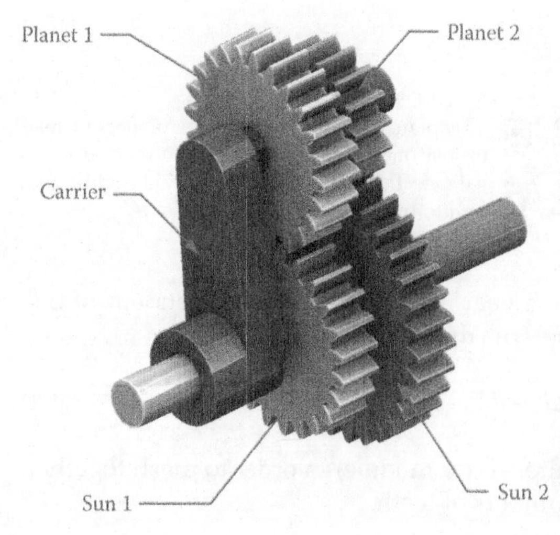

FIGURE 9.4
This planetary gearset has two suns and two planets. It is not necessary to have a ring gear for epicyclic motion.

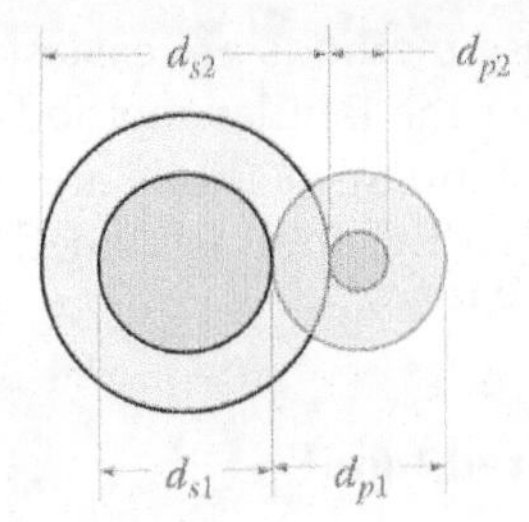

FIGURE 9.5
The gear diameters are not independent of one another.

Since the second sun/planet pair does not mesh with the first, the tooth numbers can be independent. If all gears have the same diametral pitch (or module) then we must have

$$N_{s1} + N_{p1} = N_{s2} + N_{p2} \tag{9.4}$$

9.1.4 The Differential

The gearset shown in Figure 9.6 is different from the preceding gearsets in that it is composed of miter gears rather than spur (or helical) gears. The "sun" gears are those that do not undergo the epicyclic motion experienced by the planets. As we will see in the next section, the differential can be used to measure the *difference* in speed between two shafts for the purpose of synchronization. In addition, the differential is often used in automotive drivetrains to overcome the difference in wheel speed when a car goes around a corner.

9.2 Analysis of Planetary Gearsets—The Table Method

Planetary gearsets can behave in counterintuitive ways, which makes their analysis quite interesting. However, we can develop a few simple methods that we can apply to any gearset, no matter how complicated. We will first discuss two methods that employ tables for

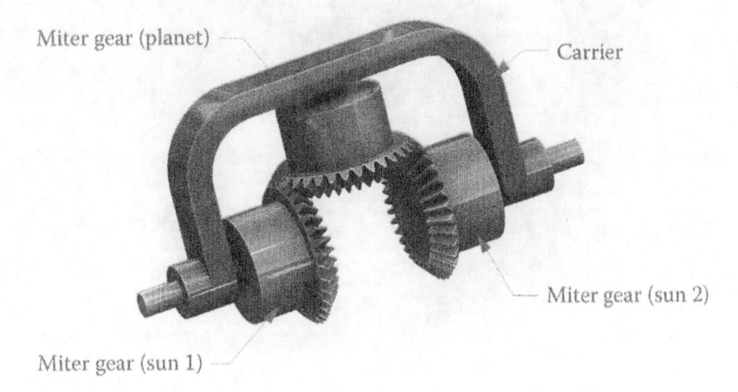

FIGURE 9.6
The differential is often used to allow the drive wheels of a car to rotate independently.

conducting the analysis, since these methods are easiest to grasp intellectually. Once our intuition has been developed using the tabular methods we will proceed to a more analytical method in Section 9.3. The tabular methods are often used when a quick analysis of a particular gearset is needed, but the analytical methods are more suited for use with modeling software such as MATLAB®.

9.2.1 Table Method with One Fixed Input

We start with the simplest case: the basic planetary gearset (shown in Figure 9.7) with one input fixed. To make things really simple, let us first assume that the arm is fixed. If the sun spins at 100 rpm cw, let us find the speed of the ring. Since the arm is fixed, no gear experiences epicyclic motion, and we have an ordinary gearset that we can analyze using the methods of Chapter 8. The planet spins at

$$n_p = -\frac{N_s}{N_p} n_s = -\frac{32}{16} 100 = -200 \text{ rpm}$$

or 200 rpm ccw. The ring spins in the same direction as the planet since it is an internal gear

$$n_r = \frac{N_p}{N_r} n_p = -\frac{16}{64} 200 = -50 \text{ rpm}$$

or 50 rpm ccw. This example is fairly straightforward, since none of the gears experience epicyclic motion. We will now develop a more general procedure that will account for epicyclic motion.

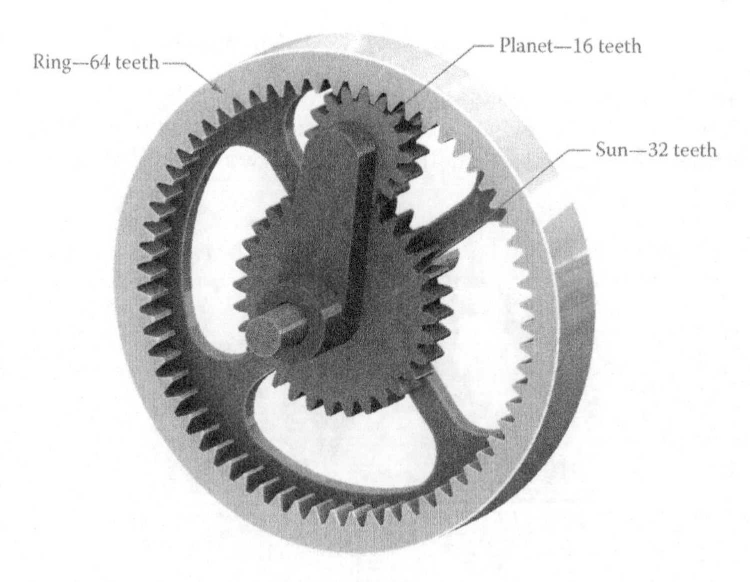

FIGURE 9.7
A basic planetary gearset with a sun, planet, and ring gear.

Example 9.1

In the gearset of the preceding example let the ring be fixed and the arm free to move. Find the speed of the arm if the sun moves at 100 rpm cw.

Solution

We will introduce a simple tabular method to conduct the analysis*. On a piece of engineering paper (or graph paper) construct a table similar to Table 9.3. The table should have six rows (including the labels at the top) and four columns, one for the speed of each of the gears in the set.

To begin, pretend that the arm is fixed and release the ring. Now, turn the ring one revolution. If the ring moves one revolution cw, then the planet moves

$$n_p = \frac{N_r}{N_p} n_r = \frac{64}{16} 1 = 4 \text{ rev}$$

and the sun moves

$$n_s = -\frac{N_p}{N_s} n_p = -\frac{16}{32} 4 = -2 \text{ rev}$$

Enter these "speeds" in the first row of the table, as shown in Table 9.4. We are using revolutions per unit time as a unit for the entries, since this method will work for any speed unit.

Next, imagine gluing the entire assembly together into a single, rigid body. If we rotate the rigid body, the kinematic relationships between the gears and the carrier are unchanged. Rotate the entire rigid body one revolution counterclockwise, and enter this operation in rows 2 and 3 of the table, as seen in Table 9.5. Row 2 consists of four entries

TABLE 9.3

Table for Analysis of Simple Planetary Gearset

n_r	n_p	n_s	n_a

TABLE 9.4

Enter the Speeds of the Gears with the Arm Fixed in Row 1

n_r	n_p	n_s	n_a
1	4	–2	0

* This method was introduced to us by Tirupathi R. Chandrupatla in private communication.

TABLE 9.5

Rotate the Entire Assembly Backwards by One Revolution

n_r	n_p	n_s	n_a
1	4	–2	0
–1	–1	–1	–1
0	3	–3	–1

of –1 – this symbolizes the counterclockwise rotation of the entire assembly. Row 3 is the sum of Rows 1 and 2.

In Row 3, we can see that the ring has moved 0 revolutions; that is, it is fixed as was given in the problem statement. The sun has moved 3 revolutions ccw, but we can scale this to 100 rpm cw by multiplying by (–100/3). In Row 4 of the table enter a row of (× –100/3). The products of Rows 3 and 4 are entered in Row 5. Your completed table should be the same as Table 9.6.

As seen in Row 5, the arm moves 33.3 rpm cw when the sun moves 100 rpm cw. The ring is still fixed, as required by the problem statement.

The following steps summarize the procedure we have followed for Example 9.1.

1. Hold the arm fixed. Rotate the fixed gear (sun or ring) by one revolution. Calculate the resulting rotations of the other gears. This completes Row 1 of the table.
2. Glue the gearset together as a rigid body, and rotate the assembly backwards by one revolution. Enter the results in Row 3 of the table.
3. Scale the results in Row 3 so that the input gear matches the problem statement. Enter the results in Row 5.

Example 9.2

Let us now examine the same gearset in Example 9.1, but with the sun fixed and the ring allowed to move. If the ring rotates at 100 rpm cw, what is the speed of the arm?

Solution

As before, we start by pretending that the arm is fixed and rotating the (fixed) sun gear one revolution. The planet rotates

$$n_p = -\frac{N_s}{N_p} n_s = -\frac{32}{16} 1 = -2 \text{ rev}$$

TABLE 9.6

The Completed Speed Analysis Table for the Simple Planetary Gearset

n_r	n_p	n_s	n_a
1	4	–2	0
–1	–1	–1	–1
0	3	–3	–1
×(–100/3)	×(–100/3)	×(–100/3)	×(–100/3)
0 rpm	–100 rpm	100 rpm	33.3 rpm

and the ring rotates

$$n_r = \frac{N_p}{N_r} n_p = \frac{16}{64}(-2) = -\frac{1}{2}\text{rev}$$

The results are entered in the first row of the table, as shown in Table 9.7.

Glue the entire assembly together and rotate backwards one revolution, as shown in Table 9.8. The problem statement gives the ring speed as 100 rpm cw, so we must multiply the entire table by –200/3 to achieve this, as shown in Table 9.9.

Thus, the arm moves at 66.6 rpm cw when the ring rotates 100 rpm cw. Or, alternatively, the arm rotates 2/3 of a revolution when the ring moves a full revolution, and both ring and arm rotate in the same direction. This is not always the case, as will be seen in the next example.

Example 9.3: A Construction Brick Planetary Gearset

Our next example, planetary gearset is made from construction bricks, as shown in Figure 9.8. The model is simple to build, requiring only a few beams and a set of gears

TABLE 9.7

Hold the Arm Fixed and Rotate the Sun One Revolution

n_r	n_p	n_s	n_a
–1/2	–2	1	0

TABLE 9.8

The Result of Rotating the Entire Assembly Backward by One Revolution

n_r	n_p	n_s	n_a
–1/2	–2	1	0
–1	–1	–1	–1
–3/2	–3	0	–1

TABLE 9.9

Final Results for Example 9.2

n_r	n_p	n_s	n_a
–1/2	–2	1	0
–1	–1	–1	–1
–3/2	–3	0	–1
×(–200/3)	×(–200/3)	×(–200/3)	×(–200/3)
100 rpm	200 rpm	0 rpm	66.6 rpm

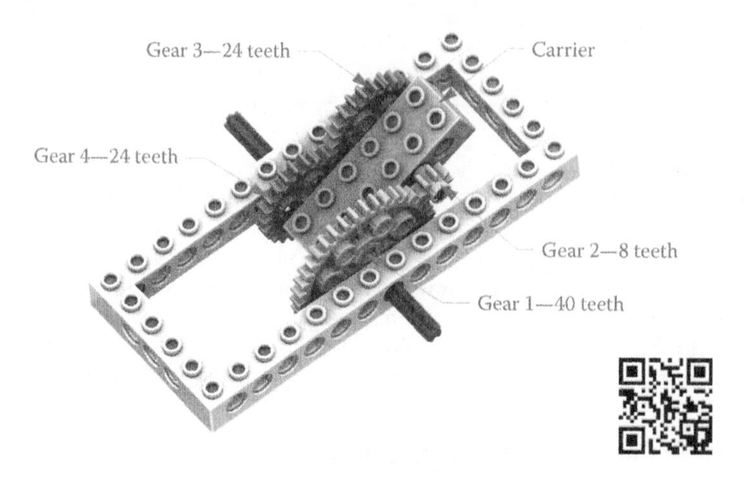

FIGURE 9.8
This construction brick planetary gearset has two suns and two planets, like the type (g) seen in Table 9.2.[1]

(8 tooth, 24 tooth, and 40 tooth) and axles. Building the model is a worthwhile exercise if the construction bricks are available, since it serves as a valuable aid to intuition and physical understanding. It is worth noting that the numbers of teeth follow Equation (9.4), since

$$40+8=24+24$$

The gearset has two suns (gears 1 and 4) and two planets (gears 2 and 3), but we can still use the tabular method to analyze it. For each case, we will hold one of the suns fixed and rotate the other sun one revolution clockwise. The goal is to determine the number of revolutions made by the carrier. To begin, let us assume that sun gear 1 is fixed, and gear 4 rotates one revolution. First, pretend that the carrier is fixed, and rotate gear 1 (the fixed gear) one revolution. Then gear 2 rotates

$$n_2=-\frac{N_1}{N_2}n_1=-\frac{40}{8}1=-5\text{ rev}$$

Gears 2 and 3 are fixed to the same shaft, so they rotate in the same direction and at the same speed

$$n_4=-\frac{N_3}{N_4}n_3=-\frac{24}{24}(-5)=5\text{ rev}$$

This completes Row 1 of the table. Next, rotate everything backwards by one revolution, and enter the result in Row 3. Since the resulting number of revolutions for gear 4 is four, we multiply everything in Row 3 by ¼ to obtain the final result, as shown in Table 9.10. Thus, the carrier rotates *backwards* by ¼ revolution when gear 4 rotates 1 revolution forward.

Now, let us use gear 1 as the input and hold gear 4 fixed. Following the same procedure as before, we obtain the results in Table 9.11. As seen in the final row of the table the carrier now rotates *forward* by 5/4 revolutions when gear 1 makes one revolution. This is an example of the sometimes counterintuitive behavior of planetary gearsets – it is often impossible to predict the rotational direction of the carrier by inspection alone. Try building the model to see for yourself.

TABLE 9.10

Solution for Construction Brick Planetary Gearset When Gear 1 Is Fixed and Gear 4 Rotates One Revolution

n_1	n_2	n_3	n_4	n_c
1	–5	–5	5	0
–1	–1	–1	–1	–1
0	6–	–6	4	–1
×(1/4)	×(1/4)	×(1/4)	×(1/4)	×(1/4)
0	–3/2 rev	–3/2 rev	1 rev	–1/4 rev

TABLE 9.11

Gear 4 Is Fixed and Gear 1 Is the Input. The carrier now rotates forward 5/4 revolutions.

n_1	n_2	n_3	n_4	n_a
1/5	–1	–1	1	0
–1	–1	–1	–1	–1
–4/5	–2	–2	0	–1
×(–5/4)	×(– 5/4)	×(–5/4)	×(–5/4)	×(–5/4)
1	5/2 rev	5/2 rev	0	5/4 rev

Example 9.4: The Differential

In the differential shown in Figure 9.9 all three gears have 32 teeth. For this example the input is through the carrier and the outputs are the shafts attached to gears 1 and 3. As a first example, let us determine the number of rotations of gear 3 if gear 1 is fixed.

Proceed in the usual manner by pretending first that the carrier is fixed and rotating gear 1 by one revolution. Since gear 2 has the same number of teeth, it also rotates one revolution, but since it is oriented perpendicular to gear 1, the choice of sign is somewhat arbitrary. This has been indicated in Table 9.12 by enclosing the 1 in parentheses. Gear 3, on the other hand, rotates backward one revolution. Completing the table as usual results in Table 9.12 where it is seen that gear 3 makes two revolutions for every one revolution of the carrier. Thus, if one of the outputs is fixed, the other output moves at twice the speed of the carrier (input). We can repeat the analysis by fixing the other input as shown in Table 9.13.

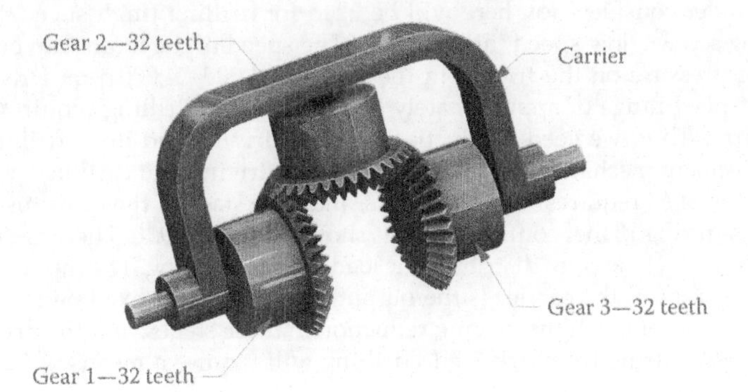

FIGURE 9.9
This differential has two suns and one planet.

TABLE 9.12

Gear 1 Is Fixed and the Carrier Makes One Revolution

n_1	n_2	n_3	n_c
1	(1)	−1	0
−1	−1	−1	−1
0	(−1)	−2	−1
×(−1)	×(−1)	×(−1)	×(−1)
0	(1)	2	1

Gear 3 moves at twice the speed of the carrier.

TABLE 9.13

Gear 3 Is Fixed and the Carrier Makes One Revolution

n_1	n_2	n_3	n_c
−1	—	1	0
−1	—	−1	−1
−2	—	0	−1
×(−1)	×(−1)	×(−1)	×(−1)
2	—	0	1

Gear 1 moves at twice the speed of the carrier.

In both cases, the input (the carrier) is the *average* of the two outputs (gears 1 and 3). This is what makes it useful as a differential. If a car is cornering, the outside wheels must move faster than the inside wheels, but the average of the two wheel speeds should be the same as the output of the transmission.

The name *differential* arises from another, very different application. If two pieces of machinery are rotating in opposite directions and attached to gears 1 and 3, the carrier will spin at the *difference* in their speeds. Thus, the carrier speed can be used as a feedback mechanism in synchronizing the speeds of the machines.

Example 9.5: Design of a Drill/Driver Transmission

For this example, we will use the tabular method to *design* a planetary speed reducer. Battery-powered drills are a common household tool, and planetary gearboxes are often used to reduce the speed and increase the torque produced by the DC motor. The drill under consideration here will be used for drilling (high speed/low torque) and driving screws (low speed/high torque). The speed of the motor can be controlled by varying pressure on the trigger in the range of 4,000–24,000 rpm. Driving screws requires a speed range of approximately 60–350 rpm and drilling requires a range of 250–1,500 rpm. Thus, we need a reduction of 64 for driving and 16 for drilling. We also need a convenient mechanism for shifting between driving and drilling speeds.

A reduction of 64 requires multiple stages, and each stage of the transmission will be the simple sun-ring-planet configuration as shown in Figure 9.10. There are five planets in the gearset which helps to distribute the load around the ring. The input to each stage will be the sun gear, the carrier is the output and the ring is fixed. Since $64 = 4^3$ and $16 = 4^2$, we can accomplish the driving reduction in three stages, and the drill reduction in two stages. Shifting from driving to drilling will require a means of "locking out" one of the stages.

Figure 9.11 shows a side view of the entire transmission. The motor drives the sun gear of stage A and the carrier of stage A drives the sun of stage B, and so on. Since we

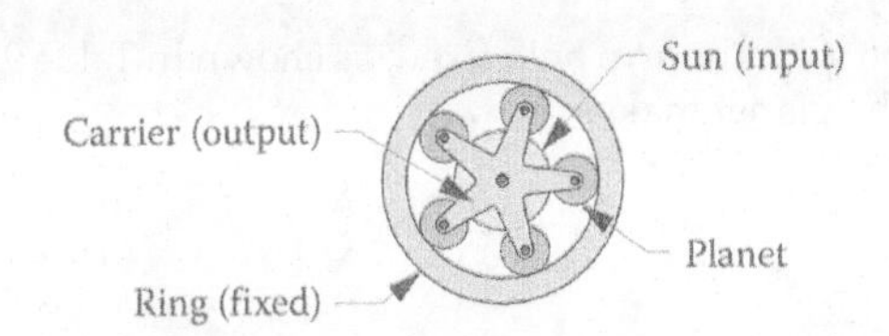

FIGURE 9.10
The simple sun-ring-planet configuration will be used for each stage of the drill transmission.

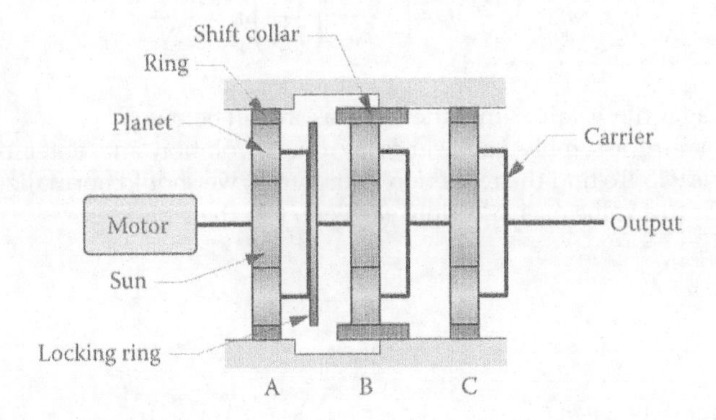

FIGURE 9.11
A side view of the drill/driver transmission. The ring gears are fixed to the body of the drill, the input to each stage is the sun gear, and the carrier serves as the output for each stage.

require a reduction of 4 in each stage, we can use three identical sun gears, three identical rings and five identical sets of planets.

We will take advantage of the versatile nature of planetary gearsets to achieve the shifting from driving to drilling speeds. As shown in Figure 9.12 a *shift collar* is used as the ring for stage B. In driving mode, the collar serves as the (fixed) ring for stage B. To switch to drilling speed, the collar is shifted to the left until it locks the planets of stage B to a locking ring on the carrier of stage A. Since the carrier of stage A is attached to the sun of stage B the shift collar has the effect of locking all of stage B into a single, rigid body. In this way, the speed reduction in stage B has been eliminated, as required.

We will now determine the number of teeth required in each stage to achieve the reduction of 4. In each stage, the sun is the input and the ring is fixed, so begin by giving

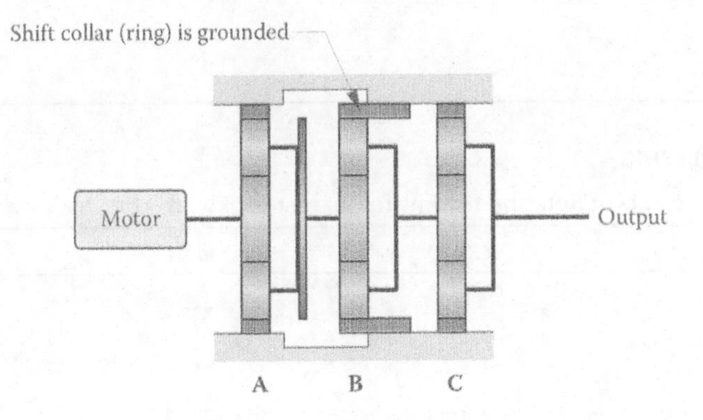

FIGURE 9.12
Shifting from drilling to driving is achieved through a shift collar. In driving mode, the collar acts as a simple ring gear fixed to ground. In drilling mode, the collar locks the planets (and carrier) of stage B to the carrier of stage A, which eliminates the reduction of stage B.

the ring one revolution with the arm held fixed, as shown in Table 9.14. If the ring makes one revolution then the planet makes

$$n_p = \frac{N_r}{N_p} n_r = \frac{N_r}{N_p}$$

revolutions. Similarly, the sun makes

$$n_s = -\frac{N_p}{N_s} n_p = \left(-\frac{N_p}{N_s}\right)\left(\frac{N_r}{N_p}\right) n_r = -\frac{N_r}{N_s}$$

revolutions. Enter these values into the table, as shown below.

Rotate the entire assembly backwards by one revolution and enter the results as shown in Table 9.15. To find the reduction in the stage, we should normalize the rotation of the sun gear to a value of: 1. Multiply each entry in the table by

$$\frac{1}{-\frac{N_r}{N_s} - 1}$$

as shown in Table 9.16. The final row of the table shows that if the sun makes one revolution, the carrier makes

$$n_a = \frac{N_s}{N_r + N_s} \tag{9.5}$$

TABLE 9.14

Step 1 in Calculating the Speed Reduction in One Stage of the Drill/Driver Transmission

n_r	n_p	n_s	n_a
1	$\frac{N_r}{N_p}$	$-\frac{N_r}{N_s}$	0
–1	–1	–1	–1

TABLE 9.15

The Results of Rotating the Entire Gearset Backwards by One Revolution

n_r	n_p	n_s	n_a
1	$\frac{N_r}{N_p}$	$-\frac{N_r}{N_s}$	0
–1	–1	–1	–1
0	$\frac{N_r}{N_p} - 1$	$-\frac{N_r}{N_s} - 1$	–1

TABLE 9.16
The Last Row of the Table Shows the Overall Speed Reduction Between Sun Gear and Carrier

n_r	n_p	n_s	n_a
1	$\frac{N_r}{N_p}$	$-\frac{N_r}{N_s}$	0
–1	–1	–1	–1
0	$\frac{N_r}{N_p}-1$	$-\frac{N_r}{N_s}-1$	–1
$\times\frac{1}{-\frac{N_r}{N_s}-1}$	$\times\frac{1}{-\frac{N_r}{N_s}-1}$	$\times\frac{1}{-\frac{N_r}{N_s}-1}$	$\times\frac{1}{-\frac{N_r}{N_s}-1}$
0	$\frac{N_s}{N_p}\left(\frac{N_p-N_r}{N_r+N_s}\right)$	1	$\frac{N_s}{N_r+N_s}$

This is a single stage in the drill/driver transmission.

revolutions. To achieve a reduction of 4, we require that $n_a = 1/4$. Here we have one equation and two unknowns, N_s and N_r. We may freely choose the number of teeth on the sun, and then use Equation (9.5) to find the number of teeth on the ring. Let us choose $N_s = 18$ teeth in order to avoid the possibility of interference or tooth undercut by choosing too few teeth. Solving for the number of teeth on the ring gives $N_r = 54$ teeth. Finally, using Equation (9.2) to solve for the number of teeth on the planet we have $N_p = 18$ teeth. This is fortunate, since it means that we can use the same tooling to make the planet and sun gears, which will reduce tooling costs in manufacturing the gearbox. Thus, our final design for each stage has $N_s = 18$ teeth, $N_p = 18$ teeth and $N_r = 54$ teeth.

As a final step, we should choose the module of the gears so that the final size of the transmission can be calculated. A commonly used module for gears of this size is 0.5 mm/tooth. Thus, the sun and planet gears will have a pitch diameter of 9 mm and the ring has pitch diameter of 27 mm. This is a reasonable diameter to fit inside the body of a handheld drill, so we conclude that our design is feasible. Of course, a thorough stress analysis of the gear train must be conducted, but that is beyond the scope of this text.

It is interesting to calculate the overall size of a set of spur gears that achieve the same reduction of 4 that was achieved with the planetary gearset. If we choose a pinion with 18 teeth, as was done for the sun in the planetary, then the gear needs 72 teeth, with a pitch diameter of 36 mm. As shown in Figure 9.13, the overall packaging size for the pinion and gear is

$$w = 9 + 36 = 45\text{ mm}$$

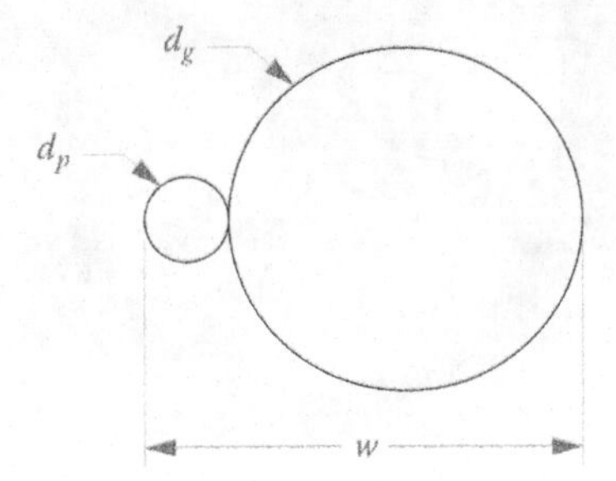

FIGURE 9.13
Overall packaging size for spur gears.

which is much larger than the ring size of 27 mm found earlier. While it is much easier and cheaper to manufacture the simple spur gear arrangement shown in Figure 9.13, the compact size and versatility of the planetary set has led to its adoption in many battery-powered drills.

9.3 Analysis of Planetary Gearsets—The Generalized Table Method

In all of the preceding examples, one of the inputs to the planetary gearset was fixed while the other input and the output were allowed to rotate. But one of the chief virtues of the planetary gearset is its ability to combine inputs from two sources of rotational power. The second tabular method will provide a means of analyzing planetary gearsets with two inputs, and we will also demonstrate the use of this method in design. Of course, we can always take one of the inputs to be zero and obtain the same results as the first tabular method. We will take as an example the planetary gearset shown in Figure 9.14. In our example, we will let the speed of the sun be 1,000 rpm, the speed of the ring be –500 rpm and our goal is to find the resulting speed of the carrier.

The new tabular method begins the same way as before, by assuming the arm to be fixed. We first rotate the sun gear through one revolution and determine the speed of the other gears. To keep the units consistent throughout the table we will work in number of revolutions, rather than speed (rpm).

$$n_{p1} = -\frac{N_s}{N_{p1}} n_s = -\frac{24}{24}(1\text{ rev}) = -1\text{ rev}$$

The second planet rotates on the same shaft as the first, so that

$$n_{p2} = n_{p1} = -1\text{ rev}$$

The number of rotations made by the third planet is

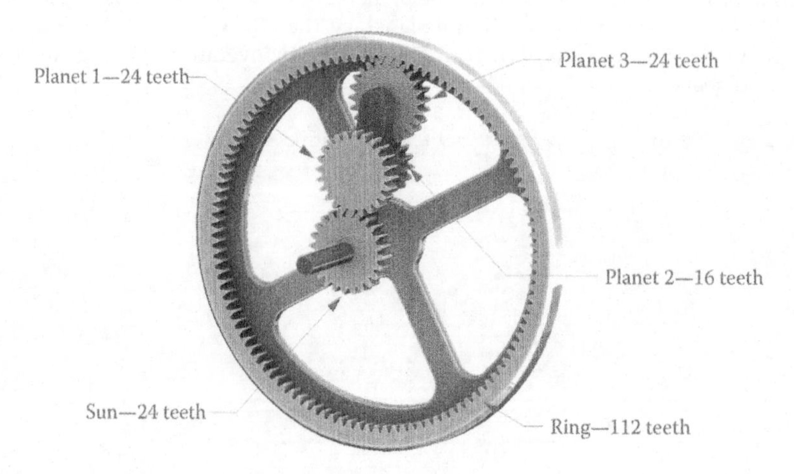

FIGURE 9.14
This planetary gearset has a sun, a ring, and three planets.

$$n_{p3} = -\frac{N_{p2}}{N_{p3}} n_{p2} = -\frac{16}{24}(-1 \text{ rev}) = \frac{2}{3} \text{ rev}$$

Finally, the ring rotates

$$n_r = \frac{N_{p3}}{N_r} n_{p3} = \frac{24}{112}\left(\frac{2}{3} \text{ rev}\right) = \frac{1}{7} \text{ rev}$$

The results are shown in Row 1 in Table 9.17.

Now multiply each row by a constant factor α. This is equivalent to increasing (or decreasing) the number of rotations of each gear by a scaling factor α. The result of this operation is shown in Row 2 in Table 9.18.

Finally, glue the entire assembly together (temporarily) and rotate it as a unit through β revolutions.

The final result is shown in Row 3 of Table 9.19. This row gives the kinematic relationship between all the gears and the carrier and can be used to determine the output speed of the planetary gearset. Note that there are two arbitrary constants, α and β, in the table.

TABLE 9.17

Turn the Sun Gear One Revolution, and Determine the Number of Revolutions of Each Other Gear

	n_r	n_{p3}	n_{p2}	n_{p1}	n_s	n_a
	1/7	2/3	−1	−1	1	0
$\times\alpha$						
$+\beta$						

All entries in the table are given in number of revolutions.

TABLE 9.18

Multiply Each Entry in Row 1 by the Constant Scaling Factor α

	n_r	n_{p3}	n_{p2}	n_{p1}	n_s	n_a
	1/7	2/3	−1	−1	1	0
$\times\alpha$	$\frac{1}{7}\alpha$	$\frac{2}{3}\alpha$	$-\alpha$	$-\alpha$	α	0
$+\beta$						

This is equivalent to scaling the number of rotations of each gear by a constant amount, and does not affect any of the kinematic relationships between gears.

TABLE 9.19

Glue the Assembly Together and Rotate it β Revolutions

	n_r	n_{p3}	n_{p2}	n_{p1}	n_s	n_a
	1/7	2/3	−1	−1	1	0
$\times\alpha$	$\frac{1\alpha}{7}$	$\frac{2\alpha}{3}$	$-\alpha$	$-\alpha$	α	0
$+\beta$	$\frac{1\alpha}{7}+\beta$	$\frac{2\alpha}{3}+\beta$	$\beta-\alpha$	$\beta-\alpha$	$\beta+\alpha$	β

The final result is shown in row 3.

These correspond neatly with the speeds of the two inputs to the gearset. In the problem statement, we have

$$n_r = -500 \text{ rpm} \qquad n_s = 1000 \text{ rpm}$$

But in Row 3 of the table we have

$$n_r = \frac{1\alpha}{7} + \beta \qquad n_s = \alpha + \beta$$

Thus, we have two equations and two unknowns. Subtracting the second equation from the first gives

$$\frac{1\alpha}{7} - \alpha = -1500$$

or

$$\alpha = 1750$$

Recall that this was used as a constant scaling factor and has no units. Next, substitute the value for α into the second equation to give

$$\beta = -750 \text{ rev}$$

This is also the speed of the arm, as shown in the final entry in Table 9.19. Thus, we have found a method for analyzing a planetary gearset with two inputs, instead of just one. To confirm that the method is producing accurate results, let us complete Example 9.2 of the previous section with the new tabular method.

Example 9.6: Planetary Gearset with One Fixed Input (Repeated from Example 9.2 of Previous Section)

In this example, the ring rotates at 100 rpm clockwise and the sun is fixed. Rotate the sun once and enter the results in Row 1 (see Table 9.20). Multiply each entry by α (Row 2) and add β (Row 3).

Since the sun is fixed, we have

$$\alpha + \beta = 0$$

or

$$\beta = -\alpha$$

TABLE 9.20

The New Tabular Method Used in Example 9.2 from the Previous Section

	n_r	n_p	n_s	n_a
	$-1/2$	-2	1	0
$\times\alpha$	$-\frac{\alpha}{2}$	-2α	α	0
$+\beta$	$-\frac{\alpha}{2}+\beta$	$-2\alpha+\beta$	$\alpha+\beta$	β

The ring rotates at 100 rpm, so that

$$-\frac{\alpha}{2}+\beta=100$$

Solving for α and β gives

$$\alpha=-\frac{200}{3} \qquad \beta=\frac{200}{3}=66.6$$

This is also the speed of the arm and is the same result as was found in Example 9.2. As you can see, the new tabular method is more general than the old one since it allows for two inputs, but we pay for this flexibility with some additional algebra at the end.

Example 9.7: Design Example

As a design example, let us reconsider the planetary gearset in Figure 9.15. We will leave the numbers of teeth on the gears unspecified for now, and try to solve for them using the required speeds of the inputs and output. In this example, we wish to combine the inputs from an electric motor (2,600 rpm) and an internal combustion engine (–1,000 rpm) so that the output speed is 200 rpm. It is not obvious which part of the planetary gearset we should connect each input and output so we must begin by guessing. Let us try attaching the electric motor (2,600 rpm) to the ring, the IC engine (–1,000 rpm) to the sun and the output (200 rpm) to the carrier. Since the modules of all three gears are the same, we have (from Equation 9.2)

$$N_s+2N_p=N_r \tag{9.6}$$

Thus, we can solve for two of the tooth numbers (say, N_s and N_r) and use Equation (9.6) to solve for the third. To begin, rotate the sun by one revolution. This will cause the planet to rotate by

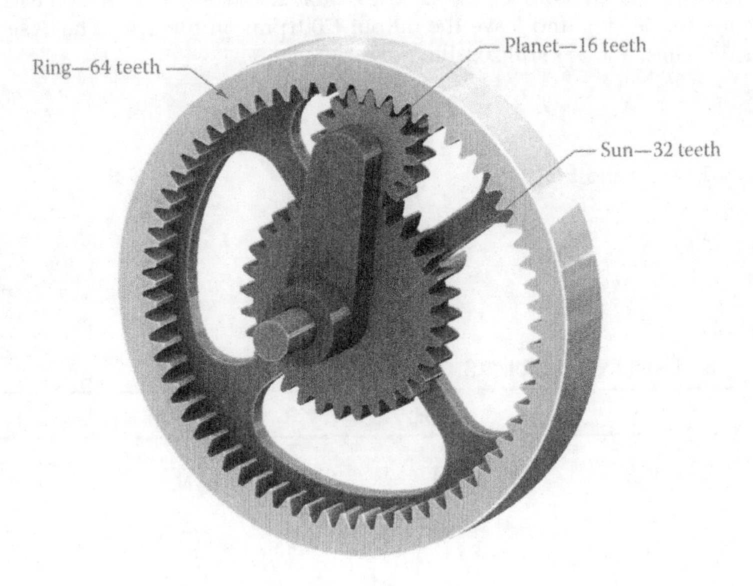

FIGURE 9.15
The planetary gearset from Example 9.2 in the previous section.

$$n_p = -\frac{N_s}{N_p} n_s = -\frac{N_s}{N_p}(1 \text{ rev}) = -\frac{N_s}{N_p}$$

rotations. The ring rotates by

$$n_r = \frac{N_p}{N_r} n_p = \left(\frac{N_p}{N_r}\right)\left(-\frac{N_s}{N_p}\right) = -\frac{N_s}{N_r}$$

Enter these values into Row 1 and then multiply by α to obtain Row 2 and add β to obtain Row 3. The results are shown in Table 9.21.

The speed of the carrier is specified in the problem statement as 200 rpm so that

$$\beta = 200$$

Similarly, the sun is specified in the problem statement as –1,000 rpm, so that

$$\alpha + \beta = -1000$$

or $\alpha = -1{,}200$. Finally, the speed of the ring is 2,600 rpm so that

$$-\frac{N_s}{N_r}\alpha + \beta = 2600$$

Solving for N_s/N_r gives

$$\frac{N_s}{N_r} = 2$$

This means that the sun must have twice as many teeth as the ring! Clearly, this cannot be true, and something has gone wrong – most likely, we have attached the inputs and output to the wrong parts of the planetary gearset.

As a second guess, let us attach the electric motor (2,600 rpm) to the sun, the IC engine (–1,000 rpm) to the ring and leave the output (200 rpm) on the arm. The design table remains the same, and β is still 200. But now

$$\alpha + \beta = 2600$$

so that $\alpha = 2{,}400$. In addition

$$-\frac{N_s}{N_r}\alpha + \beta = -1000$$

TABLE 9.21

Design Table Used for Example 9.3

	n_s	n_p	n_r	n_a
	1	$-\frac{N_s}{N_p}$	$-\frac{N_s}{N_r}$	0
$\times\alpha$	α	$-\frac{N_s}{N_p}\alpha$	$-\frac{N_s}{N_r}\alpha$	0
$+\beta$	$\alpha+\beta$	$-\frac{N_s}{N_p}\alpha+\beta$	$-\frac{N_s}{N_r}\alpha+\beta$	B

Solving for N_s/N_r gives

$$\frac{N_s}{N_r} = \frac{1}{2}$$

which is more realistic. This gives the *ratio* of the numbers of teeth on the sun and ring, but does not give the tooth numbers themselves. If we choose $N_s = 18$ to avoid interference, then $N_r = 36$. Using Equation (9.6) gives the number of teeth on the planet as $N_p = 9$. This is likely to cause interference because of the low number of teeth. If we try again with $N_s = 30$ we arrive at $N_r = 60$ and $N_p = 15$. If we use high pressure angle gears we will avoid interference.

As you have seen, designing planetary gearsets usually involves quite a bit of trial and error. In fact, this iterative approach is a normal part of the design process. Do not consider it a failure if your first (or second or third) designs are not optimal, or even feasible. The learning that accompanies this "guessing and checking" will make you a better designer.

As a final note, let us consider the entries in the bottom row of Table 9.21, leaving the input and output speeds as variables

$$n_a = \beta$$

$$n_s = \alpha + \beta$$

$$n_r = b\alpha + \beta$$

where we have replaced $-N_s/N_r$ with the variable b. This variable gives the ratio of the speed of the ring to that of the sun, and can be thought of as the most basic ratio for this particular type of planetary gearset. The observant reader will have noted that all tables made using the method described here *are identical*, with the exception of the speeds of intermediate gears (e.g. the planet) and the value of b. We can solve the equations above for α, β, and b as

$$\alpha = n_s - n_a \tag{9.7}$$

$$\beta = n_a \tag{9.8}$$

$$b = \frac{n_r - n_a}{n_s - n_a} \tag{9.9}$$

The final equation will be very important to us in the next section. Armed with this new knowledge, let us consider one final design example.

Example 9.8: Design Example of Planetary Gearset with Two Rings and Two Planets

Consider the gearset shown in Figure 9.16 with two rings and two planets. As shown in Figure 9.17 the diameters (and tooth numbers) of the gears are not independent. The diameters are related through

$$d_{r1} - d_{p1} = d_{r2} - d_{p2} \tag{9.10}$$

In order to mesh, the modules of ring 1 and planet 1 must be the same, as must the modules of ring 2 and planet 2, although the module of the first gearset need not be identical to the module of the second gearset. If the modules of all four gears are the same then we have

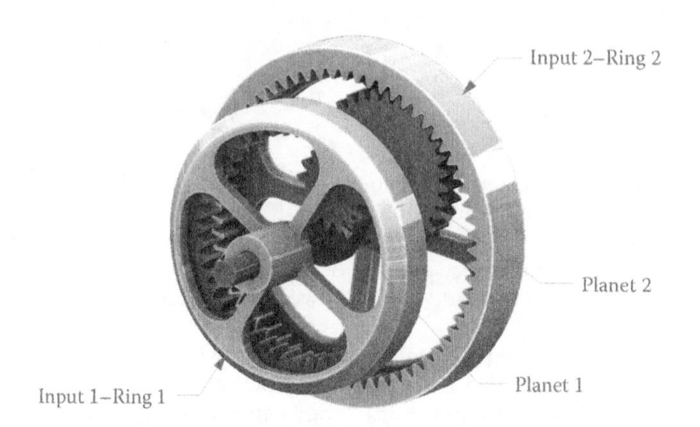

FIGURE 9.16
The planetary gearset in Example 9.4 has two rings and two planets.

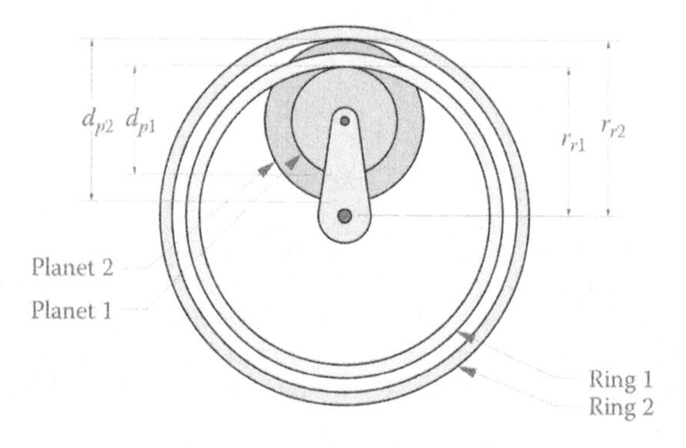

FIGURE 9.17
The diameters of the gears are not independent of one another, if all gears have the same module.

$$N_{r1} - N_{p1} = N_{r2} - N_{p2} \tag{9.11}$$

As in Example 9.7, we wish to combine the inputs from an electric motor (2,600 rpm) and an internal combustion engine (–1,000 rpm) so that the output speed is 200 rpm. Let us try attaching the 2,600 rpm input to ring 1 and the –1,000 rpm input to ring 2, with the 200 rpm output attached to the carrier. Construct a design table with one column for each of the gears and the carrier. If we give ring 1 one rotation then planet 1 spins by

$$n_{p1} = \frac{N_{r1}}{N_{p1}}(1\ \text{rev}) = \frac{N_{r1}}{N_{p1}}$$

Planet 2 is mounted on the same shaft as planet 1, so that its rotation is identical.

$$n_{p2} = n_{p1}$$

Finally, ring 2 spins by

$$n_{r2} = \frac{N_{p2}}{N_{r2}} n_{p2} = \frac{N_{r1}}{N_{p1}} \frac{N_{p2}}{N_{r2}}$$

Note that all signs are positive since both gearsets have internal gears. Scale by the factor α in the second row and then add β rotations in the third row, as seen in Table 9.22. Using the definition of b discussed earlier, we have

$$b = \frac{N_{r1}}{N_{p1}} \frac{N_{p2}}{N_{r2}} \tag{9.12}$$

This must be a positive number since the numbers of teeth are all positive and both gear pairs are internal. We can modify Equation (9.9) to account for the fact that the sun has been replaced by ring 1 and the ring has been replaced by ring 2.

$$b = \frac{n_{r2} - n_a}{n_{r1} - n_a} \tag{9.13}$$

If we substitute the speed values into this equation, we find that

$$b = \frac{-1000 - 200}{2600 - 200} = -\frac{1200}{2400}$$

Since this is a negative number, we cannot achieve these speeds with this type of planetary gearset. Instead, let us reverse the direction of the IC engine; this can be achieved by placing an idler gear between the engine and ring 2. An idler gear is a gear used to change the direction of rotation in a gear train without changing speed, as shown in Figure 9.18. The reverse gear in an automobile uses an idler gear to change the direction that the wheels rotate. If we change the rotational direction of the IC engine, we obtain

$$b = \frac{1000 - 200}{2600 - 200} = \frac{800}{2400} = \frac{1}{3}$$

TABLE 9.22

Design Table for Example 9.4

	n_{r1}	n_{p1}	n_{r2}	n_{p2}	n_a
	1	$\frac{N_{r1}}{N_{p1}}$	$\frac{N_{r1}}{N_{p1}} \frac{N_{p2}}{N_{r2}}$	$\frac{N_{r1}}{N_{p1}}$	0
$\times\alpha$	α	$\frac{N_{r1}}{N_{p1}}\alpha$	$\frac{N_{r1}}{N_{p1}} \frac{N_{p2}}{N_{r2}}\alpha$	$\frac{N_{r1}}{N_{p1}}\alpha$	0
$+\beta$	$\alpha+\beta$	$\frac{N_{r1}}{N_{p1}}\alpha+\beta$	$\frac{N_{r1}}{N_{p1}} \frac{N_{p2}}{N_{r2}}\alpha+\beta$	$\frac{N_{r1}}{N_{p1}}\alpha+\beta$	β

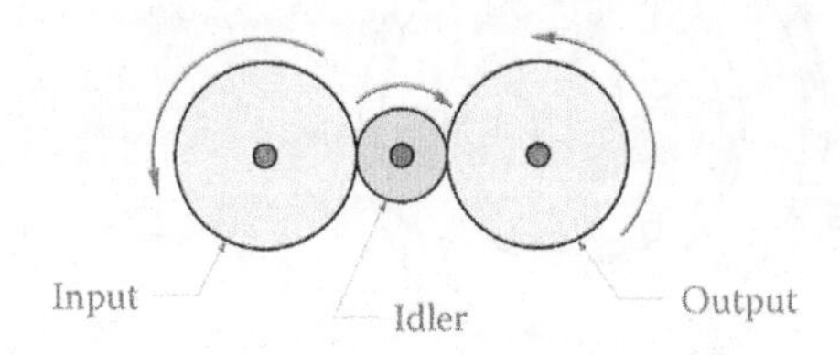

FIGURE 9.18
An idler gear is used to reverse the direction of rotation in a gear train without changing speed.

which is now a positive number. The basic ratio of the planetary gearset must be 1/3, so that

$$\frac{N_{r1}}{N_{p1}}\frac{N_{p2}}{N_{r2}} = \frac{1}{3} \tag{9.14}$$

We now have two equations (Equations (9. 12) and (9. 13)) and four unknowns, N_{r1}, N_{p1}, N_{r2}, N_{p2}. This means that we can choose two of the values somewhat arbitrarily and solve for the remaining two. Equation (9.14) states that the product of one gear ratio with the reciprocal of the other must give 1/3. Let us try

$$\frac{N_{p1}}{N_{r1}} = \frac{1}{2}$$

$$\frac{N_{p2}}{N_{r2}} = \frac{1}{6}$$

Then

$$\frac{2}{1}\cdot\frac{1}{6} = \frac{1}{3}$$

as required. Let us choose $N_{r2} = 120$ teeth, since it is divisible by 6. Then $N_{p2} = 20$ teeth. Also,

$$N_{p1} = \frac{1}{2}N_{r1}$$

Substituting this into Equation (9.11) gives

$$N_{r1} - \frac{1}{2}N_{r1} = N_{r2} - N_{p2}$$

which results in $N_{r1} = 200$ teeth and $N_{p1} = 100$ teeth.

Thus, we have found a viable solution to the problem in Example 9.8. A scaled drawing of the solution can be seen in Figure 9.19. As before, we required some amount of trial and error to find a valid solution, but our experience gained in finding the formula (and interpretation) of the basic ratio, *b*, shortened the solution time considerably. Practice makes perfect!

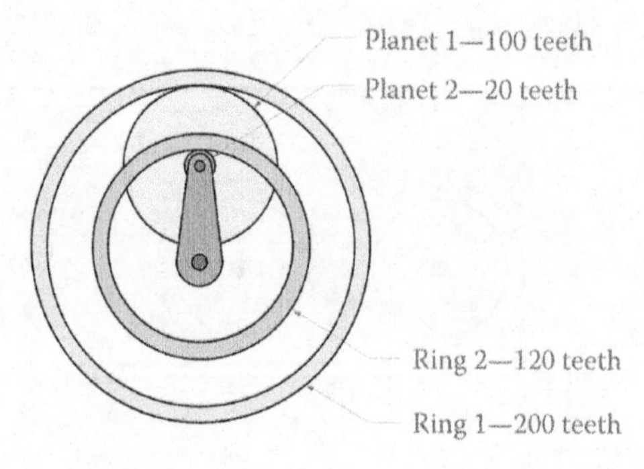

FIGURE 9.19
Scaled drawing of the solution to Example 9.4.

9.4 Analysis of Planetary Gearsets—An Algebraic Method

As we saw in the preceding section, planetary gearsets can behave in counterintuitive ways, which makes their analysis quite interesting. However, we can develop a few simple formulas, which we can apply to any gearset, no matter how complicated. We will do this by treating the planetary gearset as a "black box" with three shafts coming out of it. To begin the analysis, imagine for the moment that we hold the arm fixed. Define the *basic ratio*, *b*, as

$$b = \frac{\text{speed of high speed input shaft}}{\text{speed of low speed input shaft}} \text{ (with the arm held fixed)}$$

As an example, consider the gearset shown in Figure 9.20. Let us assume that the sun gear has $N_s = 32$, the planet has $N_p = 16$, and the ring has $N_r = 64$. As an initial guess, let us suppose that the sun gear spins faster than the ring gear, since it has fewer teeth. Hold the arm fixed and give the ring one revolution. The planet turns

$$n_p = \frac{N_r}{N_p} n_r = \frac{64}{16}(1 \text{ rev}) = 4 \text{ rev}$$

or 4 revolutions. The sign preceding the gear ratio is positive because the ring gear is internal (the ring and planet revolve in the same direction). If the planet gear makes 4 revolutions, the sun gear makes

$$n_s = -\frac{N_p}{N_s} n_p = -\frac{16}{32}(4 \text{ rev}) = -2 \text{ rev}$$

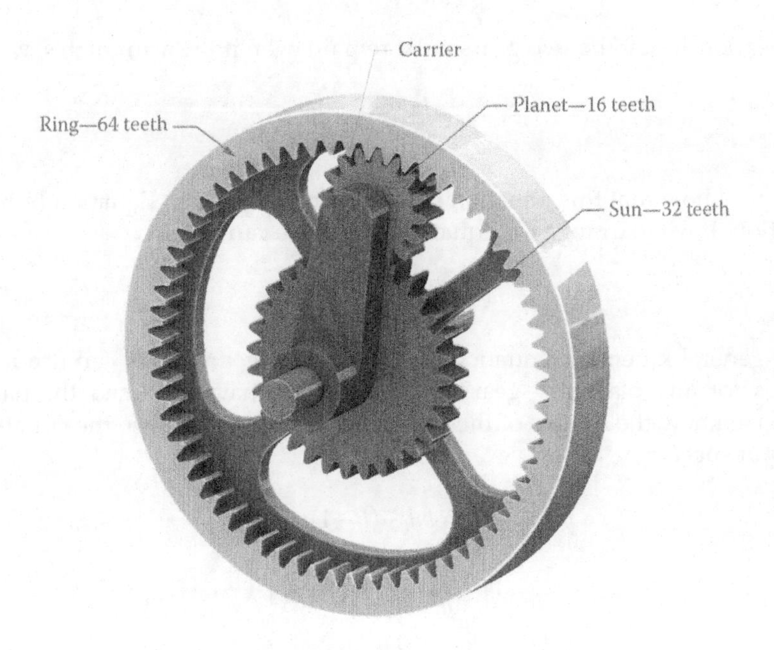

FIGURE 9.20
A typical planetary gearset with sun, planet, ring, and carrier.

Thus, for every 1 revolution of the ring, the sun makes –2 revolutions. The basic ratio is then

$$b = -2$$

While it is possible for the basic ratio to be positive or negative, its magnitude must *always* be greater than or equal to 1, since we have defined it as the ratio of the faster speed to the slower. Thus

$$|b| \geq 1$$

by definition. The efficiency formulas in the following section *will not work* if b is less than 1.

9.4.1 Overall Ratio of the Planetary Gearset

In the preceding analysis, we held the arm fixed, but the behavior of the input and output gears *relative to the arm* must hold no matter what the arm does. That is, rotation of the arm (and revolution of the planets) does not affect the ratio between sun and ring, since this ratio is determined by the numbers of teeth in each gear. Let us now repeat the preceding analysis, but this time allow the arm to rotate. Define β_h as the number of rotations that the faster (or *high speed*) gear rotates *relative to the arm*. That is

$$\beta_h = n_h - n_a \tag{9.15}$$

where the n_h and n_a give the amount of rotation relative to ground of the faster gear and arm, respectively. Also, define the β_l as the number of rotations that the slower (or *low speed*) gear makes *relative to the arm*

$$\beta_l = n_l - n_a \tag{9.16}$$

Since the relationship of the two gears must remain even if the arm rotates, we can write

$$b = \frac{\beta_h}{\beta_l} = \frac{n_h - n_a}{n_l - n_a} \tag{9.17}$$

The reader will note that this equation is remarkably similar to Equation (9.9) in the preceding section. If we rearrange the equation above, we can obtain

$$n_h - n_a - b(n_l - n_a) = 0 \tag{9.18}$$

This is the general kinematic equation for a planetary gearset. We can use it to find the speed ratios for any planetary gearset, provided we have obtained the basic ratio in advance. If we know the speeds of the two input shafts, we can use the equation to solve for the output speed.

$$n_h = bn_l - (b-1)n_a \tag{9.19}$$

$$n_l = \frac{1}{b}\left[n_h + n_a(b-1)\right] \tag{9.20}$$

$$n_a = \frac{bn_l - n_h}{b-1} \tag{9.21}$$

Example 9.9

In the gear train for the preceding example, let the ring be fixed, and the arm free to move. Find the speed of the arm if the sun moves at 100 rpm cw.

Solution

For this example, the ring has speed 0 rpm and the basic ratio is still –2. The slow gear is the ring and the fast gear is the sun. Thus

$$n_a = \frac{bn_l - n_h}{b-1}$$

$$n_a = \frac{0-100}{-2-1} = \frac{100}{3} = 33.33 \text{ rpm}$$

or 33.33 rpm cw.

Example 9.10: Construction Brick Planetary Geartrain

For the gearset in Figure 9.21, we hold gear 1 fixed. How fast does the arm rotate if gear 4 spins at 100 rpm cw?

Solution

To solve this problem, we must first find the basic ratio. As a guess, let us assume that gear 4 is the fast gear and gear 1 is the slow gear. Pretend the arm is fixed and rotate gear 1 one revolution cw.

$$n_B = -\frac{N_1}{N_2} n_1 = -\frac{40}{8} 1 = -5 \text{ rev}$$

Gears 2 and 3 rotate at the same speed, so that

$$n_D = -\frac{N_3}{N_4} n_3 = -\frac{24}{24}(-5) = 5 \text{ rev}$$

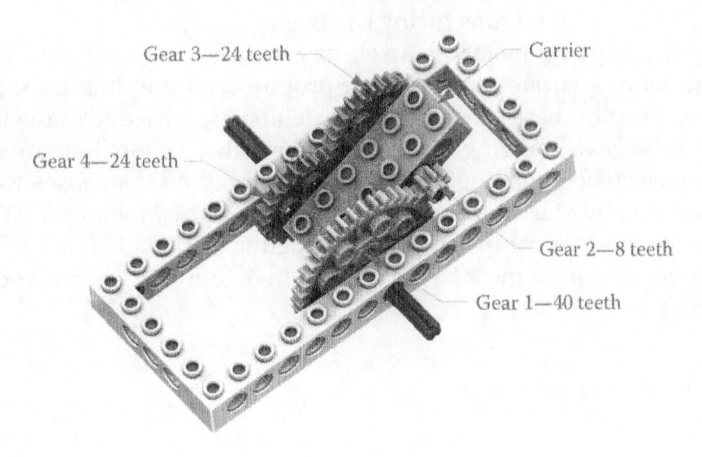

FIGURE 9.21
The construction brick planetary gearset of Example 9.2.[1]

Thus, the basic ratio is

$$b = \frac{n_4}{n_1} = 5$$

Now solve for the speed of the arm

$$n_a = \frac{bn_l - n_h}{b - 1}$$

$$n_a = \frac{5(0) - 100}{5 - 1} = \frac{-100}{4} = -25 \text{ rpm}$$

The arm rotates at 25 rpm ccw. Now, let us use gear 1 as the input and hold gear 4 fixed. Since we defined the basic ratio assuming that gear 1 is the slow gear and gear 4 is the fast, n_h still applies to gear 4 and n_l applies to gear 1, despite the fact that gear 4 is fixed in this example.

$$n_a = \frac{bn_l - n_{h\varphi}}{b - 1}$$

$$n_a = \frac{5(100) - 0}{5 - 1} = \frac{500}{4} = 125 \text{ rpm}$$

or 125 rpm cw. This is the same result as was obtained in Example 9.3.

9.5 Efficiency of Planetary Gearsets

Now that we have learned to design planetary gearsets to achieve the desired speed ratios between input and output shafts, we turn our attention to calculating the overall efficiency of a planetary gearset. In some cases, the efficiency of the planetary gearset as a whole might be higher than the basic efficiency of the gears comprising the set, while in other cases, the efficiency is so low as to render the gearset useless from a practical point of view. It is very important, therefore, to estimate the efficiency of a proposed planetary gearset before the final design and manufacturing can begin.

The topic of efficiency in planetary gearsets has received much attention in recent years, with newer and more accurate models being proposed all the time – see [1–4] for some recent examples. Most of these models require detailed knowledge of the lubricating oil, surface finish of the gears, bearing types and many other factors that are often unavailable to the engineer when creating an initial design. For our purposes we will present the method proposed by Merritt [5] and expounded upon by Molian [6]. This method can be used for the purpose of comparing two competing designs or for analyzing general trends, but laboratory testing must be carried out to obtain precise efficiency values for a particular gearset.

9.5.1 A Generic Planetary Gearset

Before embarking on the rather difficult topic of planetary gearset efficiency, we will first create a "generic" planetary gearset that we will use for our analysis. Consider the traditional sun-planet-ring gearset shown in Figure 9.22. Assume that the sun has 32 teeth, the planet has 16 teeth, and the ring has 64 teeth. If we hold the arm fixed, then the sun will spin twice as fast as the ring, but in the opposite direction. Now consider the two suns/two planets gearset shown in Figure 9.23, and assume that the sun in front has 24 teeth, the front planet has 24 teeth, the rear planet has 16 teeth, and the rear sun has 32 teeth. If we hold the arm fixed then the front sun will spin twice as fast as the rear sun, in the *same* direction. Recall that the basic ratio was defined as the ratio of the speed of the high-speed gear to that of the low-speed gear with the arm held fixed.

$$b = \frac{n_h}{n_l} \text{ (with arm fixed)}$$

where n_h is the speed of the high-speed gear and n_l is the speed of the low-speed gear. For the traditional planetary gearset in Figure 9.22 the basic ratio will be –2, since the sun spins twice as fast as the ring in the opposite direction. For the gearset in Figure 9.23, the basic ratio will be +2, since both high-speed and low-speed gears spin in the same direction with the arm fixed. Since the basic ratio is defined as the ratio of high speed to low speed, it will always have a magnitude greater than or equal to 1.

$$|b| \geq 1$$

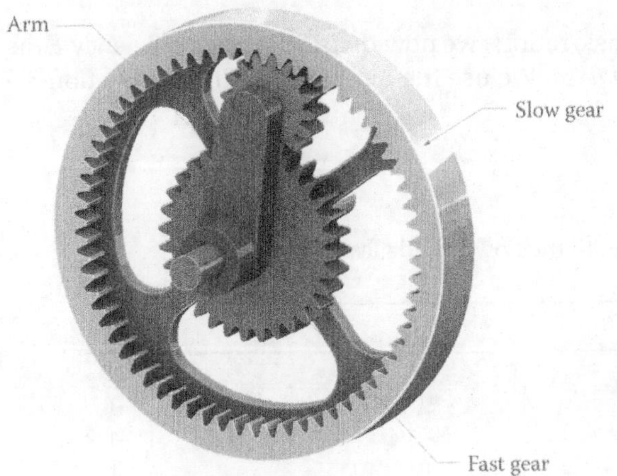

FIGURE 9.22
A planetary gearset with a sun, planet and ring. If we hold the arm fixed, then the sun spins faster than the ring.

FIGURE 9.23
A planetary gearset with two suns and two planets. If we hold the arm fixed, then the sun gear in front spins faster than the sun gear in the rear.

Now let us construct a speed table for the generic planetary gearset using the methods outlined in Section 9.2. For this case, we assume that the low-speed gear is fixed and the high-speed gear is the input. The output is taken at the arm. In Row 1 of the table we give the fixed gear (the low-speed gear) one revolution. The high-speed gear spins b times as fast, so it rotates b revolutions. Rotate the entire gearset backwards as a single entity and compute the resulting speeds in Row 3. Since we wish to find the number of rotations of the arm for a single revolution of the high-speed gear, we divide Row 3 by $(b - 1)$ to obtain Row 4. Observe that we have not made any qualifications as to the particular type of planetary gearset, and Table 9.23 is valid for the gearset shown in Figure 9.22 or Figure 9.23, or any other type of planetary gearset, as long as we take care to preserve the sign of b in our calculations. Thus, we can now think of any planetary gearset as a "black box" with two inputs and one output.

The same procedure may be used to find the speed ratios where the high-speed gear is fixed and the low-speed gear is the input. Table 9.24 gives a speed table for a generic planetary gearset with all possible combinations of input, output, and fixed shafts, neglecting those combinations where the arm is fixed (if the arm is fixed we no longer have a planetary gear train). We will make frequent use of this table in the sections that follow.

9.5.2 The Basic Efficiency

In the same manner as we defined the basic ratio, b, we now define the basic efficiency E_0 as the efficiency of the gearset *with the arm fixed*. We use the methods outlined in Section 8.5 to calculate this efficiency.

TABLE 9.23

Speed Table for a Planetary Gearset with the Low-Speed Gear Fixed and the High-Speed Gear as Input

	n_l	n_h	n_a
1	1	b	0
2	−1	−1	−1
3	0	$b-1$	−1
4	0	1	$\frac{1}{1-b}$

TABLE 9.24

Speed Table for Generic Planetary Gearset with Various Inputs, Outputs, and Fixed Shafts

Input	Output	Fixed	n_l	n_h	n_a
hi	arm	lo	0	n_h	$\frac{n_f}{1-b}$
arm	hi	lo	0	$n_a(1-b)$	n_a
lo	arm	hi	n_l	0	$\frac{bn_l}{b-1}$
arm	lo	hi	$\frac{n_a(b-1)}{b}$	0	n_a

Example 9.11: Traditional Planetary Gearset

For the traditional planetary gearset in Figure 9.22, we have $N_1 = 32$ teeth, $N_2 = 16$ teeth and $N_3 = 64$ teeth (internal). If we assume a pressure angle of 20° and a coefficient of friction of 0.1, we may consult Figure 9.36 in Section 8.5 to find that

$$E_1 = 1 - \mu L_1 \tag{9.22}$$

$$E_1 = 1 - 0.1 \cdot 0.20 = 0.98$$

Since the second gear pair has an internal gear we must multiply the tooth loss factor by $(\rho_2 - 1)/(\rho_2 + 1)$ as follows, where ρ_2 is the gear ratio for this stage of the gearset, or $\rho_2 = 64/16$.

$$E_2 = 1 - \frac{(\rho_2 - 1)}{\rho_2 + 1}\mu L_2 \tag{9.23}$$

$$E_2 = 1 - \frac{4-1}{4+1} \cdot 0.1 \cdot 0.18 = 0.9892$$

Thus, the basic efficiency of the traditional planetary gearset in Figure 9.22 is

$$E_0 = E_1 E_2 = 0.9694$$

or 96.9% efficient. Note that this is *not* the overall efficiency of the planetary gearset, it is the efficiency of the gears *within* the gearset when the arm is fixed.

Example 9.12: Two Suns, Two Planets

Assume a pressure angle of 20° and a coefficient of friction of 0.1 and calculate the basic efficiency of the planetary gearset shown in Figure 9.23.

The first gear pair has an efficiency of

$$E_1 = 1 - 0.1 \cdot 0.18 = 0.982$$

and the second gear pair has

$$E_2 = 1 - 0.1 \cdot 0.2 = 0.98$$

so that the basic efficiency is

$$E_0 = E_1 \cdot E_2 = 0.9624$$

or 96.2% efficient. At first glance, it appears as though the traditional planetary gearset might be slightly more efficient, but we must calculate the efficiency of the planetary gearset *as a whole* to determine if this is the case. We will use the basic efficiency as one input to our efficiency model of the planetary gearset, as will be seen below.

9.5.3 Torque Balance on the Gearset

Let us assume that the planetary gearset is operating at a steady speed, and with constant torques on the input and output shafts. If this is the case then the sum of the torques on all shafts must add to zero:

$$T_l + T_h + T_a = 0 \tag{9.24}$$

where T_l is the torque on the low-speed shaft, T_h is the torque on the high-speed shaft and T_a is the torque on the arm. For this equation to be true, one or two of the torques must be negative. We will assume that the *input torque* is positive, whether it occurs on the low speed, high speed, or arm shafts.

9.5.4 Power Balance of the Gearset

The power that is present at a shaft is the product of the torque on the shaft with the angular velocity of the shaft. Thus, for a given shaft we have

$$P = T \times n \tag{9.25}$$

where T is the torque present on the shaft and n is the rotational speed of the shaft. Since we have been using rpm as the unit for shaft speed, the unit for power will be (N-m)·(rev/min). We can convert this to the more standard unit of Watts by using

$$P = T \cdot \omega \tag{9.26}$$

where ω is the angular velocity in radians per second, but for now we will stick with the first equation, keeping in mind that we must multiply by a factor of $2\pi/60$ if we want the power in Watts. We will adopt the convention that a positive power represents power *supplied to the gearset* (as from a motor) and a negative power represents power *taken from the gearset* (as with a load that is being driven). A positive power can occur if torque and angular velocity are both positive or are both negative, and a negative power occurs when the torque and angular velocity have opposite signs. Thus, if the output shaft has a positive angular velocity, the load torque must be negative, and vice versa. This convention will be very important when we consider power flow within the gearset in the following sections.

Assume for the moment that the gear pairs are operating with 100% efficiency. If this is the case then the total power entering and leaving the gearbox must also add up to zero

$$T_l n_l + T_h n_h + T_a n_a = 0 \tag{9.27}$$

In our simple efficiency model, the only way that power can be lost is through the gear teeth rubbing against one another. If the entire gearset were to rotate together as a rigid unit, there would be zero power loss since the gears would not be meshing relative to one another. Thus, we must determine a method of calculating the speed of the gears *relative to one another.*

Imagine for a moment that you are sitting on the arm as it rotates inside the planetary gearset. From your viewpoint you would see the sun and planet gears meshing against each other, and you would be able to easily measure the speed at which this meshing occurs. Since the power loss is determined solely by the meshing of the gears relative to each other inside the gearset, we conclude that we should measure the gear speeds *relative to the arm* when computing power loss. Let us define

$$\beta_l = n_l - n_a \tag{9.28}$$

$$\beta_h = n_h - n_a \tag{9.29}$$

as the *relative speeds* of the low-speed and high-speed gears, respectively. If the planetary gearset is rotating as a rigid unit, then both of these relative speeds are zero, and there is zero power loss. We may tabulate the formulas for β_l and β_h for the configurations in Table 9.24 and these are shown in Table 9.25.

Note that the speeds (n_l, n_h, n_a) and relative speeds (β_l and β_h) do not depend upon the efficiency of the gearset – they are dictated solely by the numbers of teeth in the gears and the arrangement of the gears. The power loss from friction between the gears affects the *torque output* of the gearset, and not the *speed output*.

To demonstrate the method of calculating the torques on the shafts of the gearset we will first assume that there is no power loss in the gearset; that is, the basic efficiency is 100%. In this case, we have (as in Equation 9.27)

$$T_l n_l + T_h n_h + T_a n_a = 0$$

Substitute the relative velocities from Equations (9.28) and (9.29)

$$T_l(\beta_l + n_a) + T_h(\beta_h + n_a) + T_a n_a = 0 \tag{9.30}$$

collect all of the n_a terms

$$T_l\beta_l + T_h\beta_h + (T_l + T_h + T_a)n_a = 0 \tag{9.31}$$

But the sum of torques is zero, as per Equation (9.24), so that

$$T_l\beta_l + T_h\beta_h = 0 \tag{9.32}$$

This is the situation when the efficiency of the gearset is 100%. We will shortly modify this equation to account for friction losses from the rubbing of gear teeth. Let us take as an example the situation where the input to the gearset is the low-speed gear, the output is the arm, and the high-speed gear is fixed. Since the load torque is on the arm, we assume that T_a is known and we wish to solve for the other two torques T_l and T_h. Again assuming 100% efficiency, we can substitute the values for β_l and β_h from Table 9.25 into Equation (9.32) to obtain

$$\frac{T_l n_l}{1-b} + \frac{T_h b n_s}{1-b} = 0 \tag{9.33}$$

or

$$T_l = -bT_h \tag{9.34}$$

TABLE 9.25

Relative Velocities for Various Configurations of the Planetary Gearset

Input	Output	Fixed	β_l	β_h
hi	arm	lo	$\frac{n_h}{b-1}$	$\frac{n_h b}{b-1}$
arm	hi	lo	$-n_a$	$-bn_a$
lo	arm	hi	$\frac{n_l}{1-b}$	$\frac{bn_l}{1-b}$
arm	lo	hi	$-\frac{n_a}{b}$	$-n_a$

Substituting this into Equation (9.24) we obtain

$$T_h = \frac{T_a}{b-1} \tag{9.35}$$

$$T_l = -\frac{T_a b}{b-1} \tag{9.36}$$

We can use a similar procedure to find the torques for the other configurations, and the results are shown in Table 9.26.

9.5.5 Efficiency of the Overall Gearset

We are now ready to begin computing the efficiency of the overall gearset. Let us define the efficiency of a planetary gearset as

$$\eta = \frac{P_{out}}{P_{in}} \tag{9.37}$$

where P_{out} is the power of the gearset taken at the output and P_{in} is the power provided to the gearset, usually by a motor. Calculating power losses in the planetary gearset is a little tricky, since many of the gears are rotating about their own axes as well as revolving about the sun gear or gears. We must remember that the only mechanism for power loss is through the frictional rubbing of the gear teeth against one another, so that only the *relative velocities* are important for computing power loss.

To make the concepts a little bit more concrete we will return to the example gearsets in Figures 9.22 and 9.23. Recall that the traditional planetary gearset had a basic ratio of $b = -2$ and the two-suns planetary gearset had a basic ratio of $b = 2$. For both examples, let us assume that the input to the gearset is the low-speed gear and the output is the arm. The high-speed gear is fixed in both cases.

Example 9.13: Traditional Planetary Gearset

First, allow the low-speed gear to make one revolution. Then, with $b = -2$, the relative speeds of the low-speed and high-speed gears are (from Table 9.25)

$$\beta_l = \frac{n_l}{1-b} = \frac{1}{3}$$

TABLE 9.26

Shaft Torques on Planetary Gearsets with 100% Efficiency

Input	Output	Fixed	T_l	T_h	T_a
hi	arm	lo	$-\frac{b}{b-1}T_a$	$\frac{1}{b-1}T_a$	T_a
arm	hi	lo	$-bT_h$	T_h	$(b-1)T_h$
lo	arm	hi	$-\frac{b}{b-1}T_a$	$\frac{1}{b-1}T_a$	T_a
arm	lo	hi	T_l	$-\frac{1}{b}T_l$	$-\frac{b-1}{b}T_l$

$$\beta_f = \frac{bn_s}{1-b} = -\frac{2}{3}$$

The balance of power relation in Equation (9.32) tells us that, with 100% efficiency

$$T_l\beta_l + T_h\beta_h = 0$$

Since power is entering through the low-speed shaft, its torque is positive, and the product $T_l\beta_l$ is also positive. The product $T_h\beta_h$ must then be negative, and power flows from the low-speed shaft to the high-speed shaft. In order to account for frictional losses we must reduce the input power by the basic efficiency, E_0. Therefore, we can write

$$E_0 T_l\beta_l + T_h\beta_h = 0 \tag{9.38}$$

Using Equations (9.24) and (9.38) to solve for the torques again, we have

$$T_l = -\frac{b}{b-E_0}T_a \tag{9.39}$$

$$T_h = \frac{E_0 T_a}{b-E_0} \tag{9.40}$$

And the efficiency is

$$\eta = \frac{P_{out}}{P_{in}} = \frac{T_a n_a}{T_l n_l} = \frac{b-E_0}{b-1} \tag{9.41}$$

If we use the value of $E_0 = 0.9694$ found in Example 9.11, we obtain

$$\eta = \frac{-2-0.9694}{-2-1} = 0.9898$$

or 98.98% efficient. The planetary gearset is more efficient than the internal gearset itself!

Example 9.14: Two Suns, Two Planets

We will now repeat the analysis for the planetary gearset with two suns and two planets. This time the basic ratio is positive, since $b = 2$.

$$\beta_l = \frac{n_l}{1-b} = -1$$

$$\beta_h = \frac{bn_l}{1-b} = -2$$

This is where the analysis becomes tricky! The torque input to the gearset, T_l, is positive, since we assume that the speed of the input gear, n_l is positive and the product $T_l n_l$ must be positive for power to enter the gearbox. However, the product $T_l\beta_l$ is negative, which would seem to indicate that frictional power is being *consumed* by the low-speed gear.

Now consider the power used in driving the load. The speed of the arm is (from Table 9.24)

$$n_a = \frac{bn_l}{b-1} = 2$$

which is positive. Thus, the load torque, T_a, must be negative, since the load *consumes* power, rather than producing it. From Table 9.26, we see that the torque on the high-speed gear (with 100% efficiency) is

$$T_h = \frac{1}{b-1}T_a = T_a \tag{9.42}$$

which is negative, since T_a is negative. This means that the product $T_h\beta_h$ is *positive*; that is, the high-speed gear *produces* frictional power. Thus, the frictional power flow within the planetary gearset is from the high-speed gear to the low-speed gear, in spite of the fact that the overall power enters through the low-speed gear. This is quite a lengthy chain of reasoning, and it may take a few readings before it becomes clear. The end result is this: we must revise the frictional power balance equation such that the source of the power is the high-speed gear, instead of the low-speed gear as it was with the traditional planetary gearset.

$$T_l\beta_l + E_0T_h\beta_h = 0 \tag{9.43}$$

Using Equations (9.24) and (9.43), we can solve for the torques on the high-speed and low-speed shafts as

$$T_l = -\frac{E_0 b}{E_0 b - 1}T_a \tag{9.44}$$

$$T_h = \frac{1}{E_0 b - 1}T_a \tag{9.45}$$

and the efficiency is

$$\eta = \frac{P_{out}}{P_{in}} = \frac{T_a n_a}{T_l n_l} = \frac{E_0 b - 1}{E_0(b-1)} \tag{9.46}$$

Using the value of $E_0 = 0.9624$ found in Example 2, we have

$$\eta = \frac{0.9624 \cdot 2 - 1}{0.9624(2-1)} = 0.9609$$

or 96.09% efficient. This planetary gearset is *less* efficient than the gears themselves, and is probably not a very good design.

A similar procedure can be conducted for the remaining planetary configurations and the resulting torques and efficiencies are shown in Table 9.27 for positive basic

TABLE 9.27

Shaft Torques and Efficiency for Planetary Gearsets with Positive Basic Ratio

Input	Output	Fixed	T_l	T_h	T_a	η
hi	arm	lo	$-\frac{bE_0}{bE_0-1}T_a$	$\frac{1}{bE_0-1}T_a$	T_a	$\frac{bE_0-1}{b-1}$
arm	hi	lo	$-\frac{b}{E_0}T_h$	T_h	$\frac{b-E_0}{E_0}T_h$	$\frac{(b-1)E_0}{b-E_0}$
lo	arm	hi	$-\frac{bE_0}{bE_0-1}T_a$	$\frac{1}{bE_0-1}T_a$	T_a	$\frac{bE_0-1}{E_0(b-1)}$
arm	lo	hi	T_l	$-\frac{E_0}{b}T_l$	$-\frac{b-E_0}{b}T_l$	$\frac{b-1}{b-E_0}$

TABLE 9.28

Shaft Torques and Efficiency for Planetary Gearsets with Negative Basic Ratio

Input	Output	Fixed	T_l	T_h	T_a	η
hi	arm	lo	$-\frac{bE_0}{bE_0-1}T_a$	$\frac{1}{bE_0-1}T_a$	T_a	$\frac{bE_0-1}{b-1}$
arm	hi	lo	$-\frac{b}{E_0}T_h$	T_h	$\frac{b-E_0}{E_0}T_h$	$\frac{(b-1)E_0}{b-E_0}$
lo	arm	hi	$-\frac{b}{b-E_0}T_a$	$\frac{E_0}{b-E_0}T_a$	T_a	$\frac{b-E_0}{b-1}$
arm	lo	hi	T_l	$-\frac{1}{bE_0}T_l$	$-\frac{bE_0-1}{bE_0}T_l$	$\frac{(b-1)E_0}{bE_0-1}$

ratios and Table 9.28 for negative basic ratios. It must be emphasized that these efficiencies are valid for comparing between competing designs, but not for calculating absolute efficiency values. Laboratory testing must be conducted if absolute efficiency values are needed.

9.6 Design Examples for Planetary Gearsets

We will now present four design examples using planetary gearsets. In each example, we wish to achieve a specific gear reduction with the highest possible efficiency. A design is considered complete if we have determined the numbers of teeth in each gear and the overall efficiency of the gearset.

Example 9.15: Speed Reducer in a Hand Drill

The goal for the first example is to design a planetary speed reducer for a hand drill using the traditional planetary configuration shown in Figure 9.24. The load torque on

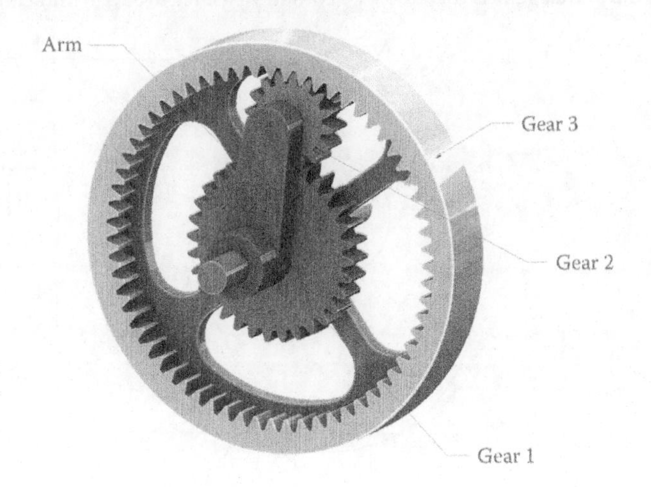

FIGURE 9.24
A traditional sun-ring-planet gearset is used to create speed reduction in a handheld drill.

the output is 5 Nm and the output speed is 1,000 rpm. The motor spins at 10,000 rpm. Find the gear tooth numbers to achieve the desired reduction and calculate the efficiency of the gearbox. Assume 20° pressure angle gears and a coefficient of friction of 0.05.

Solution

Let us use the sun as the input gear, the arm as the output gear, and leave the ring fixed. The sun is the high-speed gear and the ring is the low-speed gear, so that from Table 9.1 in Section 9.5 we find that the speed of the arm is

$$n_a = \frac{n_h}{1-b}$$

The desired speed reduction is 1/10, so that

$$\frac{n_a}{n_h} = \frac{1}{10} = \frac{1}{1-b}$$

Solving for b gives

$$b = -9$$

To calculate the basic ratio we must determine the speed of each gear with the arm held fixed. The speed of the planet is

$$n_2 = -\frac{N_1}{N_2} n_1$$

and the speed of the ring is

$$n_3 = \frac{N_2}{N_3} n_2 = -\frac{N_1}{N_3} n_1$$

or

$$\frac{n_1}{n_3} = \frac{n_f}{n_s} = -\frac{N_3}{N_1} = b$$

If we choose a sun gear with 16 teeth, the ring gear must have 144 teeth.

The size of the planet gear is dictated by geometry (see Figure 9.25) and can be found as

$$N_2 = \frac{N_3 - N_1}{2} = 64 \text{ teeth}$$

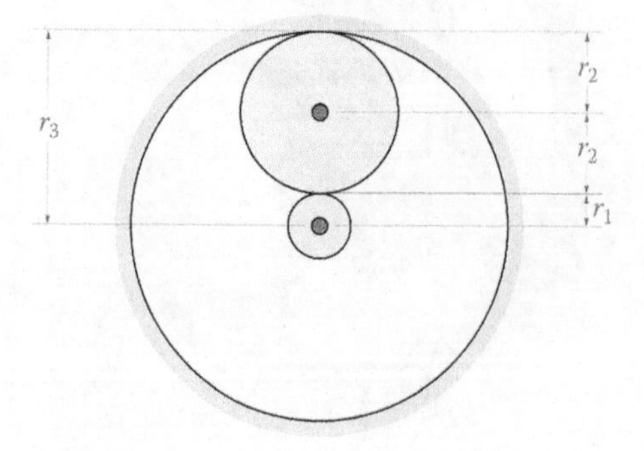

FIGURE 9.25
The number of teeth in the planet is fixed by the size of the ring and sun gears.

Consulting Figure 9.36 in Section 8.5, we see that the efficiency of the first gear pair is

$$E_1 = 1 - \mu L_1 = 1 - 0.05 \cdot 0.18 = 0.991$$

and the second pair is

$$E_2 = 1 - \mu\left(\frac{\rho_2 - 1}{\rho_2 + 1}\right)L_2 = 1 - 0.05\left(\frac{2.25 - 1}{2.25 + 1}\right)0.06 = 0.9988$$

Thus, the basic efficiency is

$$E_0 = E_1 E_2 = 0.9899$$

Using the first row of Table 9.6, we see that the efficiency of the gearset is

$$\eta = \frac{bE_0 - 1}{b - 1} = 0.9909$$

or 99.1% efficient. The required motor torque is

$$T_h = \frac{1}{bE_0 - 1}T_a = \frac{1}{-9 \cdot 0.9899 - 1}(-5\text{ Nm}) = 0.50\text{ Nm}$$

where we have used a negative load torque because the arm spins in the positive direction.

Example 9.16: A Modified Speed Reducer

Let us repeat the problem given in Example 9.15, using the configuration shown in Figure 9.26. Again, the load torque is 5 Nm at 1,000 rpm and the sun gear spins at 10,000 rpm.

Solution

There are two major differences between this gearset and Example 9.15: the numbers of teeth in the gears will be different as will the basic efficiency. We can still use the basic

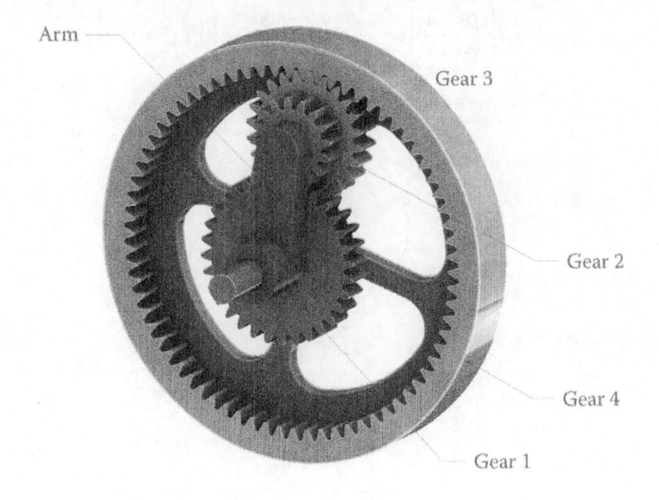

FIGURE 9.26
The planetary configuration used in Example 9.2. Here there are two planets, instead of just one.

ratio $b = -9$, since the input and output requirements are the same. If we rotate gear 1 (the sun) then the first planet spins at

$$n_2 = -\frac{N_1}{N_2} n_1$$

and the second planet spins at the same speed. The ring, gear 4, spins at

$$n_4 = \frac{N_3}{N_4} n_3 = -\frac{N_1}{N_2} \cdot \frac{N_3}{N_4} n_1$$

or

$$\frac{n_1}{n_4} = \frac{n_h}{n_l} = -\frac{N_2}{N_1} \cdot \frac{N_4}{N_3} = -9$$

To achieve this reduction, it is simplest to let each stage have a reduction of 3. If we again let the sun gear have 16 teeth, then the first planet will have 48 teeth.

The dimensions of the planetary gearset in Example 9.16 are shown in Figure 9.27. If we assume that all gears have the same diametral pitch (or module) then we have

$$N_4 = N_1 + N_2 + N_3$$

But the ratio of the ring gear to the second planet must be $N_4/N_3 = 3$, so that we have

$$N_3 = 32 \text{ teeth}$$

$$N_4 = 96 \text{ teeth}$$

Thus, the ring gear is more compact in this example than for the configuration in Example 9.15. The basic efficiency of the first gear pair is

$$E_1 = 1 - \mu L_1 = 1 - 0.05 \cdot 0.2 = 0.99$$

and the second pair is

$$E_2 = 1 - \mu\left(\frac{\rho_2 - 1}{\rho_2 + 1}\right) L_2 = 1 - 0.05\left(\frac{3-1}{3+1}\right) 0.115 = 0.9971$$

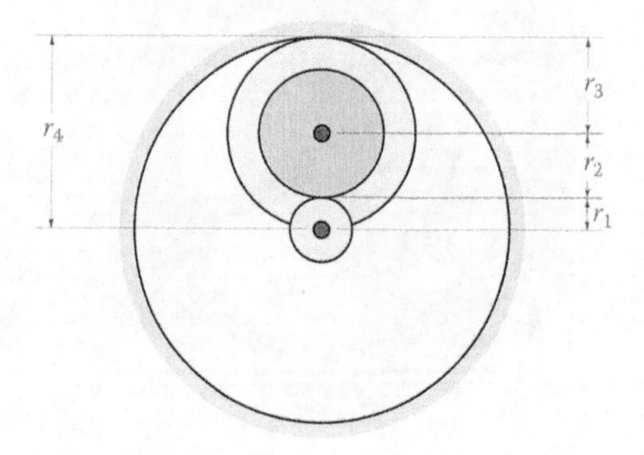

FIGURE 9.27
Dimensions of planetary gearset in Example 9.2.

so that the basic efficiency is

$$E_0 = E_1 E_2 = 0.9872$$

The overall efficiency of the gearset is then

$$\eta = \frac{bE_0 - 1}{b - 1} = 0.9885$$

or 98.85% efficient. This is slightly less efficient than the gearset in Example 9.15, but is smaller in size. There is always a tradeoff!

Example 9.17: Two Suns, Two Planets

For this example, we wish to design a planetary gearset with a very large reduction – 49:1 – in a compact space. As a first attempt, we will try the configuration shown in Figure 9.28, which has two suns and two planets. This gearset has a positive basic ratio since gear 4 rotates in the same direction as gear 1 when the arm is fixed. Let the input to the gearset be the arm and the output be the low-speed gear, with the high-speed gear fixed. From the fourth row of Table 9.1, we see that

$$\frac{n_s}{n_a} = \frac{1}{49} = \frac{b - 1}{b}$$

Solving for b gives

$$b = \frac{49}{48}$$

which is only slightly greater than unity. We must achieve this reduction in two stages, with the first stage comprised of gears 1 and 2, and the second stage comprised of gears 3 and 4. Holding the arm fixed allows us to calculate the basic ratio as

$$b = \frac{N_2}{N_1} \cdot \frac{N_4}{N_3} = \frac{49}{48}$$

FIGURE 9.28
The two suns, two planets configuration for Example 9.3.

The basic ratio has a convenient factorization

$$b = \frac{7}{8} \cdot \frac{7}{6} = \frac{49}{48}$$

so that we will use a ratio of 7/8 in the first stage and 7/6 in the second stage. Let the number of teeth in gears 2 and 4 be 28. Then $N_1 = 32$ and $N_3 = 24$. In consulting Figure 9.36, in Section 8.5, the basic efficiency of each gear pair is

$$E_1 = 1 - \mu L_1 = 1 - 0.05 \cdot 0.15 = 0.9925$$

$$E_2 = 1 - \mu L_2 = 1 - 0.05 \cdot 0.17 = 0.9915$$

so that the basic efficiency is

$$E_0 = E_1 E_2 = 0.9841$$

The overall efficiency of the gearset is

$$\eta = \frac{b-1}{b-E_0} = 0.5672$$

or 56.7% efficient. Almost half of the power in this gearset is lost to friction, so it is an unfeasible design.

Example 9.18: An Alternative Design

As an alternative design to Example 9.17, consider the gearset shown in Figure 9.29, also shown in cross-section in Figure 9.30. There are two planets and two ring gears. As in the previous example the input is the arm, the output is gear 4 and gear 2 (the high-speed gear) is fixed. Since we desire the same reduction as before, the basic ratio remains $b = 49/48$. We will most likely need larger numbers of teeth, since two of the gears are internal. The ratios at each stage can remain the same, however, let us try $N_1 = 80$, $N_2 = N_4 = 70$, $N_3 = 60$ teeth.

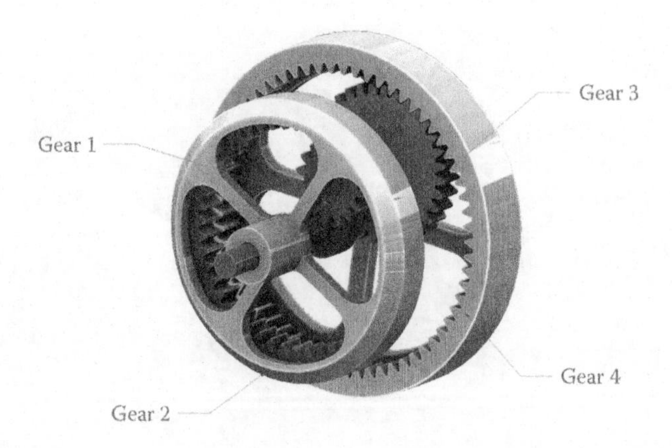

FIGURE 9.29
An alternative design based upon two ring gears and two planets. The input is the arm, the output is gear 4, and gear 2 (the high-speed gear) is fixed.

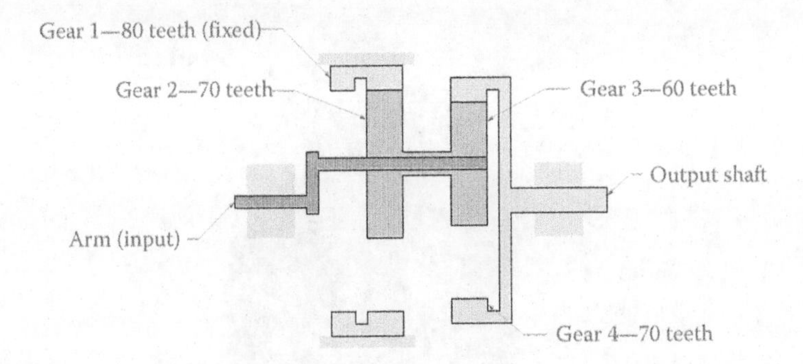

FIGURE 9.30
Cross-section of the gearset in Example 9.4. The first ring gear is fixed and the output is taken from the second ring gear. The input is the arm.

Then the basic efficiency at each stage is

$$E_1 = 1 - \mu\left(\frac{\rho_1 - 1}{\rho_1 + 1}\right)L_1 = 1 - 0.05\left(\frac{8/7 - 1}{8/7 + 1}\right)0.07 = 0.9998$$

$$E_2 = 1 - \mu\left(\frac{\rho_2 - 1}{\rho_2 + 1}\right)L_2 = 1 - 0.05\left(\frac{7/6 - 1}{7/6 + 1}\right)0.08 = 0.9997$$

The basic efficiency is

$$E_0 = E_1 E_2 = 0.9995$$

which is much higher than in Example 9.17 because we have employed internal gears. The overall efficiency is

$$\eta = \frac{b - 1}{b - E_0} = 0.9747$$

or 97.5% efficient. This is a much more feasible design than Example 9.17, and this example demonstrates the importance of performing the efficiency calculations when designing a planetary gearset. The overall efficiency of the gearset appears to be extremely sensitive to the basic efficiency in this design since the basic ratio, b, is so close to one. A very slight decrease in basic efficiency – as might be caused by a change in temperature, wear of the gears, or breakdown in the lubricant – will have a very dramatic and deleterious effect on the overall efficiency of the gearset. In practice, therefore, the high reduction should probably be achieved using multiple stages.

9.7 Practice Problems

Problem 9.1

The planetary gearset in Figure 9.31 has two suns and four planets. If all gears have module 2 mm/tooth, how many teeth are on the second sun? What are the distances from the central shaft to the first and second planetary shafts?

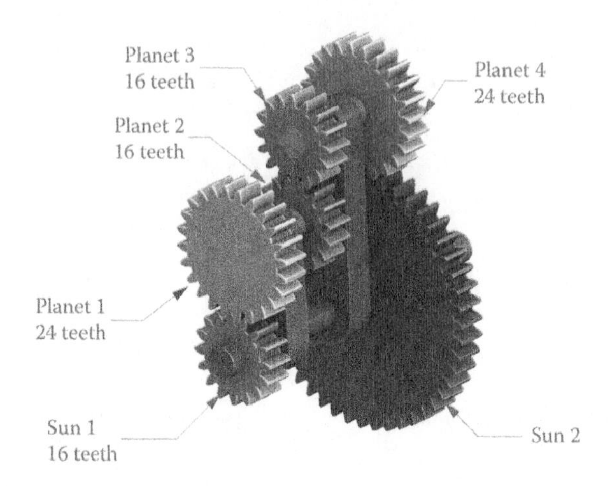

FIGURE 9.31
Problem 9.1.

Problem 9.2

The planetary gearset in Figure 9.32 has a sun, ring, and four planets. Find the number of teeth in the ring if all gears have module 5 mm/tooth. What are the distances from the central shaft to the first and second planetary shafts?

Problem 9.3

The first sun in the planetary gearset in Problem 9.1 rotates at 1,000 rpm clockwise while the second sun is fixed. What is the speed and direction of the arm? Use the first tabular method presented in Section 9.2.

Problem 9.4

The arm in the planetary gearset of Problem 9.2 rotates at 100 rpm clockwise while the sun gear is fixed. What is the speed and direction of the ring? Use the first tabular method presented in Section 9.2.

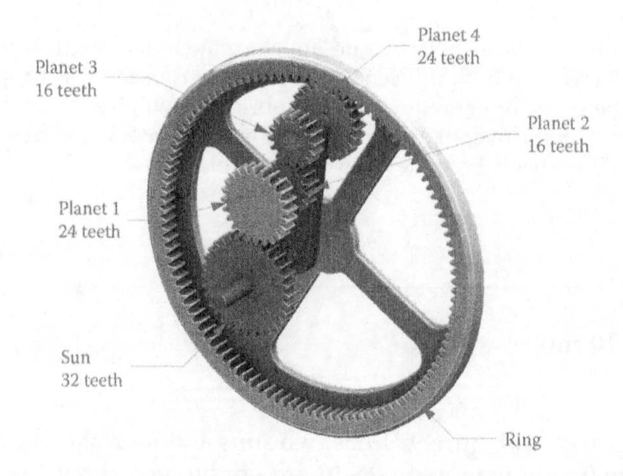

FIGURE 9.32
Problem 9.2.

Problem 9.5

In Figure 9.33, the second sun gear rotates at 500 rpm ccw while the first sun is fixed. What is the speed of the arm? Use the first tabular method presented in Section 9.2.

Problem 9.6

In Figure 9.34, the sun gear rotates at 500 rpm cw and the ring rotates at 500 rpm ccw. What is the speed of the arm? Use the second tabular method presented in Section 9.3.

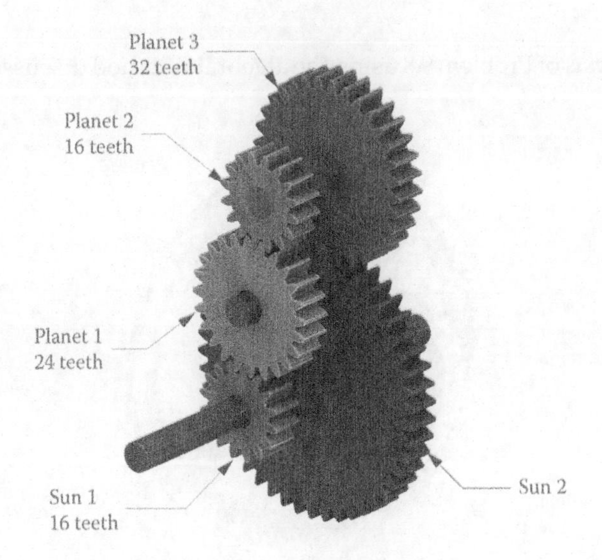

FIGURE 9.33
Problem 9.5.

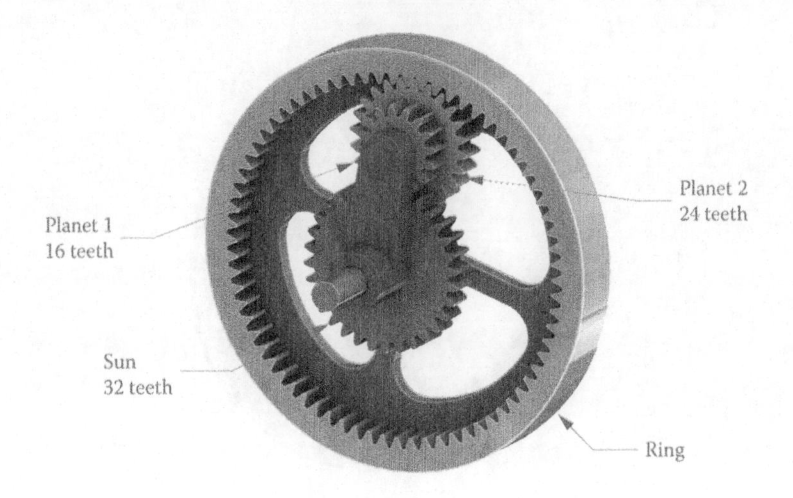

FIGURE 9.34
Problem 9.6.

Problem 9.7

In Figure 9.35, the first ring gear rotates at 200 rpm cw and the arm rotates at 100 rpm ccw. What is the speed of the second ring? Use the second tabular method presented in Section 9.3.

Problem 9.8

In Figure 9.36, the sun gear rotates at 1,000 rpm cw and the ring rotates at 1,000 rpm ccw. What is the speed of the arm? Use the second tabular method presented in Section 9.3.

Problem 9.9

Repeat the analysis of Problem 9.8 using the algebraic method discussed in Section 9.4.

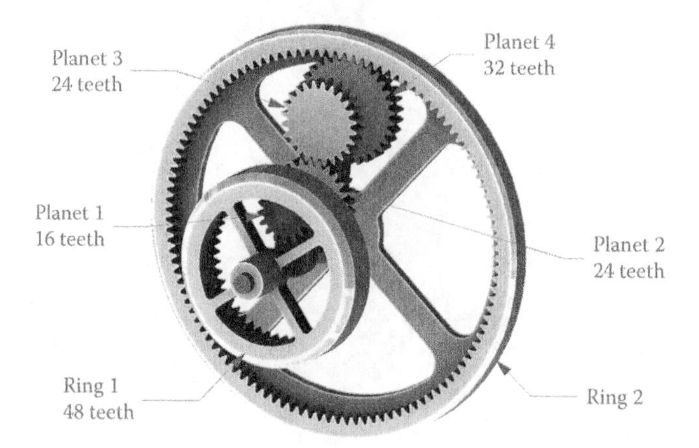

FIGURE 9.35
Problem 9.7.

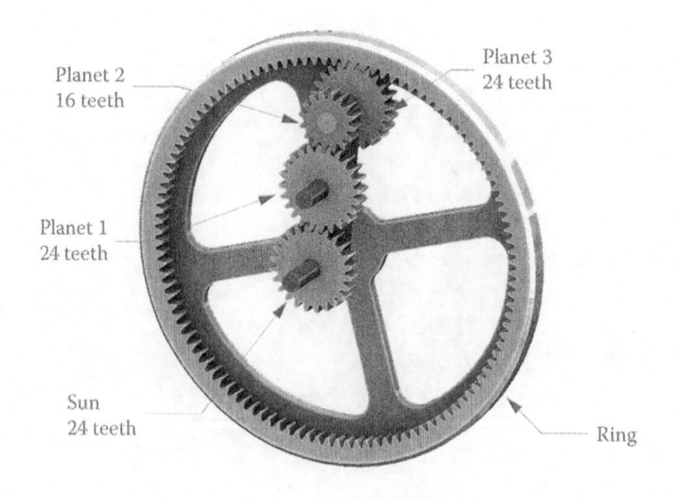

FIGURE 9.36
Problem 9.8.

Problem 9.10

Repeat the analysis of Problem 9.7 using the algebraic method discussed in Section 9.4.

Problem 9.11

The traditional planetary gearset shown in Figure 9.37 has a sun, ring, and single planet. The sun and planet both have 20 teeth. The input to the gearset is a servomotor, which can be attached to either the ring or sun, and the output is the arm. The design requires that the output rotate 135° when the servo rotates 180°. Determine which gear should be fixed and where to attach the servomotor in order to accomplish this reduction.

Problem 9.12

The planetary gearset shown in Figure 9.38 has a sun, ring, and two planets. It is desired that the arm move at 80 rpm clockwise when the sun moves at 1,680 rpm clockwise. The ring is fixed. Find the numbers of teeth on each gear to accomplish this motion using the tabular method presented in Section 9.2. Assume that the minimum allowable number of teeth on a gear is 16.

Problem 9.13

The planetary gearset shown in Figure 9.39 has a sun, ring, and two planets. The ring is attached to a shaft that spins at 500 rpm clockwise and the sun is attached to a motor spinning 2,000 rpm clockwise. The arm must spin at 1,000 rpm clockwise.

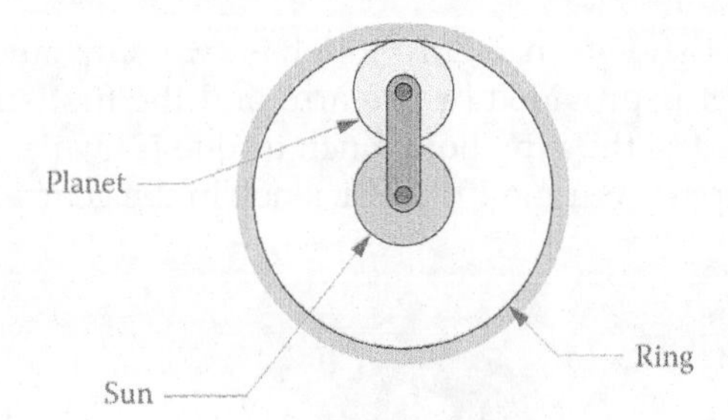

FIGURE 9.37
Problem 9.11.

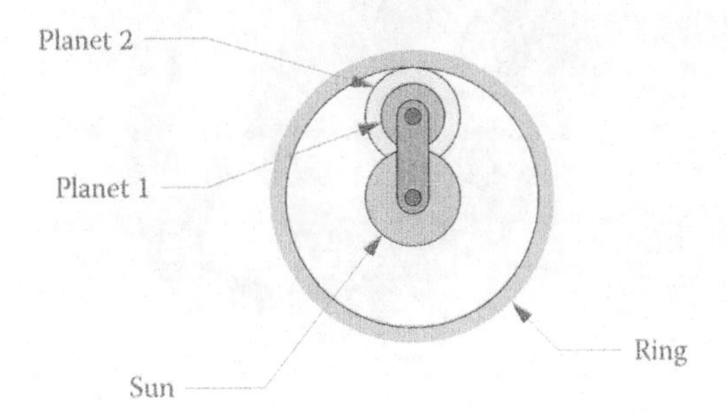

FIGURE 9.38
Problem 9.12.

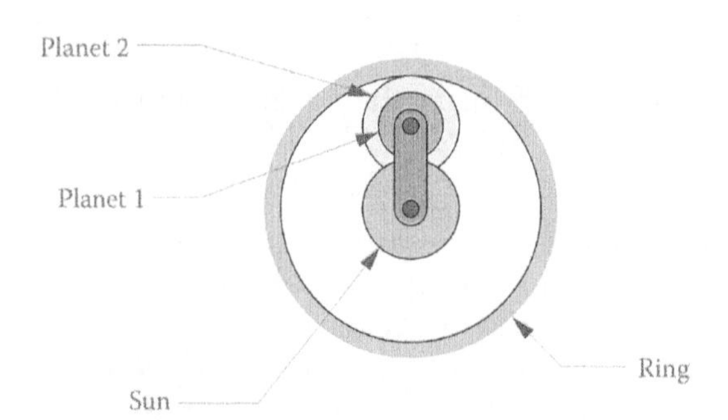

FIGURE 9.39
Problem 9.13.

Find the numbers of teeth on each gear to accomplish this motion using the tabular method presented in Section 9.2. Assume that the minimum allowable number of teeth on a gear is 16.

Problem 9.14

The planetary gearset shown in Figure 9.40 has two rings and three planets. The first ring spins at 1,000 rpm and the second ring is fixed. The arm drives a load torque of 20 Nm. Find the speed of the arm and input torque required to drive the load, assuming 100% efficiency in the gearset.

Problem 9.15

The planetary gearset shown in Figure 9.41 has two suns and three planets. The input to the gearset is provided by the arm and the first sun is fixed. If a 1,000 W motor is attached to the arm, how much torque is available at the second sun, which spins at 50 rpm? Assume 100% efficiency in the gearset.

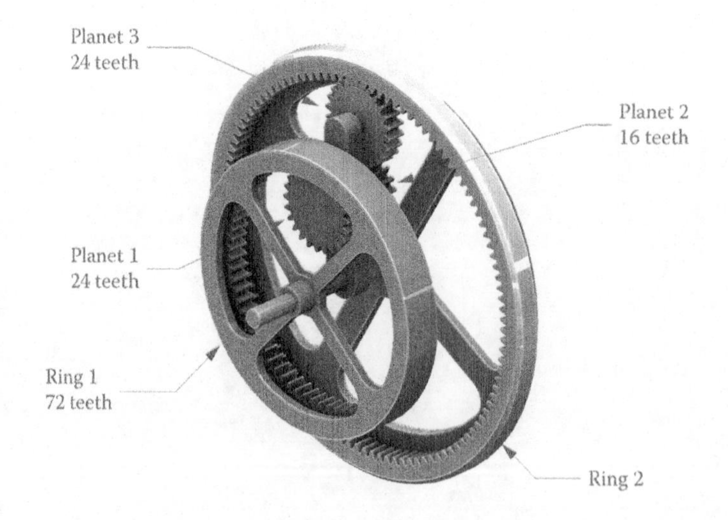

FIGURE 9.40
Problem 9.14.

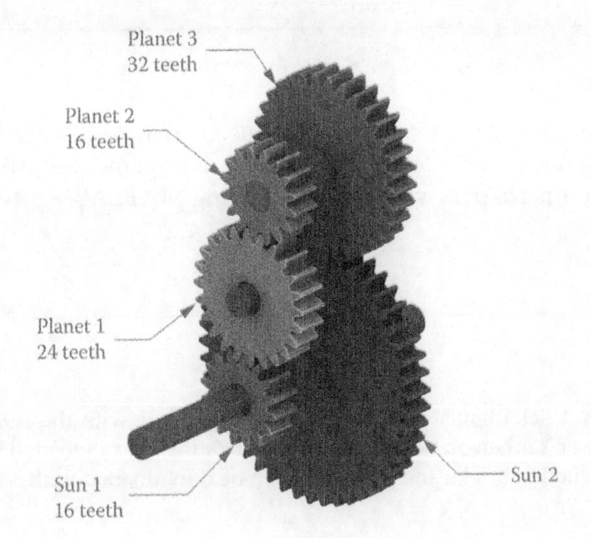

FIGURE 9.41
Problem 9.15.

Problem 9.16

Two designs are under consideration for a speed reducer, as shown in Figure 9.42. Both designs have a sun, ring, and three planets. In each case, the sun is the input and the output is taken at the arm. The desired overall speed reduction is 19, and the sun and arm move in opposite directions. Determine the number of teeth in each gear in both designs. Estimate the efficiency of each design assuming a coefficient of friction of 0.05 and a pressure angle of 20°.

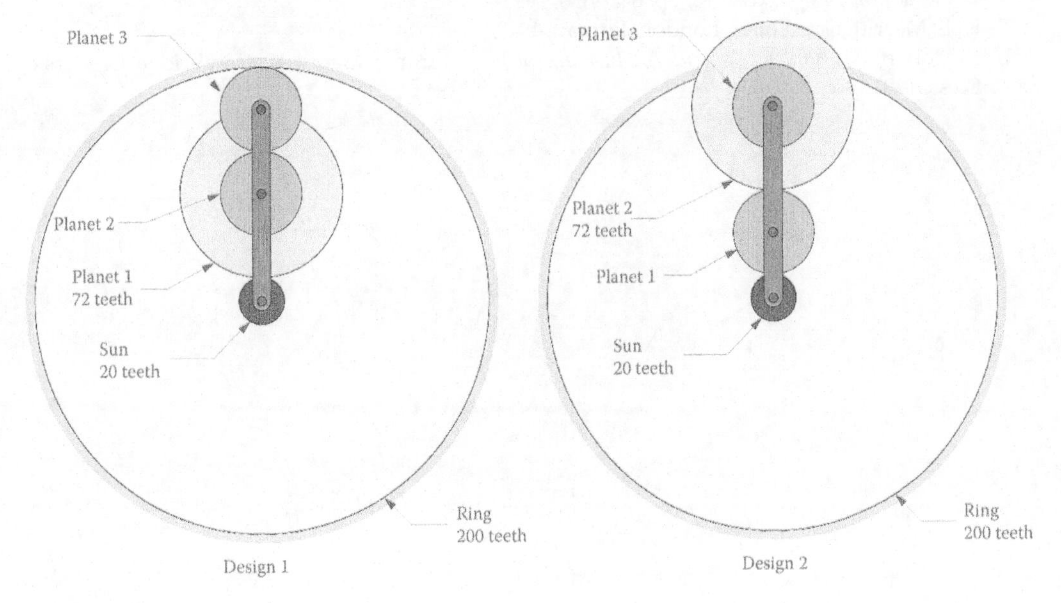

FIGURE 9.42
Problem 9.16.

Acknowledgments

Several images in this chapter were produced using SOLIDWORKS® software. SOLIDWORKS® is a registered trademark of Dassault Systèmes SolidWorks Corporation.

Several images in this chapter were produced using MATLAB® software.

Notes

1. The construction brick images used in this text were made with the SOLIDWORKS® models created by the user Yauhen on www.grabcad.com. Yauhen has uploaded hundreds of first-rate models of construction bricks and gears that can be downloaded with a free user account.

Works Cited

1. D. C. Talbot, *An experimental and theoretical investigation of the efficiency of planetary gear sets*, Ohio State University, 2012.
2. J. M. del Castillo, "The analytical expression of the efficiency of planetary gear trains," *Mechanism and Machine Theory*, vol. 37, pp. 197–214, 2002.
3. L.-C. Hsieh and H.-C. Tang, "On the meshing efficiency of 2K-2H type planetary gear reducer," *Advances in Mechanical Engineering*, vol. 2013, 2013.
4. E. Pennestri and F. Freudenstein, "The mechanical efficiency of epicyclic gear trains," *Journal of Mechanical Design*, vol. 115, pp. 645–650, 1993.
5. H. E. Merritt, *Gear Trains*, London: Pitman, 1947.
6. S. Molian, *Mechanism Design: An Introductory Text*, Cambridge, UK: Cambridge University Press, 1982.

10

Cams and Followers

10.1 Introduction to Cams

Cams are used in a very wide variety of mechanisms ranging from the valve mechanism in an automotive engine to sash locks in traditional double-hung windows. Cams are ordinarily used to convert rotary motion to linear (or almost linear) motion. The most common type of cam is the rotary cam shown in Figure 10.1. As the cam rotates, the follower is constrained to rise up and down, "following" the surface of the cam. It is much more expensive and time-consuming to manufacture a precision cam than it is to fabricate ordinary links, so their use is mostly restricted to situations where linkages cannot achieve the required motion (e.g. the valve-lifting mechanism in an automotive engine). To put this in concrete terms, you can manufacture a halfway decent linkage using a bandsaw and milling machine (or drill press, in desperation). By contrast, manufacturing a modern cam requires a CNC (computer controlled) milling machine or lathe, a CNC grinder for smoothing the profile, and heat-treating for hardening the surface. In other words, the versatility of the cam comes at a significant price.

10.1.1 Types of Cams

The most common type of cam is the *radial cam,* or *plate cam,* shown in Figure 10.1. A radial cam is a two-dimensional body and the follower moves along its outer periphery. Another common type of cam is the cylindrical cam shown in Figure 10.2. As the cylinder rotates, the follower moves in a direction parallel to the axis of the cylinder. This type of cam is common in motorcycle transmissions and lathe gearboxes. If you open up the housing of a lathe in your school's shop (with permission of the lab supervisor!) you will find a collection of cylindrical cams that serve to slide individual gears in and out of mesh for regulating the feed rate of the lathe. Another type of cam found on most lathes is the face cam, shown in Figure 10.3. This cam is a variant on the radial cam, but the follower moves along grooves (or teeth) cut into the face of the cam. Most lathe chucks use this type of cam to hold the workpiece securely. The chuck key serves to rotate the spiral, which causes the jaws of the chuck to move inward or outward. From the diagram it is easy to understand why the jaws of the chuck are numbered. As one moves around the spiral the distance to the center changes, so each jaw must be made slightly different in order to accommodate this.

Because radial cams are the most common they will be the subject of our discussion for the remainder of the chapter. Since they are often used in high-speed machinery (e.g. automotive engines) dynamic analysis of radial cams is much more important than for cylindrical or face cams, which are more often used in low-speed applications.

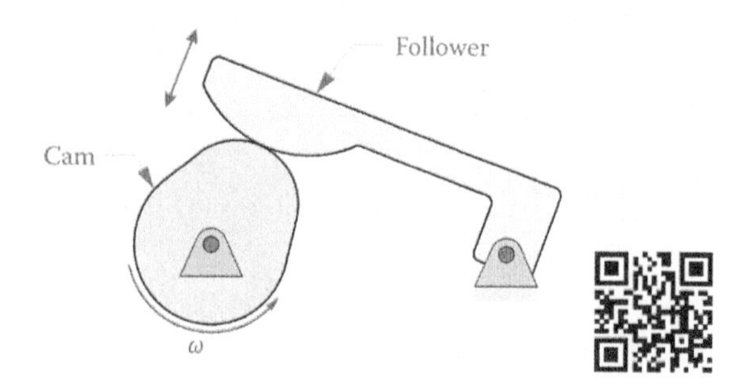

FIGURE 10.1
A basic cam-follower mechanism. As the cam rotates the follower "follows" the surface of the cam by moving up and down.

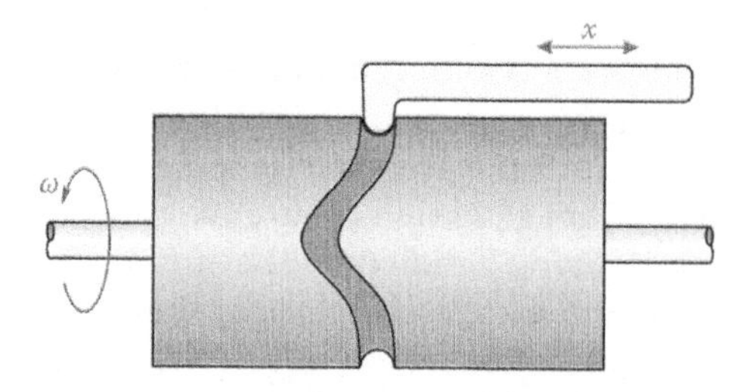

FIGURE 10.2
With the cylindrical cam the follower moves along an axis parallel to the axis of the cam.

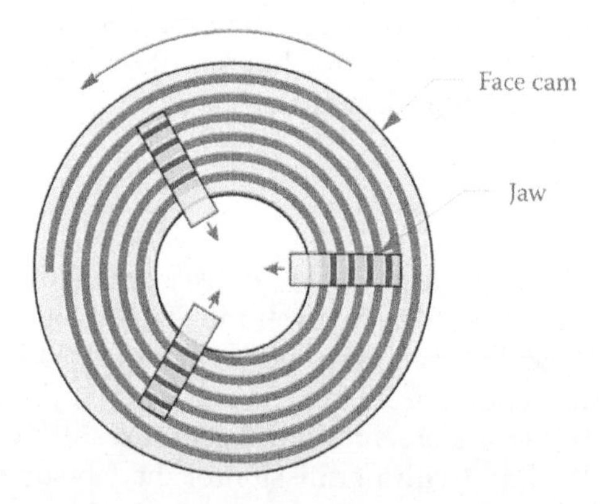

FIGURE 10.3
With a face cam the follower moves in the same plane as that of the cam, as shown in this spiral chuck.

10.1.2 Follower Motion

In general, there are two types of motion available to the follower. As shown in Figure 10.4, the follower can be configured as a rocker (left) or a slider (right). The rocker type permits mechanical advantage to be achieved through judicious design of the rocker.

Figures 10.5–10.7 show three of the more common valve-actuating mechanisms used in automobile and motorcycle engines. As shown in Figure 10.5, the direct-acting mechanism has the cam in direct contact with the valve. While this is the simplest mechanism, it limits placement of the camshaft to directly over the cylinder. This arrangement is used mainly in motorcycle engines owing to its simplicity and light weight. Observe the spring that is

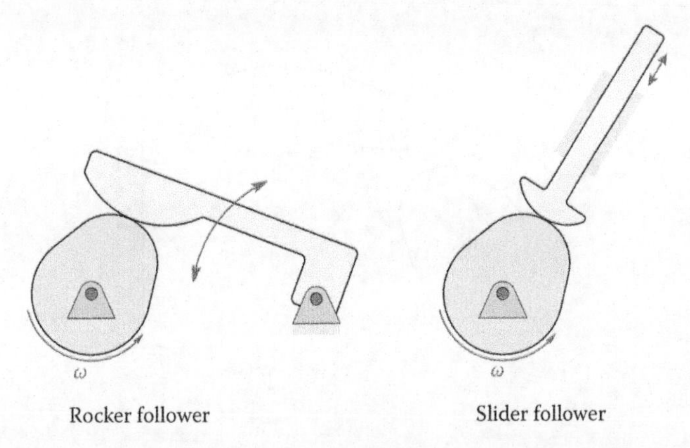

FIGURE 10.4
The follower can take the form of a rocker or a slider, depending upon the application.

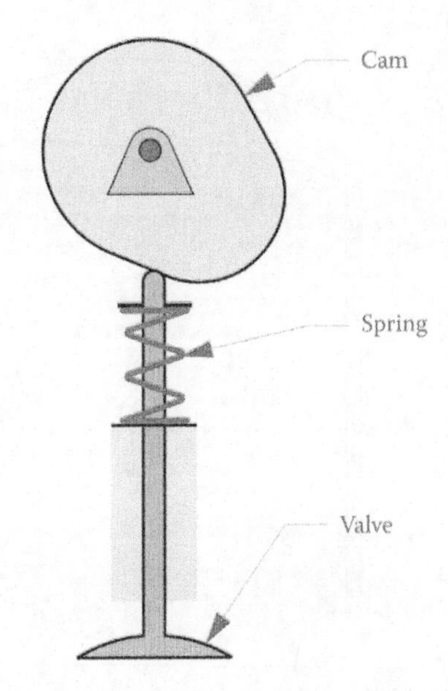

FIGURE 10.5
In a direct-acting mechanism the cam contacts the valve stem directly. The spring is used to keep the valve stem in contact with the face of the cam.

used to keep the valve in contact with the cam. Some motorcycles (e.g. Ducati) replace this spring with a second cam that pushes the valve back up when it is being closed. The two-cam system is called "desmodromic," which denotes the positive connection that the valve has with the camshaft.

Figure 10.6 shows the rocker mechanism that is used in most modern cars because it allows the placement of more than one valve per cylinder, which permits easier air-flow into and out of the cylinder. The dual overhead cam arrangement has two camshafts, one on each side of the cylinder, and each camshaft actuates a bank of rockers and valves.

Many older engines used the pushrod mechanism for actuating the valves, as shown in Figure 10.7. The pushrod permits the camshaft to be located below the cylinders, which simplifies the placement of the timing belt or chain. Since it also lowers the overall center

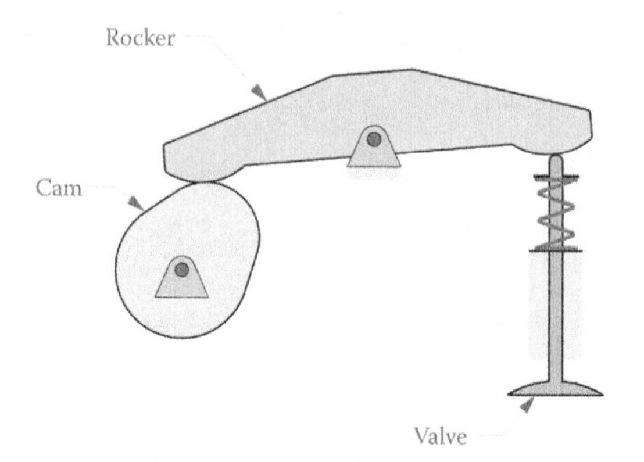

FIGURE 10.6
Most overhead valve mechanisms employ a rocker to actuate the valve. This allows more flexibility in locating the camshaft.

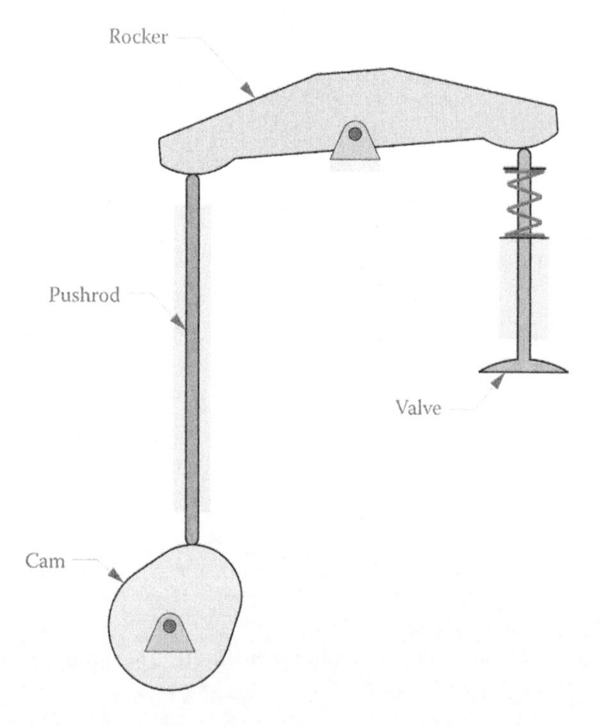

FIGURE 10.7
Many older engines (and the Corvette Z06) use pushrods to actuate the rocker. This allows the camshaft to reside below the cylinders, which may help to reduce the overall size of the engine.

of mass of the engine, it is still employed in some modern high-performance vehicles like the Z06 Corvette.

10.1.3 Types of Followers

There are three common ways for a follower to interact with a cam, as shown in Figure 10.8. The mushroom and flat-faced follower slide directly against the cam; these require continual lubrication to reduce friction and heat buildup. The roller follower avoids friction

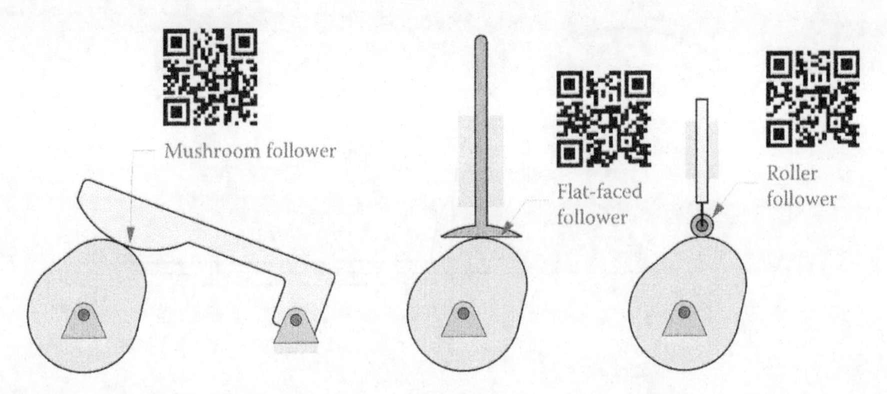

FIGURE 10.8
The three-most common types of follower. The mushroom and flat-faced followers experience sliding friction and must be well lubricated.

by rolling against the cam. Ball bearings and needle bearings are commonly used in this application. Most newer automotive engines use roller followers to reduce friction and wear on the valve stem.

10.2 Eccentric Cams

The simplest type of cam is an eccentric cam, as shown in Figure 10.9. An eccentric cam is simply a circle that is pinned to ground at a point not at its center. The distance from the center of the circle to the ground pivot is the *eccentricity*, shown as b in the diagram. If the eccentric cam is mated to a flat-faced sliding follower, the location of the face of the follower is given by

$$x = r + b\cos\theta \tag{10.1}$$

where θ is the angle of rotation of the cam. If we assume that the cam is rotating at a constant angular velocity, ω, then

$$\theta = \omega t \tag{10.2}$$

and

$$x = r + b\cos\omega t \tag{10.3}$$

We can calculate the velocity and acceleration of the follower by taking the time derivatives of the displacement

$$v = -b\omega\sin\omega t \tag{10.4}$$

$$a = -b\omega^2\cos\omega t \tag{10.5}$$

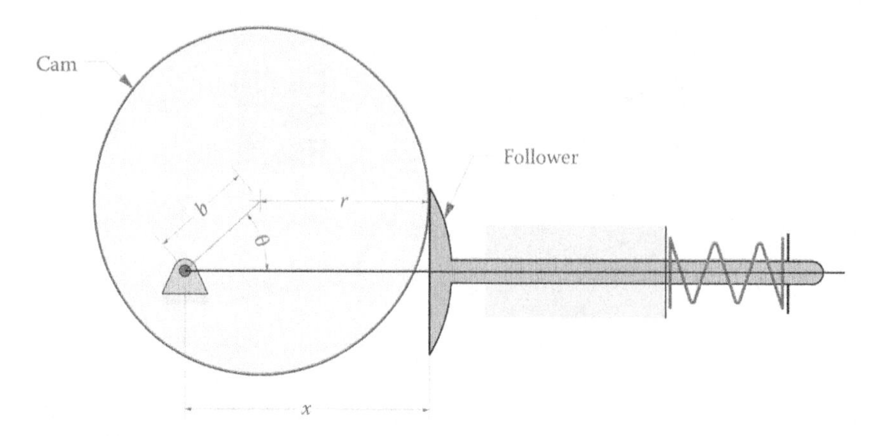

FIGURE 10.9
An eccentric cam is simply a circle that rotates about a point not at its center.

The follower experiences a smooth, continuous back and forth motion with minimum displacement $r - b$ and maximum displacement $r + b$. The motion analysis for this type of cam is obviously very simple, and a force analysis would also be quite easy. Unfortunately, this type of cam has only limited application since it is constrained to purely sinusoidal motion. In many applications, we require that the follower "dwell" at a position for a given portion of the cam rotation. The follower of an eccentric cam is in constant motion except for the instants in time where it is at the top and bottom of its stroke. For this reason, we will not further discuss the eccentric cam, and force analysis on the eccentric is left as a homework exercise.

10.3 Cams in an Automotive Engine

One of the most widespread applications of cams is in automotive engines where they are used as part of the valve-lifting mechanism. A simple schematic of a pushrod-type valve lifting mechanism is shown in Figure 10.10. The cam is used to raise and lower the pushrod, which actuates the rocker. The rocker is used to open and close the valve. Two valves are shown in the cylinder. One valve, called the *intake valve*, allows the fuel-air mixture to enter the cylinder. The *exhaust valve* lets the products of combustion escape the cylinder after combustion is complete. The valve springs have been omitted for clarity. The timing of the opening and closing of the valves is critical to proper operation of the engine; this type of timing would be difficult to achieve with a conventional linkage. The cam is attached to the *camshaft*, which has the cams for the other cylinders mounted to it (on a multicylinder engine). Similarly, the crank is mounted to the *crankshaft* along with the cranks for the other pistons.

All modern automotive engines use a variant of the four-stroke cycle shown in Figure 10.11. The first stage in the cycle is *intake*, where the fuel-air mixture is drawn into the cylinder by the partial vacuum created when the piston moves downward. The intake valve must be open during this part of the cycle to allow the mixture to flow in.

The next stage is *compression*, when the fuel-air mixture is compressed in the cylinder by the piston moving upward. Both valves are closed during this part of the cycle. The

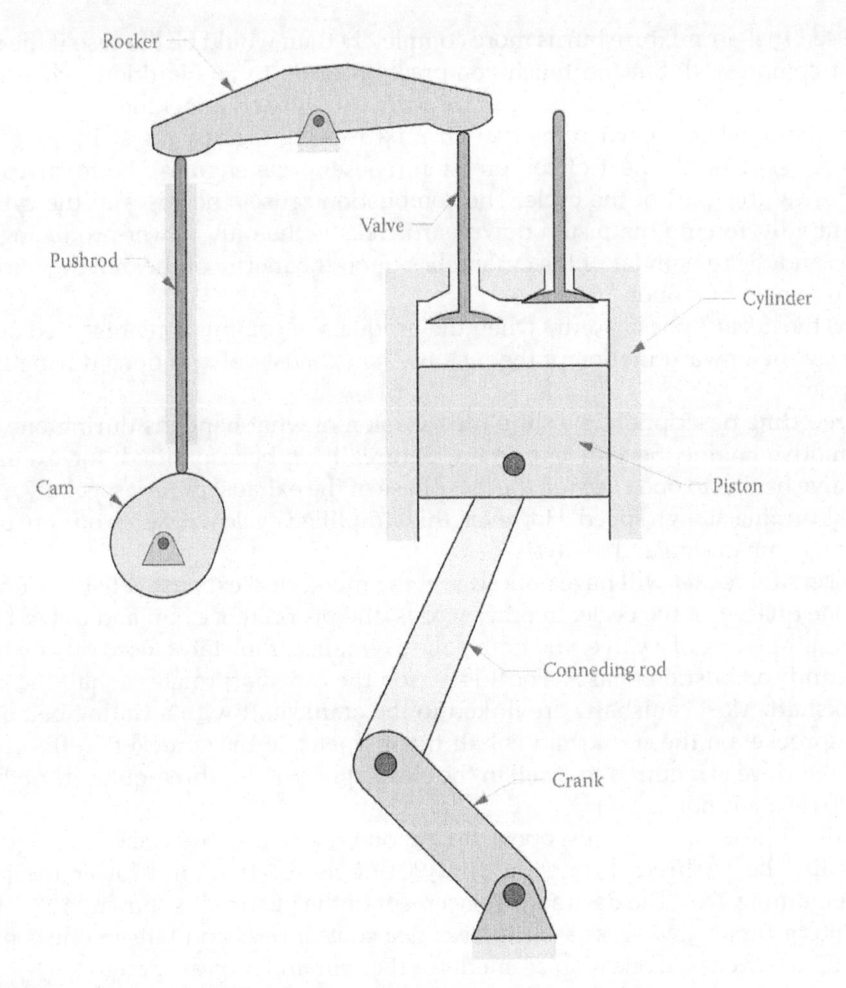

FIGURE 10.10
A very rough schematic of a pushrod-type automotive engine. Two valves (intake and exhaust) are shown, along with the valve-lifting mechanism and the slider-crank linkage.

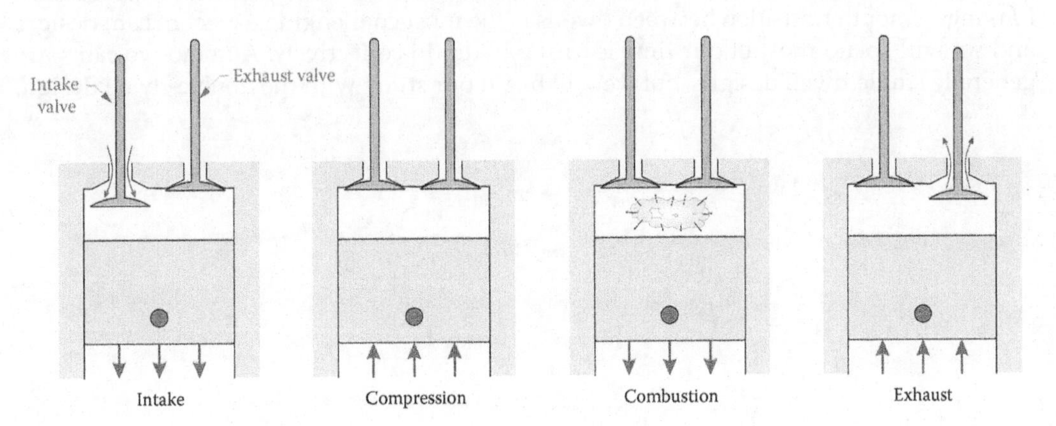

FIGURE 10.11
The four parts of the four-stroke cycle. This is the cycle used in automotive engines, most motorcycle engines and some lawnmower engines.

compressed fuel-air mixture burns more completely than would be the case if the mixture were not compressed, but too much compression leads to autoignition (when the heat created by compression ignites the mixture prematurely) and knocking.

Next comes *combustion*, when the fuel-air mixture is ignited by the spark plug (in Otto cycle engines) or by the heat of compression (Diesel cycle engines). Both valves remain closed during this part of the cycle. The combustion causes the gases in the cylinder to expand rapidly, forcing the piston downward. This is the only power-producing part of the cycle, and the remainder of the cycle relies upon the inertia of the moving parts of the engine to remain in motion.

Finally, the *exhaust* phase occurs when the products of combustion are forced out of the cylinder by the upward motion of the piston. The exhaust valve is open during this part of the cycle.

The preceding description is a simplified version of what happens during one cycle of an automotive engine. Several important features have been omitted; for example, the intake valve begins to open during the last phase of the exhaust cycle, especially when the engine is running at high speed. However, this simplified cycle will serve our purposes for introducing cam design and analysis.

The observant reader will have noticed that the intake and exhaust valves are only open during one quarter of the cycle. In other words, the piston moves up and down twice for every single opening of a valve, and both valves remain completely closed during the compression and combustion phases. For this reason the camshaft rotates at half the speed of the crankshaft. Most camshafts are linked to the crankshaft with a timing belt or chain, and the sprocket on the crankshaft is half the diameter of the sprocket on the camshaft. Because the valve is required to dwell in the closed position for three-quarters of the cycle, an eccentric cam is not suitable.

Since the intake valve is only open during one-quarter of the cycle, we require that the cam lift the pushrod during the first 90° of its rotation, and lower the pushrod for the remaining 270°. The desired displacement of the pushrod is shown in Figure 10.12. The resulting cam is known as a *single dwell cam* since it rises and falls in one continuous motion, then "dwells" at one displacement for the remainder of the cycle.

The more general case of a multiple-dwell displacement is shown in Figure 10.13. For this cam the follower remains stationary at different locations at multiple points during the cycle. The transitions between dwell states are achieved through *rise* and *fall* sections. Making a smooth transition between dwells is the most challenging aspect of cam design, and we will spend most of our time learning to do this effectively. Automotive cams are generally single-dwell designs, but we will begin our study with the double-dwell design,

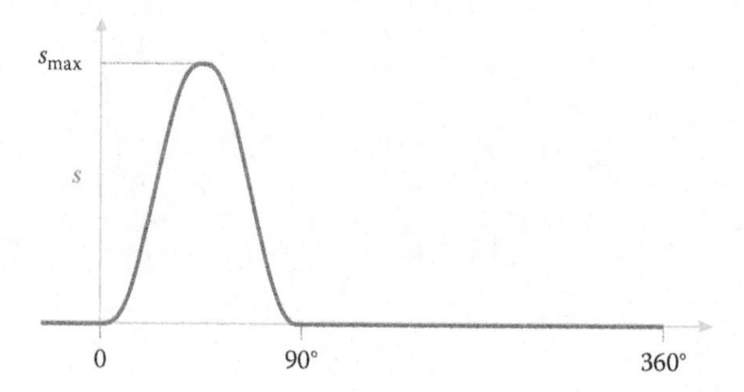

FIGURE 10.12
The valve is opened during the first 90° of the cycle and closed for the remaining 270°.

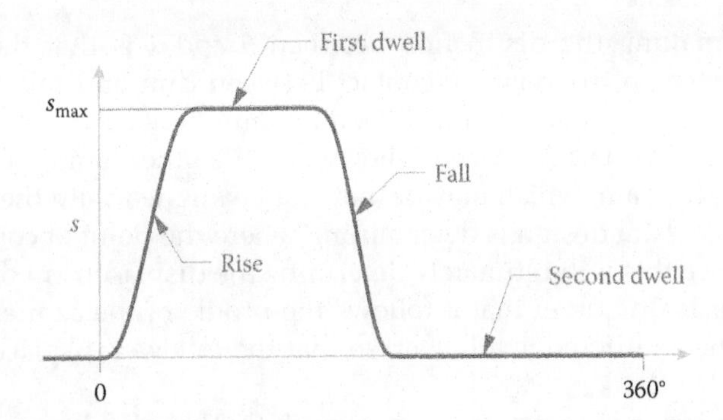

FIGURE 10.13
A double-dwell cam remains stationary at two different locations during the cycle. There is a rise section and a fall section to make the transition between dwells.

since the transition sections for these are easier to design and model. Once we have mastered the simple rise (or fall) section we will move on to the single dwell design, in which the rise and fall are achieved in a single process.

10.4 Introduction to Cam Design

Before we learn to design a cam, we must consider the interaction of the follower with the cam. As shown in Figure 10.14, we will use the angle λ to define the angular position of a point of interest, B, on the cam surface. The angle θ is used to define the rotation of the cam itself. We will define a baseline on the cam from which we will measure the angles to the points of interest on the surface of the cam. The baseline is normally located where the cam has minimum radius, as shown in Figure 10.14. The angle λ is measured *clockwise* from the baseline of the cam, since the follower engages with the profile of the cam in the opposite direction from the rotation of the cam. The angle λ is the angle you would measure if you were rotating with the cam and wanted to find the angle to a point of interest on the cam (e.g. the point of contact with the follower). The angle θ, in contrast, gives the rotation of the baseline of the cam in the global coordinate system.

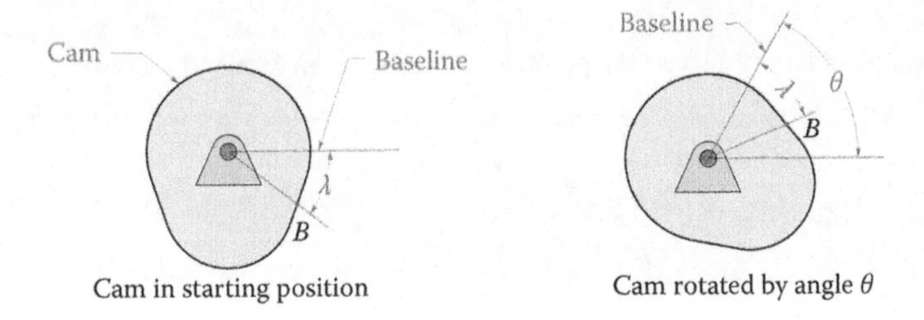

FIGURE 10.14
The angle λ is used to define a location on the cam profile and the angle θ gives the total rotation of the cam itself. Note that λ is defined as being positive clockwise, since this is the order in which the follower engages with the cam.

One reason for making the distinction between θ and λ is that the *type* of follower has a strong influence on the point of contact between cam and follower. As shown in Figure 10.15, the knife-edged follower always maintains contact at the same point, which is aligned with the x axis. The flat-faced follower, on the other hand, makes contact at the right-most point on the cam, which may or may not be aligned with the x axis. One of the most difficult parts of cam design is determining where the point of contact is for a given type of follower, since this will ultimately determine the displacement of the follower. The knife-edge follower is unique in that it follows the profile of the cam exactly. In this section, we will assume a knife-edge follower, so that the follower displacement mirrors the cam displacement.

Our goal in this section is to design the cam such that the displacement of the follower, s, is the required function of the cam profile angle, λ. Let us take an extremely simple case where the desired displacement of the follower is

$$\text{for } 0 < \lambda \leq 90^\circ \quad s = h$$

$$\text{for } 90^\circ < \lambda \leq 360^\circ \quad s = 0$$

where h, a constant, is the maximum displacement of the follower.

Let r be the radius of any point on the surface of the cam. The smallest radius of the cam is called the *base circle*, as shown in Figure 10.16. We will take the base circle as our reference point on the cam such that $s = 0$ when $r = r_b$. For the rest of the cam we have

$$s = r - r_b \tag{10.6}$$

Now imagine "unrolling" the profile of the cam so that it can be plotted along a horizontal axis. Such a plot is shown in Figure 10.17, where we have arbitrarily chosen a rise height of 0.1 units for the cam. We have taken a literal interpretation of the design requirements

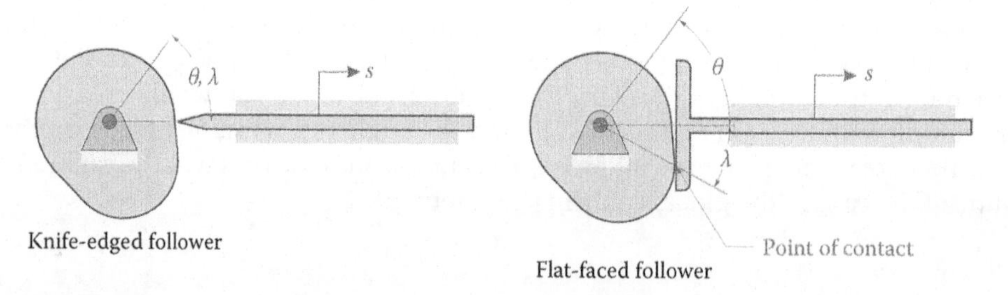

FIGURE 10.15
The type of follower has a large effect on the point of contact between cam and follower.

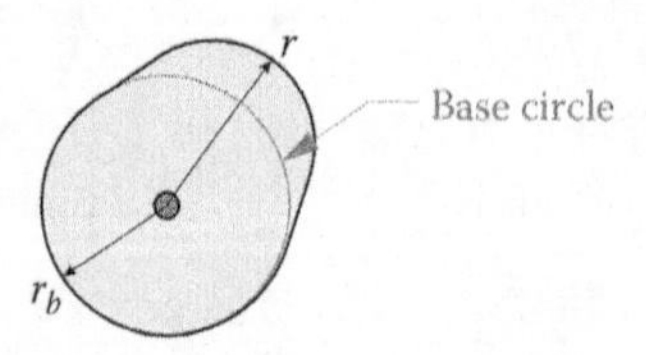

FIGURE 10.16
The smallest radius of the cam is called the base circle.

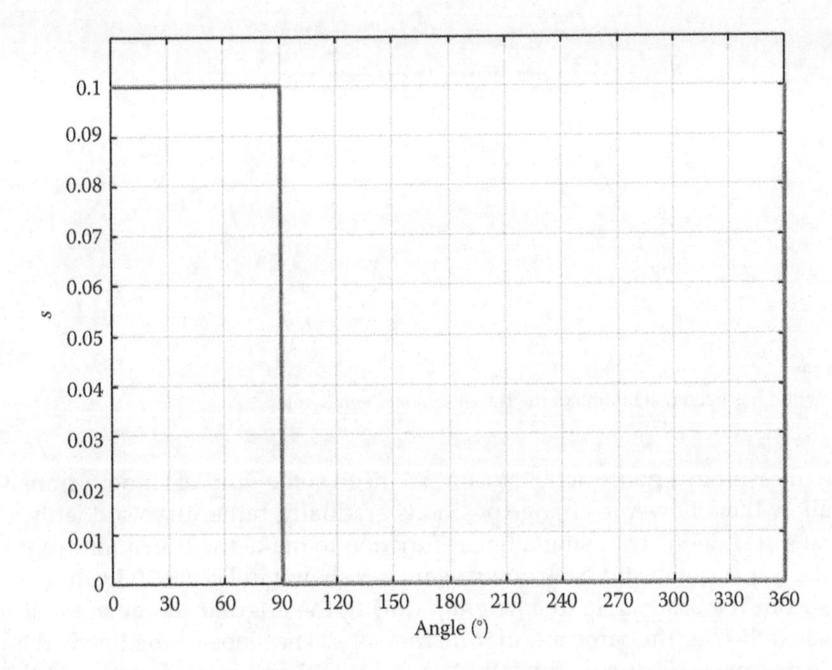

FIGURE 10.17
The displacement diagram for our initial cam design. We have arbitrarily chosen $h = 0.1$ units as the rise height for the cam.

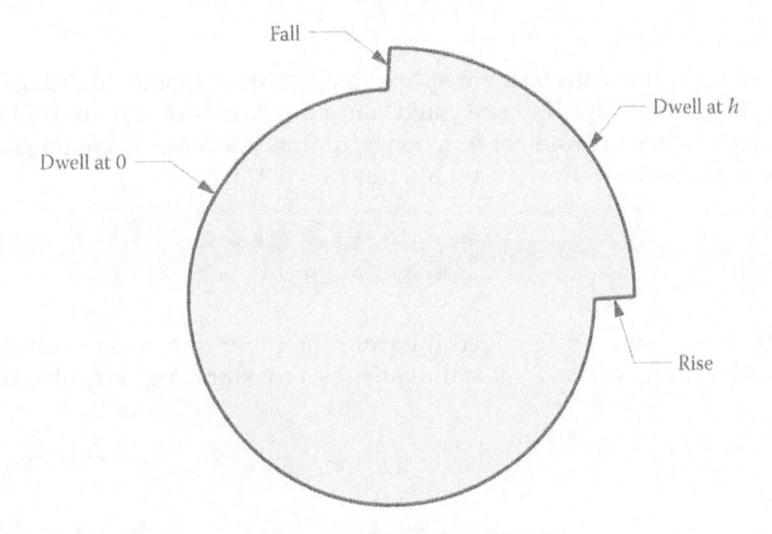

FIGURE 10.18
The cam that results from following the design requirements too literally.

(rise to h during the first 90°, then dwell at 0 for the remainder) but the sudden changes in height make it seem as though this might have been a bad idea.

If we roll up the plot in Figure 10.17 to plot the shape of the cam we obtain the monster shown in Figure 10.18. The sudden transitions from 0 to h and back again will destroy the follower very quickly. Thus, we conclude that following the design requirements too

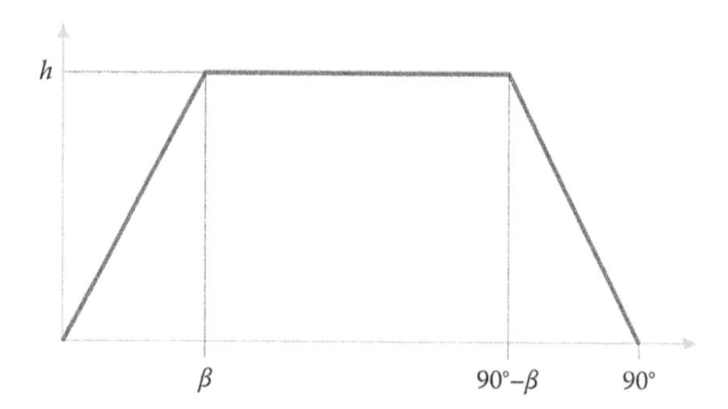

FIGURE 10.19
A linear function has been used to make the transition between dwells.

literally will result in a nonfunctional cam. We need some kind of smooth transition from 0 to h to allow the follower to change positions gradually, rather than suddenly.

As a first guess, let us try using a linear function to make the transition from the lower dwell to the upper dwell and back down again. As shown in Figure 10.19, the profile of the cam rises from the base circle to a height h during the angular increment β. It falls back to the base circle over the same angular increment β. The slope of the line is therefore h/β during the *rise* and $-h/\beta$ during the fall. We can write the rise as a function of the rotation angle of the cam profile, λ, as

$$s = \frac{h}{\beta}\lambda \tag{10.7}$$

where s is the height above the base circle. You might suspect that the sharp corners where the profile changes from dwell to rise could cause difficulties. To find out, let us take the time derivative of the displacement function, s, to find the velocity. We must employ the chain rule to take the derivative, since λ is a function of time.

$$v = \frac{ds}{dt} = \frac{ds}{d\lambda}\frac{d\lambda}{dt} = \frac{h}{\beta}\frac{d\lambda}{dt}$$

Since we are assuming a knife-edged follower, the point of contact is always aligned with the positive x axis so that $\lambda = \theta$. If the cam has a constant angular velocity, $d\theta/dt = \omega$, then

$$v = \frac{h}{\beta}\omega \tag{10.8}$$

This is the velocity of the follower during the transitions between dwells. During a dwell the velocity of the follower is (by definition) zero.

A plot of the velocity of the follower versus cam angle is shown in Figure 10.20. Because the rise is linear the velocity is constant (and finite) between dwells. But when the cam changes from dwell to rise there is a sudden change in velocity; that is, the velocity is *discontinuous* over the profile of the cam.

If we were to take the derivative of velocity to find acceleration we would find that it is undefined (or infinite) where the velocity makes the sudden change from 0 to h/β. This is

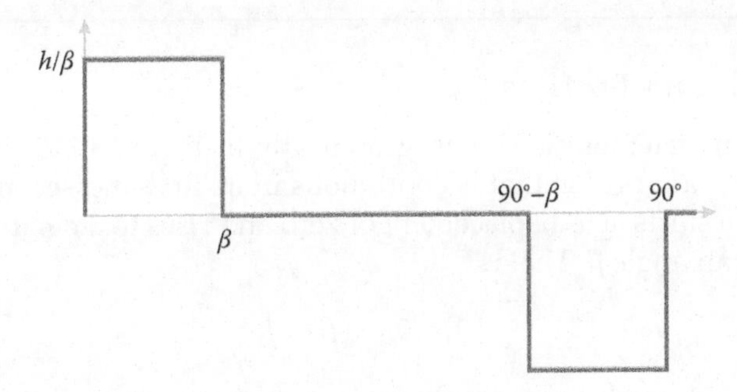

FIGURE 10.20
Velocity plot of the cam with linear rise (and fall) between dwells.

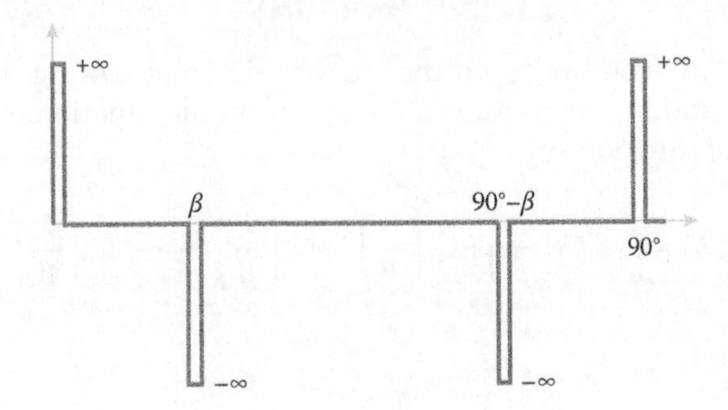

FIGURE 10.21
Acceleration of the follower with linear rise and fall. The abrupt changes in velocity create spikes of infinite acceleration.

seen in Figure 10.21, where there are spikes of infinite acceleration at each transition point. Since acceleration is proportional to force, the follower would experience "infinite" spikes of force from the cam at each transition from dwell to rise (or fall). Clearly, this is also an unacceptable design.

This discussion has led us to the *Fundamental Law of Cam Design*, which states that

> A cam must be continuous through the first and second derivatives of displacement across its entire profile.

In other words, the position, velocity, and acceleration of the follower must consist of continuous functions – no discontinuities are allowed. A corollary to the Fundamental Law of Cam Design is that

> The jerk function must be finite for the entire profile of the cam.

The jerk is the time derivative of acceleration. Since acceleration is proportional to force, the jerk gives a measure of how quickly the force on the follower changes. If the acceleration is a continuous function, then the jerk will necessarily be finite – spikes to infinity occur when we take the derivative of a discontinuous function. The problem is now to find a displacement function that is continuous through its first and second derivatives. As we will see, there are several functions that meet this criterion, and we will examine a few of the more common ones.

10.5 Polynomial Cam Profiles

The overall problem statement is shown graphically in Figure 10.22. We wish to find a rise function for the cam profile that is continuous in its first and second derivatives. We assume that the cam starts at a displacement of zero and rises to a height h by the time the cam has rotated by an angle β. That is,

$$s(0)=0 \quad s(\beta)=h$$

Since the velocity and acceleration are both zero during the dwell the first and second derivatives of the rise function must be zero at $\lambda=0$ and $\lambda=\beta$. Thus

$$v(0)=0 \quad v(\beta)=0$$
$$a(0)=0 \quad a(\beta)=0$$

We will not specify any conditions on the jerk at this point, noting only that it must be finite since the acceleration is continuous. As a first, simple function for the pushrod rise, let us try a *polynomial* function:

$$s(\lambda)=C_0+C_1\left(\frac{\lambda}{\beta}\right)+C_2\left(\frac{\lambda}{\beta}\right)^2+C_3\left(\frac{\lambda}{\beta}\right)^3+\cdots+C_n\left(\frac{\lambda}{\beta}\right)^n \tag{10.9}$$

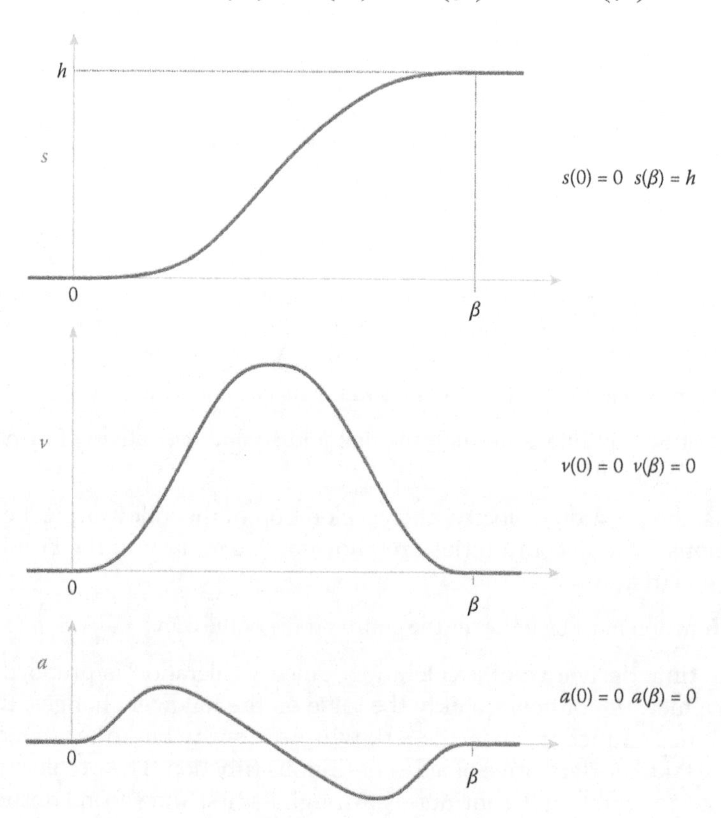

FIGURE 10.22
s-*v*-*a* diagram for the generic rise function.

We will find it much simpler to work with a dimensionless parameter x, defined as

$$x = \frac{\lambda}{\beta} \tag{10.10}$$

such that

$$s(x) = C_0 + C_1x + C_2x^2 + C_3x^3 + \cdots + C_nx^n \tag{10.11}$$

Note that $x = 0$ when $\lambda = 0$ and $x = 1$ when $\lambda = \beta$. The revised boundary conditions are then

$$s(0) = 0 \quad s(1) = h$$
$$v(0) = 0 \quad v(1) = 0$$
$$a(0) = 0 \quad a(1) = 0$$

We can employ as many terms in the polynomial as we wish, but for the moment, let us restrict ourselves to a fifth order polynomial. There are six boundary conditions, so we can only find definite values for six coefficients in the polynomial. We will use a higher-order polynomial in a later section when we create additional boundary conditions. Thus,

$$s(x) = C_0 + C_1x + C_2x^2 + C_3x^3 + C_4x^4 + C_5x^5 \tag{10.12}$$

There are six unknown coefficients in this polynomial, and we have the six boundary conditions shown in the *s-v-a* diagram of Figure 10.22. If we substitute in the first boundary condition, $s(0) = 0$, we find that $C_0 = 0$, and we have eliminated one of the coefficients. To find the velocity, take the time derivative of the displacement function using the chain rule

$$v = \frac{ds}{dt} = \frac{ds}{dx} \times \frac{dx}{d\lambda} \times \frac{d\lambda}{dt}$$

but $dx/d\lambda = 1/\beta$, $d\lambda/dt$ = omega for the knife-edged follower, so that

$$v = \left(C_1 + 2C_2x + 3C_3x^2 + 4C_4x^3 + 5C_5x^4\right)\frac{\omega}{\beta} \tag{10.13}$$

By enforcing the boundary condition $v(0) = 0$ we find that C_1 is also zero and we are left with four unknown coefficients. Differentiate once again to find the acceleration.

$$a = \left(2C_2 + 6C_3x + 12C_4x^2 + 20C_5x^3\right)\left(\frac{\omega}{\beta}\right)^2 \tag{10.14}$$

Enforcing the boundary condition $a(0) = 0$ eliminates C_2 from our polynomial, and we are left with

$$s = C_3x^3 + C_4x^4 + C_5x^5$$
$$v = \left(3C_3x^2 + 4C_4x^3 + 5C_5x^4\right)\frac{\omega}{\beta}$$
$$a = \left(6C_3x + 12C_4x^2 + 20C_5x^3\right)\left(\frac{\omega}{\beta}\right)^2$$

We must now enforce the remaining three boundary conditions that occur at $x = 1$.

$$s(1) = C_3 + C_4 + C_5 = h$$

$$v(1) = (3C_3 + 4C_4 + 5C_5)\frac{\omega}{\beta} = 0$$

$$a(1) = (6C_3 + 12C_4 + 20C_5)\left(\frac{\omega}{\beta}\right)^2 = 0$$

These are three linear equations with three unknowns, so it is easy to put them into matrix form for MATLAB® to solve. Note that the factors of (ω/β) cancel out in the velocity and acceleration equations since we have zero on the right-hand side. The matrix equation is then

$$\begin{bmatrix} 1 & 1 & 1 \\ 3 & 4 & 5 \\ 6 & 12 & 20 \end{bmatrix} \begin{Bmatrix} C_3 \\ C_4 \\ C_5 \end{Bmatrix} = \begin{Bmatrix} 1 \\ 0 \\ 0 \end{Bmatrix}$$

where we have normalized the maximum displacement h to unity for the purpose of solving for the coefficients. We will multiply the resulting functions by h after we have solved for C_3, C_4, and C_5. At the MATLAB command prompt type

```
>> A=[1 1 1; 3 4 5; 6 12 20]

A =

     1     1     1
     3     4     5
     6    12    20
```

For the **A** matrix and

```
>> b = [1; 0; 0]

b =

     1
     0
     0
```

for the **b** vector. Don't forget the semicolons between entries in the **b** vector since **b** is a column vector. Solve the matrix equation by typing

```
>> x=A\b

x =

   10.0000
  -15.0000
    6.0000
```

Thus, we conclude that

$$C_3 = 10h \quad C_4 = -15h \quad C_5 = 6h$$

Substituting these back into the *s*-*v*-*a* functions results in

$$s = h\left(10x^3 - 15x^4 + 6x^5\right) \tag{10.15}$$

$$v = \frac{30h\omega}{\beta}\left(x^2 - 2x^3 + x^4\right) \tag{10.16}$$

$$a = 60h\left(\frac{\omega}{\beta}\right)^2\left(x - 3x^2 + 2x^3\right) \tag{10.17}$$

Because of the orders of the variable x this rise function is known as the 3-4-5 polynomial. An *s*-*v*-*a* diagram of the 3-4-5 polynomial is shown Figure 10.23. As advertised, the function is continuous through its second derivative, and would seem to be a good choice. One difficulty with this function is that we have not required that the jerk be continuous. Taking the derivative of the acceleration function gives

$$j = 60h\left(\frac{\omega}{\beta}\right)^3\left(1 - 6x + 6x^2\right) \tag{10.18}$$

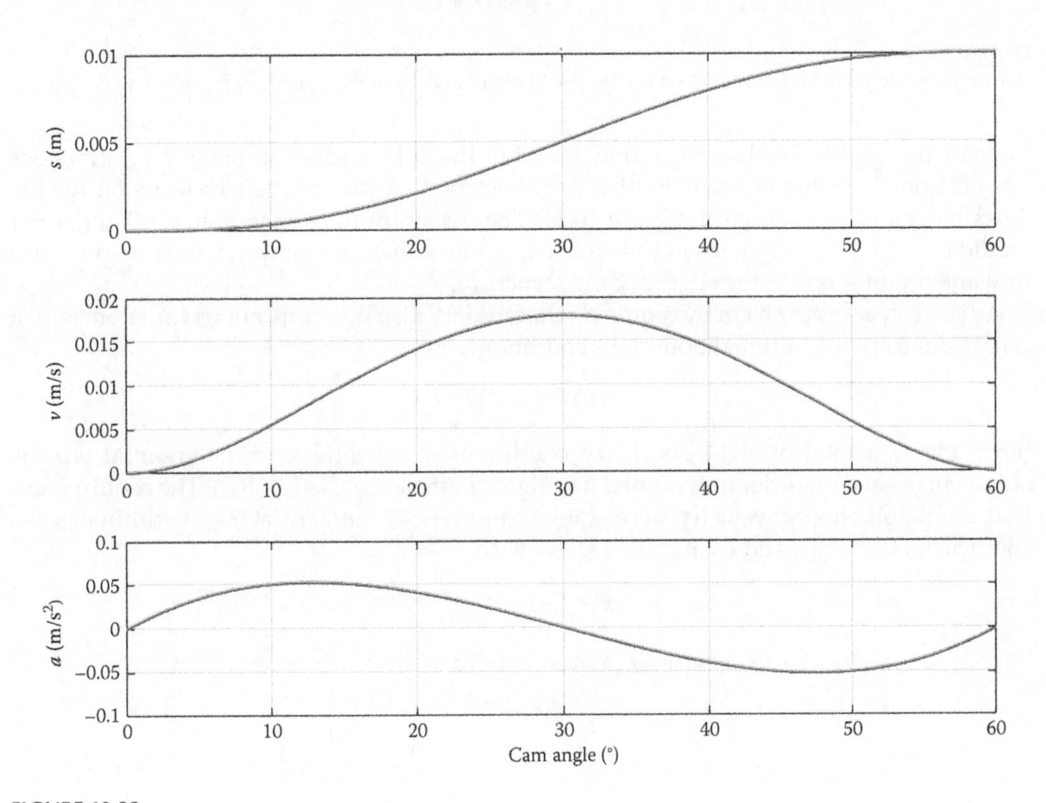

FIGURE 10.23
s-*v*-*a* diagram for 3-4-5 polynomial. The rise is 0.01 m and the angular velocity of the cam is 1 rad/s. The rise angle, β, is 60°.

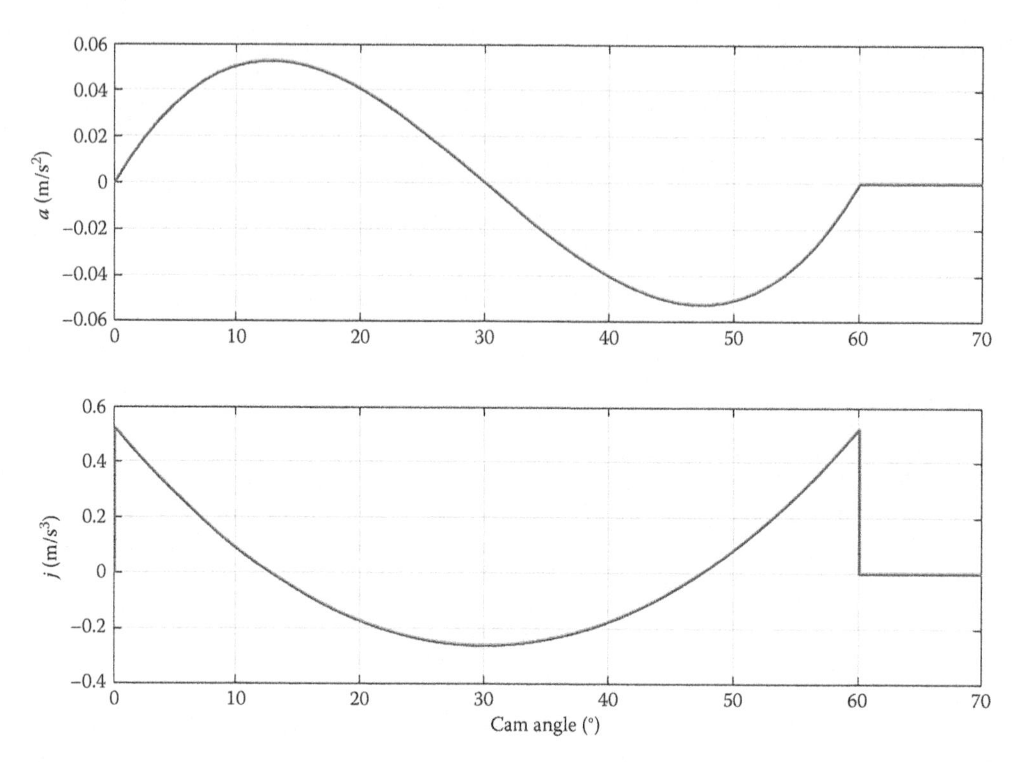

FIGURE 10.24
Plot of the acceleration and jerk function for the 3-4-5 polynomial. Note the discontinuities at 0 and β.

The jerk during the dwell is, of course, zero, but the jerk function in Equation (10.18) does not take on the value of zero at either $x = 0$ or $x = 1$. A plot of the jerk function for the 3-4-5 polynomial is shown in Figure 10.24. The discontinuities in jerk at $x = 0$ and $x = 1$ would create sudden changes in the force experienced by the follower, which might cause unwanted vibrations in the surrounding structure.

We can solve this problem by requiring that the jerk also be a continuous function, which will result in two additional boundary conditions.

$$j(0) = 0 \quad j(1) = 0$$

Since this is a total of eight boundary conditions, a seventh order polynomial will be sufficient (a seventh order polynomial has eight coefficients: C_0, C_1 … C_7). The requirement that the displacement, velocity, acceleration, and jerk all be zero at $x = 0$ eliminates the coefficients C_0, C_1, C_2, and C_3 and we are left with

$$s = C_4x^4 + C_5x^5 + C_6x^6 + C_7x^7$$

$$v = \left(4C_4x^3 + 5C_5x^4 + 6C_6x^5 + 7C_7x^6\right)\frac{\omega}{\beta}$$

$$a = \left(12C_4x^2 + 20C_5x^3 + 30C_6x^4 + 42C_7x^5\right)\left(\frac{\omega}{\beta}\right)^2$$

$$j = \left(24C_4x + 60C_5x^2 + 120C_6x^3 + 210C_7x^4\right)\left(\frac{\omega}{\beta}\right)^3$$

The remaining four boundary conditions give the set of equations

$$s(1) = C_4 + C_5 + C_6 + C_7 = h$$
$$v(1) = 4C_4 + 5C_5 + 6C_6 + 7C_7 = 0$$
$$a(1) = 12C_4 + 20C_5 + 30C_6 + 42C_7 = 0$$
$$j(1) = 24C_4 + 60C_5 + 120C_6 + 210C_7 = 0$$

We are left with four equations and the four unknown coefficients, which again seems like a good job for MATLAB. Type

```
A= [ 1  1   1   1; 4  5   6   7; 12 20  30  42; 24 60 120 210]
```

at the command prompt. The vector of knowns is

```
b = [1;0;0;0]
```

To solve the equations, type

```
x=A\b
```

and the result is

```
x =

  35.0000
 -84.0000
  70.0000
 -20.0000
```

Thus, the 4-5-6-7 polynomial functions are

$$s = 35h\left(x^4 - \left(\frac{12}{5}\right)x^5 + 2x^6 - \left(\frac{4}{7}\right)x^7\right) \tag{10.19}$$

$$v = 140h\frac{\omega}{\beta}\left(x^3 - 3x^4 + 3x^5 - x^6\right) \tag{10.20}$$

$$a = 420h\left(\frac{\omega}{\beta}\right)^2\left(x^2 - 4x^3 + 5x^4 - 2x^5\right) \tag{10.21}$$

$$j = 840h\left(\frac{\omega}{\beta}\right)^3\left(x - 6x^2 + 10x^3 - 5x^4\right) \tag{10.22}$$

A quick check of the boundary conditions reveals that the *v-a-j* functions vanish at $x = 1$ and all functions vanish at $x = 0$.

As seen in the *s-v-a-j* diagram of Figure 10.25, the jerk is now a continuous function, and we appear to have solved the problem. As you can see, polynomial functions are quite versatile and easy to model mathematically. They are among the most widely used rise and fall functions in cam design, and should be the first choice for most designers. We will examine a few other commonly used rise functions in the next sections.

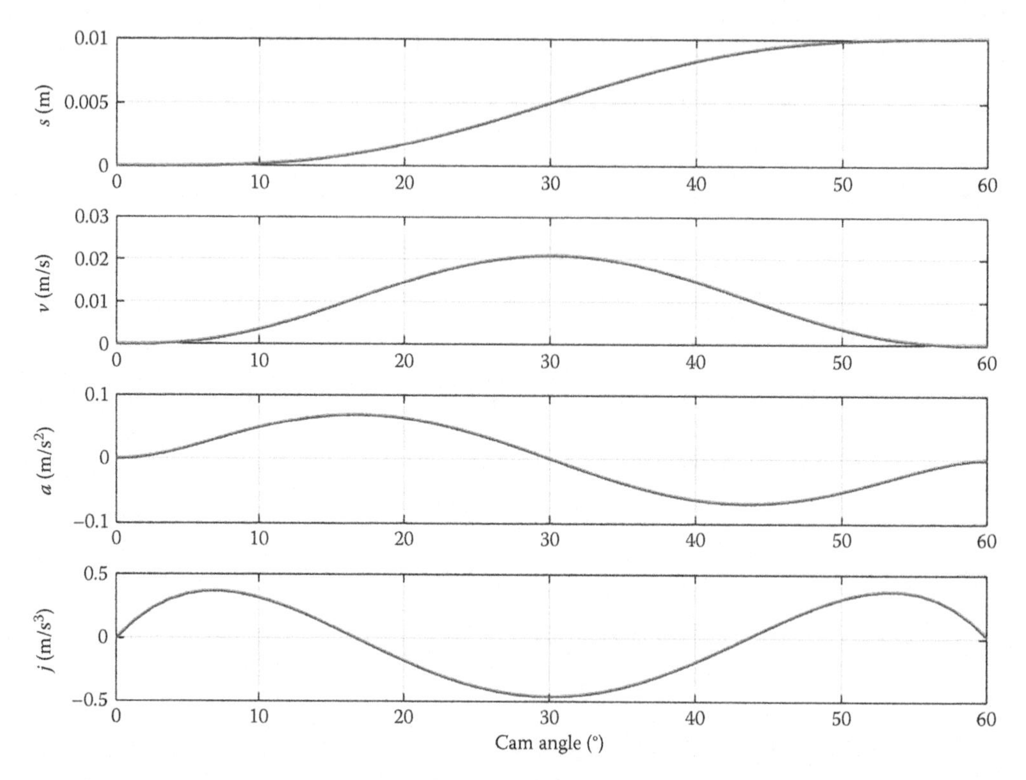

FIGURE 10.25
s-v-a-j diagram for the 4-5-6-7 polynomial with 0.01 m rise over 60° and angular velocity 1 rad/s. The jerk is now a continuous function.

10.6 Sinusoidal Cam Profiles

Now that we have mastered the polynomial rise function, we will look into another common rise function, the sinusoid. Sinusoidal functions seem like a logical choice for cam profiles, since the sine and cosine functions are smooth and allow a gradual transition from dwell to rise and back again. For the polynomial rise function we started by postulating a displacement function and worked our way through the velocity and acceleration functions by taking derivatives. But since the Fundamental Law of Cam Design states that the cam profile must be continuous through the first and second derivatives, let us try starting with the acceleration function and work our way "backwards" to displacement by integration. Since the sine function begins and ends at zero, this would seem like a good candidate for the acceleration function. As shown in Figure 10.26, our proposed acceleration function is then

$$a = C\omega^2 \sin(2\pi x) \tag{10.23}$$

where

$$x = \frac{\lambda}{\beta} \tag{10.24}$$

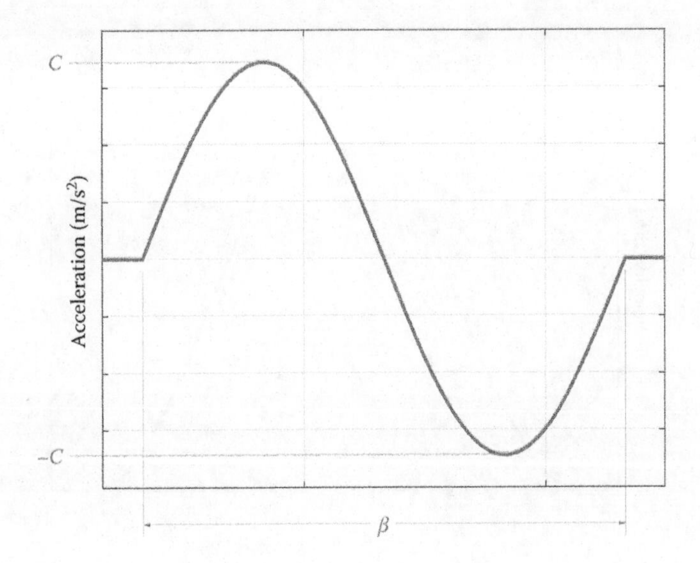

FIGURE 10.26
The acceleration function of the proposed sinusoidal cam is a sine function with amplitude C.

and ranges from 0 to 1 as before. The reader can verify that the acceleration function starts at zero when $x = 0$ and ends at zero when $x = 1$. To find velocity, we simply integrate the acceleration function with respect to time.

$$v = -\frac{C\omega}{2\pi}\cos(2\pi x) + k_1 \tag{10.25}$$

where k_1 is an unknown constant of integration.

As shown in Figure 10.27, we must have $v(0) = 0$ and $v(1) = 0$. Substituting this into Equation (10.25) above gives

$$k_1 = \frac{C\omega}{2\pi} \tag{10.26}$$

and so

$$v = \frac{C\omega}{2\pi}\left[1 - \cos(2\pi x)\right] \tag{10.27}$$

We integrate once more to find the position function

$$s = \frac{C}{2\pi}\left[x - \frac{1}{2\pi}\sin(2\pi x)\right] + k_2 \tag{10.28}$$

but $s(0) = 0$ so that $k_2 = 0$. The other boundary condition states that $s(1) = h$, and solving for C gives

$$C = 2\pi h \tag{10.29}$$

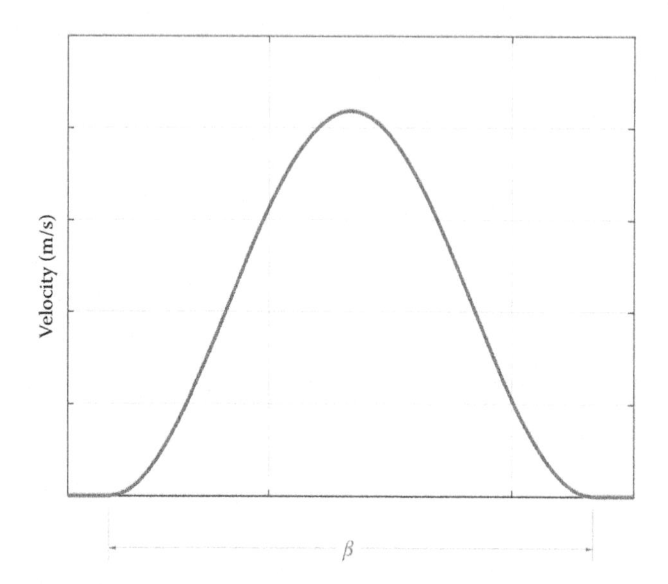

FIGURE 10.27
The velocity function must begin and end at zero to satisfy the Fundamental Law of Cam Design.

Thus, the *s*-*v*-*a*-*j* functions for our proposed cam profile are

$$s = h\left[x - \frac{1}{2\pi}\sin(2\pi x)\right] \tag{10.30}$$

$$v = h\omega\left[1 - \cos(2\pi x)\right] \tag{10.31}$$

$$a = 2\pi h\omega^2 \sin(2\pi x) \tag{10.32}$$

$$j = 4\pi^2 h\omega^3 \cos(2\pi x) \tag{10.33}$$

The displacement function is known as a *cycloidal* function.

The *s*-*v*-*a*-*j* diagram for the cycloidal rise function is shown in Figure 10.28. It is very difficult to distinguish these functions from the polynomial rise function by eye, and the differences vanish almost entirely when we couple the cam with a non-knife-edged follower. The jerk function is discontinuous for this rise function, so we could, if necessary, start with the jerk function (instead of acceleration) when deriving the cycloidal rise function.

The name "cycloid" is something of a misnomer. As shown in Figure 10.29, a true cycloid is the path followed by the wheel of a bicycle or car as it rolls without slipping over a flat surface. The coordinates of a true cycloid function are given by

$$x = r(t - \sin t) \tag{10.34}$$

$$y = r(1 - \cos t) \tag{10.35}$$

where r is the radius of the wheel and t is the angle of the wheel as it rolls along. As you can see, the rise function uses only the x component of the cycloid.

There are additional rise functions for cams including the modified sinusoid, modified trapezoidal acceleration, and many others. The interested reader can refer to [1] for more information.

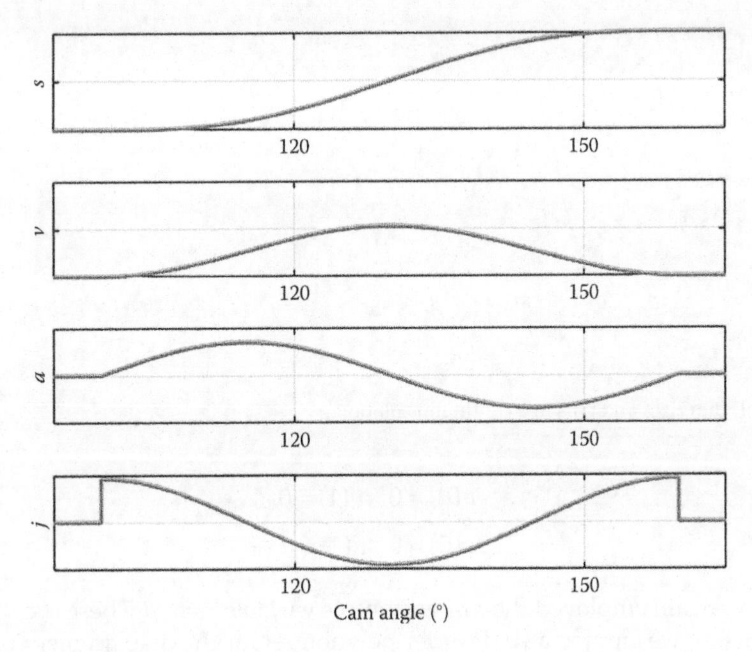

FIGURE 10.28
The s-v-a-j diagram for the cycloidal rise function.

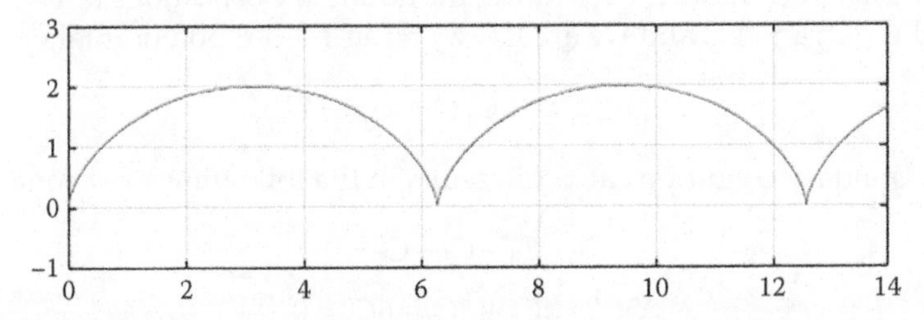

FIGURE 10.29
A cycloid is the path followed by a point on the wheel of a bicycle as it rolls without slipping.

10.7 Single-Dwell Cams

Thus far we have examined the simple rise and fall functions of multiple-dwell cams, but in many cases (such as automotive cams) we desire that the rise and fall segments be composed of a single rise-fall function. This type of cam is known as a single-dwell cam since it dwells at its resting state for most of the cycle, then rises and falls in a continuous motion. The displacement function for a typical single-dwell cam is shown in Figure 10.30. For convenience we have chosen a *symmetric* rise-fall function; that is, the displacement reaches its peak at $\beta/2$, and settles back down to zero at β. The simplest set of boundary conditions that satisfy the fundamental law of cam design are

$$s(0)=0 \quad s\left(\frac{1}{2}\right)=h \quad s(1)=0$$

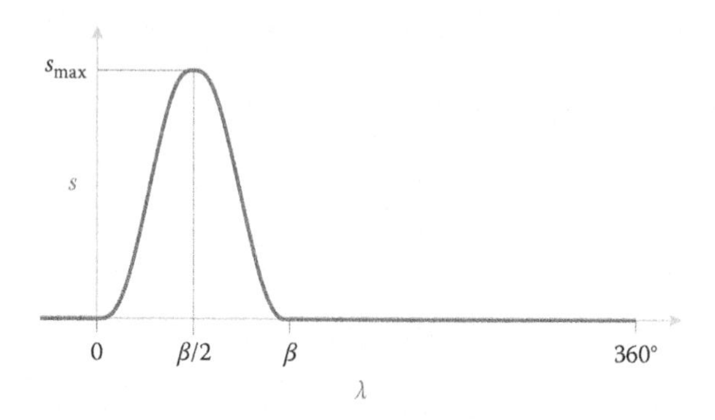

FIGURE 10.30
The single-dwell cam rises and falls in a continuous motion.

$$v(0)=0 \quad v(1)=0$$
$$a(0)=0 \quad a(1)=0$$

where we have again employed the dimensionless variable $x = \lambda/\beta$. These are seven boundary conditions, so we employ a sixth-order polynomial for the displacement function.

$$s(x) = C_0 + C_1x + C_2x^2 + C_3x^3 + C_4x^4 + C_5x_5 + C_6x^6$$

Taking the time derivatives and substituting the boundary conditions at $x = 0$ results in the elimination of C_0, C_1, and C_2, and we are left with the 3-4-5-6 polynomial

$$s(x) = C_3x^3 + C_4x^4 + C_5x_5 + C_6x^6$$

Entering the boundary conditions at $x = 1$ results in the following *s-v-a* equations

$$C_3 + C_4 + C_5 + C_6 = 0$$
$$3C_3 + 4C_4 + 5C_5 + 6C_6 = 0$$
$$6C_3 + 12C_4 + 20C_5 + 30C_6 = 0$$

and the displacement boundary condition at $x = ½$ gives

$$C_3\left(\frac{1}{8}\right) + C_4\left(\frac{1}{16}\right) + C_5\left(\frac{1}{32}\right) + C_6\left(\frac{1}{64}\right) = h$$

But this can be simplified to

$$8C_3 + 4C_4 + 2C_5 + C_6 = 64h$$

We now have four equations and four unknowns, which we can solve using MATLAB. In matrix form the equation is

$$\begin{bmatrix} 1 & 1 & 1 & 1 \\ 3 & 4 & 5 & 6 \\ 6 & 12 & 20 & 30 \\ 8 & 4 & 2 & 1 \end{bmatrix} \begin{Bmatrix} C_3 \\ C_4 \\ C_5 \\ C_6 \end{Bmatrix} = \begin{Bmatrix} 0 \\ 0 \\ 0 \\ 64h \end{Bmatrix}$$

Enter the following commands at the MATLAB command prompt

```
>> A=[1 1 1 1; 3 4 5 6; 6 12 20 30; 8 4 2 1]

A =

     1     1     1     1
     3     4     5     6
     6    12    20    30
     8     4     2     1

>> b = [0; 0; 0; 64]

b =

     0
     0
     0
    64

>> A\b

ans =

   64.0000
 -192.0000
  192.0000
  -64.0000
```

Thus, the *s-v-a-j* functions are

$$s(x) = 64h\left(x^3 - 3x^4 + 3x^5 - x^6\right) \tag{10.36}$$

$$v(x) = 192h\left(x^2 - 4x^3 + 5x^4 - 2x^5\right)z \tag{10.37}$$

$$a(x) = 384h\left(x - 6x^2 + 10x^3 - 5x^4\right)z^2 \tag{10.38}$$

$$j(x) = 384h\left(1 - 12x + 30x^2 - 20x^3\right)z^3 \tag{10.39}$$

where

$$z = \frac{\omega}{\beta} \tag{10.40}$$

An *s-v-a-j* diagram of the single-dwell cam is shown in Figure 10.31, with cam angular velocity 10 rpm and rise 0.01 m. This cam rises over 30° and falls over the next 30°. This cam fulfills all of the boundary conditions required by the fundamental law of cam design, but has a discontinuous jerk function at the beginning and end of the rise.

We can impose the boundary condition of zero jerk at the beginning and end of the rise/fall function. Since this adds two additional boundary conditions (for a total of nine) we must use an eight-order polynomial. Following the same procedure as above, we have

$$s(x) = 256h\left(x^4 - 4x^5 + 6x^6 - 4x^7 + x^8\right) \tag{10.41}$$

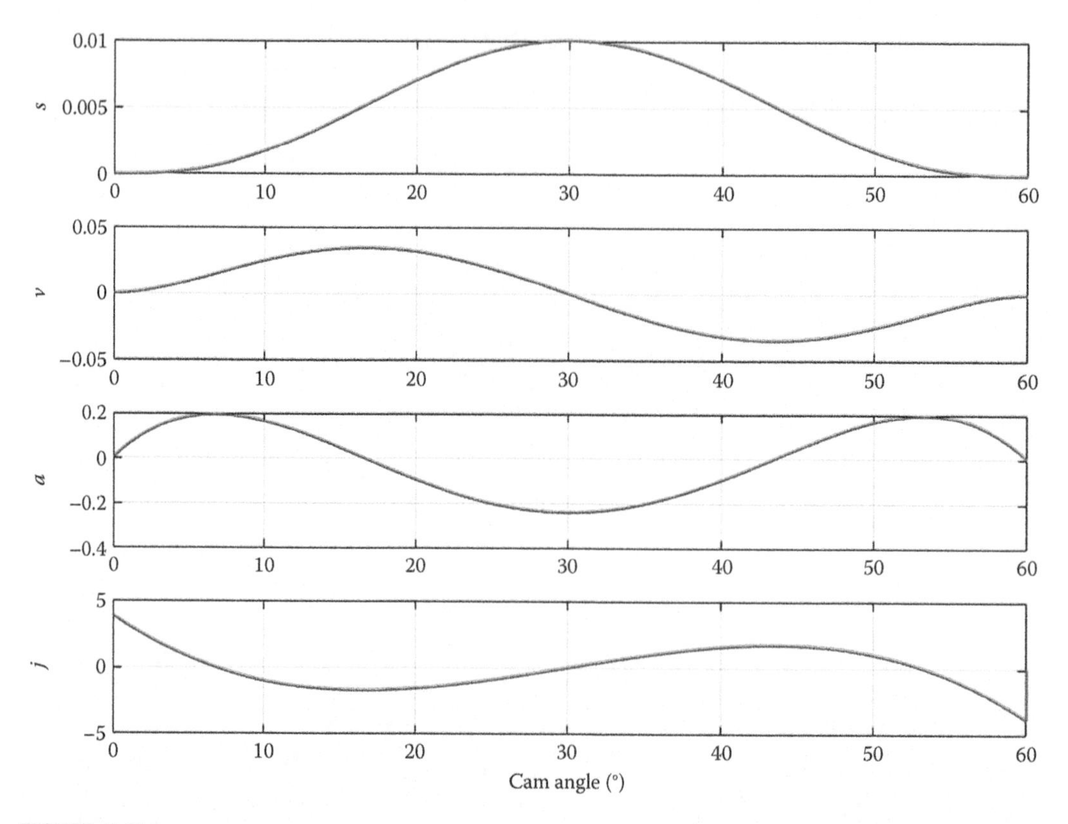

FIGURE 10.31
s-v-a-j diagram for the single-dwell cam with $\omega = 10$ rpm and rise 0.01 m. This cam fulfills all of the boundary conditions but the jerk is discontinuous.

$$v(x) = 1024h\left(x^3 - 5x^4 + 9x^5 - 7x^6 + 2x^7\right)z \tag{10.42}$$

$$a(x) = 1024h\left(3x^2 - 20x^3 + 45x^4 - 42x^5 + 14x^6\right)z^2 \tag{10.43}$$

$$j(x) = 6144h\left(x - 10x^2 + 30x^3 - 35x^4 + 14x^5\right)z^3 \tag{10.44}$$

A plot comparing the 3-4-5-6 and the 4-5-6-7-8 polynomials is shown in Figure 10.32. Note that the jerk returns to zero at each end for the 4-5-6-7-8 polynomial, but is discontinuous for the 3-4-5-6 polynomial. The 4-5-6-7-8 polynomial has higher peak velocity and slightly higher peak acceleration, which may weigh against its use in high-speed applications.

You might be wondering why we did not simply use a double-dwell cam profile with a dwell period of zero at the top of the rise, instead of specifying a completely new rise/fall function. As shown in Figure 10.33, the double-dwell cam must have zero acceleration at the top and bottom of the rise function, which results in a higher peak acceleration in the middle of the rise. While the double-dwell function would function as required, it is not an *optimal* design since its acceleration (and therefore the forces transmitted to the rest of the linkage) is higher than necessary to accomplish the required motion. Thus, you should employ the single-dwell function whenever the second dwell is not absolutely required.

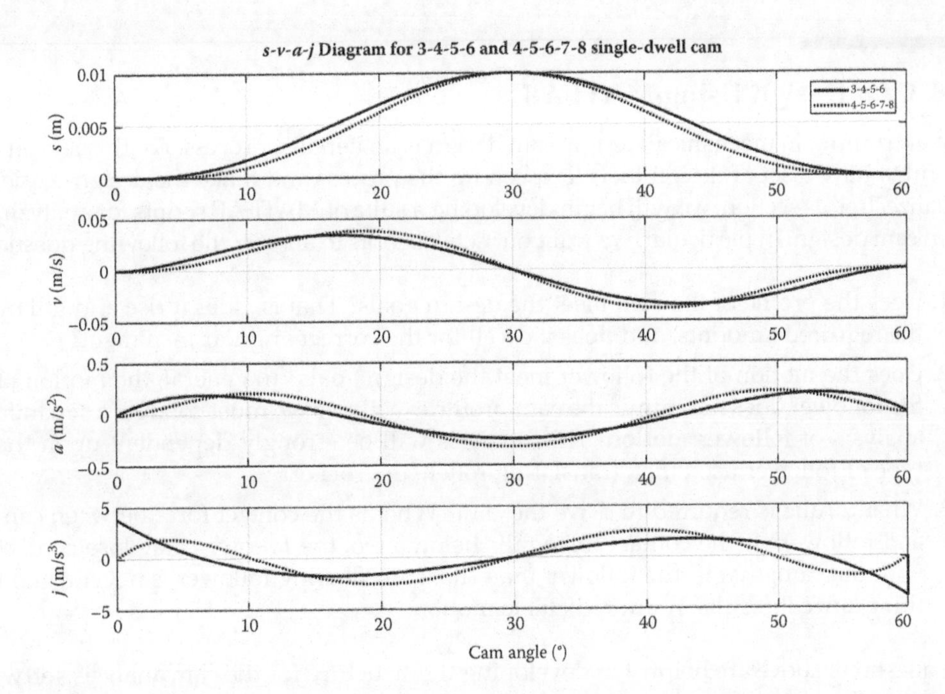

FIGURE 10.32
s-v-a-j diagram for the 4-5-6-7-8 polynomial compared with the 3-4-5-6 polynomial from before. Note the jerk returns to zero for the 4-5-6-7-8 polynomial, but is discontinuous for the 3-4-5-6 polynomial.

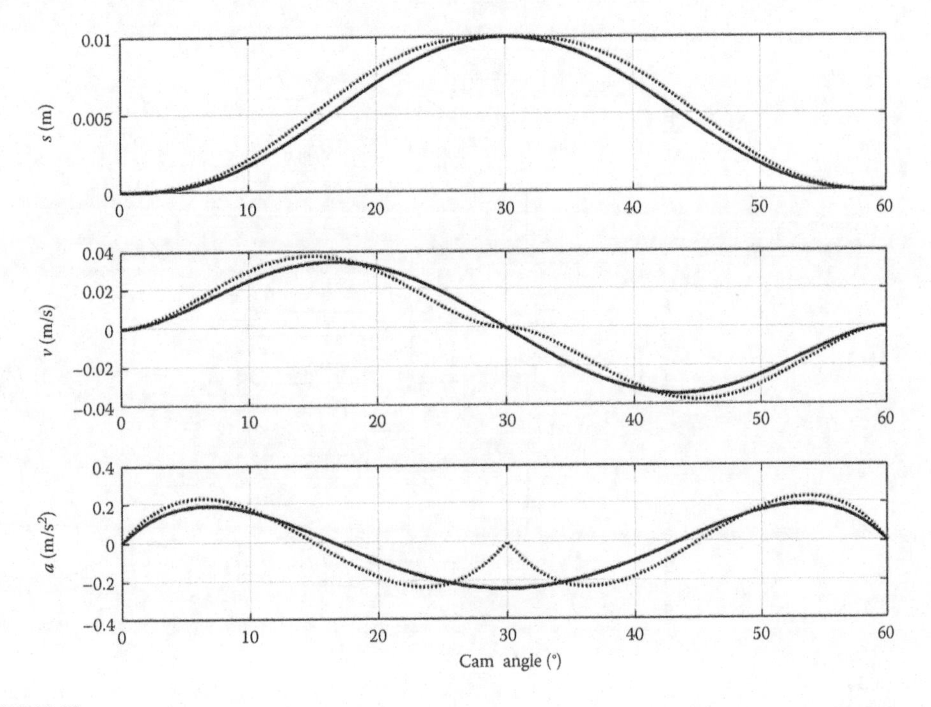

FIGURE 10.33
A comparison of the single-dwell function with a double-dwell 3-4-5 profile with zero dwell at the top of the rise. The single-dwell function is shown in the solid line and the double-dwell function (one-dwell set to zero) is shown with dashed lines.

10.8 Cam Design Using MATLAB®

Like everything in mechanical design, cam design is an iterative process. For this reason it is helpful to have a set of design tools to speed up the process and make the designs easier to visualize. In this section, we will begin developing a suite of MATLAB scripts for analyzing a given cam design. In particular, we want our set of scripts to answer the following questions:

1. Does the profile of the cam meet the design goals? That is, does it rise and fall by the required amounts, and does it dwell for the correct angular increment?
2. Does the motion of the follower meet the design goals? In general, the motion of the follower does not mimic the cam profile exactly, so we must conduct a separate analysis of follower motion. This analysis will be strongly dependent upon the type of follower used (flat, translating roller, and rocker).
3. What torque is required to drive the cam? What is the contact force between cam and follower? If the contact force falls below zero, the follower may lose contact with the cam; that is, the follower may "float." A floating follower is never a good thing since it results in unpredictable motion.

The questions above are helpful in developing the structure for the cam analysis software that we will write. A flowchart showing the program structure is shown in Figure 10.34.

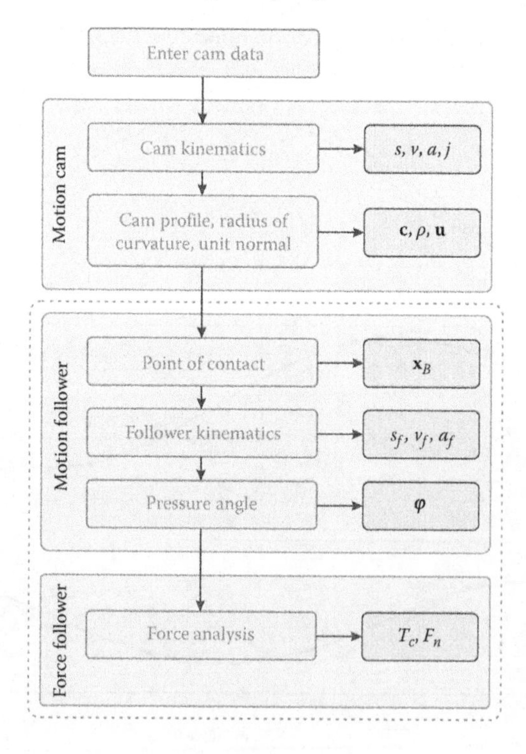

FIGURE 10.34
Flowchart of the structure of the cam analysis suite of programs. The first box is the main program and all of the computations below take place in separate functions, shown as light gray boxes. The darker gray boxes are the results of the calculations in each function, and will be plotted. The dashed box shows the computations that depend upon the type of follower (flat, roller, or rocker).

The blue box at the top symbolizes the main program where the user will enter the data for the cam, including the base radius, dwell heights, the type of rise function, the angular increments of each dwell and rise, and information about the follower. Each subsequent light gray box represents a separate function that we will write. For example, the `MotionCam.m` function will calculate the *s-v-a-j* functions for the given cam data, and will also return a vector of points around the cam profile, **c**. It will also calculate the radius of curvature and unit normal around the cam profile. We haven't discussed these yet, but they will be important for ensuring smooth motion of the follower and for conducting force analysis.

The darker gray boxes on the right represent the quantities that are returned from each of our functions. Each of these quantities may be plotted during a given point in the design cycle, but we will include commands that can turn off individual plots so that we don't end up with a screen filled with 10 plots.

The dashed blue box encloses the functions that are dependent upon the type of follower used. The functions above this box are common to all types of follower, but the flat-faced follower requires a different force analysis function than the translating roller-follower, and so on. However, the overall structure of the main program will be the same. We will write the software for designing multiple-dwell cams. Modifying the software for single-dwell cams is relatively easy and is left as an exercise for the reader.

10.8.1 The Main Program

Without further ado, let us begin writing the main program. Open a new MATLAB script and type the following header:

```
% CamDesigner.m
% Plots the profile of a cam and its s-v-a-j diagram, along with
% the pitch curve, radius of curvature, pressure angle and follower
% motion.  A force/torque analysis is completed at the end.
% By Eric Constans, August 15, 2017
%
% The user must enter the number of dwells, the height of each dwell,
% and the angular increment of each dwell and its rise (or fall).
% The angular increments of the dwells and rises must add to 360 degrees
% or less.

clear variables; close all; clc;
```

At the beginning of the program the user should enter the required data for the cam. The most important piece of information is the number of dwell segments. Figure 10.35 shows a displacement (*s*) diagram for a cam with three dwell segments. Each dwell segment is associated with a rise or fall segment, so that the number of rises and falls equals the number of dwells. Note that we have started the cam profile on a dwell segment so the sequence runs (dwell-rise-dwell-fall-dwell-fall). After entering the number of dwells, the user should enter the height of each dwell. In Figure 10.35 the first dwell has a height of zero, but this need not be the case. The number of heights entered by the user should be the same as the number of dwells, obviously. After this, the user must enter the angular increment of each dwell segment, followed by the angular increment of each rise (or fall). Type in the following set of data for a triple-dwell cam.

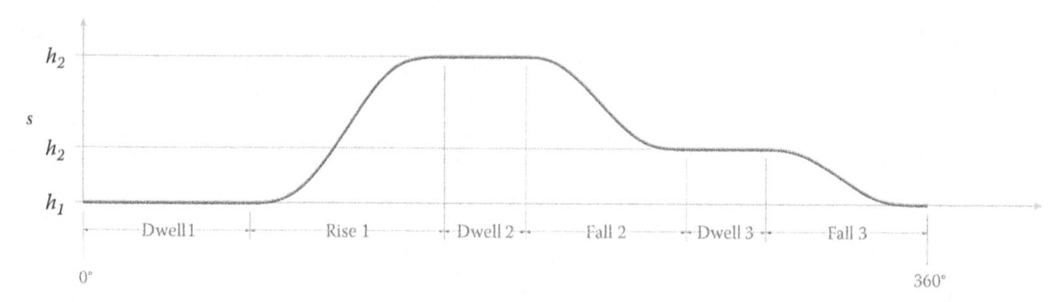

FIGURE 10.35
The *s* diagram for a three-dwell cam. Each dwell has a rise or fall attached to it. The profile after the final fall must have the same height as the initial dwell.

```
% ***** Parameters for cam profile (user-defined) *****
cam.Nd = 3;                        % number of dwells
cam.h     = [0 0.005 0.010]; % height of each dwell above base circle (m)
cam.betad = [100 40 30];           % angular increment of each dwell (degrees)
cam.betar = [ 60 50 80];           % angular increment of each rise (degrees)
```

What is the meaning of all the "`cam.`" statements? Here, we have introduced a very important concept in programming: the *structure*. A structure is used to keep a collection of different kinds of data in one object, in this case, `cam`. The quantities after the periods are called *fields*, and each field contains a separate kind of data (dwell heights, dwell angles, etc.). The really handy feature of structures is that we can pass all of the data associated with the cam to one of our functions by simply passing the `cam` structure, instead of passing each argument separately. That is, if we send `cam` to the `CamMotion` function, we automatically give it `Nd`, `h`, `betad` and `betar`. As we progress through our set of cam functions, we will add other important quantities (s, v, a, j, etc.) to the `cam` structure. We will later create a structure called `follower` that will contain all of the important parameters associated with the follower.

You might be wondering why we didn't create a structure called `link` in our earlier programs that contains the parameters associated with a given link (length, mass, moment of inertia, etc.) Good question! Creating a `link` structure would have been good programming practice, but we elected not to introduce too many abstract concepts all at once. Rewriting our suite of linkage analysis programs to include structures would be a simple, exercise.

Now back to our `CamDesigner` program. The variable `betad` is a vector containing the angular increments of each dwell. It must have as many members as the number of dwell segments, as must the vector `betar`. Here we have chosen a profile that dwells at zero for 100°, rises to 5 mm over 60°, dwells at 5 mm for 40°, rises to 10 mm over 50°, dwells at 10 mm for 30°, and finally falls back down to zero over 80°. Note that the angular increments are entered in degrees; this is for user convenience. We will convert these angles to radians later in the program. The next important pieces of data for the user to enter are the base radius and angular velocity of the cam, for calculating velocities and accelerations.

```
cam.rb = 0.025;                % radius of base circle (m)
cam.omega = 1000*2*pi/60;      % angular velocity of cam (rad/sec)
```

Here, we have chosen a cam base diameter of 50 mm and an angular velocity of 1000 rpm. We will use the data for this triple-dwell cam in all of our subsequent calculations so that you can easily check your progress as we move along. The final important piece of information for the user to enter is the rise function used in the cam profile. In the previous section, we learned several useful rise functions including the 3-4-5 polynomial, 4-5-6-7 polynomial and cycloid. We will write a separate MATLAB function for each type of rise, and let the user specify the rise type by calling the appropriate function. The syntax for doing this is

```
cam.camfunc = @svaj345;        % function handle for rise profile
```

The term "function handle" is a little cryptic. If you think of a handle as something used to grab onto an object, then a function handle is a way of grabbing a function in order to use it. Basically, the statement above defines a variable called `camfunc` that can be used to "grab" a function called `svaj345`. The "@" syntax indicates that `svaj345` is a function, and not simply a variable. Any time later in the program that we want to use the `svaj345` function we can type `cam.camfunc` instead, and obtain the same results. This makes the program much easier to use for design – if the user wishes to use a 4-5-6-7 polynomial, for example, she would enter

```
cam.camfunc = @svaj4567;        % function handle for rise profile
```

and the program would perform all calculations using the 4-5-6-7 profile instead.

Best programming practices provide error checking of user-entered parameters to ensure the program will run properly. If any errors are found, the user should be informed what corrections they should make. Open a new script for our error checking function and start with the following header, function name, and initialize a variable stating no errors found:

```
% The function ErrorCheckCam checks for errors in user input variables.
%
% ***** Inputs *****
% cam = cam parameters
%
% ***** Output *****
% errorFound = boolean value indicating if error was detected

function errorFound = ErrorCheckCam(cam)
  errorFound = false;
end
```

The most common source of error for this program is the set of cam profile variables. The sum of all angular increments must add to 360°, so this will be the first check we make. MATLAB makes this truly simple with the `sum` function:

```
anglesum = sum(cam.betad) + sum(cam.betar);
```

The `sum` function adds up all of the quantities in a vector. It will be helpful to the user to display the result of this calculation in the command window. This can be accomplished with the `disp` command.

```
disp(['Angular segments add up to: ' num2str(anglesum) ' degrees'])
```

The `disp` command displays a string of text to the command window. Anything inside the parentheses in the `disp` command must be a string (a set of characters, rather than numbers), so we have used `num2str` to convert the numerical value of `anglesum` into a string.

The three strings: "`Angular segments add up to:`", `anglesum`, and "`degrees`" are sandwiched together into a single vector, since the `disp` command can only display one vector of characters at a time.

Finally, let us check to make sure that `anglesum` is indeed 360°.

```
if (anglesum ~= 360)
  disp('ERROR: Angular segments do not add up to 360 degrees.')

  errorFound = true;
end
```

The "~=" syntax means the same as "is not equal to." If the user has entered values that do not add up to 360° then a message is displayed in the command window and we update our `errorFound` variable to true.

The next check we will make ensures the vectors defining the cam profile have the correct number of entries. For each dwell we require its height, dwell start angle, and a corresponding rise angle to the next dwell.

```
if (cam.Nd ~= length(cam.h) || cam.Nd ~= length(cam.betad)||...
    cam.Nd ~= length(cam.betar))
  disp(['Error: Cam profile parameters not of equal lengths. ',...
       'Cam parameters h, betad and betar must contain Nd ',...
       'number of entries.'])
  errorFound = true;
end
```

In writing our condition for the `if` statement, we use MATLAB's `length` command to determine the number of entries in a vector. We compare the length of each vector to the number of dwells, `Nd`, and connect each condition with "||" syntax, which is a logical OR. The OR statement causes the code with in the `if` statement to run if any of the conditions are true, meaning the length of one of the vectors does not match the number specified in `Nd`.

Our completed error checking function should look like:

```
function errorFound = ErrorCheckCam(cam)
  errorFound = false;

 % check if cam profile angles sum to 360 degrees
  anglesum = sum(cam.betad) + sum(cam.betar);
  disp(['Angular segments add up to: ' num2str(anglesum) ' degrees'])
 if (anglesum ~= 360)
    disp('Error: Angular segments do not add up to 360 degrees.')
    errorFound = true;
 end

 % check if h, betad and betar have proper number of entries
  if (cam.Nd ~= length(cam.h) || cam.Nd ~= length(cam.betad)||...
      cam.Nd ~= length(cam.betar))
    disp(['Error: Cam profile parameters not of equal lengths. ',...
    'Cam parameters h, betad and betar must contain Nd ',...
    'number of entries.'])
    errorFound = true;
 end
end
```

With our error checking function complete, the final remaining step is to return to our `CamDesigner` program and insert the error check function after the user parameters to stop the program if an error is found.

```
% ***** Error checking - validate user input *****
if ErrorCheckCam(cam);
  return;
end;
```

This completes the first part of the main program. The following sections will describe each of the many cam analysis procedures in detail, and we will keep adding functionality to our program as we move along. The full (short) text of the main program should read:

```
% CamDesigner.m
% Plots the profile of a cam and its s-v-a-j diagram, along with
% the pitch curve, radius of curvature, pressure angle and follower
% motion.  A force/torque analysis is completed at the end.
% By Eric Constans, August 15, 2017
%
% The user must enter the number of dwells, the height of each dwell,
% and the angular increment of each dwell and its rise (or fall).
% The angular increments of the dwells and rises must add to 360 degrees
% or less.

clear variables; close all; clc;

% ***** Parameters for cam profile (user-defined) *****
cam.Nd = 3;                    % number of dwells
cam.h      = [0 0.005 0.010]; % height of each dwell above base circle (m)
cam.betad = [100 40 30];       % angular increment of each dwell (degrees)
cam.betar = [ 60 50 80];       % angular increment of each rise (degrees)
cam.rb = 0.025;                % radius of base circle (m)
cam.omega = 1000*2*pi/60;      % angular velocity of cam (rad/sec)
cam.camfunc = @svaj345;        % function handle for rise profile

% ***** Error checking - validate user input *****
if ErrorCheckCam(cam);
  return;
end;
```

10.8.2 The Cam Motion Function

The first of our suite of cam analysis functions will calculate and plot the cam profile, as well as its *s-v-a-j* diagram. The coordinates of the cam can be exported to a CAD program such as SOLIDWORKS® for subsequent manufacturing, although we will not discuss this process here. This script will be one of the more complicated functions we've written so far, so we'll proceed carefully.

First, note that the cam profile can be divided into its individual rise, fall, and dwell segments, as shown in Figure 10.36. Dwell segments have constant radius, which means

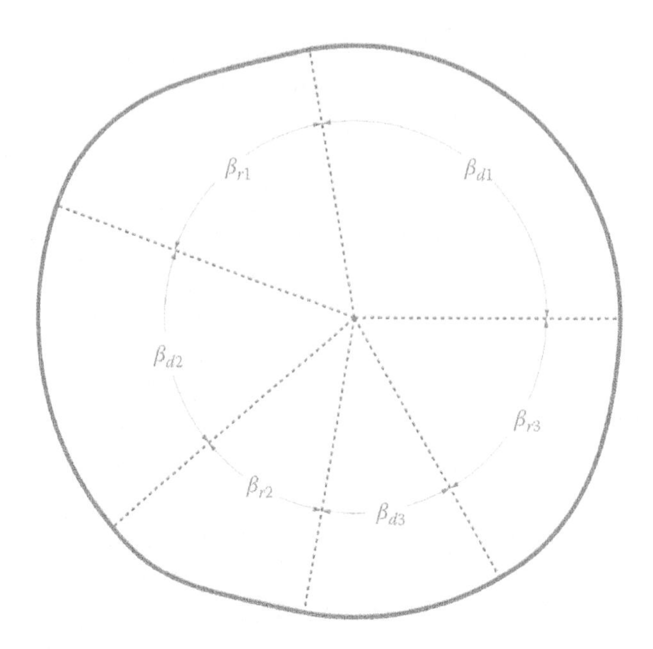

FIGURE 10.36
The angular segments of the three-dwell cam. Each dwell segment is a circular arc and each rise segment follows the specified rise function.

that they are circular arcs with radius $r_b + h$, where h is the dwell height. Rise (or fall) segments follow the rise functions described in earlier sections and have radius $r_b + s(\lambda)$, where s is the rise function. This division of the cam into its individual segments provides a hint as to how to structure the `CamMotion` function. At the end of the `CamDesigner` main program, call the `CamMotion` function.

```
% ***** Calculate cam profile and plot motion *****
cam = CamMotion(cam,showPlot);
```

Next, open up a new script, and save it as `CamMotion.m`. Then type the header for the function as shown below.

```
% The function CamMotion calculates and plots the coordinates of a
% multiple-dwell cam profile.  It also calculates and plots cams the
% s-v-a-j functions.
%
% ***** Inputs *****
% cam = parameters of cam profile
% showPlot = plotting options
%
% ***** Outputs *****
% cam = cam profile with kinematic functions

function cam = CamMotion(cam,showPlot)
```

As you can see, the `cam` structure allows the list of input and output arguments to be quite small. Before we execute the `CamMotion` function the `cam` structure contains only the dwell heights and dwell/rise/fall angles. After the function is complete the `cam` structure will also contain the kinematic variables (s, v, a, j, etc.)

We next enter the important parameters for calculating each segment of the cam profile. First is the variable `N`, which tells how many increments to divide each dwell or rise segment. The finer the increments, the smoother the curves will appear and the slower the program will execute. Since there are `N` points in each segment and $2N_d$ segments overall (each dwell has an associated rise segment) the total number of points to calculate will be

$$M = 2N_d \cdot N + 1 \tag{10.45}$$

Since the cam has been divided into $2N_d \cdot N$ divisions, we will calculate a total of $2N_d \cdot N + 1$ points on the cam, as shown in Figure 10.37. The last point on the cam will be identical to the first point, meaning that we will have one redundant point. This will not cause a problem when plotting in MATLAB, but the redundant point may need to be eliminated if the profile is to be exported to CAD software. This is entered in the code below.

```
% calculation parameters
N = 100;                % no. of points to calculate on each dwell/rise/fall
M = 2*cam.Nd*N + 1;     % total number of points to calculate
```

We have arbitrarily chosen to calculate 100 points per segment, but you can try different values if you wish. We next allocate space in memory for the variables we are about to calculate

```
% allocate space for motion variables
[lambda,s,v,a,j,rho] = deal(zeros(1,M));
[c,u]                = deal(zeros(2,M));
```

The variable `lambda` stores the value of each angle around the cam, and `c` stores the x and y coordinates of each point on the cam profile – it corresponds to the **c** vector discussed earlier. We will calculate the value of `s`, `v`, `a`, and `j` for each point around the cam, along with the radius of curvature, `rho`. In the main program, we initialized the values for the angular increments in degrees, so we shouldn't forget to convert them to radians for our calculations.

```
% convert angular segments to radians
cam.betad = pi*cam.betad/180;
cam.betar = pi*cam.betar/180;
```

We are now ready to begin calculating the cam profile and the associated s, v, a, and j functions.

We first initialize the variable k, which will store the index of the set of values we are currently calculating and will range from 1 to M. This is necessary because the loops for

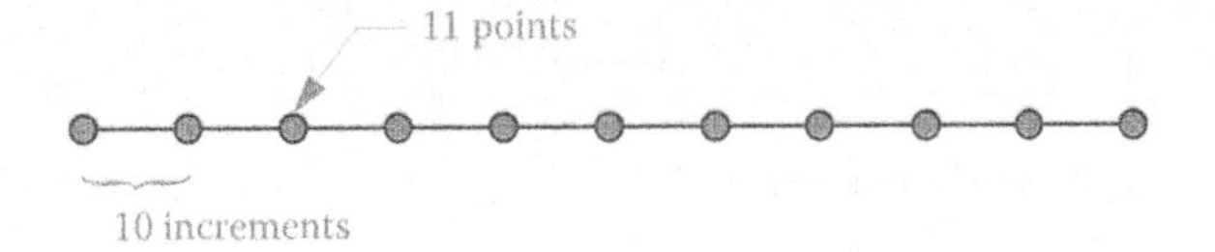

FIGURE 10.37
If we divide a segment into N increments, then we must plot $N+1$ points.

calculating dwell and rise segments each range from 1 to N, and we need to keep track of where we are on the overall cam profile. Similarly, we will initialize the variable `lambda0` that will store the starting angle of each segment.

```
% ***** calculate cam profile *****
k = 1;          % start index counter at 1
lambda0 = 0; % starting angle for each segment
```

After this, begin the main loop, which ranges from 1 to the number of dwells, N_d. Be sure to enter the end statement now, so that you do not forget it later on.

```
% main loop
for nd = 1:cam.Nd     % loop through each dwell

end
```

The looping variable `nd` tells us which dwell (or rise) segment we are currently working on. We will perform the calculations for the dwell segment first, and its associated rise (or fall) next. The dwell segment is easy to calculate since everything is constant. The displacement, s, is simply equal to the height of the current dwell

$$s(\lambda) = h_{nd}$$

and the velocity, acceleration, and jerk are all zero during the dwell.

$$v(\lambda) = 0 \quad a(\lambda) = 0 \quad j(\lambda) = 0$$

Since these were initialized to zero during the memory allocation, we can leave them as they are.

Now consider the set of vectors shown in Figure 10.38. The vector **c** starts at the center of the cam (point A) and ends at a point B on the surface. The unit vector **e** points in the same direction as **c**, and is found using the familiar formula

$$\mathbf{e} = \begin{Bmatrix} \cos\lambda \\ \sin\lambda \end{Bmatrix} \tag{10.46}$$

We learned how to find the radius of the cam at a given angle λ in the previous section:

$$r(\lambda) = r_b + s(\lambda) \tag{10.47}$$

where r_b is the base radius (a constant) and $s(\lambda)$ is the value of the displacement function at the angle λ. The pair (r, λ) gives the polar coordinates of point B, but we can convert this to Cartesian coordinates using the unit vector **e**

$$\mathbf{c} = r\mathbf{e} \tag{10.48}$$

The dwell segment of the code is then

```
% ***** dwell segment *****
  beta = cam.betad(nd);                 % angle of dwell
```

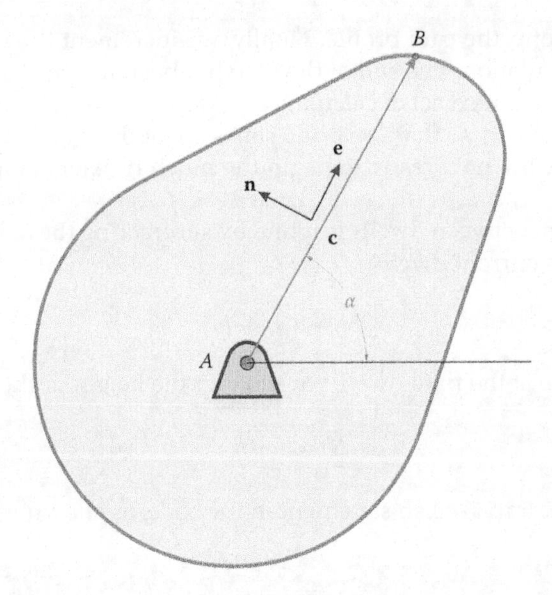

FIGURE 10.38
The vector **c** is directed from the center of the cam to a point *B* on the surface.

```
for i = 1:N                                 % loop through dwell segment
  x = (i-1)/N;                              % x ranges from 0 to 1
  lambda(k) = x*beta + lambda0;             % current angle within dwell
  s(k) = cam.h(nd);                         % current displacement
  [e,~] = UnitVector(lambda(k));            % unit vector in radial dir.
  r = cam.rb + s(k);                        % radius to point B
  c(:,k) = r*e;                             % coordinates of cam profile

  k = k + 1;                                % increment index counter
end
lambda0 = lambda0 + cam.betad(nd);  % move angle to end of current dwell
```

There are a few important things going on in this loop. First, we define the variable `beta`, which is the angle over which the current dwell occurs. The loop runs from 1 to *N*, since the angular segment was divided into *N* parts. Within the loop, we make use of the variable *x*, which was defined in Section 10.6 as

$$x = \frac{\lambda}{\beta} \tag{10.49}$$

and takes on the value of 0 at the beginning of the dwell and 1 at the end. This makes it easy to calculate the current angle on the cam, since

$$\lambda = x\beta + \lambda_0 \tag{10.50}$$

The reader can verify that this starts at λ_0 when $x = 0$ and ends at $\beta + \lambda_0$ when $x = 1$. Of course, x does not quite reach 1, but it comes very close (99/100 when $N = 100$).

The displacement function, *s*, is simply the height of the current dwell and we add this to the base circle radius to find the total radius of the cam at this angle. We next calculate the unit vector, **e**, associated with `lambda` and use this (along with the current radius) to find the coordinates of **c**, the cam profile. Finally, we increment the index counter k to be ready for the next calculation. The end of the loop has been reached, and we go back to the beginning to perform the next set of calculations. After the dwell loop is complete, we add the dwell angle to `lambda0` so that `lambda0` starts out at the angle of the beginning of the next rise segment. We are now ready to begin the much trickier computations within the rise segment.

The change in height between dwells is found by subtracting the height of the next dwell from the height of the current dwell

$$\Delta h = h_{nd+1} - h_{nd} \tag{10.51}$$

but if we happen to be at the final dwell, we subtract the height of the first dwell

$$\Delta h = h_1 - h_{nd} \tag{10.52}$$

This logic will require an `if-else` statement in the code, as shown.

```
% calculate the change in height from this dwell to the next
  if (nd == cam.Nd)                    % if we are at final dwell, then height
    dh = cam.h(1) - cam.h(nd);         % changes back to initial dwell,
  else                                 % otherwise height changes to next
    dh = cam.h(nd+1) - cam.h(nd); % dwell.
  end
```

The variable `dh` stores the change in height between one dwell and the next. Its value will be negative for a fall segment and positive for a rise segment. The next step is to calculate the values of *s-v-a-j* for the rise segment. Since it is likely that we will wish to try more than one type of function for the rise segments of our cam, it is appropriate to define a separate function for each type of rise. Open a new MATLAB script and enter the following simple function.

```
% The function svaj345 computes the current values of s-v-a-j for the
% 345 polynomial rise function
%
% ***** Inputs *****
% x  = angle parameter ranging from 0 to 1
% dh = height change from preceding dwell to next dwell
% h  = height of the preceding dwell
% z  = omega/beta for calculating v, a, j
%
% ***** Outputs *****
% s = displacement
% v = velocity
% a = acceleration
% j = jerk

function [s,v,a,j] = svaj345(x,dh,h,z)

s = dh*(10*x^3 - 15*x^4 + 6*x^5) + h; % displacement
v = 30*dh*(x^2 -  2*x^3 +   x^4)*z;    % velocity
a = 60*dh*(x   -  3*x^2 + 2*x^3)*z^2; % acceleration
j = 60*dh*(1   -  6*x   + 6*x^2)*z^3; % jerk
```

Save the function as svaj345.m in the same folder as the main program. Make sure that you use the same name as the variable camfunc in the main program or MATLAB won't be able to find the correct function. This function has been written with enough flexibility to change the *s-v-a-j* formulas, for example, the 4-5-6-7 polynomial or other functions. Go back to the main program and enter the loop for the rise segment.

```
% ***** rise (or fall) segment *****
  beta = cam.betar(nd);                  % angle of rise or fall
  h = cam.h(nd);                         % starting height of rise or fall
  z = cam.omega/beta;                    % omega/beta
  for i = 1:N
    x = (i-1)/N;                         % x ranges from 0 to 1
    lambda(k) = x*beta + lambda0;        % current angle within rise
    [s(k),v(k),a(k),j(k)] = cam.camfunc(x,dh,h,z);
    [e,~] = UnitVector(lambda(k));       % unit vector in radial dir.
    r = cam.rb + s(k);                   % radius to point B
    c(:,k) = r*e;                        % coordinates of cam profile

    k = k + 1;                           % increment index counter
  end
  lambda0 = lambda0 + cam.betar(nd); % move angle to end of current rise
```

Everything in the rise loop is similar to the dwell loop that we have already entered except for the rise function. This completes the main loop of our script. Right after the main loop we must enter the cam profile parameters for $\lambda = 360°$, which are the same as those for $\lambda = 0°$.

```
% parameters at 360deg are the same as at 0deg
lambda(M) = 2*pi;       % final angle is 360 degrees
c(:,M) = c(:,1);        % profile ends where it started
s(M) = s(1);            % final displacement is same as initial
```

10.8.3 Interpolating the Cam Profile Using the Spline Function

The calculations above were relatively straightforward, but now things will become a little trickier. You may have noticed that the angular increment between each point on the profile is different depending upon which segment of the profile we are currently evaluating. For example, in the dwell segment *i* the angular increment is

$$\Delta\lambda = \frac{\beta_{di}}{N} \tag{10.53}$$

but in the rise segment *j* the angular increment is

$$\Delta\lambda = \frac{\beta_{rj}}{N} \tag{10.54}$$

This unevenness will make plotting more complicated than it needs to be, and will also create unnecessary challenges when conducting force analysis since the time intervals for each calculation will also be different. It would be much simpler if we had a set of values for *s*, *v*, *a*, *j*, and **c** at evenly spaced angles around the cam; say every 1°. It is probably no surprise by now to learn that MATLAB has a convenient function for accomplishing this: the spline function.

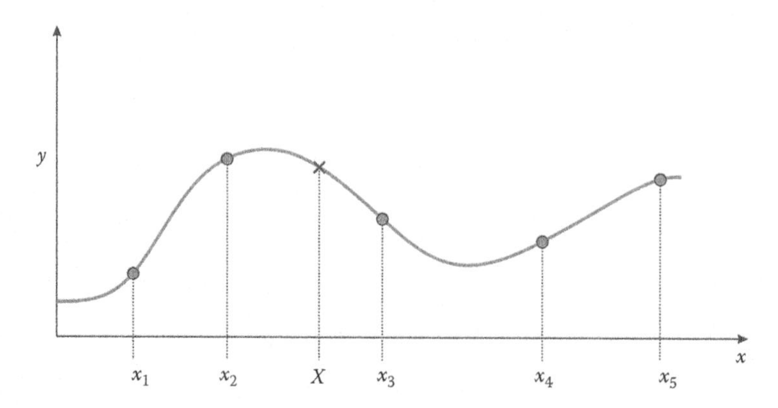

FIGURE 10.39
The value of the function y is known at (x_1, x_2, x_3, x_4, and x_5) and we wish to estimate its value at X.

A typical scenario for a spline interpolation is shown in Figure 10.39. Assume that y is a function of x, and that we know the values of y at $x = (x_1, x_2, x_3, x_4, x_5)$. With sufficient cleverness, we can construct a set of cubic polynomial functions that pass through each of the known points exactly. Because the polynomials are cubic, their slope and curvature are continuous at each of the points, which make them nice and smooth. We call the set of cubic polynomial functions a *cubic spline* and we can use the spline to estimate the value of y at an arbitrary point $x = X$. For the best accuracy the point X should be within the range defined by $x_1 \ldots x_5$, in which case it is known as an interpolation. If X lies outside the range $x_1 \ldots x_5$, then it is an extrapolation and its accuracy may be doubtful.

In MATLAB a cubic spline is constructed using the spline function, and the syntax is

```
Y = spline(x, y, X)
```

where x is a vector of data points where we know the values of the function y. The vector X is the set of x coordinates where we wish to estimate the function y, and the vector Y is the set of these estimates. Let us now try using the spline function with a simple example. Pretend for the moment that the function y is a sine function (of course, we don't know the actual form of the function y, we only know its value at a few data points). At the MATLAB command line type

```
>> x = 2*pi*(0:4)/4
x =
         0    1.5708    3.1416    4.7124    6.2832

>> y = sin(x)
y =
         0    1.0000    0.0000   -1.0000   -0.0000
```

Here we have defined x to be a five element vector spanning between 0 and 2π. Remember that the statement 0:4 returns the vector `[0 1 2 3 4]`. Now type

```
>> X = 2*pi*(0:100)/100;
>> Y = spline(x,y,X);
```

We have used the semicolon to suppress the output, but you can leave it off if you want to see the calculated values. The pairs (x, y) are the known data points, and the pairs (X, Y)

are the estimates using the spline function. Enter the following plot commands to see the result of the calculations

```
>> plot(x,y,'o')
>> hold on
>> plot(X,Y)
>> plot(X,sin(X))
```

The result of estimating the sine function with only four data points is shown in Figure 10.40. It's not too bad, considering how few data points we used. If you increase the number of known (*x*,*y*) pairs to 10, the result should resemble Figure 10.41, which is an almost exact match. Since we have calculated 100 points for each segment on the cam profile, we conclude that the cubic spline should do a reasonable job of estimating the rise, fall and dwell functions between these points. Before we use the `spline` function, we need to define a few important plotting parameters for the cam profile. Enter the following parameters at the end of the `CamMotion` function.

```
% ***** plotting parameters for the cam profile *****
cam.Nc = 361;                          % number of points around cam profile
dLambda = 2*pi/(cam.Nc-1);             % angular increment
cam.lambda = dLambda*(0:cam.Nc-1);     % angles on cam to calculate profile
cam.dt = dLambda/cam.omega;            % time increment
```

The variable `cam.Nc` gives the number of points to plot around the cam profile, which is one point for every degree. The angular spacing between plotted points is calculated as `dLambda`, and `cam.lambda` gives the angle for each plotted point on the cam profile. Finally, the variable `cam.dt` is the time increment between points on the cam profile,

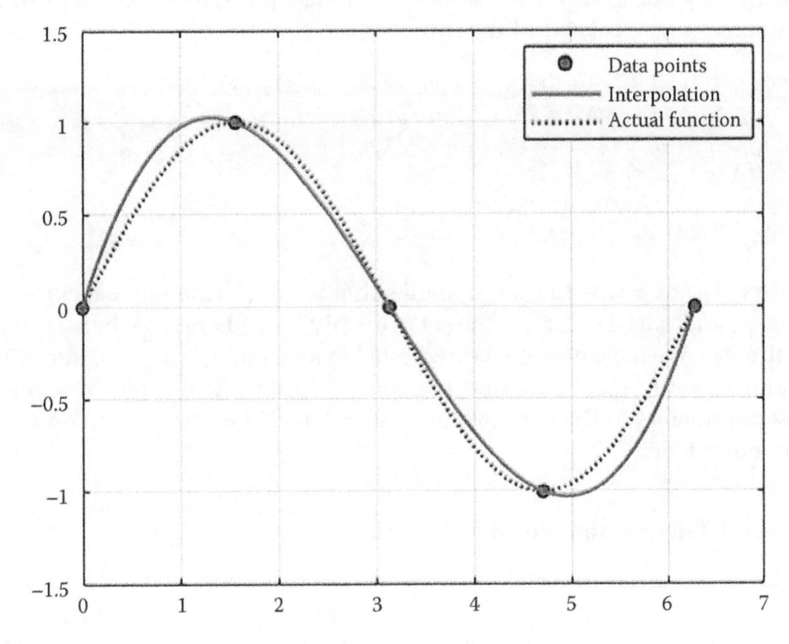

FIGURE 10.40
Interpolated values for the function y using only four known data points.

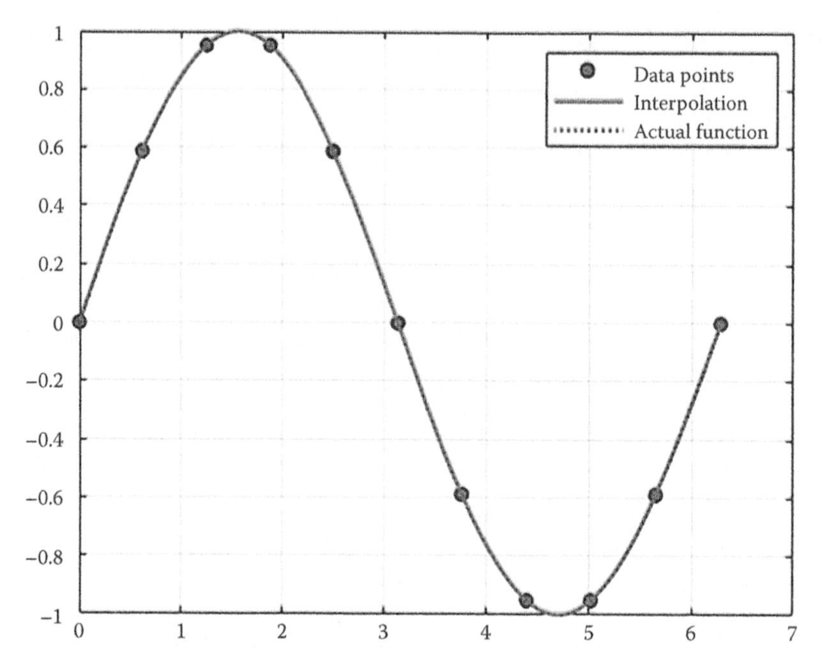

FIGURE 10.41
Interpolated values for the function *y* using 10 known data points.

assuming a constant angular velocity. Note that if we had decided to stick with 100 points per segment, we would need to calculate a different `dt` for each segment. Carrying the different `dt`'s through all of the subsequent kinematic and force calculations would be cumbersome, to say the least. We are now ready to use the spline function to estimate the profile parameters every 1° around the cam.

```
% use cubic spline to interpolate parameters on Nc points around cam
cam.s     = spline(lambda, s,cam.lambda);
cam.v     = spline(lambda, v,cam.lambda);
cam.a     = spline(lambda, a,cam.lambda);
cam.j     = spline(lambda, j,cam.lambda);
cam.c     = spline(lambda, c,cam.lambda);
```

Remember that `lambda` was the set of angles that we performed an exact calculation for each parameter, and `cam.lambda` is the set of evenly spaced angles where spline will generate the estimates. We have now converted our M values for s, v, a, j, etc. into 361 estimates for these values, evenly spaced around the cam. Before we begin plotting, there are two more useful parameters of the cam profile that we should calculate: the unit normal and the radius of curvature.

10.8.4 The Unit Tangent and Normal Vectors

Consider the set of vectors shown in Figure 10.42. The radial vector **c** stretches from the center of the cam at point A to a point B on the surface. The vector **t** is tangent to the surface of the cam at B, and the vector **u** is normal to the surface (i.e. perpendicular to **t**) pointing inward.

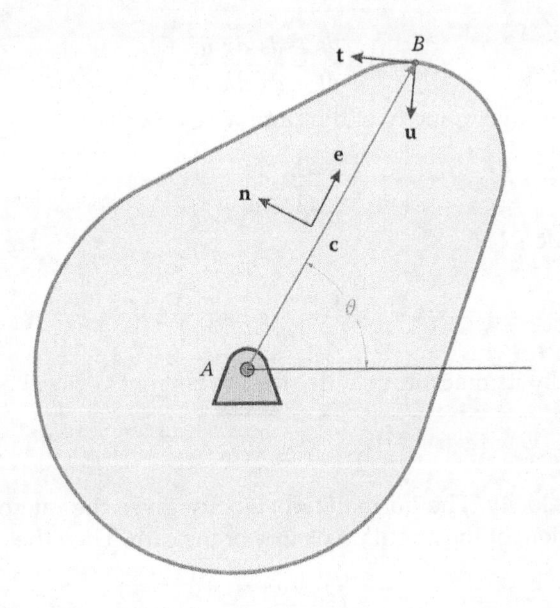

FIGURE 10.42
The unit vector **t** is tangent to the cam at point *C*, and the unit vector **u** is normal to the surface, pointing inward.

The tangent vector **t** can be thought of as the "slope" of the cam at a particular point. The formula for this vector can be found in any textbook on vector calculus; it is simply the derivative of the vector **c** with respect to the angular coordinate λ.

$$\mathbf{t} = \frac{\dfrac{d\mathbf{c}}{d\lambda}}{\left|\dfrac{d\mathbf{c}}{d\lambda}\right|} \tag{10.55}$$

where we have divided by the magnitude in order to convert **t** to a unit vector. We must use the product rule to find the derivative $d\mathbf{c}/d\lambda$

$$\frac{d\mathbf{c}}{d\lambda} = \frac{d}{d\lambda}(r_b + s)\mathbf{e} + (r_b + s)\frac{d\mathbf{e}}{d\lambda} \tag{10.56}$$

The derivative $d\mathbf{e}/d\lambda$ is simply equal to **n**, the unit normal. Since r_b is a constant, it vanishes from the first term, and we are left with

$$\frac{d\mathbf{c}}{d\lambda} = \frac{ds}{d\lambda}\mathbf{e} + (r_b + s)\mathbf{n} \tag{10.57}$$

If the point *B* happens to be in a dwell section then the displacement function, s, is constant and $ds/d\lambda = 0$. If, instead, *B* is on a rise or fall section we must employ the chain rule to find $ds/d\lambda$

$$\frac{ds}{d\lambda} = \frac{ds}{dx}\frac{dx}{d\lambda}$$

where x is the dimensionless parameter that we used to define the rise and fall sections. Now, recall the definition of velocity that we found in Section 10.4.

$$v = \frac{ds}{dt} = \frac{ds}{dx}\frac{dx}{d\lambda}\frac{d\lambda}{dt}$$

Since $d\lambda/dt = \omega$, the angular velocity of the cam, we can write

$$\frac{ds}{d\lambda} = \frac{v}{\omega}$$

Thus, the derivative $d\mathbf{c}/d\lambda$ is

$$\frac{d\mathbf{c}}{d\lambda} = \frac{v}{\omega}\mathbf{e} + r\mathbf{n} \tag{10.58}$$

Dividing this vector by its magnitude will give the tangent vector. Let us define

$$V = \frac{v}{\omega} \tag{10.59}$$

as the normalized velocity. The normalized velocity gives the rate of rise (or fall) of the cam profile independent of the angular velocity of the cam. Then the unit tangent vector is

$$\mathbf{t} = \frac{V\mathbf{e} + r\mathbf{n}}{|V\mathbf{e} + r\mathbf{n}|} \tag{10.60}$$

The unit normal vector $\mathbf{u}$ is perpendicular to the tangent vector and is found in the usual manner.

$$\mathbf{u} = \mathbf{t}^{\perp} = \begin{Bmatrix} -t_y \\ t_x \end{Bmatrix} \tag{10.61}$$

We are now able to find the unit tangent vector and unit normal vector on any point on the profile of the cam. These vectors are crucial in finding the resulting motion of the follower. We will now modify the dwell and rise segments of our code to find the unit tangent and normal vectors.

First, as shown in Figure 10.43, the unit normal on a dwell section is directed toward the center of the cam, since a dwell section is a circular arc centered at A. Thus, in the dwell section of the `CamMotion` function, add the following line of code.

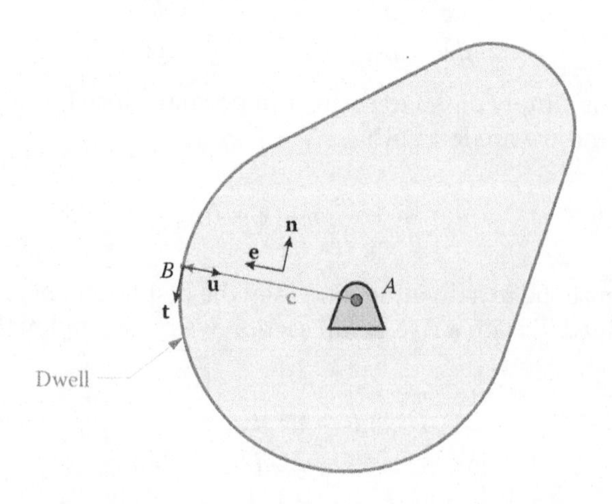

FIGURE 10.43
The unit normal on a dwell section is directed toward the center of the cam.

```
    u(:,k) = -e;            % unit normal points to center
```

The unit normal in the rise section is a little trickier, but not too terrible. Add the following lines to the rise segment of `CamMotion`.

```
% calculate unit normal to cam profile
    V = v(k)/cam.omega;                   % normalized velocity
    A = a(k)/(cam.omega^2);               % normalized acceleration
    t = V*e + r*n;                        % tangent to cam profile
    t = t/norm(t);                        % convert to unit vector
    u(:,k) = [-t(2); t(1)];               % unit normal to cam profile
```

We didn't use the normalized acceleration for the unit normal, but we will use it to calculate radius of curvature in the next section.

10.8.5 Radius of Curvature of the Cam Profile

We will next present a formula for finding the radius of curvature at any point on the cam. The concept for the radius of curvature is relatively simple, although deriving the formula is a little tricky. As seen in Figure 10.44, the radius of curvature gives the radius of a circle that "turns the corner" at the same rate as the cam profile at a given point. If the profile is turning rapidly, then the radius of curvature is small, and a gradually turning profile has a large radius of curvature. A straight line, of course, has an infinite radius of curvature.

An *inflection point* is a point on the cam profile where the curvature makes the transition from convex to concave, as shown in Figure 10.45. Since the curvature is zero here, the radius of curvature is infinite at an inflection point.

Now imagine that you are an insect traveling around the profile of the cam, as shown in Figure 10.46. Your rate of progress is such that your angular velocity, $\omega = d\lambda/dt$, is constant. Thus, your angular position can be found at any time t by

$$\lambda = \omega t \tag{10.62}$$

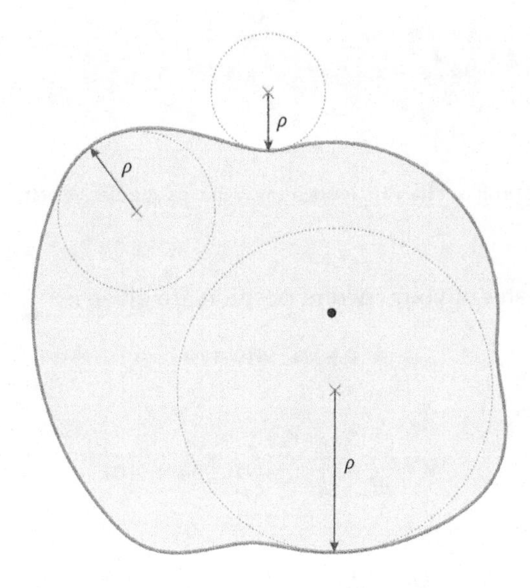

FIGURE 10.44
The radius of curvature gives the radius of a circle that has the same curvature as a point on the cam profile.

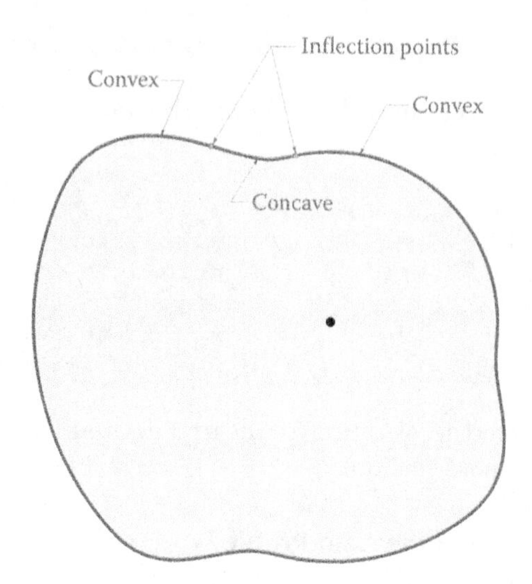

FIGURE 10.45
An inflection point occurs where the curvature makes the transition from convex to concave.

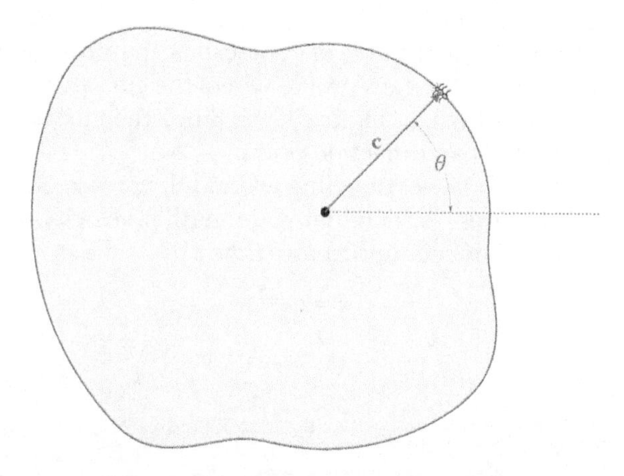

FIGURE 10.46
An insect travels around the profile of the cam in such a way that its angular velocity, $\omega = d\lambda/dt$, remains constant.

The Cartesian coordinates of your current position are given by

$$\mathbf{c} = (r_b + s)\mathbf{e} = r\mathbf{e} \tag{10.63}$$

Your current velocity is

$$\mathbf{v} = \frac{d\mathbf{c}}{dt} = \frac{ds}{dt}\mathbf{e} + \omega r\mathbf{n} = v\mathbf{e} + \omega r\mathbf{n} \tag{10.64}$$

and your current acceleration is

$$\mathbf{a} = \frac{d\mathbf{v}}{dt} = \frac{dv}{dt}\mathbf{e} + v\omega\mathbf{n} + \omega\frac{ds}{dt}\mathbf{n} - \omega^2 r\mathbf{e} \tag{10.65}$$

or

$$\mathbf{a} = a\mathbf{e} + 2v\omega\mathbf{n} - \omega^2 r\mathbf{e} \tag{10.66}$$

In the preceding equations, we have made use of the fact that

$$\frac{d\mathbf{e}}{dt} = \omega\mathbf{n} \quad \frac{d\mathbf{n}}{dt} = -\omega\mathbf{e}$$

If we separate the velocity and acceleration into x and y components, we have

$$\begin{Bmatrix} v_x \\ v_y \end{Bmatrix} = \begin{Bmatrix} v\cos\omega t - \omega r\sin\omega t \\ v\sin\omega t + \omega r\cos\omega t \end{Bmatrix}$$

and

$$\begin{Bmatrix} a_x \\ a_y \end{Bmatrix} = \begin{Bmatrix} \left(a - \omega^2 r\right)\cos\omega t - 2v\omega\sin\omega t \\ \left(a - \omega^2 r\right)\sin\omega t + 2v\omega\cos\omega t \end{Bmatrix}$$

The equation for the radius of curvature can be found in any textbook on vector calculus:

$$\rho = \frac{\left(v_x^2 + v_y^2\right)^{\frac{3}{2}}}{\left|v_x a_y - v_y a_x\right|} \tag{10.67}$$

After some algebra, we find that

$$\rho = \frac{\left(V^2 + r^2\right)^{\frac{3}{2}}}{2V^2 + r(r - A)} \tag{10.68}$$

It is remarkable that all of the sine and cosine terms have cancelled out, leaving only the position, velocity, and acceleration functions. During a dwell cycle, the velocity and acceleration are zero so that we are left with

$$\rho = r \tag{10.69}$$

that is, the radius of curvature during a dwell is simply the current radial position, as expected. If we are so foolish as to design a cam with an abrupt change in velocity (thus disobeying the fundamental law of cam design) the infinite acceleration will cause the radius of curvature to be zero – an undesirable situation.

The customary mathematical definition of radius of curvature is the absolute value of ρ found in Equation (10.67). We will use this definition when plotting the radius of curvature, but we will allow ρ to take on positive or negative values for the rest of the calculations. A positive radius of curvature means that the cam profile is *convex* at the current location and a negative radius of curvature means that the cam profile is *concave*.

We have thus found a simple formula for the radius of curvature for a cam at any point on the profile. When using a roller follower, the radius of curvature should be 2–3 times as large as the follower radius at all points on the profile, which we can check by plotting it in MATLAB. In the dwell segment of `CamMotion` add the following line

```
    rho(k)  = r;              % rad of curv is total radius
```

and in the rise segment add

```
% calculate radius of curvature of cam profile
    num =  (V^2 + r^2)^1.5;
    den = 2*V^2 + r*(r - A);
    rho(k)  = num/den;                    % radius of curvature
```

That's it! Of course, we should also use the `spline` function at the end of the code to estimate the values of **u** and ρ at 1° increments.

```
cam.rho  = spline(lambda,rho,cam.lambda);
cam.u    = spline(lambda,   u,cam.lambda);
```

If you execute the `CamDesigner` script, it will appear that nothing has happened because we haven't asked MATLAB to plot anything yet. We did, however, calculate several important cam functions, which you can see if you type `cam` at the command line.

```
>> cam

cam =
  struct with fields:
         Nd: 3
          h: [0 0.0050 0.0100]
      betad: [1.7453 0.6981 0.5236]
      betar: [1.0472 0.8727 1.3963]
         rb: 0.0250
      omega: 104.7198
    camfunc: @svaj345
         Nc: 361
    dLambda: 0.0175
     lambda: [1x361 double]
         dt: 1.6667e-04
          s: [1x361 double]
          v: [1x361 double]
          a: [1x361 double]
          j: [1x361 double]
          c: [2x361 double]
        rho: [1x361 double]
          u: [2x361 double]
```

Here you can see all of the variables that are stored in the cam structure. Vectors that are small in size are displayed in full, as are the scalars. Large vectors (such as λ) are given as a size. You can type `cam.lamba` at the command line to see the individual values. We will next create a function that will plot several of the cam parameters that we have just calculated. These will be used to verify that the *s-v-a-j* functions are the ones we intended, to check the cam profile to see that it is realistic and to verify that the radius of curvature remains at least 2–3 times the radius of the roller follower. The full text of the `CamMotion` function is shown below. At 103 lines of code, it is the longest function in our suite of cam analysis programs.

```
% The function CamMotion calculates and plots the coordinates of a
% multiple-dwell cam profile.  It also calculates and plots cams the
% s-v-a-j functions.
%
% ***** Inputs *****
% cam = parameters of cam profile
% showPlot = plotting options
%
% ***** Outputs *****
% cam = cam profile with kinematic functions

function cam = CamMotion(cam,showPlot)

% calculation parameters
N = 100;              % no. of points to calculate on each dwell/rise/fall
M = 2*cam.Nd*N + 1;   % total number of points to calculate

% allocate space for motion variables
[lambda,s,v,a,j,rho] = deal(zeros(1,M));
[c,u]                = deal(zeros(2,M));

% convert angular segments to radians
cam.betad = pi*cam.betad/180;
cam.betar = pi*cam.betar/180;

% ***** calculate cam profile *****
k = 1;                % start index counter at 1
lambda0 = 0;          % starting angle for each segment

% main loop
for nd = 1:cam.Nd     % loop through each dwell

% ***** dwell segment *****
  beta = cam.betad(nd);                % angle of dwell
  for i = 1:N                          % loop through dwell segment
    x = (i-1)/N;                       % x ranges from 0 to 1
    lambda(k) = x*beta + lambda0;      % current angle within dwell
    s(k) = cam.h(nd);                  % current displacement
    [e,~] = UnitVector(lambda(k));     % unit vector in radial dir.
    r = cam.rb + s(k);                 % radius to point B
    c(:,k) = r*e;                      % coordinates of cam profile
    rho(k) = r;                        % rad of curv is total radius
    u(:,k) = -e;                       % unit normal points to center

    k = k + 1;                         % increment index counter
  end
  lambda0 = lambda0 + cam.betad(nd); % move angle to end of current dwell

% calculate the change in height from this dwell to the next
  if (nd == cam.Nd)                % if we are at final dwell, then height
    dh = cam.h(1) - cam.h(nd);     % changes back to initial dwell,
```

```
  else                               % otherwise height changes to next
    dh = cam.h(nd+1) - cam.h(nd);    % dwell.
  end

% ***** rise (or fall) segment *****
  beta = cam.betar(nd);               % angle of rise or fall
  h = cam.h(nd);                      % starting height of rise or fall
  z = cam.omega/beta;                 % omega/beta
  for i = 1:N
    x = (i-1)/N;                      % x ranges from 0 to 1
    lambda(k) = x*beta + lambda0;     % current angle within rise
    [s(k),v(k),a(k),j(k)] = cam.camfunc(x,dh,h,z);
    [e,n] = UnitVector(lambda(k));    % unit vector in radial dir.
    r = cam.rb + s(k);                % radius to point B
    c(:,k) = r*e;                     % coordinates of cam profile

% calculate unit normal to cam profile
    V = v(k)/cam.omega;               % normalized velocity
    A = a(k)/(cam.omega^2);           % normalized acceleration
    t = V*e + r*n;                    % tangent to cam profile
    t = t/norm(t);                    % convert to unit vector
    u(:,k) = [-t(2); t(1)];           % unit normal to cam profile

% calculate radius of curvature of cam profile
    num = (V^2 + r^2)^1.5;
    den = 2*V^2 + r*(r - A);
    rho(k) = num/den;                  % radius of curvature

    k = k + 1;                         % increment index counter
  end
  lambda0 = lambda0 + cam.betar(nd);  % move angle to end of current rise
end

% parameters at 360deg are the same as at 0deg
lambda(M) = 2*pi;    % final angle is 360 degrees
c(:,M) = c(:,1);     % profile ends where it started
s(M) = s(1);         % final displacement is same as initial
u(:,M) = u(:,1);     % final unit normal is same as initial
rho(M) = rho(1);     % final radius is same as initial

% ***** plotting parameters for the cam profile *****
cam.Nc = 361;                         % number of points around cam profile
dLambda = 2*pi/(cam.Nc-1);            % angular increment
cam.lambda = dLambda*(0:cam.Nc-1); % angles on cam to calculate profile
cam.dt = dLambda/cam.omega;           % time increment

% use cubic spline to interpolate parameters on Nc points around cam
cam.s   = spline(lambda,  s,cam.lambda);
cam.v   = spline(lambda,  v,cam.lambda);
cam.a   = spline(lambda,  a,cam.lambda);
cam.j   = spline(lambda,  j,cam.lambda);
cam.c   = spline(lambda,  c,cam.lambda);
cam.rho = spline(lambda,rho,cam.lambda);
cam.u   = spline(lambda,  u,cam.lambda);
```

10.9 Plotting the Cam Profile, the *s-v-a-j* Diagram, and Other Interesting Functions

Now that we have computed the cam profile with its associated surface normals, radius of curvature and *s-v-a-j* functions, it is time to begin plotting. We will generate the following figures to help visualize our cam profile and motion:

1. The outline of the cam, **c**.
2. The *s-v-a-j* diagram for the cam.
3. The radius of curvature, ρ, versus angle on the cam, λ.
4. The profile of the cam overlaid with a circle tangent to a given point on the cam. This is used as a double-check on our radius of curvature calculations and also to help visualize where any "trouble-spots" may occur.

Every one of the cam functions that we have calculated is worthy of being plotted, and our screen will quickly become cluttered if we plot them all. For this reason, we will pass several logical variables to the `CamMotion` function. A logical variable is one that can only take on the values of `true` or `false`. In this instance, we will generate the cam profile plot if `showPlot.c` is `true`, otherwise we won't, with similar logical variables being used for the other plots. This means that we must surround each set of plotting commands with an `if`...end condition. Modify the `CamMotion` function definition so that we can pass a plotting structure called `showPlot` containing the logical variable controlling each plot. The updated first line of our `CamMotion` function should read:

```
function cam = CamMotion(cam,showPlot)
```

For completeness, we must be sure to update the function documentation as well. The new header should now read:

```
% ***** Inputs *****
% cam = parameters of cam profile
% showPlot = plotting options
```

At the bottom of the user defined variables section of our main program, `CamDesigner`, we will create our user defined plotting options `showPlot`. Populate it with the appropriate logical variables:

```
% ***** Parameters for plotting (user-defined) *****
showPlot.c = true;          % plot the profile of the cam?
showPlot.svajC = true;      % plot the s-v-a-j functions for cam?
showPlot.rho = true;        % plot the radius of curvature?
showPlot.rhoCheck = 225;    % plot radius of curvature at a point? (-1=false)
```

We can now update our function call in the CamDesigner to include our plotting options.

```
% ***** Calculate cam profile and plot motion *****
cam = CamMotion(cam,showPlot);
```

With our main program modified to accept user plotting options and our `CamMotion` function setup to receive these plotting options, let's begin generating the plots. Insert our custom color definitions at the end of `CamMotion`.

```
% Define colors for plotting
cBlu = DefineColor([  0 110 199]); % Pantone 300C
cBlk = DefineColor([  0   0   0]); % grayscale
```

Make sure that you create a copy of the `DefineColor` function in the current working directory so that the plotting colors will be defined.

10.9.1 Plotting the Cam Profile

Plotting the cam profile is very simple because we can use many of the plotting commands learned in earlier chapters. We will have many occasions to plot the cam profile as a basis for other figures, so it is useful to define it as a separate function that we can reuse as many times as we want. Open up a new script in MATLAB and type in the following function:

```
% The function PlotCamProfile plots the profile of a cam. The cam can be
% plotted in its original orientation or rotated by an angle theta2.
%
% ***** Inputs *****
% c     = 2D vector of cam profile coordinates
% theta = angle of rotation of cam
% cBlu  = shades of blue defined in the DefineColor function
% cBlk  = shades of gray defined in the DefineColor function

function PlotCamProfile(c,theta,cBlu,cBlk)

hold on
A = [cos(theta) -sin(theta);   % rotation matrix for rotating cam
     sin(theta)  cos(theta)];
cr = A*c;                      % rotate coordinates of cam profile

% plot outline of cam
fill(cr(1,:),cr(2,:),cBlu(10,:),'EdgeColor',cBlu(1,:),'LineWidth',2)

% plot cam pivot
plot(0,0,'o','MarkerSize',8,'MarkerFaceColor',cBlk(5,:),'Color',cBlk(1,:))

% plot baseline on cam
plot([0 cr(1,1)],[0 cr(2,1)],':','Color',cBlu(3,:))

axis equal; grid on
xlabel('x (m)'); ylabel('y (m)'); title('Cam Profile')
set(gcf,'Position',[300 100 800 700])
```

Most of the function is fairly self-explanatory. The first part of the function performs a coordinate transformation on the cam profile in case we wish to rotate the cam. For the present, the rotation angle `theta` will be zero, but we will need to rotate the cam when we perform motion and force analysis on the follower. Note that we have used the `fill` command, rather than `plot`. This shades within the cam boundaries to make it easier to see. The first `plot` command places a small circle at the center of the cam so that we can see the axis of rotation. The second `plot` command draws the `lambda` = 0 line on the cam so that the angle of rotation will be easier to see when we rotate the cam in later sections. Back in the `CamMotion` function enter the commands that execute the `PlotCamProfile` function.

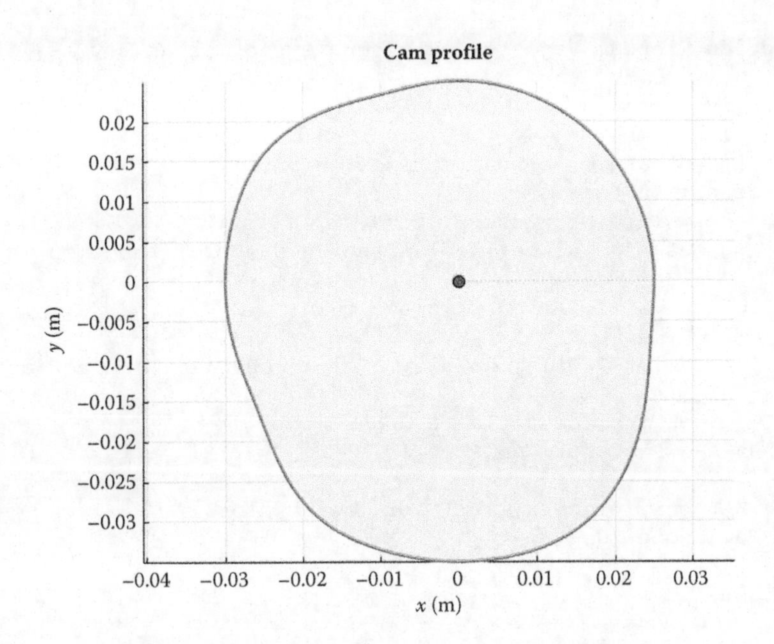

FIGURE 10.47
Cam profile as plotted by the MATLAB script with the parameters given. There are three dwells on this cam, but they are not immediately obvious by looking at the cam profile.

```
% plot profile of cam
if showPlot.c
  figure
  PlotCamProfile(cam.c,0,cBlu,cBlk)
end
```

Now hit F5 to save all of the open scripts and execute the main program.

If everything has been entered correctly, you should obtain the plot shown in Figure 10.47. The cam profile looks like a lumpy circle and it might not be immediately obvious where the three dwell sections are. Luckily, they will be more apparent in the *s-v-a-j* plot.

10.9.2 The *s-v-a-j* Diagram for the Cam

In plotting the *s-v-a-j* diagram your first thought might be to construct four separate plot windows, one for each function. It is much nicer, however, to have all four plots in one window so that they can easily be compared with one another. Luckily, MATLAB has a command `subplot` that can place several different plots in the same window. The syntax for subplot is

```
subplot(# of plot rows, # of plot columns, plot number)
```

That is, if you wanted to access the fifth plot in a window containing three rows and two columns of plots you would type

```
subplot(3, 2, 5)
```

Since each plot in the *s-v-a-j* diagram is substantially the same, we should define another function to create a single plot with axis labels, grid lines, etc. Open a new script and type the following (very short) function.

```
% The function PlotCamSvaj creates a single plot on an s-v-a-j diagram
%
% ***** Inputs *****
% i      = plot number (1-4)
% lambda = set of angles on cam (x axis on plot)
% y      = function to plot
% ytitle = the y label for the plot (s, v, a or j)
% cBlu   = shades of blue defined in the DefineColor function

function PlotCamSvaj(i,lambda,y,ytitle,cBlu)

subplot(4,1,i), plot(180*lambda/pi,y,'Color',cBlu(1,:),'LineWidth',2)

if (i == 1)  % only place title on upper plot
  title('s-v-a-j Plot for Cam Profile')
end
if (i == 4)  % only place x label for lowest plot
  xlabel('Cam angle (deg)')
end
ylabel(ytitle)

grid on
xlim([0 360])
set(gca,'xtick',0:60:360)
```

We have included an `if` statement so that the title is only placed on the uppermost plot and the x label is only placed on the bottom plot, since it is the same for all four figures. At the bottom of the `CamMotion` function, type the following

```
% create s-v-a-j diagram for cam
if showPlot.svajC
  figure
  PlotCamSvaj(1,cam.lambda,cam.s,     'Disp (m)',cBlu)
  PlotCamSvaj(2,cam.lambda,cam.v,    'Vel (m/s)',cBlu)
  PlotCamSvaj(3,cam.lambda,cam.a, 'Acc (m/s^2)',cBlu)
  PlotCamSvaj(4,cam.lambda,cam.j,'Jerk (m/s^3)',cBlu)

  set(gcf,'Position',[200 50 1000 750])
end
```

This has made the `CamMotion` program much shorter, and much less repetitive than it would have been if we had set the parameters of each plot individually. If you change `showPlot.svajC` to `true` in the main program, you will obtain the *s-v-a-j* diagram shown in Figure 10.48. It is likely that the default MATLAB plot window is too small to see the details in the *s-v-a-j* plot. This is the reason for the statement

```
  set(gcf,'Position',[100 50 1300 700])
```

in the set of plot commands. This sets the size parameters for the current plot window. The syntax for this command is

```
set(gcf,'Position',[x-coordinate y-coordinate width height])
```

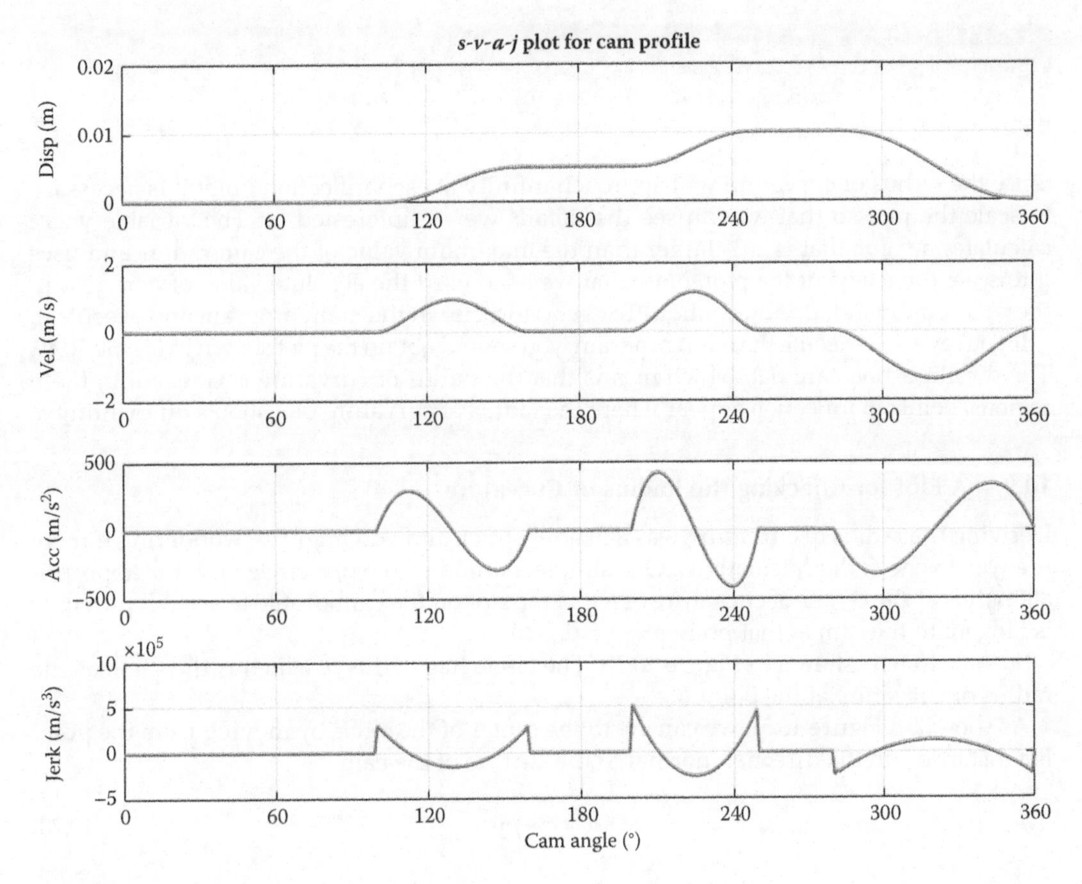

FIGURE 10.48
s-v-a-j diagram of the cam as plotted by the MATLAB script. The locations of dwells, rises, and falls are now obvious.

where `x-coordinate` and `y-coordinate` give the distance (in pixels) from the upper left corner of the screen. The values given above work well on a 14-inch laptop screen, but you may need to 'tweak' them for your own screen.

10.9.3 Plotting the Radius of Curvature

The plotting commands for radius of curvature are familiar to us by now. Enter these lines in the `CamMotion` function:

```
% Plot radius of curvature versus lambda
if showPlot.rho
  figure
  ymax = 1.2*(cam.rb + max(cam.h));   % maximum radius of cam
  plot(180*cam.lambda/pi,abs(cam.rho),'LineWidth',2,'Color',cBlu(1,:))

  xlabel('Cam Angle (deg)'); ylabel('Radius of Curvature (m)')
  title('Radius of Curvature vs. Cam Angle')
  grid on
```

```
    axis([0 360 0 ymax])
    set(gca,'xtick',0:60:360)
    set(gcf,'Position',[100 100 1000 600])
end
```

Since the radius of curvature will approach infinity at each inflection point it is necessary to scale the plot so that we can see the details we are interested in. The variable `ymax` calculates a value that is 20% larger than the maximum value of the cam radius and uses it to scale the y axis of the plot. Note that we have used the absolute value of `cam.rho` in the `plot` command. Change `showPlot.rho` to `true` in the main program and execute it.

If you save and execute the main program, you should obtain the plot shown in Figure 10.49. The dwell portions are flat, which means that the radius of curvature is constant in these regions. Points of inflection appear where the radius of curvature plot shoots off to infinity.

10.9.4 A Plot for Checking the Radius of Curvature

Deriving the radius of curvature was a little abstract, and you might be wondering if there is a way to check the calculations. One simple method is to plot a circle with the appropriate radius at the center of curvature of a given point on the cam profile to see if it the circle is tangent to the cam at that point.

Such a circle is shown in Figure 10.50. The circle has radius ρ, which is the same as the radius of curvature at the point B.

As shown in Figure 10.51, we can get to the center of the circle by moving from the point B a distance ρ in the direction normal to the surface of the cam.

$$\mathbf{g} = \mathbf{c} + \rho\mathbf{u} \tag{10.70}$$

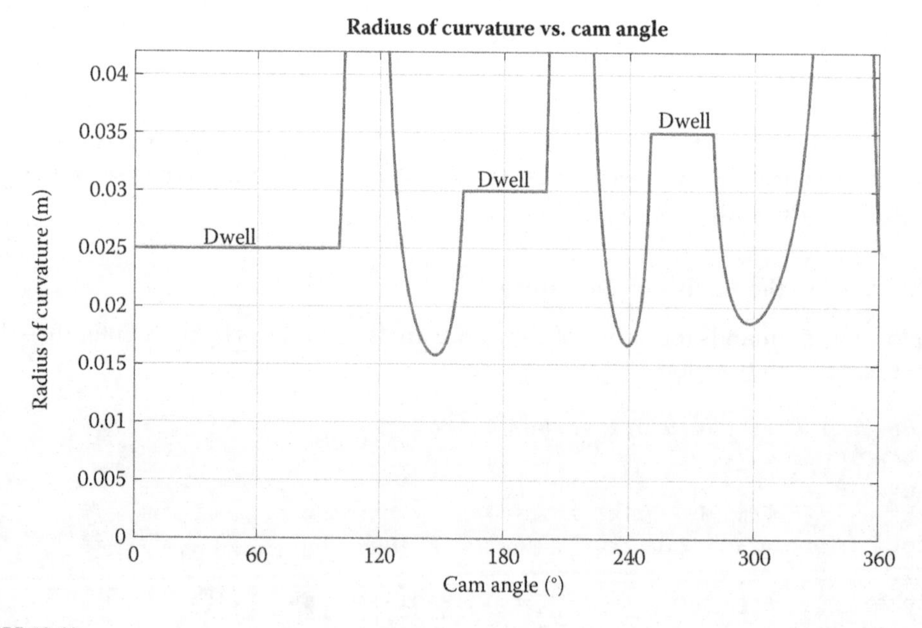

FIGURE 10.49
Radius of curvature for our example cam. Note that the radii match the outer radius of the cam during the dwell portions.

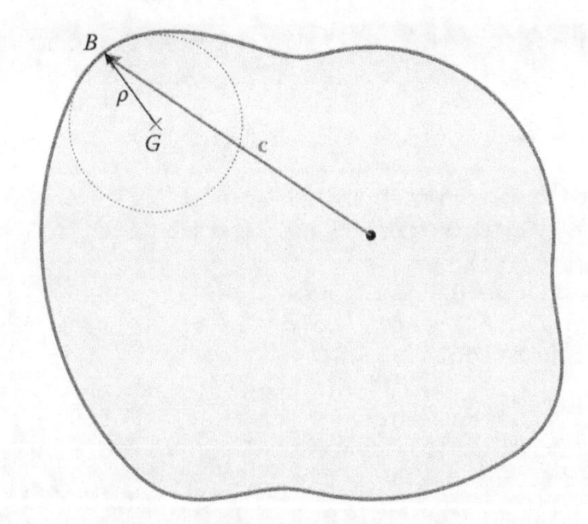

FIGURE 10.50
The circle centered at G has the same radius of curvature as the cam at point B and is also tangent to the cam at this point.

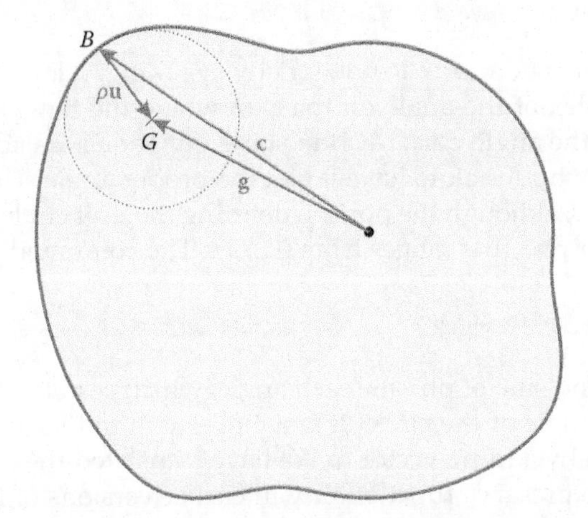

FIGURE 10.51
To get to the center of the circle move from the point B a distance ρ along the normal to the cam profile.

At the bottom of the `CamMotion` function, enter the following plotting commands:

```
% Plot tangent circle and normal to cam surface
if (showPlot.rhoCheck > -1)
  figure
  ymax = 1.2*(cam.rb + max(cam.h));          % maximum radius of cam
  CamProfilePlot(cam.c,0,cBlu,cBlk);         % plot profile of cam

  phi = pi*(0:360)/180;                      % define unit circle
  circ = [cos(phi); sin(phi)];               % centered at origin

  i = rhoCheck + 1;                          % angle of tangent point
  g = cam.c(:,i) + cam.rho(i)*cam.u(:,i); % center of circle
 circ(1,:) = cam.rho(i)*circ(1,:) + g(1); % scale and move circle x-coords
```

```
circ(2,:) = cam.rho(i)*circ(2,:) + g(2); % scale and move circle y-coords

% plot the circle
  plot(circ(1,:),circ(2,:),':','Color',cBlk(5,:),'LineWidth',2)

% plot the center of the circle
  plot(g(1),g(2),'+','Color',cBlk(5,:),'MarkerSize',8)

% plot the point B on the cam surface
  plot(cam.c(1,i),cam.c(2,i),'p','Color',cBlk(5,:),'MarkerSize',8,...
       'MarkerFaceColor',cBlk(5,:))

% plot the normal to the cam surface
  plot([g(1) cam.c(1,i)],[g(2) cam.c(2,i)],':','Color',cBlk(5,:),...
       'LineWidth',2)

  title(['Contact at ' num2str(180*cam.lambda(i)/pi) 'deg. ' ...
         ' rho = ' num2str(cam.rho(i)) 'm'])
  axis([-ymax ymax -ymax ymax])
end
```

The variable `showPlot.rhoCheck` is not, strictly speaking, a logical variable. Instead, it is used to give the index of the angle on the cam where the tangent circle should touch the cam profile. Since the angle `cam.lambda` starts out at `cam.lambda(1) = 0`, we must add one to `showPlot.rhoCheck` to have it plot the proper angle. The rest of the syntax is mostly self-explanatory, although the portion defining the unit circle may be a little subtle. We first define a vector `phi` that ranges from 0 to 2π. The command

```
  circ = [cos(phi); sin(phi)];
```

calculates the cosine and sine of `phi` for each angle, which results in a set of x and y coordinates, one for each angle of the circle. It is a circle with radius 1. We translate the unit circle to its new origin by adding vector `g`. We have translated the x and y components of the circle separately to preserve functionality in earlier versions of MATLAB. In the 2017 release this translation can be completed in one step, using

```
circ = cam.rho(i)*circ + g; %scale and translate circle
```

Remember that `showPlot.rhoCheck` gives the *index* of the current position on the cam profile, not the angle itself. We have included the angle λ in the title of the plot to make life easier for the user. Since we have done our calculations every 1°, there will be a one-to-one correspondence between `showPlot.rhoCheck` and λ, but this would not be the case if we had done our calculations every half-degree, for instance.

To execute the radius of curvature check, change the value of `showPlot.rhoCheck` in the main `CamDesigner` program to read

```
showPlot.rhoCheck = 75; % plot radius of curvature at a point? (-1=false)
```

Example plots for three different values of `showPlot.rhoCheck` can be seen in Figures 10.52–10.54. This completes our set of functions for calculating and plotting the cam profile and its associated features.

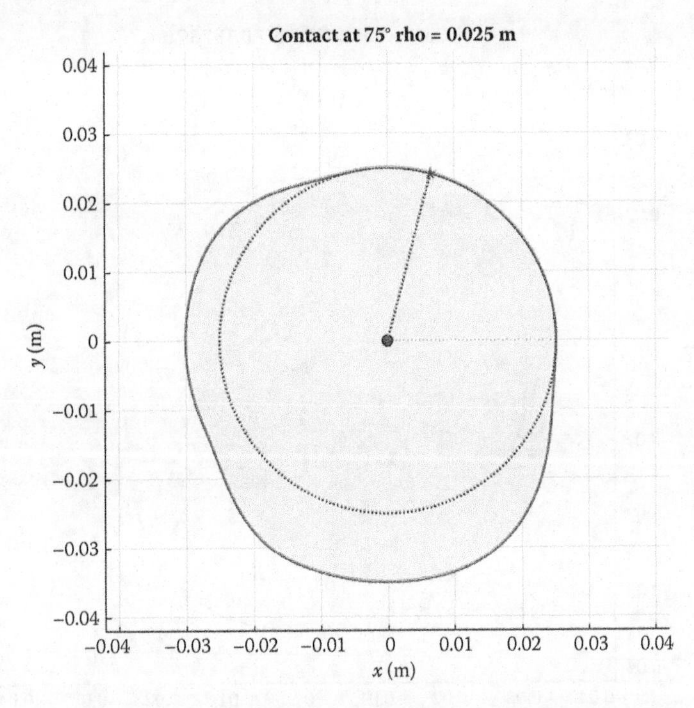

FIGURE 10.52
Tangent circle where $\lambda = 75°$. This is the first dwell segment with $s = 0$, so the circle is the same as the base circle.

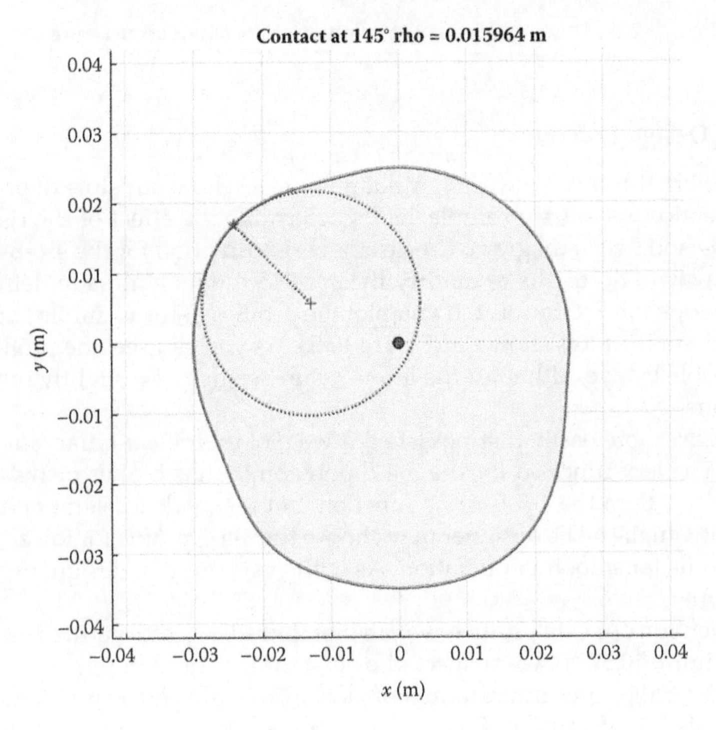

FIGURE 10.53
Tangent circle where $\lambda = 145°$. This is part of the second rise function and the radius of curvature is relatively small.

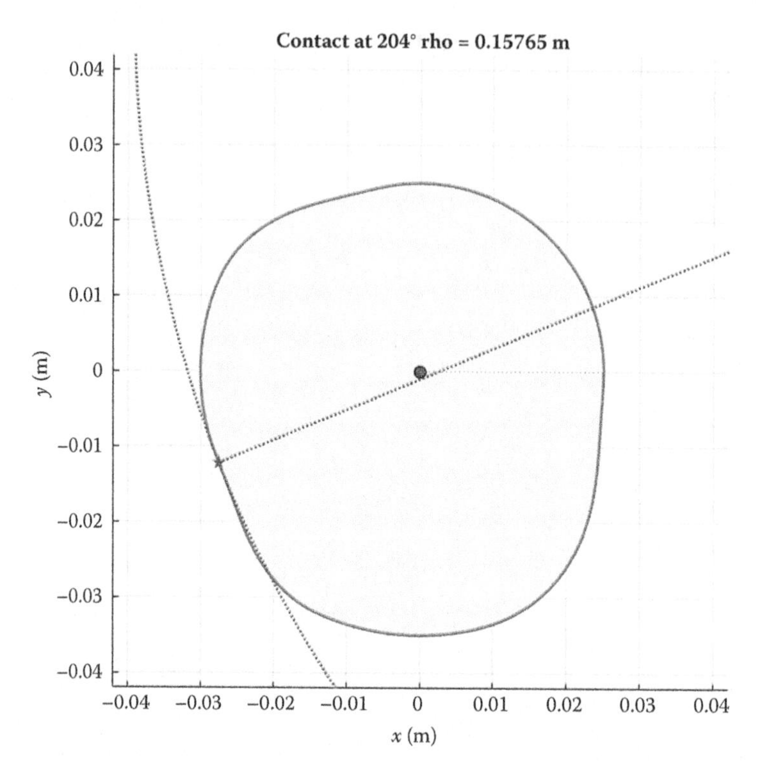

FIGURE 10.54
Tangent circle where $\lambda = 204°$. This is near a point of inflection so the radius of curvature is very large.

10.9.5 Some Design Examples

We will now work through a few design examples to see how our suite of programs might be used in practice. As a first example, let us determine the effect of the rise function on the cam profile and *s-v-a-j* diagram. Create a new rise function for the 4-5-6-7 polynomial. The simplest way to do this is to modify the svaj345 function already defined using the formulas developed in Section 10.4. If you plot the profiles of the example triple-dwell cam you should obtain the plots shown in Figure 10.55. As you can see, the profiles are almost indistinguishable by eye, although the 4-5-6-7 cam seems to be slightly more concave at the rise sections.

The profiles are more easily distinguished if we plot their *s-v-a-j* diagrams, as shown in Figure 10.56. The jerk function for the 3-4-5 polynomial has higher peaks (and is generally more "jerky") than the 4-5-6-7 rise function, but the peak acceleration for the 4-5-6-7 profile is slightly higher. The designer may choose the 3-4-5 profile for lower peak forces or the 4-5-6-7 profile for smoother operation. As with everything in design, there is no single correct solution!

As a final check on our cam profiles, we can plot the radius of curvature versus cam angle, as shown in Figure 10.57. As we suspected by examining the cam profiles, the 4-5-6-7 rise function has a smaller minimum radius of curvature – around 13 mm versus the 16 mm or so for the 3-4-5 rise function. If we choose a roller follower with roller diameter of 1/3 the minimum radius of curvature of the cam, this would require a 8.6 mm diameter roller for the 4-5-6-7 cam and a 10.7 mm diameter roller for the 3-4-5 cam. As a practical matter,

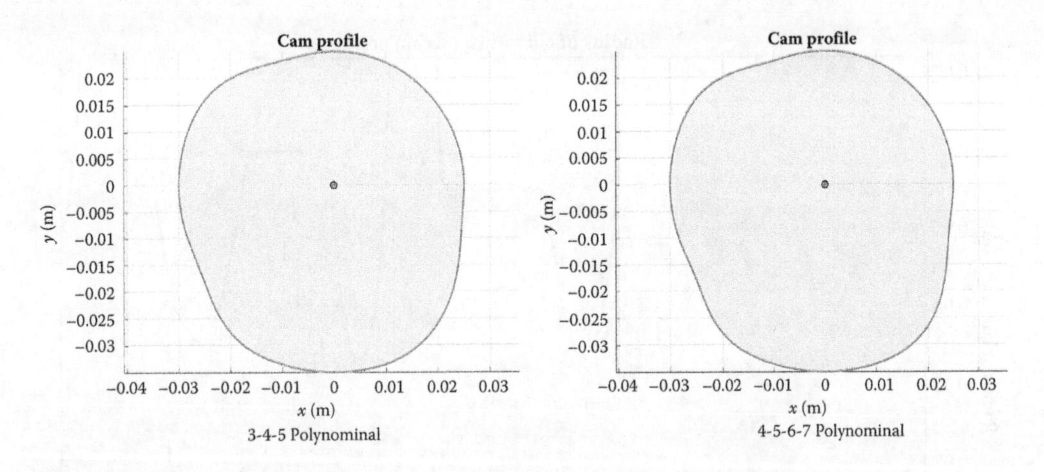

FIGURE 10.55
Comparison of cam profiles generated with the 3-4-5 polynomial and 4-5-6-7 polynomial rise functions. The profiles are almost indistinguishable.

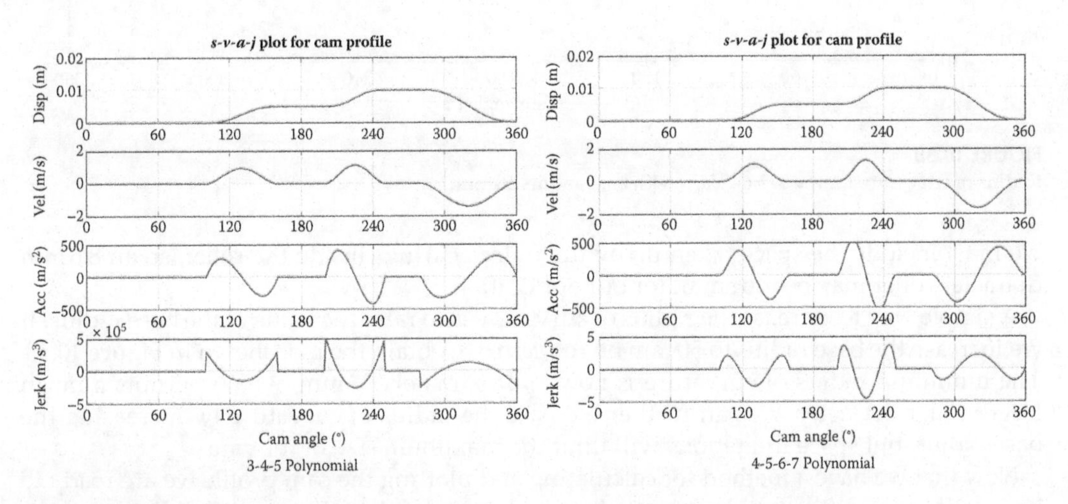

FIGURE 10.56
s-v-a-j diagrams for the 3-4-5 and 4-5-6-7 rise functions for the triple-dwell cam.

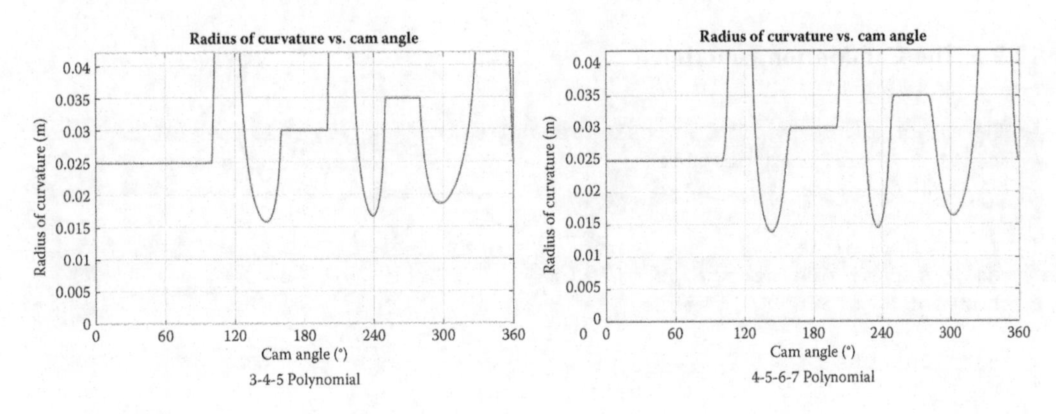

FIGURE 10.57
Radius of curvature plots for the 3-4-5 and 4-5-6-7 polynomial rise functions for the triple-dwell example cam.

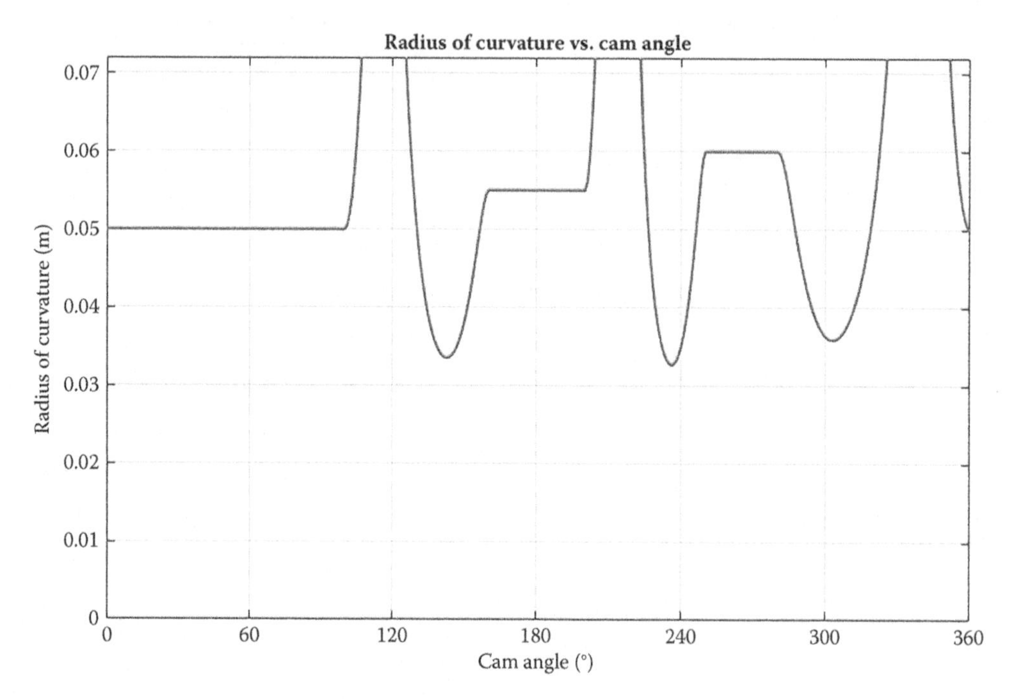

FIGURE 10.58
Radius of curvature for the 4-5-6-7 cam with base radius 50 mm.

we must include the space taken up by the roller bearings inside the roller, so an 8.6 mm diameter roller may be too small for our application.

A simple way to increase the radius of curvature is to raise the value of the base radius. If we increase the base radius to 50 mm on the cam, we obtain the plot shown in Figure 10.58. The minimum radius of curvature is now approximately 33 mm, which permits a much larger roller diameter. We can further increase the radius of curvature by increasing the base radius, but space limitations will limit the maximum size of the cam.

Now that we have a method for calculating and plotting the cam profile we are ready to begin the more difficult task of modeling the motion of the follower. A full listing of the `CamMotion` function that we have developed in this section is given below.

10.9.6 The CamMotion Function

```
% The function CamMotion calculates and plots the coordinates of a
% multiple-dwell cam profile.  It also calculates and plots cams the
% s-v-a-j functions.
%
% ***** Inputs *****
% cam = parameters of cam profile
% showPlot = plotting options
%
% ***** Outputs *****
% cam = cam profile with kinematic functions

function cam = CamMotion(cam,showPlot)
```

```
% calculation parameters
N = 100;              % no. of points to calculate on each dwell/rise/fall
M = 2*cam.Nd*N + 1; % total number of points to calculate

% allocate space for motion variables
[lambda,s,v,a,j,rho] = deal(zeros(1,M));
[c,u]                = deal(zeros(2,M));

% convert angular segments to radians
cam.betad = pi*cam.betad/180;
cam.betar = pi*cam.betar/180;

% ***** calculate cam profile *****
k = 1;               % start index counter at 1
lambda0 = 0;         % starting angle for each segment

% main loop
for nd = 1:cam.Nd   % loop through each dwell

  % ***** dwell segment *****
  beta = cam.betad(nd);                % angle of dwell
  for i = 1:N                          % loop through dwell segment
    x = (i-1)/N;                       % x ranges from 0 to 1
    lambda(k) = x*beta + lambda0;      % current angle within dwell
    s(k) = cam.h(nd);                  % current displacement
    [e,~] = UnitVector(lambda(k));     % unit vector in radial dir.
    r = cam.rb + s(k);                 % radius to point B
    c(:,k) = r*e;                      % coordinates of cam profile
    rho(k) = r;                        % rad of curv is total radius
    u(:,k) = -e;                       % unit normal points to center

    k = k + 1;                         % increment index counter
  end
  lambda0 = lambda0 + cam.betad(nd); % move angle to end of current dwell

  % calculate the change in height from this dwell to the next
  if (nd == cam.Nd)                 % if we are at final dwell, then height
    dh = cam.h(1) - cam.h(nd);      % changes back to initial dwell,
  else                              % otherwise height changes to next
    dh = cam.h(nd+1) - cam.h(nd); % dwell.
  end

  % ***** rise (or fall) segment *****
  beta = cam.betar(nd);                % angle of rise or fall
  h = cam.h(nd);                       % starting height of rise or fall
  z = cam.omega/beta;                  % omega/beta
  for i = 1:N
    x = (i-1)/N;                       % x ranges from 0 to 1
    lambda(k) = x*beta + lambda0;      % current angle within rise
    [s(k),v(k),a(k),j(k)] = cam.camfunc(x,dh,h,z);
    [e,n] = UnitVector(lambda(k));     % unit vector in radial dir.
    r = cam.rb + s(k);                 % radius to point B
    c(:,k) = r*e;                      % coordinates of cam profile
```

```
    % calculate unit normal to cam profile
    V = v(k)/cam.omega;                 % normalized velocity
    A = a(k)/(cam.omega^2);             % normalized acceleration
    t = V*e + r*n;                      % tangent to cam profile
    t = t/norm(t);                      % convert to unit vector
    u(:,k) = [-t(2); t(1)];             % unit normal to cam profile

    % calculate radius of curvature of cam profile
    num = (V^2 + r^2)^1.5;
    den = 2*V^2 + r*(r - A);
    rho(k) = num/den;                    % radius of curvature

    k = k + 1;                           % increment index counter
  end
  lambda0 = lambda0 + cam.betar(nd);    % move angle to end of current rise
end

% parameters at 360deg are the same as at 0deg
lambda(M) = 2*pi;   % final angle is 360 degrees
c(:,M) = c(:,1);    % profile ends where it started
s(M) = s(1);        % final displacement is same as initial
u(:,M) = u(:,1);    % final unit normal is same as initial
rho(M) = rho(1);    % final radius is same as initial

% ***** plotting parameters for the cam profile *****
cam.Nc = 361;                       % number of points around cam profile
dLambda = 2*pi/(cam.Nc-1);          % angular increment
cam.lambda = dLambda*(0:cam.Nc-1); % angles on cam to calculate profile
cam.dt = dLambda/cam.omega;         % time increment

% use cubic spline to interpolate parameters on Nc points around cam
cam.s   = spline(lambda,  s,cam.lambda);
cam.v   = spline(lambda,  v,cam.lambda);
cam.a   = spline(lambda,  a,cam.lambda);
cam.j   = spline(lambda,  j,cam.lambda);
cam.c   = spline(lambda,  c,cam.lambda);
cam.rho = spline(lambda,rho,cam.lambda);
cam.u   = spline(lambda,  u,cam.lambda);


% ***** Plotting Code *****

% define colors for plotting
cBlu = DefineColor([  0 110 199]); % Pantone 300C
cBlk = DefineColor([  0   0   0]); % grayscale

% plot profile of cam
if showPlot.c
  figure
  PlotCamProfile(cam.c,0,cBlu,cBlk)
end

% create s-v-a-j diagram for cam
```

```
if showPlot.svajC
  figure
  PlotCamSvaj(1,cam.lambda,cam.s,     'Disp (m)',cBlu)
  PlotCamSvaj(2,cam.lambda,cam.v,    'Vel (m/s)',cBlu)
  PlotCamSvaj(3,cam.lambda,cam.a, 'Acc (m/s^2)',cBlu)
  PlotCamSvaj(4,cam.lambda,cam.j,'Jerk (m/s^3)',cBlu)

  set(gcf,'Position',[200 50 1000 750])
end

% plot radius of curvature versus lambda
if showPlot.rho
  figure
  ymax = 1.2*(cam.rb + max(cam.h));  % maximum radius of cam
  plot(180*cam.lambda/pi,abs(cam.rho),'LineWidth',2,'Color',cBlu(1,:))

  xlabel('Cam Angle (deg)'); ylabel('Radius of Curvature (m)')
  title('Radius of Curvature vs. Cam Angle')
  grid on
  axis([0 360 0 ymax])
  set(gca,'xtick',0:60:360)
  set(gcf,'Position',[100 100 1000 600])
end

% plot tangent circle and normal to cam surface
if (showPlot.rhoCheck > -1)
  figure
  ymax = 1.2*(cam.rb + max(cam.h));    % maximum radius of cam
  PlotCamProfile(cam.c,0,cBlu,cBlk);   % plot profile of cam

  phi = pi*(0:360)/180;                % define unit circle
  circ = [cos(phi); sin(phi)];         % centered at origin

  i = showPlot.rhoCheck + 1;                 % angle of tangent point
  g = cam.c(:,i) + cam.rho(i)*cam.u(:,i); % center of circle
  circ(1,:) = cam.rho(i)*circ(1,:) + g(1); % scale and move circle x-coords
  circ(2,:) = cam.rho(i)*circ(2,:) + g(2); % scale and move circle y-coords

  % plot the circle
  plot(circ(1,:),circ(2,:),':','Color',cBlk(5,:),'LineWidth',2)

  % plot the center of the circle
  plot(g(1),g(2),'+','Color',cBlk(5,:),'MarkerSize',8)

  % plot the point B on the cam surface
  plot(cam.c(1,i),cam.c(2,i),'p','Color',cBlk(5,:),'MarkerSize',8,...
       'MarkerFaceColor',cBlk(5,:))

  % plot the normal to the cam surface
  plot([g(1) cam.c(1,i)],[g(2) cam.c(2,i)],':','Color',cBlk(5,:),...
       'LineWidth',2)
```

```
  title(['Contact at ' num2str(180*cam.lambda(i)/pi) 'deg. ' ...
        ' rho = ' num2str(cam.rho(i)) 'm'])
  axis([-ymax ymax -ymax ymax])
end
```

10.10 Motion of the Follower

In order to be useful, every cam must transmit its motion to another link or body in its environment. In this section, we will consider the motion that results when a cam rolls against one of the three types of followers shown in Figure 10.59. The simplest follower motion to model is the flat-faced translational follower and the other two types are a little more complicated. Both the first and second types move in translation only, and the third moves rotationally. To simplify things, we will refer to the translational roller-follower as the *roller-follower,* and the oscillating roller-follower we will call the *rocker-follower.* By the end of this section, we will have developed another set of functions that perform the position, velocity, and acceleration analysis for each type of follower.

We assume that the design of the cam has already been carried out using the techniques of the previous sections, and we wish to analyze the motion of the follower as the cam rotates. Do not be deceived by the apparent simplicity of the geometry shown in Figure 10.59 – the motion of the follower is surprisingly complex and will require us to learn a few new facts about vector geometry. Once we have done the math, however, it will prove to be relatively simple to write the MATLAB scripts to perform the analysis for us.

Given the complicated nature of the cam profile functions, the position analysis equations for the follower are nonlinear and very difficult (or impossible) to solve analytically. We must, therefore, employ a nonlinear equation solving routine to find the exact position of the follower. For the same reason we must use numerical differentiation to find the velocity and acceleration of the follower. A flowchart of our solution technique is shown in Figure 10.60. We first rotate the cam through one revolution in 1° increments. At each step we make an estimate of the point of contact with the follower based on the cam profile information we obtained in the `CamMotion` function. This will only be an estimate, since

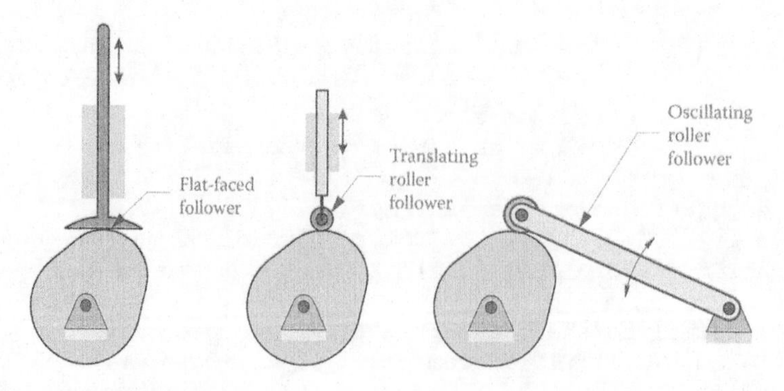

FIGURE 10.59
The motion of the cam is transmitted to a follower. The motion of the follower can be translational or rotational.

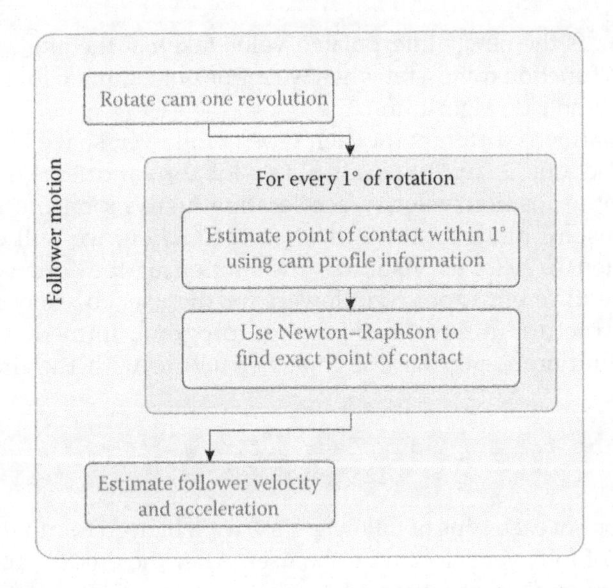

FIGURE 10.60
Flowchart for the follower motion functions. We first estimate the point of contact based on the cam profile information we have already computed, then we refine the estimate using the Newton–Raphson algorithm.

we computed the cam profile in (relatively coarse) 1° increments. It is a good enough estimate, however, to serve as a starting "guess" for the Newton–Raphson algorithm that will find the exact point of contact. Once we know the exact point of contact, we can easily find the position of the follower to a high degree of precision, along with its time derivatives (velocity and acceleration).

10.10.1 Spline Interpolation—Part 2

We can use the cam profile information already computed to find an estimate of the point of contact, but we will need to refine this information to a very fine degree when we use Newton–Raphson to solve for the exact point of contact. Recall that we used cubic spline interpolation to estimate the cam parameters at 1° increments around the profile. MATLAB allows us to "store" its values that it used in creating the cubic spline for later use if we wish to calculate interpolated values at other angles. Just before the plotting section of `CamMotion` function, type the following three lines.

```
% define spline functions for follower calculations
cam.splineU = spline(lambda,u);
cam.splineC = spline(lambda,c);
```

The field `cam.splineU` contains the coefficients of the cubic polynomials used to calculate the cubic interpolation of the surface normals to the cam. By giving the `spline` function only two arguments (as opposed to the three arguments we gave it before) we are telling MATLAB that we want the coefficients of the cubic polynomials for later use. We can use MATLAB's `ppval` function to interpolate the surface normal at any other angle by typing

```
ustar = ppval(cam.splineU,lambdaStar)
```

The variable ustar is the newly interpolated value taken at the angle lambdaStar. We will use the ppval function quite a bit when we refine our estimate for the point of contact using the Newton–Raphson algorithm.

The position equation is different for each type of follower, so we will need to develop three different functions: FlatMotion, RollerMotion, and RockerMotion. Each of these will perform a position/velocity/acceleration analysis on the follower and will generate a set of useful plots. At the end of each function, we will create a primitive MATLAB animation to help us visualize the motion of the follower. Knowing that we will have three different types of followers for the user to select from we will create a follower structure in our CamDesigner program. Immediately following our cam structure definition create the follower structure with the first field being the follower type:

```
% ***** Parameters for follower (user-defined) *****
follower.type = 'flat';     % follower type ['flat' | 'roller' | 'rocker']
```

Since the calculations of each type of follower vary, we will need to run the correct function based on the type of follower selected by the user. After the CamMotion function is run, which is the same regardless of follower type, let us use a switch-case statement to select the correct follower code to run. After the CamMotion function in our CamDesigner program add

```
% ***** Perform calculations based on follower type specified *****
switch follower.type
  case 'flat'
    % Calculate and plot motion of the follower
    follower = FlatMotion(cam,follower,showPlot);

  case 'roller'
    % Calculate and plot motion of the follower
    follower = RollerMotion(cam,follower,showPlot);

  case 'rocker'
    % Calculate and plot motion of the follower
    follower = RockerMotion(cam,follower,showPlot);

  otherwise
    disp('ERROR: Invalid follower type specified, choose flat, roller ',...
         'or rocker')
end
```

The switch reads the value stored in follower.type and then executes the code in the case that matches. If it cannot find a correct match then the otherwise block of code is executed. In our case, we use the otherwise section to throw an error and alert the user they have made an invalid selection of follower types.

10.10.2 Motion of the Flat-Faced Follower

The basic geometry of the cam and flat-faced follower is shown in Figure 10.61. We assume that the follower extends to infinity in the positive and negative y directions so that it always makes contact with the cam at its maximum x coordinate. This gives us a hint as

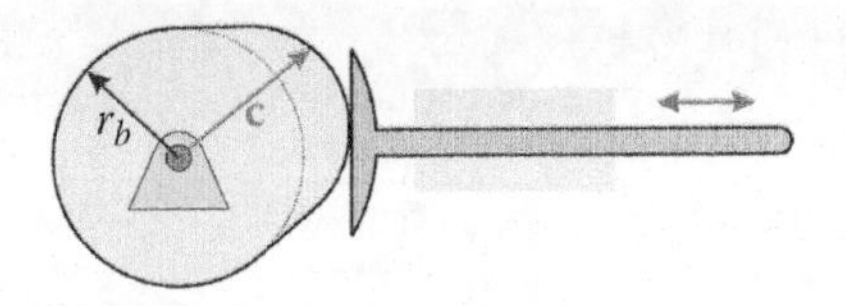

FIGURE 10.61
The flat-faced follower makes contact at the maximum x coordinate on the cam.

to how we might proceed in obtaining an estimate of the point of contact. First, rotate the cam to the desired angle. Next, search through all x coordinates of the rotated cam profile, **c**, until the maximum x value is found. This will give the point of contact to within 1°, since we solved for **c** at every 1° on the cam. Open a new script and type the following header:

```
% The function FlatMotion calculates and animates the position, velocity
% and acceleration of a flat follower.
%
% ***** Inputs *****
% cam      = cam parameters
% follower = follower parameters
% showPlot = plotting parameters
%
% ***** Outputs *****
% follower = position, velocity and acceleration of the follower

function follower = FlatMotion(cam,follower,showPlot)
```

We begin the function by rotating the cam one revolution in 1° increments. Recall that the cam profile was defined by moving around the cam in the clockwise direction – in the s diagram a positive movement along the λ axis corresponds to a clockwise rotation around the cam. Recall this λ was defined as positive in the clockwise direction so when we rotate the *cam* itself in the traditional counterclockwise angle definition direction, the follower will encounter the cam profile in the opposite direction than we designed, as shown in Figure 10.62. Enter the for loop below in the `FlatMotion` function.

```
% ***** rotate cam one revolution *****
for i = 1:cam.Nc
  theta = cam.lambda(i);          % rotation angle of cam
  A = [cos(theta) -sin(theta);    % rotation matrix
```

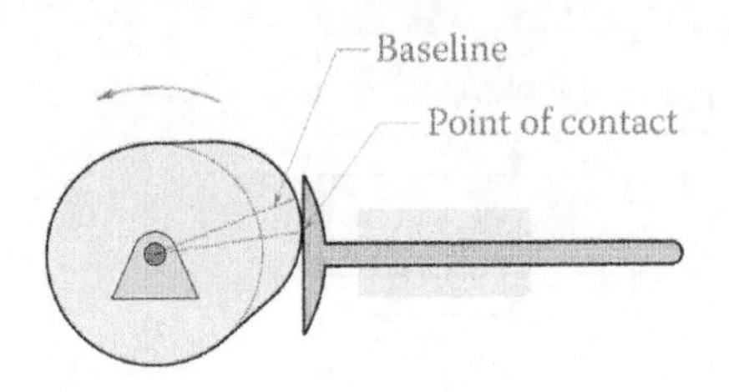

FIGURE 10.62
The follower encounters the cam profile in the opposite direction from the way we designed it.

```
        sin(theta)   cos(theta)];
    cr = A*cam.c;                          % rotated cam profile
end
```

To rotate the cam, we simply multiply the cam profile, `cam.c`, by the rotation matrix `A`. To find the maximum x coordinate of the rotated cam profile, we can use the built-in `max` function in MATLAB.

```
    [~,maxIndex] = max(cr(1,:));      % find max x coordinate of cam profile
```

Note that we are not particularly interested in the maximum *value* of the profile at this point, we just want to know which *index* of `cr` contains the maximum value, since we'll use this point as a starting value for the Newton–Raphson algorithm.

A simplified flowchart of the Newton–Raphson algorithm is shown in Figure 10.63; see Chapter 4 for a full explanation of the algorithm. The purpose of the algorithm is to find the value of the variable q that makes the function Φ take on the value of zero. In our case the variable q is the angle λ on the cam. Since we are trying to find the *maximum* value of the x coordinate on the cam profile, it is not immediately obvious which function we should set to zero.

A closer look at the cam/follower interface, shown in Figure 10.64, can give us a hint as to an appropriate function. Assuming that the cam function is continuous, the surface normal to the cam **u** at the point of contact is perpendicular to the face of the follower; that is, it is horizontal. The vertical component of **u** should be zero at the point of contact, and we have found an appropriate function.

$$\Phi(q) = u_y \tag{10.71}$$

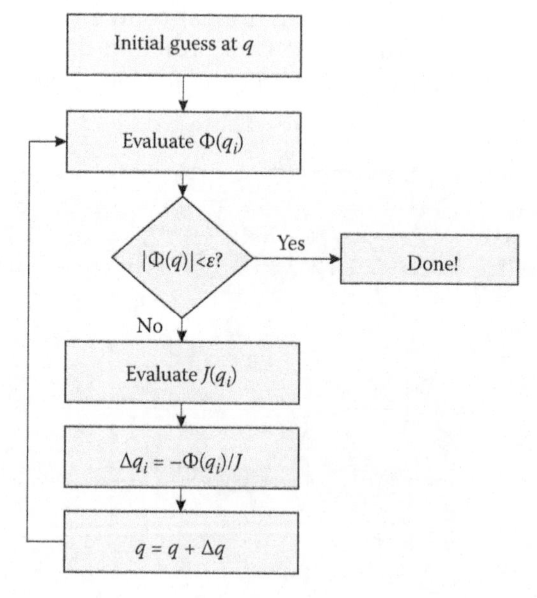

FIGURE 10.63
The Newton–Raphson algorithm for solving nonlinear equations.

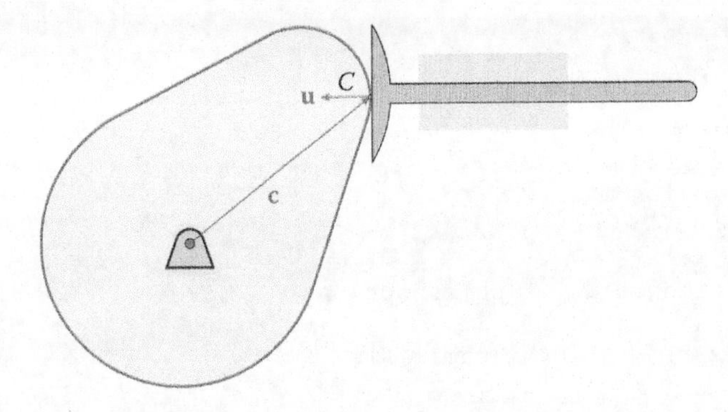

FIGURE 10.64
The surface normal at the point of contact is horizontal.

Thus, we wish to find the value of λ on the cam that makes the vertical component of the surface normal zero. The same Newton–Raphson algorithm will be used for each type of follower, but each follower will have its own function Φ that we will set to zero. We are now ready to enter the Newton–Raphson algorithm into the `FlatMotion` function.

```
% ***** use Newton-Raphson to find where cam normal is horizontal *****
  q = cam.lambda(maxIndex);   % max x coordinate is starting point for N-R
  for j = 1:100
    [phi,xB] = FlatPhi(cam,q,A);
    if (abs(phi) < 1e-10)          % have we found a solution?
      follower.s(i) = xB(1) - cam.rb;
      follower.xB(:,i) = xB(:);    % point of contact
      break
    end
    [phi1,~] = FlatPhi(cam,q-0.0001,A);   % calculate Jacobian
    [phi2,~] = FlatPhi(cam,q+0.0001,A);   % numerically
    jac = (phi2 - phi1)/0.0002;
    dq = -phi/jac;
    q = q + dq;
  end
```

You should be able to match each of the statements in the code above to the boxes in the flowchart. Remember that the Jacobian is defined as

$$J = \frac{d\Phi}{dq} = \frac{du_y}{d\lambda} \tag{10.72}$$

or the derivative of the contact function with respect to the angle. Since we lack an analytical expression for the contact function we must take the derivative numerically. The `FlatPhi` function is function we wish to minimize. Enter this function in a separate script.

```
% The function FlatPhi generates the current value of phi for the
% Newton-Raphson method while calculating point of contact between a flat
```

```
% follower and a cam
%
% ***** Inputs *****
% Cam = cam parameters
% q = current angle
% A = transformation matrix
%
% ***** Output *****
% phi = calculated value of phi function

function [phi,xBr] = FlatPhi(cam,q,A)

q = mod(q,2*pi);              % keep q under 2*pi
xB = ppval(cam.splineC,q);   % point of contact
uB = ppval(cam.splineU,q);   % surface normal
xBr = A*xB;                   % rotate point of contact
uBr = A*uB;                   % rotate surface normal

phi = uBr(2);                 % y coordinate of surface normal
```

We have used the mod function to ensure that the current value of λ remains under 2π. Note that we also calculate the point of contact at the same time as we calculate the y component of the surface normal. The variable `xBr` gives the point of contact with the follower and will enable us to compute the follower position function. This completes the follower position analysis loop.

10.10.3 Calculating Velocity and Acceleration of the Follower

We found the position of the follower numerically, so we must also use numerical differentiation to find velocity and acceleration. Since we will need to calculate derivatives for each type of follower, it makes sense to create a separate function for differentiation.

```
% The function FollowerDerivative calculates the numerical derivatives of
% follower motion
%
% ***** Inputs *****
% cam = cam parameters
% s = starting values
%
% ***** Outputs *****
% v = derivative of starting values

function v = FollowerDerivative(cam,s)

v = zeros(1,cam.Nc);
for i = 1:cam.Nc

  if (i == 1)          % if at beginning
    k1 = cam.Nc;       % loop back to end
  else
    k1 = i - 1;
  end
```

```
  if (i == cam.Nc)   % if at end
    k2 = 1;          % loop back to beginning
  else
    k2 = i + 1;
  end

  v(i) = (s(k2) - s(k1))/(2*cam.dt);
end
```

Here we are using *symmetric differentiation* with the formula

$$v_i = \frac{s_{i+1} - s_{i-1}}{2dt} \tag{10.73}$$

This works because the motion of the follower is cyclical, and $s_1 = s_{361}$. Type in the lines that call the `FollowerDerivative` function in the `FlatMotion` function.

```
% calculate velocity and acceleration of the follower numerically
follower.v = FollowerDerivative(cam,follow.s);
follower.a = FollowerDerivative(cam,follow.v);
```

We are now ready to begin plotting and animating the motion of the follower. The first figure will plot the *s-v-a* diagram for the follower overlaid with the *s-v-a* diagram for the cam. We'll need a new function to overlay both traces on the same plot.

```
% The function PlotFollowerSva creates a single plot on an s-v-a diagram
% for the follower overlaid with the same plot for the cam.
%
% ***** Inputs *****
% i      = plot number (1-3)
% lambda = angle on cam (x axis on plot)
% y1     = follower function to plot
% y2     = cam function to plot
% ytitle = the y label for the plot (s, v, a)

function PlotFollowerSva(i,lambda,y1,y2,ytitle,cBlu,cBlk)

subplot(3,1,i); plot(180*lambda/pi,y1,'Color',cBlu(1,:),'LineWidth',2)
hold on
subplot(3,1,i); plot(180*lambda/pi,y2,'Color',cBlk(3,:),'LineWidth',1)

if (i == 1)  % only place legend for highest plot
  legend('Follower','Cam','Location','Northwest')
  title('s-v-a Plot for Cam and Follower')
end

if (i == 3)  % only place x label for lowest plot
  xlabel('Cam angle (deg)')
end
ylabel(ytitle)

grid on
xlim([0 360])
set(gca,'xtick',0:60:360)
```

This function is very similar to the `PlotCamSvaj` function developed earlier, and you can use copy and paste to create most of it. In the `FlatMotion` function, use the following lines to create the plot:

```
% Define colors for plotting
cBlu = DefineColor([  0 110 199]); % Pantone 300C
cBlk = DefineColor([  0   0   0]); % grayscale

% create s-v-a diagram for cam and follower
if showPlot.svaCF
  figure
  PlotFollowerSva(1,cam.lambda,follower.s,cam.s,'Disp (m)',cBlu,cBlk)
  PlotFollowerSva(2,cam.lambda,follower.v,cam.v,'Vel (m/s)',cBlu,cBlk)
  PlotFollowerSva(3,cam.lambda,follower.a,cam.a,'Acc (m/s^2)',cBlu,cBlk)
  set(gcf,'Position',[100 50 1300 700])
end
```

Be sure to add the boolean `svaCF` field to the `showPlot` structure in the `CamDesigner` program

```
showPlot.svaCF = true;    % plot s-v-a diagram for cam and follower?
```

When you execute the main program you should obtain the *s-v-a* diagram shown in Figure 10.65. There is a surprisingly large difference between the cam functions and the follower functions, and all the work we did in carefully designing the cam seems to have been for nothing. The reason for this surprising behavior will become more clear when we

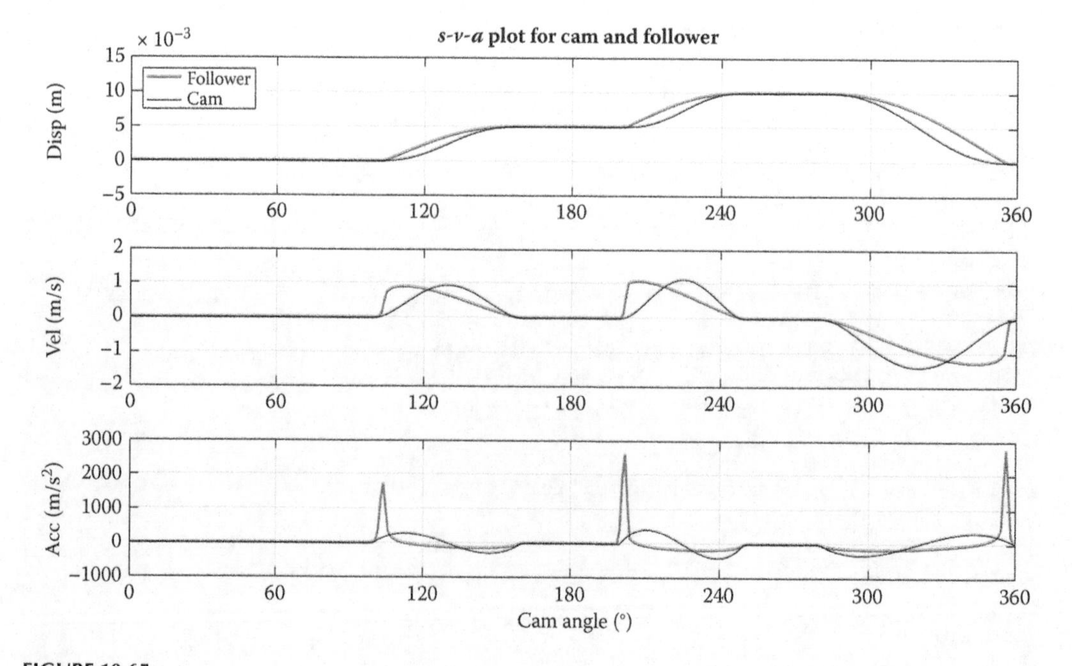

FIGURE 10.65
s-v-a diagram for the example cam. Note the large difference between the *s-v-a* functions for the cam and the follower.

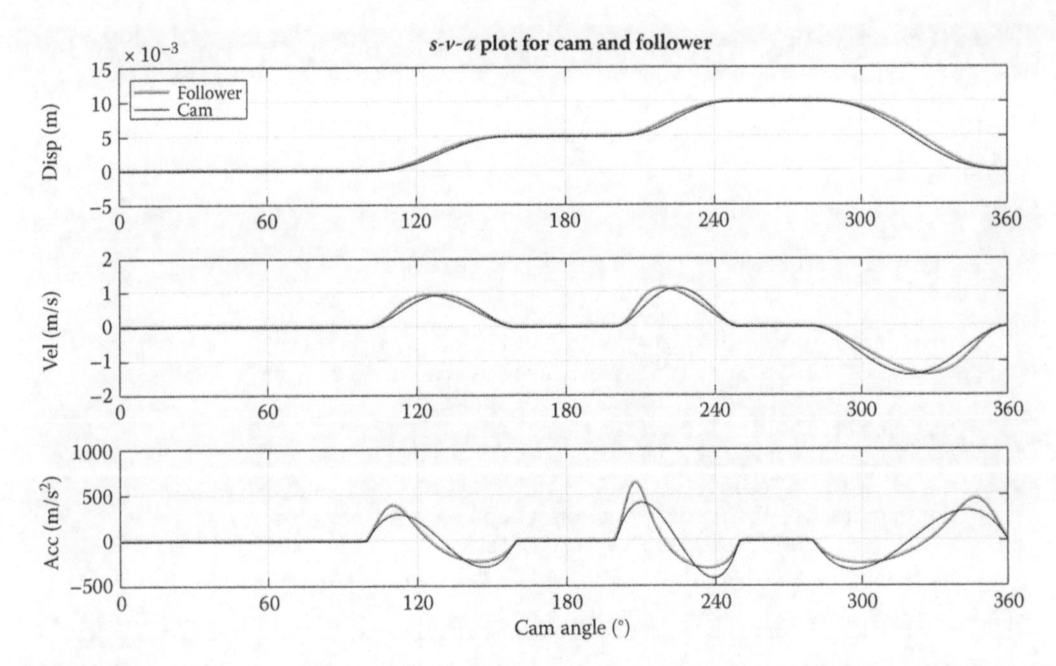

FIGURE 10.66
s-v-a diagram for the flat-faced follower and cam with base radius 0.1 m.

animate the cam/follower system, but for now we will simply state that the base radius of the cam is too small relative to the dwell heights we have specified.

Try changing the base radius on the cam to 0.1 m and you should obtain the plot shown in Figure 10.66. Here the follower and cam functions are much closer to each other. It is clear that the larger we make the base radius of the cam, the closer the follower functions will be to the cam functions. A cam with infinite base radius will have perfect follower motion! Of course, size and cost limitations will always prevent the cam from being as large as we would like, and a compromise solution must be used in practice.

The remainder of the `FlatMotion` function consists of relatively simple plotting commands, and we will not give a detailed explanation here. Remember to add appropriate fields to `showPlot` in `CamDesigner` to control the plotting options.

```
% Define coordinates of rectangle for plotting follower
rx = cam.rb;         % follower starts at base radius of cam
ry = -0.75*cam.rb; % follower is 1.5 times base radius
rw =  0.30*cam.rb; % follower is 0.3 times base radius
rh = -2*ry;
ymax = 1.5*(cam.rb + max(cam.h)); % scaling for profile plots

% Plot follower in contact with cam
if (showPlot.check > -1)
  figure
  i = showPlot.check + 1;
  PlotCamProfile(cam.c,-cam.lambda(i),cBlu,cBlk);
  hold on
  rectangle('Position',[follower.s(i) + rx, ry, rw, rh],...
            'FaceColor',cBlu(8,:),'LineWidth',2,'EdgeColor',cBlu(1,:))
```

```
  axis([-ymax ymax -ymax ymax])
end

% create animation of cam and follower for one revolution
if showPlot.anim
  figure
  for i = 1:cam.Nc
    cla
    PlotCamProfile(cam.c,-cam.lambda(i),cBlu,cBlk);
    hold on

    % plot follower rectangle and a vertical projection line of face
    rectangle('Position',[follower.s(i) + rx, ry, rw, rh],...
              'FaceColor',cBlu(8,:),'LineWidth',2,'EdgeColor',cBlu(1,:))
    plot(ones(1,101)*follower.s(i)+rx,...
         (-50:50)*(2*ymax/101),':k','linewidth',2)

    % determine if point of contact touches follower face
    if (abs(follower.xB(2,i)) > abs(ry+rh))
        plot(follower.xB(1,i),follower.xB(2,i),'xr','MarkerSize',10);
    else
        plot(follower.xB(1,i),follower.xB(2,i),'xk','MarkerSize',10);
    end

    axis([-ymax ymax -ymax ymax])
    axis manual
    drawnow
  end
end
```

When you run the animation (by setting `showPlot.anim` to true in the main program) you should be rewarded with a "cartoon" of the cam rotating and the follower maintaining contact. If you set the base radius of the cam back to 0.025 m you will clearly see why the cam and follower kinematic functions are so different: the point of contact between cam and follower, shown by the "X," jumps around because of concavities in the cam profile. Making the base radius larger reduces, and then eliminates these concavities. Watching the point of contact jump around in its discontinuous fashion is a good reminder of why we must use numerical techniques to solve the contact problem: imagine trying to develop an analytical function that would model these jumps! The general contact problem – determining when and where to bodies of general shape make contact with each other – is one of the more challenging problems in modeling, and we have obtained a small taste of this with the cam/follower system.

10.10.4 The Translating Roller-Follower

A diagram of the translating roller-follower is shown in Figure 10.67. The base circle is familiar from our previous analysis, and its radius is given by r_b. The follower is assumed to be circular with a radius r_f. The path taken by the center of the follower is called the *pitch curve*; it is offset from the cam profile by the radius of the follower. The distance from the center of the cam to the pitch curve is r_p, and is a function of the angle on the cam. The cam and follower make contact at point C. We assume that they make contact at a single point, so that the *line of transmission* is tangent to both cam and roller at this point. The *pressure*

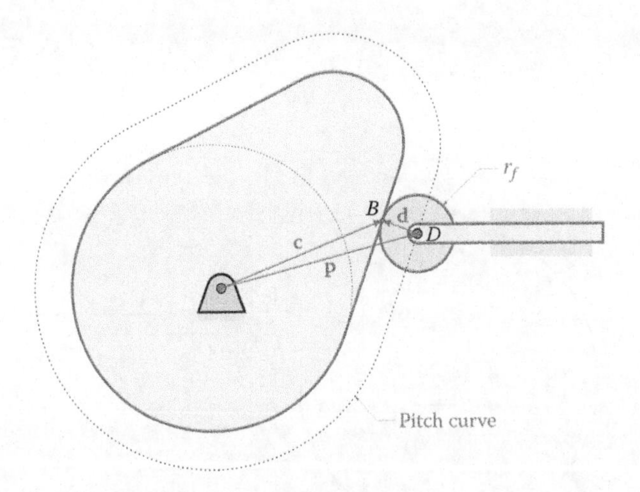

FIGURE 10.67
The translating roller-follower has vertical offset b. The line of common tangents between cam and roller is the transmission line, and the pressure line is perpendicular to this.

line is normal to both the cam surface and follower at the point C. Without loss of generality we assume that the follower translates horizontally, and the angle φ between the horizontal and the pressure line is known as the *pressure angle*. All of the useful force from the cam to the follower must be directed along this line, since any tangential component will serve only to spin the follower. We will have much to say about the pressure angle in the section on force analysis.

The follower in Figure 10.67 is shown with a vertical offset b, which is sometimes called the *eccentricity*. In most situations, b will be zero, in which case the follower is *aligned* with the axis of the cam. To keep the analysis general, we will assume that b is not zero.

The path followed by the center of the roller is known as the *pitch curve*, and is found by offsetting the cam profile by the radius of the follower, as shown in Figure 10.68. The term *parallel curve* has a strict mathematical definition: it is the locus of points that results from moving a perpendicular distance ρ from the original curve. The simplest parallel curve is a parallel line, which is found by offsetting the original line by a constant distance. A parallel curve to a circle is another circle, and a parallel curve to a square is another square.

Figure 10.69 shows a parabola that has been offset a distance ρ in the outward direction. While it seems obvious that the parallel curve is larger than the original parabola it is not obvious that the parallel curve is *no longer a parabola*. While the parallel curves of some shapes (e.g. lines and circles) are the same shape as the original curve, the parallel curves of most other shapes (e.g. parabolas and polynomial functions) are not. That is, the parallel curve of a 3-4-5 polynomial is *not* another 3-4-5 polynomial. Remarkably, the parallel curves of involutes of a circle (e.g. gear teeth) are also involutes of the same circle. The proof of the preceding discussion is beyond the scope of this text, but see [2] for an excellent explanation of the subject. The reason for including this topic in the section on follower motion is that the path traveled by the follower will not be a simple function like a 3-4-5 polynomial, which means that the follower motion will not be an exact duplicate of the cam profile, just as we found with the flat follower.

Referring back to Figure 10.67, we can write a vector $\mathbf{p}$ to the point D at the center of the follower as

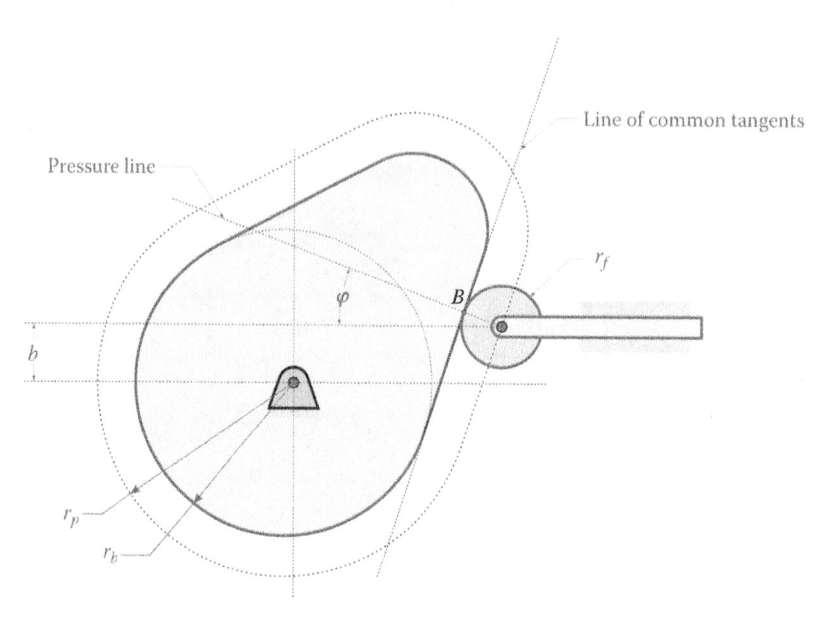

FIGURE 10.68
The pitch curve is found by offsetting the cam profile by the radius of the follower.

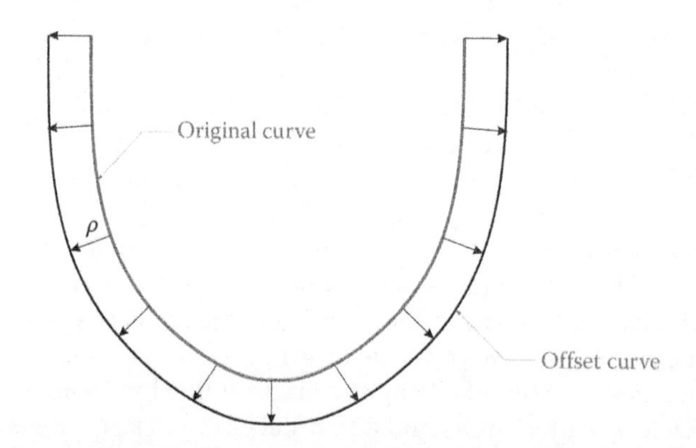

FIGURE 10.69
A parallel curve is formed by displacing the original curve a distance ρ in the perpendicular direction.

$$\mathbf{p} = \mathbf{c} - \mathbf{d} \tag{10.74}$$

Because the vector **d** points along a radial line on the follower, it is normal to the surface of the cam at that point. Thus

$$\mathbf{d} = r_f \mathbf{u} \tag{10.75}$$

where r_f is the radius of the follower. Thus, all points on the pitch curve can be traced out by computing

$$\mathbf{p} = \mathbf{c} + r_f \mathbf{u} \tag{10.76}$$

as λ varies from 0° to 360°. We can plot the pitch curve **p** in the same manner as we plotted the cam profile **c**. Let us try this out now. First, we must allow the user to enter the important parameters for the follower, which include its radius and vertical offset. We do this by defining the fields in the `follower` structure in the main program.

```
follower.r = 0.01;   % roller radius
follower.b = 0.01;   % roller vertical offset
```

Here we have used a roller radius of 10 mm and a vertical offset of 10 mm. Now, create a new function called `RollerMotion` with the header:

```
% function RollerMotion calculates and animates the position, velocity
% and acceleration of a translating roller follower.
%
% ***** Inputs *****
% cam      = cam parameters
% follower = follower parameters
% showPlot = plotting options
%
% ***** Outputs *****
% follower = position, velocity and acceleration of the follower

function follower = RollerMotion(cam,follower,showPlot)
```

The pitch curve is easy to calculate using Equation (10.76).

```
cp = cam.c - follower.r*cam.u; % calculate pitch curve for follower
```

and we can use a variation on the `PlotCamProfile` function to plot the cam profile and pitch circle.

```
if (showPlot.check > -1)
  i = showPlot.check + 1;
  theta = -cam.lambda(i);
  A = [cos(theta) -sin(theta); % rotation matrix for rotating cam
       sin(theta)  cos(theta)];
  cpr = A*cp;                  % rotate coordinates of pitch curve

  figure
  PlotCamProfile(cam.c,-cam.lambda(i),cBlu,cBlk);
  hold on
  xC = follower.xC(:,i); % center of follower circle
  fill(circ(1,:)+xC(1),circ(2,:)+xC(2),cBlu(8,:),...
   'EdgeColor',cBlu(2,:),'LineWidth',2)
  plot(cpr(1,:),cpr(2,:),':','Color',cBlk(3,:),'LineWidth',2)
end
```

If everything has been entered correctly, you should obtain the pitch curve shown in Figure 10.70.

Figure 10.71 shows what can go wrong if the roller is too large, or if the base circle is too small. Try increasing the roller radius to 30 mm and the height of the third dwell to 20 mm. On the left side of the pitch curve a discontinuity has appeared. A quick check will confirm that all points on the pitch curve are parallel to the cam profile, but the

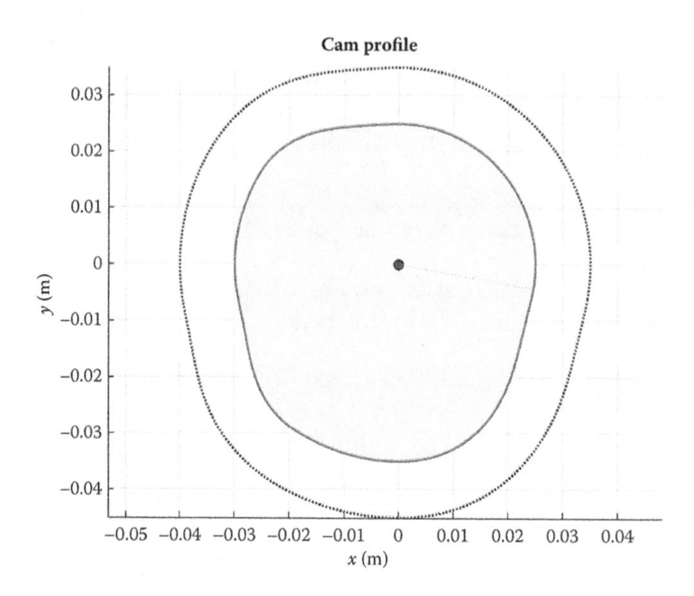

FIGURE 10.70
The cam profile and pitch curve for the example cam with roller radius 10mm.

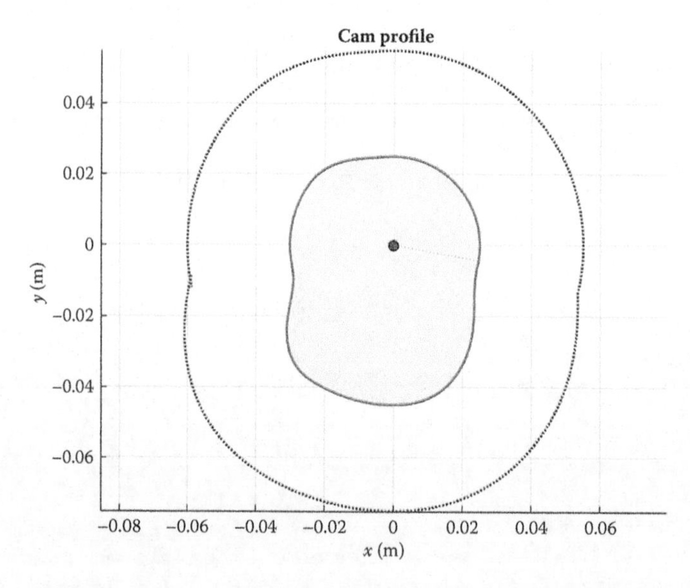

FIGURE 10.71
If the roller is too large, there may exist discontinuities in the pitch curve. In this example, the roller radius has been increased to 30mm and the third dwell height is 20mm.

path seems to reverse direction briefly at the discontinuity. Note that the discontinuity occurs where the cam profile is concave. The radius of curvature is smaller here than the radius of the roller, so the roller makes contact with the cam at two points, instead of the single point that was assumed in the analysis. In short, the radius of the roller must always be smaller (by a factor of 2 or 3, ideally) than the minimum radius of curvature of the cam profile.

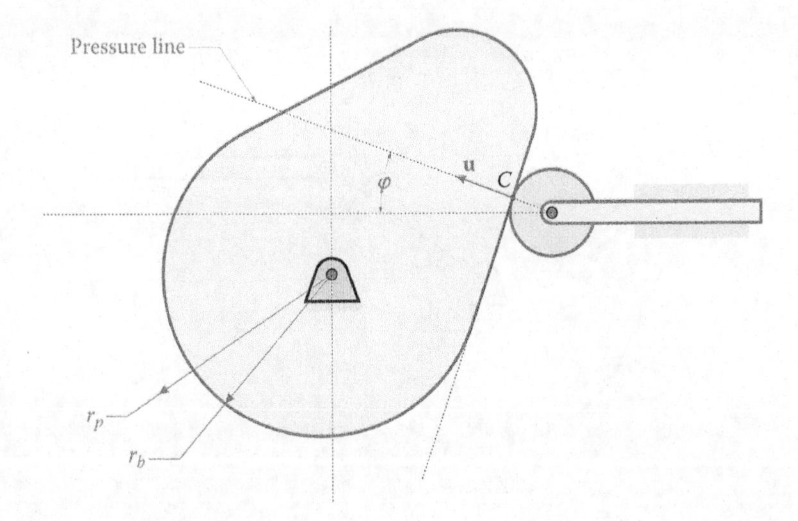

FIGURE 10.72
The pressure angle for the roller-follower is the angle between the surface normal and the negative x axis.

The contact force between the cam and follower is directed along the surface normal **u**, but the desired motion of the follower is purely horizontal, as seen in Figure 10.72. Thus, any vertical component of the contact force is "wasted" and must be taken up by the slider that the follower rides in. The angle between the contact force and the horizontal is known as the *pressure angle*. If this angle is too large, it may create excessive frictional forces in the slider, which may even result in the follower "binding up" inside the slider. Since we know the surface normal at the point of contact, the pressure angle is easy to calculate. Recall that the dot product of two unit vectors gives the cosine of the angle between the two vectors. The pressure angle is then

$$\cos\varphi = \mathbf{u} \cdot -\hat{i} \tag{10.77}$$

We use the negative of $\hat{i}$ because the unit normal points generally in the negative x direction.

Another useful quantity to compute is the neutral position of the roller; that is, the x coordinate of the roller when it is resting on the base radius of the cam. As seen in Figure 10.73, we can use the Pythagorean theorem to find this minimal x coordinate, s_0.

$$s_0 = \sqrt{\left(r_b + r_f\right)^2 - b^2} \tag{10.78}$$

Near the top of the `RollerMotion` function, add the following two lines of code.

```
rf = follower.r;
s0 = sqrt((cam.rb + rf)^2 - follower.b^2); % neutral pos of follower
```

Examining Figure 10.68 we see that the position of the center of the roller is located at the intersection of the pitch curve and the horizontal line at $y = b$. Recall that the vector to a point on the pitch curve is

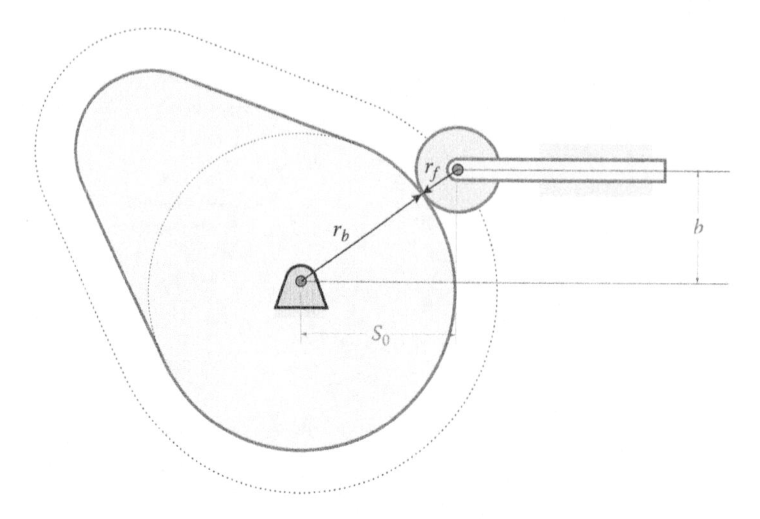

FIGURE 10.73
The minimum displacement of the roller occurs when it rolls along the base radius of the cam.

$$\mathbf{p} = \mathbf{c} - r_f\mathbf{u} \tag{10.79}$$

If the y coordinate of **p** is equal to the vertical offset b then we have found the location of the roller. We must also require that the x coordinate of **p** be positive, since there are two locations where the y coordinate of the pitch curve is equal to b. Thus, our contact function Φ is

$$\Phi(\lambda) = p_y - b \tag{10.80}$$

To find the point of contact, we will first go around the profile of the cam in 1° steps and find the angle where $dy = p_y - b$ is closest to zero. The second `for` loop in the code below accomplishes this task.

```
for i = 1:cam.Nc  % rotate cam one revolution
  theta = -cam.lambda(i);           % angle of rotation for cam
  A = [cos(theta) -sin(theta);      % rotation matrix for cam
       sin(theta)  cos(theta)];
  cpr = A*cp;                       % rotated pitch curve

  % find approximate point of contact
  minDiff = 1000;
  for j = 1:cam.Nc
    dy = abs(cpr(2,j) - follower.b);
    if ((dy < minDiff) && cpr(1,j) > 0)
      minDiff = dy;
      minIndex = j;
    end
  end

end
```

This provides our initial guess for the Newton–Raphson algorithm. The code for the algorithm is almost identical to that for the flat-faced follower, as shown below.

```
% use Newton-Raphson to find exact point of contact
  q = cam.lambda(minIndex);
  for j = 1:10
    [phi,xB,uB] = RollerPhi(cam,follower,q,A);
    if (abs(phi) < 1e-10)
      xC = xB - follower.r*uB;
      follower.s(i) = xC(1) - s0;
      follower.xB(:,i) = xB;            % point of contact
      follower.uB(:,i) = uB;            % surface normal at point of contact
      follower.xC(:,i) = xB - rf*uB; % center of roller
      follower.phi(i) = acos(dot(uB,[-1;0]));
      break
    end
    [phi1,~] = RollerPhi(cam,follower,q-0.0001,A);
    [phi2,~] = RollerPhi(cam,follower,q+0.0001,A);
    jac = (phi2 - phi1)/0.0002;
    dq = -phi/jac;
    q = q + dq;
  end
```

Once we have found the correct position for the follower, we can modify the `showPlot.check` plotting routine to plot the roller as a circle.

```
%create circle for follower
phi = pi*(0:360)/180;
circ = rf*[cos(phi); sin(phi)];
if (showPlot.check > -1)
  i = showPlot.check + 1;
  theta = -cam.lambda(i);
  A = [cos(theta) -sin(theta); % rotation matrix for rotating cam
       sin(theta)  cos(theta)];
  cpr = A*cp;                   % rotate coordinates of pitch curve

  figure
  PlotCamProfile(cam.c,-cam.lambda(i),cBlu,cBlk);
  hold on
  xC = follower.xC(:,i); % center of follower circle
  fill(circ(1,:)+xC(1),circ(2,:)+xC(2),cBlu(8,:),...
   'EdgeColor',cBlu(2,:),'LineWidth',2)
  plot(cpr(1,:),cpr(2,:),':','Color',cBlk(3,:),'LineWidth',2)
end
```

It is left as an exercise for the reader to create the necessary `RollerPhi` function, as well as the plotting code for the *s-v-a* diagram and roller animation.

10.10.5 The Oscillating Rocker-Follower

The geometry associated with the oscillating rocker-follower is shown in Figure 10.74. We assume that the ground pin for the rocker is horizontally aligned with the ground pin for the cam and the distance between ground pins is d. The rocker has length c, and its angle with the horizontal is θ_4.

The rocker reaches its lowest position when the roller is in contact with the base radius of the cam, as shown in Figure 10.75. Using the Law of Cosines we can calculate this angle as

$$\theta_0 = \pi - \cos^{-1}\left(\frac{c^2 + d^2 - \left(r_f + r_b\right)^2}{2cd}\right) \tag{10.81}$$

This is the neutral position of the rocker and we will measure its displacement from this angle.

To conduct the position analysis for the rocker-follower we first draw a vector loop diagram as shown in Figure 10.76. Adding vectors around the loop gives

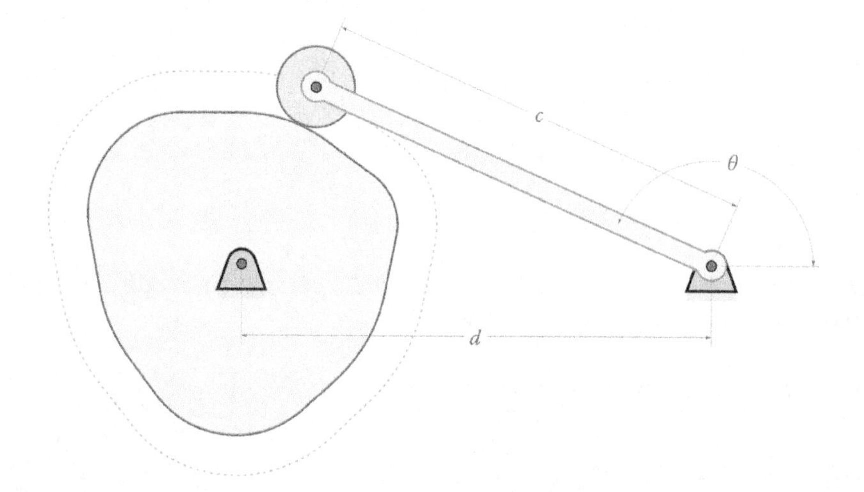

FIGURE 10.74
Geometry of the oscillating rocker-follower. The rocker makes angle θ_4 with the horizontal and has length c. The distance between ground pins is d.

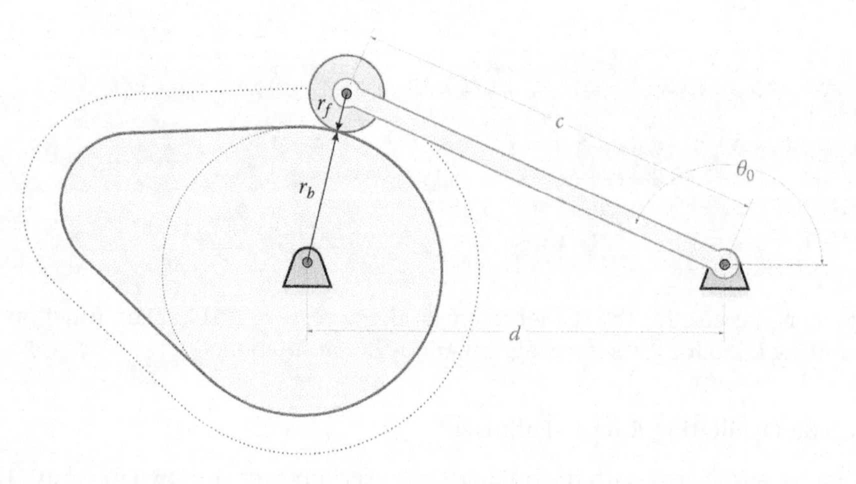

FIGURE 10.75
The neutral position of the rocker occurs when the roller touches the base radius of the cam.

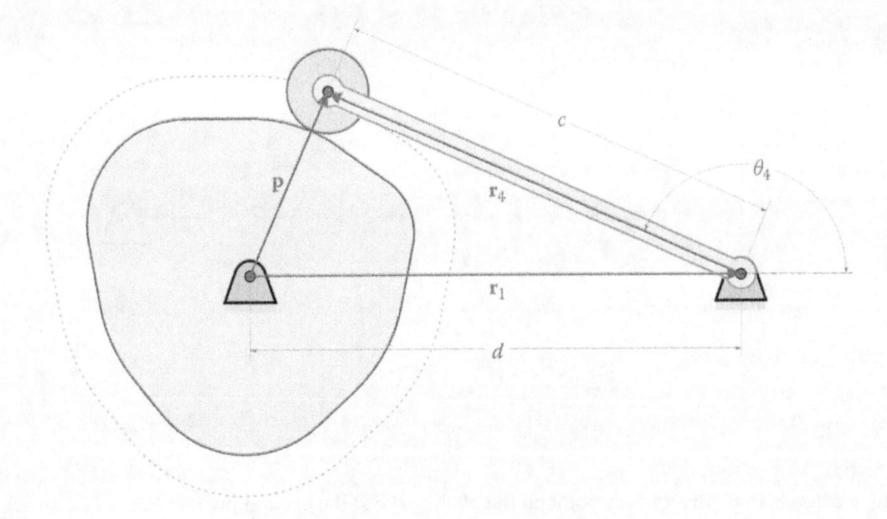

FIGURE 10.76
Vector loop diagram for the rocker follower. The vector **p** extends from the center of the cam to the pitch curve.

$$\mathbf{p} - \mathbf{r}_4 - \mathbf{r}_1 = 0 \tag{10.82}$$

where

$$\mathbf{p} = \mathbf{c} - r_f \mathbf{u} \tag{10.83}$$

is a point on the pitch curve of the cam/follower system. Expanding the vector loop equation into its components gives

$$\left(r + r_f\right)\mathbf{u} + c\mathbf{e}_4 + d\mathbf{e}_1 = 0 \tag{10.84}$$

The vector loop gives two equations and two unknowns: the angle of the vector **u** and θ_4. We can simplify this somewhat if we note that the rocker must trace out a circular arc of radius c that is centered at the ground pivot D. Any point on the pitch curve that lies a distance c from the ground pivot D gives a valid position solution for the rocker.

Let us define the vector **g** as extending between an arbitrary point on the pitch curve and the ground pivot at D as shown in Figure 10.77.

$$\mathbf{g} = \mathbf{p} - \mathbf{r}_1 \tag{10.85}$$

Then we must have

$$|\mathbf{g}| - c = 0 \tag{10.86}$$

Thus, our contact function is

$$\Phi(\lambda) = |\mathbf{g}| - c \tag{10.87}$$

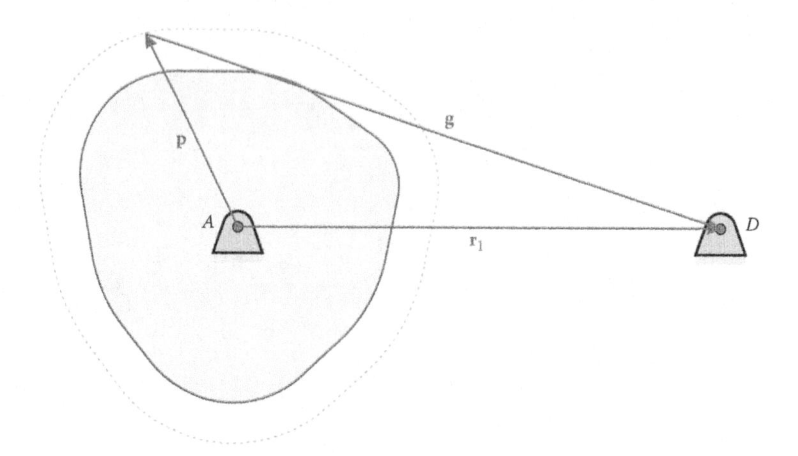

FIGURE 10.77
The vector **g** extends from an arbitrary point on the pitch curve to the ground pivot at D.

Once the vector **g** has been found, the rocker angle can be calculated as

$$\tan\theta = \frac{g_y}{g_x} \tag{10.88}$$

There will be two valid solutions for **g**: one with a positive θ and one negative. We will adopt the positive solution as shown in the diagrams.

The contact force is directed along the surface normal, as before (see Figure 10.78). When the contact force is aligned with the normal to the rocker, $\mathbf{n}_f$, then all of the contact force is used to rotate the rocker. When this is not the case, then some of the contact force is wasted in "stretching" the rocker, which may result in excessive wear on the rocker pivot. The pressure angle between cam and rocker can be computed by taking the dot product of **u** with $\mathbf{n}_f$.

$$\cos\varphi = \mathbf{u}\cdot\mathbf{n}_f \tag{10.89}$$

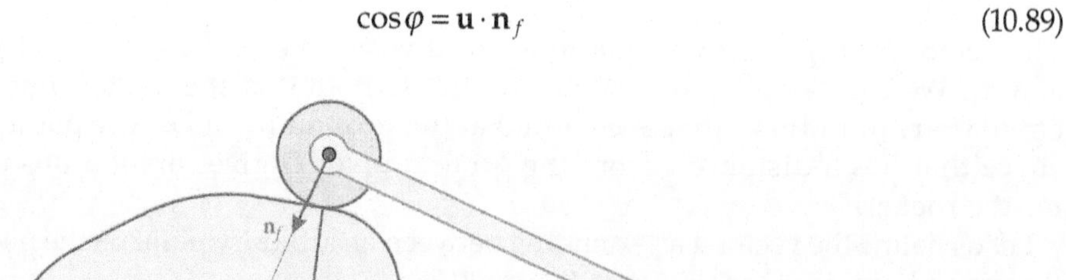
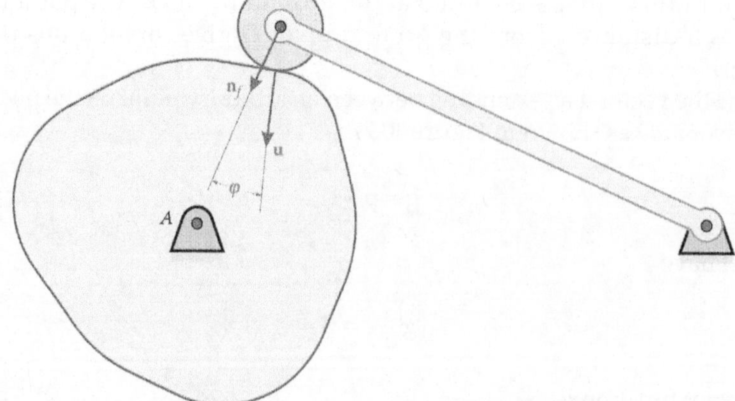

FIGURE 10.78
Pressure angle between rocker and cam.

This concludes the section on follower motion. We will use the results from this section to perform a force analysis on the cam/follower system. Try writing your own function for calculating, plotting, and animating the rocker follower system. If you get stuck, a full listing of the `MotionRock` script is shown below.

10.10.6 The RockerMotion Function

```
% function RockerMotion calculates and animates the position, velocity
% and acceleration of an oscillating roller follower.
%
% ***** Inputs *****
% cam      = cam parameters
% follower = follower parameters
% showPlot = plotting options
%
% ***** Output *****
% follower = position, velocity and acceleration of the follower

function follower = RockerMotion(cam,follower,showPlot)

% calculate angle of follower when roller is on base radius
rb = cam.rb;         % base radius of cam
rf = follower.r;     % follower radius
c = follower.c;      % rocker length
d = follower.d;      % length btw ground pins
follower.x0 = pi - acos((c^2+d^2-(rb+rf)^2)/(2*c*d));

cp = cam.c - rf*cam.u; % pitch curve for follower
xD = [d; 0];           % coordinates of ground pivot of rocker

for i = 1:cam.Nc                 % rotate cam one revolution
  theta2 = -cam.lambda(i);  % angle of rotation for cam
  A = [cos(theta2) -sin(theta2);
     sin(theta2)  cos(theta2)];
  cpr = A*cp;                    % rotated pitch curve (points C)

  minDiff = 1000;
  for j = 1:cam.Nc      % est contact point by going around pitch curve
    g = cpr(:,j) - xD;             % vector from C to D
    L = norm(g);                   % length of g
    dL = abs(c - L);               % diff btw rocker length and L
    theta = atan2(g(2), g(1)); % angle of g (must be positive)
    if ((dL < minDiff) && theta > 0)
      minDiff = dL;
      minIndex = j;
    end
  end

  q = cam.lambda(minIndex);  % best guess from cam profile
  for j = 1:10                   % Newton-Raphson algorithm
    [phi,xB,uB] = RockerPhi(cam,follower,q,A);
    if (abs(phi) < eps)          % found point of contact!
```

```
      xC = xB - rf*uB;           % center of roller (C)
      g = xC - xD;               % vector from C to D
      follower.x(i) = atan2(g(2),g(1)); % angle of rocker
      follower.xB(:,i) = xB;          % point of contact
      follower.uB(:,i) = uB;          % surface normal at point of contact
      follower.xC(:,i) = xB - rf*uB; % center of roller
      [~,nf] = UnitVector(follower.x(i)); % unit vector along follower
      follower.phi(i) = acos(dot(uB,nf));
      break
    end
    [phi1,~] = RockerPhi(cam,follower,q-0.0001,A);
    [phi2,~] = RockerPhi(cam,follower,q+0.0001,A);
    jac = (phi2 - phi1)/0.0002;
    dq = -phi/jac;
    q = q + dq;
  end
end

follower.s = follower.x - follower.x0;
follower.v = FollowerDerivative(cam,follower.s);
follower.a = FollowerDerivative(cam,follower.v);

% create circle for follower
phi = pi*(0:360)/180;
circ = rf*[cos(phi); sin(phi)];

cBlu = DefineColor([  0 110 199]); % Pantone 300C
cBlk = DefineColor([  0   0   0]); % grayscale

% create s-v-a diagram for follower
if showPlot.svaCF
  figure
  PlotFollowerSva(1,cam.lambda,follower.s,cam.s,'Disp (m)',cBlu,cBlk)
  PlotFollowerSva(2,cam.lambda,follower.v,cam.v,'Vel (m/s)',cBlu,cBlk)
  PlotFollowerSva(3,cam.lambda,follower.a,cam.a,'Acc (m/s^2)',cBlu,cBlk)
  set(gcf,'Position',[100 50 1300 700])
end

% plot pressure angle for cam/follower system
if showPlot.phi
  figure
  plot(180*cam.lambda/pi,180*follower.phi/pi,'LineWidth',2,'Color',c
Blu(1,:))

  xlabel('Cam Angle (deg)'); ylabel('Pressure Angle (deg)')
  title('Pressure Angle vs. Cam Angle')
  grid on
  xlim([0 360])
  set(gca,'xtick',0:60:360)
  set(gcf,'Position',[100 100 1000 600])
end

if (showPlot.check > -1)
  i = showPlot.check;
```

```
  figure

  PlotCamProfile(cam.c,-cam.lambda(i),cBlu,cBlk);
  hold on
  xC = follower.xC(:,i);
  fill(circ(1,:)+xC(1),circ(2,:)+xC(2),cBlu(8,:),...
      'EdgeColor',cBlu(2,:),'LineWidth',2)
  plot([xD(1) xC(1)],[xD(2) xC(2)],'Color',cBlk(3,:),'LineWidth',4)
  plot(xD(1),xD(2),'o','MarkerSize',8,'MarkerFaceColor',cBlk(5,:),...
      'Color',cBlk(1,:))
  plot(xC(1),xC(2),'o','MarkerSize',8,'MarkerFaceColor',cBlk(5,:),...
      'Color',cBlk(1,:))
end

ymax = 1.5*(cam.rb + max(cam.h));
if showPlot.anim
  figure
  for i = 1:cam.Nc
    cla
    PlotCamProfile(cam.c,-cam.lambda(i),cBlu,cBlk);
    hold on
    xC = follower.xC(:,i);
    fill(circ(1,:)+xC(1),circ(2,:)+xC(2),cBlu(8,:),...
        'EdgeColor',cBlu(2,:),'LineWidth',2)
    plot([xD(1) xC(1)],[xD(2) xC(2)],'Color',cBlk(3,:),'LineWidth',4)
    plot(xD(1),xD(2),'o','MarkerSize',8,'MarkerFaceColor',cBlk(5,:),...
        'Color',cBlk(1,:))
    plot(xC(1),xC(2),'o','MarkerSize',8,'MarkerFaceColor',cBlk(5,:),...
        'Color',cBlk(1,:))
    plot(follower.xB(1,i),follower.xB(2,i),'xk','MarkerSize',10);
    axis([-ymax ymax -ymax ymax])
    axis manual
    drawnow
  end
end
```

10.11 Force Analysis in Cams

After all of the hard work we've done to analyze the motion of the follower, force analysis on the cam/follower system will be surprisingly straightforward. A free-body diagram of the flat-faced follower and its cam is shown in Figure 10.79. Here we make the assumption that the interface between the cam and follower is well lubricated, so that the tangential friction force is neglected. Our primary goal in conducting the force analysis is to ensure that the cam and follower maintain contact with each other. If they lose contact then the motion of the follower will not follow the cam surface, with unpredictable (and sometimes tragic) results. Contact between cam and follower is maintained with a spring, as shown in the figure. The spring force is proportional to the motion of the follower as

$$F_k = -k\left(s_f + s_{f0}\right) \tag{10.90}$$

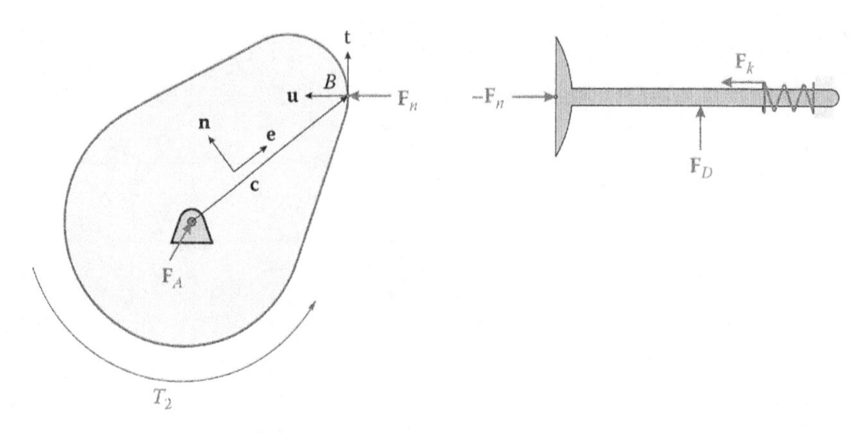

FIGURE 10.79
Free-body diagram of the flat-faced follower and cam.

where k is the spring constant, s_f is the displacement of the follower and s_{f0} is the initial compression of the spring (a constant). The minus sign is used because the spring force points in the negative x direction for a positive displacement of the follower, and vice versa. The contact force between cam and follower is denoted $\mathbf{F}_n$, and is normal to the face of the cam at the point of contact. The face of the follower is assumed to be vertical, so the normal force is horizontal. The normal force can be written as

$$\mathbf{F}_n = F_n\mathbf{u} \tag{10.91}$$

where $\mathbf{u}$ is the unit normal to the face of the cam at the point of contact. Since the unit normal points in the negative x direction, we can write

$$\mathbf{F}_n = \left\{ \begin{matrix} -F_n \\ 0 \end{matrix} \right\} \tag{10.92}$$

Using the free-body diagrams we sum forces on each body as

$$\mathbf{F}_A + \mathbf{F}_n = m_2\mathbf{a}_2 \tag{10.93}$$

$$-\mathbf{F}_n + \mathbf{F}_k = m_f\mathbf{a}_f \tag{10.94}$$

where m_f is the mass and $\mathbf{a}_f$ is the acceleration of the follower, respectively. The motion of the follower is purely horizontal, so that it has only an x component. We found the displacement, velocity, and acceleration of the follower during the motion analysis, so that we can easily solve for the contact force as

$$F_n = k\left(s_f + s_{f0}\right) + m_f a_f \tag{10.95}$$

Next, we sum moments on the cam. As we discussed at the end of Section 7.2, we can choose to sum moments about the center of mass of the cam or about its ground pivot. Since we have already solved for the vector $\mathbf{c}$ from the ground pivot to the point of contact, it is simpler to sum moments about the ground pivot. The moment created by the contact force about the ground pivot is

$$\mathbf{M}_n = (\mathbf{c} \times \mathbf{F}_n)\hat{k} \tag{10.96}$$

Using the same logic as in Section 7.2 this may be rewritten as

$$M_n = rF_n \, \mathbf{n} \cdot \mathbf{u} \tag{10.97}$$

where r is the distance from the ground pivot to the point of contact. Remember that $\mathbf{n}$ is the unit normal to the vector $\mathbf{c}$ and $\mathbf{u}$ is the unit normal to the surface of the cam at the point of contact. Since $\mathbf{u} = \{-1\ 0\}^T$ we can further simplify by writing

$$M_n = -rF_n n_x \tag{10.98}$$

Summing moments about the ground pivot gives

$$M_n + T_2 = I_{2A}\alpha_2 \tag{10.99}$$

where we have used I_{2A} to indicate that the moment of inertia is calculated about the ground pivot A. We have assumed thus far that the cam rotates at a constant angular velocity and $\alpha_2 = 0$, so that

$$T_2 = rF_n n_x \tag{10.100}$$

That's it! We have solved for the contact force and driving torque for the flat-faced follower system. We could also solve for the reaction force $\mathbf{F}_A$ at the ground pivot, but we leave that as an exercise for the reader.

As an interesting side note, we can come up with an initial "guesstimate" of the speed of the cam when the follower begins to lose contact; that is, when the normal force F_n goes to zero. As a very rough estimate, assume that the motion of the follower is sinusoidal. It's not really, as we have seen, but the motion of the follower does indeed bob up and down, much like a sine wave. If we make this assumption, we can write

$$s_f = A \sin \omega t \tag{10.101}$$

where A is the amplitude that the follower bobs up and down and ω is the angular velocity of the cam. Taking the time derivative gives the velocity

$$v_f = \omega A \cos \omega t \tag{10.102}$$

and differentiating once again gives acceleration

$$a_f = -\omega^2 A \sin \omega t \tag{10.103}$$

Substituting this into Equation (10.95) for the contact force gives

$$F_n = \left(k - m_f \omega^2\right) A \sin \omega t \tag{10.104}$$

The contact force goes to zero when

$$\omega = \sqrt{\frac{k}{m_f}} \tag{10.105}$$

This is the *natural frequency* of the spring/follower system. The natural frequency of a spring/mass system arises quite often in oscillating or vibrating systems and it is almost always an indication that something "interesting" is about to happen. In our case, the follower loses contact with the cam when the cam spins at the natural frequency.

10.11.1 Force Analysis of the Roller-Follower

A free-body diagram of the roller-follower mechanism is shown in Figure 10.80. We now have three bodies to analyze: the cam, roller, and follower. We cannot neglect the frictional force on the surface of the cam, F_t, since it causes the roller to rotate. Luckily, this force tends to be very small and unimportant relative to the contact force, F_n. In fact, the roller and follower can be combined into a single body by making a simplifying assumption: since the radius of the roller is assumed to be small compared with the cam (recall the radius of curvature discussion in Section 10.8) we may safely neglect its rotary inertia. The contact force points directly toward the center of the roller and does not create a rotational moment on it. If we sum torques about the center of the roller we have

$$-F_t r_f = I_r \alpha_r \tag{10.106}$$

Since we are neglecting the rotational inertia of the roller, $F_t = 0$ and the frictional force vanishes.

The roller and follower have the same translational motion, which we have previously calculated as s_f. By ignoring the rotation of the roller, we can combine the roller and follower into a single body, as shown in Figure 10.81. The force $\mathbf{F}_C$ is now an internal force within the roller/follower body, so we do not need to solve for it. Our analysis will,

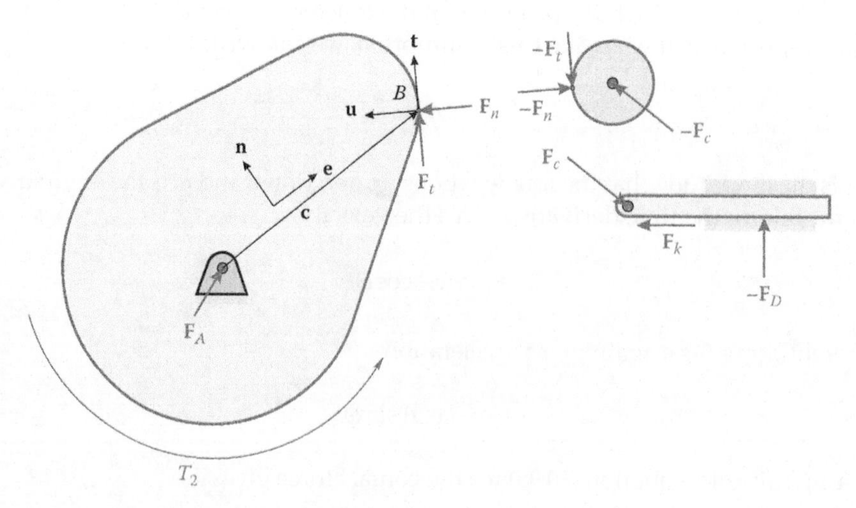

FIGURE 10.80
Free-body diagram of the roller follower.

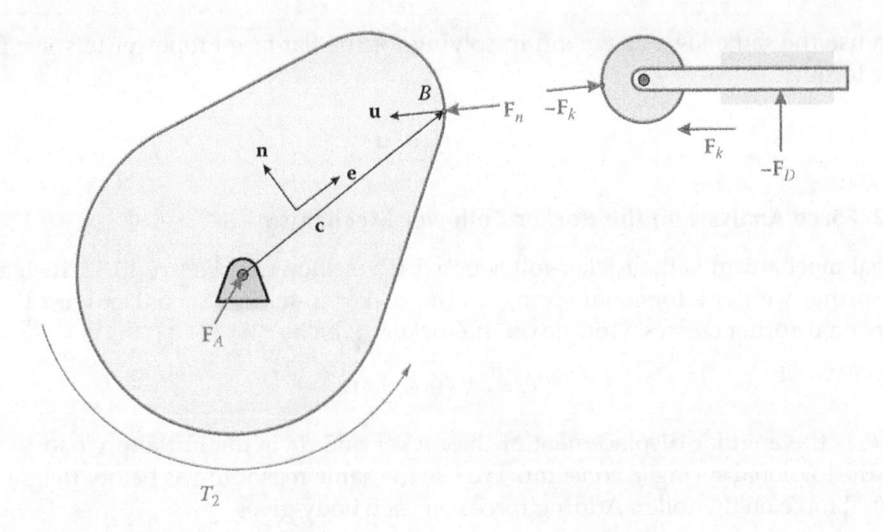

FIGURE 10.81
Free-body diagram of the roller-follower with the roller and follower combined into a single body.

therefore, be quite similar to that of the flat-faced follower. Summing forces on each body gives

$$\mathbf{F}_A + \mathbf{F}_n = m_c \mathbf{a}_c \tag{10.107}$$

$$-\mathbf{F}_n + \mathbf{F}_k - \mathbf{F}_D = m_f \mathbf{a}_f \tag{10.108}$$

here $\mathbf{F}_D$ is the vertical force that keeps the follower in its slot. The direction of the contact force varies depending upon the cam profile, unlike the flat-faced follower, whose contact force was always horizontal. The contact force points in the direction of the normal to the cam surface

$$\mathbf{F}_n = F_n \mathbf{u} \tag{10.109}$$

where F_n is the "strength" of the contact force, which can be positive or negative. If the contact force is negative then the follower loses contact with the cam. We can compute the horizontal component of the contact force by taking the dot product with the unit vector in the x direction, $\hat{i}$

$$F_{nx} = \mathbf{F}_n \cdot \hat{i} = F_n \mathbf{u} \cdot \hat{i} \tag{10.110}$$

But remember that the dot product $\mathbf{u} \cdot \hat{i}$ is the negative of the cosine of the pressure angle, so that

$$F_{nx} = -F_n \cos\varphi \tag{10.111}$$

Since $\mathbf{F}_D$ is entirely vertical, we can use the x component of Equation (10.108) to solve for the strength of the contact force

$$F_n = \frac{k\left(s_f + s_{f0}\right) + m_f a_f}{\cos\varphi} \tag{10.112}$$

We can use the same logic as we did in solving for the flat-faced follower to solve for the driving torque:

$$T_2 = -rF_n\mathbf{n}\cdot\mathbf{u} \tag{10.113}$$

10.11.2 Force Analysis on the Rocker-Follower Mechanism

Our final mechanism is the rocker-follower, which is shown in Figure 10.82. Instead of a linear spring, we use a torsional spring on the rocker to maintain contact with the cam. The torsional spring creates a torque on the rocker given by

$$T_4 = -k_t\left(\theta_4 + \theta_{40}\right) \tag{10.114}$$

where θ_f is the angular displacement of the rocker and θ_{f0} is the initial preload twist we give to the torsional spring (a constant). We use the same reasoning as before to ignore the tangential force on the roller. Adding forces on each body gives

$$\mathbf{F}_A + \mathbf{F}_n = m_c\mathbf{a}_c \tag{10.115}$$

$$-\mathbf{F}_n - \mathbf{F}_D = m_f\mathbf{a}_f \tag{10.116}$$

The torque on the rocker caused by the contact force can be written

$$\mathbf{M}_n = \left(\mathbf{r}_f \times -\mathbf{F}_n\right)\hat{k} \tag{10.117}$$

where

$$\mathbf{r}_f = c\mathbf{e}_f \tag{10.118}$$

is the vector from the rocker pivot to the center of the roller and

$$\mathbf{e}_f = \begin{Bmatrix} \cos\theta_f \\ \sin\theta_f \end{Bmatrix} \tag{10.119}$$

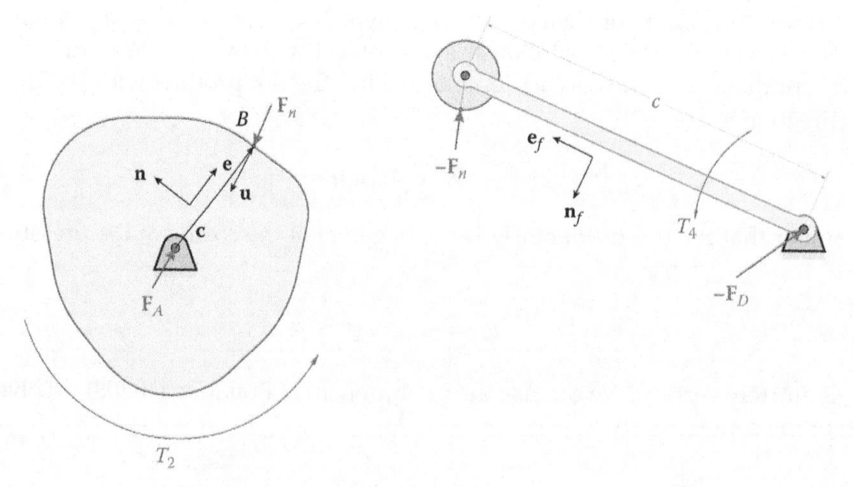

FIGURE 10.82
Free-body diagram of the rocker-follower mechanism.

is the unit vector in the direction of the rocker. Thus, the moment expression simplifies to

$$M_n = -cF_n \mathbf{n}_f \cdot \mathbf{u} \tag{10.120}$$

As before, the dot product $\mathbf{n}_f \cdot \mathbf{u}$ is equal to the cosine of the pressure angle so that

$$M_n = -cF_n \cos\varphi \tag{10.121}$$

The sum of torques on the rocker is then

$$M_n + T_4 = I_{fD}\alpha_f \tag{10.122}$$

where I_{fD} is the moment of inertia of the rocker about the ground pivot D. Solving for the contact force gives

$$F_n = -\frac{k_t(\theta_f + \theta_{f0}) + I_{fD}\alpha_f}{c\cos\varphi} \tag{10.123}$$

and the driving torque is

$$T_2 = -rF_n \mathbf{n} \cdot \mathbf{u} \tag{10.124}$$

A summary of the driving torque and contact force formulas found so far can be seen in Table 10.1. It is remarkable how similar the formulas are, given the different natures of the followers. This similarity will make programming them into MATLAB a relatively straightforward task.

10.11.3 Force Analysis of the Rocker-Follower in MATLAB®

We will develop the function for performing the force analysis on the rocker-follower, with the other two left as exercises for the reader. The rocker-follower is the most complicated, and adapting the `RockerForce` function to the other two is not too difficult. We start by defining the following new fields in the main program `CamDesigner`:

```
follower.Ip = 0.0001;          % moment of inertia of rocker about ground pivot
follower.kt = 10;              % torsional spring constant - rocker (N*m/rad)
follower.st0 = -10*pi/180;  % initial angular displacement of spring
```

TABLE 10.1

Contact Force and Driving Torque for the Three Types of Follower

	Contact Force F_n	Driving Torque T_c
Flat-faced	$k(s_f + s_{f0}) + m_f a_f$	$rF_n \times n_x$
Roller	$\dfrac{k(s_f + s_{f0}) + m_f a_f}{\cos\varphi}$	$-rF_n \mathbf{n} \cdot \mathbf{u}$
Rocker	$-\dfrac{k_t(\theta_f + \theta_{f0}) + I_{fD}\alpha_f}{c\cos\varphi}$	$-rF_n \mathbf{n} \cdot \mathbf{u}$

Note that the initial preload of the spring must be negative, since pushing the rocker up corresponds to a negative angle for the follower (i.e. it rotates clockwise). Next, call the `RockerForce` function in the appropriate case of the `CamDesigner` program.

```
% Conduct force analysis on the follower
RockerForce(cam,follower,showPlot)
```

This function will generate only two plots: contact force and driving torque. Next, open a new script and create the header for `RockerForce`

```
% function RockerForce calculates and plots the contact force and
% driving torque of an oscillating roller follower.
%
% ***** Inputs *****
% cam      = cam parameters
% follower = follower parameters
% showPlot = plotting options

function RockerForce(cam,follower,showPlot)

[Fn,Tc] = deal(zeros(1,cam.Nc)); % allocate memory
```

As usual, we loop through each angle on the cam, and calculate the contact force and driving torque at each position.

```
for i = 1:cam.Nc
  xB = follower.xB(:,i);          % point of contact
  r = norm(xB);                   % radius on cam to point of contact
  theta = atan2(xB(2),xB(1)); % angle on cam to point of contact
  [~,nc] = UnitVector(theta); % unit vector to pt of contact on cam
```

We first define the variable `xB` so that we don't need to type `follower.xB(:,i)` in each of our formulas. We use this to calculate the radius to the point of contact and the angle of the vector **c**, which points from the ground pin on the cam to the point of contact.

```
% calculate contact force
  tNet = follower.s(i) + follower.st0;  % total angle of spring
  num = -(follower.kt*tNet + follower.Ip*follower.a(i));
  den = follower.c * cos(follower.phi(i));
  Fn(i) = num/den;
```

To calculate the contact force we must first determine the total angle of twist of the spring. Remember that `follower.s(i)` stores the *angular displacement* of the spring; that is, its angle of twist relative to its neutral position. We have divided the formula for contact force into three lines to make it easier to read.

```
% calculate driving torque
  Tc(i) = -r * Fn(i) * dot(nc,follower.uB(:,i));
```

We finish up the loop by calculating the driving torque, and that concludes the force analysis portion of the program. Be sure to add the appropriate fields to `showPlot` in the main program and then add the the plotting routines familiar to us by now:

```
% Define colors for plotting
cBlu = DefineColor([  0 110 199]); % Pantone 300C

if showPlot.force
  figure
  % figure('MenuBar','none')
  plot(180*cam.lambda/pi,Fn,'Color',cBlu(1,:),'LineWidth',2)
  grid on
  xlim([0 360])
  set(gca,'xtick',0:60:360)
  set(gcf,'Position',[200 400 1200 350])
  title('Contact Force between Cam and Follower')
  xlabel('Cam Angle (degrees)'); ylabel('Force (N)')
end

if showPlot.torque
  figure
  % figure('MenuBar','none')
  plot(180*cam.lambda/pi,Tc,'Color',cBlu(1,:),'LineWidth',2)
  grid on
  xlim([0 360])
  set(gca,'xtick',0:60:360)
  set(gcf,'Position',[200 100 1200 350])
  title('Driving Torque for Oscillating Roller Follower')
  xlabel('Cam Angle (degrees)')
  ylabel('Torque (N-m)')
end
```

In the figure command we have set the property "MenuBar" to "none," which eliminates the menu and toolbar in the plot windows. This frees up valuable real estate on the screen for what we are really interested in – the plots themselves.

If everything goes well, you should obtain the plots shown in Figures 10.83 and 10.84 when you execute the code. It may be surprising that the first peaks in the driving torque are negative, but remember that we are rotating the cam in the *clockwise* direction and a negative torque is also clockwise. The contact force is positive through the entire rotation, which means that the follower remains in contact with the cam the entire time. Observe

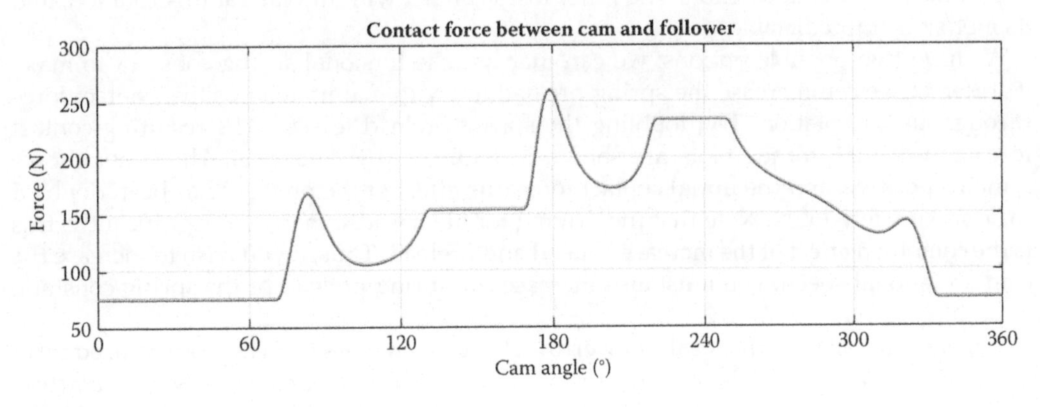

FIGURE 10.83
Contact force for the example three-dwell cam.

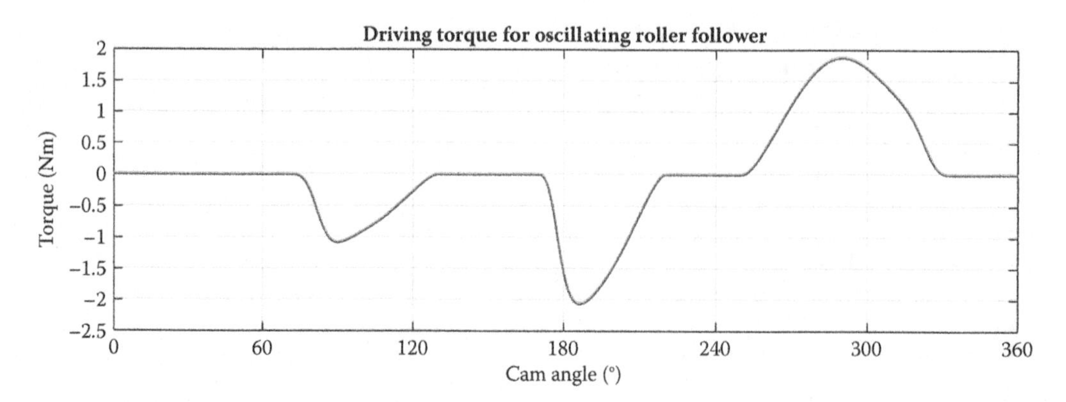

FIGURE 10.84
Driving torque for the example three-dwell cam.

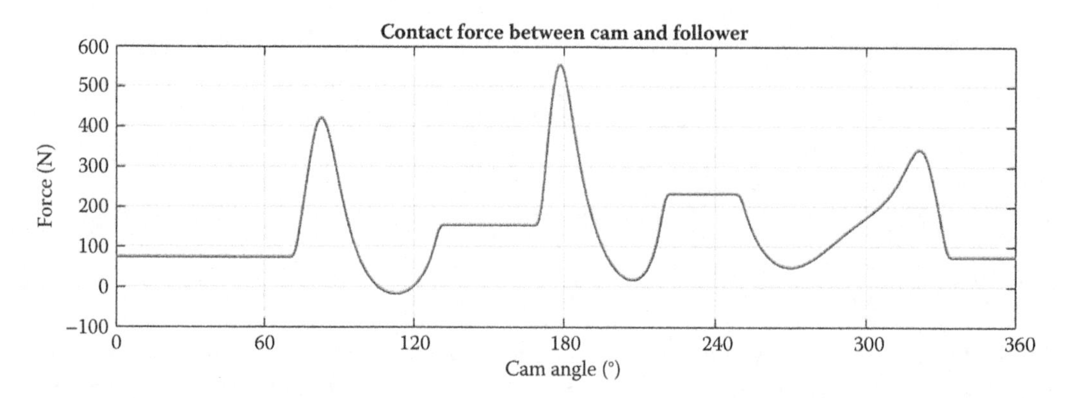

FIGURE 10.85
Contact force with cam rotational speed of 2000 rpm. Note the negative excursion at around 100° of rotation.

that the contact force begins and ends at approximately 75 N, which is the effect of preloading the torsional spring.

Now increase the speed of the cam to 2000 rpm. The resulting contact force is shown in Figure 10.85. It is positive throughout most of the rotation of the cam, but becomes negative at around 100° of cam rotation. The roller loses contact with the cam at this rotation, and its motion is unpredictable.

We have two possible options: we can increase the torsional spring constant to make it stiffer, or we can increase the spring preload to try to maintain a positive contact force throughout the rotation. Try doubling the spring preload to −20°. The resulting contact force and driving torque plots are shown in Figures 10.86 and 10.87. The contact force remains positive, and the initial contact force (the effect of the preload) has been doubled to approximately 150 N. Note that the driving torque has also increased significantly: this is the combined effect of the increased speed and preload. Thus, if you wish to increase the cam's rotational speed, you must also increase the spring preload or the spring constant, or both.

You now have at your disposal a powerful set of cam analysis tools that can be used in an iterative design process. It is clear that the nonlinear nature of the cam/follower interaction will normally prevent hitting the "optimum" design on the first try, but practicing with the design tools will tend to reduce the number of design iterations as you gain experience.

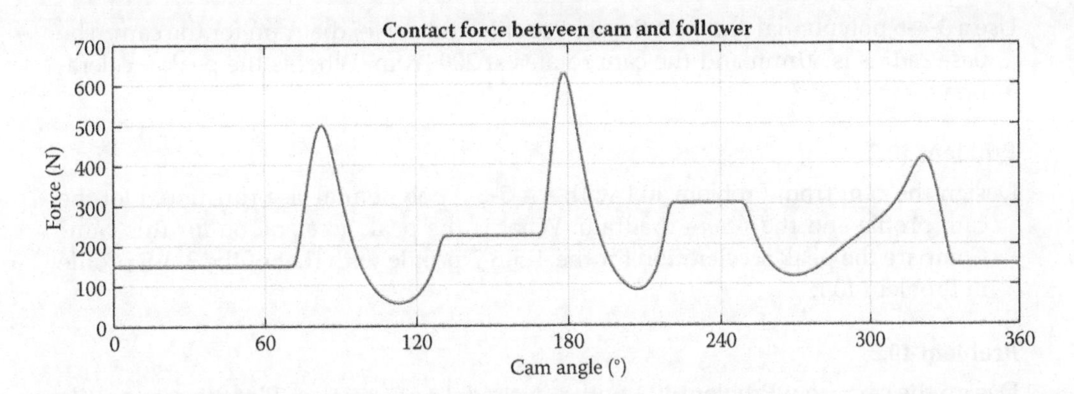

FIGURE 10.86
Contact force with the spring preload doubled to –20°. The contact force remains positive through the entire rotation.

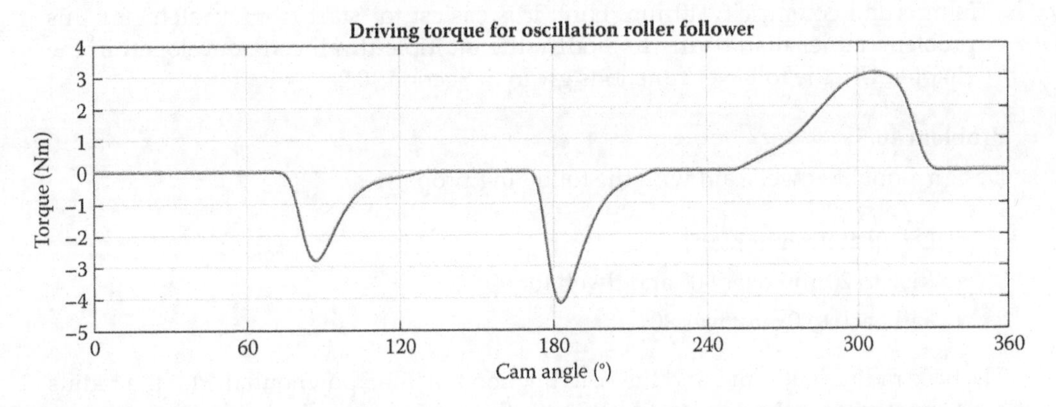

FIGURE 10.87
Driving torque with the increased spring preload. The spring preload has increased the driving torque.

10.12 Practice Problems

Problem 10.1

Design a quadruple-dwell cam with the following properties:

- Dwell at 0 mm for 30°
- Rise to 10 mm over 40° and dwell for 30°
- Rise to 15 mm over 20° and dwell for 60°
- Fall to 10 mm over 40° and dwell for 60°
- Fall back to 0 mm over 80°

Use a 3-4-5 polynomial and plot the cam profile and *s-v-a-j* diagram for the cam. The base radius is 50 mm and the cam rotates at 2000 rpm. What is the peak acceleration for this cam?

Problem 10.2

Design the cam from Problem 10.1 with a 4-5-6-7 polynomial rise function. Plot the cam profile and the *s-v-a-j* diagram. What is the peak acceleration for this cam? Compare the peak acceleration for the 4-5-6-7 profile with that of the 3-4-5 profile in Problem 10.1.

Problem 10.3

Design the cam from Problem 10.1 with a cycloidal rise function. Plot the cam profile and the *s-v-a-j* diagram. What is the peak acceleration for this cam? Compare the peak acceleration for this profile with that of the 3-4-5 profile in Problem 10.1.

Problem 10.4

Plot the *s-v-a-j* diagram for a flat-faced follower on an eccentric cam with base radius 50 mm and eccentricity 10 mm. Hint: it is easiest to "start from scratch" for this problem, rather than trying to modify the multiple-dwell code developed in the chapter. Use the follower functions given in Section 10.2.

Problem 10.5

Design a double-dwell cam with the following properties

- Dwell at 0 mm for 90°
- Rise to 20 mm over 90° and dwell for 90°
- Fall back to 0 mm over 90°

The base radius is 100 mm and the rise function is a 3-4-5 polynomial. Plot the radius of curvature for the cam profile. What is the minimum radius of curvature?

Problem 10.6

Repeat the exercise in Problem 10.5 with a 4-5-6-7 polynomial profile. What is the minimum radius of curvature?

Problem 10.7

Repeat the exercise in Problem 10.5 with a cycloidal rise profile. What is the minimum radius of curvature?

Problem 10.8

Plot the tangent circle and unit normal at the point of minimum radius of curvature in Problem 10.5.

Problem 10.9

Plot the *s-v-a* diagram of the flat-faced follower mated with the cam of Problem 10.5. The cam rotates at 1000 rpm. What is the maximum acceleration of the follower?

Problem 10.10

Plot the *s-v-a* diagram of a roller follower mated with the cam of Problem 10.5. The diameter of the roller is 40 mm and its vertical offset is 20 mm. The cam rotates at 1000 rpm. What is the maximum acceleration of the follower?

Problem 10.11

Plot the *s-v-a* diagram of a rocker follower mated with the cam of Problem 10.5. The diameter of the roller is 40 mm and the length of the rocker is 100 mm. The distance between ground pins is 100 mm and the cam rotates at 1000 rpm. What is the maximum amplitude of the angular acceleration of the rocker?

Problem 10.12

Figure 10.88 shows an eccentric cam mated with a flat-faced follower of mass m_f. Conduct a force analysis on this system and find an expression for the contact force between cam and follower as a function of cam angle θ. The spring has stiffness k, and is in its neutral position when $\theta = 180°$; that is, when the follower displacement is at a minimum. Assume that there is no friction between cam and follower, or between the follower and its guide bushing. The cam rotates at a constant angular velocity, ω. At which angular velocity will the contact force between cam and follower vanish if the cam is at $\theta = 0°$?

Problem 10.13

The double-dwell cam of Problem 10.5 is mated with a flat-faced follower, as shown in Figure 10.89. The spring has a stiffness of 1000 N/m and is compressed 10 mm when it is in contact with the base radius. The follower has a mass of 20 g. Find the angular velocity of the cam where it begins to lose contact with the follower and plot the contact force at this speed.

Problem 10.14

The double-dwell cam of Problem 10.5 is mated with a roller follower, as shown in Figure 10.90. The vertical offset of the follower is 20 mm and the radius of the roller is 10 mm. The spring has a stiffness of 1000 N/m and is compressed 10 mm when

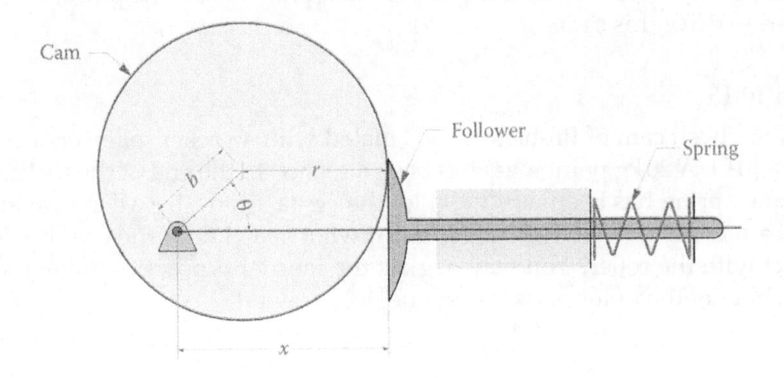

FIGURE 10.88
Problem 10.12.

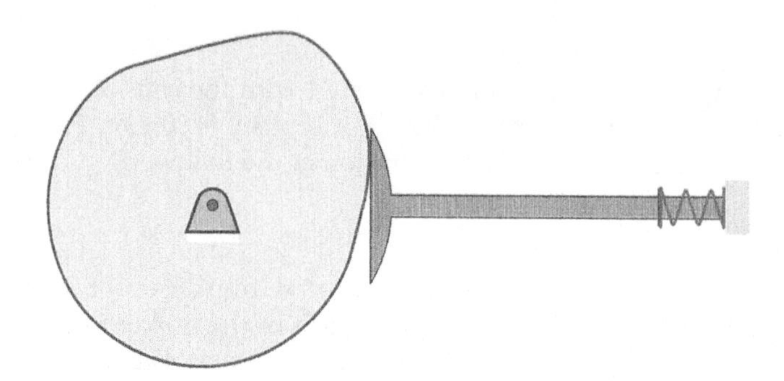

FIGURE 10.89
Problem 10.13.

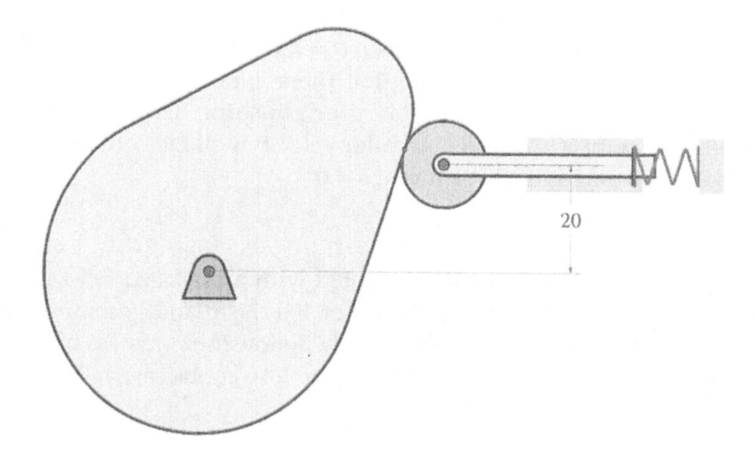

FIGURE 10.90
Problem 10.14.

it is in contact with the base radius. The follower has a mass of 20 g. Plot the driving torque for the cam when its speed is 1000 rpm. What is the power of the motor needed to drive this cam?

Problem 10.15

The double-dwell cam of Problem 10.5 is mated with a rocker follower, as shown in Figure 10.91. A 20 kg point mass has been attached to the end of the rocker and the torsional spring has been omitted. Plot the contact force between cam and roller for a cam angular velocity of 100 rpm. At what speed does the cam begin to lose contact with the roller? You may neglect the inertial properties of the rocker and roller, but not the point mass. Do not neglect gravity!

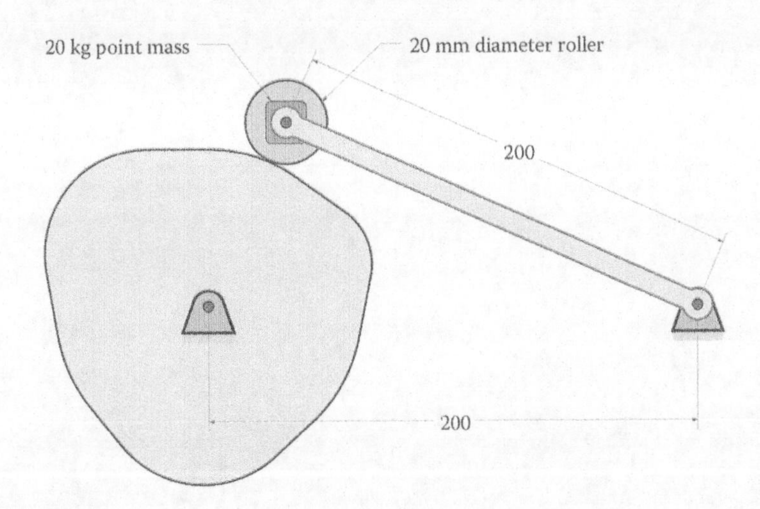

FIGURE 10.91
Problem 10.15.

Acknowledgments

SOLIDWORKS is a registered trademark of Dassault Systèmes SolidWorks Corporation.
Several images in this chapter were produced using MATLAB software.
MATLAB is a registered trademark of The MathWorks, Inc.

Works Cited

1. R. L. Norton, *Cam Design and Manufacturing Handbook*, New York: Industrial Press, 2002.
2. F. M. Stein, "The curve parallel to a parabola is not a parabola: parallel curves," *The Two-Year College Mathematics Journal*, vol. 11, no. 4, pp. 239–246, 1980.

Appendix: Inertial Properties of some Common Shapes

Variable Definitions

V = volume m = mass ρ = mass density

$I_x = \int (y^2 + z^2)dm$ = mass moment of inertia about x axis

$I_y = \int (x^2 + z^2)dm$ = mass moment of inertia about y axis

$I_z = \int (x^2 + y^2)dm$ = mass moment of inertia about z axis

Properties of Common Shapes

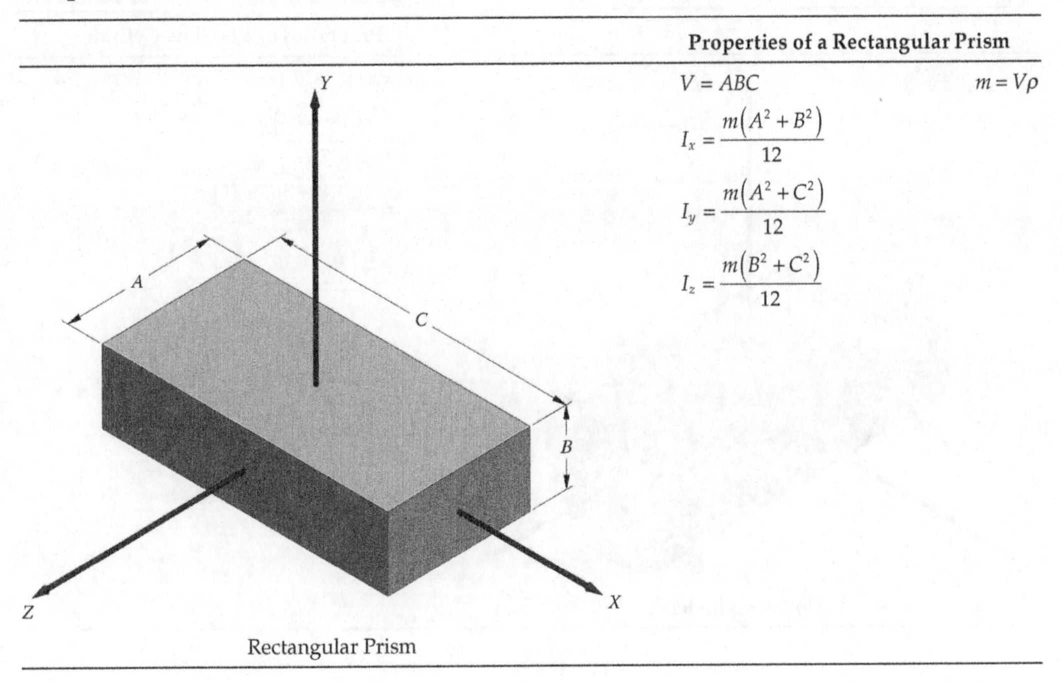

Rectangular Prism

Properties of a Rectangular Prism

$V = ABC$ $\qquad m = V\rho$

$$I_x = \frac{m(A^2 + B^2)}{12}$$

$$I_y = \frac{m(A^2 + C^2)}{12}$$

$$I_z = \frac{m(B^2 + C^2)}{12}$$

Properties of a Cylinder

$V = \pi R^2 L \qquad m = V\rho$

$$I_x = \frac{mR^2}{2}$$

$$I_y = \frac{m\left(3R^2 + L^2\right)}{12}$$

$$I_z = \frac{m\left(3R^2 + L^2\right)}{12}$$

Cylinder

Properties of a Hollow Cylinder

$V = \pi\left(R_o^2 - R_i^2\right)L \qquad m = V\rho$

$$I_x = \frac{m\left(R_o^2 + R_i^2\right)}{2}$$

$$I_y = \frac{m\left(3R_o^2 + 3R_i^2 + L^2\right)}{12}$$

$$I_z = \frac{m\left(3R_o^2 + 3R_i^2 + L^2\right)}{12}$$

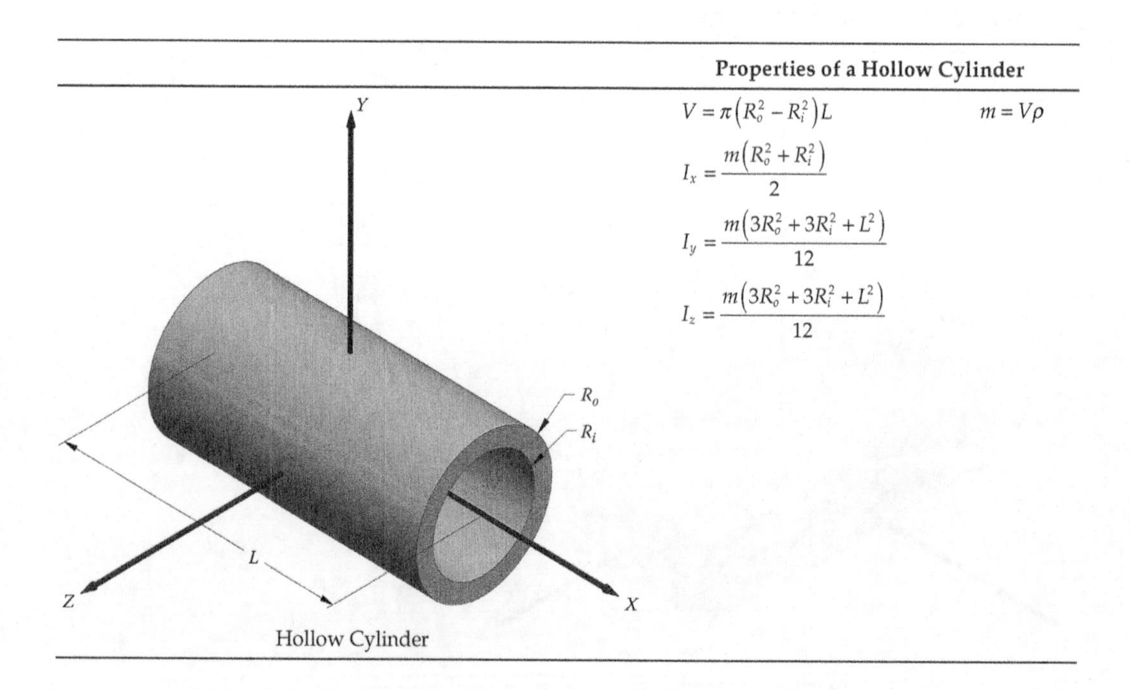

Hollow Cylinder

Properties of a Sphere

$$V = \frac{4}{3}\pi R^3 \qquad m = V\rho$$

$$I_x = \frac{2}{5}R^2$$

$$I_y = \frac{2}{5}R^2$$

$$I_z = \frac{2}{5}R^2$$

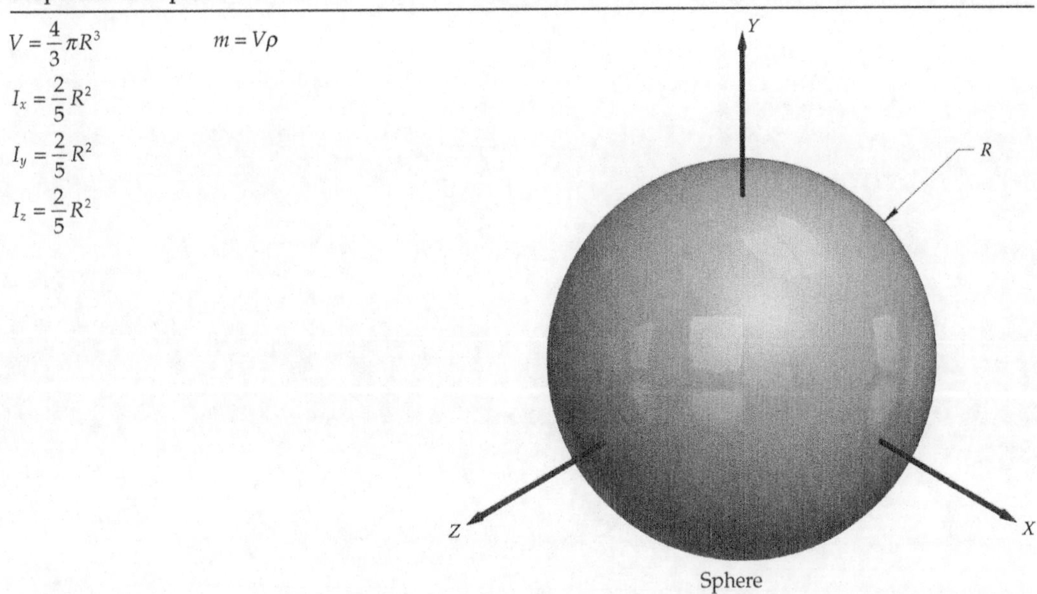

Sphere

Properties of a Cone

$$V = \frac{\pi}{3}R^2 H \qquad m = V\rho$$

$$I_x = \frac{3}{10}mR^2$$

$$I_y = \frac{m\left(12R^2 + 3H^2\right)}{80}$$

$$I_z = \frac{m\left(12R^2 + 3H^2\right)}{80}$$

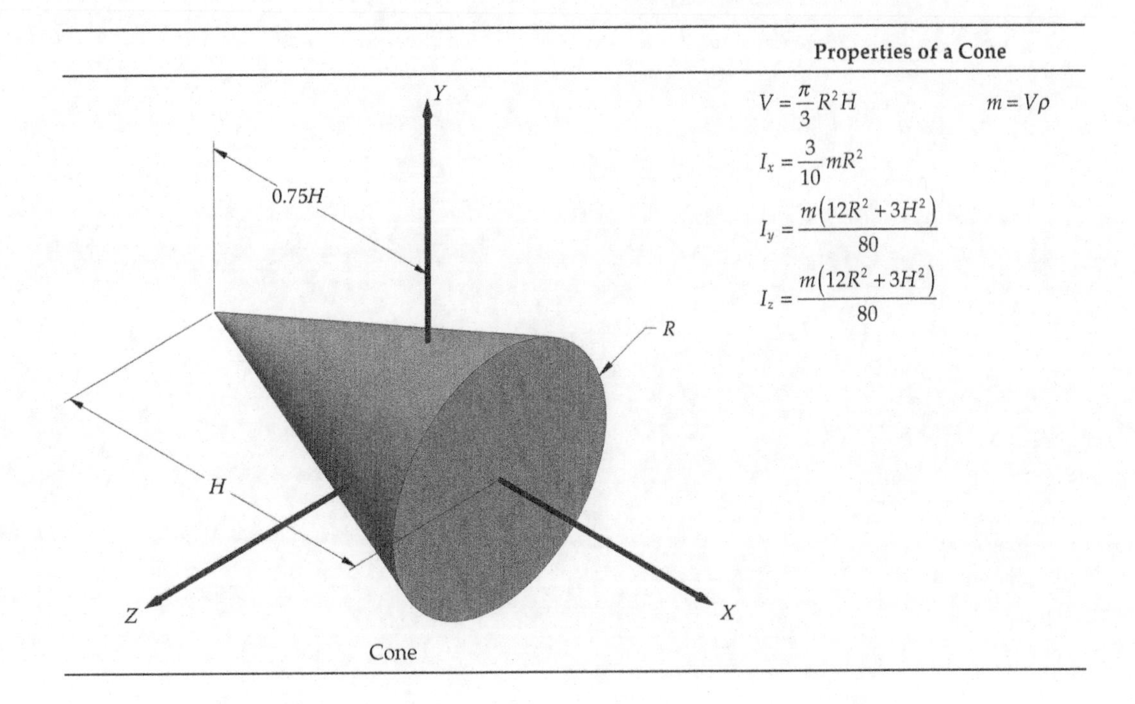

Cone

Index

R

S

T

U

V